T0141685

METEOROLOGICAL MONOGRAPHS

VOLUME 27 | DECEMBER 1998 | NUMBER 49

METEOROLOGY OF THE SOUTHERN HEMISPHERE

Edited by

David J. Karoly
Dayton G. Vincent

American Meteorological Society
45 Beacon Street, Boston, Massachusetts 02108

ISBN 1-878220-29-2
ISSN 0065-9401 CIP 98-074523

Support for this monograph has been provided by the Cooperative Research Centre for Southern Hemisphere Meteorology, Australia.

Published by the American Meteorological Society
45 Beacon St., Boston, MA 02108

Printed in the United States of America
by Cadmus Journal Services, Easton, MD

TABLE OF CONTENTS

PREFACE

Most books on meteorology have an emphasis on the Northern Hemisphere, but much can be learned about the behavior of the global atmospheric circulation by comparing and contrasting the circulation and weather systems between the two hemispheres. The different distributions of oceans and continents, and of topography, lead to differences in the circulation. Explaining these differences helps in understanding the basis for the global circulation. In addition, a description and understanding of the meteorology of the Southern Hemisphere is vital for all the people living there.

A comprehensive description of the meteorology of the Southern Hemisphere was provided for the first time by the monograph on this subject published by the American Meteorological Society in 1972, *Meteorology of the Southern Hemisphere,* edited by C. W. Newton. That monograph was, of necessity, preliminary in nature because the time series of observations from the few islands in southern middle and high latitudes were short, nearly all of the Antarctic stations had functioned only since the International Geophysical Year (IGY; July 1957–December 1958), radiosondes were concentrated on the continents, the existing climatological atlases were based on heterogeneous datasets, and the only daily analyses of the hemisphere that existed were of sea level pressure and 500-mb geopotential height for the 18 months of the IGY. Daily sea level pressure maps were indeed analyzed before this period, but they were made without the benefit of Antarctic observations and were not useful over large areas. In addition, very little data was available for the stratosphere.

In the quarter century that has passed since the first monograph, much has happened to warrant an updated edition: new observational techniques based on satellites, anchored and drifting buoys, and more ground-based stations have expanded the observational network to cover the whole hemisphere. These techniques have made it possible to provide daily objective analyses of a higher quality than previously possible. Furthermore, the better oceanic and atmospheric database that we have today owes much to the many field experiments that have taken place in the intervening years. These include the First GARP (Global Atmospheric Research Program) Global Experiment (FGGE), during which geostationary satellites and drifting buoys were deployed; the Tropical Oceans Global Atmosphere (experiment); the TOGA Coupled Ocean–Atmosphere Response Experiment; the Global Energy and Water Cycle Experiment; the World Ocean Circulation Experiment; the Australian Monsoon Experiment; and the First Regional Observing Study of the Troposphere. The time is right, therefore, for a fresh look at the circulation features over the Southern Hemisphere.

One of the objectives of this monograph is to reexamine the features presented in the first monograph, taking advantage of the better datasets and analyses. Equally or more important is to document and substantiate the many findings and ideas that have emerged since the original monograph. The present monograph deals in greater detail with regional climates and updates the knowledge of circulation features such as the Australian monsoon; the South Pacific convergence zone; the South Atlantic convergence zone; the subtropical and polar jet streams, cyclones, and storm tracks; and the circulation in the stratosphere. These topics are dealt with mainly in the first four chapters and in chapter 6. Subjects that were not mentioned in the previous monograph are found in the remaining five chapters. They include mesoscale circulations; stratospheric ozone depletion; ocean circulations; air–sea interaction; interannual variability, including the Southern Oscillation and its attending El Niño; intraseasonal variability, including the Madden–Julian oscillation; climate variability and changes; and climate modeling. There is less emphasis on synoptic meteorology in the present edition, as it was covered quite adequately in the first edition. The main topic that is lacking is numerical weather prediction (NWP). This topic was originally envisioned as chapter 11, but it was decided to forego it because NWP developments and improvements generally are not so specific to a geographical region.

Because of the increased interest in the Southern Hemisphere that followed FGGE, the American Meteorological Society formed an ad hoc committee (1981) and later a formal committee (1984) on meteorology and oceanography of the Southern Hemisphere, one of the functions of which is to plan and conduct professional scientific meetings. As a result, five international conferences on the meteorology and oceanography of the Southern Hemisphere have been held: in Brazil (1983), New Zealand (1986), Argentina (1989), Australia (1993), and South Africa (1997). The exchange of ideas at these conferences has been an important factor in the initiation of many of the studies that are summarized in this monograph.

Finally, it's important to note that projects were just recently completed at ECMWF, NCEP, and other organizations to reanalyze the global atmospheric circulation for the last few decades using the same data assimilation system. This will provide a more consistent analysis database for a longer period and will be particularly important in the Southern Hemisphere, where the sparsity of conventional meteorological observing stations means that analyses are more sensitive to changes in data assimilation systems. These reanalyses, however, were generally not available in time to be used in this monograph.

Harry van Loon
David J. Karoly
Dayton G. Vincent
Editors

METEOROLOGY OF THE SOUTHERN HEMISPHERE
Common topics covered in the different chapters

Chapter	Time-mean circulation	Transient eddies	Planetary waves	Mesoscale systems	El Niño–Southern Oscillation	Monsoons and MJO	Climate change	Air–ice interactions	Air–sea interactions
Mean state of the troposphere	*	*	*				+		
General circulation	*	*	*						
Meteorology of the Tropics	+			+	+	*			+
Meteorology of the Antarctic	+	+	+	+	+			*	+
Mesoscale meterology		*		*					+
The stratosphere in the SH	*	*	*				*		
The role of the oceans in SH climate	*	+	+		*		+	*	*
Interannual and intraseasonal variability		+	+		*	+			+
Climate change and long-term variability		+	+		*		*		+
Climate modelling	*	+	+		+	+	*	+	*

Key: * Major discussion + Minor discussion

CONTRIBUTORS

Dr. Robert J. Allan
CSIRO Atmospheric Research
Private Bag No. 1
Aspendale, Vic 3195
Australia
E-mail: rja@dar.csiro.au

Prof. David H. Bromwich
Polar Meteorology Group
Byrd Polar Research Centre
The Ohio State University
1090 Carmack Road
Columbus, OH 43210-1002
USA
E-mail: bromwich@polarmet1.mps.ohio-state.edu

Dr. J. Stuart Godfrey
CSIRO Marine Research
GPO Box 1538
Hobart, TAS 7001
Australia
E-mail: godfrey@marine.csiro.au

Dr. James W. Hurrell
NCAR
P.O. Box 3000
Boulder, CO 80307-3000
USA
E-mail: jhurrell@ucar.edu

Prof. Philip D. Jones
Climatic Research Unit
School of Environmental Sciences
University of East Anglia
Norwich NR4 7TJ
UK
E-mail: p.jones@uea.ac.uk

Prof. David J. Karoly
Meteorology CRC
Monash University
Clayton, Vic 3168
Australia
E-mail: d.karoly@sci.monash.edu.au

Dr. George N. Kiladis
CIRES, Campus Box 449
University of Colorado
Boulder, CO 80309
USA
E-mail: gkiladis@al.noaa.gov

Dr. John McBride
Bureau of Meteorology Research Centre
GPO Box 1289K
Melbourne, Vic 3001
Australia
E-mail: j.mcbride@bom.gov.au

Dr. Gerald A. Meehl
NCAR
P.O. Box 3000
Boulder, CO 80307
USA
E-mail: meeh@ucar.edu

Dr. Kingtse C. Mo
Climate Prediction Centre
NCEP/NWS/NOAA
Washington, DC 20233
USA
E-mail: wd52km@sgi44.wwb.noaa.gov

Dr. Paul A. Newman
NASA/Goddard Space Flight Center
Code 916, Greenbelt MD 20771
USA
E-mail: newman@notus.gsfc.nasa.gov

Dr. Carlos Nobre
Centro de Previsao de Tempo e Estudos Climaticos (CPTEC)
Instituto Nacional de Pesquisas Espaciais (INPE)
Km 40 Rodovia Presidente Dutra
Cachoeira Paulista
SP, CEP: 12.630-000
Brazil
E-mail: nobre@cptec.inpe.br

Dr. Thomas R. Parish
Department of Atmospheric Science
University of Wyoming
Box 3038, University Station
Laramie, WY 82071
USA
E-mail: parish@grizzly.uwyo.edu

Dr. William J. Randel
NCAR
P.O. Box 3000
Boulder, CO 80307-3000
USA
E-mail: randel@ucar.edu

Dr. Michael J. Reeder
Centre for Dynamical Meteorology and Oceanography
Monash University
Clayton, Vic 3168
Australia
E-mail: m.reeder@sci.monash.edu.au

Dr. Stephen R. Rintoul
CSIRO Marine Research
GPO Box 1538
Hobart, TAS 7001
Australia
E-mail: steve.rintoul@marine.csiro.au

Dr. Prakki Satyamurty
Centro de Previsao de Tempo e Estudos Climaticos (CPTEC)
Instituto Nacional de Pesquisas Espaciais (INPE)
Km 40 Rodovia Presidente Dutra
Cachoeira Paulista
SP, CEP: 12.630-000
Brazil
E-mail: saty@cptec.inpe.br

Dr. Jon M. Schrage
Department of Earth and Atmospheric Sciences
Purdue University
1397 Civil Engineering Building
West Lafayette, IN 47907-1397
USA
E-mail: schragej@purdue.edu

Dr. Dennis J. Shea
NCAR
P.O. Box 3000
Boulder, CO 80307-3000
USA
E-mail: shea@ucar.edu

Dr. Pedro L. Silva Dias
Department de Ciencias Atmosfericas
Instituto Astronomico e Geofisico
Universidade de Sao Paulo
Rua do Matao 1226
05508-900 Sao Paulo SP
Brazil
E-mail: pldsdias@model.iag.usp.br

Dr. Roger K. Smith
Meteorological Institute
University of Munich
Theresienstr. 37
80333 Munich
Germany
E-mail: roger@meteo.physik.uni-muenchen.de

Dr. J.J. Taljaard
South African Weather Bureau
Private Bag X97
Pretoria
South Africa

Prof. John van Heerden
Department of Civil Engineering
University of Pretoria
Pretoria
South Africa
E-mail: jheerden@postino.up.ac.za

Dr. Harry van Loon
NCAR
P.O. Box 3000
Boulder, CO 80307-3000
USA
E-mail: vanloon@ucar.edu

Prof. Dayton G. Vincent
Department of Earth and Atmospheric Sciences
Purdue University
1397 Civil Engineering Building
West Lafayette, IN 47907-1397
USA
E-mail: dvincent@purdue.edu

Chapter 1

The Mean State of the Troposphere

JAMES W. HURRELL, HARRY VAN LOON, AND DENNIS J. SHEA

National Center for Atmospheric Research, Boulder, Colorado*

1.1. Introduction

The mean state of the troposphere on broad scales in space and time is described in this chapter. Emphasis is placed on the climate of the Southern Hemisphere (SH), although global maps will be presented and some relevant comparisons with the Northern Hemisphere (NH) will be made. General descriptions of the seasonal mean distributions of many atmospheric variables will be given, while the following chapters will present discussions on topics including the variability of the SH circulation and the dominant processes that maintain the general circulation.

The primary data source for most of the diagrams is the set of global analyses produced by the European Centre for Medium-Range Weather Forecasts (ECMWF) through 1993. The original version of the monograph (van Loon et al. 1972) was based on a series of atlases that described the mean conditions from sea level to 100 mb (Taljaard et al. 1969; van Loon et al. 1971; Jenne et al. 1971; Crutcher et al. 1971). The data on which these atlases were based ended with October 1966. Daily analyses of sea level pressure and 500-mb geopotential height for the International Geophysical Year (IGY; July 1957–December 1958) were also extensively used in the original monograph.

The ECMWF analyses are believed to be the best *operational* global analyses available for general use (Trenberth and Olson 1988). A comprehensive evaluation of the ECMWF global analyses is given by Trenberth (1992), who also discusses the usefulness of the analyses for climate studies. A concern is the influence of operational changes in the analysis–forecast system employed at ECMWF to produce the analyses. The apparent climate changes that result can obscure the true short-term climate changes or interannual climate variability and are, therefore, a motivation for the reanalysis projects at the major operational centers (Trenberth and Olson 1988; Bengtsson and Shukla 1988). The relative strengths and weaknesses of the reanalysis products will, however, need to be assessed and documented. At the time of this writing, the reanalysis products were just becoming available. A few comparisons will be made in the summary (section 1.7) to reanalyses from both ECMWF (Gibson et al. 1996) and the National Centers for Environmental Prediction (NCEP; Kalnay et al. 1996). For most of the variables presented in this chapter, differences between the operational and reanalyzed products are minimal and do not affect the conclusions.

Two archived datasets from ECMWF were used to produce the circulation statistics. The first is the World Meteorological Organization (WMO) archive, which consists of twice-daily, initialized analyses at seven pressure levels in the vertical from 1979 through 1989. The other is the World Climate Research Programme's Tropical Oceans Global Atmosphere program (WCRP/TOGA) archive, which begins in 1985 and consists of four-times-daily, uninitialized analyses at 14 pressure levels in the vertical. Both archives are available on a triangular wavenumber 42 (T42) Gaussian grid (Trenberth 1992).

For fields minimally affected over time by the operational changes at ECMWF, the WMO and WCRP/TOGA archives were combined to produce 15-yr (1979–93) climatologies. Fields associated with moisture and the divergent component of the wind, however, clearly exhibit spurious trends over these years (e.g., see Figs. 35–41 in Trenberth 1992; Trenberth and Guillemot 1995). Climatologies for these variables were constructed from the WCRP/TOGA archive over relatively short periods that fall between major operational changes in May 1986 and May 1989.

Use was also made of analyses based on daily operational maps from the Australian Bureau of Meteorology, which have been processed into monthly means and are available since May 1972. These analyses have been used, for example, to document the mean state (e.g., Trenberth 1980; Karoly et al. 1986), observed eddy statistics (e.g., van Loon 1980; Trenberth 1981a, 1982), and interannual variations (e.g., Trenberth 1981b, 1984; Kidson 1988) of the SH circulation. The quality of these data has been discussed by Trenberth (1981a,b) and Swanson and Tren-

* The National Center for Atmospheric Research is sponsored by the National Science Foundation.

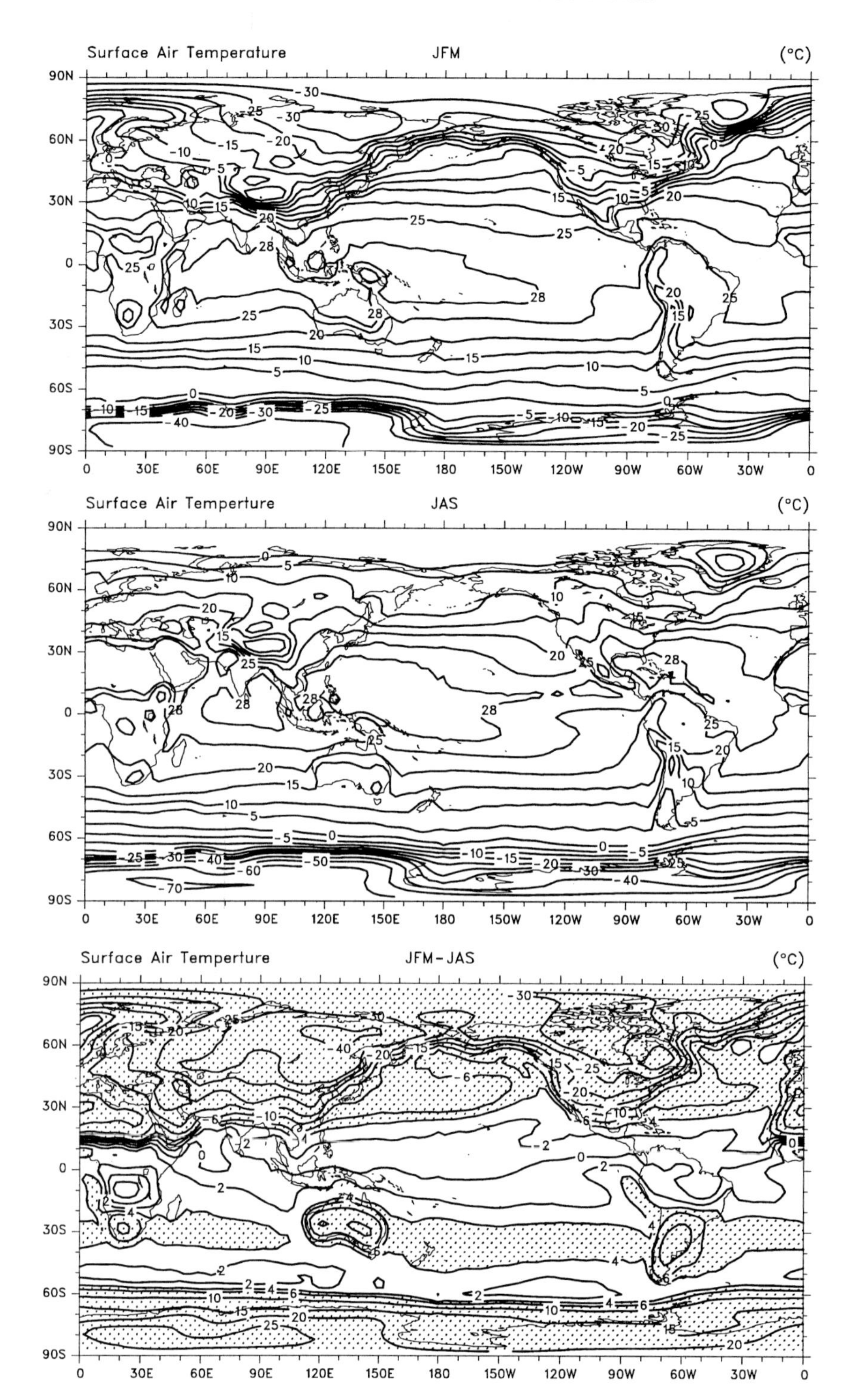

FIG. 1.1. Mean surface air temperatures from ECMWF (1985–1993) for JFM (top), JAS (middle), and JFM−JAS (bottom) in °C. Differences greater than 4°C in magnitude in the bottom panel are stippled.

berth (1981), who noted that the 0000 UTC surface pressure and 500-mb geopotential height charts are the most reliable. Consequently, our analyses are confined to those data.

For several variables, statistics obtained from the operational ECMWF analyses were supplemented with data from stations with long records throughout the SH. This not only provides a check on the analyses,

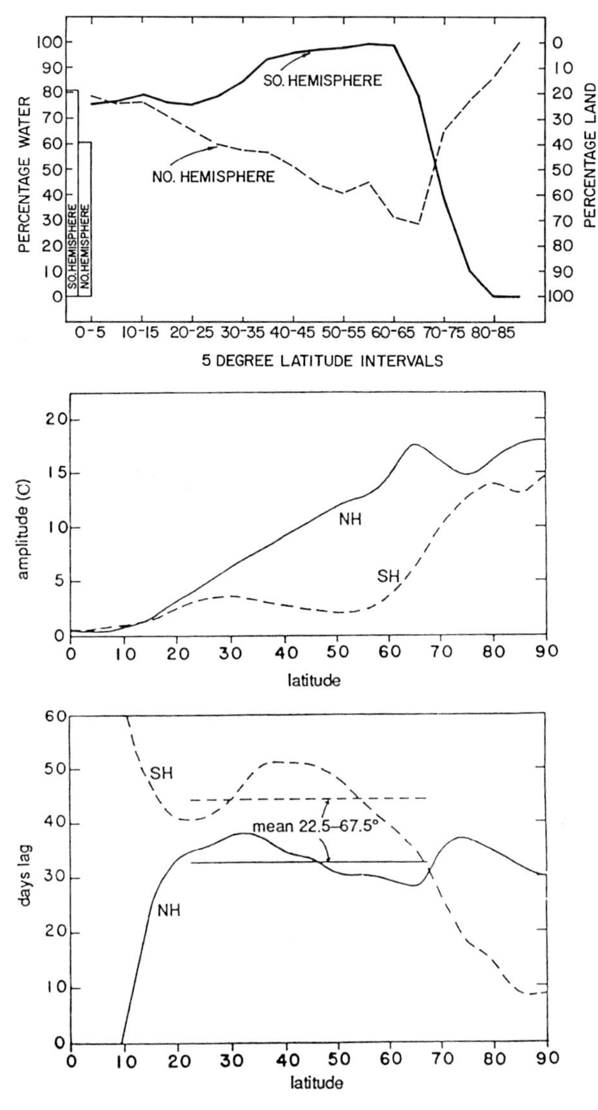

FIG. 1.2. Zonal averages for each hemisphere of percentage of land and water around each latitude circle (top), amplitude of the 12-month annual cycle harmonic of surface temperature in °C (middle), and phase of the 12-month harmonic of surface temperature in terms of lag behind the sun (bottom). (Adapted from van Loon 1972 and Trenberth 1983.)

but is useful since it is unlikely that the periods covered by the operational data provide stable statistics of variability. It is known, for instance, that a quite marked change in the SH circulation took place from the 1970s to the 1980s (van Loon et al. 1993; Hurrell and van Loon 1994), in the middle of the period covered by the Australian analyses and at the time the ECMWF data began.

Satellite data for climate monitoring, which have become increasingly available over the past decade, provide another important source of global information. To examine aspects of the radiation budget, data over the period 1985–89 from the Earth Radiation Budget Experiment (ERBE) were utilized (Barkstrom et al. 1989). Cloud-coverage estimates over the years 1983–91 were obtained from the International Satellite Cloud Climatology Project (ISCCP) (Rossow and Schiffer 1991; Rossow and Garder 1993a,b). Global gridded estimates of monthly precipitation covering the period January 1979–December 1995 were obtained from Xie and Arkin (1996). Although the periods of coverage for the ERBE and ISCCP products are comparatively short, the observations provide a reliable and representative description of the large-scale features.

The original monograph dedicated four chapters to a description of the distribution of temperature, pressure, wind, and moisture in the SH troposphere (van Loon 1972). Since that time a wealth of data has become available, and much has been learned about SH meteorology. Many topics that were not included in the original monograph are covered in the following chapters. For these reasons, our discussion of the mean state is not as comprehensive as that in the original monograph, to which we must therefore refer for discussion of some topics. We have attempted, however, to illustrate and describe those features that are most relevant to the results presented in the following chapters. The reader is also encouraged to refer to the original monograph for a thorough bibliography of earlier studies dealing with SH meteorology.

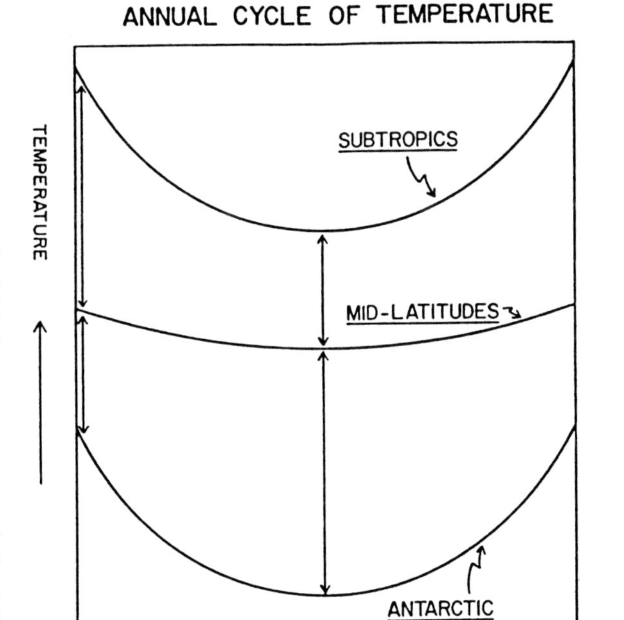

FIG. 1.3. Schematic diagram of the annual cycle of the air temperature in the Southern Hemisphere subtropics, middle latitudes, and the Antarctic. The vertical arrows show the meridional temperature contrasts in summer and winter between the latitude belts.

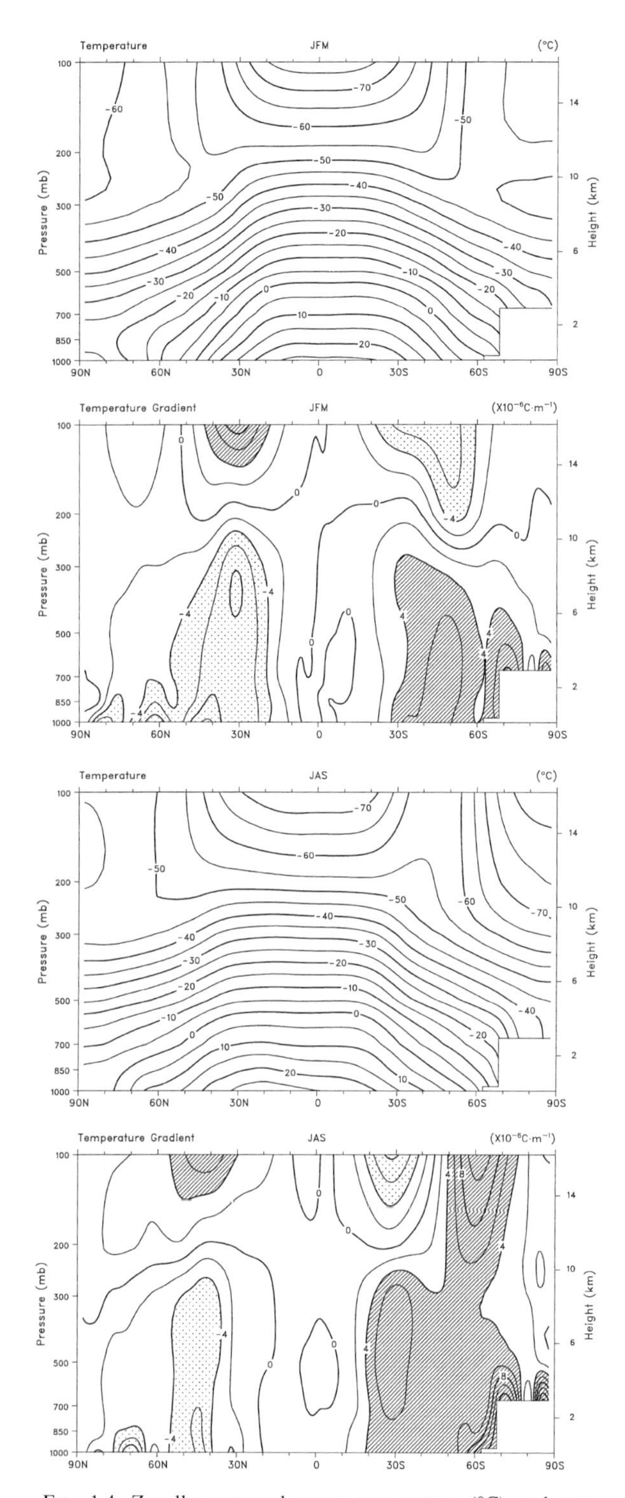

FIG. 1.4. Zonally averaged mean temperature (°C) and mean meridional temperature gradients (10^{-6}°C m^{-1}) for JFM (top two panels) and JAS (bottom two panels) over the years 1979–93 from ECMWF. Gradients greater than 4 units are hatched and gradients less than −4 units are stippled.

1.2. Temperature

One of the basic climate elements is temperature. The mean monthly maps of surface air temperature in Taljaard et al. (1969) for the SH and Crutcher and Meserve (1970) for the NH used in the original monograph were based on the best possible coverage over the greatest number of years. These maps differ little from more recent climatologies, such as those from Oort (1983) or derived from ECMWF analyses (Fig. 1.1). The surface temperature maps can be further compared with data from selected SH stations in Table 1.1a. Since the astronomical definition of the seasons (e.g., from the December solstice to the March equinox for northern winter and southern summer) is most appropriate over the oceanic regions of the SH (Trenberth 1983), the southern summer season is defined as the period January–March (JFM), while winter is defined as July–September (JAS).

The highest mean surface air temperatures are in the intertropical regions, where the largest amounts of solar radiation are received during the course of the year, with maximum temperatures in the tropical western Pacific and Indian Oceans. The meridional temperature gradients in the Tropics are very small. In the SH, summer (JFM) surface temperatures decrease from the Tropics toward the pole except over the lower-latitude continents. The rates of decrease, however, are not uniform. Over the Atlantic and Indian Oceans the temperature differences between 40° and 50°S in summer are up to four times greater than the differences between 55° and 65°S, reflecting conditions at the ocean's surface (Fig. 1.1; van Loon 1966). There are no such large discontinuities in gradient over the Pacific Ocean. At its widest extent, the region of weak gradient spans as much as 20° latitude (50°–70°S), so that subantarctic temperatures in summer vary by only modest amounts. Over the South Atlantic and Indian Oceans poleward of 54°S, 90% of the observed summer air temperatures lie between −1° and 3°C with a mean temperature of about 0.8°C (see van Loon 1966). The meridional gradients of temperature over the middle latitude oceans in southern winter (Fig. 1.1) are approximately the same as those in summer, but the gradients steepen at high latitudes over the expanded ice pack.

The isotherms are primarily zonally oriented in both seasons over the SH, except for the trade-wind regions in the South Atlantic and Pacific Oceans where the west coasts are colder than the east coasts, a pattern not observed in the Australian region. Zonal variations in surface temperature between 40° and 60°S are characterized by positive anomalies in the Pacific Ocean, with opposing negative anomalies in the Atlantic and Indian Oceans (see Fig. 3.2 of van Loon 1972; Shea et al. 1990). The temperature pattern over South America and Africa is determined not only by the latitude but also by the topography. The intensely

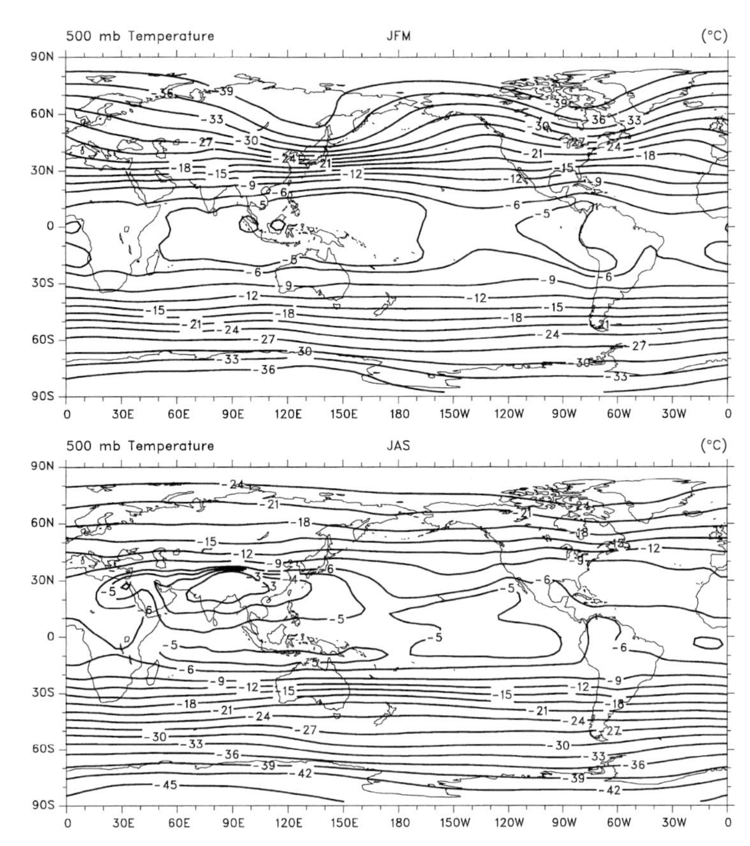

FIG. 1.5. Mean 500-mb temperatures from ECMWF (1979–93) for JFM (top) and JAS (bottom) in °C.

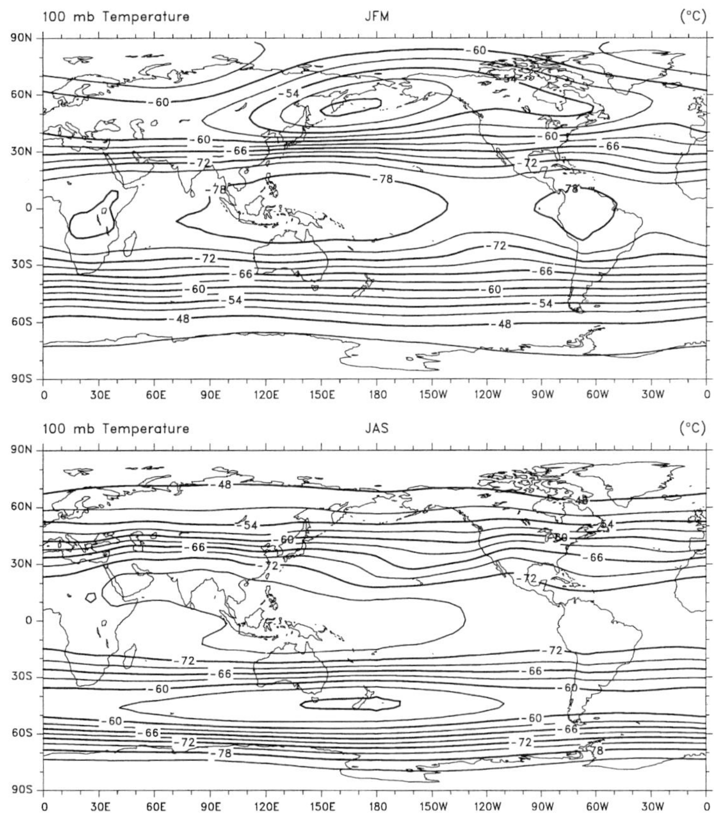

FIG. 1.6. As in Fig. 1.5, but for mean 100-mb temperatures.

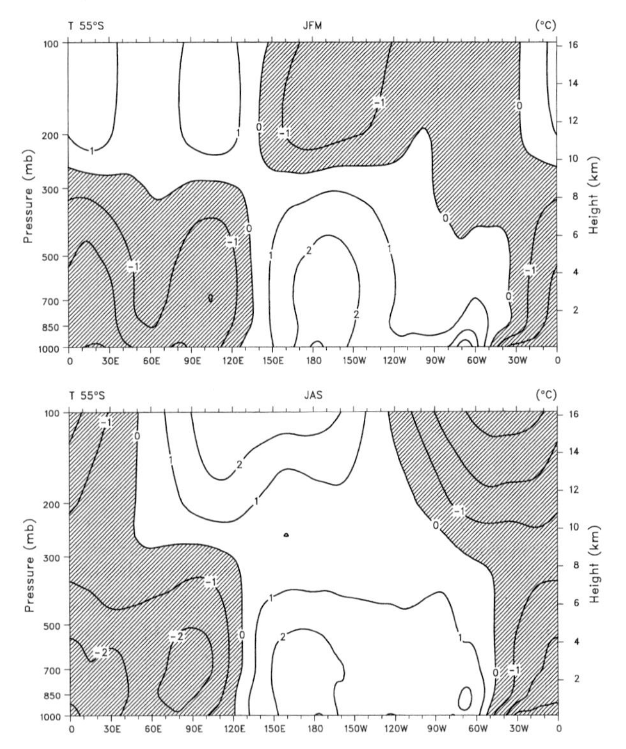

FIG. 1.7. Deviation (°C) of the mean temperature at 55°S from the mean of the latitude in JFM (top) and JAS (bottom) from ECMWF (1979–93). Negative values are hatched.

heated surface of the Australian interior in summer strongly contrasts with the cool water of the Great Australian Bight so that the isotherms are concentrated along the south coast. In winter, the southern interior of Australia is not only colder than the northern part and the east and west coasts, but it is also colder than the waters of the bight (see also Fig. 3.8 of van Loon 1972). The isotherms in this season are concentrated over northern Australia.

The annual range of surface air temperature over the globe (Fig. 1.1), measured by JFM minus JAS, highlights land–sea contrasts and relatively small annual changes over the oceans. These features reflect the very different heat capacities of ocean versus land and the depth of the layer linked to the surface. The largest ranges occur over the continents, although, except for Australia and Antarctica, values are smaller in the SH than in the NH at corresponding latitudes. An interesting feature of the SH is that, over the ocean, the annual range increases from the Equator to about 35°S, beyond which it has a smaller amplitude to about 55°S. This feature is also found over the open ocean of the NH (van Loon 1966, 1972).

The contrast between the zonally averaged mean annual cycle of surface temperature in the NH (60.7% water) versus the SH (80.9% water) further illustrates these points (Fig. 1.2). The amplitude of the first harmonic between 40° and 60°S is less than 3°C in the SH but is nearly 12°C in the NH, and the average lag in the temperature response relative to the sun for the annual cycle is 32.9 days in the NH versus 43.5 days in the SH (averaged between 22.5° and 67.5° latitude), again reflecting the difference in thermal inertia (Trenberth 1983; see also Prescott and Collins 1951). The amplitude of the first harmonic is relatively large over the subtropics and high latitudes of the SH—both regions where the harmonic explains more than 90% of the mean annual surface air temperature variance (see Fig. 3.21 of van Loon 1972).

The structure of the amplitude of the annual cycle in meridionally alternating belts of high and low values in the SH (Fig. 1.1) has interesting implications for the change of the meridional temperature gradients from summer to winter. The schematic drawing in Fig. 1.3 demonstrates that because the annual range is larger in the subtropics than in middle latitudes of the SH, the meridional gradient of temperature is steeper in summer than winter. Poleward of the middle latitudes, the annual range increases, and thus the meridional gradient is steeper in winter than summer. The same pattern exists in the 1000–500-mb thickness fields (not shown), and as a result the thermal wind (the change in the geostrophic wind with height) is stronger in summer than winter in the latitudes near 50°S, whereas the opposite occurs to the north and south of these latitudes (van Loon 1965, 1967a, 1972).

In the middle troposphere the influence of the continents and oceans is less evident, and the isotherms are much more zonally symmetric in both hemispheres (Fig. 1.5). The zonally averaged temperature contrast at 500 mb in southern summer between the Equator and the South Pole is 31°C, of which 17°C is in the interval between 30° and 60°S (Fig. 1.4). The largest gradients are in the middle latitudes of the Atlantic and Indian Oceans (Fig. 1.5). Note that over Australia and north of New Zealand the gradients are larger than elsewhere at the same latitudes, and the isotherms are deflected southward over the middle and high latitudes of the Pacific Ocean (see also Fig. 3.4 of van Loon 1972).

In southern winter the zonally averaged temperature difference between the Equator and the South Pole is about 40°C at 500 mb (Fig. 1.4). Zonally averaged temperatures exhibit large gradients over a wider area in winter than summer in the SH, and the steepest gradients move from 40°–50°S in summer to 30°S in winter. Increased subtropical gradients are especially evident in winter over Australia and the Pacific Ocean (Fig. 1.5). The isotherms bend poleward from the Indian Ocean over to the Pacific Ocean in winter, resulting in a zone of relatively weak gradient at middle latitudes from the Great Australian Bight into the central Pacific.

In the low stratosphere the horizontal gradients of temperature are reversed (Fig. 1.4), since the warmest air in summer is at the South Pole (Fig. 1.6). In winter

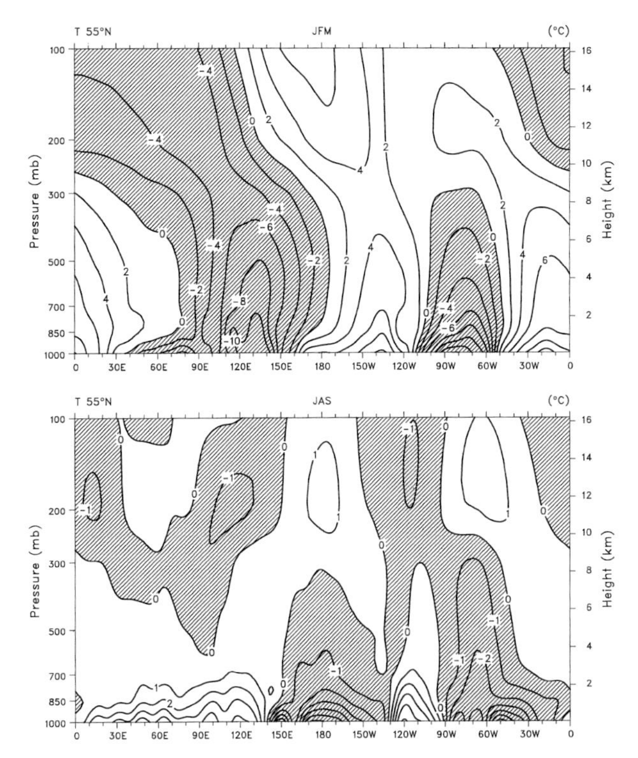

FIG. 1.8. As in Fig. 1.7, but for temperature anomalies at 55°N.

the temperature increases from the Equator to a circumpolar belt of high temperature at 40° to 50°S. From this peak the temperature drops to a polar minimum. A similar circumpolar maximum is found at 100 mb in the winter of the NH, but farther poleward and with less zonal symmetry than its counterpart in the SH.

The average temperatures for each hemisphere and for the earth are given in Table 1.2 for JFM and JAS as a function of elevation. At the surface, the SH is colder than the NH in the respective summer seasons but is warmer in the winter. The range of surface air temperature from summer to winter is thus appreciably smaller over the SH than over the NH. The area between 30° and 60° latitude, where the SH is mostly covered with water and more than half of the NH is land (Fig. 1.2), contributes the most to the hemispheric differences (see also Fig. 3.15 of van Loon 1972). The SH is also colder than the NH in summer at 500 mb, but there is little difference between the two hemispheres in winter; thus, although the temperature range is still larger over the NH in the middle troposphere, the JFM − JAS hemispheric differences are smaller than at the surface. Moreover, note that the annual range hardly changes with elevation over the SH but decreases over the NH, since the influence of the continents is primarily limited to the lowest levels of the troposphere. Globally averaged temperatures are higher in JAS than in JFM between the surface and the 300-mb level, and the annual range is greater in the lower troposphere. The difference between JAS and JFM halves between the surface and 500 mb, and at the 100-mb level it is small.

The extent to which the character of the underlying surface influences tropospheric temperatures is further illustrated in Fig. 1.7 by selecting the same parallel in the two hemispheres such that the difference in distribution of land and water is large. The temperature departures from the latitude mean along 55°S change little from summer to winter. In both seasons the parallel is divided into one half with positive anomalies across the Pacific and another half with negative anomalies. This reflects the pattern in the surface temperature, which, at this latitude, is determined by the shape and extent of Antarctica (Fig. 1.1; see also section 1.3 for further discussion). The amplitude of the anomalies decreases with height up to the tropopause, above which the amplitude increases. The phase changes at the tropopause by 180° in summer and 90° in winter. The anomalies in the NH at 55°N and their change with season (Fig. 1.8) are closely related to the distribution of land and ocean. The colder air is over the continents in winter and the oceans in summer. In both JFM and JAS, the amplitude of the anomalies is much larger in the troposphere of the NH, as well as in the stratosphere in winter.

The standard deviation of day-to-day variability (i.e., the intraseasonal variability) is well measured by the ECMWF data (Trenberth 1992). Such variations in time are associated with the alternation of different air masses at a given location. The variability at 850 mb (Fig. 1.9) is smallest in the Tropics and largest in middle and high latitudes and is larger in the NH than in the SH during both seasons. The intraseasonal variability during summer in the SH is greatest in coastal areas of the lower-latitude continents and in the middle latitude belt of high cyclonic activity. The variability is also high over the southern parts of the three continents in winter, but the largest values occur over high latitudes and are probably associated with frequent interruptions of the stable stratification over Antarctica and the sea ice in the lowest levels of the troposphere (see Fig. 16 in van Loon 1967b).

The interannual variability of temperature is more difficult to assess because operational changes at ECMWF over time have introduced spurious variability into the analyzed temperatures. These changes have apparently affected tropical temperatures the most (Hurrell and Trenberth 1992). Despite this shortcoming, the interannual variability of the seasonal 850-mb temperatures is presented in Fig. 1.10. The standard deviations of interannual changes of summer and winter surface air and 500- and 100-mb temperatures at selected SH stations are given in Table 1.1.

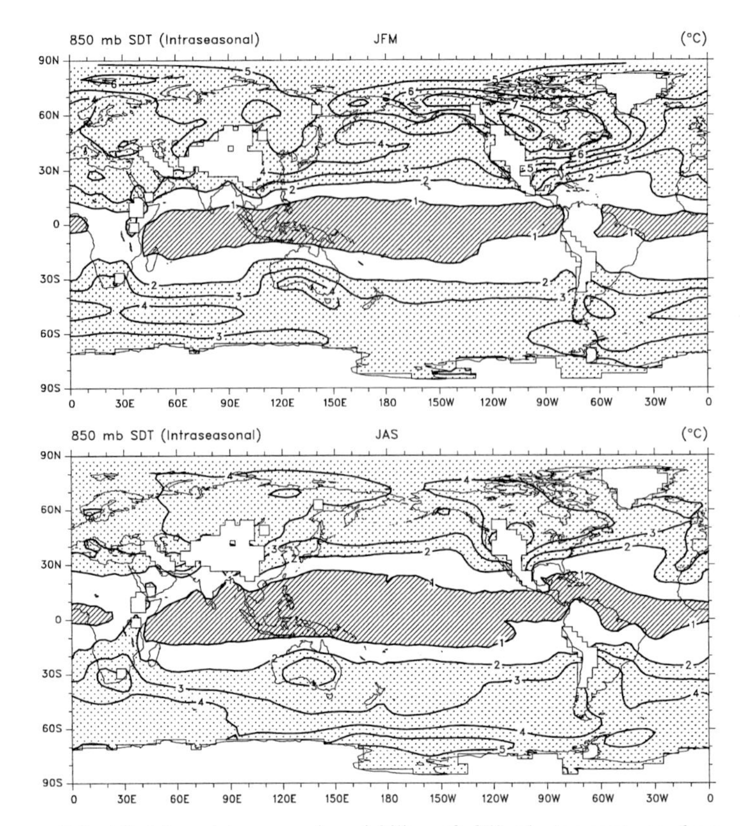

FIG. 1.9. Mean intraseasonal variability of 850-mb temperatures from ECMWF (1979–93) for JFM (top) and JAS (bottom) in °C. Values less than 1°C are hatched and values greater than 2°C are stippled.

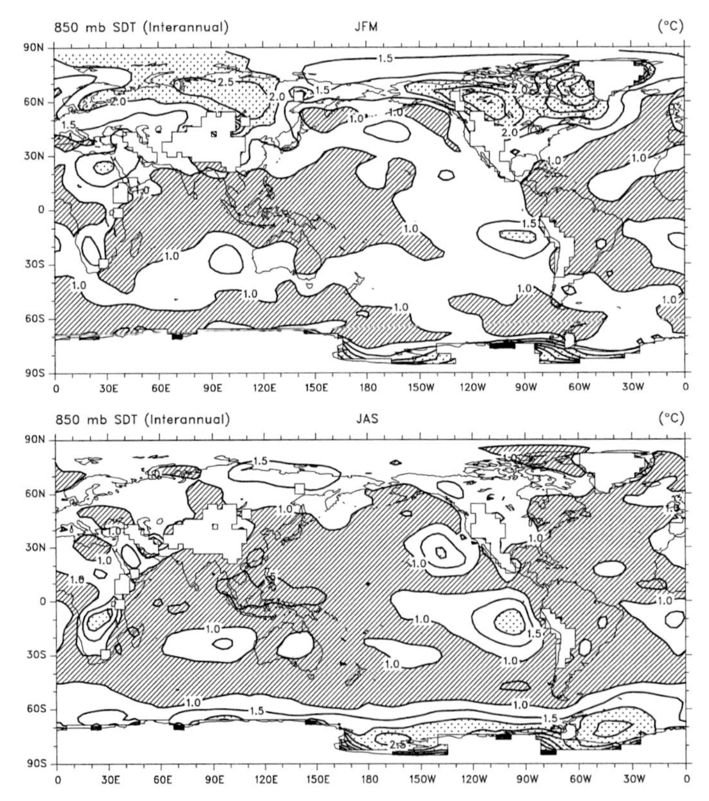

FIG. 1.10. Interannual variability of JFM (top) and JAS (bottom) 850-mb temperatures from ECMWF (1979–93) in °C. Values less than 1°C are hatched and values greater than 2°C are stippled.

TABLE 1.1. Mean surface (a), 500-mb (b), and 100-mb (c) temperatures (°C) for JFM and JAS at selected Southern Hemisphere stations. Also given are the number of years included in the mean and the standard deviation about the mean (i.e., the interannual variability). Values in parentheses are for the period 1979–93.

a. Surface temperature (°C)

Name (Location)	JFM			JAS		
	Mean	Yrs	σ	Mean	Yrs	σ
Nairobi (1.3°S, 36.9°E)	20.1 (20.1)	42 (15)	0.7 (0.6)	17.6 (17.8)	42 (14)	0.5 (0.4)
Luanda (8.8°S, 13.2°E)	26.3 (27.0)	80 (3)	0.8 (0.7)	20.7 (20.7)	79 (3)	0.7 (0.4)
Lima (12.1°S, 77.0°W)	22.0 (22.8)	47 (8)	0.8 (0.6)	15.3 (16.4)	50 (7)	0.9 (0.5)
Darwin (12.5°S, 130.9°E)	28.4 (28.3)	109 (14)	0.6 (0.4)	26.3 (26.4)	112 (15)	0.7 (0.5)
St. Helena (16.0°S, 5.7°W)	18.8 (20.9)	91 (6)	1.0 (0.5)	14.5 (16.6)	93 (7)	0.8 (0.3)
St. Brandon (16.5°S, 59.6°E)	27.8 (27.8)	42 (14)	0.4 (0.3)	23.4 (23.4)	40 (12)	0.4 (0.4)
Tahiti (17.5°S, 149.6°W)	26.9 (27.2)	59 (15)	0.6 (0.4)	24.7 (24.9)	59 (15)	0.6 (0.4)
Pretoria (25.8°S, 28.2°E)	21.5 (22.4)	42 (14)	1.1 (0.8)	14.6 (15.4)	42 (14)	1.1 (0.7)
Johannesburg (26.1°S, 28.2°E)	18.8 (19.4)	42 (14)	1.0 (1.1)	12.7 (13.1)	42 (14)	0.8 (0.8)
Isla de Pasqua (27.2°S, 109.4°W)	23.4 (23.3)	48 (14)	0.4 (0.5)	18.2 (18.0)	51 (14)	0.4 (0.5)
Durban (29.9°S, 31.0°E)	24.1 (23.9)	107 (14)	0.9 (1.9)	18.3 (18.3)	107 (14)	0.8 (0.5)
Perth (32.0°S, 115.9°E)	23.5 (24.0)	66 (14)	1.0 (0.7)	13.7 (13.7)	67 (15)	0.5 (0.6)
Sydney (33.9°S, 151.2°E)	21.6 (22.5)	113 (14)	0.5 (0.7)	13.4 (13.7)	112 (15)	0.6 (0.5)
Cape Town (34.0°S, 18.6°E)	20.0 (20.0)	42 (14)	0.4 (0.4)	12.6 (13.0)	41 (13)	0.5 (0.5)
Buenos Aires (34.6°S, 58.4°W)	22.4 (23.5)	135 (12)	0.9 (0.6)	11.5 (12.7)	133 (10)	1.0 (0.6)
Melbourne (37.8°S, 145.0°E)	19.4 (20.5)	126 (3)	0.9 (1.5)	10.9 (12.0)	127 (3)	0.5 (0.6)
Ile N. Amsterdam (37.8°S, 77.6°E)	16.5 (17.1)	38 (10)	0.8 (0.5)	11.3 (11.9)	37 (9)	0.6 (0.5)
Gough (40.3°S, 9.9°W)	14.1 (14.1)	37 (14)	0.7 (0.7)	9.1 (9.2)	35 (12)	0.5 (0.5)
Christchurch (43.5°S, 172.6°E)	15.8 (15.9)	105 (15)	0.8 (0.7)	7.3 (7.5)	105 (15)	0.6 (0.6)
Chatham Is. (44.0°S, 176.6°W)	14.3 (14.9)	90 (14)	0.8 (0.8)	8.3 (8.6)	89 (14)	0.5 (0.4)
Ile Crozet (46.5°S, 51.0°E)	7.3 (7.4)	11 (10)	0.4 (0.4)	3.2 (3.3)	13 (9)	0.7 (0.7)
Marion Is. (46.9°S, 37.8°E)	7.3 (7.8)	42 (12)	0.6 (0.3)	3.7 (4.3)	44 (13)	0.5 (0.4)
Kerguelen (49.3°S, 70.2°E)	7.2 (7.6)	37 (9)	0.6 (0.4)	2.1 (2.2)	40 (12)	0.6 (0.5)
Campbell Is. (52.5°S, 169.1°E)	9.1 (9.4)	49 (12)	0.5 (0.6)	5.2 (5.6)	51 (13)	0.5 (0.6)
Macquarie (54.5°S, 159.0°E)	6.8 (7.1)	43 (13)	0.7 (0.7)	3.3 (3.6)	45 (14)	0.5 (0.4)
Islas Orcadas (60.7°S, 44.7°W)	0.3 (0.7)	50 (7)	0.7 (0.4)	−8.1 (−7.2)	49 (6)	2.7 (4.0)
Mawson (67.6°S, 62.9°E)	−5.1 (−5.1)	5 (5)	1.1 (1.1)	−18.3 (−18.3)	5 (5)	2.3 (2.3)
Davis (68.6°S, 77.9°E)	−3.5 (−3.4)	20 (10)	0.8 (0.8)	−17.3 (−17.1)	25 (13)	2.4 (2.9)
S.A.N.A.E. (70.5°S, 2.9°W)	−9.0 (−9.0)	24 (8)	0.9 (1.3)	−26.1 (−25.1)	30 (10)	2.4 (2.3)
Halley Bay (75.5°S, 26.6°W)	−10.3 (−10.0)	29 (6)	0.9 (1.0)	−27.8 (−28.0)	28 (5)	2.0 (2.4)
McMurdo (77.8°S, 166.6°E)	−9.9 (−10.3)	24 (5)	1.5 (1.0)	−25.8 (−25.6)	24 (4)	2.2 (2.1)
Amundsen-Scott (90.0°S, 0.0°E)	−40.9 (−40.8)	28 (6)	1.3 (1.5)	−59.6 (−59.6)	28 (6)	1.5 (1.6)

b. 500 mb: Temperature (°C)

Name (Location)	JFM Mean	JFM Yrs	JFM σ	JAS Mean	JAS Yrs	JAS σ
Darwin (12.4°S, 130.9°E)	−5.0 (−5.0)	41 (13)	0.5 (0.4)	−5.8 (−5.5)	40 (12)	0.5 (0.4)
Tahiti (17.6°S, 149.6°W)	−5.6 (−5.3)	34 (14)	1.2 (1.3)	−6.0 (−6.0)	34 (14)	0.8 (0.7)
Pretoria (25.9°S, 28.2°E)	−6.8 (−6.9)	19 (15)	0.5 (0.5)	−11.6 (−11.7)	16 (12)	0.8 (0.9)
Durban (29.8°S, 31.0°E)	−7.5 (−7.5)	40 (15)	0.9 (1.3)	−13.8 (−13.8)	38 (11)	0.6 (0.8)
Perth (31.7°S, 116.1°E)	−9.9 (−10.0)	39 (12)	0.6 (0.5)	−18.7 (−18.5)	36 (11)	1.0 (0.9)
Cape Town (34.0°S, 18.6°E)	−9.6 (−10.0)	40 (15)	2.1 (2.7)	−16.6 (−16.3)	35 (11)	0.8 (0.8)
Ile N. Amsterdam (37.8°S, 77.6°E)	−11.5 (−11.2)	27 (9)	1.0 (0.6)	−19.9 (−19.4)	29 (8)	1.0 (0.9)
Gough Is. (40.4°S, 9.9°W)	−13.5 (−13.6)	35 (14)	0.7 (0.7)	−21.3 (−20.4)	34 (13)	2.8 (4.3)
Christchurch (43.5°S, 172.5°E)	−14.8 (−15.0)	33 (9)	1.1 (1.2)	−24.5 (−24.3)	33 (8)	0.7 (0.8)
Chatham Is. (44.0°S, 176.5°W)	−15.2 (−15.0)	36 (15)	2.0 (3.0)	−24.4 (−23.4)	33 (11)	2.7 (4.6)
Marion Is. (46.9°S, 37.9°E)	−18.1 (−17.6)	41 (14)	1.1 (1.1)	−25.3 (−25.2)	36 (11)	0.9 (1.1)
Kerguelen (49.3°S, 70.2°E)	−19.3 (−18.7)	19 (10)	1.5 (1.6)	−27.3 (−27.3)	16 (8)	1.1 (1.2)
Campbell (52.6°S, 169.1°E)	−20.2 (−20.3)	35 (15)	0.9 (1.0)	−27.6 (−27.7)	34 (13)	1.0 (0.6)
Islas Orcadas (60.7°S, 44.7°W)	−27.0 (−24.7)	10 (1)	1.3 (0.0)	−33.4 (−34.5)	16 (2)	0.9 (0.1)
Davis (68.6°S, 77.9°E)	−32.2 (−32.1)	13 (10)	0.8 (0.9)	−39.3 (−39.3)	17 (10)	1.4 (1.5)
S.A.N.A.E. (70.3°S, 2.4°W)	−31.7 (−30.8)	18 (5)	1.1 (0.9)	−39.8 (−39.4)	21 (4)	0.9 (0.2)
Amundsen-Scott (90.0°S, 0.0°E)	−37.7 (−37.4)	16 (3)	0.8 (0.4)	−44.7 (−44.3)	18 (4)	0.6 (0.6)

TABLE 1.1. Continued.

c. 100 mb: Temperature (°C)

Name (Location)	JFM			JAS		
	Mean	Yrs	σ	Mean	Yrs	σ
Darwin (12.4°S, 130.9°E)	−82.6 (−82.8)	33 (13)	1.1 (1.2)	−78.6 (−78.8)	34 (12)	1.1 (1.0)
Tahiti (17.6°S, 149.6°W)	−74.3 (−76.8)	33 (14)	3.2 (2.1)	−72.8 (−75.1)	34 (14)	2.8 (1.2)
Pretoria (25.9°S, 28.2°E)	−75.7 (−75.7)	19 (15)	0.8 (0.8)	−68.5 (−68.2)	16 (12)	1.4 (1.5)
Durban (29.8°S, 31.0°E)	−72.1 (−72.3)	32 (14)	1.5 (0.9)	−65.1 (−64.9)	29 (10)	1.6 (1.6)
Perth (31.7°S, 116.1°E)	−71.1 (−71.2)	33 (12)	1.6 (1.8)	−62.8 (−63.0)	30 (11)	1.6 (1.6)
Cape Town (34.0°S, 18.6°E)	−66.9 (−65.4)	35 (14)	7.7 (12.1)	−61.3 (−61.1)	32 (11)	1.5 (1.6)
Ile N. Amsterdam (37.8°S, 77.6°E)	−63.4 (−64.9)	23 (8)	1.7 (1.0)	−57.0 (−57.8)	27 (8)	1.7 (1.3)
Gough Is. (40.4°S, 9.9°W)	−61.6 (−62.3)	34 (14)	1.6 (1.2)	−57.5 (−57.6)	33 (13)	1.2 (0.8)
Christchurch (43.5°S, 172.5°E)	−59.2 (−58.9)	33 (9)	1.4 (1.3)	−52.7 (−52.5)	33 (8)	1.4 (1.2)
Chatham Is. (44.0°S, 176.5°W)	−58.9 (−59.2)	35 (14)	1.4 (1.5)	−52.8 (−52.6)	33 (11)	1.3 (1.2)
Marion Is. (46.9°S, 37.9°E)	−55.4 (−56.2)	28 (13)	1.8 (1.1)	−56.4 (−56.1)	25 (9)	1.3 (1.3)
Kerguelen (49.3°S, 70.2°E)	−54.0 (−54.4)	18 (10)	1.5 (0.8)	−56.2 (−56.8)	14 (6)	1.4 (1.0)
Campbell (52.6°S, 169.1°E)	−52.5 (−52.5)	35 (15)	1.1 (1.0)	−55.3 (−55.7)	35 (13)	1.9 (1.6)
Davis (68.6°S, 77.9°E)	−45.6 (−45.8)	13 (10)	0.9 (0.8)	−76.7 (−77.0)	14 (10)	1.1 (0.9)
S.A.N.A.E. (70.3°S, 2.4°W)	−44.7 (−45.0)	15 (5)	1.0 (0.8)	−77.3 (−77.8)	16 (2)	0.9 (1.2)
Amundsen-Scott (90.0°S, 0.0°E)	−43.3 (−44.4)	17 (3)	1.1 (0.7)	−83.1 (−84.8)	15 (1)	1.1 (0.0)

1.3. Pressure

The annually averaged surface pressure differs considerably between the hemispheres: 982.1 mb for the NH and 987.3 mb for the SH. The globally averaged pressure at the earth's surface is 984.7 mb (Trenberth and Guillemot 1994). These mean values are removed from the data to show the mean annual cycle of global and hemispheric surface pressures (Fig. 1.11). The total mass of the global atmosphere changes systematically as a result of changes in water vapor content. Because of the greater land mass and continentality of the NH, it is warmer in summer and undergoes a larger annual cycle in temperature than the SH (Table 1.2). Thus, although the mean relative humidity is generally less, latitude by latitude, in the NH than the SH (see Fig. 1.45), the mean mass of water vapor and its annual cycle are larger in the north (Trenberth 1981c).

The interseasonal difference between maps of sea level pressure (SLP) (Fig. 1.12) is most striking in the NH, where the subtropical high pressure centers are especially well-developed during northern summer and the high-latitude low pressure systems intensify during winter. The largest seasonal pressure variations in the NH are found over the Asian continent and are related to the development of the Siberian anticyclone during northern winter and the monsoon low over southeast Asia during summer.

In the SH the SLP rises from a nearly continuous low pressure zone near the Equator (the intertropical convergence zone, ITCZ) to a circumpolar peak in the subtropics where there is an anticyclone center in each ocean. The subtropical belt of high pressure lies a few degrees latitude farther poleward and is weaker during southern summer than winter (Fig. 1.13) so that the oceanic subtropical highs over both hemispheres reach their peaks during JAS (Fig. 1.12). Farther poleward the mean SLP rapidly drops to its lowest values in the circumpolar trough between 60° and 70°S. There are three minima in the trough whose positions change little with season (Fig. 1.12). The zonally averaged pressure in the circumpolar trough is about 4 mb higher in summer than winter (Fig. 1.13). The mean SLP increases south of the trough, but probably does not reach mean values much above 1000 mb over Antarctica, where the high elevations and strong temperature inversions make reduction to sea level of questionable value and accuracy.

Although there is little land between 40°S and the coast of Antarctica, there are notable deviations of mean SLP from the latitudinal average. As in the case of temperature (e.g., Fig. 1.7), wavenumber 1 dominates the mean SLP at middle and high latitudes with higher pressures over the Pacific than over the other oceans. The concentric alignment of the isobars is often interpreted as indicating a mainly zonal circulation; however, it is in reality a statistical mean that results from a multitude of moving low and high pressure systems (e.g., Taljaard and van Loon 1963). The standard deviation of daily values from the seasonal-mean SLP (Fig. 1.14) is largest in the region of the midlatitude westerlies (Fig. 1.29) and is somewhat higher in winter than summer. Values decrease toward the Tropics with a larger gradient in winter than summer, reflecting the meridional expansion of cyclonic activity over much of the middle latitudes during winter. The interannual variability of the sea-

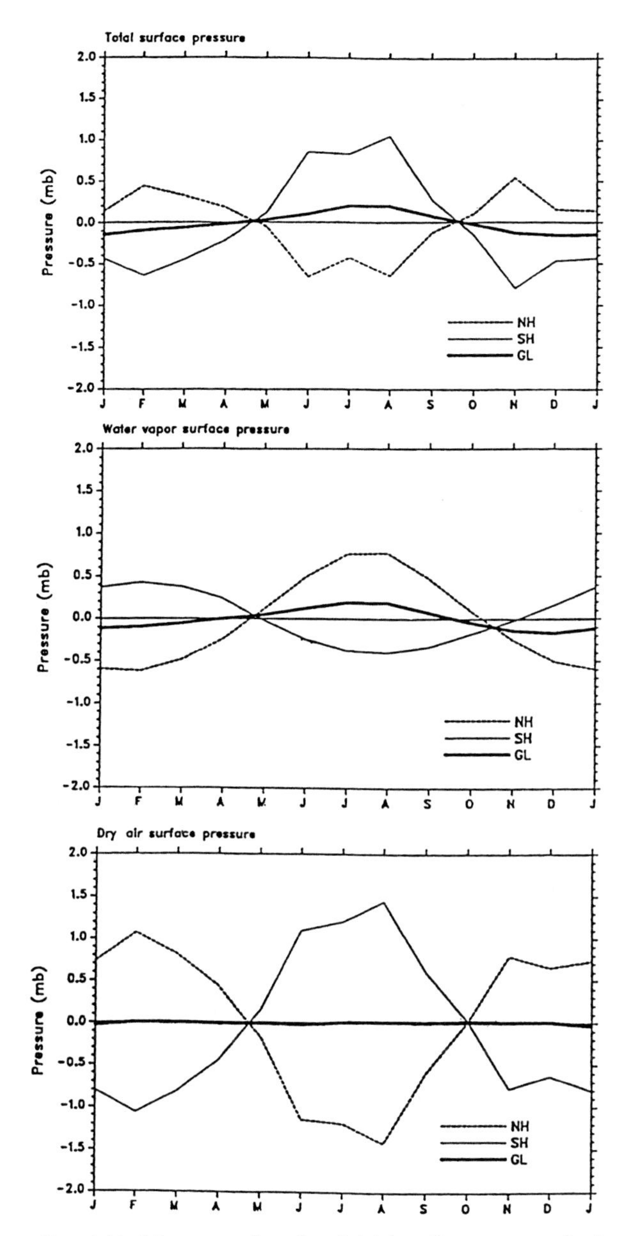

FIG. 1.11. Mean annual cycle of total surface pressure (top), water vapor surface pressure (middle), and their difference (bottom) in mb from ECMWF (1985–93). Shown are values after the removal of annual hemispheric and global means. (From Trenberth and Guillemot 1994.)

sonal mean SLP is shown in Fig. 1.15, and longer-term values from individual stations are given in Table 1.3. In general, the year-to-year variations are largest and farther equatorward in southern winter, when they contain a marked wavenumber 3 pattern. Values over Antarctica are unrealistic in Fig. 1.15.

In the subtropics, and especially over the three lower-latitude continents, the mean SLP is higher in southern winter than summer (Fig. 1.12; see also Fig. 2 of van Loon and Rogers 1984a). In middle and high latitudes, however, the annual march of SLP is less simple. At these latitudes the dominant component in the annual pressure curve is a semiannual oscillation (SAO) that is associated with the position and intensity of the subantarctic trough of low pressure (van Loon 1967b, 1972; van Loon and Rogers 1984b; Meehl 1991). Twice a year, from March to June and from September to December, the trough moves equatorward and weakens, and twice a year, from June to September and December to March, it moves toward Antarctica and deepens. The effect on the annual curve of SLP on either side of the trough appears as a double wave, which on the equatorward side has maxima in the transition seasons and minima in the extreme seasons, and which on the poleward side has the opposite phase (Fig. 1.16). The phase change results in maximum meridional pressure gradients in March and September near 60°S and, thus, a double wave in the zonal geostrophic wind and surface wind stress with a peak amplitude between 55° and 65°S (van Loon 1972; van Loon and Rogers 1984b; Large and van Loon 1989; Trenberth et al. 1990). The amplitude of the SAO in pressure decreases equatorward from its peak near 50°S but remains in the same phase, so that the oscillation in the wind changes phase north of 50°S to maxima in June and December. This structure of the SAO in pressure and wind exists throughout the troposphere (van Loon 1972).

Shown in Fig. 1.17 is the amplitude of and the percent variance explained by the second harmonic in

TABLE 1.2. Average temperatures (°C) of the hemisphere and of the earth.

SH	JFM	JAS	JFM − JAS
SFC	16.1	11.3	4.8
850	9.8	4.7	5.1
700	2.0	−2.7	4.7
500	−12.9	−17.6	4.7
300	−38.1	−42.9	4.8
200	−52.3	−56.9	4.6
100	−65.6	−67.7	2.1
NH	JAS	JFM	JAS − JFM
SFC	20.9	10.5	10.4
850	14.8	5.2	9.6
700	5.9	−2.2	8.1
500	−9.6	−17.4	7.8
300	−35.6	−41.9	6.3
200	−52.3	−54.6	2.3
100	−65.6	−67.0	1.4
Globe	JAS	JFM	JAS − JFM
SFC	16.1	13.3	2.8
850	9.7	7.5	2.2
700	1.6	−0.1	1.7
500	−13.6	−15.1	1.5
300	−39.3	−40.0	0.7
200	−54.6	−53.5	−1.1
100	−66.7	−66.3	−0.4

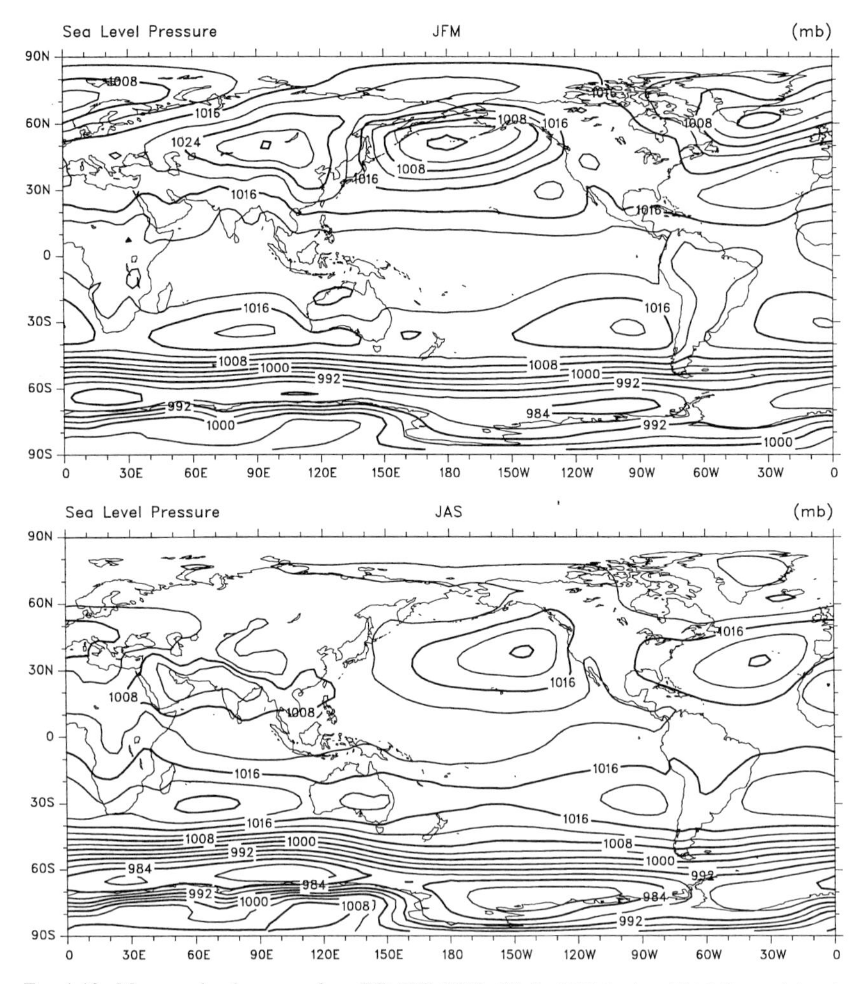

FIG. 1.12. Mean sea level pressure from ECMWF (1979–93) for JFM (top) and JAS (bottom) in mb.

the annual cycle of SLP averaged over the period 1973–79, using the operational maps from the Australian Bureau of Meteorology. It is noteworthy that the pattern of the SAO during the 1970s is very similar to earlier analyses of the semiannual wave based on ship and station data (e.g., see Figs. 4.7 and 4.8 of van Loon 1972). Peak amplitudes occur in all three oceans near 50°S, minima are found near 60°S, and a peak is noted over Antarctica. In the regions of peak amplitude, the semiannual wave accounts for more than 50% of the mean annual variance. The same plots for the period 1980–89 (Fig. 1.18) reveal, however, that the SAO weakened considerably after the late 1970s as a result of significant decadal changes in the monthly means that occurred primarily during the second half of the year (van Loon et al. 1993; Hurrell and van Loon 1994). In particular, the central pressure of the circumpolar trough was generally lower during the 1980s than the 1970s, and the trough was located farther equatorward. The SAO in the wind changed in accordance with the changes in the annual cycle of pressure. Such low-frequency variability emphasizes that a few decades of data do not necessarily yield statistics representative of other periods.

At 500 mb, the mean position of the tropical ridge during JFM is about 10° latitude poleward of its position during JAS, and during both seasons it is equatorward of its position at sea level (Fig. 1.13). The pressure in the middle troposphere decreases from the Tropics to the South Pole, but the poleward slope is weaker in the middle latitudes of the Pacific Ocean than in the same latitudes elsewhere (Fig. 1.19). The 500-mb height profile (Fig. 1.13) has a larger latitudinal gradient in the SH than in the NH during the summer seasons. The largest values during summer in both hemispheres are only 20 gpm apart, but the lowest values (at the poles) differ by about 300 gpm. The zonal means of the hemispheres differ less during the two winter seasons. While the difference in the gradient of 500-mb heights is small between summer and winter in the middle latitudes of the SH, it is somewhat steeper in summer than winter between 40° and 55°S and is steeper in winter in the subtropical and polar regions.

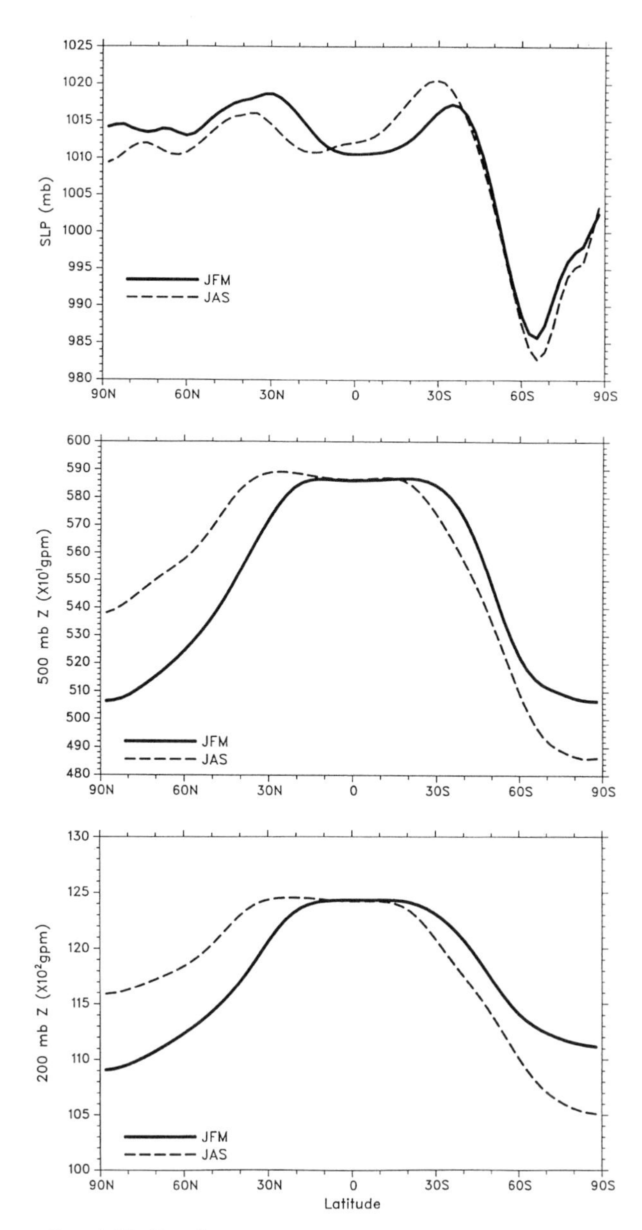

FIG. 1.13. Zonally averaged mean sea level pressure in mb (top), 500-mb geopotential height in 10^1gpm (middle), and 200-mb geopotential height in 10^2gpm (bottom) for JFM (solid) and JAS (dash) from ECMWF (1979–93).

The patterns of the zonally asymmetric 500-mb heights (Fig. 1.20) are the same in the SH winter and summer seasons, although the amplitudes of the anomalies are larger during JAS. South of 40°S the patterns are related to the fact that Antarctica is not centered on the South Pole. The larger portion of the continent lies on the eastern side of the hemisphere, and sea ice, ocean currents, and water and air temperatures display the same zonal asymmetry as the Antarctic coastline. This asymmetry, which consists principally of a wave with its crest on the western (Pacific) and a trough on the eastern (Atlantic–Indian) side of the hemisphere, is found throughout the troposphere in all seasons—not only in the geopotential heights, but also in other climate elements (e.g., Fig. 1.7). Ultimately, it is linked to the observation that steady easterly surface winds along the periphery of Antarctica drive water and ice westward until they are deflected to the north in the Atlantic by the Antarctic Peninsula and the Scotia Arc. The northward limit of the cold water is the Antarctic convergence zone (Oceanic Polar Front), which lies near 50°S in the South Atlantic and Indian Oceans. The northward deflection of ice and water at the west coast of the Ross Sea is small by comparison, and the oceanic polar front lies approximately 10° latitude farther south in the South Pacific Ocean.

The large-scale deviations over the SH evident in Fig. 1.20 are described well by the first wave in a Fourier decomposition of the geopotential heights along latitude circles. Wave 1 reaches its peak between 50° and 60°S in both winter and summer (Figs. 1.21 and 1.22). The amplitude is larger in JAS than in JFM, and during southern winter the amplitude of wave 1 increases into the stratosphere. Also note that the amplitude in winter (JAS) is comparable to the corresponding winter (JFM) value at 50°N.

The large and seasonally persistent wave 1 over the continuous southern oceans closely follows the pattern of the latitude anomalies of temperature (see, for instance, Fig. 3.4 of van Loon 1972; van Loon et al. 1973a). It is therefore likely that surface thermal forcing is an essential component in establishing the structure of the pressure wave. Throughout the year, wave 1 over the SH is quasi-barotropic; that is, its phase is the same at all levels in the troposphere (Figs. 1.21 and 1.22). On the other hand, wave 1 during winter over the NH is baroclinic, leaning westward with height. This difference is found in the higher zonal harmonic waves as well (not shown). The result is that the quasi-stationary waves in the SH cannot transport sensible heat poleward in any month. This is in contrast to the vigorous transport of sensible heat during the colder part of the year by the quasi-stationary waves in the NH (van Loon 1979, 1983; see also Fig. 2.1).

It should also be noted that the amplitude of wave 1 in the SH has a second peak between 30° and 40°S that is largest in winter but weaker than the subantarctic maximum during both seasons. The ridge of the second peak lies between 30° and 60°E and is quasi-barotropic. The phase reversal between the subtropical and subantarctic maxima occurs near 40°S in both winter and summer.

The first wave is clearly dominant in the SH (Figs. 1.21 and 1.22). In winter, wave 1 explains more than 90% of the mean spatial variance near 60°S and over 80% near 35°S. At these latitudes in summer, wave 1 explains more than 70% of the mean spatial variance.

The annual march of wave 1 at 300 mb is shown in Fig. 1.23. The wave over the subantarctic moves

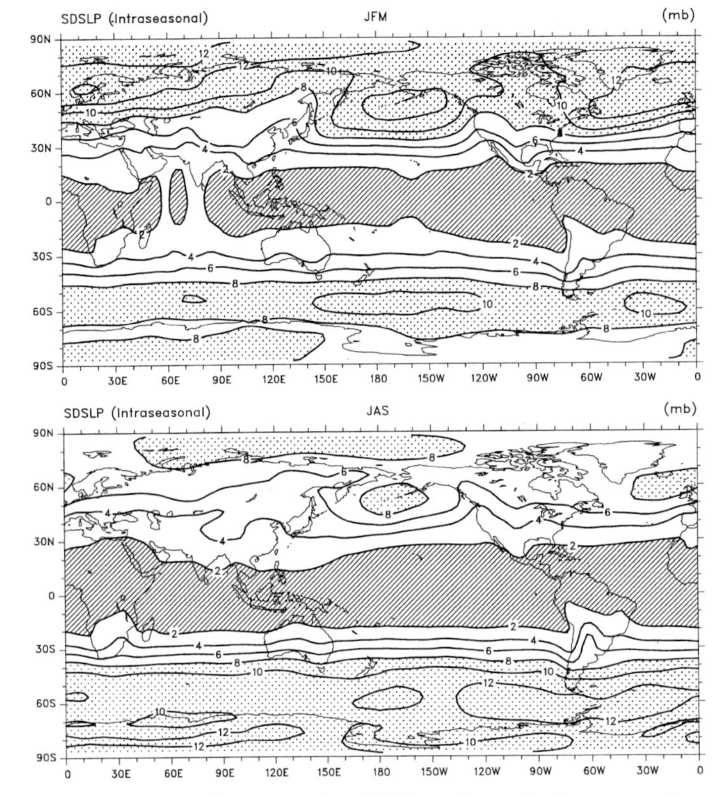

FIG. 1.14. Mean intraseasonal variability of sea level pressure from ECMWF (1979–93) for JFM (top) and JAS (bottom) in mb. Values less than 2 mb are hatched and values greater than 8 mb are stippled.

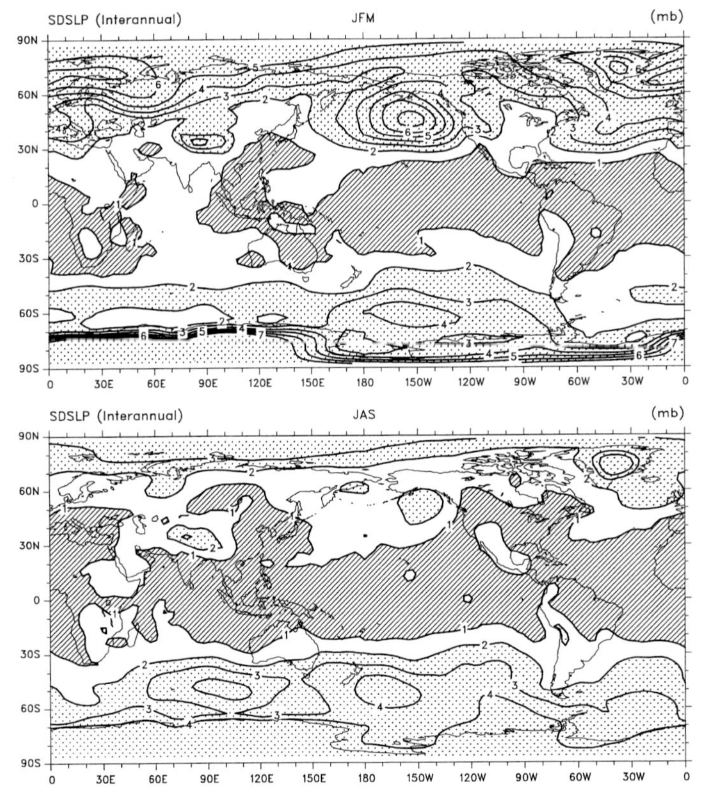

FIG. 1.15. Interannual variability of JFM (top) and JAS (bottom) sea level pressure from ECMWF (1979–93) in mb. Values less than 1 mb are hatched and values greater than 2 mb are stippled.

TABLE 1.3. As in Table 1.1, but for (a) the mean sea level pressure (mb) and (b) 500- and (c) 200-mb geopotential heights (gpm).

a. Sea level pressure (mb)

Name (Location)	JFM			JAS		
	Mean	Yrs	σ	Mean	Yrs	σ
Luanda (8.8°S, 13.2°E)	1010.2 (1010.0)	80 (3)	1.2 (0.6)	1014.8 (1014.4)	78 (3)	1.1 (0.2)
Lima (12.0°S, 77.1°W)	1011.0 (1011.0)	20 (8)	0.6 (0.5)	1014.5 (1014.7)	18 (7)	0.5 (0.4)
Darwin (12.5°S, 130.9°E)	1006.7 (1007.5)	112 (15)	1.1 (1.1)	1012.5 (1013.0)	112 (15)	0.9 (0.8)
St. Helena (16.0°S, 5.7°W)	1014.7 (1015.6)	25 (6)	1.0 (0.6)	1019.5 (1020.1)	26 (7)	0.7 (0.4)
St. Brandon (16.5°S, 59.6°E)	1010.4 (1010.3)	42 (14)	1.2 (1.1)	1017.6 (1017.8)	40 (12)	0.7 (0.8)
Tahiti (17.5°S, 149.6°W)	1011.3 (1010.9)	108 (15)	0.9 (1.1)	1014.3 (1014.3)	107 (15)	1.0 (0.8)
Isla de Pasqua (27.2°S, 109.4°W)	1019.1 (1018.4)	53 (14)	1.4 (1.6)	1021.0 (1019.8)	53 (14)	1.9 (1.9)
Durban (29.9°S, 31.0°E)	1014.9 (1013.9)	108 (14)	1.2 (0.7)	1021.5 (1020.9)	107 (14)	1.4 (1.0)
Perth (32.0°S, 115.9°E)	1013.6 (1013.8)	106 (14)	1.1 (0.7)	1018.7 (1019.4)	109 (15)	1.5 (1.3)
Sydney (33.9°S, 151.2°E)	1014.3 (1015.0)	112 (14)	1.2 (1.0)	1017.7 (1017.9)	110 (15)	1.9 (2.4)
Cape Town (34.0°S, 18.6°E)	1013.7 (1013.8)	42 (14)	0.9 (1.2)	1020.3 (1020.6)	41 (13)	0.8 (0.6)
Buenos Aires (34.6°S, 58.4°W)	1012.2 (1012.5)	50 (12)	0.8 (0.7)	1018.1 (1018.4)	48 (10)	1.3 (1.0)
Melbourne (37.8°S, 145.0°E)	1014.8 (1015.1)	81 (5)	1.1 (0.8)	1017.6 (1018.5)	80 (4)	2.3 (3.4)
Ile N. Amsterdam (37.8°S, 77.6°E)	1019.7 (1020.3)	37 (10)	1.7 (2.1)	1020.1 (1019.2)	36 (9)	2.7 (3.4)
Gough (40.3°S, 9.9°W)	1015.1 (1015.2)	37 (14)	1.4 (1.6)	1016.0 (1015.4)	35 (12)	2.2 (2.3)
Christchurch (43.5°S, 172.6°E)	1013.0 (1012.0)	88 (15)	2.2 (1.3)	1013.1 (1011.6)	88 (15)	3.4 (3.8)
Chatham Is. (44.0°S, 176.6°W)	1014.0 (1013.7)	63 (14)	2.1 (2.1)	1012.5 (1010.6)	63 (14)	4.0 (4.7)
Ile Crozet (46.5°S, 51.0°E)	1010.3 (1010.6)	11 (10)	1.7 (1.5)	1008.7 (1008.7)	13 (9)	2.6 (3.1)
Marion Is. (46.9°S, 37.8°E)	1007.4 (1008.7)	42 (12)	2.5 (2.5)	1008.5 (1008.9)	44 (13)	2.4 (3.0)
Kerguelen (49.3°S, 70.2°E)	1004.2 (1004.5)	36 (9)	2.8 (2.6)	1001.0 (1000.1)	40 (12)	3.0 (3.8)
Campbell Is. (52.5°S, 169.1°E)	1005.3 (1004.8)	49 (12)	2.4 (2.5)	1005.5 (1004.1)	51 (13)	3.9 (4.0)
Macquarie (54.5°S, 159.0°E)	1000.0 (1000.2)	41 (13)	2.7 (2.6)	1000.8 (1000.2)	43 (14)	3.8 (3.4)
Islas Orcadas (60.7°S, 44.7°W)	990.8 (992.0)	53 (7)	2.5 (1.3)	993.9 (994.4)	54 (7)	3.2 (3.7)
Mawson (67.6°S, 62.9°E)	987.9 (987.9)	5 (5)	1.2 (1.2)	985.2 (985.2)	5 (5)	2.0 (2.0)
Davis (68.6°S, 77.9°E)	988.0 (988.0)	20 (10)	2.0 (1.4)	985.0 (984.2)	25 (13)	3.1 (2.9)
S.A.N.A.E. (70.5°S, 2.9°W)	989.6 (988.6)	25 (8)	3.0 (1.7)	986.6 (985.2)	30 (10)	2.7 (3.2)
Halley Bay (75.5°S, 26.6°W)	990.5 (990.1)	28 (6)	2.6 (1.9)	988.1 (987.8)	28 (5)	3.1 (3.2)

b. 500 mb: Geopotential height (gpm)

Name (Location)	JFM Mean	JFM Yrs	JFM σ	JAS Mean	JAS Yrs	JAS σ
Diego Garcia (7.2°S, 72.4°E)	5858 (5867)	7 (4)	28 (34)	5852 (5856)	9 (5)	10 (9)
Lima (12.0°S, 77.1°W)	5854 (5858)	18 (8)	12 (12)	5848 (5853)	17 (5)	7 (5)
Darwin (12.4°S, 130.9°E)	5851 (5856)	41 (13)	13 (10)	5857 (5859)	40 (12)	12 (18)
Tahiti (17.6°S, 149.6°W)	5861 (5862)	34 (14)	14 (12)	5862 (5865)	34 (14)	15 (8)
Pretoria (25.9°S, 28.2°E)	5872 (5872)	19 (15)	9 (10)	5830 (5829)	16 (12)	12 (14)
Durban (29.8°S, 31.0°E)	5855 (5852)	40 (15)	11 (11)	5793 (5793)	38 (11)	15 (17)
Perth (31.7°S, 116.1°E)	5814 (5815)	39 (12)	10 (8)	5654 (5661)	36 (11)	21 (20)
Cape Town (34.0°S, 18.6°E)	5811 (5813)	40 (15)	15 (13)	5712 (5720)	35 (11)	22 (19)
Ile N. Amsterdam (37.8°S, 77.6°E)	5777 (5793)	27 (9)	28 (26)	5622 (5633)	29 (8)	33 (40)
Gough Is. (40.4°S, 9.9°W)	5703 (5705)	35 (14)	24 (28)	5552 (5554)	34 (13)	31 (31)
Christchurch (43.5°S, 172.5°E)	5698 (5681)	33 (9)	32 (33)	5518 (5512)	33 (8)	32 (39)
Chatham Is. (44.0°S, 176.5°W)	5675 (5663)	36 (15)	33 (38)	5503 (5496)	33 (11)	32 (31)
Marion Is. (46.9°S, 37.9°E)	5525 (5541)	41 (14)	34 (27)	5411 (5413)	36 (11)	32 (39)
Kerguelen (49.3°S, 70.2°E)	5483 (5506)	19 (10)	57 (52)	5314 (5310)	16 (8)	33 (43)
Campbell (52.6°S, 169.1°E)	5491 (5481)	35 (15)	37 (45)	5365 (5366)	35 (13)	41 (31)
Davis (68.6°S, 77.9°E)	5101 (5110)	13 (10)	55 (58)	4908 (4900)	17 (10)	52 (55)
S.A.N.A.E. (70.3°S, 2.4°W)	5108 (5106)	18 (5)	30 (31)	4911 (4919)	21 (4)	29 (18)
Amundsen-Scott (90.0°S, 0.0°E)	5021 (5059)	16 (3)	38 (9)	4841 (4852)	18 (4)	31 (22)

c. 200 mb: Geopotential height (gpm)

Name (Location)	JFM Mean	JFM Yrs	JFM σ	JAS Mean	JAS Yrs	JAS σ
Lima (12.0°S, 77.1°W)	12416 (12427)	18 (8)	30 (29)	12355 (12363)	17 (5)	19 (26)
Darwin (12.4°S, 130.9°E)	12440 (12444)	41 (13)	20 (19)	12417 (12425)	40 (12)	16 (19)
Tahiti (17.6°S, 149.6°W)	12433 (12429)	34 (14)	51 (50)	12403 (12396)	34 (14)	38 (28)
Pretoria (25.9°S, 28.2°E)	12395 (12391)	19 (15)	22 (21)	12240 (12241)	16 (12)	57 (66)
Durban (29.8°S, 31.0°E)	12341 (12334)	35 (15)	27 (25)	12138 (12128)	33 (11)	33 (33)
Perth (31.7°S, 116.1°E)	12253 (12248)	39 (12)	28 (23)	11961 (11965)	35 (11)	32 (30)
Cape Town (34.0°S, 18.6°E)	12243 (12240)	38 (15)	29 (22)	11992 (11996)	35 (11)	36 (26)
Ile N. Amsterdam (37.8°S, 77.6°E)	12190 (12197)	27 (9)	40 (33)	11851 (11854)	29 (8)	47 (56)
Gough Is. (40.4°S, 9.9°W)	12066 (12056)	35 (14)	34 (36)	11737 (11735)	35 (13)	50 (42)
Christchurch (43.5°S, 172.5°E)	12024 (12006)	33 (9)	54 (67)	11628 (11617)	33 (8)	37 (47)

TABLE 1.3. Continued.

Name (Location)	JFM			JAS		
	Mean	Yrs	σ	Mean	Yrs	σ
Chatham Is. (44.0°S, 176.5°W)	11987 (11979)	36 (15)	51 (56)	11615 (11606)	33 (11)	39 (41)
Marion Is. (46.9°S, 37.9°E)	11817 (11820)	40 (13)	42 (30)	11530 (11524)	35 (11)	48 (47)
Kerguelen (49.3°S, 70.2°E)	11742 (11755)	19 (10)	69 (79)	11511 (11392)	16 (8)	46 (47)
Campbell (52.6°S, 169.1°E)	11730 (11713)	35 (15)	51 (57)	11408 (11394)	35 (13)	43 (43)
Mawson (67.6°S, 62.9°E)	11294 (11294)	2 (2)	205 (205)	10668 (10668)	3 (3)	58 (58)
Davis (68.6°S, 77.9°E)	11179 (11184)	13 (10)	27 (28)	10673 (10665)	17 (10)	75 (78)
S.A.N.A.E. (70.3°S, 2.4°W)	11211 (11215)	18 (5)	45 (57)	10677 (10679)	21 (4)	38 (23)
Amundsen-Scott (90.0°S, 0.0°E)	11067 (11102)	16 (3)	46 (24)	10489 (10505)	18 (4)	42 (19)

slightly poleward from summer to winter, and its amplitude reaches a peak in late winter and early spring. The phase of wave 1 over the subantarctic remains the same through the year (not shown) and explains from 70% to 80% of the mean spatial variance from January through May to nearly 90% or more from August through December (see also van Loon and Jenne 1972 and Trenberth 1980). Wave 2 over the SH is comparatively small (Fig. 1.23), reaching its maximum of more than 40 m over the Antarctic during the colder part of the year. Wave 3 over the SH is also small in comparison with wave 1, but its amplitude varies considerably from one year to another (e.g., van Loon et al. 1993). The amplitude of wave 3 between 50° and 60°S is more than 40 m during southern summer and is only slightly less in winter.

The interseasonal changes in the quasi-stationary waves are large over the NH (Figs. 1.21–1.23). Wave 1 has a maximum amplitude of more than 150 m near 45°N during the cold part of the year, but the amplitude weakens from winter to summer to a maximum of about 50 m near 70°N. The subtropical maximum also changes character from winter to summer. During JAS, wave 1 has peaks in the lower and upper troposphere with a minimum near 500 mb (Fig. 1.22). The phase reverses between the two peaks, reflecting the change in the monsoon over Asia from a low pressure center at sea level to a high pressure center in the upper troposphere. The monsoonal phase reversal with height is seen in waves 2 and 3 as well (not shown). Wave 2 is much more pronounced over the NH than over the SH, and during northern winter its amplitude equals or exceeds wave 1 (Fig. 1.23). Wave 3 is also large over the NH during winter, but the amplitudes of both it and wave 2 dwindle to near 20 m in summer. More information about the zonal harmonic waves may be obtained in Anderssen (1965), van Loon and Jenne (1972), van Loon et al. (1973a), van Loon (1979, 1983), Hartmann (1977), Trenberth (1980), Karoly (1985), and Randel (1988).

The intraseasonal variability of the 500-mb height field (Fig. 1.24) shows many of the same features evident in the same maps for SLP (Fig. 1.14). Maximum values in the SH are zonally oriented and occur over a broad region of the middle latitudes during both seasons. In winter, when the day-to-day variability is the largest, the magnitude of the variability is similar in both hemispheres, but in summer the deviations are larger in the SH than in the NH. Thus, despite the zonality of the mean flow in the SH, the degree of daily deviation is not less than in the NH.

Estimates of the interannual variability of the winter and summer 500-mb heights are given in Table 1.3b for selected long-term stations throughout the SH, and maps of the variability over the period covered by the ECMWF data are shown in Fig. 1.25. In both seasons the variability is zonally asymmetric in the SH and is largest in the Pacific, where the influence of the Southern Oscillation is strongest. The variability in the NH has a strong seasonal cycle and maximizes during winter in the vicinities of the Aleutian and Iceland lows. As was the case for the intraseasonal variations, the interannual variability of the summertime flow is larger in the SH than in the NH.

The pattern of the geopotential heights at 200 mb (Fig. 1.26) is similar to that of the middle troposphere. During southern summer the heights peak over Africa and South America and a large ridge covers most of the tropical western Pacific. The troughs over the tropical oceans are related to the thermal troughs evident in Fig. 1.6. The poleward decrease of height is generally greatest at middle latitudes of the eastern hemisphere, as at 500 mb (Fig. 1.19). In southern winter, the meridional height gradient at 200 mb is markedly smaller between 40° and 50°S in the southwestern Pacific than elsewhere in the same latitudes, which was also noted of the temperature (section 1.2).

1.4. Zonal wind

The horizontal wind distribution is closely linked geostrophically to the temperature and pressure distributions. The vertical and meridional distributions of the zonally averaged mean zonal wind for winter and summer (Fig. 1.27) are in close agreement with the plots of zonally averaged geostrophic wind in van

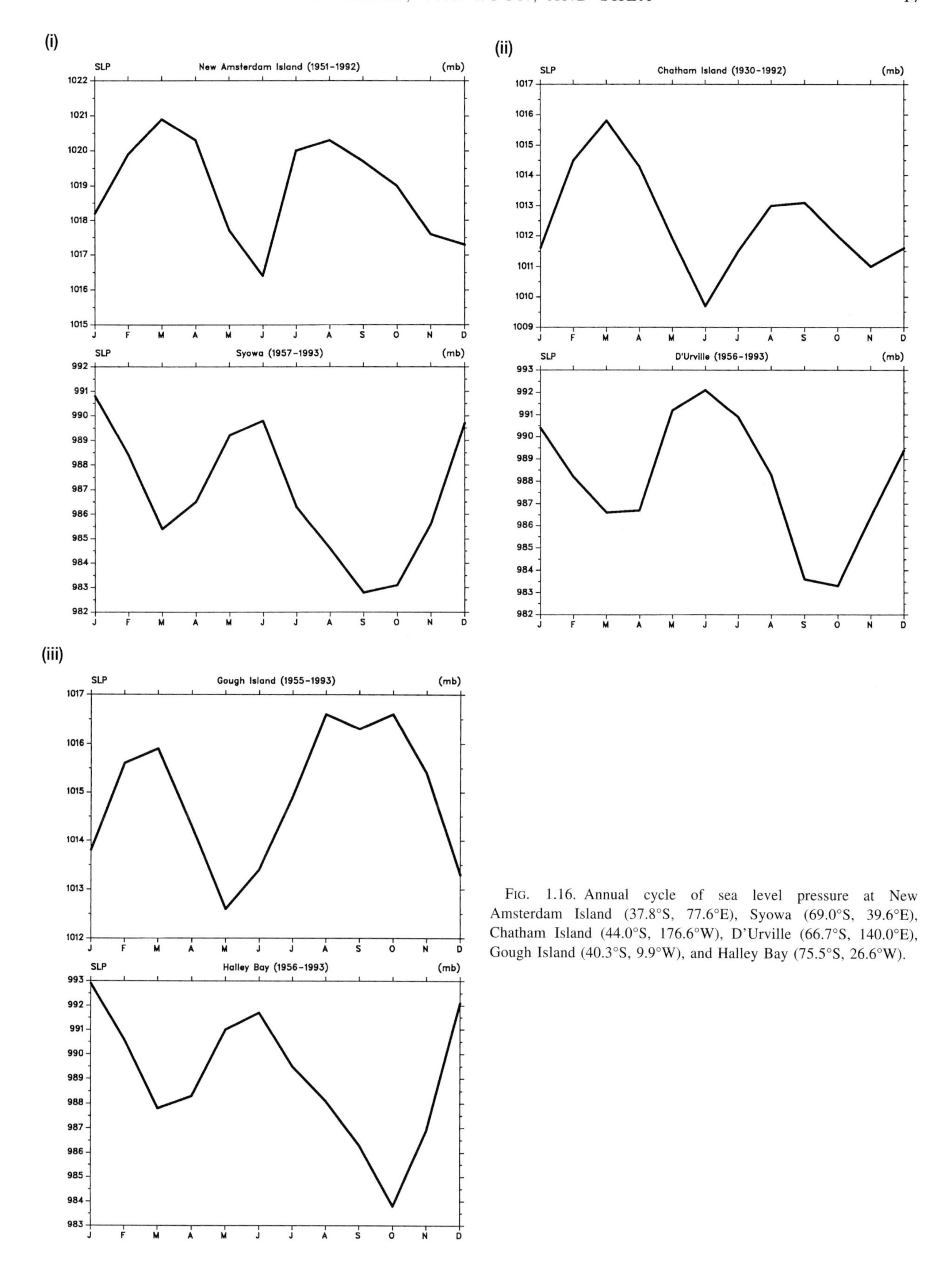

FIG. 1.16. Annual cycle of sea level pressure at New Amsterdam Island (37.8°S, 77.6°E), Syowa (69.0°S, 39.6°E), Chatham Island (44.0°S, 176.6°W), D'Urville (66.7°S, 140.0°E), Gough Island (40.3°S, 9.9°W), and Halley Bay (75.5°S, 26.6°W).

Loon (1972). The mean zonal flow in the winter hemispheres equatorward of 40° latitude is similar, with the strongest westerlies near 40 m s^{-1} at the 200-mb level. The maximum in the SH is about 2°–3° latitude nearer the Equator and is about 5 m s^{-1} weaker than the NH winter maximum. The low-level easterlies are also of nearly equal magnitude in each winter hemisphere and cover similar areas. Poleward of 40° latitude the zonal winds differ appreciably in

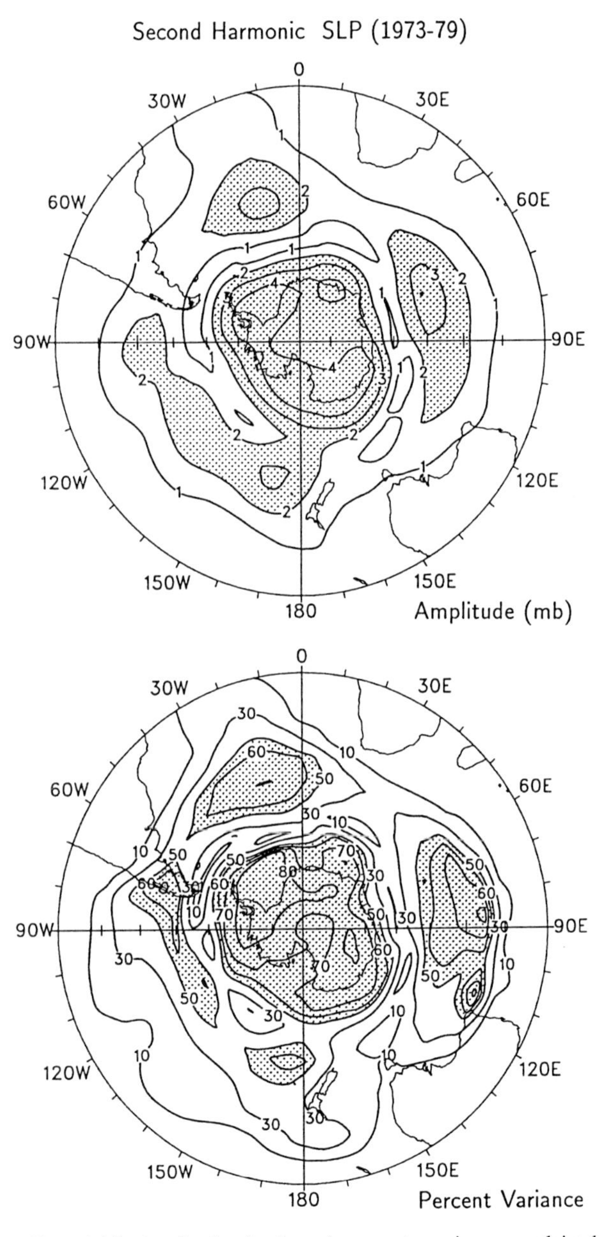

FIG. 1.17. Amplitude (top) and percent variance explained (bottom) of the second harmonic of sea level pressure averaged from 1973 to 1979 from the Australian Bureau of Meteorology analyses. Amplitudes greater than 2 mb are stippled and percent variance explained greater than 50% is hatched. (Adapted from Hurrell and van Loon 1994.)

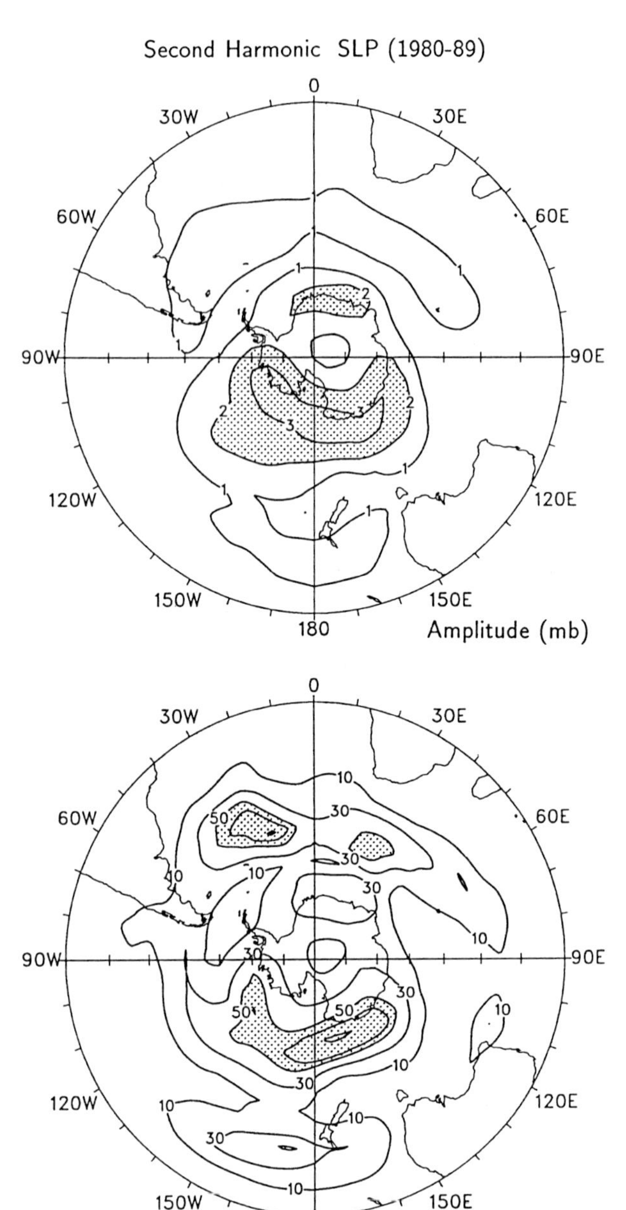

FIG. 1.18. As in Fig. 1.17, but for the period 1980–89.

winter, with stronger winds in the SH. A westerly maximum in the upper troposphere that continues into the stratosphere is evident between 50° and 60°S in accordance with the upward-increasing meridional temperature contrasts poleward of 45°S (Fig. 1.4). The distribution of wind differs considerably between the summer hemispheres. The upper-troposphere westerly maximum is nearly twice as strong in the SH and is located farther poleward than the peak in the NH. In the middle and upper troposphere the tropical easterlies are much stronger in the NH than in the SH, and in the subtropics the westerlies are much stronger in the SH.

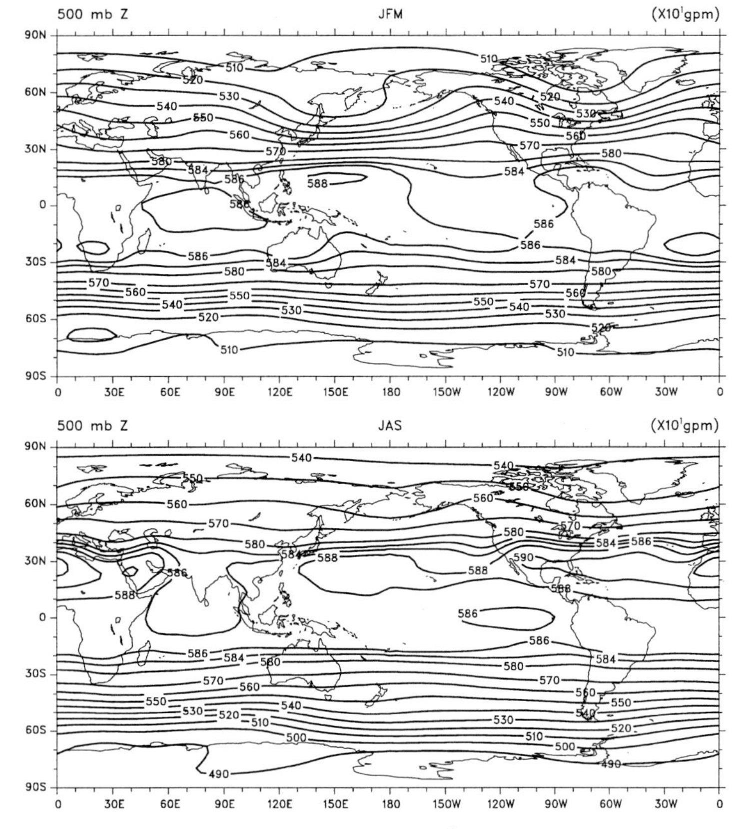

FIG. 1.19. Mean 500-mb geopotential height (10^1gpm) from ECMWF (1979–93) for JFM (top) and JAS (bottom).

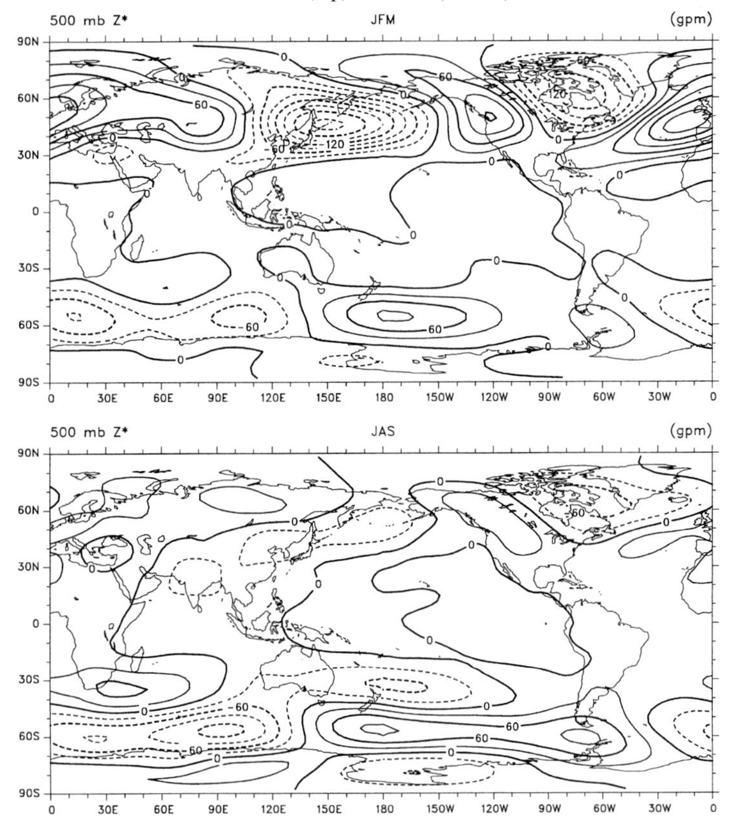

FIG. 1.20. Departures from zonally averaged mean 500-mb geopotential height (gpm) from ECMWF (1979–93) for JFM (top) and JAS (bottom).

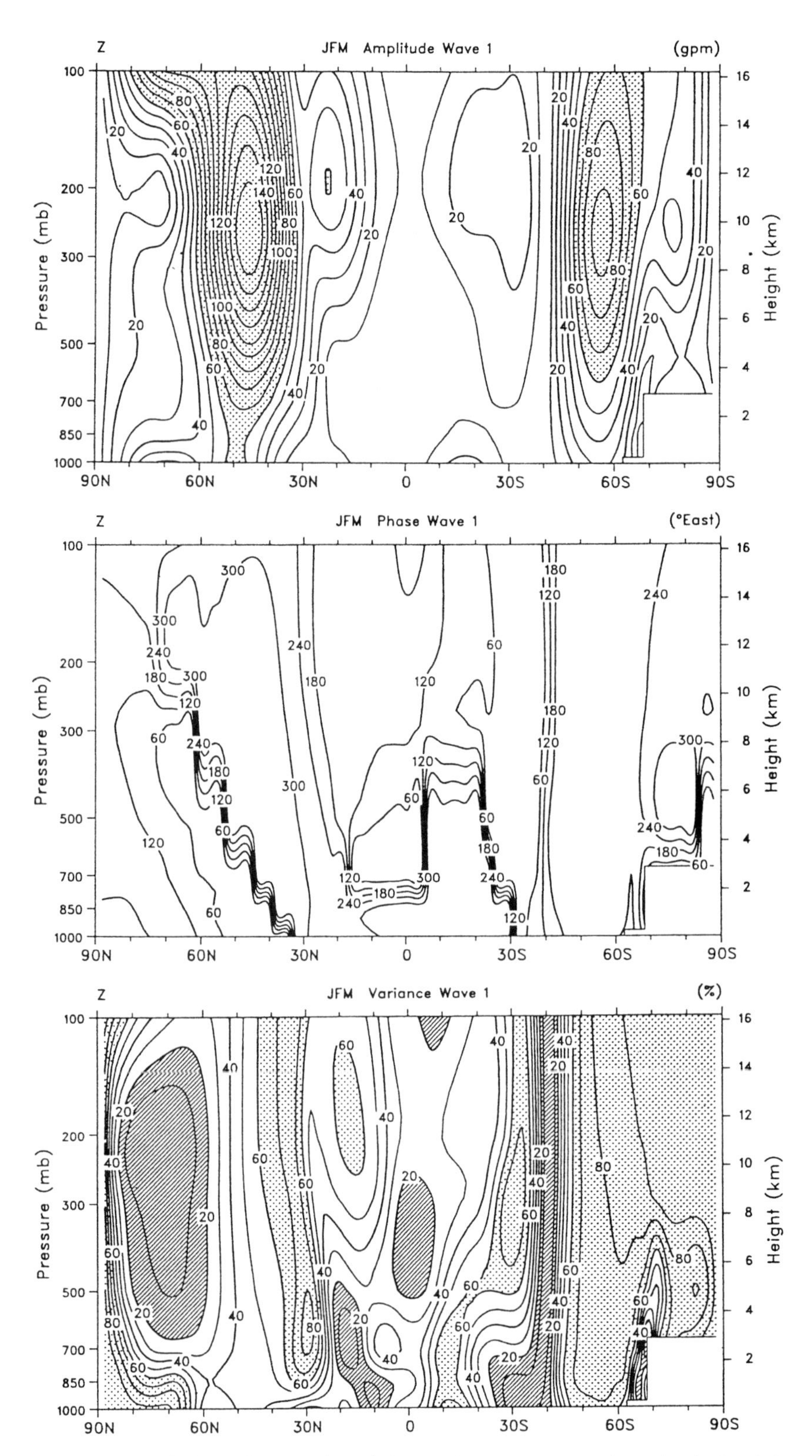

FIG. 1.21. Vertical meridional section of the amplitude (gpm, top), phase (longitude of the ridge in °E, middle), and the percentage of the mean spatial variance explained (%, bottom) of quasi-stationary wave 1 in JFM from ECMWF (1979–93). In the top panel, values greater than 60 gpm are stippled. In the bottom panel, values less than 20% are hatched and values greater than 60% are stippled.

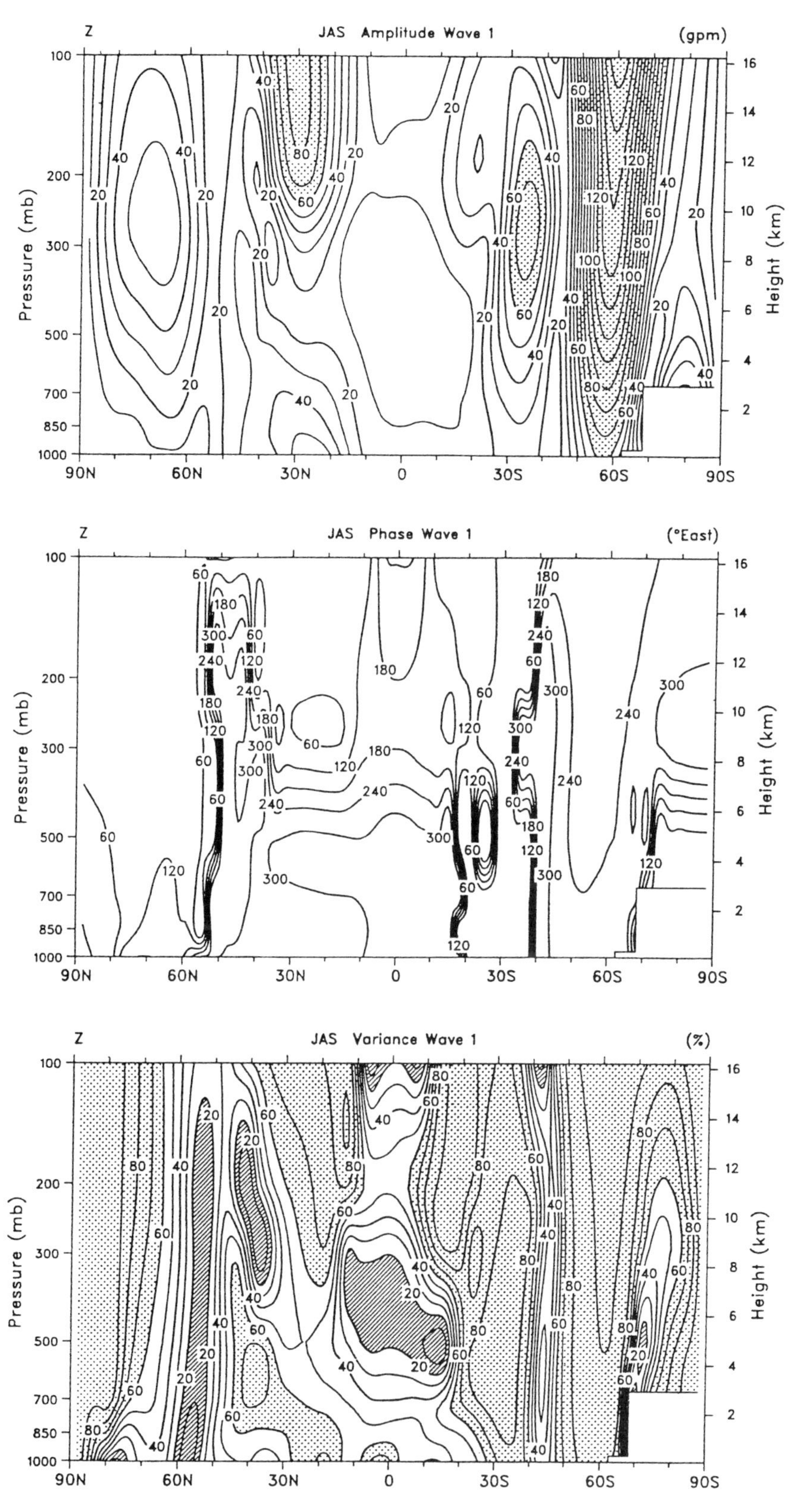

FIG. 1.22. As in Fig. 1.21 but for JAS.

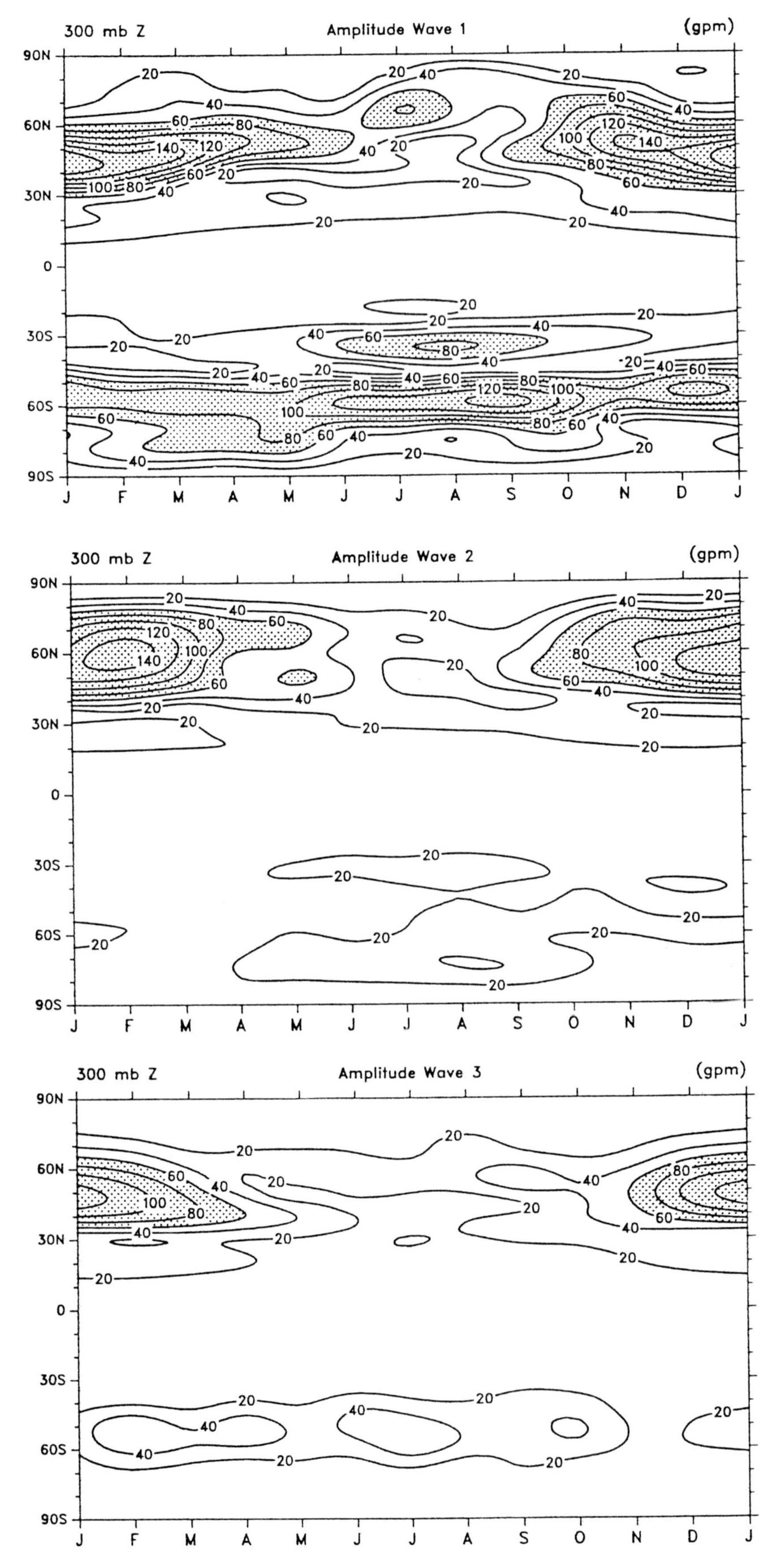

FIG. 1.23. The annual march of the amplitude (gpm) of quasi-stationary waves 1–3 at 300 mb from ECMWF (1979–93). Values greater than 60 gpm are stippled.

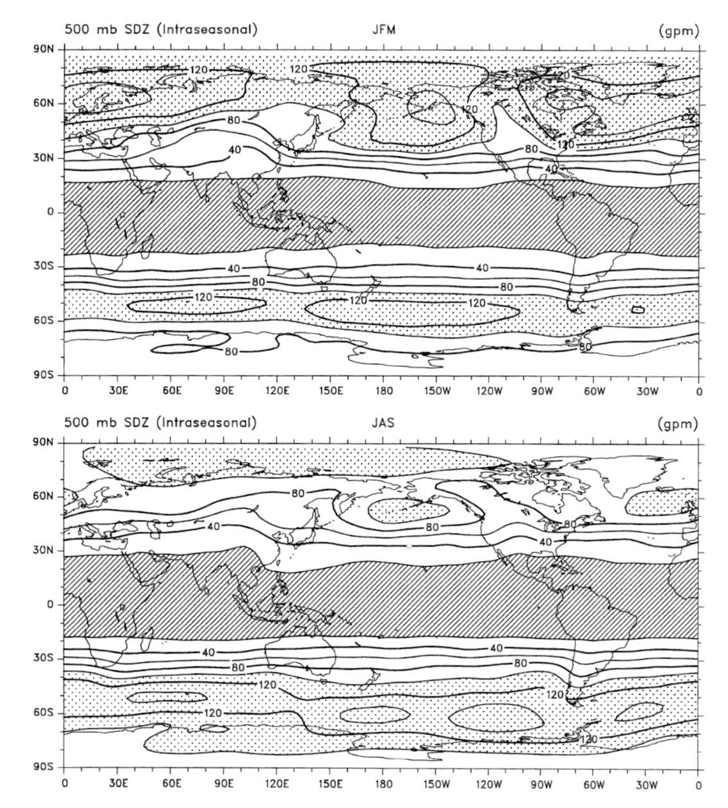

FIG. 1.24. Mean intraseasonal variability of 500-mb geopotential height (gpm) from ECMWF (1979–93) for JFM (top) and JAS (bottom). Values less than 20 gpm are hatched and values greater than 100 gpm are stippled.

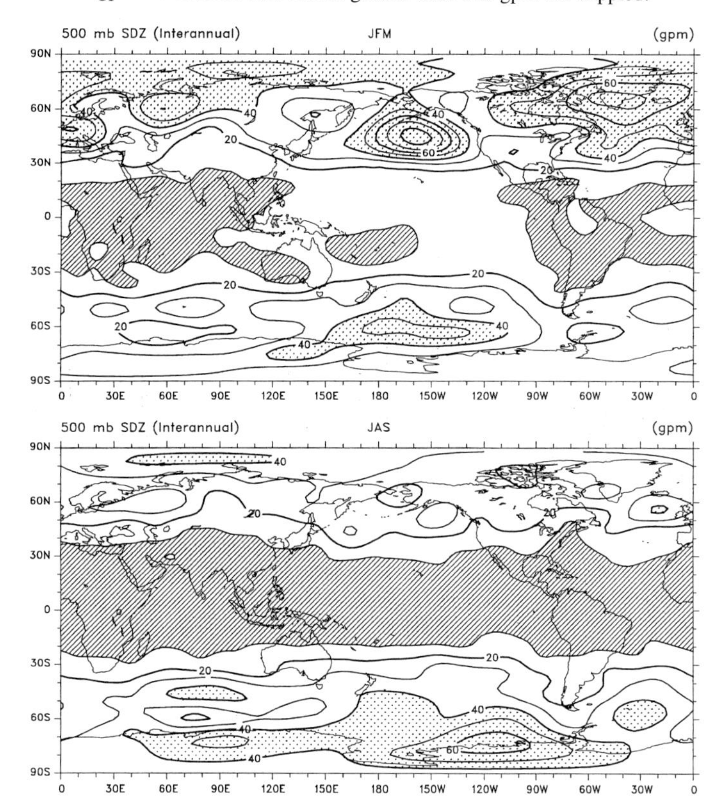

FIG. 1.25. Interannual variability of JFM (top) and JAS (bottom) 500-mb geopotential height (gpm) from ECMWF (1979–93). Values less than 10 gpm are hatched and values greater than 40 gpm are stippled.

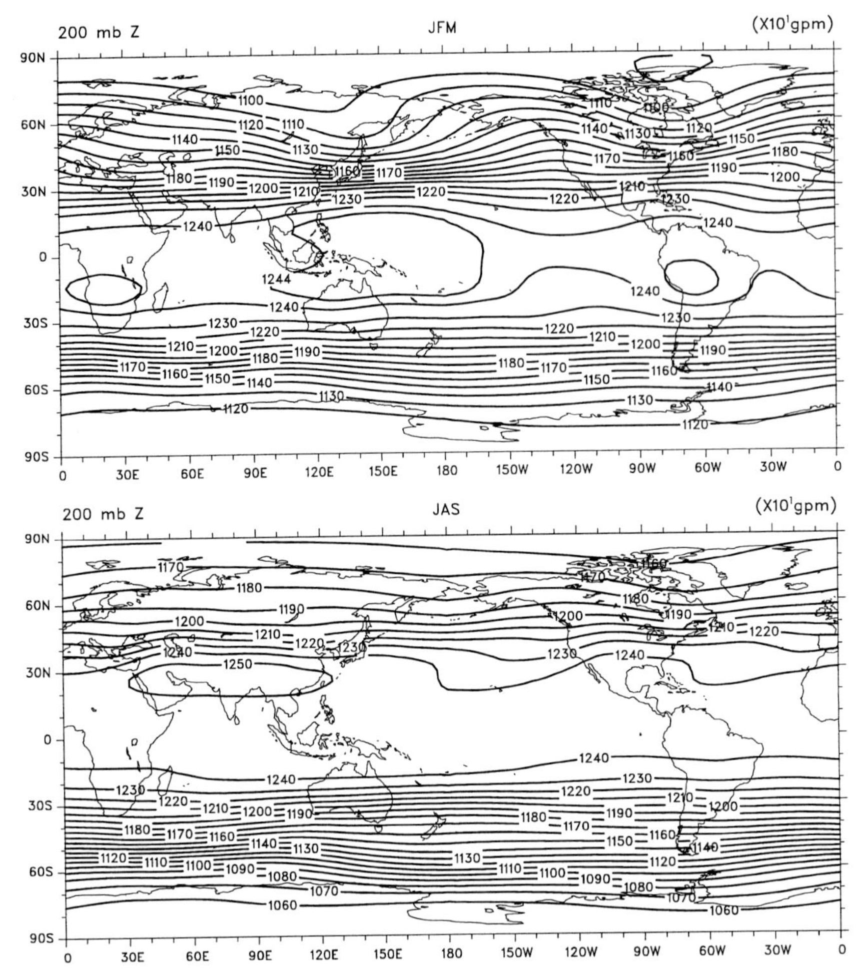

FIG. 1.26. Mean 200-mb geopotential height (10^1gpm) from ECMWF (1979–93) for JFM (top) and JAS (bottom).

Vertical profiles of the zonally averaged zonal wind (Fig. 1.28) further illustrate circulation differences between the hemispheres. At 25° latitude the zonal wind is stronger in winter than summer and the two hemispheric winter profiles are similar. In summer, however, the flow is easterly throughout the troposphere in the NH but is westerly above 700 mb in the SH. The zonal flow is stronger in winter than summer in both hemispheres at 60° latitude as well, but in the SH it is appreciably stronger than in the NH during winter. Moreover, the hemispheric wintertime differences grow larger with height due to the stronger thermal wind in the SH (see Fig. 1.4).

The largest differences in the mean zonal flow between the hemispheres occurs in middle latitudes. Not only are the westerlies stronger in the SH than in the NH during both seasons at 45° latitude (especially during summer, when the differences are nearly 10 m s^{-1} in the middle and upper troposphere), but in the SH the westerlies are several meters per second stronger in summer than winter below 200 mb, in contrast to the NH. This is a result of the stronger temperature gradients across the middle latitudes of the SH during summer, which in turn come about from the smaller annual range of temperature at these latitudes compared to the subtropics and higher latitudes (van Loon 1966; see also Figs. 1.1 and 1.3).

The regional distribution of the zonal wind at 850 mb (Fig. 1.29) illustrates that the strongest westerlies in the SH are in the latitudes near 50°S in both summer and winter. They are weakest in the Pacific Ocean and reach a peak of more than 15 m s^{-1} in the Indian Ocean, reflecting the zonal asymmetry in temperature and pressure described earlier. The area covered by the westerlies expands from summer to winter but the strength in their core changes little. Weak low-level easterlies cover most of the Tropics during both seasons and are associated with the trades, although equatorial westerlies extend from Africa to the western Pacific Ocean in southern summer. The largest interseasonal changes in the 850-mb zonal wind occur in the NH and reflect the interseasonal varia-

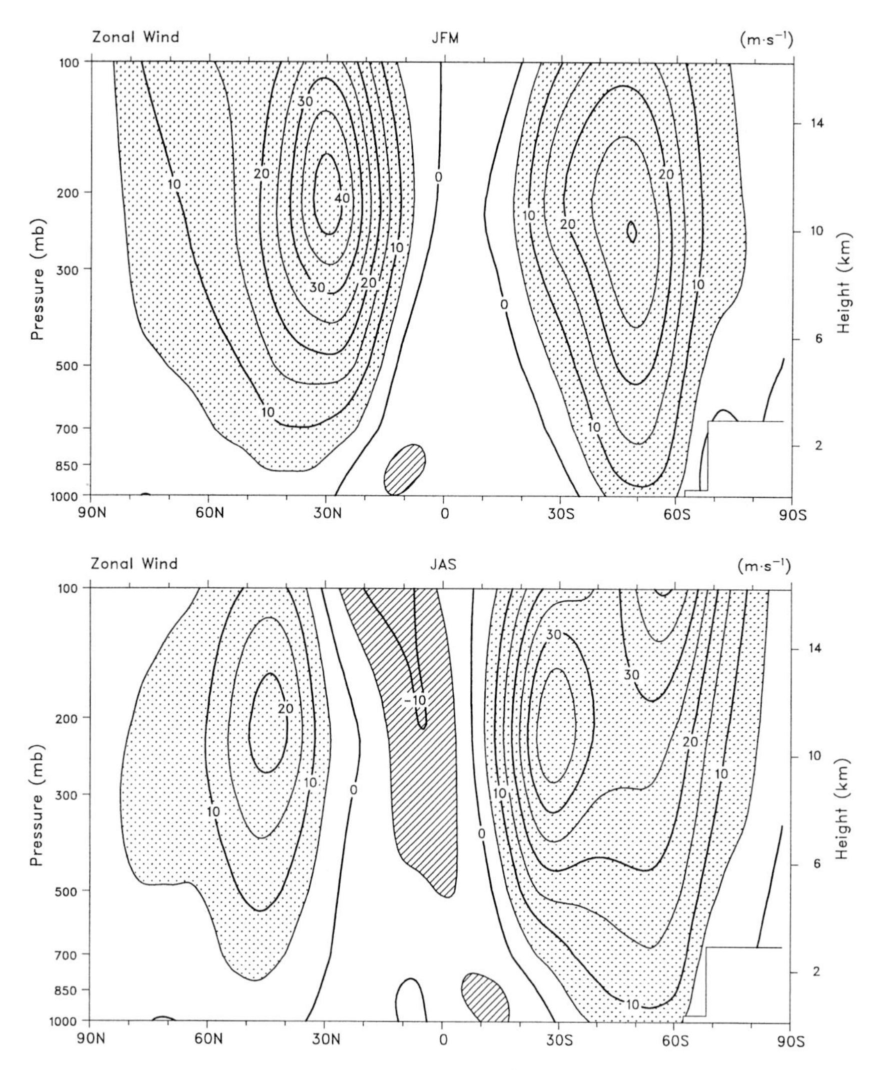

FIG. 1.27. Zonally averaged mean zonal wind (m s^{-1}) for JFM (top) and JAS (bottom) over the years 1979–93 from ECMWF. Values greater than 5 m s^{-1} are stippled and values less than −5 m s^{-1} are hatched.

tions of the semipermanent surface pressure centers (Fig. 1.12).

The tropical easterlies at 200 mb (Fig. 1.30) encircle the near-equatorial latitudes in southern winter and are strongest over the Indian Ocean (the tropical easterly jet), but westerlies extend across the Equator over the Atlantic and central Pacific Oceans during southern summer. The peak in the westerlies is reached between 40° and 50°S in summer with maxima over the Atlantic and Indian Oceans. A striking feature of the winter map in the SH is the maximum near 30°S with values of more than 45 m s^{-1} over Australia and the Pacific. The maximum westerlies during JAS over the Indian Ocean spiral poleward through the middle to higher latitudes over the Pacific Ocean. The result is that the distance between the middle- and high-latitude westerly maxima is greatest over the western Pacific, with the weakest westerlies over the South Island of New Zealand. There are easterlies equatorward of the maximum westerlies in the eastern hemisphere during winter, but westerlies extend farther toward the Equator over the Pacific and Atlantic. In the NH the strongest westerlies occur in winter (Fig. 1.30) and reach 65 m s^{-1} off the Asian coast. During summer, a closed circulation is evident over the southern part of Asia (see Fig. 1.38), accompanied by a northward shift and substantial weakening of the westerly jet stream.

The intraseasonal and interannual variability of the 850- and 200-mb zonal winds are shown in Figs. 1.31 through 1.34, and the interannual variability is supplemented by the SH station data in Table 1.4. The day-to-day variability of the 850-mb winds in the SH is broadly consistent with the day-to-day variability of SLP discussed previously (Fig. 1.14). The standard deviation of daily values (Fig. 1.31) is largest poleward of the midlatitude westerlies (Fig. 1.29) and is

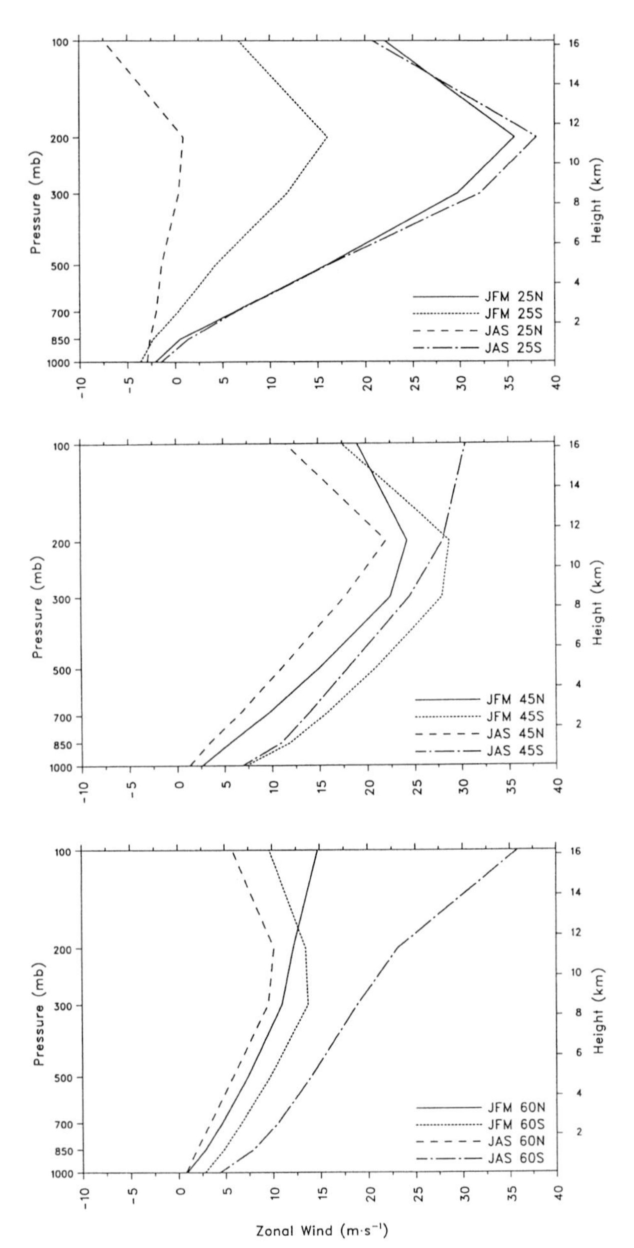

FIG. 1.28. Vertical profiles of the zonally averaged mean zonal wind (m s^{-1}) for JFM and JAS at 25° latitude (top), 45° latitude (middle), and 60° latitude (bottom).

larger in winter than summer. Values decrease toward the Tropics more rapidly in winter than summer, in accordance with the meridional expansion of cyclonic activity over much of the middle latitudes during winter. In the NH the largest values occur in winter and are associated with the storm tracks over the Pacific and Atlantic Oceans.

The intraseasonal variability of the 200-mb zonal winds in the SH reaches a maximum in a circumpolar belt between 30° and 50°S during summer (Fig. 1.33), centered a few degrees equatorward of the belt of strongest westerlies (Fig. 1.30). The day-to-day variability is stronger in winter than summer, and the largest variations occur over Australia within a circumpolar belt centered near 35°S. In the NH, maxima occur in winter near the exit regions of the east Asian and North American jets. The largest interannual variability at 200 mb in southern summer is in the area most affected by the Southern Oscillation (Fig. 1.34).

The interseasonal variation of the pressure and wind in the SH troposphere contains a pronounced semiannual oscillation at middle and high latitudes as discussed in the previous section. There is, in addition, a tropical–subtropical semiannual oscillation (TSSAO) of the zonal wind and temperature in the upper troposphere which, over large parts of the tropical SH, dominates the annual cycle (van Loon and Jenne 1969, 1970; Shea et al. 1995). In equatorial regions the temperature oscillation has maxima in April–May and October–November, whereas it peaks in January–February and July–August in the subtropics (not shown). As a result, both the thermal wind and the wind in the Tropics, especially over the southern half of the Tropics, have a marked semiannual variation. The 200-mb zonal components of the observed wind at eight tropical stations between 97° and 172°E are shown in Fig. 1.35. This is a region where the second harmonic generally exceeds 80% of the mean annual variance (see Fig. 9 of Shea et al. 1995). The TSSAO is clear: at five of the stations the long-term zonal wind is westerly in the transitional months and easterly in the extreme seasons. At the other three stations the wind is easterly throughout the year, but is weakest in the transitional months. It is noteworthy that, of the westerly maxima/easterly minima, the one in May is more marked than the one in November, which is also the case for the SAO in the stratosphere (van Loon et al. 1973b).

Van Loon and Jenne (1969) found that the TSSAO is most pronounced in the areas where the ITCZ crosses the Equator twice a year—in particular, from Africa eastward to the central Pacific Ocean. This region is also where the 40–50-day oscillation is most evident (e.g., Madden and Julian 1994). Where the southeast trades blow across the Equator throughout the year, the ITCZ stays in the NH and the TSSAO is weaker.

1.5. Rotational and irrotational flow

One way to illustrate the overall picture of the flow is in terms of the horizontal streamfunction, which is related to the rotational (i.e., nondivergent) component of the wind. The largest gradient in the 850-mb streamfunction (Fig. 1.36) is observed between 40° and 60°S and is slightly larger in winter than summer. This circumpolar belt corresponds to the latitudes of the strongest westerlies (Fig. 1.29). Wave 1 dominates the seasonal mean zonally asymmetric flow (Fig. 1.37) and is most pronounced during southern winter, with mean northerly flow over a broad

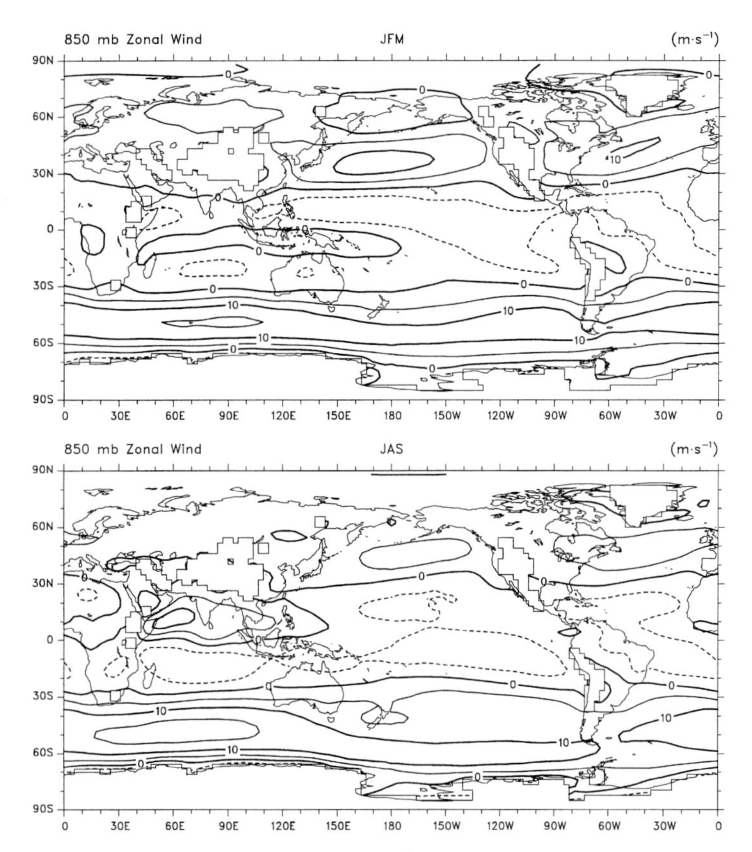

FIG. 1.29. Mean 850-mb zonal wind (m s^{-1}) from ECMWF (1979–93) for JFM (top) and JAS (bottom). Negative values are dashed.

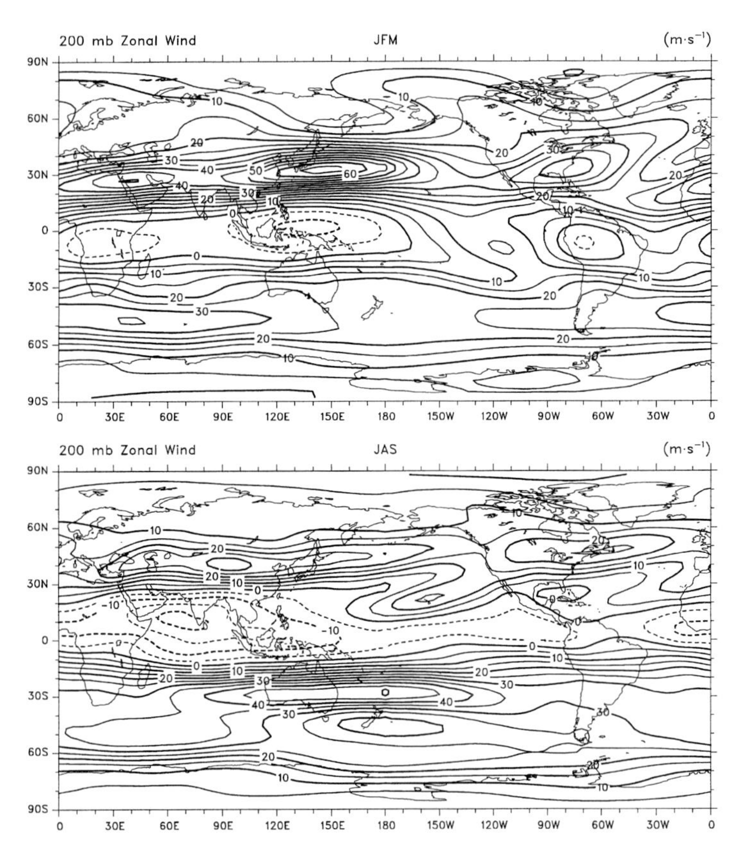

FIG. 1.30. As in Fig. 1.29, but for 200-mb zonal winds.

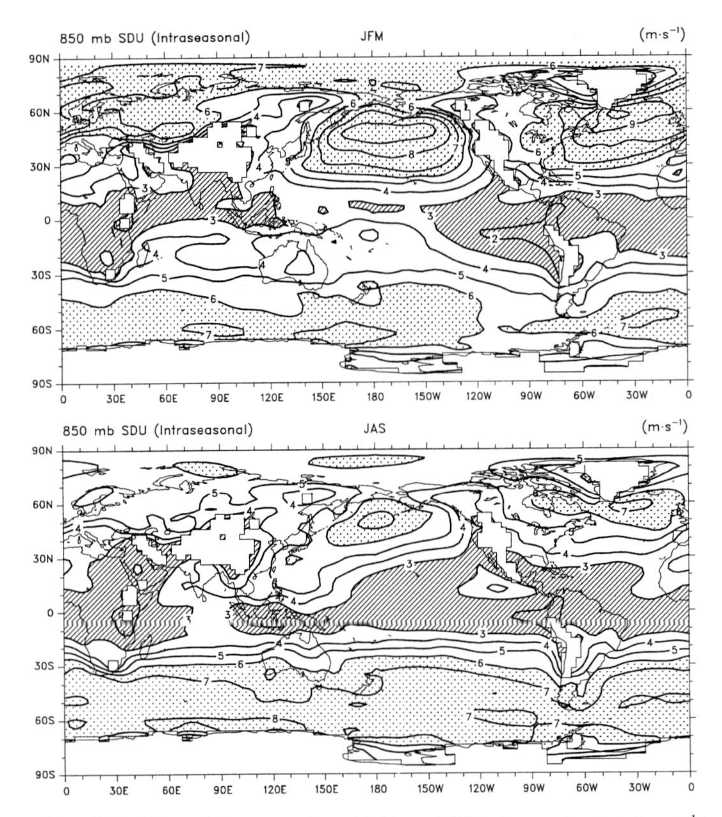

FIG. 1.31. Mean intraseasonal variability of 850-mb zonal wind (m s^{-1}) from ECMWF (1979–93) for JFM (top) and JAS (bottom). Values less than 3 m s^{-1} are hatched and values greater than 6 m s^{-1} are stippled.

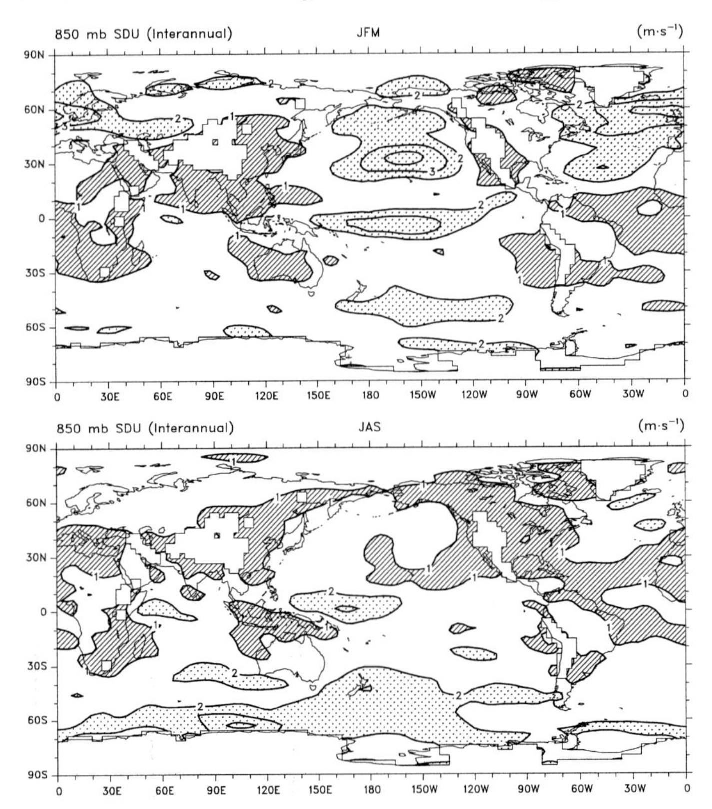

FIG. 1.32. Interannual variability of JFM (top) and JAS (bottom) 850-mb zonal wind (m s^{-1}) from ECMWF (1979–93). Values less than 1 m s^{-1} are hatched and values greater than 2 m s^{-1} are stippled.

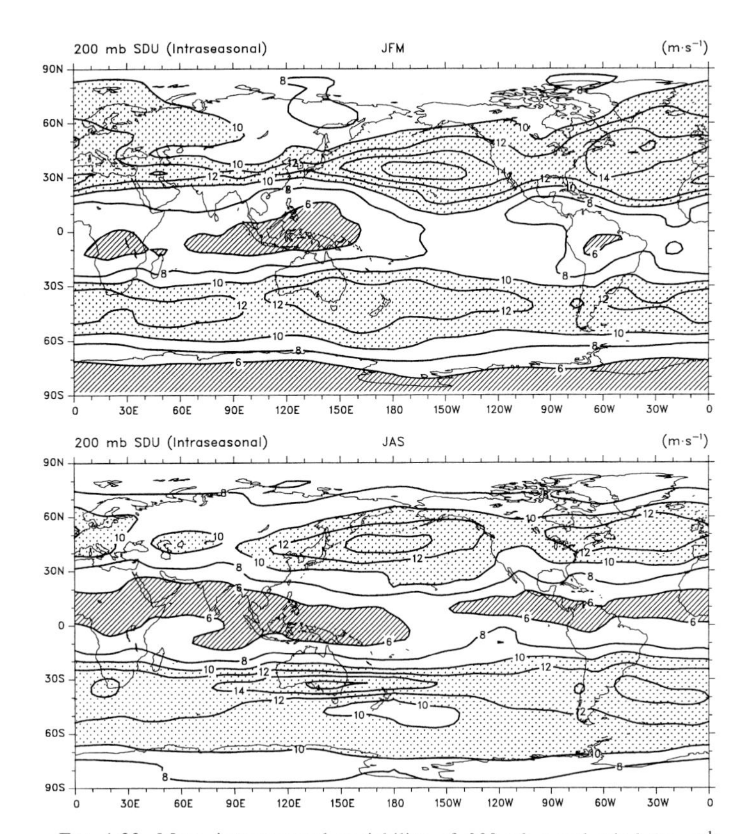

FIG. 1.33. Mean intraseasonal variability of 200-mb zonal wind (m s^{-1}) from ECMWF (1979–93) for JFM (top) and JAS (bottom). Values less than 6 m s^{-1} are hatched and values greater than 10 m s^{-1} are stippled.

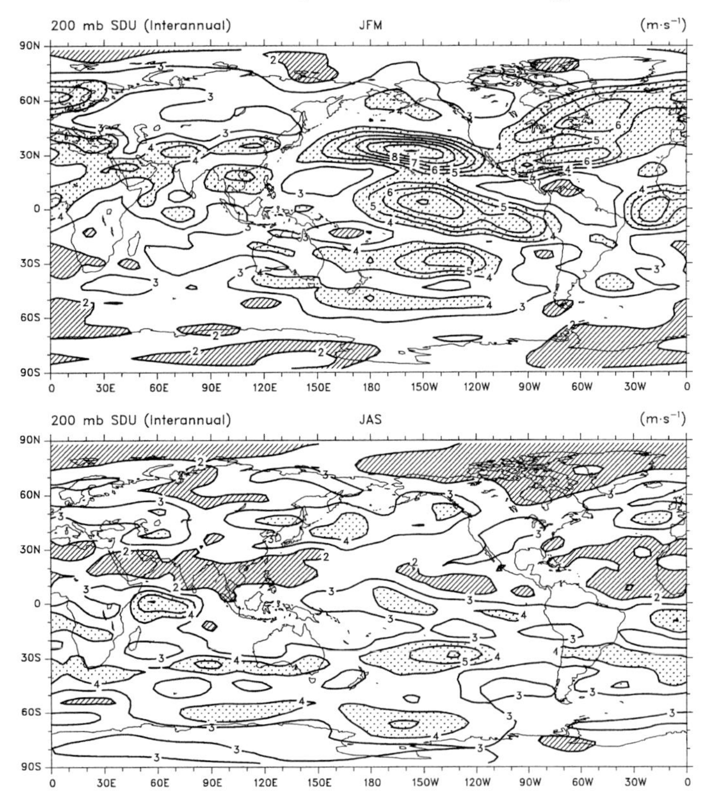

FIG. 1.34. Interannual variability of JFM (top) and JAS (bottom) 200-mb zonal wind (m s^{-1}) from ECMWF (1979–93). Values less than 2 m s^{-1} are hatched and values greater than 4 m s^{-1} are stippled.

TABLE 1.4. As in Table 1.1, but for (a) the mean 850- and (b) 200-mb zonal wind (m s^{-1}).

a. 850 mb: Zonal wind (m s^{-1})

Name (Location)	JFM			JAS		
	Mean	Yrs	σ	Mean	Yrs	σ
Lima (12.0°S, 77.1°W)	0.2 (0.1)	15 (7)	0.3 (0.4)	0.0 (0.0)	14 (4)	1.1 (0.2)
Darwin (12.4°S, 130.9°E)	0.6 (0.0)	27 (13)	2.8 (2.6)	−7.1 (−6.9)	27 (12)	1.0 (1.2)
Tahiti (17.6°S, 149.6°W)	−3.3 (−3.1)	29 (14)	1.2 (0.8)	−2.1 (−2.0)	28 (14)	1.4 (1.3)
Pretoria (25.9°S, 28.2°E)	−1.4 (−1.4)	6 (6)	0.3 (0.3)	−1.0 (−0.9)	16 (12)	0.5 (0.5)
Durban (29.8°S, 31.0°E)	0.6 (0.5)	31 (15)	0.6 (0.5)	2.1 (2.1)	29 (11)	0.8 (0.9)
Perth (31.7°S, 116.1°E)	0.1 (0.5)	25 (12)	1.1 (0.8)	5.6 (5.7)	25 (11)	2.0 (2.0)
Cape Town (34.0°S, 18.6°E)	1.6 (1.5)	33 (15)	0.8 (0.9)	4.8 (5.0)	31 (11)	1.0 (1.4)
Ile N. Amsterdam (37.8°S, 77.6°E)	6.6 (7.5)	18 (9)	1.8 (1.7)	10.8 (13.1)	21 (7)	3.0 (2.3)
Gough Is. (40.4°S, 9.9°W)	10.5 (11.2)	25 (11)	2.6 (2.3)	10.1 (10.8)	26 (13)	2.4 (1.9)
Christchurch (43.5°S, 172.5°E)	4.2 (4.9)	22 (9)	1.2 (1.3)	3.9 (4.5)	21 (8)	1.2 (1.1)
Chatham Is. (44.0°S, 176.5°W)	4.5 (4.5)	5 (5)	1.5 (1.5)	5.9 (5.9)	5 (5)	2.4 (2.4)
Marion Is. (46.9°S, 37.9°E)	13.0 (13.1)	32 (14)	2.2 (2.9)	14.5 (14.9)	28 (11)	1.9 (2.6)
Kerguelen (49.3°S, 70.2°E)	16.9 (17.4)	18 (9)	2.7 (3.1)	17.8 (18.1)	15 (8)	3.8 (4.4)
Campbell (52.6°S, 169.1°E)	14.5 (14.5)	12 (12)	4.3 (4.3)	12.8 (12.8)	10 (10)	4.4 (4.4)
Davis (68.6°S, 77.9°E)	−4.9 (−4.7)	8 (6)	1.7 (1.7)	−6.1 (−6.2)	13 (8)	1.7 (1.5)
S.A.N.A.E. (70.3°S, 2.4°W)	−3.8 (−5.7)	16 (5)	2.4 (1.7)	2.5 (4.8)	16 (4)	2.6 (1.8)

b. 200 mb: Zonal wind (m s^{-1})

Name (Location)	JFM Mean	JFM Yrs	JFM σ	JAS Mean	JAS Yrs	JAS σ
Darwin (12.4°S, 130.9°E)	−4.3 (−3.9)	27 (13)	2.5 (2.5)	2.3 (1.8)	27 (12)	2.2 (2.2)
Tahiti (17.6°S, 149.6°W)	4.4 (4.9)	29 (14)	5.2 (4.3)	20.1 (20.4)	29 (14)	3.2 (3.6)
Pretoria (25.9°S, 28.2°E)	14.8 (15.6)	19 (15)	3.6 (3.4)	30.1 (29.1)	16 (12)	3.7 (3.7)
Durban (29.8°S, 31.0°E)	19.3 (20.3)	28 (15)	3.3 (2.7)	32.7 (33.5)	19 (10)	4.4 (4.4)
Perth (31.7°S, 116.1°E)	25.4 (25.8)	26 (12)	3.6 (3.8)	46.5 (47.5)	25 (11)	5.7 (7.6)
Cape Town (34.0°S, 18.6°E)	28.3 (30.9)	33 (15)	3.9 (3.6)	34.5 (35.5)	29 (11)	6.0 (5.6)
Ile N. Amsterdam (37.8°S, 77.6°E)	18.1 (19.2)	7 (4)	3.5 (3.0)	24.9 (24.8)	10 (2)	7.0 (12.5)
Gough Is. (40.4°S, 9.9°W)	34.7 (36.7)	22 (13)	8.9 (5.2)	33.7 (35.7)	20 (12)	7.2 (5.8)
Christchurch (43.5°S, 172.5°E)	21.3 (23.4)	20 (9)	5.0 (6.1)	18.4 (20.7)	20 (8)	4.0 (5.2)
Chatham Is. (44.0°S, 176.5°W)	18.5 (18.5)	5 (5)	4.7 (4.7)	20.3 (20.3)	5 (5)	5.8 (5.8)
Marion Is. (46.9°S, 37.9°E)	36.0 (37.0)	22 (13)	7.6 (5.1)	35.7 (36.6)	20 (11)	4.7 (3.7)
Kerguelen (49.3°S, 70.2°E)	38.8 (39.0)	18 (10)	6.1 (7.4)	42.8 (44.5)	14 (8)	8.9 (9.4)
Campbell (52.6°S, 169.1°E)	29.2 (29.2)	12 (12)	8.6 (8.6)	26.2 (26.2)	10 (10)	7.8 (7.8)
Davis (68.6°S, 77.9°E)	6.3 (5.2)	6 (4)	3.8 (2.1)	7.2 (8.1)	8 (4)	2.2 (2.2)
S.A.N.A.E. (70.3°S, 2.4°W)	2.9 (3.7)	15 (5)	2.7 (1.5)	5.5 (7.7)	15 (4)	4.4 (2.7)
Amundsen-Scott (90.0°S, 0.0°E)	−0.2 (0.2)	10 (3)	1.2 (1.1)	3.2 (4.4)	12 (4)	2.2 (1.6)

area south of Australia and southerly flow over the Atlantic and western Indian Oceans. Large zonal asymmetry is also evident at the lower latitudes of the SH associated with broad regions of anticyclonic flow around the subtropical highs (Fig. 1.12). The northeast and southeast flows of the trade winds are apparent, and the circulation around the SH high pressure centers results in southerly flow along the west coasts of the continents, which is strongest off of South America and Africa. The anticyclonic regions are situated somewhat farther poleward and are weaker in the SH during summer than winter. A cyclonic deviation from the zonal mean flow extends from the Indian Ocean across northern Australia into the Pacific during southern summer, which is associated on its equatorward side with a band of weak westerlies (Fig. 1.29). A notable change in the flow pattern at 850 mb occurs in association with the Asian monsoon cycle.

In the upper troposphere (Figs. 1.38 and 1.39) the general lower-latitude pattern in each hemisphere is one of anticyclonic (cyclonic) flow above each of the regions of cyclonic (anticyclonic) flow noted in the lower troposphere (Figs. 1.36 and 1.37). A strong anticyclonic couplet straddles the Equator in the west Pacific during JFM and tropical cyclonic centers are prominent over the eastern Pacific and Atlantic Oceans of both hemispheres. The largest zonal asymmetry occurs in the middle latitudes of the NH during winter, when strong southerly flow is found to the west of major ridges over western North America and the eastern Atlantic and strong northerlies exist in the mean to the west of troughs over Japan and Hudson Bay. During JAS the main features include anticyclonic centers in the eastern hemisphere and subtropical troughs over the Pacific and Atlantic Oceans. The zonal asymmetry is markedly smaller in the middle latitudes of the SH during both seasons than at lower levels, although wave 1 is still evident, especially during winter when the meridional gradient of the streamfunction is largest. A pronounced split occurs over New Zealand, so that the westerlies there are at a minimum (Fig. 1.30), the flow has a strong southerly component from the Indian Ocean across south-

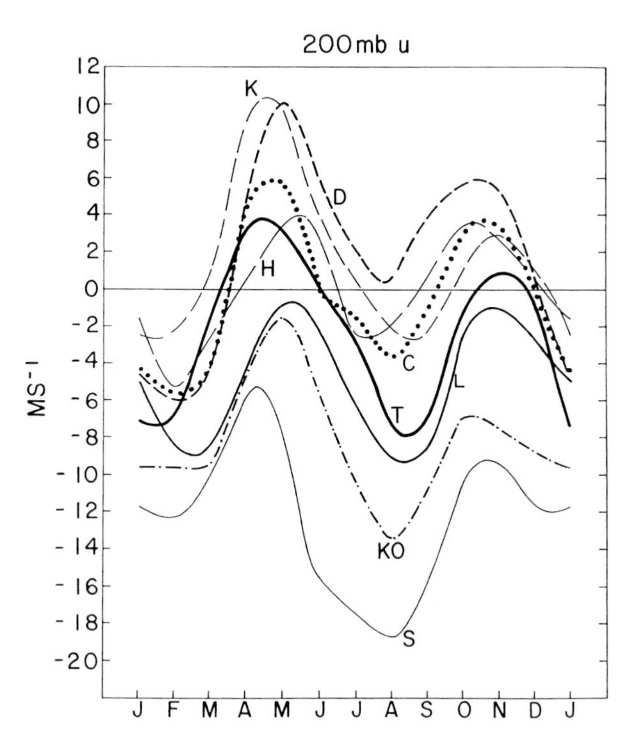

FIG. 1.35. The long-term zonal component of the wind (m s^{-1}) at 200 mb at: K, Kwajalein (6.1°N, 167.4°E); D, Darwin (12.4°S, 130.9°E); H, Honiara (9.2°S, 159.8°E); C, Cocos Island (12.0°S, 97.0°E); L, Lae (6.4°S, 147.0°E); T, Tarawa (1.2°N, 172.6°E); KO, Koror (7.2°N, 134.3°E); S, Singapore (1.2°N, 103.6°E). (Adapted from Shea et al. 1995.)

ern Australia and over the southwest Atlantic, and a strong northerly component is present in the mean 200-mb wind between 50° and 70°S south of Australia.

The divergent (i.e., irrotational) component of the wind plays a much more prominent role in the Tropics than in higher latitudes. The analysis of observed divergence is sensitive to the initialization technique, the numerical prediction model used for the data assimilation, and, in particular, the parameterizations of convection used in the assimilating model. Trenberth (1992) has documented many of the problems with the analyzed divergence fields from ECMWF. Yet it is generally true that, while the magnitude of the analyzed divergent wind (or vertical motion) has shown large changes over time, the general patterns of divergence are more robust. With this in mind, the ECMWF climatology is derived from the relatively short periods of JFM 1987–1989 and JAS 1986–1988. These periods come after a major operational change in May 1986 and before a major change in May 1989.

Seasonal plots of the zonal mean meridional streamfunction (Fig. 1.40) reveal a strong intensification and predominance of the winter Hadley cell in each hemisphere with mass fluxes of over 180×10^9 kg s^{-1}. In southern winter, the Ferrel cell in the SH is of comparable strength to the northern Hadley cell. The southern Hadley cell penetrates across the Equator and reinforces the upward motions characteristic of the mean ITCZ near 10°N. In northern winter, the maximum upward motions are centered approximately on the Equator.

In the lower troposphere during southern summer, the strongest convergence is located over New Guinea (Fig. 1.41) and divergent flow out of the subtropical anticyclones is evident in both hemispheres. At 200 mb (Fig. 1.42) a broad band of tropical outflow extends zonally from the Indian Ocean to well east of the date line with the strongest flow from the SH into the NH. Strong upper-level divergence also occurs over Africa and South America and convergence is notable over the eastern Pacific and Atlantic Oceans. During southern winter the strongest divergent flow in the upper troposphere is from the NH into the SH and emanates from a broad region of divergence centered near 20°N that stretches from India into the Pacific. Cross-equatorial flow is also present over the eastern Pacific and Atlantic Oceans (Fig. 1.42).

1.6. Clouds, moisture, and precipitation

The total cloud cover estimated from ISCCP as a function of latitude[1] (Fig. 1.43) reveals that the largest amounts of mean total cloudiness are found at subpolar latitudes from 50° to 60°S, where the mean is more than 85% during both southern summer and winter. Higher mean total cloudiness in the Tropics and south of 45°S occurs in southern summer, while total cloudiness is a few percent larger between 25° and 40°S in winter. A maximum is evident just north of the Equator in northern summer close to the mean latitude of the ITCZ, and subtropical minima are associated with Hadley circulation subsidence in each hemisphere (see Fig. 1.40). Over the middle latitudes of the NH, maximum total cloudiness reaches nearly 70% near 38°N in northern winter and 55°N in summer. The total cloudiness equatorward of 70° latitude is greater in the SH than in the NH during both seasons, 70.1% compared to 57.4% during JFM and 65.4% compared to 61.6% during JAS. The profile in Fig. 1.43 agrees well with the earlier SH estimates of van Loon (1972, Fig. 6.3), although ISCCP cloud amounts throughout the SH subtropics and middle latitudes are higher by 10%–15%. Total cloudiness from ISCCP is also 5%–10% higher than the ground-based climatology of Warren et al. (1986, 1988) over the southern oceans (see Table 3 and Fig. 6 of Rossow et al. 1993b).

The distribution of mean total cloudiness (Fig. 1.44) reveals a wide circumpolar belt of cloud cover exceeding 80% over the subpolar latitudes of the SH that is slightly more extensive in southern summer than winter. Local maxima of more than 90% cover the south

[1] The ISCCP cloud-detection algorithms are much less reliable at polar latitudes, so estimates of total cloud poleward of 70° latitude are not presented.

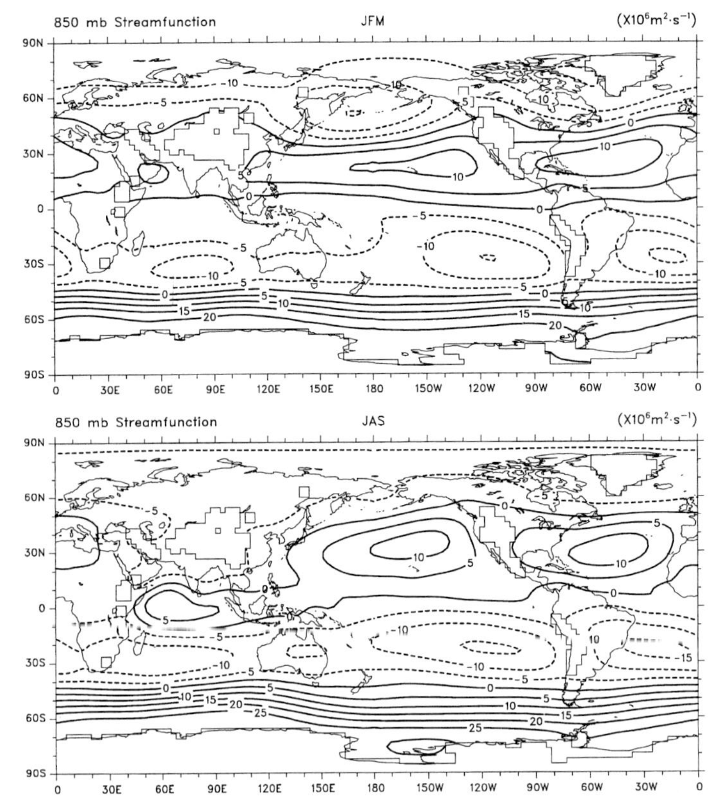

FIG. 1.36. Mean 850-mb streamfunction (10^6 m^2 s^{-1}) from ECMWF (1979–93) for JFM (top) and JAS (bottom). Negative values are dashed.

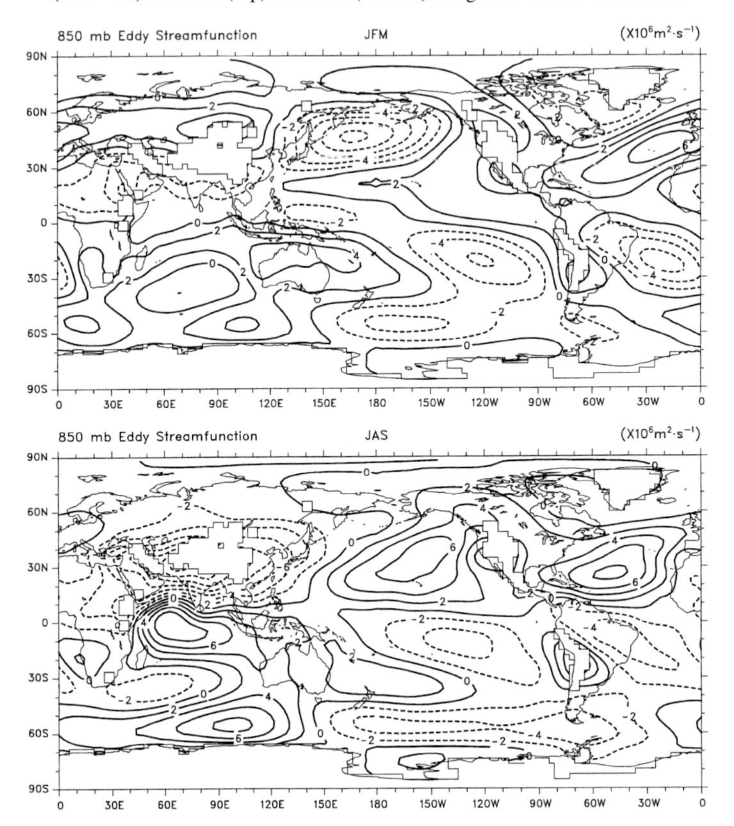

FIG. 1.37. Departures from zonally averaged mean 850-mb streamfunction (10^6 m^2 s^{-1}) from ECMWF (1979–93) for JFM (top) and JAS (bottom). Negative values are dashed.

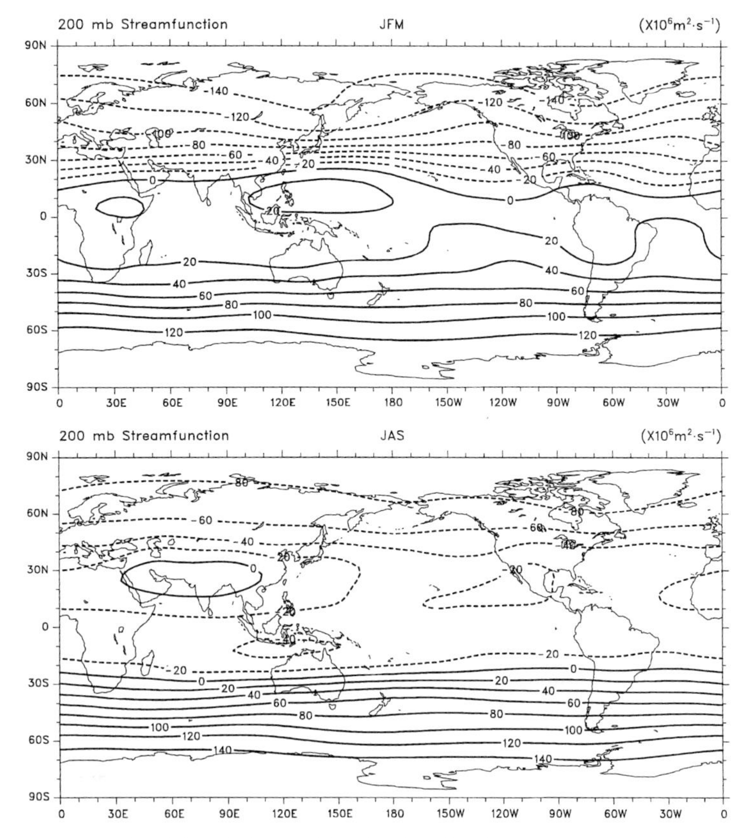

FIG. 1.38. As in Fig. 1.36, but for 200-mb streamfunction.

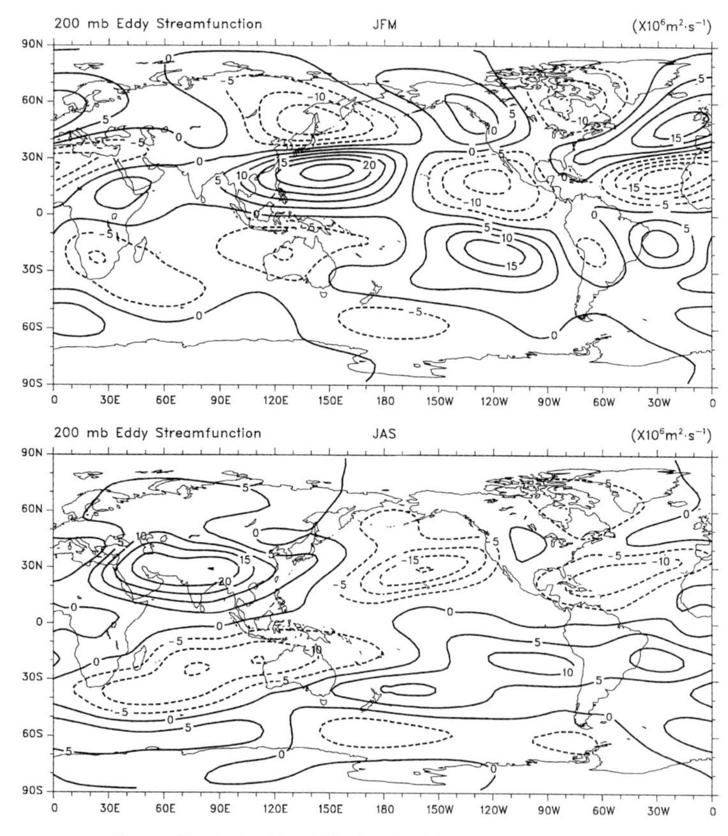

FIG. 1.39. As in Fig. 1.37, but for 200-mb streamfunction.

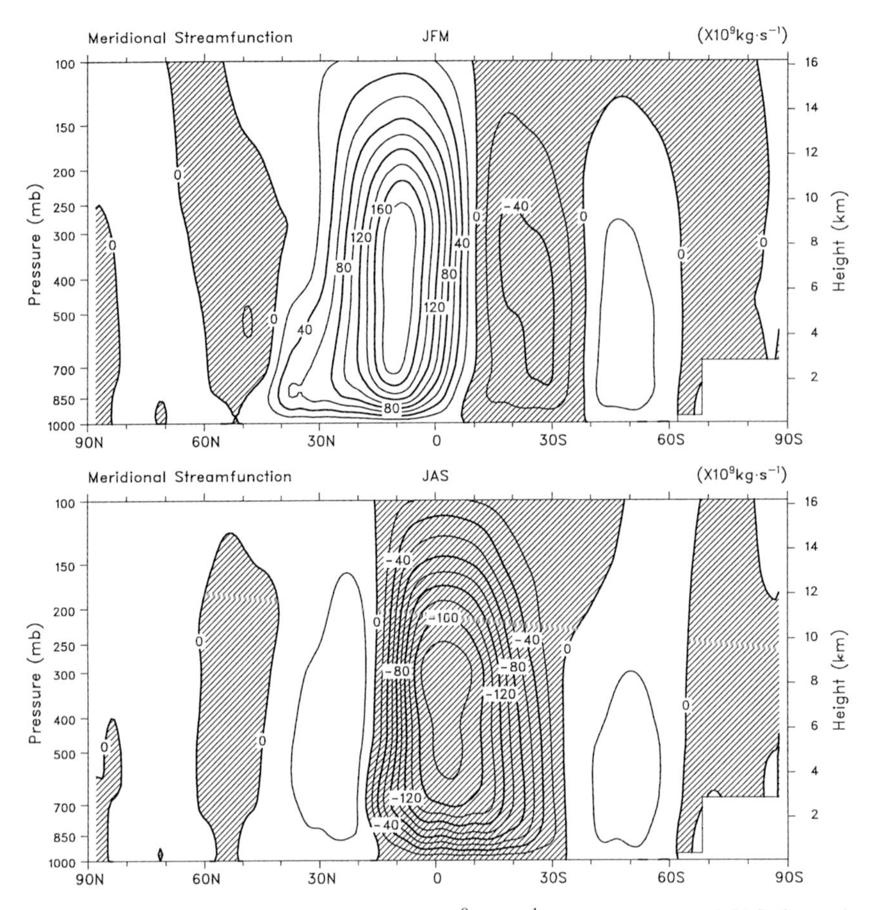

FIG. 1.40. Mean meridional streamfunction (10^9 kg s^{-1}) for JFM (top) and JAS (bottom). Negative values are hatched. ECMWF data from 1987–89 (1986–88) were used to construct the JFM (JAS) climatology.

Atlantic during both seasons and extend over parts of the Indian and Pacific Oceans during summer, in agreement with observations from whaling ships for more than 25 years (Vowinckel and van Loon 1957). The zonal eccentricity south of 45°S that exists in temperature pressure, and other elements is also evident in the total cloudiness, the maximum of which lies farther south over the Pacific than over the Indian and Atlantic Oceans. There are few clouds over the interior of Australia in both seasons, and total cloud amounts of less than 20% cover a large portion of southern Africa during winter. The consistent cloudiness over the Amazon and Congo basins in summer is almost equaled over Indonesia and the tropical coasts of Australia, from where it curves southeastward as a narrow band (known as the South Pacific convergence zone, SPCZ) that merges with the high cloudiness at middle and high latitudes. A similar connection between high and low latitudes appears in summer over the western Atlantic (associated with the South Atlantic convergence zone, SACZ), but not over the Indian Ocean.

Overall, the cloudiness in both hemispheres is higher over the oceans than over the continents (see also Table 3 of Rossow et al. 1993). The percent cloud cover is high throughout the year in the equatorial belt, as maximum cloudiness follows the interseasonal movement of the ITCZ. Low-percentage cloud amounts are notable over the subtropical belts of high pressure, which shift poleward from winter to summer in each hemisphere. Major seasonal changes are evident in the monsoon regions, and the lowest values of total cloud cover are observed over the continental deserts. Percent cloud cover is very high in the coastal stratus regions of both hemispheres.

Plots of seasonal-mean relative humidity corroborate many of the features on the maps of total cloudiness. Zonally averaged profiles of relative humidity (Fig. 1.45) reveal low values in the subtropical middle troposphere in the subsidence regions of the Hadley circulation with minima less than 30% in the winter hemispheres. The tropical troposphere is relatively moist, and midtroposphere relative humidities increase from the subtropics toward the poles. In the lower

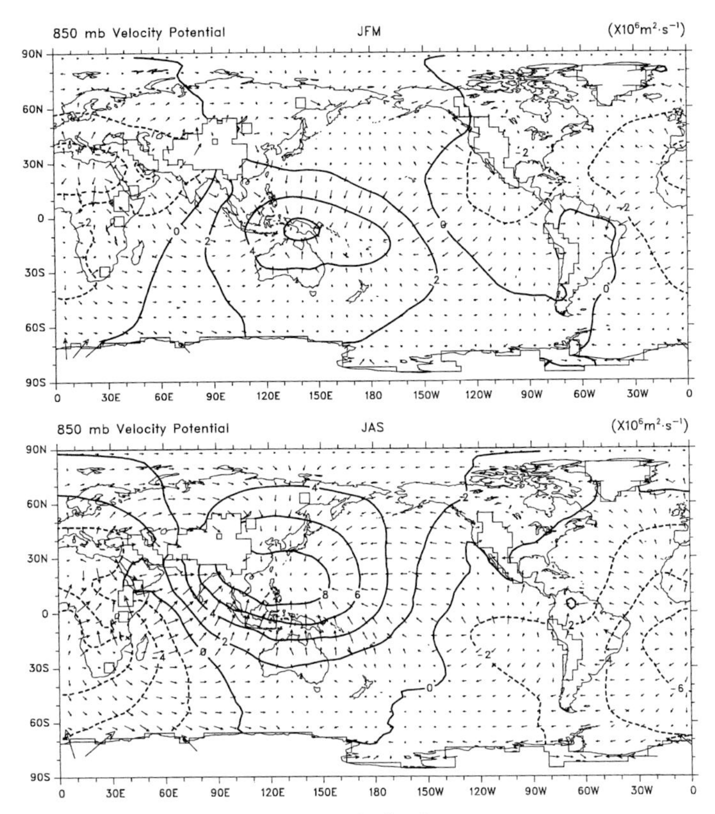

FIG. 1.41. Mean velocity potential (10^6 m^2 s^{-1}) and vector divergent wind at 850 mb for JFM (top) and JAS (bottom). Negative values of the velocity potential are dashed, and the largest vector corresponds to 8 m s^{-1}. ECMWF data from 1987–89 (1986–88) were used to construct the JFM (JAS) climatology.

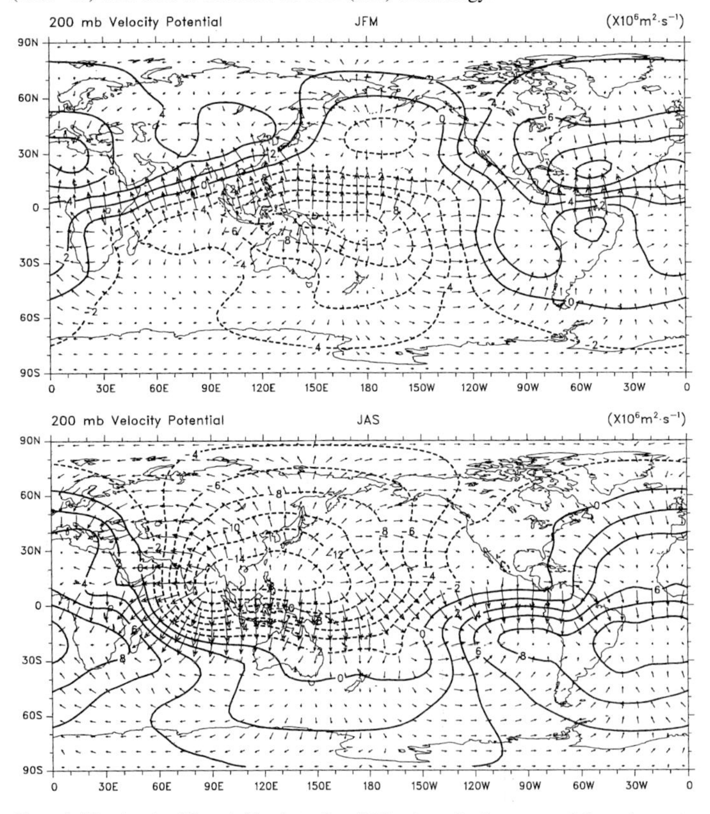

FIG. 1.42. As in Fig. 1.41, but for 200-mb velocity potential and vector divergent wind.

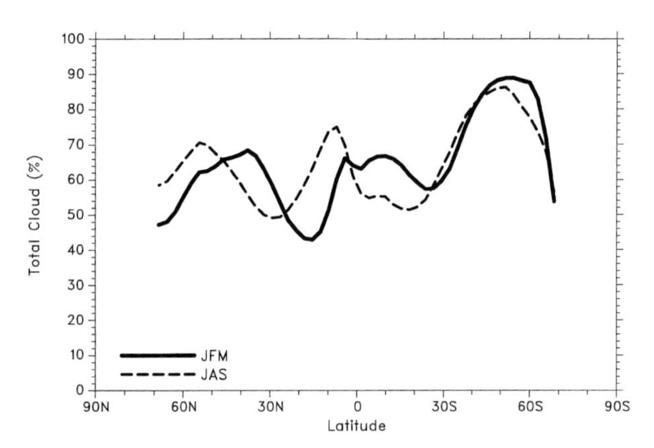

FIG. 1.43. Zonally averaged mean total cloudiness (%) from ISCCP for JFM (solid) and JAS (dash). Data over the period 1984–91 (1983–90) were used in the JFM (JAS) climatology.

troposphere the SH generally has higher relative humidities latitude by latitude than in the NH during the same season.

The relative humidity at 850 mb (Fig. 1.46) has a circumpolar minimum between 30° and 45°S in summer with the lowest humidities found over the continents and along the west coasts of South America and Africa stretching toward lower latitudes. Values of relative humidity exceed 80% at 850 mb at subpolar latitudes and in the Tropics following the interseasonal migration of the ITCZ. In southern winter the middle-latitude band of lower humidities is 5°–10° latitude farther equatorward than in summer, with lowest values still observed over the continents. At 500 mb (Fig. 1.47) the subtropical dry zones and their seasonal shifts are well defined, as are the maximum humidities associated with the ITCZ. Relative maxima extend from the Tropics to middle latitudes during southern summer over the Pacific and Atlantic, in accordance with the belts of maximum cloudiness (Fig. 1.44).

The total moisture content of the atmosphere, as given by the precipitable water (Fig. 1.48), is greater at all latitudes in summer than winter, and the largest tropical to high-latitude meridional gradients occur in the winter hemisphere. The area-averaged values from the ECMWF estimates are 22.0 kg m^{-2} for both winter hemispheres, but the summer values range from 35.3 kg m^{-2} in the NH to 29.7 kg m^{-2} in the SH, reflecting the continentality of the NH. The largest seasonal changes occur in the subtropics and over the continents, especially in the monsoon regions and over the extensive NH landmasses. Maximum values of more than 50 kg m^{-2} reside throughout the year over Indonesia and the western tropical Pacific. Comparisons between the global estimates of precipitable water from ECMWF and the global ocean estimates from the special sensor microwave imager reveal that the two datasets agree over most ocean areas, but that large differences are evident in the dry air masses over the eastern tropical and subtropical oceans, particularly in the SH (Liu et al. 1992; Trenberth and Guillemot 1994). In these regions, it appears as though the ECMWF data overestimate the total moisture content of the atmosphere.

Reasons for uncertainty in our knowledge of both the moisture content of the atmosphere and precipitation stem from the nature of the quantities and the lack of observations over vast uninhabited areas such as the oceans. One proxy of convective activity throughout the Tropics is outgoing longwave radiation (OLR). The mean OLR fields for JFM and JAS are shown in Fig. 1.49. The lowest values of tropical OLR, representing the temperature at the cloud tops, lie between the Equator and 20°S during southern summer, except for the SPCZ and SACZ, which extend into middle latitudes. The dry subtropical regions of subsidence are indicated by high values of OLR and extend into lower latitudes over the eastern Pacific and the Atlantic. In southern winter most of the tropical convective activity is located north of the Equator and is most prominent over southeast Asia.

Estimates of seasonal-mean precipitation rates (Figs. 1.50 and 1.51) come from the data of Xie and Arkin (1996), and values for selected SH stations are given in Table 1.5. The global gridded fields of monthly precipitation were constructed by combining estimates from gauge observations, three types of satellite estimates, and predictions from the ECMWF operational forecast model. The result is a product that well represents with reasonable amplitude the large-scale spatial patterns, both over the Tropics and the extratropics as evidenced by its agreement with the independently derived climatologies of Dorman and Bourke (1978), Legates and Wilmott (1990), and Spencer (1993). As Xie and Arkin (1996) caution, however, the actual quality of the merged product depends highly on the uncertain error structures of the individual data sources.

The large rainfall rates in the equatorial latitudes are associated with strong convection in the ITCZ, which migrates north and south in phase with the insolation and is the factor that determines the marginal climates bordering the arid subtropical regions. The low precipitation rates over the subtropical regions with minima over the eastern parts of the oceans are well defined in both hemispheres and are associated with the large semipermanent anticyclones (Fig. 1.12). Rainfall rates are less than 1 mm day^{-1} over large regions of the SH tropical and subtropical eastern Pacific and Atlantic during both seasons. During summer in the SH, large rainfall rates associated with convection in the SPCZ and SACZ extend through the subtropics toward higher latitudes (Fig. 1.51), creating a broad maximum evident in the zonal averages (Fig. 1.50) that tapers to a minimum near 30°S. Rainfall rates are much smaller during southern winter over these latitudes, when a clear and sharp maximum associated with the ITCZ occurs near 10°N, and the subtropical minimum in the SH is near 20°S. Southern winter is the dry season

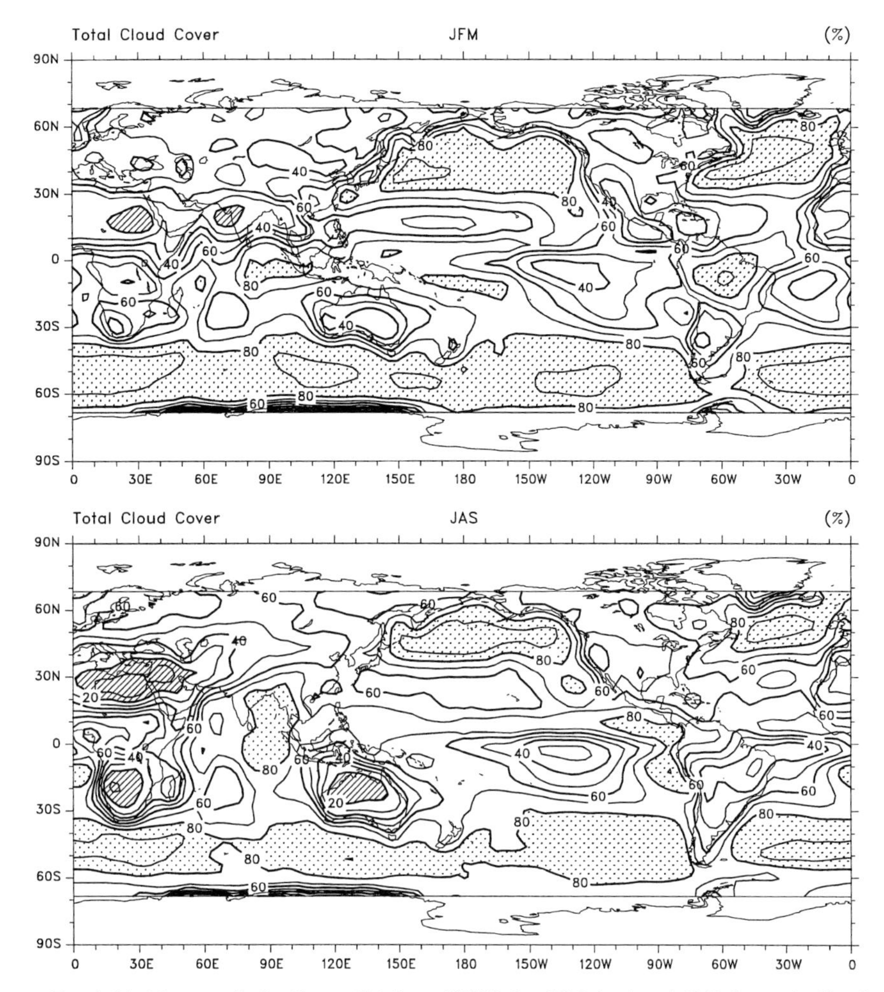

FIG. 1.44. Mean total cloudiness (%) from ISCCP for JFM (top) and JAS (bottom). Cloud coverage greater than 80% is indicated by stippling, and coverage less than 20% by hatching. Data over the period 1984–91 (1983–90) were used to construct the JFM (JAS) climatology.

over the subtropical continents, when rainfall rates less than 1 mm day^{-1} cover large areas (Fig. 1.51). Secondary maxima in precipitation occur over midlatitudes where the polar fronts and the associated disturbances predominate. The more zonal orientation in the SH than in the NH broadly reflects the different patterns of cyclone frequency. Over the polar regions the moisture content of the atmosphere is very low (e.g., Fig. 1.48) and the amounts of precipitation are small during both seasons.

1.7. Summary

Since the publication of *Meteorology of the Southern Hemisphere* by van Loon et al. (1972), a great deal of new data has led to a better understanding and description of the mean state of the global atmosphere and the dominant processes that maintain the general circulation. The global analyses produced specifically for the Global Weather Experiment in 1979 and the ensuing global operational analyses, in addition to the many satellite-derived products over the past several decades, are now mainstays of atmospheric research. The more recent data also confirm, however, the accuracy of the analyses of the different climate variables presented in the four-volume atlas, *Climate of the Upper Air: Southern Hemisphere* (Taljaard et al. 1969; van Loon et al. 1971; Jenne et al. 1971; Crutcher et al. 1971), upon which the original monograph was based.

The second volume (van Loon et al. 1971), for instance, presents a comprehensive account of the wind over the SH. Because there was much less information about wind than about pressure heights, van Loon et al. showed maps of the mean and variability of the geostrophic winds at different pressure levels. We have combined these data with geostrophic wind analyses derived from the NH height data of Crutcher and Meserve (1970) and compared the result-

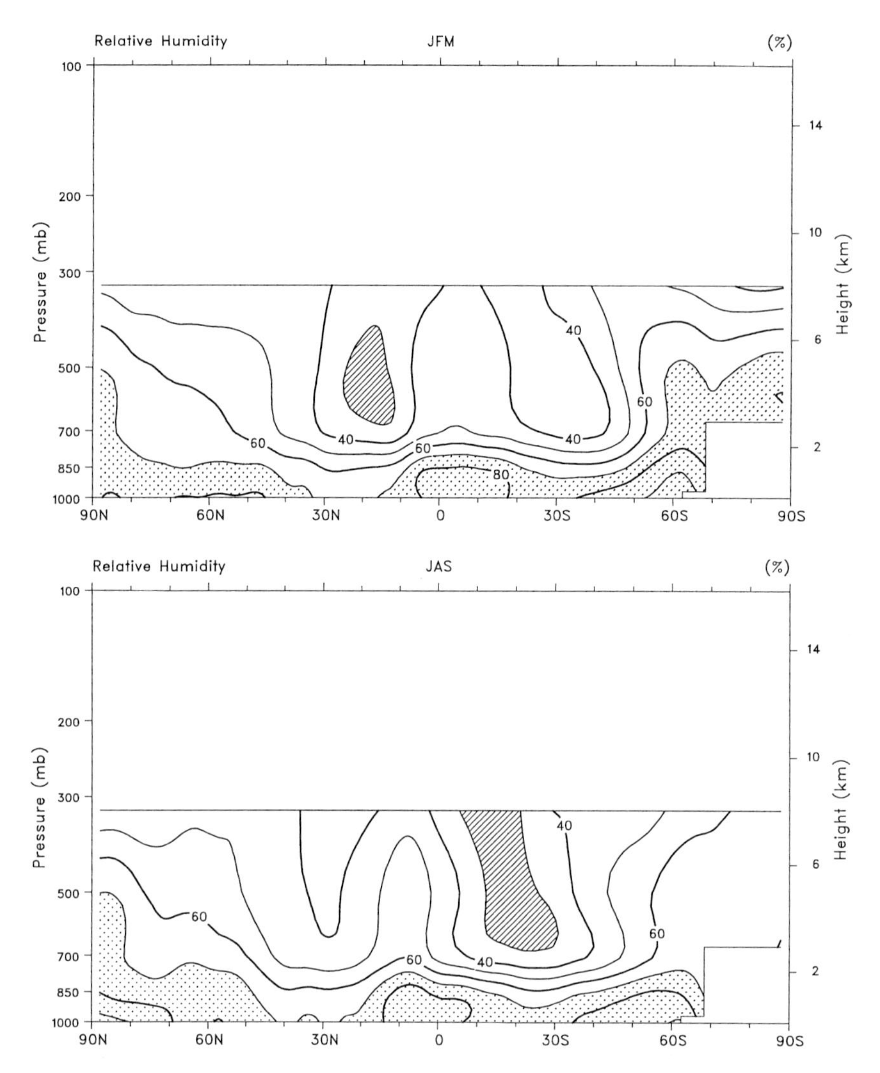

FIG. 1.45. Zonally averaged mean relative humidity (%) for JFM (top) and JAS (bottom) from ECMWF. Values greater than 70% are stippled and values less than 30% are hatched. Data from 1987–89 (1986–88) were used to construct the JFM (JAS) climatology.

ing global geostrophic winds to zonal winds from the 15-yr ECMWF climatology (section 1.4). Zonal wind values equatorward of 20° latitude are excluded from the comparison since, at these latitudes, the geostrophic wind is a poor approximation of the real wind.

The agreement between the two climatologies for both southern summer (Fig. 1.52) and winter (Fig. 1.53) is remarkable, with differences generally less than 5 m s^{-1} at 200 mb. Over the NH, larger differences are found over the Pacific and Atlantic Oceans during winter (Fig. 1.52). These are consistent, however, with observed pronounced changes in the atmospheric circulation over the northern oceans since the mid-1970s. In particular, the Aleutian low pressure system was deeper and shifted eastward throughout the late 1970s and 1980s compared to preceding decades, resulting in stronger westerlies over the subtropical Pacific (e.g., Trenberth and Hurrell 1994). Similarly, over the North Atlantic, the Icelandic low and the Azores high pressure systems have been intensified during winter since the late 1970s, resulting in a persistent positive phase of the North Atlantic oscillation with increased westerlies between 50° and 60°N and increased easterlies over the subtropical Atlantic (Hurrell 1995). Equally impressive agreement between the atlas and ECMWF climatologies is evident in the zonal winds at other pressure levels as well, so differences in the vertical and meridional distributions of the zonally averaged mean JFM and JAS zonal winds are less than 2 m s^{-1} poleward of 30° latitude in both hemispheres (not shown). Similarly, the mean distributions of temperature and pressure from ECMWF (sections 1.2 and 1.3) compare well to the atlas data of Taljaard et al. (1969) presented in the original monograph (van Loon 1972), and the largest differences (not shown) are consistent with observed decadal

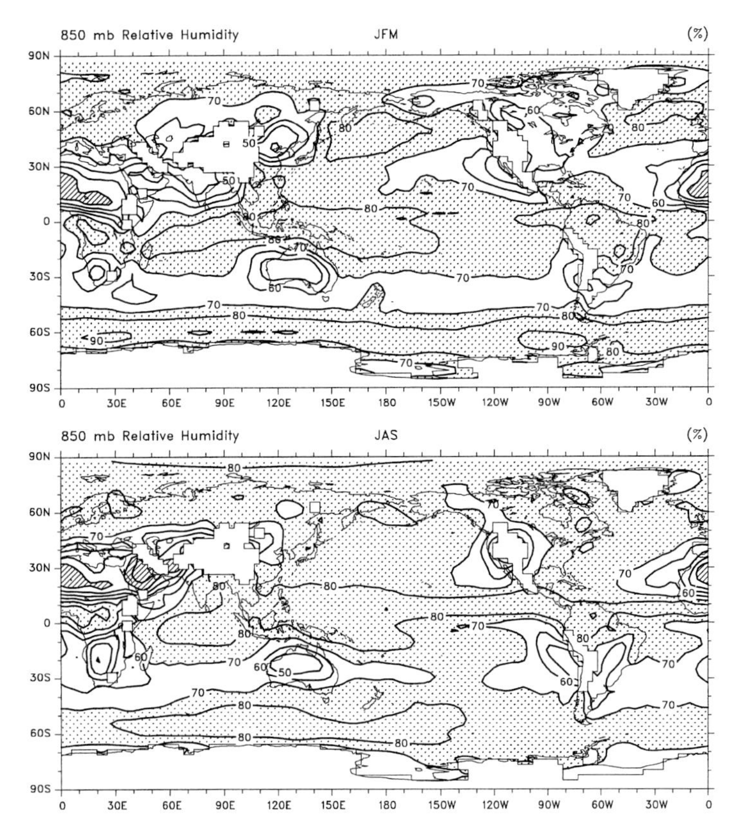

FIG. 1.46. Mean relative humidity (%) at 850 mb for JFM (top) and JAS (bottom). Values greater than 70% are stippled and values less than 30% are hatched. Data from 1987–89 (1986–88) were used to construct the JFM (JAS) climatology.

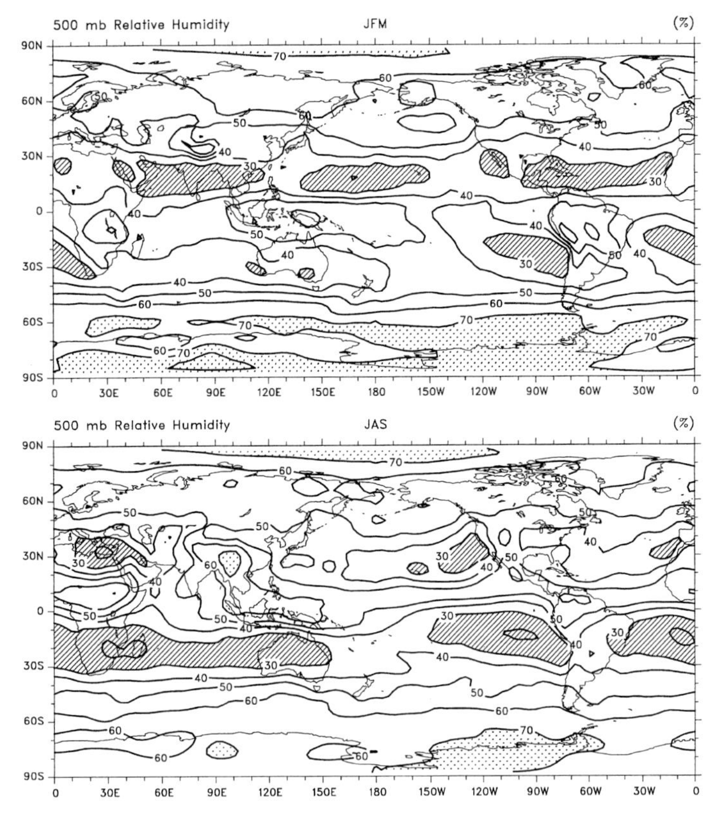

FIG. 1.47. As in Fig. 1.46, but for 500-mb relative humidity.

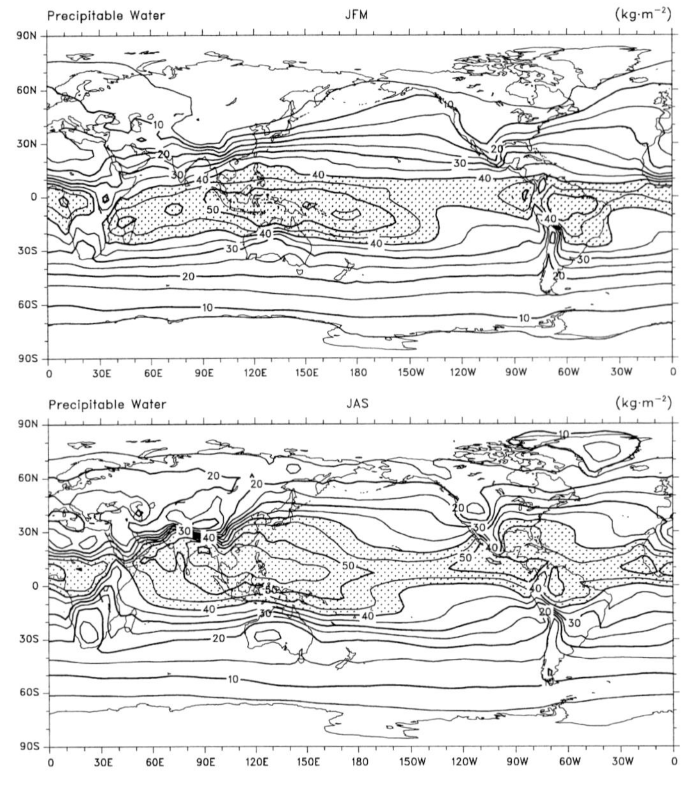

FIG. 1.48. Mean precipitable water (kg m^{-2}) for JFM (top) and JAS (bottom) from ECMWF. Values greater than 40 kg m^{-2} are stippled. Data from 1987–89 (1986–88) were used to construct the JFM (JAS) climatology.

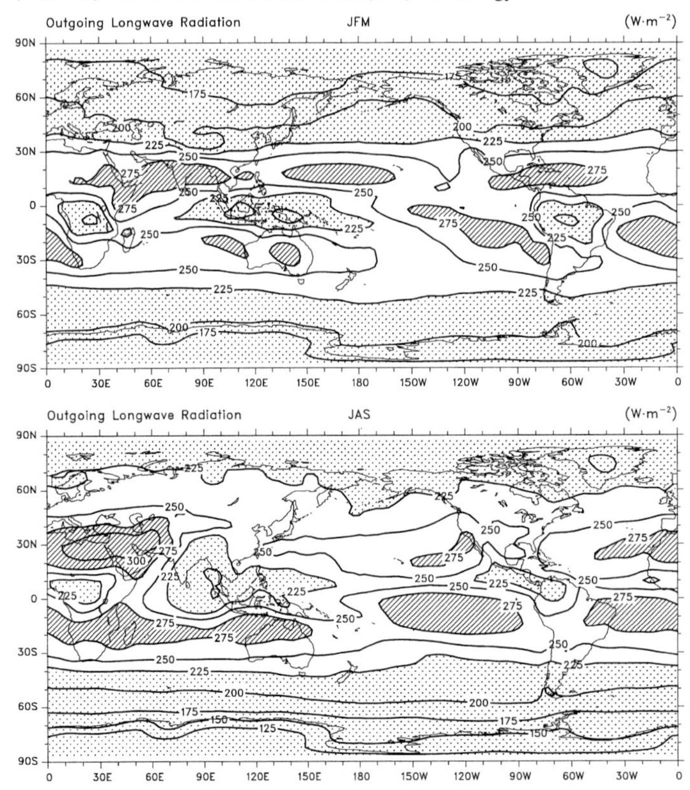

FIG. 1.49. Mean outgoing longwave radiation (W m^{-2}) from ERBE for JFM (top) and JAS (bottom). Values greater than 275 W m^{-2} are hatched and values less than 225 W m^{-2} are stippled. Data over the period 1986–89 (1985–88) were used to construct the JFM (JAS) climatology.

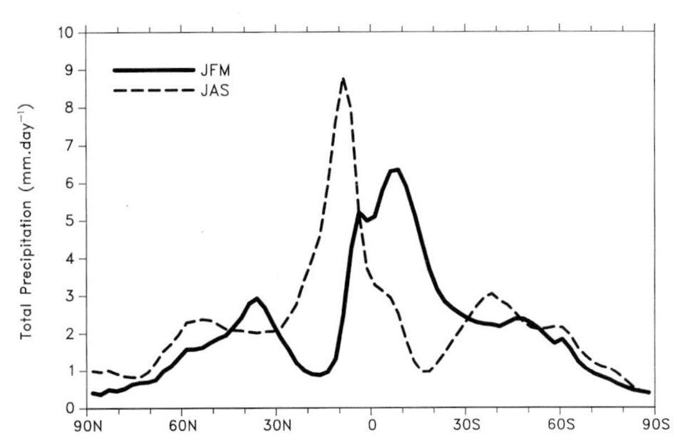

FIG. 1.50. Zonally averaged mean total precipitation (mm day^{-1}) for JFM (solid) and JAS (dash) from the data of Xie and Arkin (1996). The climatologies were constructed over the period 1979–95.

changes in the SH circulation (e.g., van Loon et al. 1993; Hurrell and van Loon 1994; see also Figs. 1.17 and 1.18). The salient point is that, on broad scales in space and time, the mean state of the troposphere is very well described in the original monograph, which remains an authoritative treatise on the climate of the SH.

The reanalysis efforts at both ECMWF (Gibson et al. 1996) and NCEP (Kalnay et al. 1996) involve the recovery of many different sources of data and the quality control and assimilation of these data with data-assimilation systems that remain unchanged over the reanalysis periods. This eliminates perceived climate jumps associated with changes in the data-assimilation systems, which might obfuscate interannual variability or true short-term climate changes, although other discontinuities associated with changes in the input data will still be present. Undoubtedly, the global reanalyses will provide vastly superior datasets for carrying out retrospective analyses. Moreover, both ECMWF and NCEP plan to analyze the archived sets of observations back to at least the IGY, which will result in improved climate statistics as well as better understanding of climate processes and climate change.

Most of the fields presented in this chapter were minimally affected over time by the operational changes at ECMWF, and for these fields differences with the reanalyses are minimal. This point is illustrated in the top panel of Fig. 1.54, which shows differences between the ECMWF operational analyses and reanalyses (ERA) for the JFM 200-mb zonal wind component averaged over 1979–1993. Similar differences are evident during JAS (not shown). During both seasons, the largest discrepancies are only ~2 m s^{-1} and occur over low latitudes where uncertainties in both products are greatest. This point is further illustrated in the lower panel, which, for the same period and pressure level, shows differences between the ERA and NCEP reanalyzed winds. The two reanalyses show excellent agreement at most locations, although the NCEP winds are locally ~4 m s^{-1} stronger than the ERA winds over the tropical Indian and eastern Pacific Oceans. Similarly, comparisons between the ECMWF operational and reanalyzed measures of the intraseasonal and interannual variability of 200-mb zonal winds reveal very small differences during JFM (Fig. 1.55) and JAS (not shown). Differences in measures of the interannual variability are most pronounced over the Tropics, where operational changes in the analysis–forecast system have their largest influence (Trenberth 1992).

The divergent wind components exhibit more significant differences, including differences between the two reanalyses. Such products are highly dependent on the assimilating model and, in particular, to the parameterization of convection in the model. Consequently, while it is clear the analyzed divergence fields in the operational analyses suffer from large changes over time, it is not easy to interpret differences between the reanalyses and the shorter ECMWF climatologies (constructed from data between major operational changes in May 1986 and May 1989) shown in this chapter.

REFERENCES

Anderssen, E. C., 1965: A study of atmospheric long waves in the Southern Hemisphere. *Notos,* **14,** 57–65.

Barkstrom, B. R., E. Harrison, G. Smith, R. Green, J. Kibler, R. Cess, and ERBE Science Team, 1989: Earth Radiation Budget Experiment (ERBE) archival and April 1985 results. *Bull. Amer. Meteor. Soc.,* **70,** 1254–1262.

Bengtsson, L., and J. Shukla, 1988: Integration of space and in situ observations to study global climate change. *Bull. Amer. Meteor. Soc.,* **69,** 1130–1143.

Crutcher, H. L., and J. M. Meserve, 1970: Selected-level heights, temperatures, and dewpoint temperatures for the Northern Hemisphere. NAVAIR 50-1C-52 rev., 17 pp. plus charts.

———, R. L. Jenne, J. J. Taljaard, and H. van Loon, 1971: Climate of the Upper Air: Southern Hemisphere. Vol. IV, Selected Meridional Cross Sections of Temperature, Dew Point, and Height. NAVAIR 50-1C-58, 62 pp. [Available from National Climatic Center, Asheville, North Carolina.]

Dorman, C. E., and R. H. Bourke, 1978: Precipitation over the Pacific Ocean, 30°S to 60°N. *Mon. Wea. Rev.,* **107,** 896–910.

Gibson, R., P. Kållberg, and S. Uppala, 1996: The ECMWF Re-Analysis (ERA) Project. *ECMWF Newsletter,* **73,** 7–17. [Available from ECMWF, Shinfield Park, Reading, Berkshire, RG2 9AX, U.K.]

Hartmann, D. L., 1977: Stationary planetary waves in the Southern Hemisphere. *J. Geophys. Res.,* **82,** 4930–4934.

Hurrell, J. W., 1995: Decadal trends in the North Atlantic Oscillation: Regional temperatures and precipitation. *Science,* **269,** 676–679.

———, and K. E. Trenberth, 1992: An evaluation of monthly mean MSU and ECMWF global atmospheric temperatures for monitoring climate. *J. Climate,* **5,** 1424–1440.

———, and H. van Loon, 1994: A modulation of the atmospheric annual cycle in the Southern Hemisphere. *Tellus,* **46A,** 325–338.

Jenne, R. L., H. L. Crutcher, H. van Loon, and J. J. Taljaard, 1971: Climate of the Upper Air: Southern Hemisphere. Vol. III,

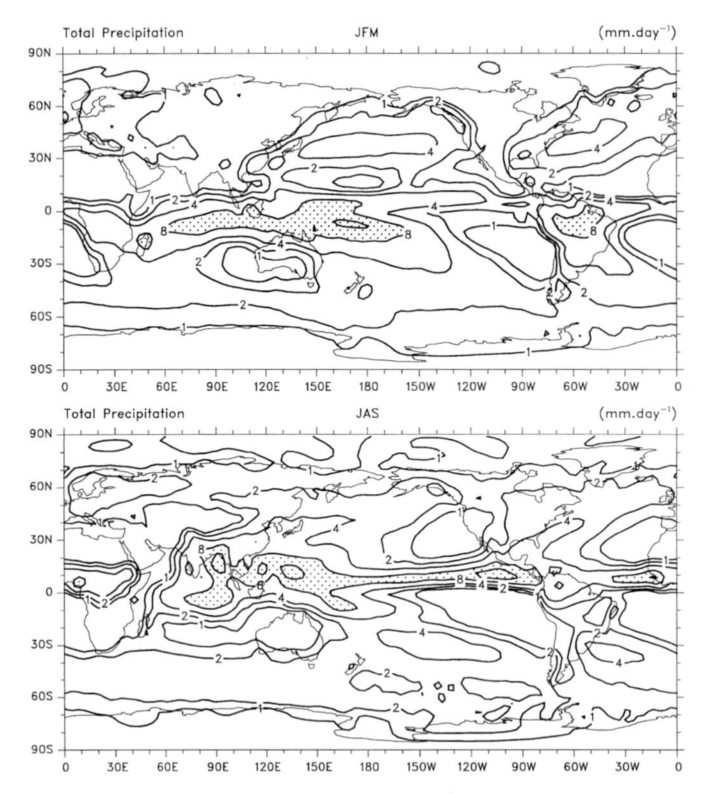

FIG. 1.51. Mean total precipitation (mm day^{-1}) for JFM (top) and JAS (bottom). Rainfall rates greater than 8 mm day^{-1} are stippled.

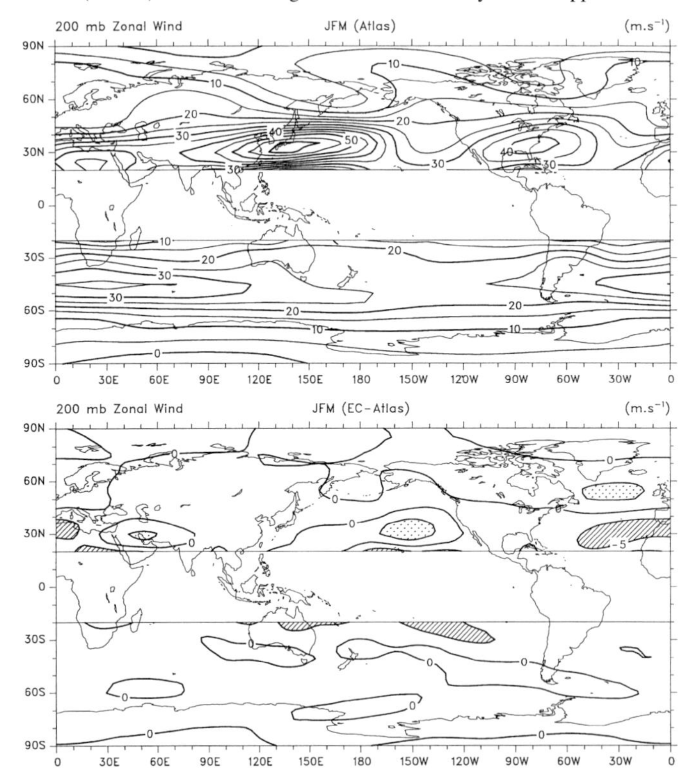

FIG. 1.52. Mean JFM 200-mb zonal geostrophic wind (m s^{-1}) from the atlas data of Crutcher and Meserve (1970) and van Loon et al. (1971) contoured every 10 m s^{-1} (top), and the differences from the ECMWF zonal wind climatology (1979–93) contoured every 5 m s^{-1} (bottom). Geostrophic values between 20°S and 20°N have been excluded, and differences greater (less) than 5 m s^{-1} (-5 m s^{-1}) are stippled (hatched).

TABLE 1.5. As in Table 1.1, but for the mean total precipitation rate (mm day^{-1}).

Name (Location)	Precipitation (mm day^{-1}) JFM Mean	JFM Yrs	JFM σ	JAS Mean	JAS Yrs	JAS σ
Nairobi (1.3°S, 36.9°E)	1.9 (2.3)	41 (14)	1.7 (2.6)	0.6 (0.6)	41 (13)	0.6 (1.1)
Luanda (8.8°S, 13.2°E)	1.7 (1.2)	79 (2)	1.3 (1.1)	0.0 (0.0)	78 (3)	0.1 (0.0)
Lima (12.1°S, 77.0°W)	0.0 (0.0)	46 (2)	0.0 (0.0)	0.1 (0.0)	49 (2)	0.1 (0.0)
Darwin (12.5°S, 130.9°E)	11.3 (12.3)	123 (14)	2.8 (3.2)	0.2 (0.4)	124 (15)	0.3 (0.6)
St. Helena (16.0°S, 5.7°W)	2.8 (1.4)	92 (6)	1.0 (0.6)	2.6 (1.4)	94 (7)	0.7 (0.3)
St. Brandon (16.5°S, 59.6°E)	5.4 (5.4)	42 (14)	2.2 (1.8)	1.4 (1.4)	40 (12)	0.7 (0.6)
Tahiti (17.5°S, 149.6°W)	8.4 (8.2)	81 (15)	3.5 (3.4)	2.0 (2.0)	82 (15)	1.0 (0.8)
Pretoria (25.8°S, 28.2°E)	3.5 (2.9)	78 (14)	1.1 (0.9)	0.4 (0.4)	77 (14)	0.4 (0.3)
Johannesburg (26.1°S, 28.2°E)	3.4 (3.4)	42 (14)	1.0 (1.0)	0.4 (0.5)	42 (14)	0.5 (0.5)
Isla de Pasqua (27.2°S, 109.4°W)	3.0 (2.8)	39 (13)	1.2 (1.0)	3.1 (3.8)	42 (14)	1.4 (1.8)
Durban (29.9°S, 31.0°E)	4.2 (3.9)	120 (14)	1.6 (1.8)	1.7 (2.2)	119 (13)	1.0 (1.3)
Perth (32.0°S, 115.9°E)	0.4 (0.5)	117 (14)	0.4 (0.5)	4.1 (3.5)	118 (15)	1.0 (0.5)
Sydney (33.9°S, 151.2°E)	3.8 (3.3)	133 (4)	2.0 (0.3)	2.7 (1.4)	133 (4)	1.5 (1.6)
Cape Town (34.0°S, 18.6°E)	0.5 (0.7)	42 (14)	0.3 (0.3)	2.1 (2.0)	41 (13)	0.7 (0.7)
Sydney (34.0°S, 151.2°E)	3.7 (3.3)	63 (14)	2.1 (1.9)	2.3 (2.4)	64 (15)	1.3 (1.5)
Buenos Aires (34.6°S, 58.4°W)	3.2 (3.9)	130 (12)	1.4 (1.8)	2.2 (2.2)	128 (10)	1.0 (0.7)
Melbourne (37.8°S, 145.0°E)	1.7 (1.2)	127 (4)	0.8 (0.1)	1.7 (1.3)	128 (4)	0.5 (0.4)
Ile N. Amsterdam (37.8°S, 77.6°E)	2.8 (2.6)	38 (10)	1.0 (1.2)	3.1 (3.4)	37 (9)	0.7 (0.8)
Gough (40.3°S, 9.9°W)	7.2 (7.2)	37 (14)	1.4 (1.2)	9.6 (9.7)	35 (12)	1.8 (1.5)
Christchurch (43.5°S, 172.6°E)	1.6 (1.6)	116 (15)	0.7 (0.9)	1.8 (1.7)	117 (15)	0.7 (0.8)
Chatham Is. (44.0°S, 176.6°W)	2.3 (2.0)	96 (14)	0.9 (0.5)	2.7 (2.7)	98 (14)	1.2 (0.7)
Ile Crozet (46.5°S, 51.0°E)	5.9 (5.8)	10 (9)	0.9 (0.9)	7.1 (7.2)	13 (9)	1.4 (1.3)
Marion Is. (46.9°S, 37.8°E)	6.8 (5.8)	42 (12)	1.5 (1.0)	6.1 (5.6)	44 (13)	1.0 (1.0)
Kerguelen (49.3°S, 70.2°E)	1.8 (1.7)	37 (9)	0.6 (0.6)	2.4 (2.2)	40 (12)	0.9 (0.6)
Campbell Is. (52.5°S, 169.1°E)	3.9 (3.8)	49 (12)	0.7 (0.9)	3.8 (3.6)	51 (13)	1.9 (0.7)
Macquarie (54.5°S, 159.0°E)	3.1 (3.7)	42 (13)	1.6 (2.8)	2.2 (2.5)	45 (14)	0.5 (0.5)
Islas Orcadas (60.7°S, 44.7°W)	1.7 (1.5)	54 (8)	0.8 (0.6)	1.2 (1.3)	54 (7)	0.8 (0.7)
Davis (68.6°S, 77.9°E)	0.4 (0.4)	13 (9)	0.6 (0.6)	0.2 (0.2)	15 (9)	0.1 (0.1)
McMurdo (77.8°S, 166.6°E)	0.5 (0.5)	23 (5)	0.4 (0.2)	0.4 (0.7)	21 (4)	0.3 (0.4)

Vector Mean Geostrophic Winds. NCAR Tech. Note NCAR/TN-STR58 and NAVAIR 50-1C-57, 68 pp.

Kalnay, E., and Coauthors, 1996: The NCEP/NCAR Reanalysis Project. *Bull. Amer. Meteor. Soc.*, **77,** 437–471.

Karoly, D. J., 1985: An atmospheric climatology of the Southern Hemisphere based on ten years of daily numerical analyses (1972–1982), Part II: Standing wave climatology. *Aust. Meteor. Mag.*, **33,** 105–116.

———, G. A. M. Kelly, J. F. Le Marshall, and D. J. Pike, 1986: An atmospheric climatology of the Southern Hemisphere based on ten years of daily numerical analyses (1972–1982). WMO Long-Range Forecasting Research Rep. 7, WMO/TD 92, 73 pp.

Kidson, J. W., 1988: Indices of the Southern Hemisphere zonal wind. *J. Climate,* **1,** 183–194.

Large, W. G., and H. van Loon, 1989: Large scale, low frequency variability of the 1979 FGGE surface buoy drifts and winds over the Southern Hemisphere. *J. Phys. Oceanogr.*, **19,** 216–232.

Legates, D. R., and C. J. Wilmott, 1990: Mean seasonal and spatial variability in gauge-corrected, global precipitation. *Int. J. Climatol.*, **10,** 111–128.

Liu, W. T., W. Tang, and F. J. Wentz, 1992: Precipitable water and surface humidity over global oceans from special sensor microwave imager and European Center for Medium-Range Weather Forecasts. *J. Geophys. Res.*, **97,** 2251–2264.

Madden, R. A., and P. R. Julian, 1994: Observations of the 40–50-day Tropical Oscillation—A review. *Mon. Wea. Rev.*, **122,** 814–837.

Meehl, G. A., 1991: A reexamination of the mechanism of the semiannual oscillation in the Southern Hemisphere. *J. Climate,* **4,** 911–926.

Oort, A. H., 1983: Global atmospheric circulation statistics, 1958–1973. NOAA Professional Paper 14, 180 pp. plus 47 microfiches.

Prescott, J. A., and J. A. Collins, 1951: The lag of temperature behind solar radiation. *Quart. J. Roy. Meteor. Soc.*, **77,** 121–126.

Randel, W. J., 1988: The seasonal evolution of planetary waves in the Southern Hemisphere stratosphere and troposphere. *Quart. J. Roy. Meteor. Soc.*, **14,** 1385–1409.

Rossow, W. B., and R. A. Schiffer, 1991: ISCCP cloud data products. *Bull. Amer. Meteor. Soc.*, **72,** 2–20.

———, and L. C. Garder, 1993a: Cloud detection using satellite measurements of infrared and visible radiances for ISCCP. *J. Climate,* **6,** 2341–2369.

———, and ———, 1993b: Validation of ISCCP cloud detections. *J. Climate,* **6,** 2370–2393.

———, A. W. Walker, and L. C. Garder, 1993: Comparison of ISCCP and other cloud amounts. *J. Climate,* **6,** 2394–2418.

Shea, D. J., K. E. Trenberth, and R. W. Reynolds, 1990: A global monthly sea surface temperature climatology. NCAR Tech. Note NCAR/TN-345+STR, 167 pp.

———, H. van Loon, and J. W. Hurrell, 1995: The tropical–subtropical semi-annual oscillation in the upper troposphere. *Int. J. Climatology,* **15,** 975–983.

Spencer, R. W., 1993: Global oceanic precipitation from the MSU during 1979–1991 and comparison to other climatologies. *J. Climate,* **6,** 1301–1326.

Swanson, G. S., and K. E. Trenberth, 1981: Trends in the Southern Hemisphere tropospheric circulation. *Mon. Wea. Rev.*, **109,** 1879–1889.

Taljaard, J. J., and H. van Loon, 1963: Cyclogenesis, cyclones and anticyclones in the Southern Hemisphere during summer 1957–1958. *Notos,* **12,** 37–50.

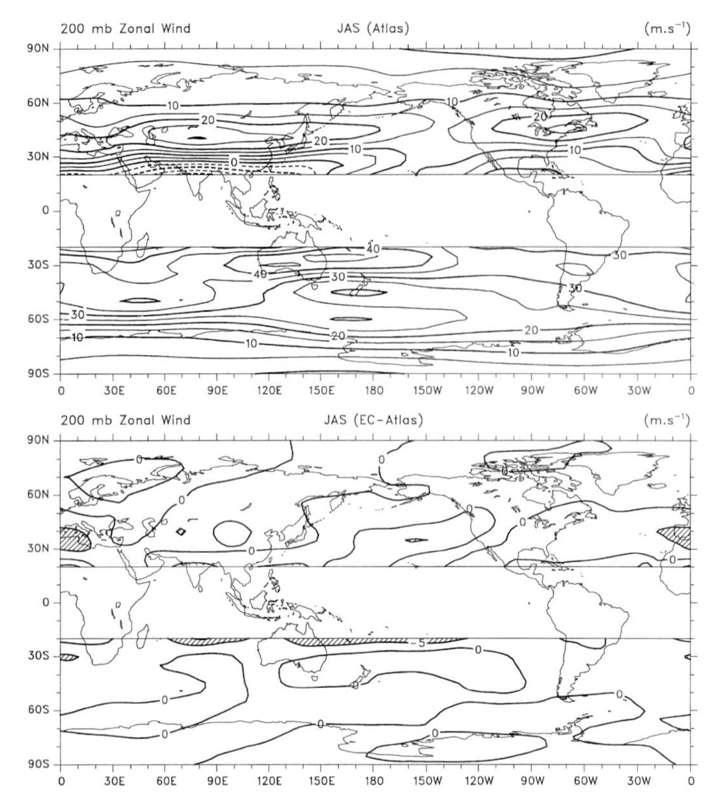

FIG. 1.53. As in Fig. 1.52 but for JAS.

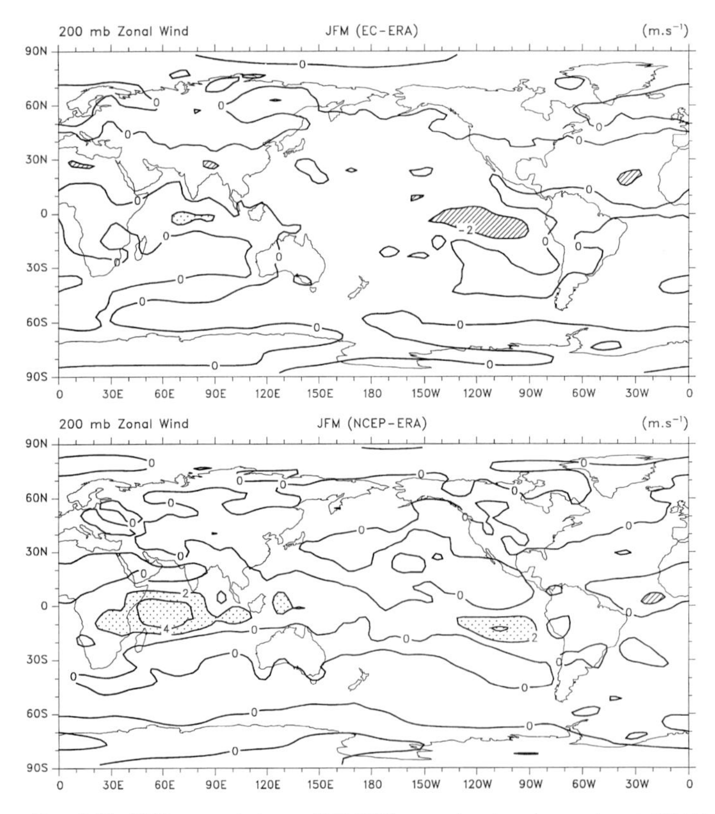

FIG. 1.54. Differences between ECMWF operational and reanalyzed (ERA) mean JFM 200-mb zonal winds (top) and between ERA and NCEP reanalyzed winds (bottom). The contour increment is 2 m s^{-1}, and differences greater (less) than 2 m s^{-1} (-2 m s^{-1}) are stippled (hatched). The climatologies were constructed over the period 1979–93.

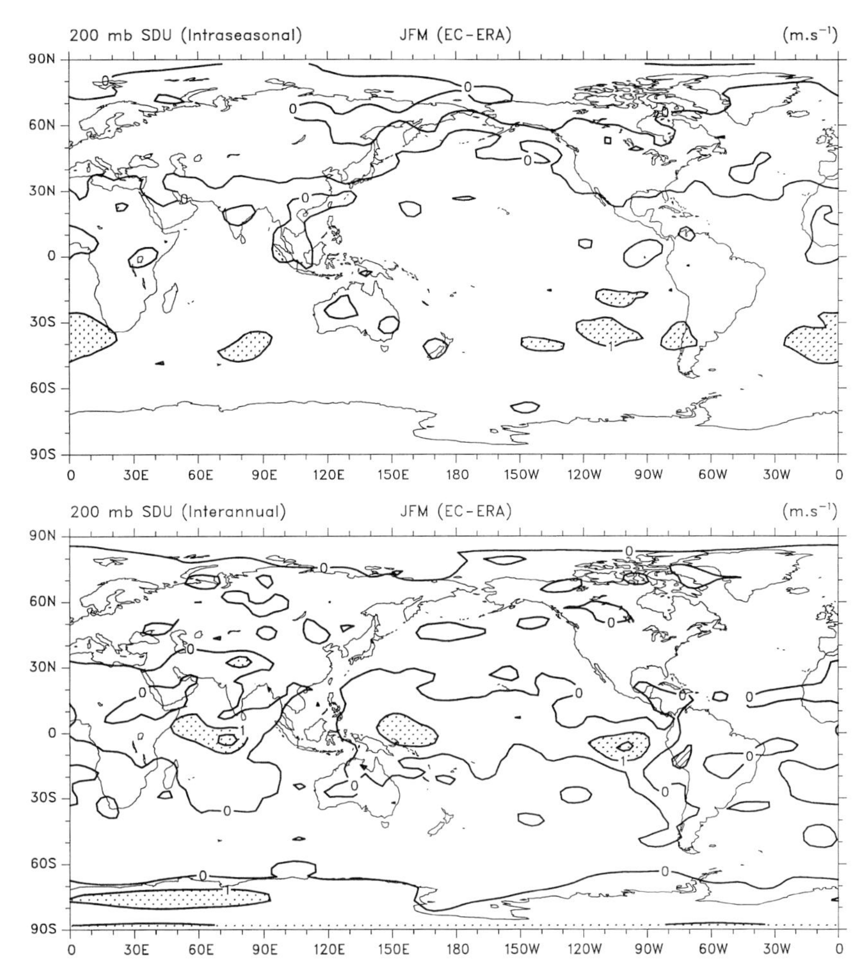

FIG. 1.55. Differences between ECMWF operational and reanalyzed (ERA) measures of the intraseasonal (top) and interannual (bottom) variability of 200-mb JFM zonal winds over the period 1979–93. The contour increment is 1 m s^{-1}, and differences greater (less) than 1 m s^{-1} (-1 m s^{-1}) are stippled (hatched).

———, ———, H. L. Crutcher, and R. L. Jenne, 1969: Climate of the Upper Air: Southern Hemisphere. Vol. I, Temperatures, Dew Points, and Heights at Selected Pressure Levels. NAVAIR 50-1C-55, 135 pp.

Trenberth, K. E., 1980: Planetary waves at 500 mb in the Southern Hemisphere. *Mon. Wea. Rev.,* **108,** 1378–1389.

———, 1981a: Observed Southern Hemisphere eddy statistics at 500 mb: Frequency and spatial dependence. *J. Atmos. Sci.,* **38,** 2585–2605.

———, 1981b: Interannual variability of the Southern Hemisphere 500 mb flow: Regional characteristics. *Mon. Wea. Rev.,* **109,** 127–136.

———, 1981c: Seasonal variations in global sea level pressure and the total mass of the atmosphere. *J. Geophys. Res.,* **86,** 5238–5246.

———, 1982: Seasonality in Southern Hemisphere eddy statistics at 500 mb. *J. Atmos. Sci.,* **39,** 2507–2520.

———, 1983: What are the seasons? *Bull. Amer. Meteor. Soc.,* **64,** 1276–1282.

———, 1984: Interannual variability of the Southern Hemisphere circulation: Representativeness of the year of the Global Weather Experiment. *Mon. Wea. Rev.,* **112,** 108–123.

———, 1992: Global analyses from ECMWF and atlas of 1000 to 10 mb circulation statistics. NCAR Tech. Note NCAR/TN-373+STR, 191 pp. and 24 fiche.

———, and J. G. Olson, 1988: An evaluation and intercomparison of global analyses from NMC and ECMWF. *Bull. Amer. Meteor. Soc.,* **69,** 1047–1057.

———, and J. W. Hurrell, 1994: Decadal atmosphere–ocean variations in the Pacific. *Clim. Dyn.,* **9,** 303–319.

———, and C. J. Guillemot, 1994: The total mass of the atmosphere. *J. Geophys. Res.,* **99D,** 23 079–23 088.

———, and ———, 1995: Evaluation of the global atmospheric moisture budget as seen from analyses. *J. Climate,* **8,** 2255–2272.

———, W. G. Large, and J. G. Olson, 1990: The mean annual cycle in global ocean wind stress. *J. Phys. Oceanogr.,* **20,** 1742–1760.

van Loon, H., 1965: A climatological study of the atmospheric circulation in the Southern Hemisphere during the IGY, Part I. *J. Appl. Meteor.,* **4,** 479–491.

———, 1966: On the annual temperature range over the southern oceans. *Geogr. Rev.,* **58,** 497–515.

———, 1967a: A climatological study of the atmospheric circulation in the Southern Hemisphere during the IGY, Part II. *J. Appl. Meteor.,* **6,** 803–815.

———, 1967b: The half-yearly oscillation in middle and high southern latitudes and the coreless winter. *J. Atmos. Sci.,* **24,** 472–486.

———, 1972: Temperature, pressure, wind, cloudiness and precipitation in the Southern Hemisphere. *Meteorology of the Southern Hemisphere, Meteor. Monogr.,* No. 35, Amer. Meteor. Soc., 25–111.

———, 1979, The association between latitudinal temperature gradient and eddy transport. Part I: Transport of sensible heat in winter. *Mon. Wea. Rev.,* **107,** 525–534.

———, 1980: Transfer of sensible heat by transient eddies in the atmosphere of the Southern Hemisphere: An appraisal of the data before and during FGGE. *Mon. Wea. Rev.,* **108,** 1774–1781.

———, 1983: A comparison of the quasi-stationary waves on the Northern and Southern Hemispheres. *Proc. First International Southern Hemisphere Conf.,* Sao Jose dos Campos, Brazil, Amer. Meteor. Soc. and Cosponsors, 77–84.

———, and R. L. Jenne, 1969: The half-yearly oscillations in the tropics of the Southern Hemisphere. *J. Atmos. Sci.,* **26,** 218–232.

———, and ———, 1970: On the half-yearly oscillations in the tropics. *Tellus,* **22,** 391–398.

———, and ———, 1972: The zonal harmonic standing waves in the Southern Hemisphere. *J. Geophys. Res.,* **77,** 992–1003.

———, and J. C. Rogers, 1984a: The yearly wave in pressure and zonal geostrophic wind at sea level on the Southern Hemisphere and its interannual variability. *Tellus,* **36A,** 348–354.

———, and ———, 1984b: Interannual variations in the half-yearly cycle of pressure gradients and zonal wind at sea level on the Southern Hemisphere. *Tellus,* **36A,** 76–86.

———, J. J. Taljaard, R. L. Jenne, and H. L. Crutcher, 1971: Climate of the Upper Air: Southern Hemisphere. Vol. II, Zonal Geostrophic Winds. NCAR Tech. Note NCAR/TN-STR57 and NAVAIR 50-1C-56, 43 pp.

———, ———, T. Sasamori, J. London, D. V. Hoyt, K. Labitzke, and C. W. Newton, 1972: *Meteorology of the Southern Hemisphere, Meteor. Monogr.* No. 35, C. W. Newton, Ed., Amer. Meteor. Soc., 263 pp.

———, R. L. Jenne, and K. Labitzke, 1973a: Zonal harmonic standing waves. *J. Geophys. Res.,* **78,** 4463–4471.

———, K. Labitzke, and R. L. Jenne, 1973b: A note on the annual temperature wave in the stratosphere. *J. Geophys. Res.,* **78,** 2672–2678.

———, J. W. Kidson, and A. B. Mullan, 1993: Decadal variation of the annual cycle in the Australian dataset. *J. Climate,* **6,** 1227–1231.

Vowinckel, E., and H. van Loon, 1957: Das Klima des Antarktischen Ozeans: III Die Verteilung der Klimaelemente und ihr Zusammenhang mit der allgemeinen Zirkulation. *Arch. Meteor. Geophys. Bioklim.,* **B8,** 75–102.

Warren, S. G., C. J. Hahn, J. London, R. M. Chervin, and R. L. Jenne, 1986: *Global Distribution of Total Cloud Cover and Cloud Type Amounts over Land.* NCAR Tech. Note NCAR/TN-273+STR, 29 pp. and 199 maps.

———, ———, ———, ———, and ———, 1988: *Global Distribution of Total Cloud Cover and Cloud Type Amounts over Ocean.* NCAR Tech. Note NCAR/TN-317+STR, 42 pp. and 170 maps.

Xie, P., and P. A. Arkin, 1996: Analyses of global monthly precipitation using gauge observations, satellite estimates, and numerical model predictions. *J. Climate,* **9,** 840–858.

Chapter 2

General Circulation

DAVID J. KAROLY

Meterology CRC, Monash University, Clayton, Victoria, Australia

DAYTON G. VINCENT, AND JON M. SCHRAGE

Department of Earth and Atmospheric Sciences, Purdue University, West Lafayette, Indiana

2.1. Introduction

The mean state of the troposphere has been described in the previous chapter, with emphasis on features in the SH. In this chapter, we seek to explain the origin of these features in terms of the dominant processes that maintain the general circulation of the SH, primarily in the summer and winter seasons, and particularly in middle latitudes. More detailed discussions of the features in the Tropics and in high southern latitudes, and of the variability of the SH circulation, are given in chapters 3, 4, and 8, respectively.

The major factor determining the general circulation of the atmosphere is the distribution of diabatic heating, with net heating in the Tropics and net cooling in middle and high latitudes. These net heating differences are balanced by a poleward transport of energy in the atmosphere and ocean. Major constraints on the atmospheric circulation are imposed by the rotation of the earth and the global requirements to conserve angular momentum, mass, moisture, and total energy. Other factors that determine the general circulation, particularly differences between the SH and the NH, include the distributions of land and ocean, the topography and other land surface properties, and the sea surface temperature distribution. As discussed in chapter 1 and shown in Fig. 1.2, the fraction of the surface covered by land in the SH is much less than in the NH, which leads to a smaller-amplitude seasonal cycle in the SH. Also, the topography and land–ocean thermal contrasts are smaller in SH middle latitudes, leading to smaller zonal asymmetries than in the NH. The strong westerly flow in the SH middle latitudes is interrupted by the regular passage of weather systems in storm tracks. These storm tracks are more zonally uniform in the SH than the NH.

In the next section, we describe the maintenance of the zonal-mean circulation in the SH and the relationship between the zonal-mean flow and the mean transports by the weather systems (transient baroclinic eddies). Following this, the maintenance of the general circulation is described using the budgets of angular momentum, heat, moisture, and total energy. The discussion of the SH general circulation in relation to global momentum, moisture, and energy-balance requirements follows that in the earlier SH monograph (van Loon et al. 1972) but makes use of recent global observational datasets. Finally, the maintenance of the zonal variations of the SH circulation is discussed in terms of the forcing of the zonal asymmetries of the mean flow and the storm tracks.

As in chapter 1, the atmospheric distributions presented will be global in order to place the results for the SH into perspective, but the focus is on the SH. Most diagrams are based on the same dataset used in chapter 1 (i.e., analyses for the 1979–93 period from the ECMWF). It is argued in chapter 1 that the ECMWF analyses are among the best available. Exceptions to the use of ECMWF analyses, or to the 15-yr period of these analyses, will be stated when they arise. We concentrate on the extreme seasons, January–March (JFM) and July–September (JAS), for the same reasons as described in chapter 1.

2.2. Maintenance of the zonal-mean circulation

The simplest view of the general circulation of the atmosphere can be obtained by considering the maintenance of the zonal-mean distributions of temperature, wind, and moisture, as described in chapter 1 and shown in Figs. 1.4, 1.27, and 1.45. These zonal-mean fields can be understood in terms of the differential diabatic heating between the Tropics and polar regions and the constraints on the poleward energy transport in the atmosphere. There is an excess of latent heat release over radiative cooling in the Tropics, as shown by the zonal-mean diabatic heating in Fig. 2.1. This is primarily balanced by ascent and adiabatic cooling in the Hadley cell. In middle and high latitudes, there is mean radiative cooling as the temperatures are higher

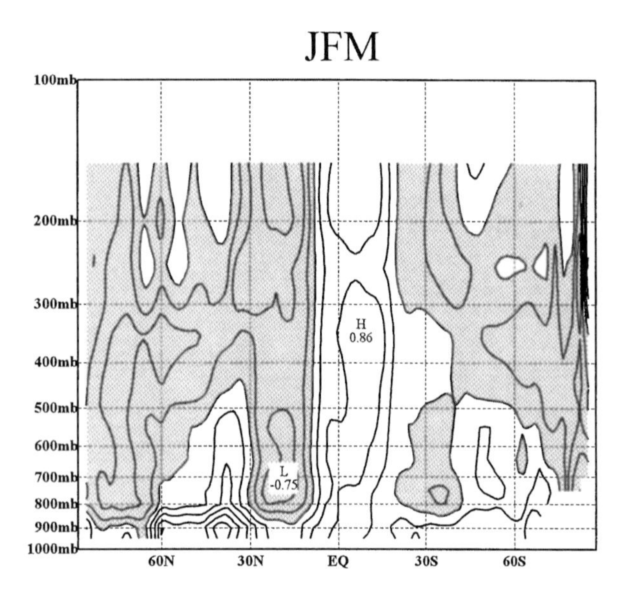

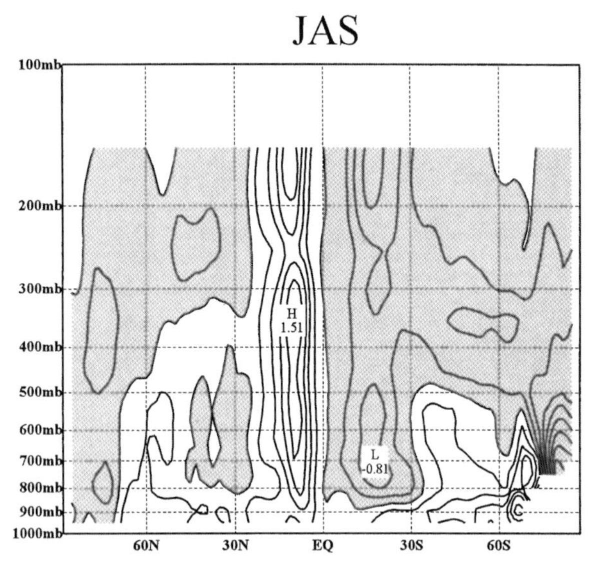

FIG. 2.1. Zonal-mean net diabatic heating for January–March (JFM) and July–September (JAS) calculated as the residual in the heat budget [term (a) in (2.11)], based on ECMWF analyses for the period 1979–93. Contour interval is 0.3 K day^{-1}, and negative values are shaded.

than for a local radiative equilibrium, due to the energy transport from lower latitudes. A discussion of the atmospheric heat budget, as well as an explanation for how the diabatic heating was obtained, is given in section 2.3b. The largest meridional temperature gradients are in middle latitudes (Fig. 1.4), with the mean westerly flow (Fig. 1.27) in thermal wind balance with the meridional temperature gradient. The strong westerly jet in the upper troposphere and the strong meridional temperature gradient in middle latitudes provide an energy source for the growth of disturbances in the mean zonal flow due to baroclinic instability. These disturbances grow to finite amplitude and decay. They are the traveling weather systems, which are often called transient eddies.

a. Zonal-mean eddy transports

We use a simple mathematical framework for describing the role of the mean meridional flow and the transports by transient and stationary eddies in the budgets of momentum, heat, moisture, and energy in the atmosphere. This framework is introduced here, but the budget equations are considered in more detail in section 2.3. Let s' be the departure of a variable s from its time mean $\bar{s}$, and let s^* be the departure from its zonal mean $[s]$. The transient eddies are associated with time variations, s', and the stationary eddies are associated with time-mean departures from the zonal mean, $\bar{s}^*$. The time-mean, zonal-mean northward transport of a variable, s, per unit mass $[\overline{vs}]$ can thus be represented as the contributions from three terms,

$$[\overline{vs}] = [\bar{v}][\bar{s}] + [\overline{v's'}] + [\bar{v}^*\bar{s}^*]. \qquad (2.1)$$

The terms on the right side of (2.1) are the transport by the mean meridional circulation, transport by the transient eddies, and transport by the stationary eddies, respectively. Except where noted, time means will be for a seasonal period within a year and eddies will be based on departures calculated from the seasonal mean. Long-term means, therefore, will be the average of each year's seasonal statistics.

The distributions of the zonal-mean transports of momentum, heat, and moisture per unit mass by the transient variations and stationary waves are shown in Figs. 2.2–2.4 for JFM and JAS. We compare these with earlier results on the zonal-mean transports presented by Newell et al. (1972, 1974) and Peixoto and Oort (1992) from analyses of upper-air station data and by Trenberth (1992) from a shorter period of ECMWF analyses than used here. Karoly and Oort (1987) have compared the estimates of eddy transports in the SH obtained from station data and from numerical analyses and found that the sparse network of upper-air stations in the SH can lead to a poor representation of the meridional variations of the transport in analyses that use only station data. It was shown by van Loon (1980) that early numerical analyses in the SH could be biased by the distribution of stations (i.e., too much weight placed on sparsely distributed observations). More recent data-assimilation products do not appear to suffer as much from this problem.

Vertical cross sections of transient and stationary eddy momentum fluxes are shown in Fig. 2.2 for JFM and JAS. They show that the transports by the transient eddies are much greater than those by the stationary eddies, except in winter in the NH. The reason for the relatively strong poleward transports of upper-tropospheric momentum by stationary eddies at 30°N

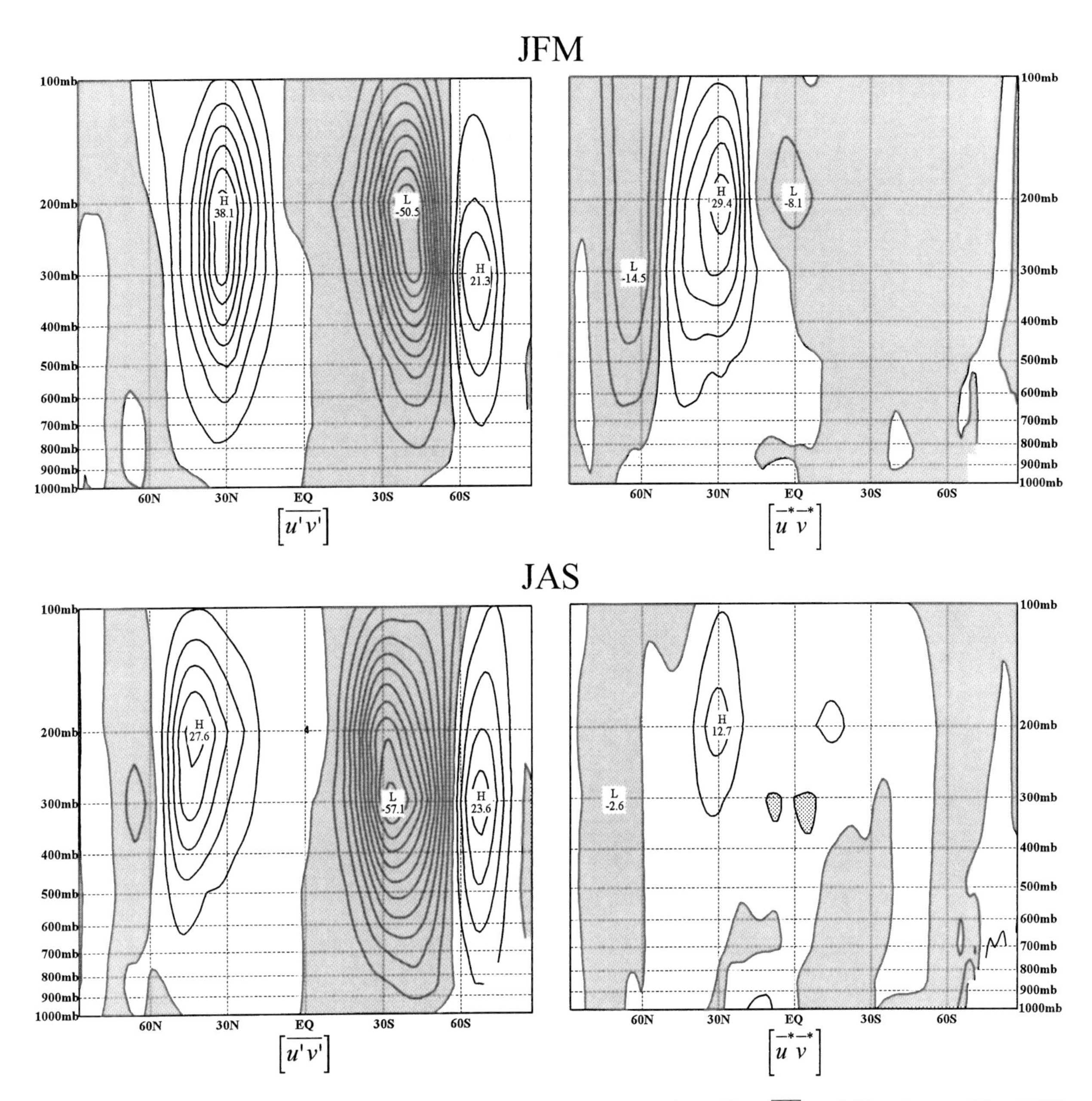

FIG. 2.2. Time–zonal-mean eddy momentum flux for JFM and JAS due to (a) transient eddies, $[\overline{u'v'}]$, and (b) stationary eddies, $[\bar{u}^*\bar{v}^*]$, based on ECMWF analyses for 1979–93. Contour interval is 5 m^2s^{-2}. Positive values are northward, negative values are shaded.

in JFM is, most likely, due to the persistent influence that the major continents have on causing NE–SW-oriented troughs to be located just east of these large land masses (Starr 1948; Newell et al. 1972). There is a poleward momentum flux by the transient eddies from near the Equator to approximately 55° latitude in both hemispheres. Maximum values occur at 30°–35° (40°–45°) in the winter (summer) hemisphere, reflecting the seasonal migration of the jet streams and associated storm tracks. At high latitudes, there is equatorward transport by the transient eddies, leading to substantial flux convergence by the transient eddies in middle latitudes. The momentum flux maxima (both positive and negative) occur in the upper troposphere between 200 and 300 hPa, regardless of season, but the seasonal cycle consists of stronger poleward transports in winter than in summer in both hemispheres. In the SH, as is well known, the transient eddy fluxes are greater than those in the NH during corresponding seasons. In general, the patterns of eddy momentum flux presented by earlier investigators (Newell et al. 1970, 1972; Peixoto and Oort 1992; and Trenberth 1992) agree with those presented here, but there are some differences in details. For example, the Newell

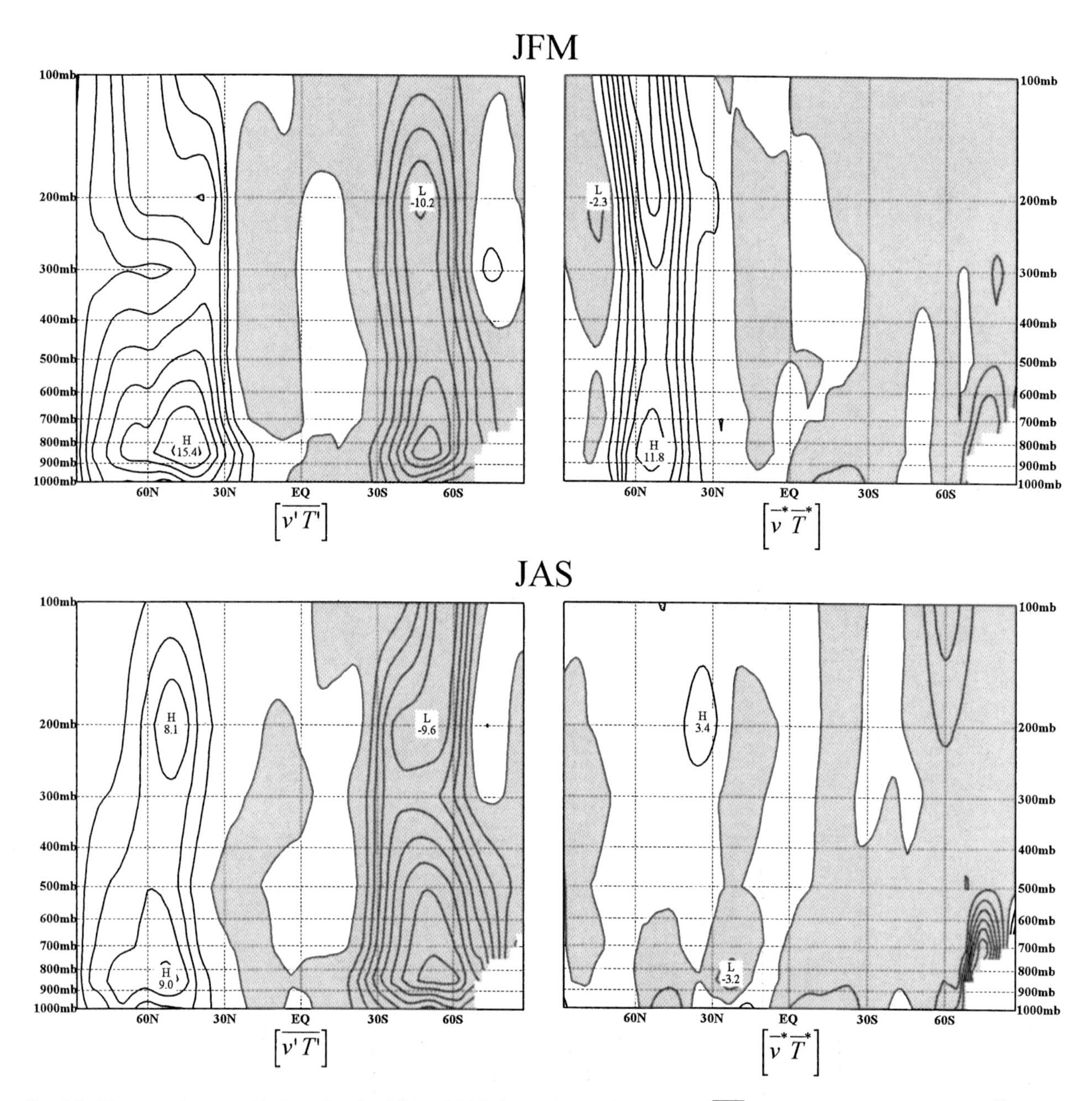

FIG. 2.3. Time–zonal-mean eddy heat flux for JFM and JAS due to (a) transient eddies, $[\overline{v'T'}]$, and (b) stationary eddies, $[\bar{v}^*\bar{T}^*]$, based on ECMWF analyses for 1979–93. Contour interval is 2 K m s^{-1}. Positive values are northward, negative values are shaded.

et al. references show a reasonably strong cross-equatorial transport of momentum by transient eddies in the upper troposphere from the winter to the summer hemisphere that is not shown in the present results or those of Trenberth (1992). In addition, both Newell et al. and Trenberth show cross-equatorial transports by stationary eddies in the upper troposphere from winter to summer that are about twice as large as the values shown here. These discrepancies are due to differences in data sources, analysis methods, or the different time periods considered.

Vertical cross sections of sensible heat flux by the transient and stationary eddies for JFM and JAS are shown in Fig. 2.3. In middle and high latitudes, the transient eddy flux is large and predominantly poleward in both hemispheres and both seasons. Maximum values occur in the lower troposphere, near 850 hPa, with a secondary maximum near 200 hPa, and they shift slightly with season between about 40° and 55° latitude. The maximum at 850 hPa is caused primarily by closed cyclonic/anticyclonic disturbances and fronts, while the maximum at 200 hPa is a result of

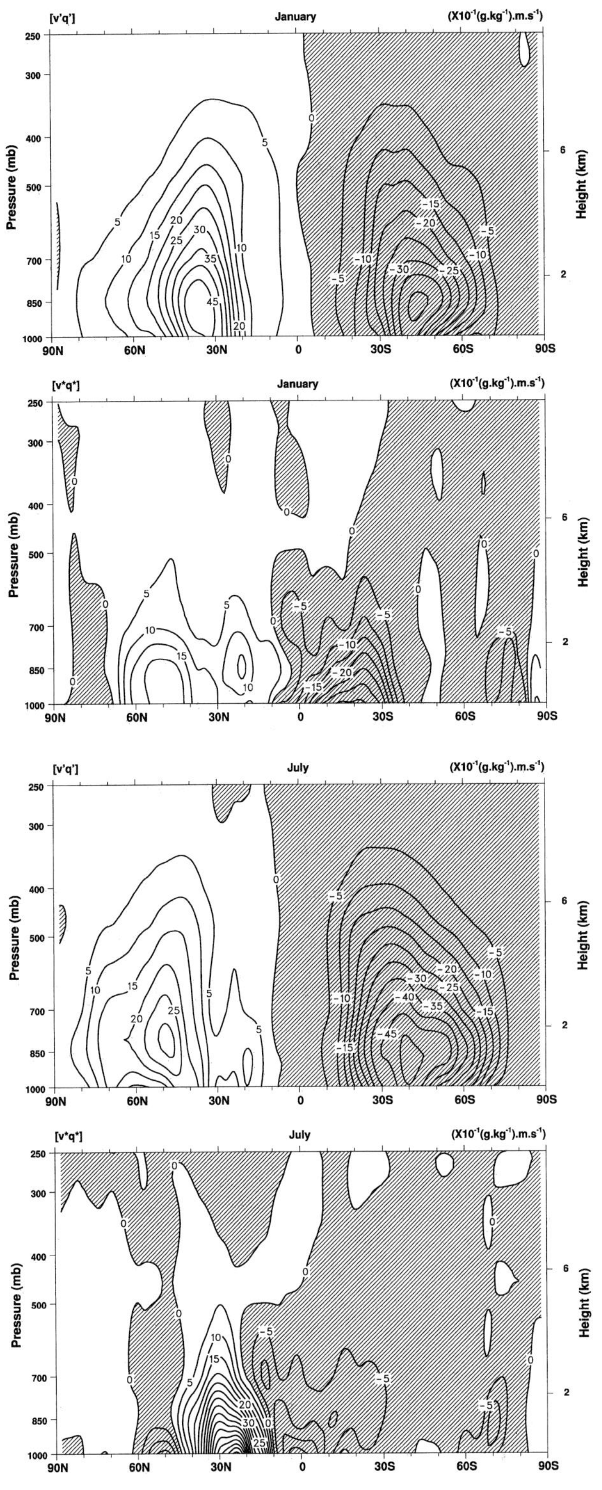

FIG. 2.4. Time–zonal-mean eddy moisture flux for January and July due to transient eddies, $[\overline{v'q'}]$, and stationary eddies, $[\bar{v}^*\bar{q}^*]$, based on ECMWF analyses for 1987–89. Contour interval is 7.5×10^{-1} (g kg^{-1}) (m s^{-1}). Positive values are northward, negative values are shaded. (Reproduced from Trenberth 1992.)

open waves. In the tropical regions of both hemispheres, the transient eddy flux is small, but primarily equatorward. The stationary eddy heat flux is significant in the SH only in the lower troposphere over Antarctica, where there is poleward transport (see chapter 4 for further discussion), and in the NH in JFM, where there is substantial poleward heat transport by the stationary waves in the troposphere and lower stratosphere in middle and high latitudes. These distributions are similar to those given by Trenberth (1992), even though his results were based only on the years 1986–89 and generally show similar distributions, but larger values, than the station-based analyses of Newell et al. (1974) and Peixoto and Oort (1992).

The vertical cross sections of eddy moisture fluxes in January and July are shown in Fig. 2.4. They have been extracted from Trenberth (1992). Because of the problems noted in chapter 1 with regard to ECMWF moisture analyses, only the three years 1987–89 were used to construct this figure. Since the specific humidity decreases rapidly with height above the surface, the largest moisture transports are in the lower atmosphere. As for the eddy heat flux, the eddy moisture flux is primarily poleward in both hemispheres throughout the year and is dominated by the transient eddy transport. In the tropical to subtropical latitudes in the summer season, stationary eddies also play a substantial role in the moisture transport. This is due primarily to monsoonal-type, quasi-stationary circulations, as well as to the locations of subtropical high pressure centers. In the SH, there is little latitudinal shift with season in the location of the transient eddy flux maxima, but in the NH the peak poleward transports occur near 35°N in January and from 50° to 60°N in July. The latter is due to the larger percentage of land masses in the NH.

b. Baroclinic eddy life cycles

The zonal-mean distributions of the mean flow and daily variability presented in chapter 1 can be combined with the zonal-mean eddy flux distributions above to give a picture of the general circulation of the SH. The major baroclinic zone in middle latitudes shows relatively little seasonal variation and is associated with strong transient eddy activity in the SH storm track. The transient eddies play the dominant role in the eddy transports in the SH, with the stationary eddies playing a very minor role. The transient eddies extract energy from the mean flow, transporting heat poleward and reducing the baroclinicity of the mean flow. The momentum (and vorticity) transport by the transient eddies, however, acts to strengthen the barotropic component of the mean flow in middle latitudes (e.g., see Hurrell 1995 and references therein). The relationship between the meridional distributions of the zonal-mean flow and the eddy transports

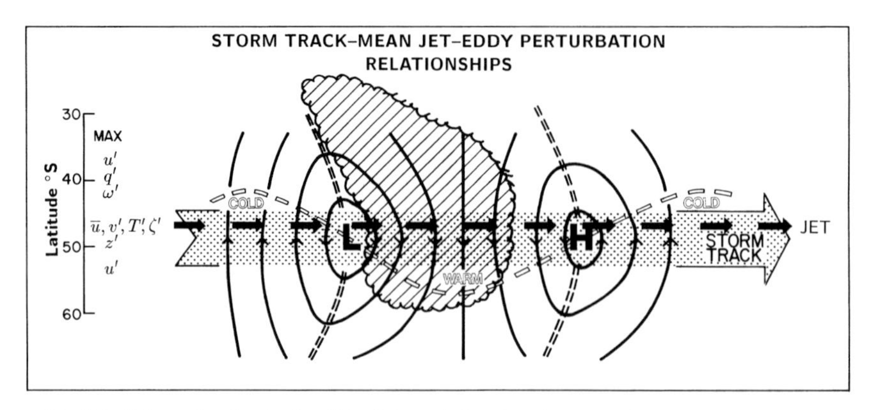

FIG. 2.5. Schematic plot in the SH of the relationships between the mean jet (solid heavy arrow) and storm tracks (broad stippled arrow) and associated eddy statistics. Perturbation geopotential heights are greatest in the storm track, producing a maximum $\overline{z'^2}$. Maxima in all quantities are indicated at left. Vorticity perturbations are a maximum just north of the storm track, owing to the Coriolis parameter variations and consistent with the geostrophic relation. Therefore, meridional wind perturbations are similarly shifted north while zonal wind perturbations are a maximum north and south of the storm track. Moreover, the perturbations are elongated north–south so that $\overline{v'^2} < \overline{u'^2}$, and they have a characteristic shape with trough and ridge lines (shown by the double line) such that there is convergence of momentum by the eddies into the storm track ($\overline{v'\zeta'} > 0$). Temperature perturbations (outlined dashed line) are a maximum in the lower troposphere and highly correlated with $-v'$. Maximum perturbations in moisture and vertical motion are closely related and occur in lower latitudes in close correspondence with v' and T', as indicated by the hatched cloud zone. (Reproduced from Trenberth 1991b.)

has been described by Trenberth (1991b) and is shown schematically in Fig. 2.5.

Numerical modeling studies of baroclinic eddy lifecycles by Hoskins (1983) and Hoskins et al. (1983) have shown that they provide a good explanation for the observed transient eddy fluxes in the SH and their relationship with the mean flow. The relatively smaller role for stationary eddies in the SH means that the observed SH transient eddy statistics are generally more similar to these idealized model simulations than those in the NH.

The major contribution to the transient eddy heat and momentum flux comes from eastward-moving waves with zonal wavenumbers of around 3 to 7 and periods of 2 to 8 days (Randel and Held 1991). Coherent wave–zonal flow interactions for these high-frequency transients have been revealed using a cross-correlation analysis by Randel (1990). Consistent with theoretical models of nonlinear baroclinic wave lifecycles, the poleward eddy heat flux maximizes in the lower troposphere as the wave is growing, while the eddy momentum flux maximizes at upper levels as it matures and decays. There is a close relationship between the meridional temperature gradient and the location of the storm track (Trenberth 1991b). The baroclinic eddies reduce the meridional temperature gradient through their heat fluxes, but it is maintained by solar heating, surface heat fluxes, and latent heating within the baroclinic eddies. "Baroclinic waves are very active throughout the year in the SH and continually undergo lifecycles of growth to maturity and decay, at the same time influencing the zonal-mean flow" (Trenberth 1991b).

c. *Alternative views of the zonal-mean circulation*

So far, we have presented a view of the zonal-mean general circulation from an Eulerian perspective in pressure or height coordinates. A somewhat different view of the role of the mean meridional circulation and eddy transports is obtained if the zonal averaging is carried out in isentropic coordinates (along constant potential temperature surfaces) rather than in height coordinates. This approach has been reviewed by Johnson (1989) and may be a more natural way to approach the general circulation, since diabatic heating in the atmosphere forces motion across isentropic surfaces. The zonal-mean meridional mass-transport streamfunction computed on isentropic surfaces and then mapped back onto pressure coordinates (Karoly et al. 1997) is shown in Fig. 2.6 [computed from only three years (1993–95) of ECMWF analyses]. This shows the Hadley cells in the Tropics in much the same way as in the Eulerian mean meridional circulation shown in Fig. 1.40, but the mean meridional circulation in middle latitudes is quite different. The indirect Ferrel cells in middle latitudes do not exist in the zonal-mean meridional circulation in isentropic coordinates and have been replaced by an extension of the direct Hadley cell into high latitudes, with a

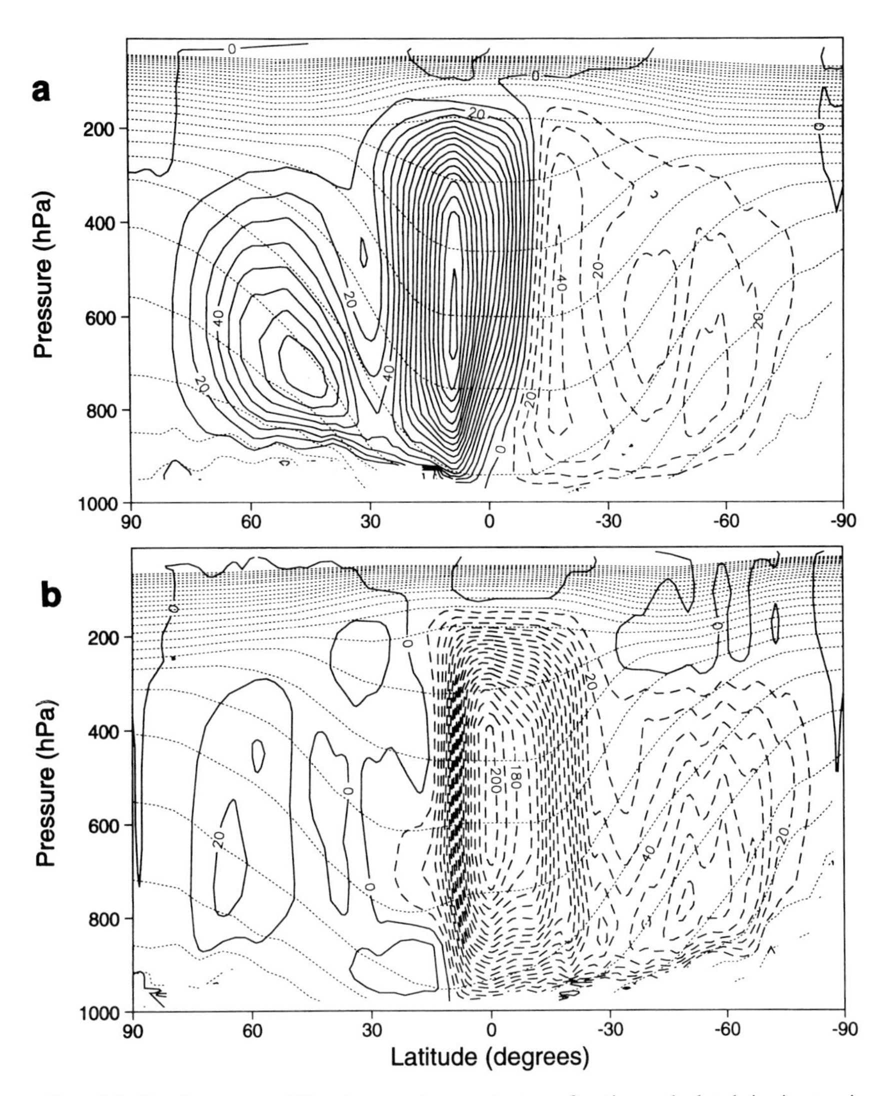

FIG. 2.6. Zonal-mean meridional mass transport streamfunction calculated in isentropic coordinates and remapped to isobaric coordinates using the mean pressure of the isentropic surfaces. Computed from ECMWF analyses for 1992–95 for (a) December–February and (b) June–August. Dashed lines indicate negative values of the streamfunction and dotted lines are mean isentropes. (Reproduced from Karoly et al. 1997.)

second region of ascent in middle latitudes. The cross-isentropic flow shown in the isentropic meridional circulation in Fig. 2.6 is more consistent with forcing by the zonal-mean diabatic heating in Fig. 2.1 than is the Eulerian mean meridional circulation in Fig. 1.40.

It is possible to link the pressure coordinate and isentropic coordinate views of the zonal-mean general circulation by considering the transformed Eulerian mean flow, introduced by Edmon et al. (1980). The zonally averaged momentum and thermodynamic equations in pressure coordinates for large-scale quasi-geostrophic flow can be written:

$$\frac{\partial}{\partial t}[u] + [v]\frac{\partial[u]}{\partial y} - f[v] = -\frac{\partial[v^*u^*]}{\partial y} - [X]$$

$$\frac{\partial}{\partial t}[T] + \Gamma[\omega] = -\frac{\partial[v^*T^*]}{\partial y} + [Q], \quad (2.2)$$

where the Coriolis parameter, $f = 2\Omega\sin(\phi)$, is twice the local vertical component of the earth's rotation rate at latitude ϕ, X is the x-component of frictional forces, Γ is the static stability parameter defined by $\Gamma = (R_d/c_p p)[T] - (\partial[T]/\partial p)$, and Q is the diabatic heating. Here, the zonal-mean adiabatic motion balances the diabatic heating and the eddy heat flux convergence.

Readers should note that (2.2), as well as (2.3–2.5), which follow, have been formulated without time averaging and with the eddy fluxes due to the departures from the zonal mean. After time averaging, the total eddy transport terms in these equations are essen-

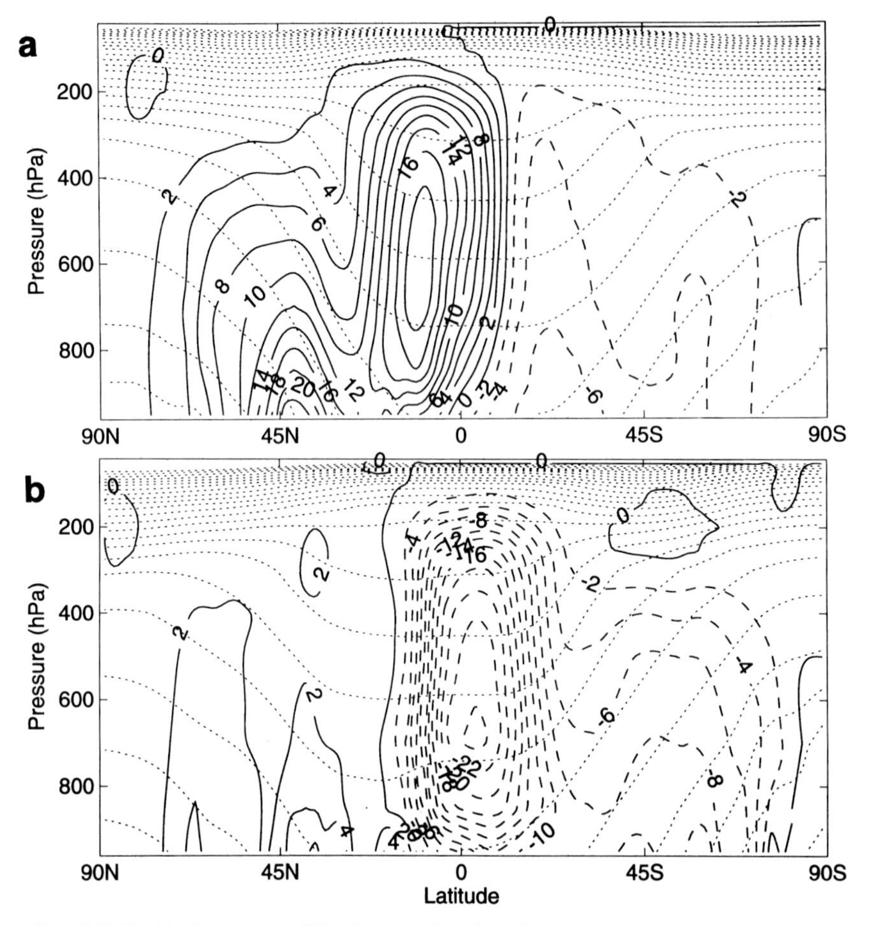

FIG. 2.7. Residual mean meridional streamfunction for (a) December–February and (b) June–August computed from eight years (1986–93) of operational analyses from the U.K. Meteorological Office. (Reproduced from Karoly et al. 1997.)

tially identical to the sums of the transient and stationary eddy fluxes (i.e., $\overline{[v^*u^*]} \cong \overline{[v'u']} + [\bar{v}^*\bar{u}^*]$).

An alternative form can be obtained by introducing the residual mean meridional circulation (0, $[v_T]$, $[\omega_T]$) (Edmon et al. 1980), such that

$$[v_T] = [v] - \frac{\partial([v^*T^*]/\Gamma)}{\partial p}$$

$$[\omega_T] = [\omega] + \frac{\partial([v^*T^*]/\Gamma)}{\partial y}. \tag{2.3}$$

When we use advection by the residual mean circulation, we obtain the transformed Eulerian mean equations (Edmon et al. 1980),

$$\frac{\partial}{\partial t}[u] - f[v_T] = \nabla \cdot F + [X]$$

$$\frac{\partial}{\partial t}[T] + \Gamma[\omega_T] = [Q]. \tag{2.4}$$

Now, the zonal-mean residual vertical motion balances only the mean diabatic forcing, and the residual meridional advection of vorticity balances the nonconservative eddy forcing, represented by the Eliassen–Palm (EP) flux,

$$F = (0, -[v^*u^*], f[v^*T^*]/\Gamma). \tag{2.5}$$

The use of these transformed Eulerian mean equations simplifies the form of the mean momentum and thermodynamic equations.

The residual mean meridional mass-transport streamfunction is shown in Fig. 2.7 and is very similar to the isentropic mean meridional circulation in Fig. 2.6. This similarity has been noted for the ocean by McIntosh and McDougall (1996), who showed that, to leading order, the mean meridional circulation on isentropic surfaces is the same as the residual mean meridional circulation. This will be discussed again later in chapter 7 on the meridional circulation in the Southern Ocean. The effects of the heat flux convergence by the eddies in the Eulerian perspective can be

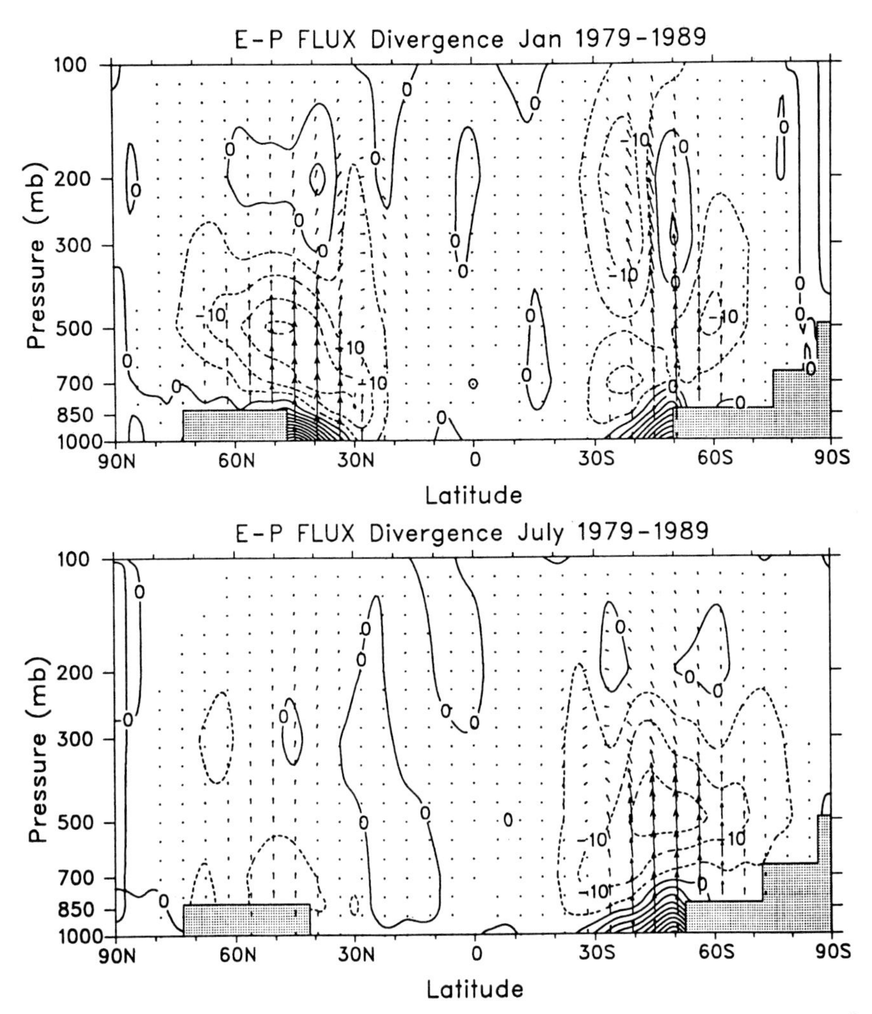

FIG. 2.8. Zonal-mean EP flux vectors scaled by $1/p$ and the EP flux divergence (in 10^{-6} m s^{-2}) for January and July 1979–89 for the 2–8-day eddies. Negative divergences are dashed and regions influenced by topography (low pressure surfaces) are stippled. (Reproduced from Trenberth 1991b.)

effectively removed using the residual mean circulation, leading to a meridional circulation that is more directly related to mean density forcing. Hence, the Ferrel cells are primarily the Eulerian meridional circulations required to balance the zonal-mean eddy heat and vorticity flux convergences on constant height surfaces. In the SH, the eddy flux convergences are almost solely due to transient eddies, and the Ferrel cells in the SH are of similar strength in both seasons. The eddy transport of heat in isentropic coordinates is very small, as geostrophic advection of dry static energy is precluded in isentropic coordinates (Johnson 1989).

The EP flux has a second use—namely, as a diagnostic of the meridional propagation of wave-like disturbances in the atmosphere. It can be shown that the EP flux is locally in the direction of the group velocity of small-amplitude Rossby waves (Edmon et al. 1980). Figure 2.8 shows the EP flux plotted as vectors and contours of the EP flux divergence, computed from the eddy flux statistics shown earlier. It can be seen that there is predominately upward and then equatorward propagation of the transient eddies from a source region in the midlatitude baroclinic zone in the lower troposphere. From (2.4), the EP flux divergence represents the mean forcing of the zonal-mean flow by the eddies. Hence, the region of EP flux divergence in the lower troposphere and convergence in the upper troposphere represents a reduction in the vertical shear of the mean zonal flow but an overall increase in the barotropic component of the zonal-mean flow. Therefore, the transient eddies reduce the mean meridional temperature gradient and mean vertical shear of the zonal flow in the middle latitudes (the baroclinicity of the mean flow), but increase the vertically averaged zonal flow.

2.3. Atmospheric budgets

For the earth–atmosphere system as a whole, there are conservation laws that dictate the four-dimensional distributions of mass, momentum, heat, water vapor, and energy. These fundamental laws are frequently expressed in the form of budget equations, which consist of letting the individual rate of change (*d/dt*) of a specified variable be determined by the sources and sinks of that variable. In this section, four atmospheric budgets are examined. They are the budgets of 1) angular momentum, including maintenance of the zonal wind; 2) sensible heat, including maintenance of the temperature; 3) water vapor; and 4) total energy. The approach used will be primarily based on observations and is similar to other studies of the general circulation over the past three decades (e.g., Lorenz 1967; Newell et al. 1972, 1974; Oort and Peixoto 1983; Peixoto and Oort 1992; and Trenberth 1992). For each budget, the contribution by advection and/or divergence of the primary budget variable will be illustrated and discussed. This contribution is, of course, equal to the local rate of change, as well as the net sources and sinks of the budget variable.

Within each budget, there is more than one source or sink term. When the budgets are vertically integrated, these terms, in the angular momentum budget, for example, are the mountain (or pressure) and frictional torques. In the sensible heat budget, they comprise the various contributions from diabatic heating, while in the water vapor budget, they are evaporation at the ground and precipitation. In general, these sources and sinks are not directly available from observations, at least not on a global basis; therefore, they are not part of the analyses used here that are generally the same as those used in chapter 1 (e.g., a combination of WMO and WCRP versions of ECMWF gridpoint analyses). For this reason, the source/sink terms will be computed as budget residuals, which will make it difficult to sort out the various physical processes associated with them. The main exception to this is the mountain torque, which was calculated from the ECMWF analyses. Nevertheless, an attempt will be made, wherever possible, to discuss the significance of the different physical processes.

Another equally important problem to that of assessing sources/sinks within the budgets is related to the accuracy of the time- and zonally averaged meridional wind component, $[\bar{v}]$. As is well known, the meridional transports of budget variables are primarily influenced by the mean meridional circulations at low latitudes (i.e., by Hadley-type circulations). Unfortunately, the ECMWF analyses used here do not conserve mass (Trenberth 1991a; Trenberth et al. 1995) and, as pointed out by Trenberth (1991a), budget residuals may therefore contain large errors. Obviously, this will impact on the results presented in this section, and the reader should be aware of this problem.

With all of the above in mind, we decided to proceed in the following manner. At low latitudes, values of $[\bar{v}]$ were calculated directly from our ECMWF dataset. At latitudes poleward of 25°N and 25°S, values were computed based on momentum balance considerations. All values were then mass adjusted so that the vertical integral of $[\bar{v}]$ vanished within each latitude band. The indirect method was necessary at middle and high latitudes because zonally averaged, vertically integrated mass imbalances were found to occur frequently. It is important to emphasize that the procedure of meshing direct and indirect values of $[\bar{v}]$ is a common, apparently successful, practice (e.g., Kuo 1956; Gilman 1965; Holopainen 1967; Lorenz 1967; Vincent 1968; Newell et al. 1972; Rosen 1976; Oort and Peixoto 1983; and Peixoto and Oort 1992).

a. Earth–atmosphere momentum budget

The discussion in this section consists of two parts. First, the angular momentum budget for the earth–atmosphere system is presented; then, results are given for the linear momentum balance equation. The latter is simply a statement of the maintenance of the time- and zonally averaged zonal wind equation.

1) ABSOLUTE ANGULAR MOMENTUM

The form of the absolute angular momentum equation used here is essentially the same as that given by several previous authors (e.g., Lorenz 1967; Newell et al. 1972; Oort and Peixoto 1983; and Peixoto and Oort 1992). It can be written in spherical isobaric coordinates as

$$\frac{\partial[\bar{M}]}{\partial t} = \underset{\text{(a)}}{-\frac{1}{a\cos\phi}\frac{\partial}{\partial\phi}[\bar{J}_\phi]\cos\phi} \underset{\text{(b)}}{-\frac{\partial}{\partial p}[\bar{J}_p]} + \underset{\text{(c)}}{g\frac{\partial[\bar{z}]}{\partial\lambda}} \underset{\text{(d)}}{- a\cos\phi g\frac{\partial[\bar{\tau}_\lambda]}{\partial p}}, \qquad (2.6)$$

where

$$[\bar{M}] = \Omega a^2\cos\phi + [\bar{u}]a\cos\phi; \qquad (2.7)$$

$$[\bar{J}_\phi] = \Omega a^2\cos^2\phi[\bar{v}] + a\cos\phi([\bar{u}][\bar{v}] + [\overline{u'v'}] + [\bar{u}^*\bar{v}^*]); \qquad (2.8)$$

$$[\bar{J}_p] = \Omega a^2\cos^2\phi[\bar{\omega}] + a\cos\phi([\bar{u}][\bar{\omega}] + [\overline{u'\omega'}] + [\bar{u}^*\bar{\omega}^*]); \qquad (2.9)$$

and Ω = earth's angular velocity, a = earth's radius, g = gravity, τ_λ = frictional stress, λ = longitude, and

ϕ = latitude. The remaining symbols (overbars, brackets, primes, and asterisks) are as defined in section 2.2.

The terms in (2.6) that can be computed from observations are the left side and the terms (a), (b), and (c) on the right side. Since the results discussed here are centered close to the climatological solstice seasons, the left side of (2.6) should be small compared to the remaining terms in the equation (Newton 1972). Thus, the left side was not calculated. On the other hand, it has been shown by Newton (1971a,b) and others that the left side can be comparable to the mountain torque during the equinox seasons. In this context, as will be seen, the mountain torque is considerably less than the frictional torque on seasonal time scales.

The terms on the right side of (2.6) represent mechanisms that can change the absolute angular momentum, M, with time, from an atmospheric perspective. Terms (a) and (b) are the meridional and vertical convergence (signs included) of the earth's momentum plus the air's momentum relative to the earth. In our sign convention, convergence adds westerly momentum to the atmosphere. These two terms contain contributions by mean motions, and transient and stationary eddies, as defined in section 2.2. Terms (c) and (d) in (2.6) are known as the mountain (or pressure) torque and the frictional torque, respectively. Positive values, including the sign, represent a gain of westerly momentum by the atmosphere that occurs with regard to the frictional torque—for example, at the earth–atmosphere interface where low-level easterly winds are observed. As mentioned above, only terms (a), (b), and (c) were calculated. The sum of these terms was calculated as the residual and assumed to equal term (d). Of course, the latter not only contains the frictional torque, but also the effects of subgrid-scale contributions to the budget, as well as errors in any of the computed terms and the local rate of change of $[\bar{M}]$. It is worth mentioning that in some current studies that consider daily values of global M, term (d) is being obtained from the output of 4-dimensional data-assimilation systems. Also, even though the length of the day (earth's rotation rate) is affected by angular momentum exchanges between the earth and atmosphere (e.g., Hide et al. 1980; Barnes et al. 1983; Rosen and Salstein 1983; Dickey et al. 1991; and Rosen 1993), this impact should be slight on the long-term seasonal means of momentum budget terms considered here.

Latitudinal distributions of vertically integrated transports of relative angular momentum across latitude circles are shown in Fig. 2.9 for JFM and JAS. The transports by transient and stationary eddies are the same as those used earlier in this chapter. These transports are shown first because they make important contributions both to (2.6) and to the maintenance of the zonal wind, discussed later. Figure 2.9 shows that the contribution to the total momentum transport

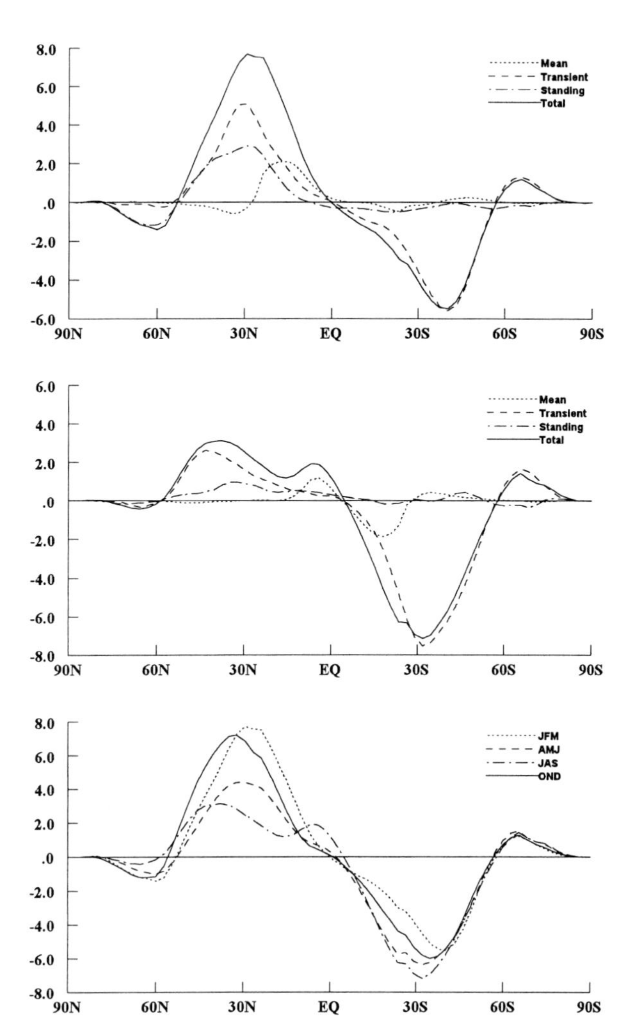

FIG. 2.9. Vertical integrals of time–zonal-mean momentum transports by mean motions (dotted), transient eddies (dashed), stationary eddies (dot–dash), and total (solid) for JFM (top) and JAS (center). Total momentum transport by season (bottom). Units are 10^{12} kg (m s^{-2}). Positive values are northward, negative are southward.

by the mean motions is greatest in the region occupied by the wintertime Hadley cell. Elsewhere, it is negligible, except just north of the Equator in JAS. The mean motion transport by the Hadley cells is poleward and is due to the vertical integral being dominated by an equatorward transport of easterly momentum at low levels and a poleward transport of westerly momentum at upper levels. In either case, the role of the mean meridional circulations, with regard to relative angular momentum, is to remove zonal momentum from the equatorial region (where it is accumulated in the surface easterlies) and initiate its poleward transport (where ultimately it is deposited in the surface westerlies). An important component of this Hadley cell transport is the earth's angular momentum (often referred to as Ω-momentum). In the rising branch, Ω-momentum is greater than in the sinking branch due to the longer momentum arm (hence, torque) at low latitudes. Thus, there is a requirement for a poleward transport of absolute angular momentum in the upper branches of the Hadley cells. The transport of relative angular momentum by transient eddies takes over as the dominant mechanism in the poleward transport at subtropical and middle latitudes. This contribution is actually poleward from near the Equator to approximately 55° latitude in both hemispheres, with maximum values occurring at 30°–35° (40°–45°) in the winter (summer) hemisphere. A seasonal cycle, consisting of stronger poleward transports in winter than in summer, exists in both hemispheres, and the maximum values both in the winter and summer seasons occur in the SH. The stationary eddy contribution is generally quite small, particularly in the SH. In the NH, it is significant only in middle latitudes where, in both seasons, it supports the poleward transport of momentum provided by the transient eddies. Finally, at high latitudes there is an equatorward transport by eddy fluxes that carries easterly momentum from the polar regions toward midlatitudes.

Figure 2.9 also shows the latitudinal distributions of the total momentum transport for all four seasons. It is seen that momentum is transported poleward, away from the Equator in both hemispheres, to about 55° latitude. At that location, it is met by equatorward transports that originated over the polar caps. The seasonal variation of both the poleward and equatorward transports is much greater in the NH than in the SH. Particularly noteworthy is the similarity of the seasonal curves poleward of 40°S. The strong poleward transports each season, between about 35° and 55°S, are due to the year-round storm tracks (hence, dominated by transient eddies) in that latitude belt (Trenberth 1991b; Sinclair 1994, 1995). The annual average transport (not shown) depicts the maxima of poleward (equatorward) transports between latitudes 30° and 35° (60° and 65°) in both hemispheres. The SH maxima are slightly greater than their counterparts in the NH.

There have been several estimates of momentum transports over the last few decades with which the present results can be compared. Most estimates since about 1970 have been based on global datasets, including the present, and there are only minor differences among them (e.g., Newell et al. 1972; Oort and Peixoto 1983; and Peixoto and Oort 1992). Estimates prior to 1970, however, do not agree as well with the current ones, except over the NH middle latitudes, where a majority of the routine upper-air observations were located. The main problem with the earlier values is that they underestimated the effects of the wintertime Hadley cell and middle latitude transient waves in the SH (e.g., see results of Buch 1954 and Obasi 1963, presented in Lorenz 1967).

We now turn our attention to the terms in the angular momentum budget. We begin with the content, M, as defined in (2.7). The first term is not very interesting to examine because it varies only as the cosine of latitude. Moreover, seasonal values of this Ω-momentum contribution have been shown to be less than the second term (e.g., Rosen 1993). Thus, we focus on the second term in (2.7). Table 2.1 compares values of the relative angular momentum contents between the NH and SH for each of the four seasons. Since this quantity depends primarily on the zonal wind, it is not surprising that it shows maxima in the winter hemisphere and minima in the summer hemisphere. In three of the four seasons, it is seen that the SH has the greater content of angular momentum. It is also seen that the global value is lowest in JAS, when easterly winds occupy a large portion of the NH. For comparison, several authors provide seasonal values of M in both hemispheres. The earlier values by Newell et al. (1972) and Oort and Peixoto (1983) showed the same seasonal cycle as the current values, but their magnitudes were much lower than the present ones. More recent results, which appear in Peixoto and Oort (1992), but are credited to Rosen and Salstein (1983) with a verbal update from Salstein, are in excellent agreement with those given here.

TABLE 2.1. Values of relative angular momentum contents in 10^{23} kg m^2 s^{-1} computed from $\int\int(a^2\cos\phi/g)\int ua\cos\phi\ dpd\lambda d\phi$.

	JFM	AMJ	JAS	OND
NH	952	514	109	665
SH	610	936	1020	840
Globe	1562	1450	1129	1505

Terms in the vertically integrated form of (2.6) are now discussed. First, a time–latitude diagram of the mountain torque [term (c)] in the SH is shown (Fig. 2.10). The units used are Hadleys (HU), where 1 HU = 10^{18} kg m^2 s^{-2}. As for the other computed budget terms in (2.6), the mountain torque was based on averages of ECMWF analyses over the 15-yr period, 1979–93. The pattern depicted in Fig. 2.10 arises from the torque caused by the horizontal component of the pressure stress normal (perpendicular) to the inclined surface that occurs due to pressure (mass) differences across topography (Johnson 1980). Thus, it is primarily impacted by the major north–south mountain ranges, which consist mainly of the Andes (0°–50°S) and the various mountain groups over southern Africa (0°–35°S). The continent of Antarctica also affects the mountain torque. Before looking at results of the mountain torque, it should be stressed that since this quantity is based on ECMWF analyses and topography, it can contain errors due to a poor resolution of pressure relative to orography.

Figure 2.10 shows that middle latitudes are dominated by negative values, which are caused primarily by near-surface westerlies piling up mass on the windward side of the Andes, thus creating a higher pressure than that on the leeward side. This east–west pressure differential causes the mountains to "push back" and,

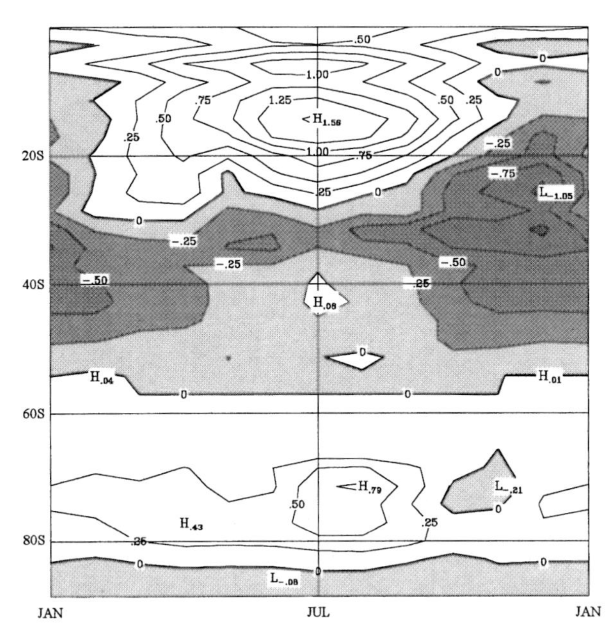

FIG. 2.10. Annual cycle of the mountain torque [term (c) in (2.6)], in Hadleys, for the Southern Hemisphere, with negative values shaded (1 Hadley = 10^{18} kg m^2 s^{-2}).

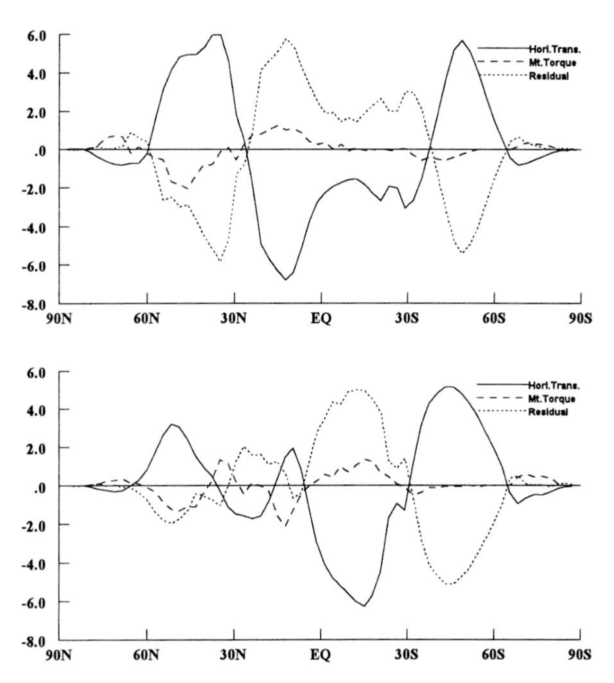

FIG. 2.11. Vertically integrated terms of the angular momentum budget, (2.6), for JFM (top) and JAS (bottom). Units are Hadleys, where 1 Hadley = 10^{18} kg m^2 s^{-2}. Term (a) is the meridional convergence of absolute angular momentum (solid) and term (c) is the mountain torque (dashed). The residual (dotted line) should be approximately equal to the frictional torque, term (d).

therefore, extract angular momentum from the atmosphere. In the Tropics, and at high latitudes, the pattern reverses to one where the atmosphere gains momentum. The Andes most likely play the dominant role in the Tropics; the mountainous regions over southern Africa, however, may also make a contribution. At high latitudes, the only influence is the topography of Antarctica. The positive values in Fig. 2.10 appear to be due to near-surface easterlies (which predominate at low and high latitudes) piling up mass on the eastern flanks of mountains compared to the western flanks. In this case, the reaction by the mountains is to reduce the easterly momentum or, in effect, increase or add westerly zonal momentum, as observed. In this context, it was shown in Fig. 1.27 that low-level easterlies are a maximum near 10°S in winter. Figure 2.10 clearly shows this to be the location where, and time of year when, positive values are at a maximum.

Latitudinal distributions of the vertically integrated terms in (2.6) are presented in Fig. 2.11 for JFM and JAS. As noted earlier, the left side of (2.6) was not computed. Additionally, the vertical convergence of total momentum, term (b), vanishes upon integration, and term (d) becomes the torque due to the surface frictional stress. It is assumed that the local rate of change of $[\bar{M}]$ is small during the solstice seasons (see earlier discussion); thus, the residual of (2.6) should be approximately equal to the surface frictional torque. In general, it is seen that the mountain torque is much less than the frictional torque (i.e., the residual); the main exception is at high latitudes, where both torques are comparable. Nonetheless, the two torques usually complement each other. Similar results have been found by several investigators (e.g., Lorenz 1967; Newton 1971a,b; Newell et al. 1972; Oort and Peixoto 1983; Wahr and Oort 1984; Swinback 1985; Oort 1989; and Peixoto and Oort 1992). The SH mountain torque has already been discussed. At low latitudes in the NH, the mountain torque supplies momentum to the atmosphere in JFM and extracts it in JAS. This reflects the reaction of topography to the pressure differentials across mountains set up by the changing low-level wind distribution from easterlies in winter to more monsoonal westerlies in summer. In the winter hemisphere, the value of the mountain torque is approximately 20%–25% of that for the frictional torque. In the middle latitude NH, angular momentum is extracted from the atmosphere by both torques. In JFM, the mountain torque reaches a maximum (negative) value of −2 HU, while the frictional torque peaks at almost −6 HU. In JAS, the respective values are −1 and −2 HU. In the SH, as already noted, the mountain torque at midlatitudes is nearly absent; the only contributors are the Andes, which extract a relatively small amount of momentum from the atmosphere. The frictional torque, on the other hand, is quite large and reaches maximum (negative) values of −5 HU in both JFM and JAS.

Term (a), the convergence of meridional eddy momentum flux, shows negative values throughout the Tropics in JFM, as well as in JAS, except for a narrow band of positive values from 5° to 15°N. Thus, divergence of momentum transfer dominates at low latitudes and is balanced primarily by the frictional torque. In middle latitudes, convergence of momentum takes place and acts to accelerate the westerly flow found there, which is continuously losing its momentum due primarily to the frictional torque. Figure 2.11 shows that in JFM the maxima of convergence have nearly the same value in both hemispheres, whereas in JAS the SH value is considerably greater than its counterpart in the NH. This is because the poleward eddy momentum fluxes at southern midlatitudes exhibit less seasonal variation than those in the NH (Fig. 2.9).

Appropriate comparisons (i.e., seasonal means using similar units) with the results shown in Fig. 2.11 can be found in Oort and Peixoto (1983), Oort (1989), and Peixoto and Oort (1992). These authors show that their maxima and minima of total torque (i.e., sum of the mountain and frictional torques) are generally greater than the corresponding values found here. This is particularly true in the SH, where their middle latitude minima are −8 and −7 HU for JJA and DJF, compared to −5 HU for JAS and JFM in the present study. Moreover, their tropical peak at 10°–15°S in JJA is 10 HU, compared to about 6 HU in our JAS results. Although some progress appears to have been made in the past several years, the strength of the seasonal mean values of the earth's torques still seems to be an unresolved issue, especially at southern latitudes.

Before advancing to the next section, it is worth mentioning that recent studies have shown that the relative magnitudes of the mountain and frictional torques can vary substantially among temporal scales. Rosen (1993) summarizes the literature on this topic and provides a schematic diagram that suggests that the frictional torque is three to four times greater than the mountain torque when seasonal-to-annual averages are compared. This is compatible with results presented here, as well as those cited elsewhere in this section. When averages are computed over shorter or longer time periods, however, Rosen suggests that the mountain torque becomes increasingly important compared to the frictional torque. In fact, the former becomes greater than the latter on weekly and El Ninõ–Southern Oscillation timescales. Madden and Speth (1995) examined the behavior of both torques on intraseasonal time cycles (~50 days) during 1988 and found that they complemented one another, but that the mountain torque's contribution to changes in absolute angular momentum during two of these intraseasonal oscillations exceeded the frictional torque's contribution by a factor of three.

2) MAINTENANCE OF THE ZONAL WIND

The time- and zonally averaged zonal-wind equation used here is written in isobaric coordinates as:

$$\frac{\partial[\bar{u}]}{\partial t} = \underset{(a)}{-\frac{\partial}{\partial y}([\overline{u'v'}] + [\bar{u}^*\bar{v}^*])} \underset{(b)}{- \frac{\partial}{\partial p}([\overline{u'\omega'}] + [\bar{u}^*\bar{\omega}^*])} \tag{2.10}$$

$$\underset{(c)}{-[\bar{v}]\frac{\partial[\bar{u}]}{\partial y}} \underset{(d)}{- [\bar{\omega}]\frac{\partial[\bar{u}]}{\partial p}} \underset{(e)}{+ f[\bar{v}]} \underset{(f)}{+ [\bar{F}_x]},$$

where the left side is essentially zero or at least small compared to the major contributors to (2.10). Spherical coordinates were applied; results will be presented in terms of 10^{-5} m s^{-2}. Furthermore, each term includes the sign; therefore, a positive value represents an increase locally, in the zonal wind. Terms (e) and (f) could also be thought to represent a local acceleration of the zonal wind. While all terms except (f) on the right side of (2.10) were computed, terms (a), (c), and (e) were found to dominate, together with the residual, which was assumed to represent term (f), the frictional force. Thus, only these four terms are discussed. As for the momentum budget, the 15-yr dataset from ECMWF was used in all calculations.

Vertical cross sections of the four terms mentioned above are displayed in Figs. 2.12 and 2.13 for JFM and JAS. Figure 2.12 shows that the zonal wind speed is decreased throughout most of the tropical troposphere by momentum flux divergence [term (a)]. Maximum losses occur at subtropical latitudes near 200 hPa, with the winter (NH) value being about three times greater than the summer value. Middle latitudes in both hemispheres are dominated by convergence, again with maximum values occurring near 200 hPa; the magnitudes in each hemisphere, however, are comparable. At high latitudes, flux divergence returns to both hemispheres, and maxima of these negative values, which are comparable, occur at 300 hPa. Term (c), shown in Fig. 2.12, is negative over most of the globe, indicating that advection of zonal momentum by the mean meridional motions acts to reduce the strength of the zonal flow. Most values of term (c), however, are small, except in the upper tropical troposphere of the Northern (winter) Hemisphere. There, the maximum (negative) value occurs at 200 hPa near 15°N. This location is where the poleward branch of the strong wintertime Hadley cell is advecting easterly (or weak westerly) momentum against the zonal-wind gradient, thus attempting to reduce its speed. In this context, it supports the role played by the eddies. The primary force that acts in opposition to the momentum flux convergence

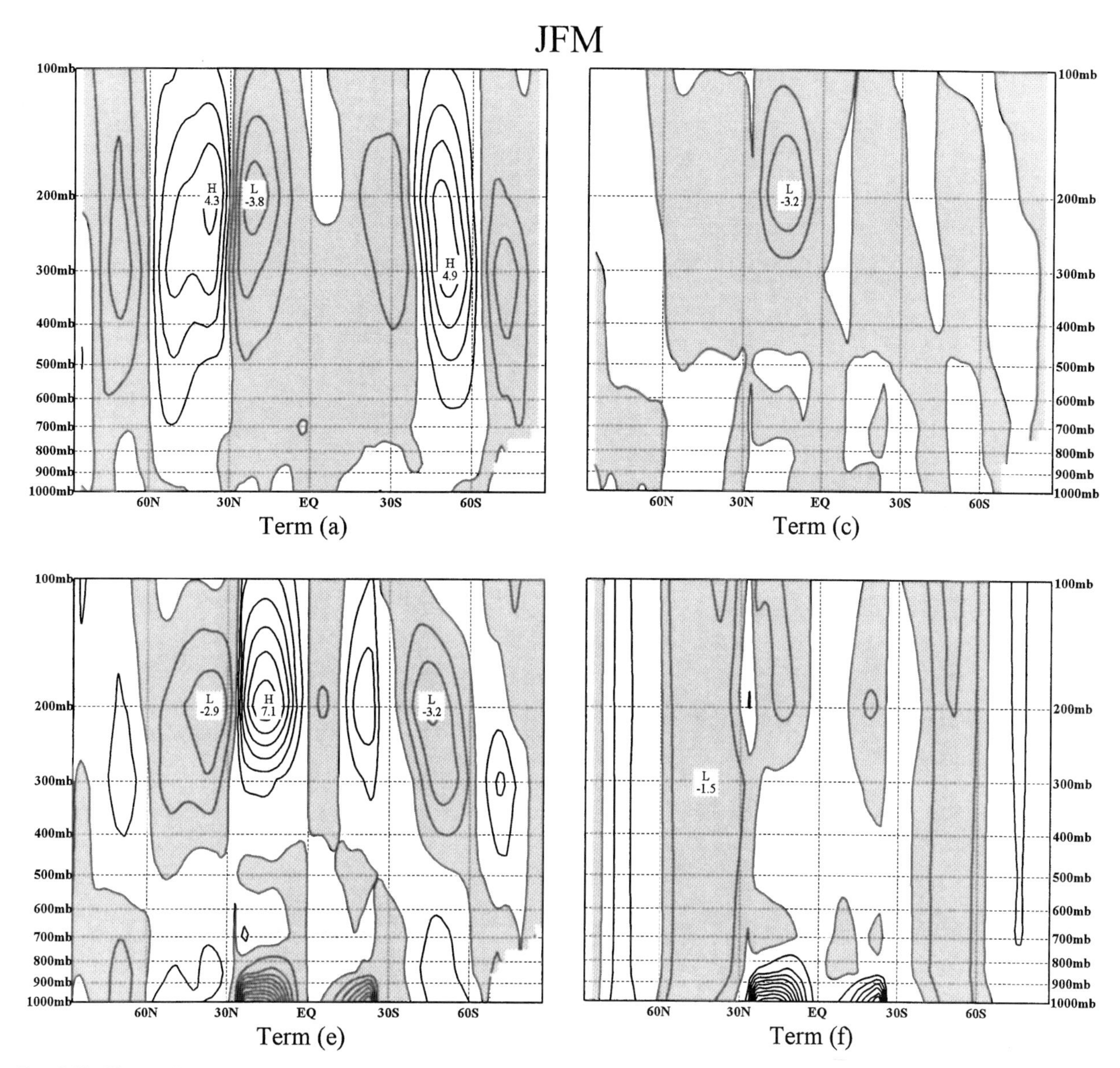

FIG. 2.12. Time and zonally averaged terms of the maintenance of the zonal wind for JFM, according to (2.10). Contour interval is 1×10^{-5} m s^{-2}. Shaded areas represent a decrease of the westerly wind.

[term (a)] in the upper troposphere is the Coriolis force [term (e)]. Figure 2.12 shows that, above approximately 500 hPa, the Coriolis force generally acts to increase or locally accelerate the zonal wind in the Tropics, decrease or decelerate it in the middle latitudes, and increase or accelerate it again at the high latitudes. Maxima of both positive and negative values occur at 200 hPa, except at high latitudes, where maxima are at 300 hPa. In the lower half of the troposphere, the values reverse sign, since the distribution of this term responds to the mean meridional wind, $[\bar{v}]$ (see Fig. 1.40). As expected, lower tropospheric maxima occur in the boundary layer. Last, the residual term [term (f)] is shown in Fig. 2.12. It represents the impact of the frictional force plus any effects caused by subgrid-scale processes, the local rate-of-change term, and errors in other terms. The pattern is very encouraging in that it provides a qualitative picture of what the frictional force might be expected to look like. For example, it tends to decrease or decelerate the westerly momentum of the air throughout the midlatitude troposphere in both hemispheres. At high latitudes and throughout much of the Tropics, where easterly winds dominate, this term is positive. This implies that easterly momentum (wind) is being decreased (decelerated).

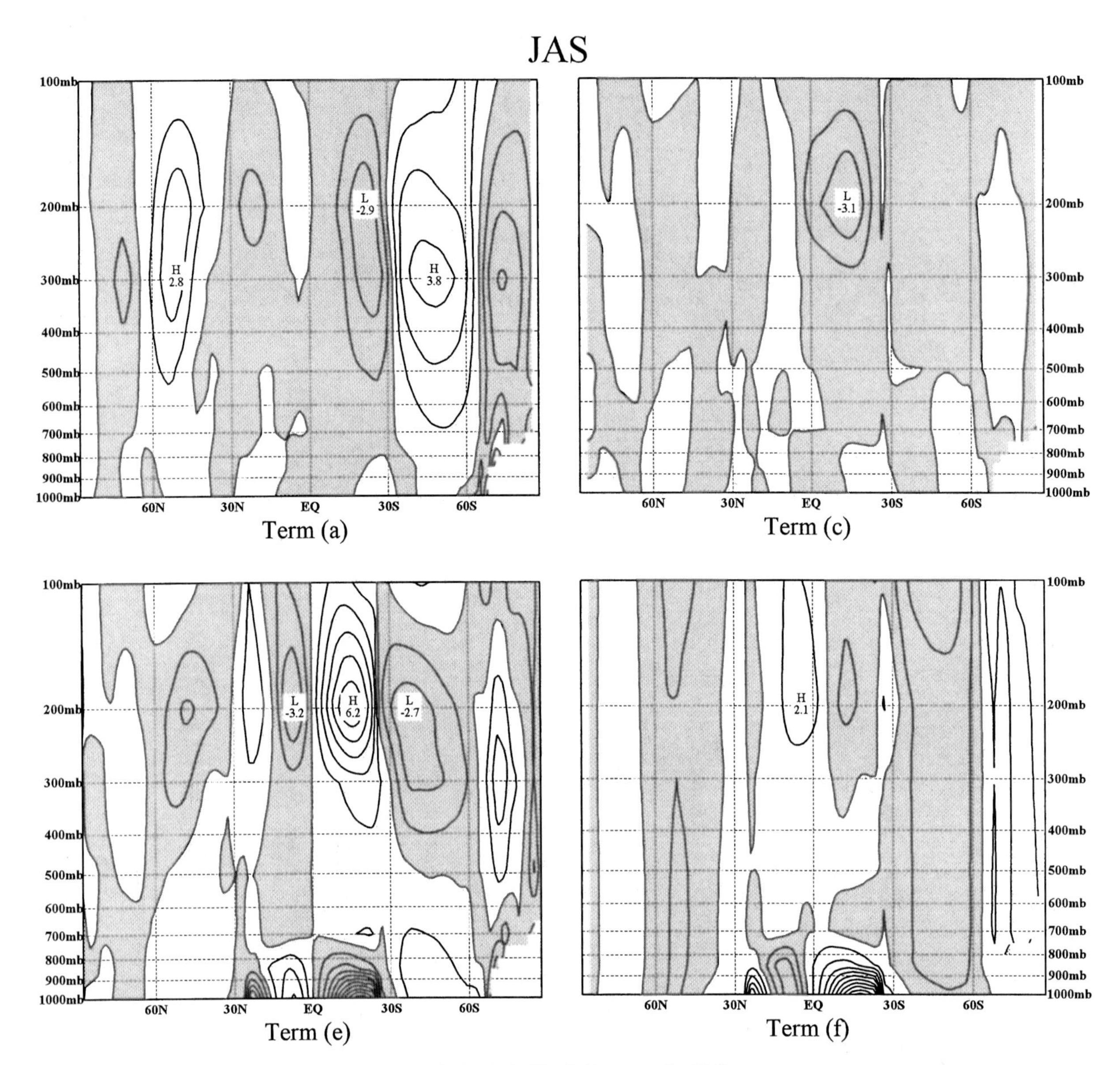

FIG. 2.13. As in Fig. 2.12, except for JAS.

The same four terms are illustrated in Fig. 2.13 for JAS. For the most part, the patterns are mirror images of those for JFM. As expected, the maximum magnitudes of term (a), the momentum flux convergence, are less in JAS than they were in JFM for the NH. The SH Tropics also show a greater negative value for the maximum in winter (JAS) compared to summer. In the middle latitude SH, however, the maximum value of term (a) is greater in summer than it is in winter, even though the vertical integrals are of similar magnitude (Fig. 2.11). As for JFM, convergence maxima in JAS occur between 200 and 300 hPa. Results of term (c), the horizontal advection by mean meridional motions, for JAS are nearly identical to those for JFM, in that the only significant contribution occurs in the upper branch of the wintertime Hadley cell where weak westerly (or sometimes easterly) momentum is transported poleward toward the westerly jet. The Coriolis force [term (e)] shows the expected pattern, dictated by the three-cell mean meridional circulations in both hemispheres. As for JFM, the more intense Hadley cell in JAS occurs in the winter hemisphere. The magnitude of this term in the upper troposphere at 15°S is slightly less in JAS than it was at 15°N in JFM. At other latitudes, the SH shows little seasonal variation in term (e), whereas the maxima of this term in the middle- and high-latitude cells of the NH are much greater in the winter season than they are in the summer season. The frictional force [term (f)] shows the same pattern as it did in JFM. Values are negative

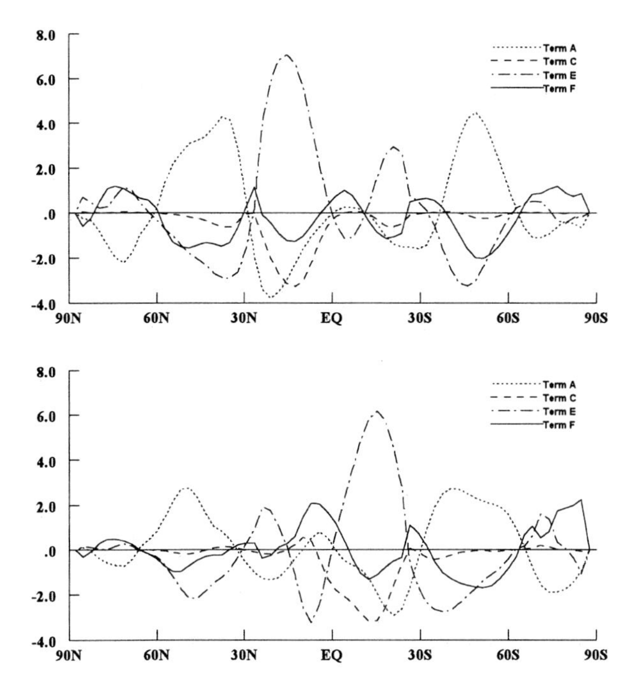

FIG. 2.14. Time and zonally averaged terms of the maintenance of the zonal wind at 200 hPa, in 10^{-5} m s^{-2}, for JFM (top) and JAS (bottom).

(positive) where westerly (easterly) winds tend to dominate, and maxima (minima) in this term are located where maximum easterly (westerly) winds occur. Thus, as expected, friction acts to decrease (increase) westerly (easterly) momentum everywhere.

The only appropriate reference with which to compare our results is Newell et al. (1972). They present vertical cross sections of the momentum flux convergence [term (a)] and Coriolis force [term (e)] for all four seasons that show excellent agreement with the present results in the NH. In the SH, their patterns are similar to those given here, but the magnitudes of the present results are greater for both terms, especially in the upper troposphere. Thus, the intensity of the circulation by both eddies and mean motion is probably stronger in the SH than had been thought more than two decades ago. No doubt, this is due to better sources of data over the southern oceans with regard to wave activity. Furthermore, the middle- and high-latitude estimates of $[\bar{v}]$ presented here were based on momentum distributions (see introduction of section 2.3).

We close this section with a graph showing the latitudinal distributions of the four major terms just discussed at 200 hPa for JFM and JAS (Fig. 2.14). This level is at or near the location of maximum values for each term. It is seen that in the tropical winter hemisphere the zonal wind is maintained by a strong acceleration due to the Coriolis force, which is opposed by moderate forcings due to eddy momentum flux divergence and mean meridional advection, plus a weaker deceleration due to friction. In the tropical summer hemisphere, the balance among forces is more complex, and most forces, except friction, are weaker than their counterpart across the Equator. In middle latitudes, the zonal wind is maintained by a strong convergence of westerly momentum in both hemispheres and in both seasons. This acceleration is offset by decelerations due to the Coriolis force in the Ferrel cells, as well as to friction. The contribution by mean meridional circulations [term (c)] is negligible at middle and high latitudes. In the latter region, the roles of the three remaining terms are reversed from what they were in middle latitudes.

b. Sensible heat budget

The equation used to diagnose the sensible heat budget is a form of the First Law of Thermodynamics, written in terms of contributions due to time- and zonally averaged mean motions and transient and stationary eddies. The form of the equation resembles that given by Newell et al. (1974) and is expressed as follows:

$$\frac{\partial[\bar{T}]}{\partial t} = \left[\frac{\overline{dh}}{dt}\right] + \Gamma[\bar{\omega}] - [\bar{v}]\frac{\partial[\bar{T}]}{\partial y} - \frac{\partial}{\partial y}([\overline{v'T'}] + [\bar{v}^*\bar{T}^*])$$

(a) (b) (c) (d)

(2.11)

$$-\frac{\partial}{\partial p}([\overline{\omega'T'}] + [\bar{\omega}^*\bar{T}^*]) + \frac{R_d}{c_p p}([\overline{\omega'T'}] + [\bar{\omega}^*\bar{T}^*]).$$

(e) (f)

Seasonal means of the left side of (2.11) are small compared to other terms and are nearly zero for the solstice seasons discussed here. Contributions to the right side, expressed as temperature increases, are (a) total diabatic heating, which includes shortwave and longwave radiation, latent heat release, sensible heat exchange with the underlying surface, and conduction; (b) adiabatic subsidence, where Γ is a stability parameter defined as

$$\Gamma = \frac{R_d}{c_p p}[\bar{T}] - \frac{\partial[\bar{T}]}{\partial p};$$

(c) meridional advection of warm air by mean motions; (d) meridional eddy heat-flux convergence; (e) vertical eddy heat-flux convergence; and (f) vertical eddy heat flux.

The units in (2.11) are Kelvins per day, and the dataset and symbols used are the same as for the momentum budget. All terms were calculated, except for term (a). It was determined as the residual. Additionally, term (f) was found to be negligibly small and results for it will not be shown. As for momentum, the discussion begins with meridional transports by mean

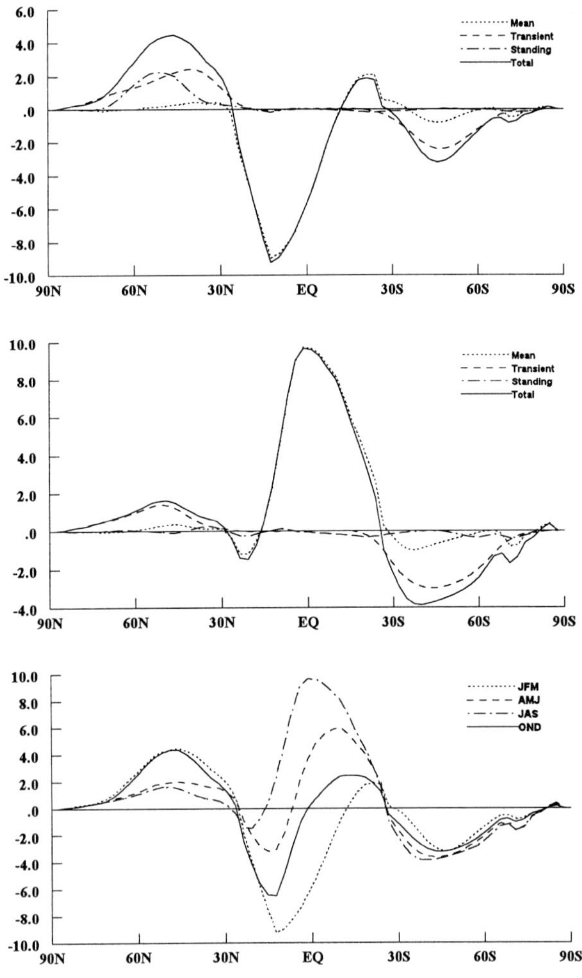

FIG. 2.15. As in Fig. 2.9, except for sensible heat transport. Units are 10^{15} watts.

and eddy motions. The eddy heat transports, which appear in term (d) of (2.11), were introduced in section 2.2. They will also appear later, together with the transport by mean motions, in discussions of the total energy transports (section 2.3d). Values of $[\bar{v}]$ in term (c) were determined as described in the introductory remarks of section 2.3.

Latitudinal distributions of the vertically integrated, zonally averaged transports of sensible heat by the mean meridional circulations and the transient and stationary eddies for JFM and JAS are shown in Fig. 2.15. Also shown is the total transport for all four seasons. It is seen that the only significant contribution to the total heat transport at low latitudes, in both JFM and JAS, is that made by the mean motions. The seasonal shift of the Hadley circulation dictates the direction of the transport. Since the temperatures decrease with height, it is the lower branch(es) of the Hadley cell(s) that determines (determine) whether the vertically integrated transport will be northward or southward. During the solstice seasons, there is one dominant cell centered in the winter hemisphere that transports heat in the deep Tropics southward in JFM and northward in JAS. This results in a cross-equatorial flow of heat from the winter to the summer hemisphere. In middle latitudes, transient eddies generally become the dominant mechanism for transporting heat, and the direction of the transport is poleward in both hemispheres. The one exception occurs in the NH winter season, when a poleward flux by stationary eddies becomes more important and supports the transient eddy flux. Maximum transports by eddies meander between 40° and 50° latitude and, as expected, change location more with season in the NH. Of secondary, but not negligible, importance in middle latitudes are the mean motion transports by Ferrel cells, which are also poleward.

The total transport for all four seasons is shown in the bottom panel of Fig. 2.15. The distributions reflect the contributions by mean motions at low latitudes, and primarily those by transient eddies at higher latitudes. The seasonal shift of the transport due to the changing Hadley circulation causes an appreciable cross-equatorial flow of heat from the winter hemisphere to the summer hemisphere. Furthermore, the annual average transport of sensible heat (not plotted) is from the thermal equator toward the polar regions in both hemispheres, as required for maintenance of the global temperature distribution.

Attention now shifts to the terms in the heat budget equation given in (2.11). As indicated earlier, only terms (b) through (e) will be presented. Vertical cross sections of these terms, from 90°N to 90°S, are shown in Figs. 2.16 and 2.17 for JFM and JAS. It is seen that adiabatic subsidence [term (b)] and meridional and vertical heat-flux convergence [terms (d) and (e)] are the major contributors toward maintaining the temperature and that their order of magnitude is several tenths of a Kelvin per day. Term (c), on the other hand, which measures the meridional advection by mean motions, is an order of magnitude smaller and only approaches the other terms in the upper tropical troposphere. Thus, no further comments regarding this term will be given. Term (b) contributes to temperature increases when subsidence occurs and to decreases when rising air cools adiabatically. Since the zonally averaged stability parameter, Γ, is positive, the sign of this term is determined by the rising and sinking branches of the mean meridional cells. Figure 2.16 clearly shows the effects of the three-cell pattern in each hemisphere during JFM. There is strong adiabatic cooling just south of the Equator associated with the rising motion in the Hadley circulation. Maxima of warming occur in the sinking branches of the Hadley cells, with the greatest warming taking place in the NH in response to the stronger wintertime circulation. The rising motion in the poleward branches of the Ferrel cells in both hemispheres contributes to temper-

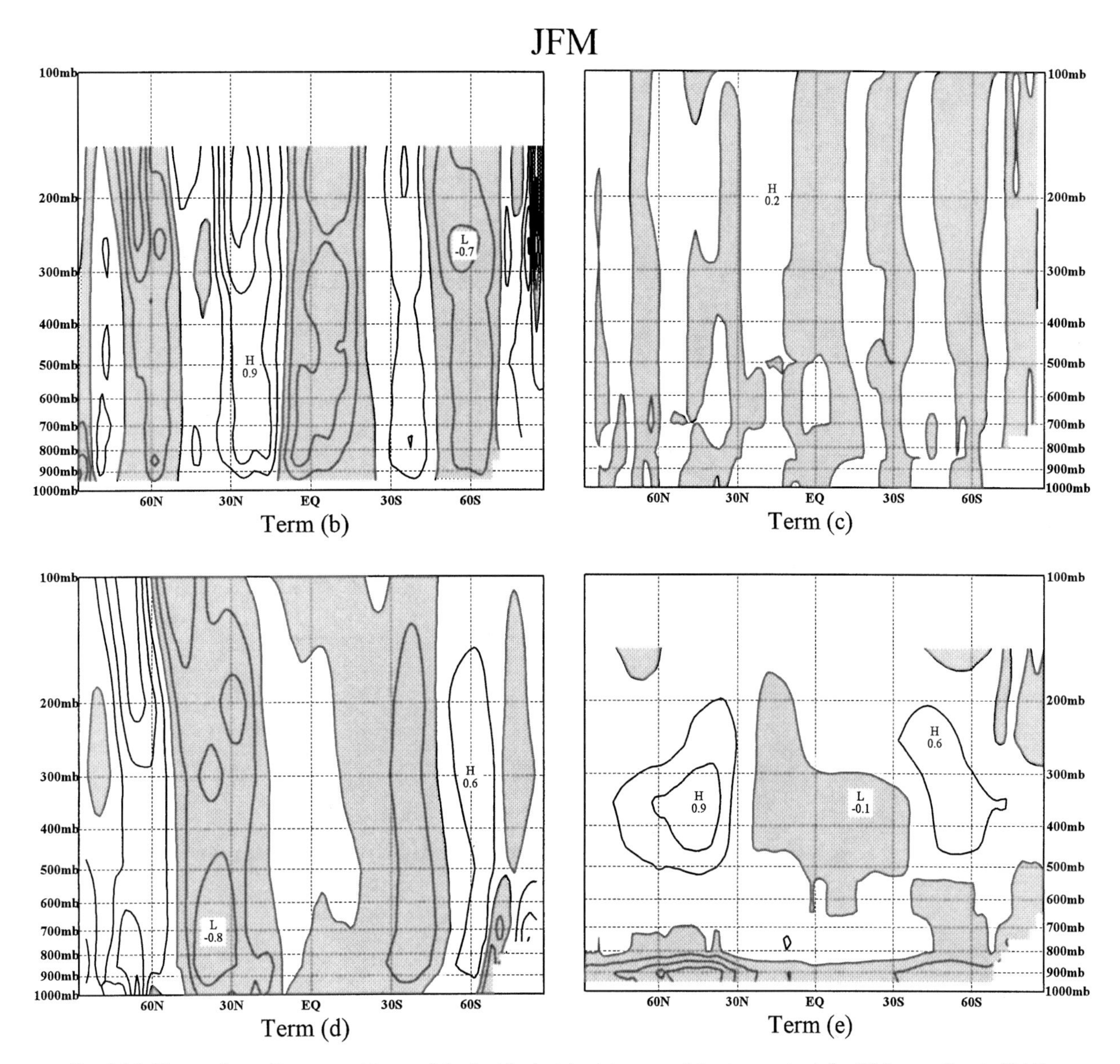

FIG. 2.16. Time and zonally averaged terms of the heat budget (maintenance of the temperature) for JFM, according to (2.11). Contour interval is 0.3 K day^{-1}. Shaded areas represent a decrease in heating rate.

ature decreases, while the return flow, due in part to the thermally direct high-latitude cells, causes warming over much of the polar regions. As for the Hadley cell, the Ferrel cell in the NH exhibits a stronger circulation than its counterpart in the SH.

Term (d) in (2.11) represents the heating (cooling) of the air due to meridional heat-flux convergence (divergence) caused by transient and stationary eddies. Figure 2.16 shows that there is a weak narrow band of convergence in the equatorial region and a much stronger band of convergence at higher latitudes in both hemispheres. Located between these zones of convergence are wide bands of heat-flux divergence that span from subtropical to middle latitudes. Thus, the role of the eddies is to remove heat from the temperate latitudes and deposit it, through poleward transports, at higher latitudes, where solar input is minimal. In this manner, the heat transport tends to destroy the temperature gradient that differential diabatic heating continues to create. Term (e), the vertical heat-flux convergence by transient and stationary eddies, makes an important contribution to the heat budget at middle and higher latitudes. At low latitudes, it is negligible. Figure 2.16 shows that this term reaches maximum values of heating in the upper troposphere, both at 45°N and at 45°S. The greater heating rate, as well as a lower level of maximum heating, occurs in the winter hemisphere. Below about 800

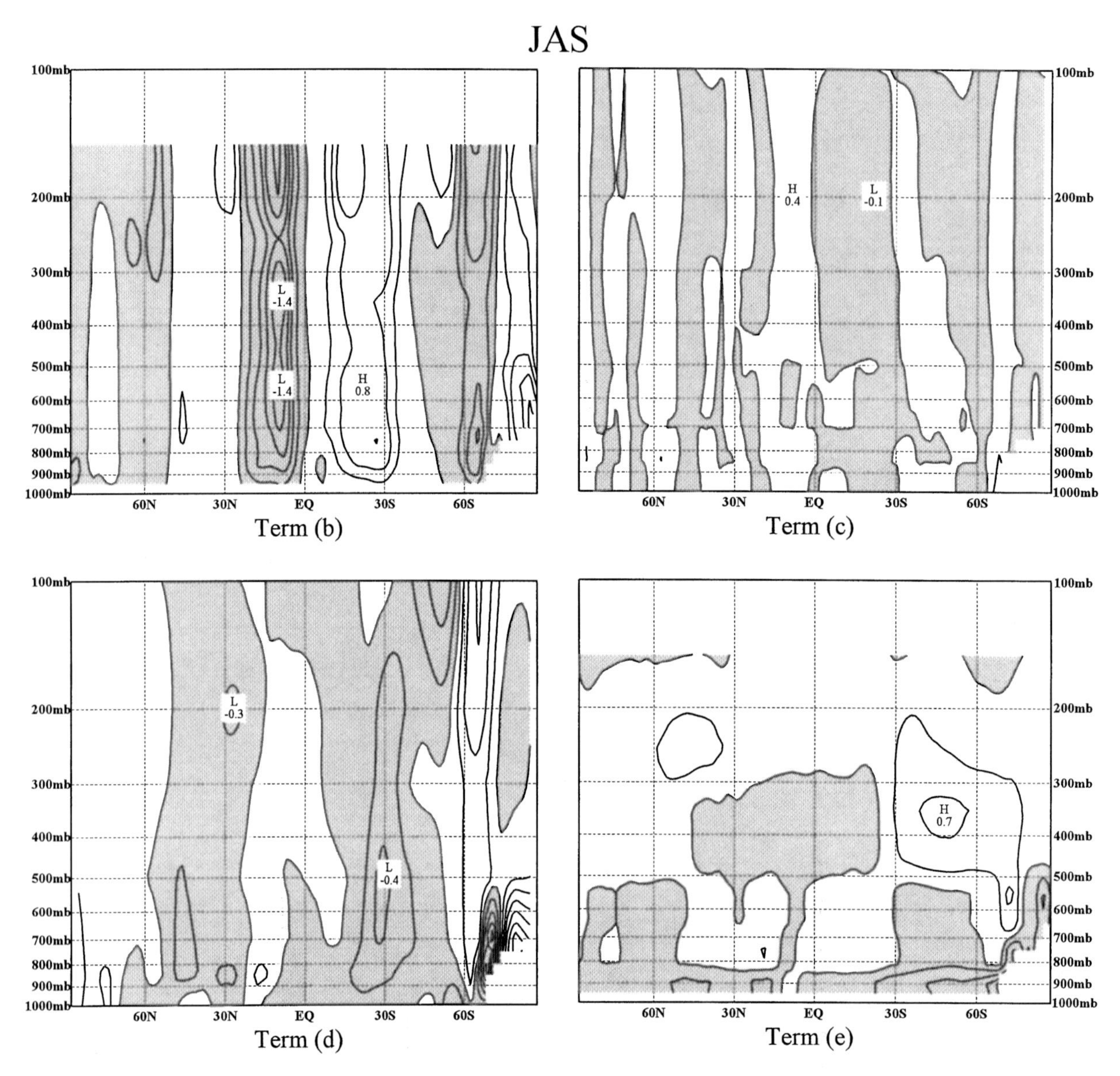

FIG. 2.17. As in Fig. 2.16, except for JAS.

hPa, strong divergence (cooling) takes place, particularly in the NH. Thus, the role of this term is to remove heat from the boundary layer and deposit it in the middle and upper troposphere and, as for term (d), this heat transport is down the temperature gradient. This transport is primarily accomplished by lower tropospheric disturbances, poleward of the main jet streams.

The results of the heat budget terms for JAS (Fig. 2.17) are only briefly summarized, since they generally represent a mirror image (with respect to the thermal equator) of the JFM results. Term (b) shows the expected adiabatic heating and cooling distributions that are associated with the three-cell patterns of sinking and rising air in the mean meridional circulations in both hemispheres. As for JFM, the winter hemisphere circulations in JAS are strongest. The patterns of meridional and vertical eddy heat-flux convergence [terms (d) and (e)] are similar to their counterparts in JFM if small seasonal shifts are taken into account. Again, the greater contributions occur in the winter hemisphere, although, not surprisingly, the seasonal change is much less in the SH than in the NH for both terms.

The patterns described above with regard to term (e) are easily seen in Fig. 2.18, which shows cross sections of the vertical flux of heat by transient eddies for JFM and JAS. It is clear that there is a strong upward transport of heat in middle latitudes in both seasons

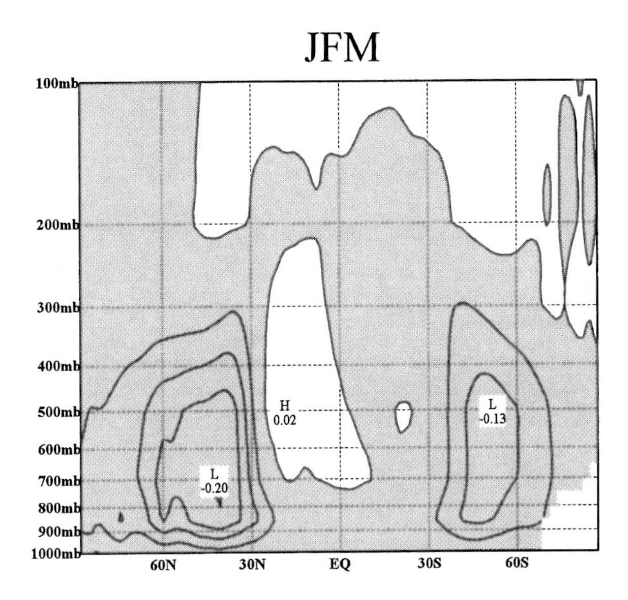

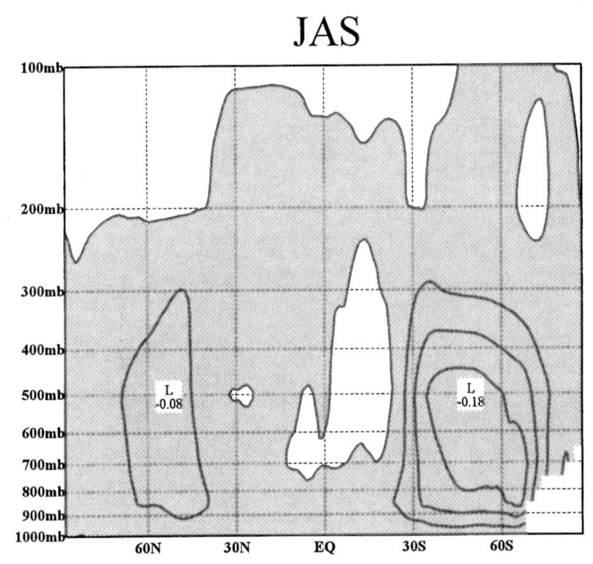

FIG. 2.18. Time and zonally averaged vertical flux of sensible heat by transient eddies, $[\overline{\omega' T'}]$, for JFM (top) and JAS (bottom). Contour interval is 0.05 K Pa s^{-1}. Shaded areas represent temperature decreases due to warm air rising or cold air sinking.

and both hemispheres, which helps to offset the radiative cooling that occurs there. Trenberth (1992) shows cross sections of $[\overline{\omega' T'}]$, but for January and July. As for meridional fluxes, his vertical fluxes are based only on the years 1986–89. Nevertheless, there is very good agreement between his results and those presented here.

The sum of terms (b) through (e) were already shown for JFM and JAS in Fig. 2.1. If there were no errors in these terms, as well as no other important contributions to the heat budget, the distributions seen in that figure would represent total diabatic heating. In this regard, the results in Fig. 2.1 seem quite reasonable. For example, near the Equator, centered slightly into the summer hemisphere, are bands of positive heating rates. These extend from 10°N to 25°S in JFM, and from 0° to 25°N in JAS, and reach maxima in the middle and upper troposphere. No doubt this pattern is due to latent heat release dominating over adiabatic cooling, as well as net radiative cooling, in the ascending branch of this Hadley circulation (Figs. 2.16 and 2.17). In middle latitudes, there are also bands of heating; however, they do not extend above 400–500 hPa, since latent heat release associated with extratropical precipitating systems does not often reach as high a level as it does in the Tropics.

Except for the two regions discussed above, Fig. 2.1 shows that diabatic cooling is generally present. This implies that net radiative cooling dominates over latent heat release, which seems reasonable, especially since the values are on the order of several tenths of a Kelvin per day. At subtropical and lower midlatitudes, this cooling is supported by the meridional flux divergence of eddy heat. The primary heat source that balances these heat losses is subsidence warming in the sinking branch of the Hadley circulation. At higher latitudes, the primary sink for heat is probably longwave radiation, while the main source is eddy heat-flux convergence. The polar direct cells also play a role, but their contribution is secondary. The role of the vertical fluxes, as noted earlier, is to remove heat from the boundary layer, where sensible heat exchange from the underlying surface has undoubtedly provided a supply, and carry it to the middle and upper layers of the middle-latitude troposphere to help counterbalance some of the radiative cooling that occurs there. To highlight the physical processes just described, terms (b), (d), and (e), as well as the residual [which approximately represents term (a) in (2.11)], are illustrated at selected pressure levels in Figs. 2.19 and 2.20 for JFM and JAS.

To conclude this section, it is worth mentioning that Newell et al. (1974) also presented vertical cross sections and latitudinal plots of the major terms in (2.11). Their general findings were similar to those described here. For example, they parameterized the diabatic heating components and combined them with observed values for terms (b), (c), and (d) in (2.11). They conjectured, therefore, that their residual primarily consisted of the contribution by term (e), the vertical eddy heat-flux convergence, plus any errors in other terms. In the present set of results, term (e) was calculated, and term (a), the diabatic heating, was derived as the residual. A careful examination of Newell et al.'s residual pattern shows that it requires a distribution of term (e), which is compatible with the distribution found here.

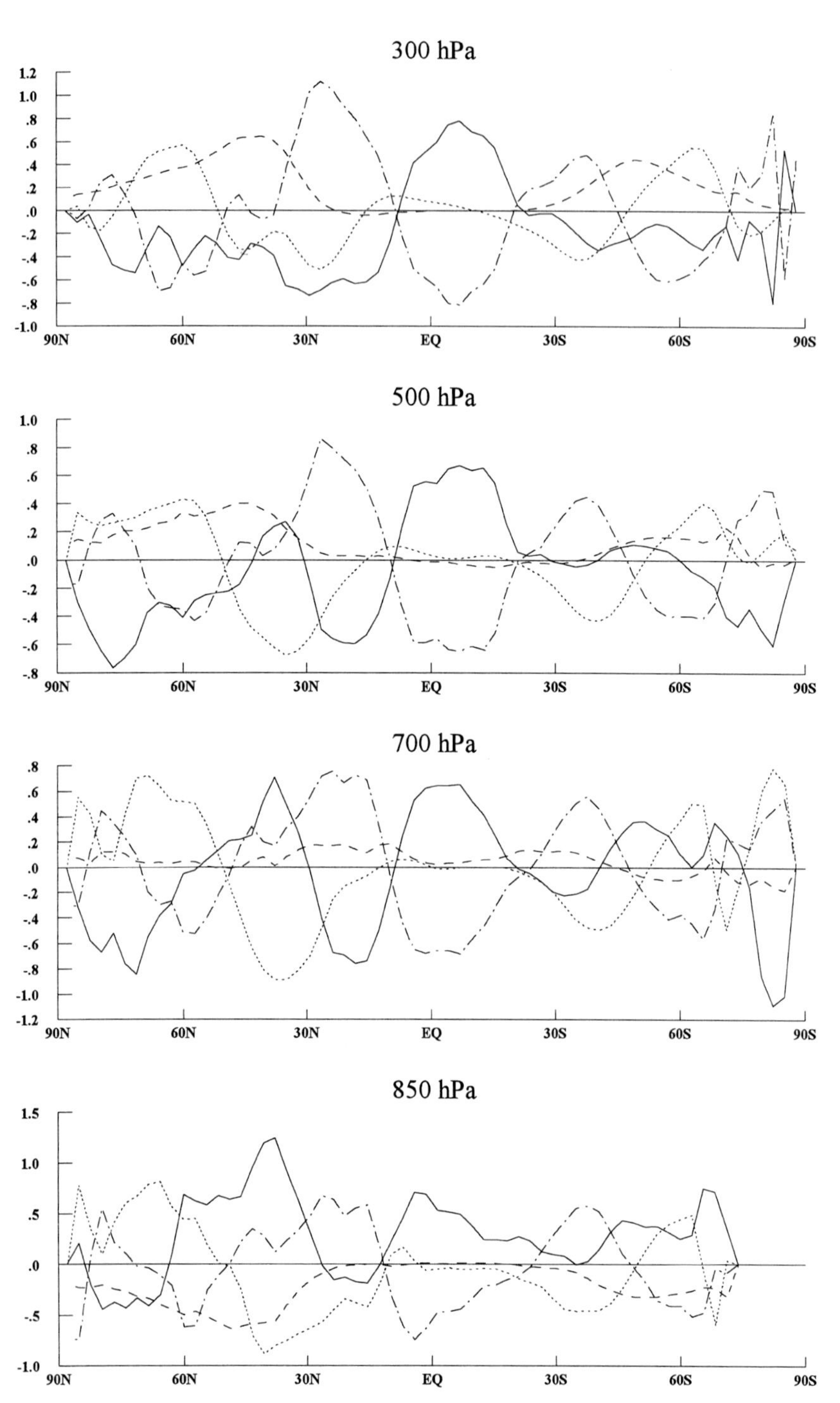

FIG. 2.19. Time and zonally averaged values of horizontal eddy heat divergence (dotted), vertical eddy heat divergence (dashed), subsidence heating (dash–dot), and diabatic heating (solid) for JFM at 300 (top), 500 (middle top), 700 (middle bottom), and 850 hPa (bottom). Units are K day^{-1}.

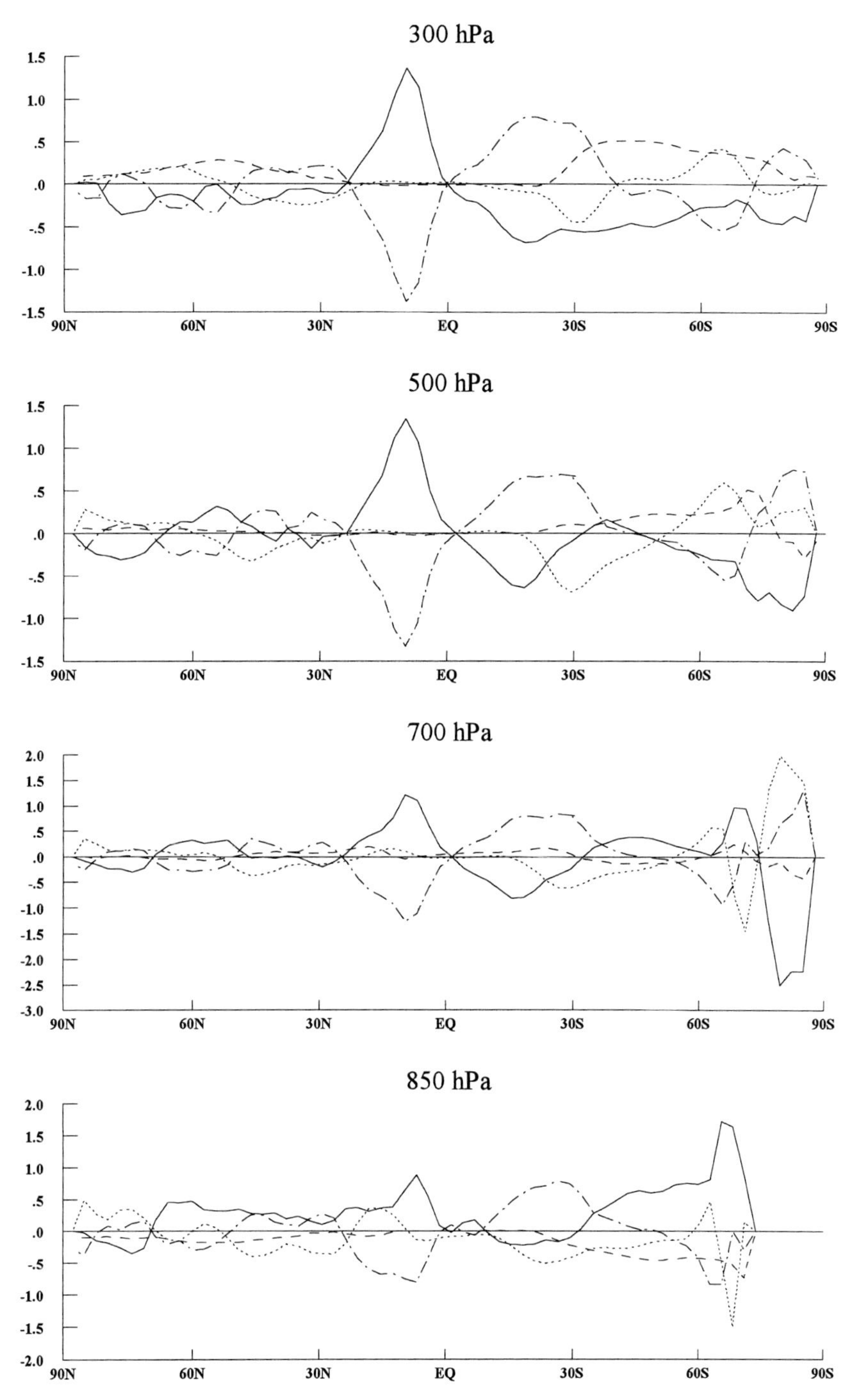

FIG. 2.20. As in Fig. 2.19, except for JAS.

c. Water vapor budget

The conservation equation for water vapor (specific humidity) is:

$$\frac{dq}{dt} = S_q, \qquad (2.12)$$

where S_q represents the sources and sinks of vapor. By using the equation of continuity in isobaric coordinates, (2.12) can be written as

$$\frac{\partial q}{\partial t} = -\nabla_p \cdot q\mathbf{V} - \frac{\partial}{\partial p}(q\omega) + S_q. \qquad (2.13)$$

When (2.13) is vertically integrated, the result is

$$\frac{\partial W}{\partial t} + \nabla \cdot \mathbf{Q} = E - P, \qquad (2.14)$$

where

$$W = \frac{1}{g}\int_{p_o}^{p_s} q dp = \text{precipitable water}; \qquad (2.15)$$

$$\mathbf{Q} = Q_\lambda \mathbf{i} + Q_\phi \mathbf{j};$$

$$Q_\lambda = \frac{1}{g}\int_{p_o}^{p_s} (uq) dp = \text{zonal transport of vapor}; \qquad (2.16)$$

$$Q_\phi = \frac{1}{g}\int_{p_o}^{p_s} (vq) dp = \text{meridional transport of vapor}; \qquad (2.17)$$

E is the evaporation at the surface; and P is the precipitation. As for the previous budget equations, the moisture transports in (2.16) and (2.17) can be partitioned into time- and zonally averaged mean and eddy components. When (2.14) is zonally averaged, the contribution from (2.16) vanishes.

Global maps of precipitable water, W, for JFM and JAS were presented and discussed in chapter 1 (Fig. 1.48). In this chapter, therefore, although monthly mean maps of W were produced, they are not shown. Instead, Fig. 2.21 is presented; it shows a time–latitude diagram of the annual cycle of W. To maintain consistency with the mean state moisture variables presented in chapter 1 and elsewhere in this chapter, ECMWF analyses for only the three-year period, May 1986 through April 1989, were used to compile the budget results discussed in this section. The same period was used by Vincent and Schrage (1995) in their atlas of TOGA Coupled Ocean–Atmosphere Response Experiment (COARE) moisture climatology and appeared to give acceptable results over much of the Tropics. Further explanation and justification for using this 3-yr period can be found in Trenberth (1992), as well as in Vincent and Schrage (1995) and chapter 1 herein. Figure 2.21 shows that, in both hemispheres, maximum values of W occur in midsummer at middle and high latitudes, whereas they peak about one month later in the Tropics. It is seen that the seasonal change is considerably less in the SH than it is in the NH. For example, at 30°S, the value of W ranges only from a maximum of 3.0 cm in February to a minimum of 1.9 cm in August, while at 30°N the values for the same two months are 1.7 and 3.5 cm. The seasonal difference is even more dramatic at 60° latitude, where the SH values for January and July are 1.2 and 0.7 cm and those for the NH are 0.5 and 2.2 cm. The axis of maximum values meanders from about 5°S in southern summer to approximately 10°N in the northern summer and early fall. Maxima slightly in excess of 5 cm occur at 5°N in May and again at 10°N from July to September. The overall patterns and magnitudes are very similar to those in Newell et al. (1972), except values at low latitudes are slightly greater in the present analysis. Finally, it is seen that the high-latitude SH is much drier than its counterpart in the NH, especially during their respective summer seasons. This was also noted by Peixoto and Oort (1992), who reasoned that the higher altitudes and lower temperatures over Antarctica were responsible for the reduced values of W over the southern polar region. Their Fig. 12.20 is reproduced here as Fig. 2.22, and it shows that the moisture content is greater over the NH polar cap than over the SH polar cap in all months. Moreover, in northern summer, the NH value is three to four times greater than the SH value. A more detailed discussion of high-latitude moisture distributions in the SH can be found in chapter 4 of this monograph.

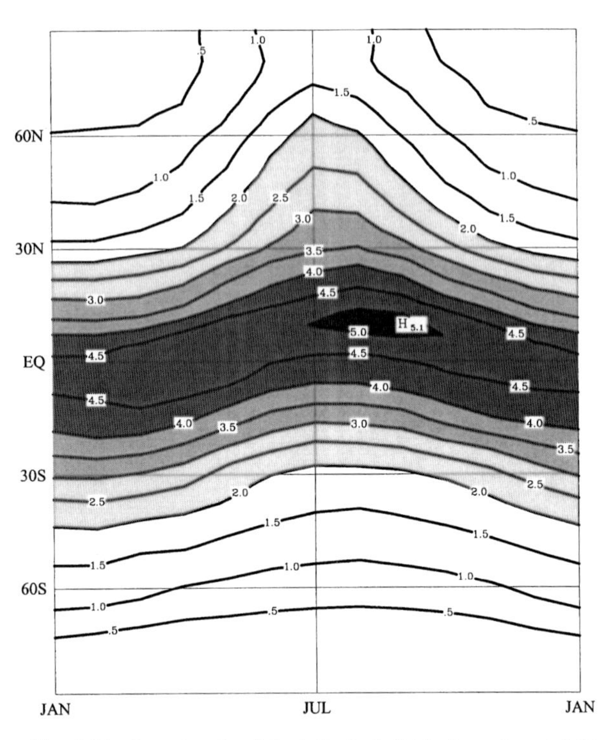

FIG. 2.21. Annual cycle of the latitudinal distribution of precipitable water, W, in cm. Contour interval is 0.5 cm. Shaded regions have $W \geq$ 2.0 cm.

The latitudinal distribution of the vertically integrated meridional moisture transport terms are shown in Fig. 2.23 for the JFM and JAS seasons. Also shown are the distributions of the total transport for all four seasons. The seasonal transports are dominated by the mean meridional circulations at low latitudes, whereas,

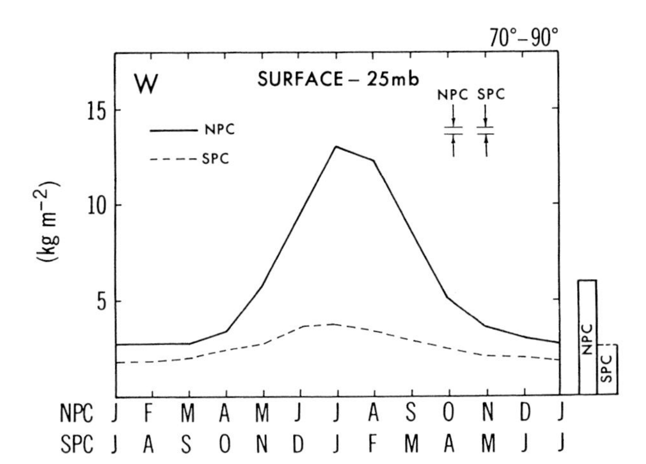

FIG. 2.22. Annual cycle of precipitable water, W, for the north (solid) and south (dashed) polar caps in kg m^{-2}. (Reproduced from Peixoto and Oort 1992.)

at middle and high latitudes, the transient eddy flux term dominates. Since the lower branch of the Hadley circulation contains the majority of the moisture, the maxima (absolute values) of the integrated transport occur near 10°N in JFM and near the Equator in JAS. This results in a cross-equatorial transport from the winter into the summer hemisphere, which is in agreement with many previous results (e.g., Chen et al. 1996).

With regard to transient eddies, Fig. 2.23 shows that their role is to transport moisture away from the Equator toward higher latitudes. In the NH, there is a 20° latitudinal shift in the location of the maximum flux between the winter and summer seasons. In contrast, there is very little latitudinal shift with season in the location of transient flux maxima in the SH. In all four seasons, the maximum flux occurs near 40°S. The primary reason for the large latitudinal shift in the NH compared to the SH is, of course, that the former has a higher percentage of land, and therefore its annual cycle is more influenced by land–sea contrast. As seen in Fig. 2.23, the contribution by the stationary eddies is generally small, especially in the SH. Nevertheless, they typically complement the transient eddies and transport moisture poleward.

The bottom panel in Fig. 2.23 illustrates the latitudinal distributions of the total integrated moisture transport for all four seasons. As indicated earlier, the major fluctuations occur at low latitudes due to seasonal changes in the Hadley circulation. It is seen that, throughout most of the year, the SH, with its vast oceanic reservoir, is supplying moisture to the NH, again in agreement with other studies (e.g., Chen et al. 1996). This cross-equatorial transport is greatest in JAS and helps provide fuel for many of the NH's summer monsoon circulations (e.g., the southwest African and Indian monsoons, which account for more than one half of the low-level transport in northern summer). The only season in which the transport is into the SH is in the southern summer season. This was true for sensible heat transport as well, which is not too surprising since these two variables often complement each other. The bottom panel also shows that in all four seasons the vapor transports are poleward from about 25°N to the North Pole and 25°S to the South Pole. In addition, they are equatorward between 25°N and 25°S in all seasons except summer. It is well known that one of the main sources of water vapor for the global atmosphere occurs in the regions occupied by oceanic subtropical highs (e.g., Lorenz 1967 and Newell et al. 1972).

In the original SH monograph (van Loon et al. 1972), the moisture transports were determined as residuals from the $E–P$ budget (Newton 1972). Newton used values of evaporation and precipitation based on the work of prior authors to compute the right side of (2.14), which yielded $\nabla \cdot \mathbf{Q}$ from balance considerations. Then, he performed a pole-to-pole integration to arrive at his transports. When Newton compared his

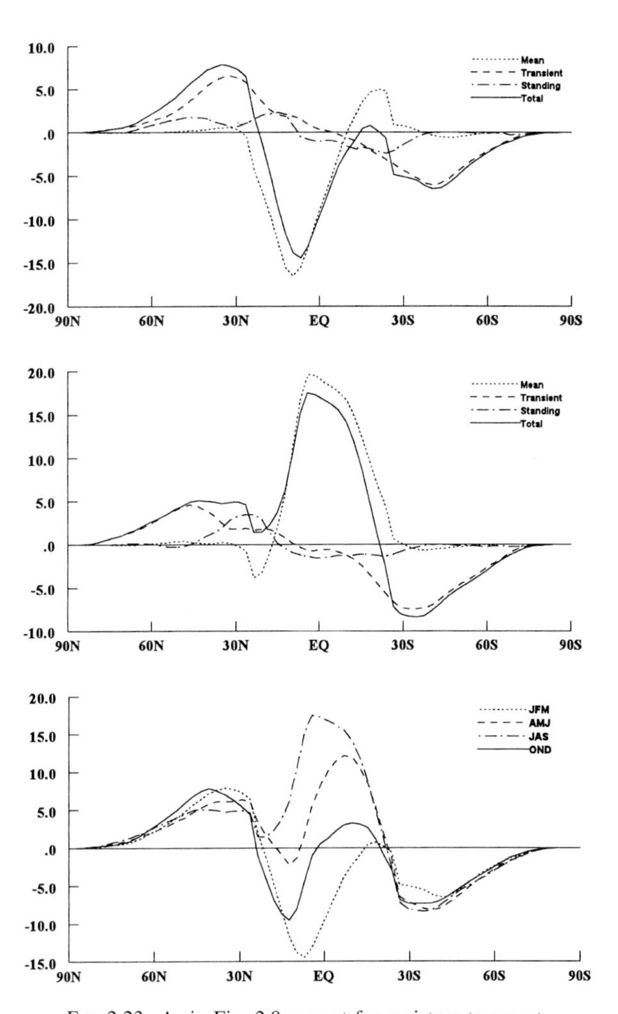

FIG. 2.23. As in Fig. 2.9, except for moisture transport. Units are 10^8 kg s^{-1}.

residual estimates to those from one of the observational studies available at that time (Starr et al. 1969), he found good agreement between the two patterns; his values, however, were generally lower in the NH and higher in the SH than the ones in Starr et al. The latitudinal distributions given by Newton also compare favorably with the present results, especially at low latitudes, where magnitudes of the maximum transports are nearly the same in both studies. Poleward of the subtropical high pressure belts, however, the present results for the winter hemisphere are 20%–30% greater than those given in the original monograph. This seems reasonable, since the transport at these latitudes is mainly performed by eddy motions, and there has been considerable improvement in the documentation of moisture transports by middle latitude disturbances in the last two decades.

Since the date of the original monograph, there have been several papers written about latitudinal distributions of moisture transports (e.g., Rosen et al. 1979; Peixoto et al. 1981; Salstein et al. 1983; Chen 1985; and Peixoto and Oort 1992). Differences in units between these results and the ones given here (Fig. 2.23), as well as lack of global coverage or shortness of time period examined, make it difficult to do quantitative comparisons. Since the latter reference is global, a few remarks are offered regarding how that study compares with the present one. For some reason, the results of Peixoto and Oort are much smoother than the present ones. Nevertheless, there is one location where a noticeable difference between the two sets occurs. This is in the SH between about 20° and 40°S, where Peixoto and Oort's results show almost no seasonal fluctuations in any of the contributions to the total transport. The present results, on the other hand, exhibit identifiable seasonal changes between these latitudes, especially with regard to transient eddies, which show a stronger poleward transport in JAS than in JFM. This trend in the seasonal cycle is well known in the NH and is easily seen in Fig. 2.23, as well as in the results of Peixoto and Oort. The present results, however, reveal that there is also a stronger poleward flux by transient eddies in the SH winter season than was previously thought, at least in lower midlatitudes.

The seasonal shifts in the total vertically integrated moisture transport are illustrated in Fig. 2.24. All of the features discussed earlier are clearly evident in this figure. Due to the meandering Hadley circulation, the maximum northward transport occurs just south of the Equator in July, while the maximum southward transport takes place in January at about 10°N. In middle latitudes, where eddy transports dominate, the axis of maximum flux in the NH shifts from 35°–40° latitude in late winter to 45°–50° in late summer. Values are also much less in summer than in winter. In contrast, the SH transport shows less seasonal variation with latitude. The values, however, do vary with season, as noted above, and fluctuate from maxima in the winter half of the year to minima in the summer half.

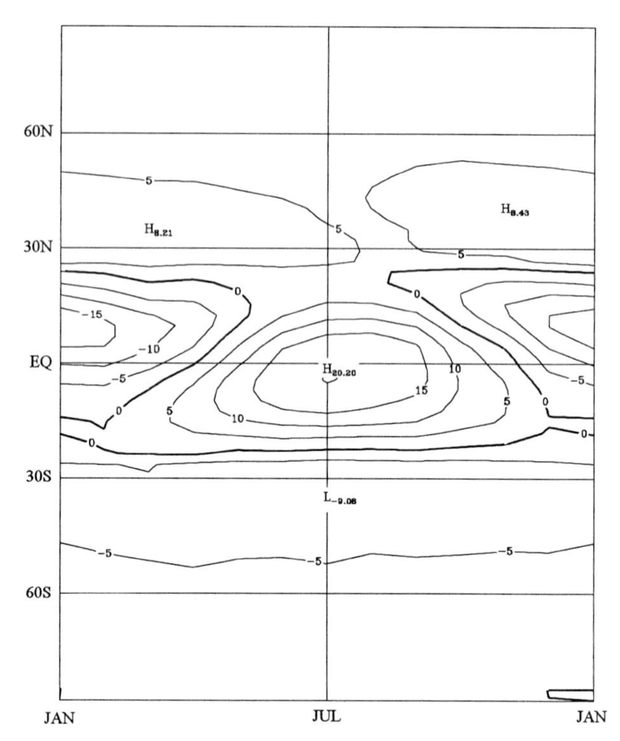

FIG. 2.24. Annual cycle of the latitudinal distribution of total moisture transport. Contour interval is 5×10^8 kg s^{-1}.

Attention now is focused on the water vapor budget as described by (2.14). Figure 2.25 shows the latitudinal distributions of the divergence of the total vapor transport, $\nabla \cdot \mathbf{Q}$, for each of the four seasons. This quantity is essentially equivalent to E–P, especially when all variables have been temporally and zonally averaged. It is seen that a zone of convergence, about 20°–25° latitude in width, exists near the Equator throughout the year. This band meanders with season and is centered (i.e., has its greatest magnitude) near 10°N in JAS and just south of the Equator in JFM, which yields $P > E$ for the equatorial belt. Poleward

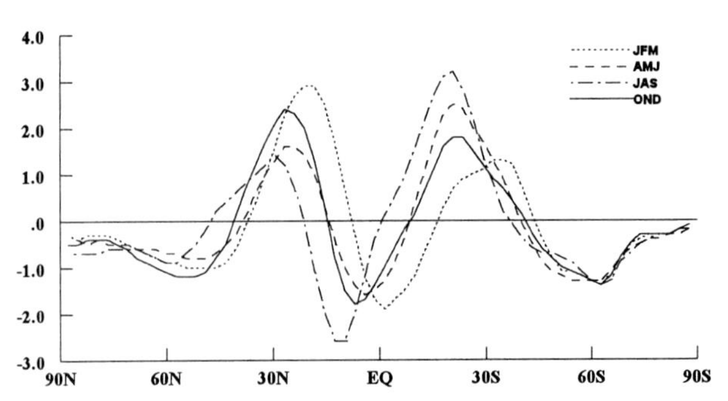

FIG. 2.25. Seasonal distributions of the vertically integrated total moisture divergence [approximately equal to E–P from the moisture budget equation, (2.14)]. Units are mm day^{-1}.

of this convergence zone, there are bands of divergence that extend to about 40°–45° latitude in both hemispheres. In these regions, which include the subtropical high pressure belts, $E > P$. Maxima of divergence meander from approximately 30° in the winter hemisphere to about 20° in the summer hemisphere. Furthermore, wintertime values are twice as large as summertime values. At higher latitudes, convergence ($P > E$) occurs throughout the year, and there is much less variation in the SH, both with season and magnitude, than in the NH. These latitudes are where disturbances associated with the polar jet occur, and the jet in the SH is known to be a more quasi-stationary, semipermanent feature than its counterpart in the NH.

The latitudinal distribution for the annually averaged moisture divergence in the present set of results (not shown) is in good agreement with that derived from balance considerations by Newton (1972). The maxima and minima in both sets of results occur at the same latitudes. The primary difference between the two estimates lies in the strength of the divergence at subtropical latitudes in both hemispheres. In those regions, the present values are much greater than the ones offered in the original monograph. Peixoto and Oort (1992) also show annual averages of the latitudinal distributions of E–P, as well as results for the solstice seasons. Overall, their patterns agree quite well with the present ones; there are, however, two differences worth noting. First, in the northern winter season, the present results show maximum divergence at 20°N, while Peixoto and Oort show it at 15°N. Second, at SH high latitudes (i.e., poleward of 45°S), the convergence computed in the present work is much greater than that given by Peixoto and Oort. They show some annual mean values, however, by Baumgartner and Reichel (1975), which agree very well with the current values at southern high latitudes. Finally, Chen and Pfaendtner (1993) show two sets of latitudinal distributions of $\nabla \cdot \mathbf{Q}$ for the globe and for the northern winter and summer seasons. One set is from ECMWF and is only for the First Global Atmospheric Research Program Global Experiment (FGGE) year; the other is a 10-yr average based on data from the Massachusetts Institute of Technology (MIT) Planetary Circulation Project. In the SH, there are large differences (frequently approaching a factor of two) between their two sets of results, which would suggest that the FGGE year was quite different from normal. On the other hand, the present results (Fig. 2.25) show values that are consistently greater in magnitude (about 25%–50%) than the long-term MIT values. Thus, it appears that SH fluxes and divergence of these fluxes are greater than previously thought.

d. Total energy transports

This section addresses the total (atmospheric plus oceanic) energy transport and, therefore, extracts and combines variables that were discussed in sections 2.4a, b, and c. In this context, the results provide a convenient summary of the atmospheric budgets presented in this chapter. The time- and zonally averaged transport of total energy across latitude circles has been examined by several different authors using a variety of forms for the equation (e.g., Starr 1951; Sellers 1965; Lorenz 1967; Palmen and Newton 1969; van Loon et al. 1972; Newell et al. 1974; Oort and Peixoto 1983; Peixoto and Oort 1992; Trenberth and Solomon 1994; and Gleckler et al. 1995). The form of the equation used here is vertically integrated and can be written as

$$[\bar{H}] = [\bar{h}] + \frac{2\pi a \cos\phi}{g} \int_{p_o}^{p_s} \overline{[v(c_pT + Lq + \Phi + k)]}dp, \qquad (2.18)$$

where $[\bar{H}]$ = total energy transport; $[\bar{h}]$ = heat transport by the oceans; c_pT = sensible heat energy (c_p = specific heat at constant pressure); Lq = latent energy (L = latent heat of condensation); Φ = potential energy; k = kinetic energy; and other variables and symbols are as previously defined. The divergence of $[\bar{H}]$ across each latitude band must meet the radiative balance requirement for the column of air above that band (i.e., net incoming minus net outgoing radiation at the top of the atmosphere). Since this study is based on atmospheric observations, the sources and sinks due to the net radiation were not calculated. Furthermore, ocean heat transports could not be calculated; Trenberth and Solomon (1994) present some values for 1988, however, that they claim were the best available at the time of their article. Thus, even though their values are only for one year, they will be used for comparison to our total atmospheric transports. Finally, in (2.18), the present study neglects the contribution from the kinetic energy transport, which is a common practice (e.g., Newell et al. 1974 and Oort and Peixoto 1983). Results for the three important atmospheric contributions are presented in Fig. 2.26 for JFM and JAS. Also shown in the figure is the sum of these three contributions (which equals the total atmospheric energy transport) for each of the four seasons. The sensible heat and moisture transports have already been thoroughly discussed in sections 2.2 and 2.3 and will be discussed further in section 2.4a. Thus, no additional discussion is necessary here. The transport of potential energy is essentially accomplished by the mean motions; that is, eddy transports of potential energy are negligible (e.g., Newell et al. 1974). Before proceeding, it should be noted that sensible heat and potential energy transports were calculated using analyses from 1979 through 1993, while the latent energy transports were based on 1986–89 analyses. This was done to maintain continu-

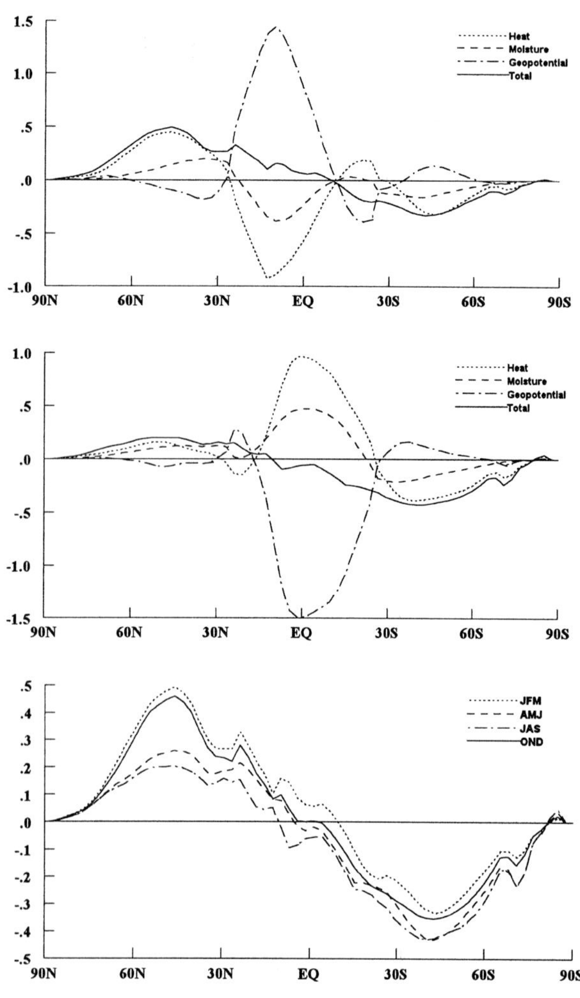

FIG. 2.26. Vertical integrals of time–zonal-mean terms in the total energy transport equation, (2.18), for JFM (top) and JAS (middle). Terms included are meridional transports of total sensible heat (dotted), total moisture (dashed), potential energy (dot–dash), and the sum of these three (solid). The bottom panel gives the seasonal distributions of the sum. Units are 10^{16} watts.

ity among datasets by using the same analyses in this section as were used in the previous sections.

Figure 2.26 shows the well-known latitudinal distribution of energy transports, consisting of heat and moisture transports that complement each other and a potential energy transport that acts in opposition to both of these. At low latitudes, where mean motions dominate all three components, the transport of Φ is greater than the transport of $(c_pT + Lq)$. Thus, the direction of total energy transport is determined by the transport of potential energy in the upper branch of the Hadley circulation, which generally is poleward. The main exception occurs in the solstice seasons, when the winter Hadley cell dominates and there is a cross-equatorial transport from the summer into the winter hemisphere. In middle latitudes, the eddy components of sensible heat and latent energy transports become much more important than the mean motion transports, and, even though the transport of Φ is equatorward in the upper branch of the Ferrel cells, the total transport is poleward. The high-latitude transports are weak but generally poleward. Thus, there is a net poleward transport of total atmospheric energy across all latitudes during all seasons (third panel in Fig. 2.26), except for the aforementioned cross-equatorial transports primarily in JFM and JAS. This leads to a divergence of total atmospheric energy away from the (thermal) equator in both hemispheres, in which the energy-rich tropical air redistributes the energy to higher latitudes. The primary physical mechanisms at work are the low-level transports of heat and moisture at low latitudes toward the Equator, which cause convergence in the ITCZ; a lifting of this energy into the upper troposphere in the form of latent heat and an accompanying increase in potential energy; a poleward transport of total atmospheric energy by the upper branches of the Hadley cells; and a subsequent transport poleward at higher latitudes by wave systems. The latitudes of maximum total energy transport by the atmosphere are between 40° and 45° in each hemisphere, with the largest transport occurring in the winter hemisphere and the greatest seasonal variation in the NH.

It is worth noting that our latitudinal distribution of total annually averaged atmospheric energy transport (not shown) is in excellent agreement with that of Trenberth and Solomon (1994). For this reason, it is appropriate to comment on their ocean heat transport distribution and compare it to our atmospheric transports. Before doing this, however, it should be mentioned that several previous authors have provided latitudinal annually averaged estimates of atmospheric and oceanic energy transports. These include Bryan (1962), Rakipova (1966), Emig (1967), van Loon et al. (1972), Newell et al. (1974), Oort and Vonder Haar (1976), Trenberth (1979), Bryan (1982), Carissimo et al. (1985), and Peixoto and Oort (1992). Furthermore, a detailed discussion of the heat transport by the oceans is given in chapter 7. In general, all of these studies, including Trenberth and Solomon, have shown that the oceanic heat transport complements the poleward atmospheric transport, and, in fact, the two transports are comparable within about 20° of the Equator. Maximum annual transport of heat by the oceans in the Trenberth and Solomon reference occurs at 15°N and 20°S. Moreover, the values of ocean transport in the subtropics and lower midlatitudes, given by Trenberth and Solomon, are about 25% greater in the NH and more than 50% greater in the SH than the values that appeared in the original SH monograph (Newton 1972). The net result of combining the annual mean oceanic and atmospheric transports is to cause the latitude of maximum poleward energy transport by the earth–atmosphere system ($[\bar{H}]$) to be at approximately 40° latitude in each hemisphere.

At stated earlier, it is the divergence of $[\bar{H}]$ that must be balanced by the solar energy impinging at the top of the atmosphere minus the outgoing net radiation emitted to space at that latitude band. If an imbalance exists, the energy surplus (or deficit) in a column will cause the temperature of the air, land, or ocean to increase (decrease). For the annual mean, this imbalance should be small; however, on a seasonal basis, it can become quite large. It is well known that the oceans account for most of the heat storage (chapter 7). Thus, in JFM, for example, the southern oceans, with their vast expanse, would accumulate a large amount of heat.

2.4. Maintenance of the zonal variations of the mean circulation

After concentrating on the zonal-mean aspects of the general circulation of the SH in the previous sections, we shall focus now on the zonal variations of the mean flow and eddy statistics. The mean fields presented in chapter 1 show substantial zonal asymmetries in the geopotential height (Fig. 1.20) and circulation (Figs. 1.37 and 1.39), primarily associated with large zonal and small meridional scale variations. These zonal variations in the mean flow are equivalent barotropic in middle and high latitudes, having increasing amplitude but little phase tilt with increasing height. The Fourier decomposition of the zonal variations in the SH in chapter 1 (Figs. 1.21–1.23) shows large amplitudes for zonal wavenumber 1 in both summer and winter, which are comparable with the amplitudes in the NH. The amplitudes of zonal wavenumbers 2, 3, and 4, however, are weaker in the SH (Fig. 1.23). In addition, there is little westward phase tilt with height of the stationary waves in the SH, indicative of small heat transport and weak vertical propagation.

The zonal flow at 200 hPa (Fig. 1.30) shows enhanced westerly jets near 45°S in JFM, stretching across most of the southern oceans, and near 30°S in JAS, over Australia and the western Pacific Ocean. These westerly jets are often collocated with regions of enhanced daily variability, which manifests itself in the transient eddy activity associated with storm tracks. In the SH, the storm tracks are much more zonally extended than the similar storm tracks in the NH. Further discussion of the storm tracks is given later, but first the horizontal distributions of the transient eddy fluxes of momentum, heat, and moisture are presented, since these variables comprise the physical quantities within the storm systems.

a. Horizontal distributions of transient eddy transports

Maps of the transient eddy momentum flux ($\overline{u'v'}$) at 200 hPa, the level where it is strongest, are shown in Fig. 2.27 for JFM and JAS. In JFM, there is a band of poleward transport extending around the SH between about 30° and 50°S with little zonal variation. In the NH in middle latitudes, there are three centers of maximum poleward transport: one over the central Pacific, another over the central Atlantic, and a third over the Mediterranean and eastern Europe. These maxima are located east of the major trough systems, which tilt northeast–southwest with latitude. This orientation is favorable for poleward momentum transport in the NH (Starr 1948). In JAS, the poleward fluxes in the NH weaken considerably, while those in the SH strengthen a little. In the SH in JAS, there are regions of maximum poleward transport near the southern parts of the three continents.

Maps of the transient eddy heat flux ($\overline{v'T'}$) at 850 and 200 hPa are shown in Fig. 2.28 for JFM and JAS. In JFM, there are poleward fluxes in middle latitudes of both hemispheres, which are stronger at 850 hPa than at 200 hPa. The latitudinal extent of these poleward fluxes at both levels is much greater in the NH than in the SH. In the SH, for example, the maxima of the poleward flux lie between 40° and 50°S, whereas in the NH they meander from 30°N to inside the Arctic Circle. Regional maxima at 850 hPa are usually found over the western oceans in each hemisphere. The variations with longitude in the SH, particularly at the lower level, are considerably less than in the NH. In JAS, there are several differences in the heat-flux distributions from those in JFM. In the SH, the transport at 850 hPa is still stronger than at 200 hPa, but this is not true in the NH. The latitudinal extent of the regions of strong poleward transport is greater in JAS in the SH than in the NH. In JAS, most regional transport maxima in the NH occur over land masses, rather than over the oceans as in JFM.

Maps of transient eddy moisture flux at 850 hPa are shown in Fig. 2.29. As for moisture results in chapter 1 and in Fig. 2.4, only three years of analyses were used. The 850-hPa surface most clearly represents the level of maximum transport. It is seen that poleward transports dominate the patterns in each season in both hemispheres. The largest values in the NH occur in middle latitudes over the Pacific Ocean, stretching from the eastern part of North America to the central Atlantic. As expected, these bands are farther north in JAS than in JFM, and the magnitudes of the maxima are greatest in JFM. In the SH, maximum moisture transports are primarily located over the middle-latitude portions of the main ocean basins, and the latitudinal change of these maxima, with season, is less than it is in the NH. The figure also shows that the magnitudes of the maxima over the southern oceans undergo a substantial seasonal variation, but one that is slightly different from its counterpart in the NH. For instance, although the moisture fluxes over the Pacific and Atlantic Oceans are greater in winter than summer (as for the NH), the flux over the SH Indian Ocean is greater during the summer season.

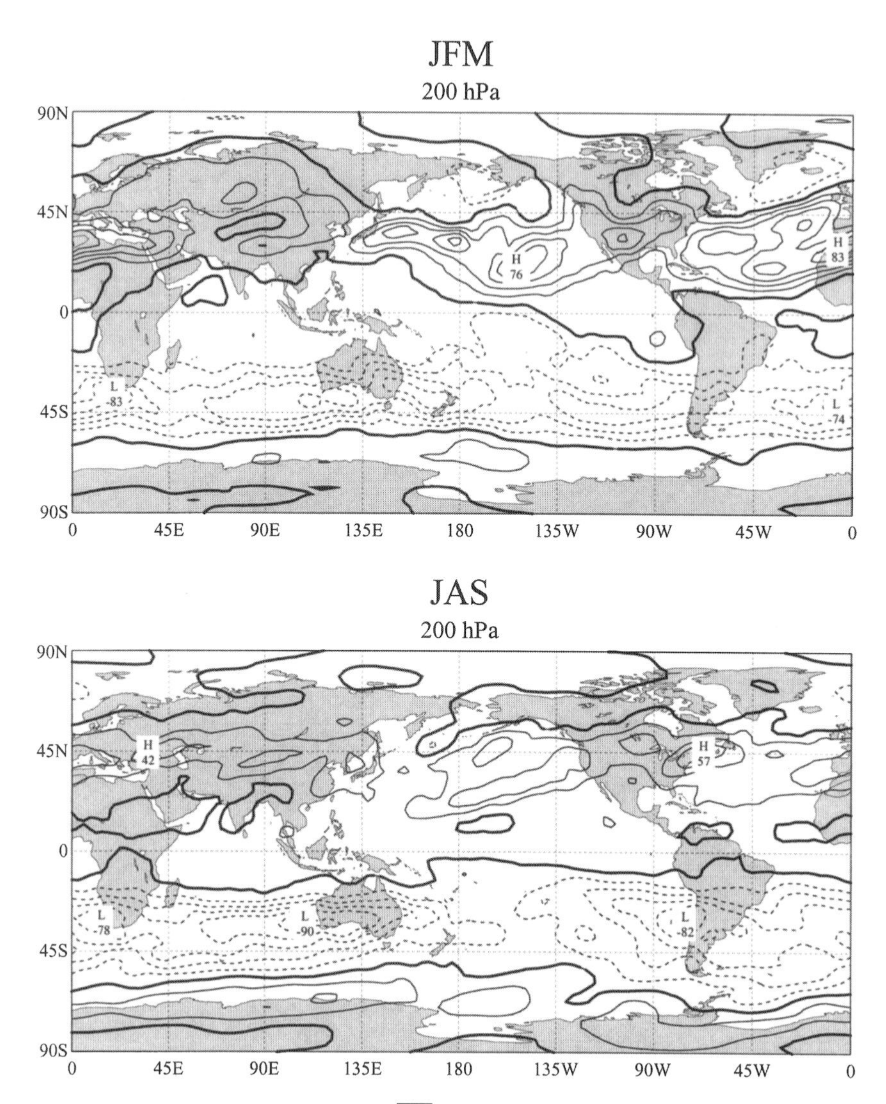

FIG. 2.27. Distributions of the transient eddy momentum flux, $\overline{u'v'}$, at 200 hPa for JFM (top) and JAS (bottom). Contour interval is 15 $m^2 s^{-2}$ and positive values are northward. Values $\geq$ 30 are highlighted in small hatching and values ≤ -30 by large hatching.

b. Storm tracks

Overall, the regions of strongest eddy activity in the SH, shown by the largest values of transient eddy fluxes in Figs. 2.27–2.29, are located between 30° and 60°S. In addition, the zonal variations of eddy activity in the SH are smaller than in the NH. These regions of large transient eddy activity are often called storm tracks, because of the dominant role of baroclinic waves in producing these eddy variations. The local meridional temperature gradient in the lower troposphere is a good indicator of the location of these storm tracks (Trenberth 1991b). James and Anderson (1984) were unable to account for the location of the SH storm tracks from the baroclinicity of the mean flow and argued that regional moisture transport was a factor in the storm track over the South Atlantic Ocean. Frederiksen (1985), however, used linear baroclinic theory to show that the distribution of storm tracks was entirely consistent with the mean flow and thermal distribution and that the main storm track was in the eastern hemisphere, poleward and downstream of the jet maximum.

A simple measure of the baroclinicity of the mean flow is the Eady growth rate $0.31\, f/N|du/dz|$ (James and Anderson 1984). This depends mainly on the vertical shear of the zonal wind or the meridional temperature gradient, as the mean flow is in thermal wind balance. The climatological distribution in the lower troposphere for January and July is shown in Fig. 2.30, computed from 10 years of ECMWF analyses. In January, there is a zonal band of high baroclinicity around 50°S, with the largest values in the Indian Ocean region. In July, the high-latitude belt of large baroclinicity is located around 55°–60°S, and

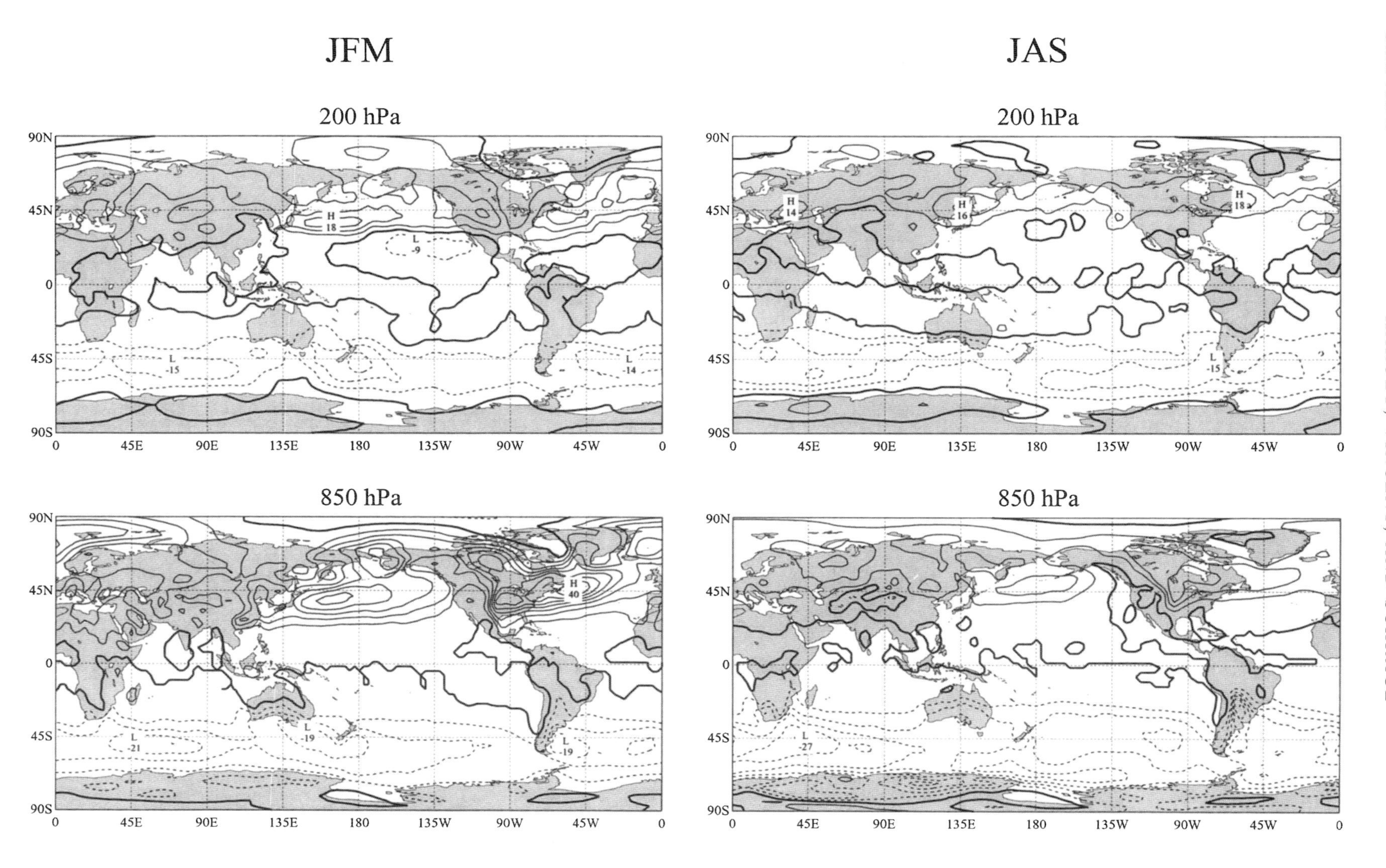

FIG. 2.28. Distributions of the transient eddy heat flux, $\overline{v'T'}$, at 200 (top two panels) and 850 hPa (bottom two panels) for JFM (left) and JAS (right). Contour interval is 5 K m s^{-1}, and positive values are northward. Values ≥ 5 are highlighted by small hatching and values ≤ -5 by large hatching.

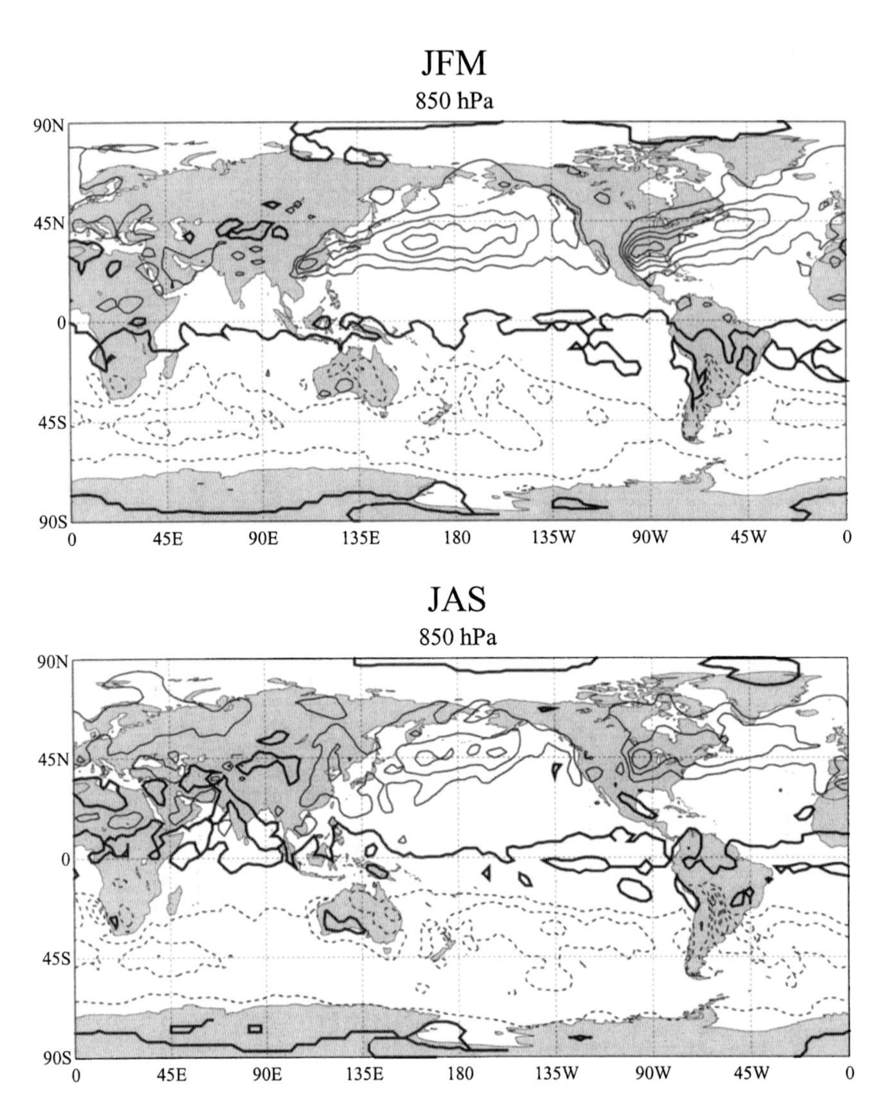

FIG. 2.29. Distributions of the transient eddy moisture flux, $\overline{v'q'}$, at 850 hPa, for JFM (top) and JAS (bottom). Contour interval is 2.5 (g kg^{-1}) (m s^{-1}), and positive values are northward. Values ≥ 5 are highlighted by small hatching and values ≤ -5 by large hatching.

there is a second band of high baroclinicity associated with the subtropical jet, which spirals from Australia across the Pacific. The regions of large baroclinicity are consistent with the regions of large transient eddy activity shown in Figs. 1.24, 1.31, and 1.33, as well as the transient eddy fluxes described above.

The storm tracks referred to above should be compared with the distributions of cyclone and anticyclone centers and their tracks. These distributions are a classical part of synoptic climatology and were presented by Taljaard (1972) in the chapter on synoptic meteorology in the original monograph. Those distributions were based on analyses for the IGY and involved manual identification and tracking of centers from synoptic charts. More recently, automated procedures have been developed for identifying and tracking centers from gridpoint numerical analyses or model output. These have been applied in the SH to obtain climatologies of cyclone and anticyclone centers and tracks (Jones and Simmonds 1993, 1994; Sinclair 1994, 1996). The distribution of cyclone track density in Fig. 2.31 closely follows the region of largest poleward eddy heat flux but is shifted slightly poleward, with a maximum around 55°S. The distribution of anticyclone track density is located equatorward of the storm track, between 25° and 45°S. The variability of these storm tracks and eddy activity is described in other chapters (see chapters 4 and 5).

c. Forcing of the mean flow variations

We now discuss the forcing of these zonal variations of the mean flow and eddy activity and the

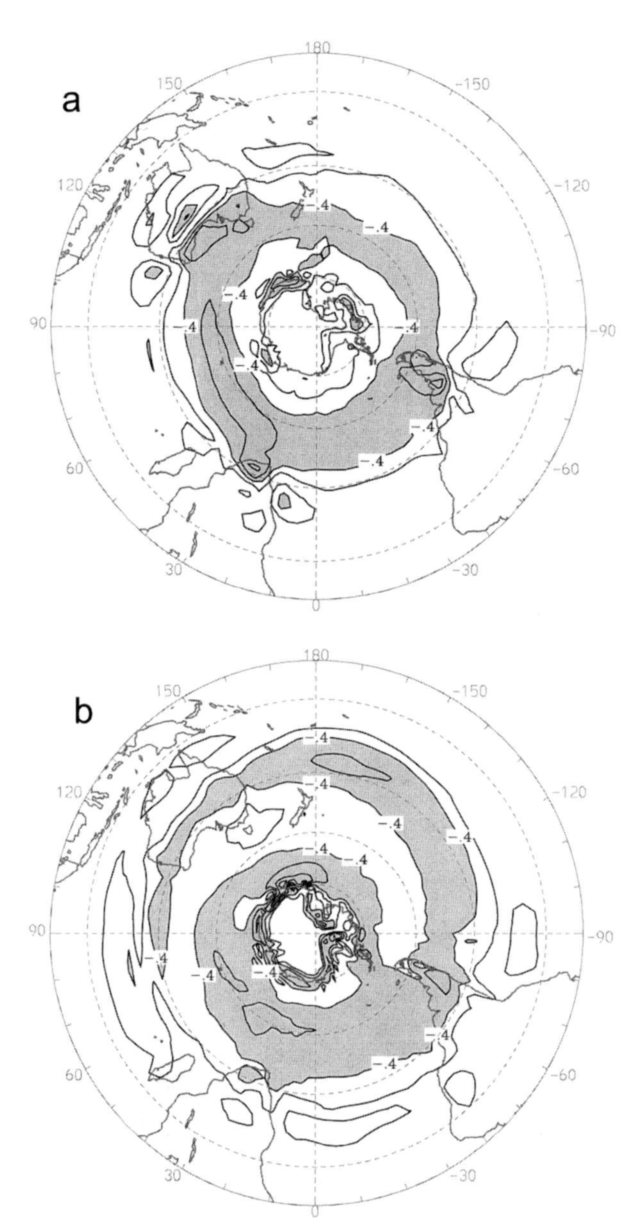

FIG. 2.30. Eady growth rate 0.31 $f/N|du/dz|$ for (a) January and (b) July, computed between 850 and 700 hPa from 10 years (1979–88) of ECMWF analyses.

feedbacks between the two. First, we consider the forcing of the stationary waves in the SH. For the time-mean stationary waves, Plumb (1985) introduced a diagnostic of three-dimensional wave propagation, analogous to the zonal mean EP flux described earlier. Regions of divergence of this stationary wave flux indicate sources of stationary waves, while regions of convergence indicate dissipation. The horizontal stationary wave activity flux is defined from the zonally asymmetric part of the time-mean flow by

$$\mathbf{F}_s = (F_\lambda, F_\phi) = \alpha\cos\phi\left[\bar{v}^{*2} - \frac{1}{2a\Omega\sin 2\phi}\frac{\partial}{\partial\lambda}(\bar{v}^*\bar{\Phi}^*), -\bar{u}^*\bar{v}^* + \frac{1}{2a\Omega\sin 2\phi}\frac{\partial}{\partial\lambda}(\bar{u}^*\bar{\Phi}^*)\right], \quad (2.19)$$

where α is p/p_0 (p_0 is the 1000 hPa) and (u, v) are the horizontal geostrophic velocities calculated from the geopotential Φ. Maps of the stationary wave flux for the SH mean flow from Karoly et al. (1989) (shown in Fig. 2.32) and Quintanar and Mechoso (1995a) show mainly zonal and equatorward propagation of stationary wave activity, with an apparent source over the Indian Ocean and other sources at high latitudes, as well as from the Tropics. We shall seek to interpret these apparent sources in terms of different forcing processes.

Since the stationary waves are close to equivalent barotropic in the SH (as described in chapter 1) and have their largest amplitudes in the upper troposphere, we shall use the time-mean barotropic vorticity equation

$$\left(\frac{\partial}{\partial t} + \bar{\mathbf{v}}_\psi \cdot \nabla\right)(\overline{\zeta + f}) = -(\zeta + f)\nabla \cdot \bar{\mathbf{v}}_\chi - \bar{\mathbf{v}}_\chi \cdot \nabla(\overline{\zeta + f}) - \nabla \cdot (\overline{\mathbf{v}'_\chi \zeta'}) - \nabla \cdot (\overline{\mathbf{v}'_\psi \zeta'}) - X \quad (2.20)$$

applied in the upper troposphere to understand the different forcing processes. The forcing of extratropical Rossby waves by the anomalous tropical heating can be best understood by partitioning the horizontal flow, $\mathbf{v}$, into irrotational and nondivergent components, such that $\mathbf{v} = \mathbf{v}_\psi + \mathbf{v}_\chi = \mathbf{k} \times \nabla\Psi + \nabla\chi$ with streamfunction, Ψ, and velocity potential, χ. Equation (2.20) has been written so that the left side retains all terms necessary to support Rossby wave propagation, while the right side involves terms associated with the forcing of stationary waves. This form of barotropic vorticity equation was discussed by Sardeshmukh and Hoskins (1988), who described the first three terms on the right side as the Rossby wave source, $\bar{S} = -\nabla \cdot \overline{\mathbf{v}_\chi(\zeta+f)}$, associated with the divergent flow. The other forcing terms on the right side of (2.20) are associated with the convergence of vorticity by transient eddies and friction. Hence, the forcing of stationary waves is through (i) divergent flow generated by flow over topography, by gradients of sensible heating, or by latent heat release in the Tropics or in the midlatitude storm tracks; and (ii) transient eddy forcing associated with eddy flux convergence of vorticity. No thorough diagnostic analysis of the upper-troposphere vorticity budget in the SH has been carried out to analyze the relative magnitudes of the different

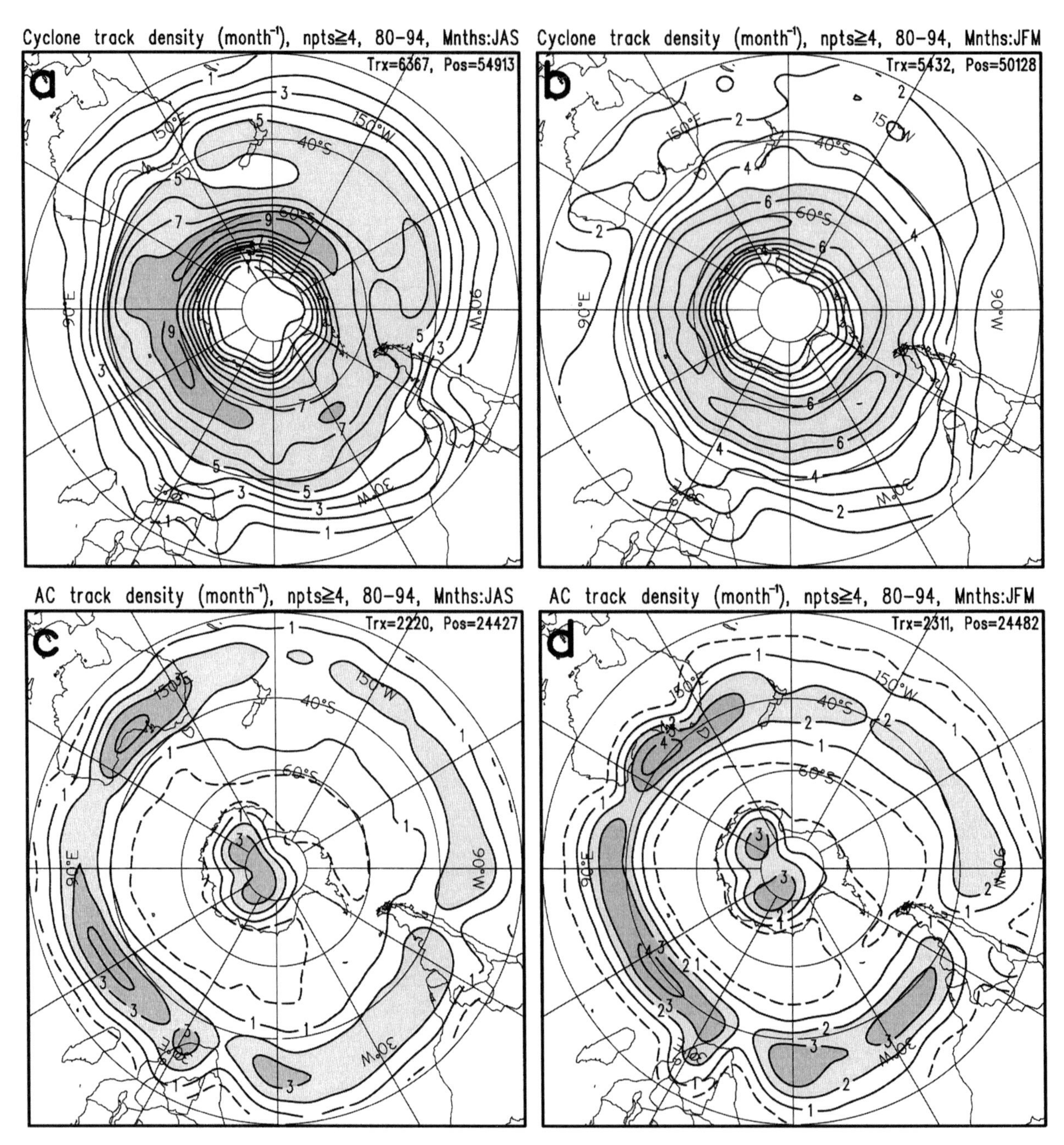

FIG. 2.31. Mean monthly cyclone track density for cyclones lasting two or more days for (a) winter (July–September) and (b) summer (January–March), drawn every 1 cyclone per 5°-latitude-radius circle per month, with values > 5 and 8 shaded. Obtained from twice-daily ECMWF 1000-hPa analyses during 1980–94 using automated center finding and tracking algorithms. (c) and (d) As for (a) and (b), except for anticyclone tracks, with values > 2 and 3 shaded, and the 0.3 contour added (dashed) to delineate regions of low anticyclone occurrence. [Provided by M. Sinclair using procedures described by Sinclair et al. (1997).]

terms in (2.20). The forcing of the mean flow by transient eddies is discussed later.

The sources of stationary wave activity shown in Fig. 2.32 suggest that there is significant forcing from high latitudes, probably associated with topography or zonal asymmetries in sea surface temperature and sea ice, and from the Tropics, probably associated with latent heat release in convection over the tropical land masses of Indonesia, South America, and Africa. Barotropic model simulations by Grose and Hoskins (1979) and James (1988) show that the observed SH stationary waves at middle and high latitudes in winter can be simulated using forcing by idealized large-scale orography centered off the South Pole. Using the full

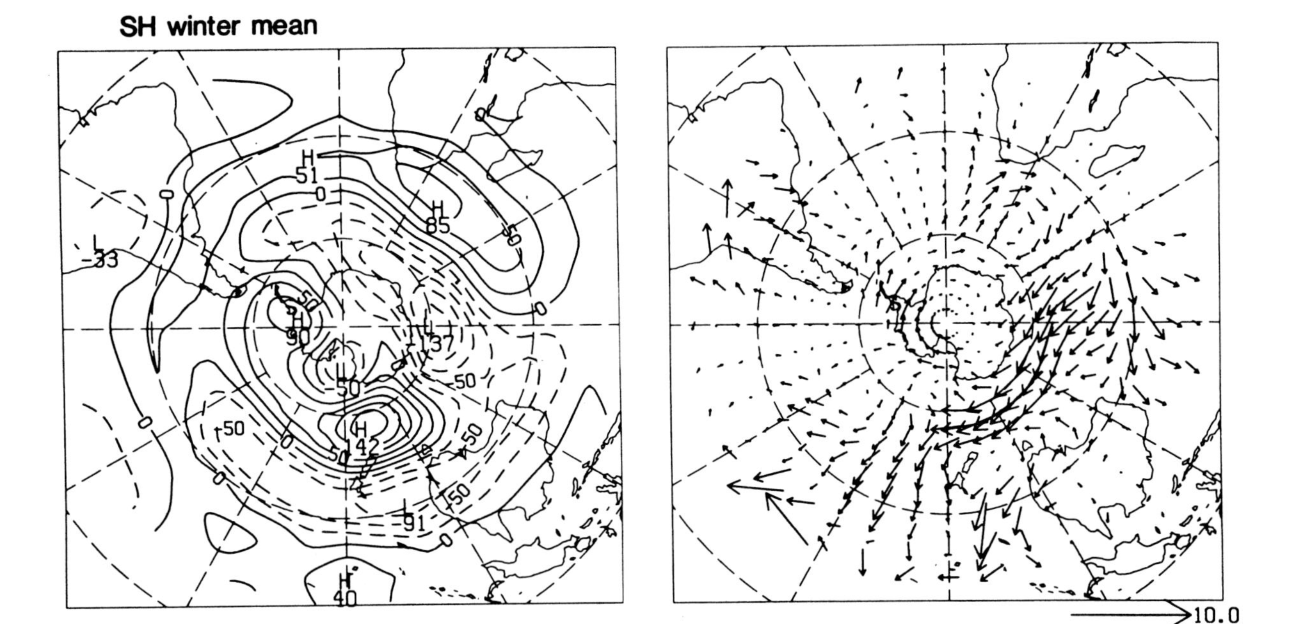

FIG. 2.32. Distributions of stationary wave activity flux vector, $\mathbf{F}_s$, in the SH for June–August from Australian analyses. (Reproduced from Karoly et al. 1989.)

SH orography gave some improvement to the simulation of the stationary waves, but the essential characteristics in the SH of large zonal-scale and small meridional-scale zonal asymmetries were simulated reasonably well by simple large-scale forcing centered just off the pole, representing the asymmetries of Antarctica. A recent general circulation model study by Quintanar and Mechoso (1995b), however, has shown that it is possible to simulate the SH stationary waves even after removing the Antarctic topography and the zonal sea surface temperature asymmetries poleward of 45°S. They found that remote forcing from low latitudes was mainly responsible for the stationary waves in SH middle latitudes. This ambiguity between the roles of forcing from Antarctica and from low latitudes appears to be due to the very small phase difference between the responses to the two different forcings. Further study is required to quantify the relative importance of forcing from Antarctica and from low latitudes for the SH stationary waves in middle latitudes.

The feedback between the transient eddies and the mean flow is equally important for the three-dimensional variations of the general circulation as it was for the zonal-mean circulation discussed in section 2.2. Hoskins et al. (1983) developed the **E** vector to relate the shape and propagation of eddies to the feedback of these eddies onto the time-mean flow. Defining

$$\bar{\mathbf{E}}_u = \left[\frac{1}{2}(\overline{v'^2} - \overline{u'^2}),\ -\overline{u'v'},\ \frac{f}{\Gamma}\overline{v'T'}\right]\cos\phi$$

from Trenberth (1991b), the effect of the eddies on the mean flow is such that

$$\frac{d\bar{u}}{dt} = \frac{1}{\cos\phi}\nabla\cdot\mathbf{E}_u.$$

At 300 hPa, the horizontal distribution of the **E** vectors in Fig. 2.33 shows local divergence in the storm tracks, consistent with transient eddy forcing of the mean flow through the momentum flux convergence. One exception is in the New Zealand region in winter, where there is often a split jet and weak convergence of **E**, suggesting that the eddies act to reinforce this split.

An alternative approach to identifying the forcing of the zonal asymmetries of the mean flow by the transient eddies was used by Lau and Holopainen (1984). They calculated the mean tendencies of height and temperature due to eddy flux convergences and discussed the distributions of these tendencies and the zonal asymmetries of height and temperature in terms of forcing by the transient eddies. The quasigeostrophic forcing of the height tendency by transient eddies is

$$\nabla^{-2}\left[\frac{-f_0}{g}\nabla\cdot(\overline{\mathbf{v}'\zeta'})\right],$$

and that of the temperature tendency is $-\nabla\cdot(\overline{\mathbf{v}'T'})$. For the SH, Cuff and Cai (1995) have shown that the barotropic feedback of the transient eddies due to the vorticity

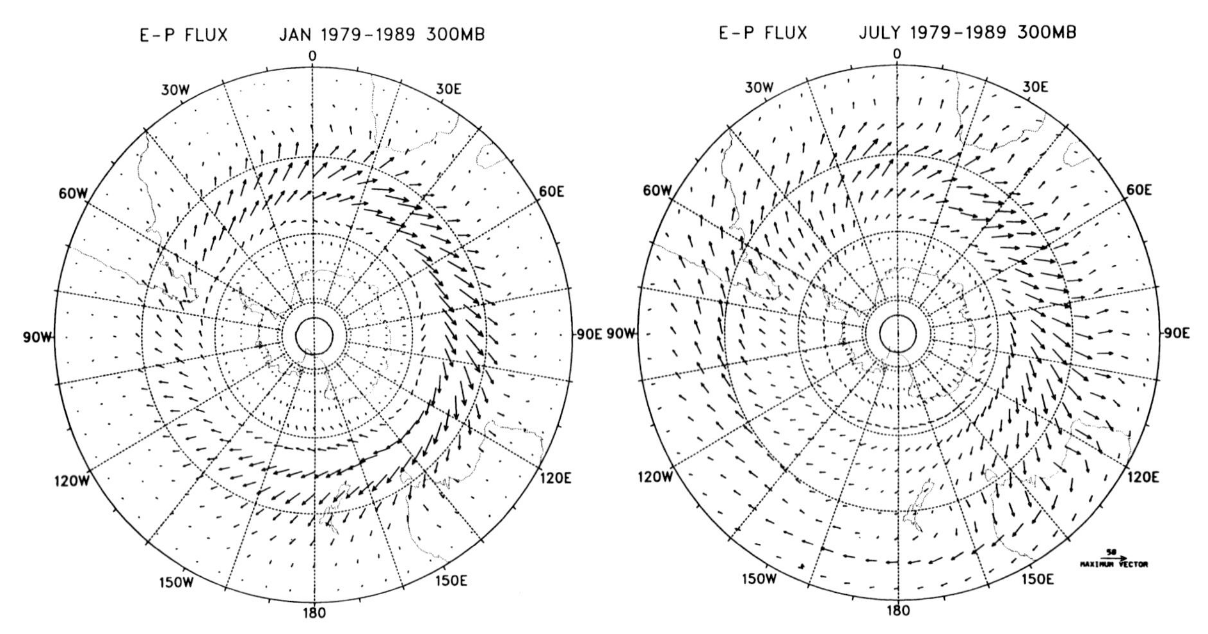

FIG. 2.33. EP flux vectors at 300 mb with the arrow at lower right indicating 50 m^2 s^{-2} for January and July 1979–89 for the 2–8-day eddies.

flux (Fig. 2.34b) shows a nearly in-phase relationship with the stationary wave field, while the baroclinic feedback due to the transient eddy heat flux (Fig. 2.34) is mostly out of phase with the temperature wave field. Hence, the forcing of the mean flow asymmetries by the transient eddies is much the same as for the zonal-mean flow; the vorticity flux by the high-frequency transients acts to reinforce the barotropic component of the stationary waves, while the heat flux associated with high-frequency transients acts to diminish the baroclinic component of the stationary wave (Cuff and Cai 1995). This relationship is similar to that found in the NH by Lau and Holopainen (1984).

2.5. Summary

The general circulation of the SH has been described in terms of the interactions between the mean flow and the high-frequency transient eddy activity. It is well known that the general circulation is constrained by the global requirements to conserve energy, mass, moisture, and angular momentum; by the distribution of diabatic heating; and by the distribution of land masses and oceans. In this chapter, the atmospheric budgets that account for these requirements were examined, and similarities and differences between the NH and SH were noted. It is also known, and has been demonstrated in this chapter, that jet streams and storm tracks in the SH are more zonally elongated than in the NH. This is, of course, because of the very different distribution of topography in the two hemispheres. It was shown that the evolution and structure of transient eddies in the SH is described well by nonlinear baroclinic eddy lifecycles. The location of the storm tracks was found to be closely related to the regions of largest meridional temperature gradient at low levels. Furthermore, the heat transport by the transient eddies acted to reduce the baroclinicity of the mean flow, while the momentum transport acted to reinforce the mean flow. In addition, it was shown that quasi-stationary waves in the SH, which have an equivalent barotropic vertical structure, contributed little to the eddy transports. Finally, it was noted that there are significant zonal asymmetries to the circulation in the SH that have very long zonal scales, since they are primarily associated with zonal wavenumber 1.

REFERENCES

Barnes, R. T. H., R. Hide, A. A. White, and C. A. Wilson, 1983: Atmospheric angular momentum fluctuations, length-of-day changes and polar motion. *Proc. R. Soc. London,* **A387,** 31–73.

Baumgartner, A. and E. Reichel, 1975: *The World Water Balance.* Elsevier, 179 pp.

Bryan, K., 1962: Measurements of meridional heat transport by ocean currents. *J. Geophys. Res.,* **67,** 3403–3414.

———, 1982: Poleward heat transport by the ocean: Observation and models. *Ann. Rev. Earth Planetary Sci.,* **10,** 15–38.

Buch, H., 1954: Hemispheric wind conditions during the year 1950. Final Rep., Part 2, Contract No. AF19(122)-153, Planetary Circulations Project, Dept. of Meteorology, Massachusetts Institute of Technology, 126 pp.

Carissimo, B. C., A. H. Oort, and T. H. Vonder Haar, 1985: Estimating the meridional energy transports in the atmosphere and oceans. *J. Phys. Oceanogr.,* **15,** 82–91.

Chen, T.-C., 1985: Global water vapor flux and maintenance during FGGE. *Mon. Wea. Rev.,* **113,** 1801–1819.

———, and J. Pfaendtner, 1993: On the atmospheric branch of the hydrological cycle. *J. Climate,* **6,** 161–167.

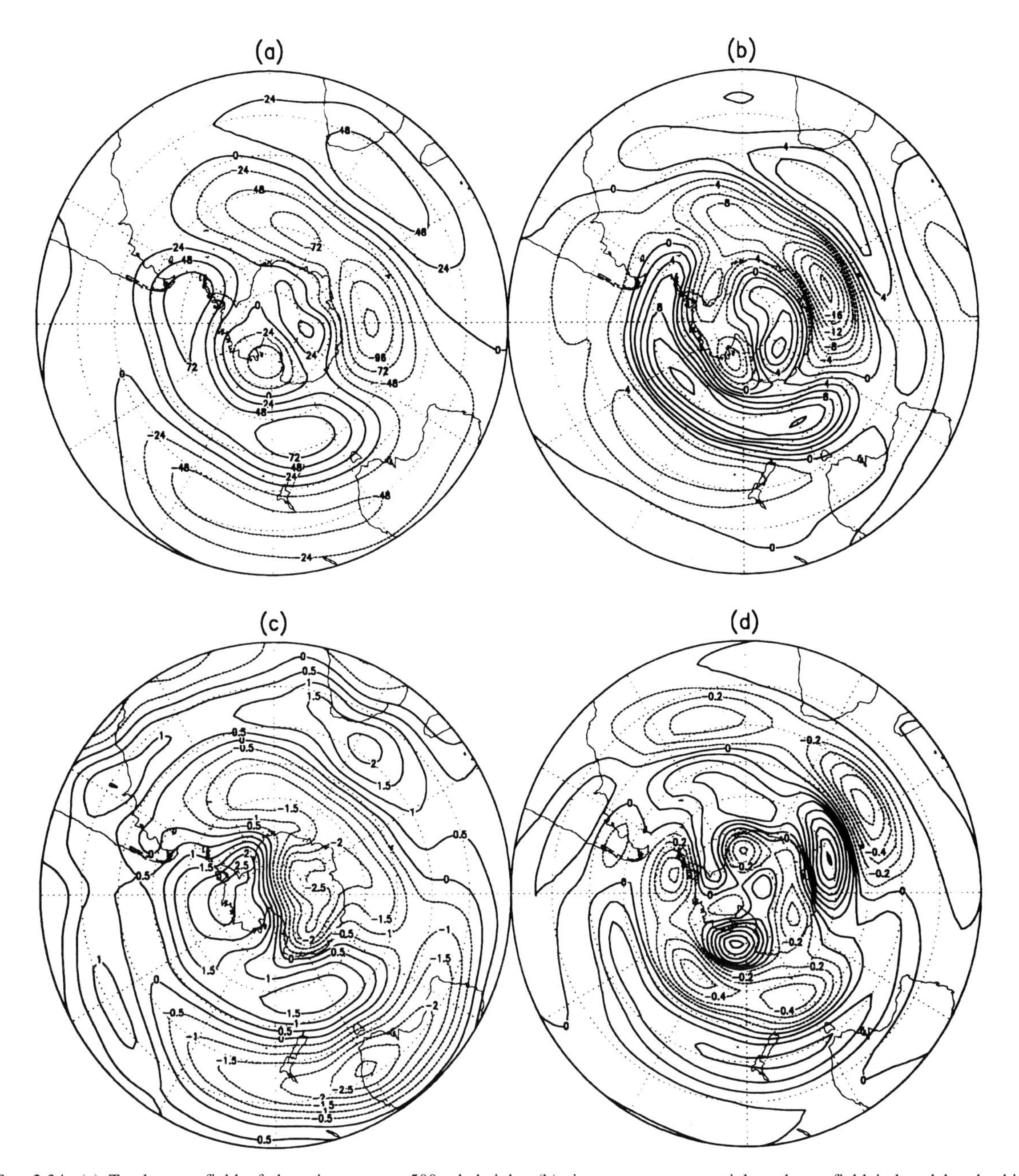

FIG. 2.34. (a) Total wave field of the winter mean 500-mb height; (b) time-mean geopotential tendency field induced by the high-frequency transients; (c) as in (a), except for the 500-mb temperature field; and (d) as in (c), except for the temperature tendency. Contour interval for (a), (b), (c), and (d) is 24 m, 2×10^{-5} m s^{-1}, 0.5 °C, and 0.1×10^{-5} °C s^{-1}, respectively. (Reproduced from Cuff and Cai 1995.)

———, M.-C. Yen, J. Pfaendtner, and Y. C. Sud, 1996: Annual variation of the global precipitable water and its maintenance: Observation and climate simulation. *Tellus,* **48A,** 1–16.

Cuff, T. J., and M. Cai, 1995: Interaction between the low- and high-frequency transients in the Southern Hemisphere winter circulation. *Tellus,* **47A,** 331–350.

Dickey, J. O., M. Ghil, and S. L. Marcus, 1991: Extratropical aspects of the 40–50 day oscillation in length-of-day and atmospheric angular momentum. *J. Geophys. Res.,* **96,** 22 643–22 658.

Edmon, H. J., B. J. Hoskins, and M. E. McIntyre, 1980: Eliassen–Palm cross sections for the troposphere. *J. Atmos. Sci.,* **37,** 2600–2616.

Emig, M., 1967: Heat transport by ocean currents. *J. Geophys. Res.,* **72,** 2519–2529.

Frederiksen, J. S., 1985: The geographic locations of Southern Hemisphere storm tracks: Linear theory. *J. Atmos. Sci.,* **42,** 710–723.

Gilman, P. A., 1965: The mean meridional circulation of the Southern Hemisphere inferred from momentum and mass balance. *Tellus,* **17,** 277–284.

Gleckler, P. J., and Coauthors, 1995: Cloud-radiative effects on implied oceanic energy transports as simulated by atmospheric general circulation models. *Geophys. Res. Lett.,* **22,** 791–794.
Grose, W. L., and B. J. Hoskins, 1979: On the influence of orography on large-scale atmospheric flow. *J. Atmos. Sci.,* **36,** 223–234.
Hide, R., N. T. Birch, L. V. Morrison, D. J. Shea, and A. A. White, 1980: Atmospheric angular momentum fluctuations and changes in the length of day. *Nature,* **286,** 114–117.
Holopainen, E. O., 1967: On the mean meridional circulation and the flux of angular momentum over the Northern Hemisphere. *Tellus,* **19,** 1–13.
Hoskins, B. J., 1983: Modelling of the transient eddies and their feedback on the mean flow. *Large-Scale Dynamical Processes in the Atmosphere*, G. J. Hoskins and R. P. Pearce, Eds., Academic Press, 397 pp.
———, I. N. James, and G. H. White, 1983: The shape, propagation, and mean-flow interaction of large-scale weather systems. *J. Atmos. Sci.,* **40,** 1595–1612.
Hurrell, J. W., 1995: Transient eddy forcing of the rotational flow during northern winter. *J. Atmos. Sci.,* **52,** 2286–2300.
James, I. N., 1988: On the forcing of planetary-scale Rossby waves by Antarctica. *Quart. J. Roy. Meteor. Soc.,* **114,** 619–637.
———, and D. L. T. Anderson, 1984: The seasonal mean flow and distribution of large-scale weather systems in the Southern Hemisphere: The effects of moisture transports. *Quart. J. Roy. Meteor. Soc.,* **110,** 943–966.
Johnson, D. R., 1980: A generalized transport equation for use with meteorological coordinate systems. *Mon. Wea. Rev.,* **108,** 733–745.
———, 1989: The forcing and maintenance of global monsoon circulations: An isentropic analysis. *Adv. Geophys.,* **31,** 43–316.
Jones, D. A., and I. Simmonds, 1993: A climatology of Southern Hemisphere extratropical cyclones. *Clim. Dyn.,* **9,** 131–145.
———, and ———, 1994: A climatology of Southern Hemisphere anticyclones. *Clim. Dyn.,* **10,** 131–145.
Karoly, D. J., and A. H. Oort, 1987: A comparison of Southern Hemisphere circulation statistics based on GFDL and Australian analyses. *Mon. Wea. Rev.,* **115,** 2033–2059.
———, R. A. Plumb, and M. F. Ting, 1989: Examples of the horizontal propagation of quasi-stationary waves. *J. Atmos. Sci.,* **46,** 2802–2811.
———, P. C. McIntosh, P. Berrisford, T. J. McDougall, and A. C. Hirst, 1997: Similarities of the Deacon cell in the Southern Ocean and the Ferrel cells in the atmosphere. *Quart. J. Roy. Meteor. Soc.,* **122,** 519–526.
Kuo, H. L., 1956: Forced and free meridional circulations in the atmosphere. *J. Meteor.,* **13,** 561–568.
Lau, N. C., and E. O. Holopainen, 1984: Transient eddy forcing of the time-mean flow as identified by geopotential tendencies. *J. Atmos. Sci.,* **41,** 313–328.
Lorenz, E. N., 1967: *The Nature and Theory of the General Circulation of the Atmosphere.* WMO Pub. 218, 161 pp.
Madden, R. A. and P. Speth, 1995: Estimates of atmospheric angular momentum, friction, and mountain torques during 1987–1988. *J. Atmos. Sci.,* **52,** 3681–3694.
McIntosh, P. C., and T. J. McDougall, 1996: Isopycnal averaging and the residual mean circulation. *J. Phys. Oceanogr.,* **26,** 1655–1660.
Newell, R. E., D. G. Vincent, T. G. Dopplick, D. Ferruzza, and J. W. Kidson, 1970: The energy balance of the global atmosphere. *The Global Circulation of the Atmosphere,* G. A. Corby, Ed., Royal Meteorological Society, 42–90.
———, J. W. Kidson, D. G. Vincent, and G. J. Boer, 1972: *The General Circulation of the Tropical Atmosphere and Interactions with Extratropical Latitudes, Vol. 1.* Dept. of Meteorology, Massachusetts Institute of Technology, 258 pp.
———, ———, ———, and ———, 1974: *The General Circulation of the Tropical Atmosphere and Interactions with Extratropical Latitudes, Vol. 2.* Dept. of Meteorology, Massachusetts Institute of Technology, 371 pp.
Newton, C. W., 1971a: Global angular momentum balance: Earth torques and atmospheric fluxes. *J. Atmos. Sci.,* **28,** 1329–1341.
———, 1971b: Mountain torques in the global angular momentum balance. *J. Atmos. Sci.,* **28,** 623–628.
———, 1972: Southern Hemisphere general circulation in relation to global energy and momentum balance requirements. *Meteorology of the Southern Hemisphere, Meteor. Monogr.,* No. 35, Amer. Meteor. Soc., 215–246.
Obasi, G. O. P., 1963: Atmospheric momentum and energy calculations for the Southern Hemisphere during the IGY. Rep. No. 6, Planetary Circulations Project, Dept. of Meteorology, Massachusetts Institute of Technology, 353 pp.
Oort, A. H., 1989: Angular momentum cycle in the atmosphere–ocean–solid earth system. *Bull. Amer. Meteor. Soc.,* **70,** 1231–1242.
———, and J. P. Peixoto, 1983: Global angular momentum and energy balance requirements from observations. *Adv. Geophys.,* **25,** 355–490.
———, and T. H. Vonder Haar, 1976: On the observed annual cycle in the ocean–atmosphere heat balance over the Northern Hemisphere. *J. Phys. Oceanogr.,* **6,** 781–800.
Palmen, E., and C. W. Newton, 1969: *Atmospheric Circulation Systems: Their Structure and Physical Interpretation.* International Geophysics Series 13, Academic Press, 603 pp.
Peixoto, J. P., and A. H. Oort, 1992: *Physics of Climate.* American Institute of Physics, 520 pp.
———, D. A. Salstein, and R. D. Rosen, 1981: Intra-annual variation in large-scale moisture fields. *J. Geophys. Res.,* **86**(C2), 1255–1264.
Plumb, R. A., 1985: On the three-dimensional propagation of stationary waves. *J. Atmos. Sci.,* **42,** 217–229.
Quintanar, A. I., and C. R. Mechoso, 1995a: Quasi-stationary waves in the Southern Hemisphere. Part I: Observational data. *J. Climate,* **8,** 2659–2672.
———, and ———, 1995b: Quasi-stationary waves in the Southern Hemisphere. Part II: Generation mechanisms. *J. Climate,* **8,** 2673–2690.
Rakipova, L. R., 1966: Heat transfer and general circulation of the atmosphere. *Izv. Atmos./Oceanic Phys.,* **2,** No. 9, 983–986.
Randel, W. J., 1990: Coherent wave-zonal mean flow interactions in the troposphere. *J. Atmos. Sci.,* **47,** 439–456.
———, and I. M. Held, 1991: Phase speed spectra of transient eddy fluxes and critical layer absorption. *J. Atmos. Sci.,* **48,** 688–697.
Rosen, R. D., 1976: The flux of mass across latitude walls in the atmosphere. *J. Geophys. Res.,* **81,** 2001–2002.
———, 1993: The axial momentum balance of earth and its fluid envelope. *Surv. Geophys.,* **14,** 1–29.
———, and D. A. Salstein, 1983: Variations in atmospheric angular momentum on global and regional scales, and the length of day. *J. Geophys. Res.,* **88,** 5451–5470.
———, ———, and J. P. Peixoto, 1979: Variability in the annual fields of large-scale atmospheric water vapor content. *Mon. Wea. Rev.,* **107,** 26–37.
Salstein, D. A., R. D. Rosen, and J. P. Peixoto, 1983: Modes of variability in annual hemispheric water vapor and transport fields. *J. Atmos. Sci.,* **40,** 788–803.
Sardeshmuhk, P. D., and B. J. Hoskins, 1988: The generation of global rotational flow by steady tropical divergence. *J. Atmos. Sci.,* **45,** 1228–1251.
Sellers, W. D., 1965: *Physical Climatology.* University of Chicago, 272 pp.
Sinclair, M. R., 1994: An objective cyclone climatology for the Southern Hemisphere. *Mon. Wea. Rev.,* **122,** 2239–2256.
———, 1995: A climatology of cyclogenesis for the Southern Hemisphere. *Mon. Wea. Rev.,* **123,** 1601–1619.
———, 1996: A climatology of anticyclones and blocking for the Southern Hemisphere. *Mon. Wea. Rev.,* **124,** 245–263.

———, J. A. Renwick, and J. W. Kidson, 1997: Low-frequency variability of Southern Hemisphere sea level pressure and weather system activity. *Mon. Wea. Rev.,* **125,** 2531–2543.

Starr, V. P., 1948: An essay on the general circulation of the atmosphere. *J. Meteor.,* **5,** 39–43.

———, 1951: Applications of energy principles to the general circulation. *Compendium of Meteorology,* Amer. Meteor. Soc., 568–574.

———, J. P. Peixoto, and R. G. McKean, 1969: Pole-to-pole moisture conditions for the IGY. *Pure Appl. Geophys.,* **75,** 300–331.

Swinback, R., 1985: The global atmospheric angular momentum balance inferred from analyses made during the FGGE. *Quart. J. Roy. Meteor. Soc.,* **111,** 977–992.

Taljaard, J. J., 1972: Synoptic meteorology of the Southern Hemisphere. *Meteorology of the Southern Hemisphere, Meteor. Monogr.,* No. 35, Amer. Meteor. Soc., 139–213.

Trenberth, K. E., 1979: Mean annual poleward energy transports by the oceans in the Southern Hemisphere. *Dyn. Atmos. Oceans,* **4,** 57–64.

———, 1991a: Climate diagnostics from global analyses: Conservation of mass in ECMWF analyses. *J. Climate,* **4,** 707–722.

———, 1991b: Storm tracks in the Southern Hemisphere. *J. Atmos. Sci.,* **48,** 2159–2178.

———, 1992: Global analyses from ECMWF and atlas of 1000 to 10 mb circulation statistics. NCAR/TN-373+STR, 191 pp.

———, and A. Solomon, 1994: The global heat balance: Heat transports in the atmosphere and ocean. *Clim. Dyn.,* **10,** 107–134.

———, J. W. Hurrell, and A. Solomon, 1995: Conservation of mass in three dimensions in global analyses. *J. Climate,* **8,** 692–708.

van Loon, H., 1980: Transfer of sensible heat by transient eddies in the atmosphere on the Southern Hemisphere: An appraisal of the data before and during FGGE. *Mon. Wea. Rev.,* **108,** 1774–1781.

———, J. J. Taljaard, T. Sasamori, J. London, D. V. Hoyt, K. Labitzke, and C. W. Newton, 1972: *Meteorology of the Southern Hemisphere, Meteor. Monogr.,* No. 35, C. W. Newton, Ed., Amer. Meteor. Soc., 263 pp.

Vincent, D. G., 1968: Mean meridional circulations in the Northern Hemisphere lower stratosphere during 1964 and 1965. *Quart. J. Roy. Meteor. Soc.,* **94,** 333–349.

———, and J. M. Schrage, 1995: *Climatology of the TOGA-COARE and Adjacent Regions (1985–1990), Vol. 2: Thermodynamic and Moisture Variables.* Dept. of Earth and Atmospheric Sciences, Purdue University, West Lafayette, IN 47907, 123 pp.

Wahr, J. M. and A. H. Oort, 1984: Friction and mountain-torque estimates from global atmospheric data. *J. Atmos. Sci.,* **41,** 190–204.

Chapter 3

Meteorology of the Tropics

This chapter is divided into four sections, according to geographical regions as a function of longitude. The regions progress eastward from the maritime continent/Australia sector, to the Pacific, to South America, and finally to southern Africa. The three continental regions also include some discussion of their adjacent oceanic areas. Each section is written by different authors and, by agreement between the editors and authors, was designed to stand alone; an attempt has been made, however, to provide some continuity among the sections. For example, section authors were encouraged to cross-reference each other. In addition, reviewers were asked to read and comment on two or more sections, and, in most cases, they agreed. Despite these attempts by the editors, authors, and reviewers, the attention given to the interaction among the four regions remains limited. Presumably, some of the chapter references provide information concerning regional interactions. Moreover, chapter 2 provides some discussion of the Tropics on a global scale, which includes the interaction among tropical regions as well as between the Tropics of the Northern and Southern Hemispheres.

Chapter 3A

Indonesia, Papua New Guinea, and Tropical Australia: The Southern Hemisphere Monsoon

JOHN MCBRIDE
Bureau of Meteorology Research Center, Melbourne, Victoria, Australia

3A.1. Overview

In this section we discuss the tropical atmosphere over the longitudes of Indonesia, Papua New Guinea, and Australia. This area is described variously as the location of the SH summer monsoon (Murakami and Sumi 1982; McBride 1983), the Australian summer monsoon (Troup 1961; McBride 1987), and the maritime continent (Ramage 1968). It is the location of one of the three global tropical heat sources and is the primary region of latent heat release associated with both the Southern Oscillation and the Madden–Julian oscillation (MJO; see chapter 8 for more details). Thus, it plays an important role in the dynamics of the global climate.

Culturally and historically, the region is divided into the Indonesian sector, defined as including the Indonesian archipelago, the island of New Guinea, and the large equatorial islands of Sumatra, Borneo, and Sulawesi (Fig. 3A.1); and the north Australian sector, defined as the tropical part of the Australian continent. The latter region has been the focus of much research in recent decades, particularly in the context of a number of tropical field experiments, such as the Australian Monsoon Experiment (AMEX; McBride and Holland 1989); the Equatorial Mesoscale Experiment (Webster and Houze 1991); the Island Thunderstorm Experiment (Keenan et al. 1989); the Down Under Doppler and Electricity Experiment (Rutledge et al. 1992); and the Maritime Continent Thunderstorm Experiment (Keenan et al. 1994). The meteorology of the north Australian sector, its monsoonal structure, and transient behavior, including active and break phenomena, have been reviewed extensively by McBride (1987) and Manton and McBride (1992).

The monsoonal character of tropical Australia has been recognized for the last hundred years. Troup and Berson (Berson 1961; Troup 1961; Berson and Troup 1961) first documented the similarities with the Northern Hemisphere (viz. Indian) monsoon with the summer rainfall episodes being associated with low-level westerlies extending from the Equator to about 15°S. These authors also showed that the onset occurs rapidly over an interval of 5 to 10 days and is associated with a transition to both low-level westerlies and upper-level equatorial easterlies over a wide longitudinal span. One of the key synoptic features in the region is the *monsoon trough,* or *monsoon shear line,* defined as the separation at low levels (e.g., the 850-hPa level) between a zone of westerlies extending from the Equator and a zone of trade-wind easterlies at higher latitude (McBride and Keenan 1982; McBride 1995).

During an active monsoon period, the monsoon shearline usually extends across the northern portion of the Australian continent. The convection is organized into mesoscale weather systems embedded within a *supercluster,* or envelope, of convective activity on a scale of 2000–3000 km (McBride 1983; Keenan and Brody 1988). Gunn et al. (1989) and Mapes and Houze (1993) demonstrated that during AMEX the active-break cycle of the monsoon system consisted of the spin up of tropical cyclones and monsoon depressions within the monsoon shear line, followed by their subsequent dissipation as they moved to higher latitudes. These important subcomponents of the Australian monsoon system are discussed further in chapter 5.

In contrast, the Indonesian sector has been studied much less in recent decades, and there are many gaps in knowledge and unanswered fundamental questions concerning the meteorology of this region. The most comprehensive study during the colonial period was that of Braak (1929). This massive work of over 1000 pages documented and discussed in detail the climate and weather of the whole region. Besides the detailed treatment of parameters such as wind, temperature, rain, humidity, and pressure, volume 1 also provided chapters on more specialized phenomena like evaporation, haze, and hail. The second volume contained separate chapters on the climate of individual islands and regions. Important aspects of the meteorology documented by Braak include the strong increase in rainfall and decrease of temperature with altitude

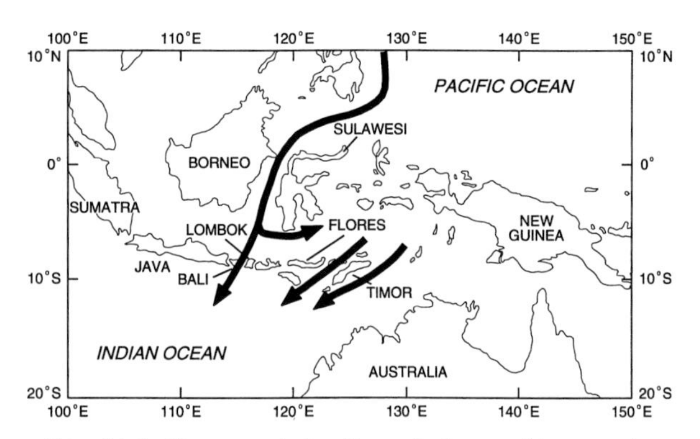

FIG. 3A.1. Names and location of the maritime continent countries involved in the Southern Hemisphere (or Australian) summer monsoon. The thick arrows depict the location of the main transport associated with the Indonesian throughflow described in section 6.

throughout the region and the large interregional variations in seasonal and diurnal distributions of rainfall. He also documented and discussed a number of important and interesting mesoscale phenomena in the region—for example, the Bohorok, a damaging warm, dry föhn wind in Sumatra, and the Sumatras, a coast-parallel squall line in the Malacca Strait.

After Indonesian independence, one of the most important studies was that of Schmidt and Ferguson (1951). Following the principles derived by Mohr (1933), Schmidt and Ferguson developed a new system of climatic classification based on the ratio of dry (less than 60 mm) to wet (greater than 100 mm) months in individual years. This system was developed specifically for application to agriculture in a region where temperatures are high year-round and depend linearly on altitude. Detailed maps were presented for the whole of Indonesia (Fig. 3A.2). The main features were the division of the country into two distinct climatic zones; the all-year high-rainfall region of Sumatra, western Java, Kalimantan, Sulawesi, and Irian Jaya, and the summer-wet/winter-dry zone of the eastern end of the archipelago.

These principles have been developed further by Oldeman and coauthors (Oldeman 1975; Oldeman et al. 1979). According to these papers, an agro-climatic classification should describe the existing environment in relation to crop requirements. Accordingly, their definitions of dry and wet months are based on consecutive months exceeding the minimum rainfall requirements for the growth of lowland rice (wet months) and on consecutive months not reaching the minimum rainfall for growth of most economic upland crops (dry months).

The interannual variability of the circulation over both the north Australian and the Indonesian sectors are strongly affected by the Southern Oscillation (Braak 1919; Nicholls 1981; McBride and Nicholls 1983). For the Australian monsoon, the relationship is strongest during the premonsoon or transition season from September to November. Rainfall in that season is strongly correlated to indices of the Southern Oscillation (McBride and Nicholls 1983; Nicholls et al. 1982). Following earlier authors back to Troup (1961), Holland (1986), and Hendon and Liebmann (1990a), Drosdowsky (1996) defined the onset of the monsoon as the beginning of sustained periods of westerly wind (at Darwin) throughout the lower and middle troposphere. He found that the onset occurs on 28 December, with a range from 22 November to 25 January. Drosdowsky also found the onset date to be significantly related to the Southern Oscillation Index (over the preceding September to October) with a correlation coefficient of −0.56 over a 35-yr time series (in the direction that low pressure over Darwin corresponds to an early monsoon onset).

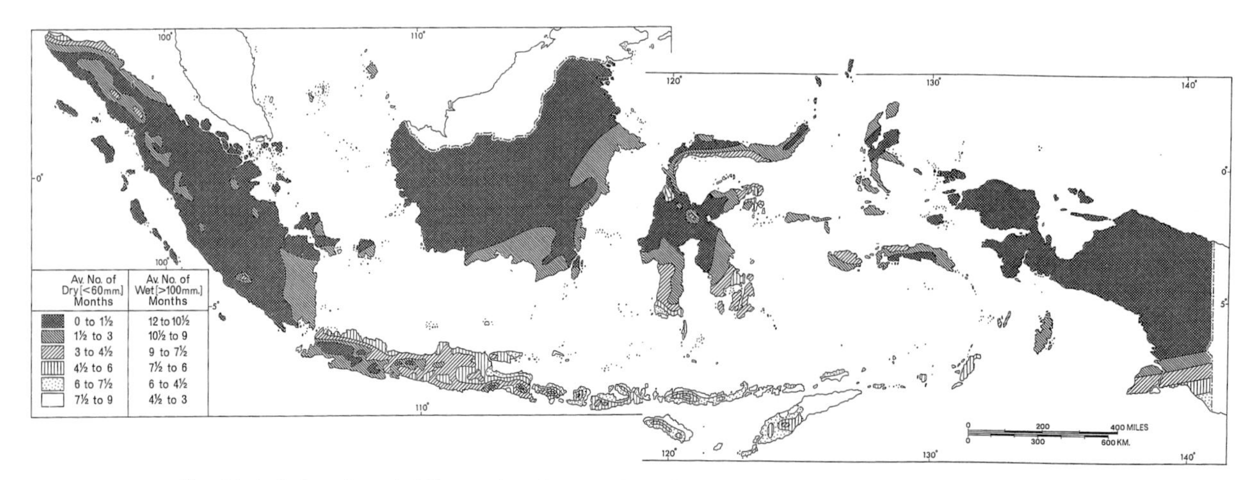

FIG. 3A.2. Indonesian rainfall types based on numbers of wet (>100 mm) and dry (<60 mm) months. (From Missen 1972—adapted from Schmidt and Ferguson 1951.)

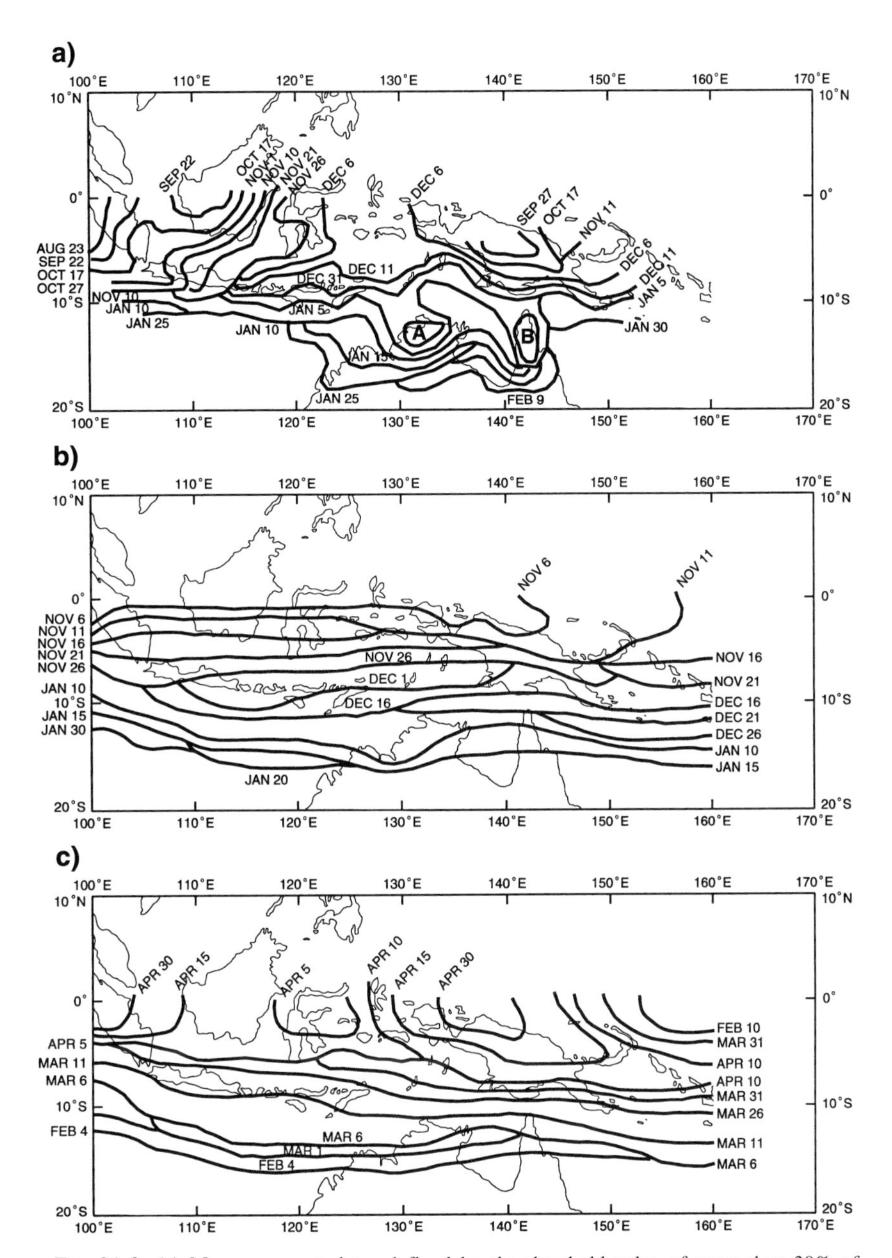

FIG. 3A.3. (a) Monsoon onset dates defined by the threshold value of more than 30% of the mean high cloud amount for the monsoon season (data from 1978–92). The region marked "A" has onset dates prior to 15 December; the region marked "B" had onset prior to 26 December. (Adapted from Tanaka 1994.) (b) Monsoon onset dates defined by dates of simultaneous observations of westerly wind at 850 hPa and easterly wind at 200 hPa. (Data from ECMWF analyzed grids, 1980–88; adapted from Tanaka 1994.) (c) Monsoon retreat dates following the wind criteria of panel b. (Adapted from Tanaka 1994.)

Tanaka (1992, 1994) has extended the definition of monsoon onset and retreat to include the Indonesian sector. Based on GMS imagery, he defined onset as the first pentad with high cloud amount exceeding the seasonal mean for that location by more than 30%. Tanaka showed that the onset date advances eastward and southward but with convective activity enhanced earlier over the land areas (Fig. 3A.3a). For example, the onset dates over New Guinea and northern Australia are 15 to 60 days earlier, compared to the adjacent seas in the same latitudes. Using a wind-based definition, Tanaka found a much simpler pattern (Fig. 3A.3b), whereby a southward advance of the summer monsoon is zonally oriented, as is the retreat (Fig. 3A.3c).

As discussed above, various authors have demonstrated that, over northern Australia and the southeastern portion of the Indonesian archipelago, interannual variations of rainfall during the September to November transition season are well related to the Southern Oscillation. The relationship breaks down, however, during the peak of the monsoon season (McBride and Nicholls 1983; Drosdowsky and Williams 1991).

The relationship between monsoon activity and the Madden–Julian (or 30–60-day) oscillation is less clear cut. As documented in chapter 8, the eastward-moving outgoing longwave radiation (OLR) anomaly associated with the MJO moves slowly across equatorial Indonesia. Thus, presumably, the monsoon rainfall will have a modulation on this periodicity. Farther south in the northern Australian region, however, spectral studies of rainfall by Hendon and Liebmann (1990b) and Drosdowsky (1996) found no significant deviation from a red-noise spectrum. Hendon and Liebmann (1990a) do, however, suggest that monsoon *onset* over northern Australia is linked to the MJO. As the 40–50-day wave moves eastward, the region of convection expands southward over the Australian continent before retreating northward as the wave continues to the east.

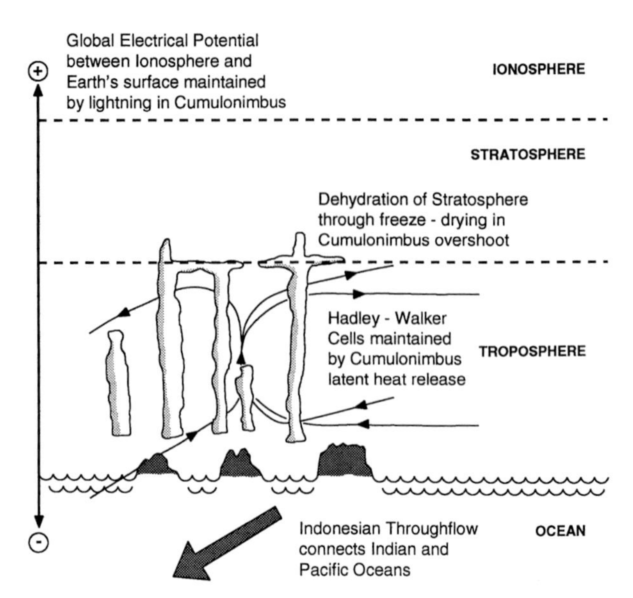

FIG. 3A.4. Schematic representation of some contributions of phenomena in the maritime continent to global-scale circulations.

3A.2. Source of latent heat release for the planetary-scale monsoon circulation

In this and the following subsections, we discuss various contributions to global climate made by tropical meteorological and oceanic phenomena located in the Indonesia–New Guinea–tropical Australia region. These are summarized in a schematic in Fig. 3A.4.

The major energy release in the tropical troposphere occurs in the form of latent heat in deep cumulonimbus convection and large areas of upper-tropospheric stratiform clouds associated with that convection. Diagnostically, this heat release can be considered to be the forcing mechanism for the observed large-scale circulation of the entire Tropics, as was originally proposed and modeled by Matsuno (1966), Webster (1972), and Gill (1980). The regions of this latent heat release can be seen on global maps of outgoing longwave radiation, where low values of OLR can be taken as a proxy for convective activity. Referring to the OLR maps in Fig. 1.49, it is seen that the tropical rainfall is concentrated into three global tropical heat sources. The first two of these are located over the equatorial regions of Africa and South America. The third is located over the monsoon longitudes, between approximately 70° and 150°E, and undergoes a seasonal migration to encompass the Northern Hemisphere summer monsoon in July–September (Fig 1.49, bottom panel) and the Southern Hemisphere summer monsoon in January–March, located over equatorial Indonesia (Fig. 1.49, top panel).

Given that the other two global heat sources occur over continents, the monsoonal region encompassing Indonesia, Malaysia, the Philippines, and New Guinea is referred to as the maritime continent (Ramage 1968). The reasons for the global convection being concentrated in the maritime continent are complex, but they are related to the very warm sea surface temperatures (SSTs) in the region. The convection is also forced by the diurnal heating of the land areas over the many islands in the region. The diurnally heated high surface temperatures, combined with the available moisture from the surrounding sea, lead to air with high values of equivalent potential temperature (θ_e), which can rise easily, through moist convective processes, to the upper troposphere.

Referring to the velocity potential and divergent wind-component maps in chapter 1 (Figs. 1.41, 1.42), the Indonesian/monsoon longitudes heat source appears to be the most dominant in terms of the world's divergent circulations. In fact, the strong winter Hadley circulation that appears on zonal mean meridional streamfunction maps (Fig. 1.40) reflects mainly the divergent flow over these central Asian longitudes, with a much smaller contribution from the western hemisphere. This was previously noted by Lorenz (1970). This divergent flow constitutes the secondary circulation associated with the planetary-scale monsoon. The primary (rotational) circulation is the accompanying zonal flow with low-level monsoon westerlies and upper-level equatorial easterlies. The upward branch is located near the seasonally heated Euro–Asian continent in the northern summer, over the maritime continent during the southern summer, and migrates southward, as depicted in Fig. 3A.5, adapted from Lau and Chan (1983). Meehl (1987) shows a similar pattern for the annual cycle of maximum precipitation.

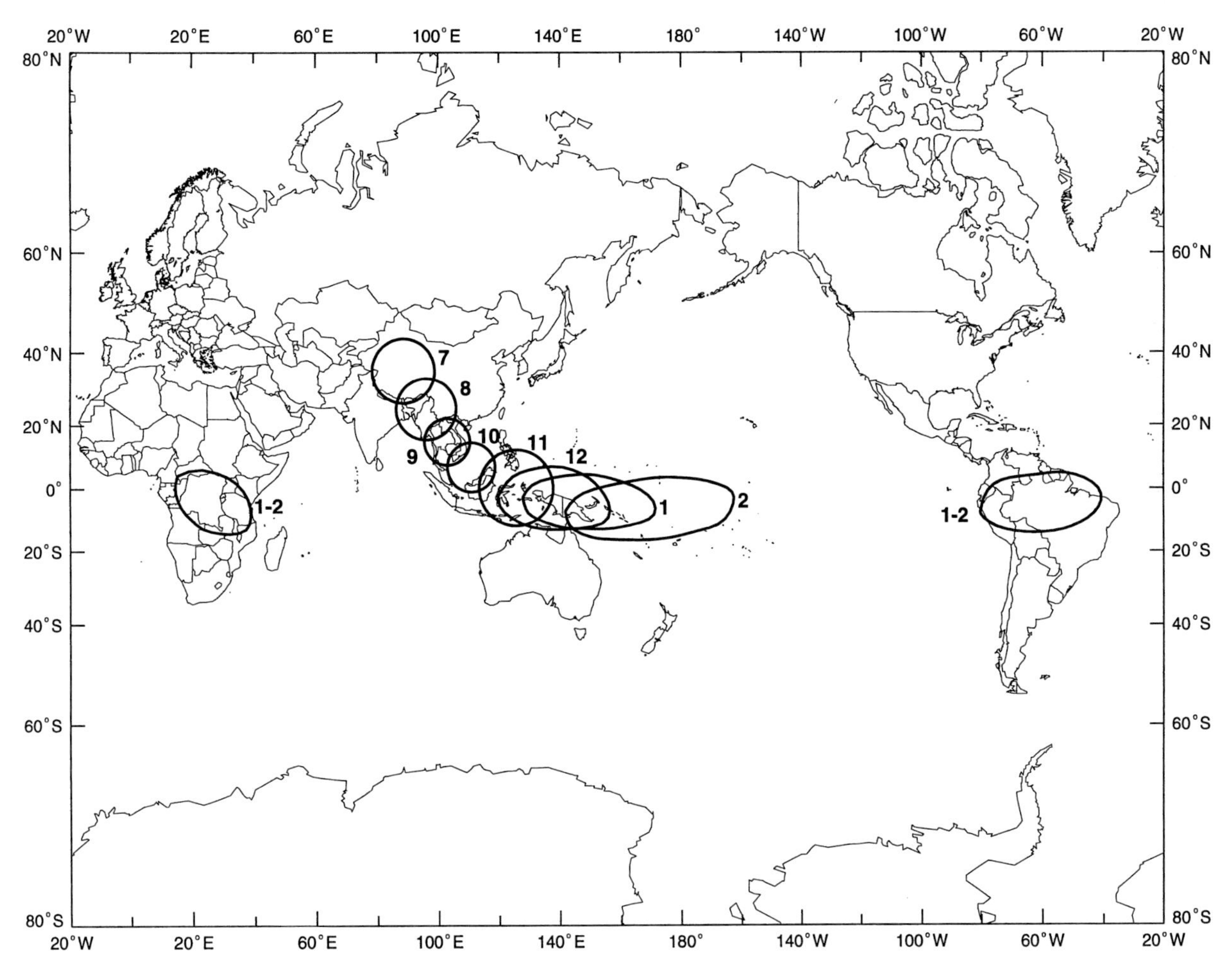

FIG. 3A.5. Seasonal migration of the monsoon diabatic heat sources during the latter half of the year (July–February, denoted by matching numerals). The extent of the diabatic heat sources is determined from the area with OLR values <225 W m^{-2} from monthly OLR climatology and is approximately proportional to the size and orientation of the schematic drawings. (Adapted from Lau and Chan 1983.)

During the southern summer, the rainfall from cumulonimbus clouds over Indonesia and tropical Australia constitutes the primary energy source for the global tropical circulation. There are two distinct modes of convection existing in this region (Keenan and Carbone 1992; Williams et al. 1992). During the early season prior to monsoon onset, and during "breaks" in the monsoon, convection occurs mainly over land (e.g., islands and the Australian continent). This *break-period convection* occurs in the form of isolated thunderstorms extending through the depth of the troposphere (up to 20 km). These have very high values of convective available potential energy (CAPE) and vigorous vertical velocities in the updrafts. Heavy rain occurs over a small area and there is vigorous electrical activity.

In contrast, convection during active monsoon periods occurs in a low CAPE regime. The monsoon convection is organized more into synoptic-scale clusters and bands and is associated with monsoonal rain over a large areal extent. The electrical activity in this second type of system is much less. The monsoon-type convection also tends to be embedded in mesoscale areas of upper-level stratiform cloud. The different properties of these two convection types are sketched schematically in Fig. 3A.6.

3A.3. Major activity center for the global electrical circuit

The monsoon-type convection gives large rainfall amounts over a large area; thus, it would be reasonable to conclude that this convection type is the major contributor to the global heat source and the upward branch of the Hadley circulation discussed earlier. In contrast, the break-type convection produces an order of magnitude more lightning ground flashes per unit of precipitation (Williams et al. 1992). As with precipitation, global lightning activity has three major source regions, one of which is the maritime continent (Whipple 1929; Orville and Henderson 1986). In particular, it has been suggested that half of all global lightning occurs within 10° of the Equator (Rut-

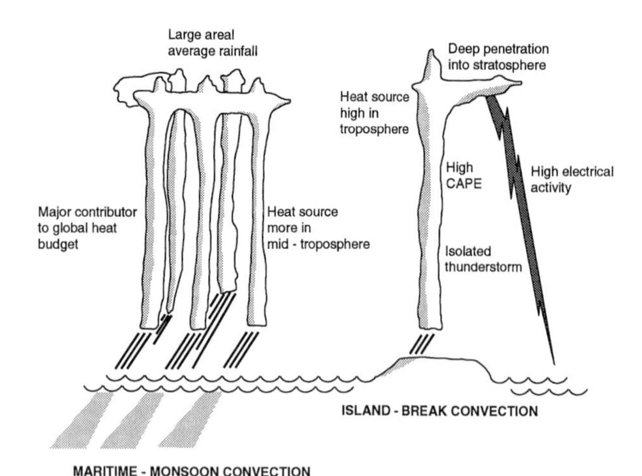

FIG. 3A.6. Schematic representation of the properties of monsoon–maritime convection versus break-island convection over the maritime continent.

ledge et al. 1992); thus, the break-type convection may well be the principle contributor to the maintenance of the net negative charge on the earth and the ionospheric potential across the depth of the troposphere (Markson 1986).

There are a large number of observational unknowns, however, associated with this scenario. In particular, during the large-scale monsoon outbreaks a considerable portion of the rainfall occurs over the large and small islands within the maritime continent. Various authors have documented the existence of a diurnal trade-off, with convection occurring over land through the afternoon and local evening and over the sea during the local morning (Holland and Keenan 1980; Murakami 1983; Williams and Houze 1987; Hendon and Woodberry 1993; McBride et al. 1995). The relative proportions of land-based rain versus sea-based rain, however, are unknown in the region.

Based on all available surface reports, Wyrtki (1956) documented very large differences in annual rainfall. He found the average value on land to be 1.5 times the average value over the sea. The area of land is small, however, compared to that of sea, so that the ratio of the average rain over the total region to that over the sea is reduced to 1.16. The extent to which these differences between land and sea are supported by satellite-derived precipitation estimates is unclear. Even if it were known, it is also unclear whether the land-based monsoon convection is of the high CAPE, enhanced electrical break-convection type, or whether it is the low-CAPE monsoonal type.

3A.4. Vertical distribution of heating

Returning to the heating (as distinct from the electrical) effect of convection in the region, the large-scale quasi-steady monsoon structure can be, to first order, understood as a dynamical response to the cumulonimbus heating, according to the classical Matsuno–Webster–Gill model. In particular, the low-level monsoon westerlies and upper-level easterlies are associated with equatorially trapped Rossby modes to the west of the region of large-scale heating. The low-level easterlies and upper-level westerlies in the Pacific Ocean are associated with an equatorial Kelvin mode generated to the east of the monsoon heating. Various authors have demonstrated that the horizontal scale and vertical structure of this solution is critically sensitive to the shape function distributing the heat in the vertical (Hartmann et al. 1984; Webster and Houze 1991; Johnson 1992; Mapes and Houze 1995). In addition, the transient components of the monsoon have been shown to depend fundamentally on the interaction between convection and the surrounding wind fields (Lau et al. 1989; Yano and McBride 1998). Thus, the transient structure is also sensitive to the vertical heating distribution. Vertical distributions of convective heating are also discussed in the next section of this chapter (chapter 3B).

Accordingly, there has been considerable energy directed toward determining the vertical distribution of heating in the monsoon region (see the review by Johnson 1992). Noting that measurements of heating are almost always inferred from measurements of divergence, Mapes and Houze (1993) compared airborne Doppler radar measurements on the cloud and mesoscale with the larger-scale, radiosonde-based measurements of Frank and McBride (1989). They found the divergence and heating to have an upscale simplification: the flow on larger horizontal scales being increasingly dominated by deeper, simpler vertical structure in the heating. Through a rotated empirical orthogonal function analysis of the vertical heating, Alexander et al. (1993) also demonstrated a large, high-frequency temporal variation of the vertical heating profile, the main variables being the stage of evolution of the life cycle of the parent cloud cluster and the diurnal signal.

Frank et al. (1997) calculated the cumulonimbus heat source based on observations from a radiosonde array surrounding the large island of New Guinea during TOGA COARE. They found that, compared to other tropical maritime datasets, the heating over New Guinea has a maximum at a higher level in the troposphere. They attributed this to the diurnal heating of the boundary layer over land producing higher values of CAPE over the island rather than over the surrounding oceans. This effect is increased by the elevated nature of the boundary layer over the New Guinea highlands. These processes favor the growth of convection that has higher tops and more heating in the upper troposphere than those that occur over the nearby maritime continent. For vertical distribution of heating, therefore (as with net rain amount and electrical properties), it seems that the land- or island-based convection has different properties from the convection over the adjoining maritime regions (Fig. 3A.6). This difference in vertical heating profile between oceanic and island/continental convection is supported by the consistent differences in radar structure (vertical profiles of reflectivity) found for the two types of system (Keenan and Carbone 1992; Keenan et al. 1994).

3A.5. Source region for stratospheric drying

Another important role played by the maritime continent region is its contribution to the meridional cell governing the exchange of mass and chemicals between the troposphere and stratosphere. Global scale aspects of tropospheric–stratospheric exchange have been reviewed by Holton et al. (1995). The meridional cells consist of upward motion in the Tropics, through the troposphere into the lower stratosphere, and equivalent downward motion into the troposphere in the extratropics and polar regions. The timescale of the cell is of the order of several years, but the strength of the flow varies both interannually and seasonally. The forcing mechanism is believed to be through eddies in the extratropical upper stratosphere and lower mesosphere, associated with breaking Rossby waves. Thus, the upwelling through the tropical tropopause is forced remotely by the extratropical mesospheric wave-driven pumping.

This picture for the slow global circulation is consistent with evidence from chemical tracers. There is a particular problem, however, with the water-vapor budget of the stratosphere. As described in a classical paper by Newell and Gould-Stewart (1981), observations of tropical stratospheric water-vapor mixing ratio in tropical South America reveal very low values of the order of 3 ppmv at altitudes several kilometers above the tropopause. Such values are too low to be explained by freeze-drying of air passing through the local tropopause, so Newell and Gould-Stewart proposed the existence of "stratospheric fountains"—regions where the local tropopause is cold enough for air passing through the tropopause to be dehydrated to the observed values. Based on monthly analyses of 100-hPa temperatures, those authors proposed that the major regions for tropical upwelling are the monsoon heat sources, specifically over the Bay of Bengal–India region in the northern summer and over the Australian–Indonesian monsoon area in the southern summer. Newell and Gould-Stewart further suggested that the upward transport in these fountains be slow, mean diabatically forced ascent through the tropopause. It was suggested by Robinson (1980), however, that the upward flow across the tropopause occurs primarily in the updrafts from deep convection penetrating into the stratosphere. Danielsen (1982) proposed a mechanism whereby the observed dehydration could occur by small-scale turbulent entrainment into overshooting cumulonimbus turrets, followed by particle sedimentation into anvils, leaving sheets of dry air behind (also see Holton et al. 1995).

To evaluate these ideas, two field experiments were conducted with high-altitude ER2 aircraft: one over tropical Panama during the northern summer (Margozzi 1983) and the other over Darwin, Australia, during the southern summer (Russell et al. 1993). Both experiments found evidence of tropospheric air penetrating high into the stratosphere within the cumulonimbus anvils. In the northern summer case, however, the net effect was to hydrate rather than dehydrate the local stratosphere (Kley et al. 1982). In the case of the Southern Hemisphere summer experiment [Stratosphere–Troposphere Exchange Project (STEP/Tropical)], water-vapor mixing ratios were observed in the overshooting cumulonimbus clouds, with values ranging between 1.8 and 3.2 ppmv (Kelly et al. 1993). Simultaneous summertime observations of radon (a tracer of tropospheric air), ice-crystal sedimentation, and the air being near saturation provide convincing evidence that the overshooting cumulonimbus clouds near Darwin in the Southern Hemisphere provide a mechanism by which tropospheric air can be transferred to the stratosphere and dehydrated to prevailing stratospheric minimum water-vapor mixing ratios. (Danielsen 1993; Russell et al. 1993).

Similar mechanisms may be operating over the Northern Hemisphere location of the monsoon heat source during the northern summer. The extratropically forced mass transport into the tropical stratosphere, however, is believed to be twice as large during the Southern Hemisphere summer (Holton et al. 1995); thus, it seems the major location for dehydration of the global stratosphere is over the maritime continent during the southern summer monsoon.

Inspection of data from STEP/Tropical reveals several different modes of transfer of tropospheric air into the stratosphere. The two most efficient modes occur in cumulonimbus convection in either break-monsoon/continental-convection conditions or tropical cyclones. In both these situations, the data showed that overshooting towers deposited tropospheric air as high as 18 km and had saturation mixing ratios as low as 1.1 ppmv. By comparison, the monsoon convection had anvil tops rarely above 17 km and saturation mixing ratios only down to 2.25 ppmv. Thus, as with the other global roles played by convection in the maritime continent, the role in stratospheric drying seems highly dependent on the break or island convection rather than the main monsoon convection (Fig. 3A.6).

3A.6. Indonesian throughflow

An interesting role played by the maritime continent in global climate is through the phenomenon known as the "Indonesian throughflow." This is a system of currents flowing from the Pacific Ocean to the Indian

Ocean through the seas and ocean channels between the major islands constituting Indonesia and surrounding countries. As sketched in Fig. 3A.1, the major component of the flow occurs in the Makassar Strait between Borneo and Sulawesi, with secondary flows in the eastern seas (dashed arrows). Farther south, the flow enters the Indian Ocean through the Lombok Strait between Bali and Lombok and through the Sawu and Timor Seas on either side of the island of Timor.

Despite the small area of this current, the throughflow is thought to play a major role in global climate. This is because the source of the flow, the equatorial western Pacific, is a region of very deep thermocline; thus, the transport-averaged temperature of the throughflow is of the order of 24°C (Fieux et al. 1994). The mass-balancing return flow occurs south of the Australian continent, so the transport-averaged temperature is of the order of 6°–8°C (Godfrey et al. 1993). It is thought that the net heat transport in the throughflow is a substantial fraction of the total heat absorbed by the equatorial Pacific Ocean. Consequently, in the heat budget of the Southern Hemisphere, the Indian Ocean may actually transport more heat poleward than is transported by the Pacific. Further discussion of the role of oceanic currents can be found in chapter 7.

The net throughflow transport is believed to be westward at all times. It occurs as a western boundary current (along Mindanao and Borneo—see Fig. 3A.1) and is made up of two components: the surface current driven by the monsoons and the deeper-reaching interoceanic flow (Tomczak and Godfrey 1994). The wind-driven flow actually opposes the throughflow during May to September, yet the throughflow is in the same direction all year and reaches a maximum in August, with a minimum in February. This is because the main driving force is the interoceanic pressure gradient (and, thus, relative sea levels between oceans), rather than the local monsoonal winds.

Estimates of the magnitude of the annual mean throughflow range between 2 and 16 Sv (1 Sv = 10^6 $m^3\ s^{-1}$) (Qu et al. 1994), and there has been considerable research activity aimed at the measurement and monitoring of it (Fieux et al. 1994; Meyers et al. 1995; Gordon 1995). Several recent measurements have varied enormously, reflecting the large seasonal and interannual variability though, in the mean, estimates are toward the middle of the above-mentioned range (Wjiffels et al. 1996). This is consistent with the magnitude suggested by experimentation with global ocean models (e.g., Masumoto and Yamagata 1996). Latest observational results suggest an average throughflow during the past several years of 5–8 Sv only, but winds have been rather anomalous during this time, both in the equatorial Pacific and apparently in the midlatitude South Pacific, which is important for throughflow forcing.

Currently, there is also considerable activity with ocean-modeling experiments aimed at determining the importance of the throughflow to global and regional ocean circulations (Hughes et al. 1992; Hirst and Godfrey 1993, 1994; Wajsowicz 1995). A number of authors have also studied its interannual variability and, in particular, its relationship to the Southern Oscillation (Wajsowicz 1994; Meyers 1996). These studies find a maximum throughflow during a La Niña event and a minima during an El Niño, with a peak to trough amplitude (of the ENSO signal) of 5 Sv (Meyers 1996).

3A.7. Summary

The meteorology of the region has been described briefly in the context of traditional monsoon meteorology. Various studies have been summarized concerning the morphology of monsoon onset and retreat, as well as active and break patterns.

The major emphasis of this section has been on the role of the Indonesian–Australian region in global dynamics. As noted originally by Ramage (1968), during the southern summer the major global tropical heat source is located over this region; for this reason, the region is referred to as the maritime continent. The associated latent heat release can be considered, from energetic considerations, to be the forcing mechanism for the observed large-scale circulation of the global Tropics. In particular, at this time of year, the maritime continent is the location of the major center of upward motion associated with the Hadley and Walker circulations, which in turn are major components of both the ENSO phenomenon and the MJO. The latent heat release occurs primarily in cumulonimbus convection, extending through the depth of the troposphere. The effect of the heating on large-scale dynamics is believed to be crucially dependent on the vertical shape function of net heat release. This has led to a large number of observational studies and field programs directed toward the determination of this vertical distribution.

In addition to being the major source of vertical mass transport in the Tropics, the maritime continent is also probably the major contributor to maintenance of the global electrical circuit. As with latent heat release, lightning activity is concentrated largely into three tropical activity centers, one of which is the Indonesian–Australian region.

The role of the region has also been described in the global meridional circulation of mass between the troposphere and the stratosphere. The water vapor budget of the stratosphere is such that it must be dried through mechanisms involving overshoot of deep tropical cumulonimbus cells. Analysis of data from field experiments directed at this problem suggest that, once again, the major location for dehydration of the global stratosphere may be the maritime continent during the Southern Hemisphere summer.

As discussed above, the maritime deep cumulonimbus convection in the maritime continent plays a major role in the global-scale Hadley and Walker cells, the global electrical circuit, and the meridional cell of mass transport through the tropopause. In recent decades, field experiments have been carried out to study all three phenomena. A major finding emerging in all three cases is that the cumulonimbus activity in the region has fundamentally different properties depending on whether it is land based or sea based and whether the seasonal conditions are monsoonal or in a break transition. Specifically, maritime and monsoon convection tends to be organized on the synoptic scale, is associated with large areal average magnitudes of rainfall and latent heat release, and is presumably the major contributor to vertical motions in the Hadley and Walker cells. In contrast, the island and break conditions convection is associated with vigorous electrical activity and extends through a greater depth, such that it is the principle contributor to stratospheric drying. In addition, the vertical distribution of heating is believed to be quite different in the two types of convection.

Given the mix of islands and ocean within the maritime continent, the internal structure of the global source region for mass circulation, electrical activity, and troposphere–stratosphere exchange is still largely undocumented. Specifically, the relative contributions of land and sea to areal average rainfall, the vertical distributions of heating during active versus break monsoon conditions, and the type of convection over land during monsoon active conditions are not well known.

This section also briefly summarized some aspects of the Indonesian throughflow phenomenon. This is a transport of water from the Pacific Ocean to the Indian Ocean through the seas and channels in the maritime continent region. The throughflow is believed to play a major role in global climate, as the associated net heat transport is a substantial fraction of the total heat absorbed by the Pacific Ocean. Consequently, seasonal and interannual variations in throughflow can have a major influence on large-scale climate.

REFERENCES

Alexander, G. D., G. S. Young, and D. V. Ledvina, 1993: Principal component analysis of vertical profiles of Q_1 and Q_2 in the tropics. *Mon. Wea. Rev.,* **121,** 535–548.

Berson, F. A., 1961: Circulation and energy balance in a tropical monsoon. *Tellus,* **13,** 472–485.

———, and A. J. Troup, 1961: On the angular momentum balance in the equatorial trough zone of the eastern hemisphere. *Tellus,* **13,** 66–78.

Braak, C., 1919: Atmospheric variations of short and long duration in the Malay archipelago and neighbouring regions, and the possibility to forecast them. *Verhandelingen,* No. 5, Koninklijk Magnetisch en Meteorologisch Observatorium te Batavia, 57 pp.

———, 1929: The climate of the Netherland Indies, Volumes I and II. *Verhandelingen,* No. 8, Koninklijk Magnetisch en Meteorologisch Observatorium te Batavia, 1602 pp.

Danielsen, E. F., 1982: A dehydration mechanism for the stratosphere. *Geophys. Res. Lett.,* **9,** 605–608.

———, 1993: In situ evidence of rapid, vertical, irreversible transport of lower tropospheric air into the lower tropical stratosphere by convective cloud turrets and by larger-scale upwelling in tropical cyclones. *J. Geophys. Res.,* **98(D5),** 8665–8681.

Drosdowsky, W., 1996: Variability of the Australian summer monsoon at Darwin: 1957–1992. *J. Climate,* **9,** 85–96.

———, and M. Williams, 1991: The Southern Oscillation in the Australian region. Part I: Anomalies at the extremes of the oscillation. *J. Climate,* **4,** 619–638.

Fieux, M., C. Andrie, P. Delecluse, A. G. Ilahude, A. Kartavtseff, F. Mantisi, R. Molcard, and J. C. Swallow, 1994: Measurements within the Pacific–Indian Oceans throughflow region. *Deep Sea Res.,* **41,** 1091–1130.

Frank, W. M., and J. L. McBride, 1989: The vertical distribution of heating in AMEX and GATE cloud clusters. *J. Atmos. Sci.,* **46,** 3464–3478.

———, Wang, H.-J., and J. L. McBride, 1997: Rawinsonde budget analyses during the TOGA COARE IOP. *J. Atmos. Sci.,* **53,** 1761–1780.

Gill, A. E., 1980: Some simple solutions for heat-induced tropical circulation. *Quart. J. Roy. Meteor. Soc.,* **106,** 447–462.

Godfrey, J. S., A. C. Hirst, and J. Wilkin, 1993: Why does the Indonesian throughflow appear to originate from the North Pacific? *J. Phys. Oceanogr.,* **23,** 1087–1098.

Gordon, A. L., 1995: When is appearance reality? A comment on why does the Indonesian throughflow appear to originate from the North Pacific. *J. Phys. Oceanogr.,* **25,** 1560–1567.

Gunn, B. W., J. L. McBride, G. J. Holland, T. D. Keenan, and N. E. Davidson, 1989: The Australian summer monsoon circulation during AMEX phase II. *Mon. Wea. Rev.,* **117,** 2554–2574.

Hartmann, D. L., H. H. Hendon, and R. A. Houze Jr., 1984: Some implications of the mesoscale circulations in tropical cloud clusters for large-scale dynamics and climate. *J. Atmos. Sci.,* **41,** 113–121.

Hendon, H. H., and B. Liebmann, 1990a: A composite study of onset of the Australian summer monsoon. *J. Atmos. Sci.,* **47,** 2227–2240.

———, and ———, 1990b: The intraseasonal (30–50-day) oscillation of the Australian summer monsoon. *J. Atmos. Sci.,* **47,** 2909–2923.

———, and K. Woodberry, 1993: The diurnal cycle of tropical convection. *J. Geophys. Res.,* **98,** 16 623–16 637.

Hirst, A. C., and J. S. Godfrey, 1993: The role of the Indonesian throughflow in a global ocean GCM. *J. Phys. Oceanogr.,* **23,** 1057–1086.

———, and ———, 1994: The response to a sudden change in the Indonesian throughflow in a global ocean GCM. *J. Phys. Oceanogr.,* **24,** 1895–1910.

Holland, G. J., 1986: Interannual variability of the Australian summer monsoon at Darwin: 1952–82. *Mon. Wea. Rev.,* **114,** 594–604.

———, and T. D. Keenan, 1980: Diurnal variations of convection over the "maritime continent." *Mon. Wea. Rev.,* **108,** 223–225.

Holton, J. R., P. H. Haynes, M. E. McIntyre, A. R. Douglass, R. B. Rood, and L. Pfister, 1995: Stratosphere–troposphere exchange. *Rev. Geophys.,* **33** (4), 403–439.

Hughes, T. M. C., A. J. Weaver, and J. S. Godfrey, 1992: Thermohaline forcing of the Indian Ocean by the Pacific Ocean. *Deep-Sea Res.,* **39,** 965–995.

Johnson, R. H., 1992: Heat and moisture sources and sinks of Asian monsoon precipitating systems. *J. Meteor. Soc. Japan,* **70,** 223–372.

Keenan, T. D., and L. R. Brody, 1988: Synoptic-scale modulation of convection during the Australian summer monsoon. *Mon. Wea. Rev.*, **116,** 71–85.

———, and R. E. Carbone, 1992: A preliminary morphology of precipitation systems in tropical northern Australia. *Quart. J. Roy. Meteor. Soc.*, **118,** 283–326.

———, B. R. Morton, M. J. Manton, and G. J. Holland, 1989: The Island Thunderstorm Experiment (ITEX)—A study of tropical thunderstorms in the maritime continent. *Bull. Amer. Meteor. Soc.*, **70,** 152–159.

Keenan, T. D., and Coauthors, 1994: Science Plan Maritime Continent Thunderstorm Experiment (MCTEX). BMRC Research Rep. 44, Australian Bureau of Meteorology, 61 pp.

Kelly, K., M. H. Profit, K. R. Chan, M. Loewenstein, J. R. Podolske, S. E. Strahan, J. C. Wilson, and D. Kley, 1993: Water vapour and cloud water measurements over Darwin during the STEP 1987 tropical mission. *J. Geophys. Res.*, **98(D5),** 8713–8723.

Kley, D., A. L. Schmeltekopf, K. Kelly, R. H. Winkler, T. L. Thompson, and M. McFarland, 1982: Transport of water through the tropical tropopause. *Geophys. Res., Lett.*, **9,** 617–620.

Lau, K. M., and P. H. Chan, 1983: Short-term climate variability and atmospheric teleconnections from satellite-observed outgoing longwave radiation. Part II: Lagged correlations. *J. Atmos. Sci.*, **40,** 2751–2767.

———, L. Peng, C. H. Sui, and T. Nakazawa, 1989: Dynamics of supercloud clusters, westerly wind bursts, 30–60-day oscillation, and ENSO: A unified view. *J. Meteor. Soc. Japan,* **67,** 205–219.

Lorenz, E. N., 1970: The nature of the global circulation of the atmosphere: A present view. *The Global Circulation of the Atmosphere,* G. A. Corby, Ed., Royal Meteorological Society, 3–23.

Manton, M. J., and J. L. McBride, 1992: Recent research on the Australian Monsoon. *J. Meteor. Soc. Japan,* **70,** 275–285.

Mapes, B., and R. A. Houze Jr., 1993: An integrated view of the 1987 Australian monsoon and its mesoscale convective systems. Part I: Vertical structure. *Quart. J. Roy. Meteor. Soc.*, **119,** 733–754.

———, and ———, 1995: Diabatic divergence profiles in western Pacific mesoscale convective systems. *J. Atmos. Sci.*, **52,** 1807–1828.

Margozzi, A. P., Ed., 1983: The 1980 Stratospheric–Tropospheric Exchange Experiment. NASA Tech. Memo. NASA TM 84297, 418 pp.

Markson, R., 1986: Tropical convection, ionospheric potential and global circuit variation. *Nature,* **320,** 588–594.

Masumoto, Y., and T. Yamagata, 1996: Seasonal variations of the Indonesian throughflow in a general ocean circulation model. *J. Geophys. Res.*, **101,** 12 287–12 295.

Matsuno, T., 1966: Quasi-geostrophic motions in the equatorial area. *J. Meteor. Soc. Jap.*, **44,** 25–43.

McBride, J. L., 1983: Satellite observations of the Southern Hemisphere monsoon during winter MONEX. *Tellus,* **35A,** 189–197.

———, 1987: The Australian summer monsoon. *Monsoon Meteorology,* C. P. Chang and T. N. Krishnamurti, Eds., Oxford University Press, 203–231.

———, 1995: Tropical cyclone formation. *Global Perspectives on Tropical Cyclones,* WMO Tech. Doc. WMO/TD 693, World Meteorological Organization, 63–105.

———, and T. D. Keenan, 1982: Climatology of tropical cyclone genesis in the Australian region. *J. Climatol.*, **2,** 13–33.

———, and N. Nicholls, 1983: Seasonal relationships between Australian rainfall and the Southern Oscillation. *Mon. Wea. Rev.*, **111,** 1998–2004.

———, and G. J. Holland, 1989: The Australian Monsoon Experiment (AMEX): Early results. *Aust. Meteor. Mag.*, **37,** 23–35.

———, N. E. Davidson, K. Puri, and G. C. Tyrrell, 1995: The flow during TOGA COARE as diagnosed by the BMRC tropical analysis and prediction system. *Mon. Wea. Rev.*, **123,** 717–736.

Meehl, G. A., 1987: The annual cycle and interannual variability in the tropical Pacific and Indian Ocean regions. *Mon. Wea. Rev.*, **115,** 27–50.

Meyers, G., 1996: Variation of Indonesian throughflow and the El Niño–Southern Oscillation. *J. Geophys. Res.*, **101,** 12 255–12 263.

———, R. J. Bailey, and A. P. Worby, 1995: Geostrophic transport of Indonesian throughflow. *Deep-Sea Res.*, **42,** 1163–1174.

Missen, G. J., 1972: *Viewpoint on Indonesia: A Geographical Study.* Thomas Nelson Limited, 359 pp.

Mohr, E. C., 1933: De Bodem der tropen in het algemeen en die van Ned.-Indië in the bijzonder. Meded. Kon. Ver. Kol. Inst., afd. Handelsmuseum, no. 31, D1. I le Stuk.

Murakami, M., 1983: Analysis of the deep convective activity over the western Pacific and southeast Asia. Part I: Diurnal variation. *J. Meteor. Soc. Japan,* **61,** 60–76.

Murakami, T., and A. Sumi, 1982: Southern Hemisphere monsoon circulation during the 1978–79 WMONEX. Part II: Onset, active and break monsoons. *J. Meteor. Soc. Jap.*, **60,** 649–671.

Newell, R. E., and S. Gould-Stewart, 1981: A stratospheric fountain? *J. Atmos. Sci.*, **38,** 2789–2796.

Nicholls, N., 1981: Air–sea interaction and the possibility of long-range weather prediction in the Indonesian Archipelago. *Mon. Wea. Rev.*, **109,** 2435–2443.

———, J. L. McBride, and R. J. Ormerod, 1982: On predicting the onset of the Australian wet season at Darwin. *Mon. Wea. Rev.*, **110,** 14–17.

Oldeman, L. R., 1975: An agro-climatic map of Java. *Contributions from the Central Research Institute for Agriculture,* No. 17, Central Research Institute for Agriculture (Indonesia), 22 pp.

———, I. Las, and S. N. Darwis, 1979: An agro-climatic map of Sumatra. *Contributions from the Central Research Institute for Agriculture,* No. 52, Central Research Institute for Agriculture (Indonesia), 35 pp.

Orville, R. E., and R. W. Henderson, 1986: Global distribution of midnight lightning: September 1977 to August 1978. *Mon. Wea. Rev.*, **114,** 2640–2653.

Qu, T., G. Meyers, and J. S. Godfrey, 1994: Ocean dynamics in the region between Australia and Indonesia and its influence on the variation of sea surface temperature in a global general circulation model. *J. Geophys. Res.*, **99,** 18 433–18 445.

Ramage, C. S., 1968: Role of a tropical "maritime continent" in the atmospheric circulation. *Mon. Wea. Rev.*, **96,** 365–369.

Robinson, G. D., 1980: The transport of minor atmospheric constituents between troposphere and stratosphere. *Quart. J. Roy. Meteor. Soc.*, **106,** 227–253.

Russell, P. B., L. Pfister, and H. B. Sleeker, 1993: The tropical experiment of the Stratosphere–Troposphere Exchange Project (STEP): Science objectives, operations, and summary findings. *J. Geophys. Res.*, **98,** 8563–8589.

Rutledge, S. A., E. A. Williams, and T. D. Keenan, 1992: The Down Under Doppler and Electricity Experiment (DUNDEE): Overview and preliminary results. *Bull. Amer. Meteor. Soc.*, **73,** 3–16.

Schmidt, F. H., and J. H. A. Ferguson, 1951: Rainfall types based on wet and dry period ratios for Indonesia with Western New Guinea. *Verhandelingen,* No. 42, Kementerian Pehubungan Djawatan Meteorologi dan Geofisik, 77 pp. plus figures.

Tanaka, M., 1992: Intraseasonal oscillation and the onset and retreat dates of the summer monsoon over east, southeast Asia and the western Pacific region using GMS high cloud amount data. *J. Meteor. Soc. Japan,* **70,** 613–629.

———, 1994: The onset and retreat dates of the Austral summer monsoon over Indonesia, Australia and New Guinea. *J. Meteor. Soc. Japan,* **72,** 255–267.

Tomczak, M., and J. S. Godfrey, 1994: *Regional Oceanography: An Introduction.* Pergamon Press, 422 pp.

Troup, A. J., 1961: Variations in upper tropospheric flow associated with the onset of the Australian summer monsoon. *Indian. J. Meteor. Geophys.*, **12,** 217–230.

Wajsowicz, R. C., 1994: A relationship between interannual variations in the South Pacific wind stress curl, the Indonesian throughflow, and the west Pacific warm water pool. *J. Phys. Oceanogr.*, **24,** 2180–2187.

———, 1995: The response of Indo–Pacific throughflow to interannual variations in the Pacific wind stress. Part I: Idealized geometry and variations. *J. Phys. Oceanogr.*, **25,** 1805–1826.

Webster, P. J., 1972: Response of the tropical atmosphere to local, steady forcing. *Mon. Wea. Rev.*, **100,** 518–541.

———, and R. A. Houze, Jr., 1991: The Equatorial Mesoscale Experiment (EMEX): An overview. *Bull. Amer. Meteor. Soc.* **72,** 1481–1505.

Whipple, F. J. W., 1929: On the association of the diurnal variation of electrical potential gradient in fine weather with the distribution of thunderstorms over the globe. *Quart. J. Roy. Meteor. Soc.*, **55,** 1–17.

Williams, E. R., S. A. Rutledge, S. G. Geotis, N. Renno, E. Rasmussen, and T. Rickenbach, 1992: A radar and electrical study of tropical "hot towers." *J. Atmos. Sci.*, **49,** 1386–1395.

Williams, M., and R. A. Houze Jr., 1987: Satellite-observed characteristics of winter monsoon cloudclusters. *Mon. Wea. Rev.*, **115,** 505–519.

Wjiffels, S. E., N. Bray, S. Hautala, G. Meyers, and W. M. L. Morawitz, 1996: The WOCE Indonesian throughflow repeat hydrology sections: I10 and IR6. International WOCE Newsletter No. 24, 25–30. [Available from WOCE IPO, Southampton Oceanography Centre, Empress Dock, Southampton SO14 3ZH, United Kingdom.]

Wyrtki, K., 1956: The rainfall over the Indonesian waters. *Verhandelingen,* No. 49, Lembaga Meteorologi dan Geofysik, 24 pp.

Yano, J.-I., and J. L. McBride, 1998: An aquaplanet monsoon. *J. Atmos. Sci.*, **79,** 1373–1399.

Chapter 3B

Pacific Ocean

DAYTON G. VINCENT
Purdue University, West Lafayette, Indiana

The equatorial and southern low-latitude region of the Pacific Ocean contains many of the world's most interesting and significant circulation features. These occur on a variety of temporal scales. For example, the El Niño takes place there. The Madden–Julian (1971, 1972) oscillation and other intraseasonal fluctuations, which account for much of the variability in convection and related circulation patterns, also occur in this region. These topics, however, are discussed in detail in chapter 8 and, thus, are not emphasized in this chapter. Instead, the focus here will be on other important large-scale features that occur in the region, such as the ITCZ, SPCZ, and subtropical jet. The distributions of several variables, including SSTs, convective heating, precipitation, mean SLP (MSLP), and kinematic properties of the flow field will be related to these features. In addition, because of the enhanced scientific interest during the last decade in the "warm pool" region of the western Pacific, a special section is devoted to the discussion of the climatological features in that region. This is particularly appropriate because of its relevance to many of the recently published results from TOGA COARE, held from 1 November 1992 to 28 February 1993. Finally, the last part of this section will be devoted to an examination of the vertical distributions of total convective heating over the South Pacific. Smaller-scale and shorter-term phenomena (e.g., mesoscale features) are delegated to chapter 5 and, thus, will not be discussed here.

The dataset used in this section, unless otherwise specified, is for the 6-yr period, 1985–90, and consists of the WCRP TOGA archive II ECMWF gridpoint analyses. An excellent description and assessment of this dataset is given by Trenberth (1992). The same 6-yr period has been used in the TOGA COARE atlases by Vincent and Schrage (1994, 1995) and Hennon and Vincent (1996), who note that this period offers an acceptable representation of the climatology of the tropical South Pacific and adjacent regions because it contains one El Niño and one La Niña.

The SST distributions for the tropical Pacific Ocean are shown in Fig. 3B.1 for the solstice seasons. The values, which were extracted from the aforementioned 6-yr period of global data given in the TOGA CD-ROM archive, are at grid increments of 2.0° lat/long. Discussions of SSTs occur elsewhere in this monograph (e.g., chapter 7); however, they are shown here because of their importance to many of the circulation features discussed below. The SSTs in Fig. 3B.1 are based on Climate Analysis Center monthly mean analyzed SST fields, as well as climatological SST fields provided by the Comprehensive Ocean Atmosphere Data Set and climatological ice data (Reynolds 1988). According to Finch (1994), this SST dataset is intended for those interested in air–sea interaction climatological studies, which makes it ideal for the present study. Figure 3B.1 shows that a large portion of the equatorial western Pacific has SSTs in excess of 29°C. This area is referred to in the opening paragraph as the warm pool region, which was selected as the location for the TOGA COARE field program. Note that the position of maximum SST shifts from about 5°–10°S in December–February (DJF) to about 10°N in June–August (JJA). In austral summer, an axis of maximum SST stretches southeastward across the South Pacific toward subtropical latitudes. As will be seen later, this axis of warm water is an integral part of the SPCZ, which is most intense in the DJF season. In JJA, the axis of maximum SST in the SH extends eastward along about 10°S and does not penetrate into higher latitudes. Also present is an axis of maximum SST in both seasons, located along 5°–10°N across the entire Pacific. This axis is related to the NH ITCZ, which is strongest in the summer and early fall seasons. More will be said later about the relationship between SSTs, the SPCZ, and the NH and SH ITCZs.

A variable that frequently has been used to indicate tropical convection and rainfall is the OLR. Figure 3B.2 shows the distributions of OLR for DJF and JJA. OLR data tapes, with values at 2.5° lat/long grid points for 1974–92, were kindly supplied to us by George Kiladis. Details concerning this dataset can be found in Liebmann and Smith (1996). In compiling Fig. 3B.2, only the years 1985–90 were used to be compatible with the SST maps shown in Fig. 3B.1. A typical threshold value used to indicate convective

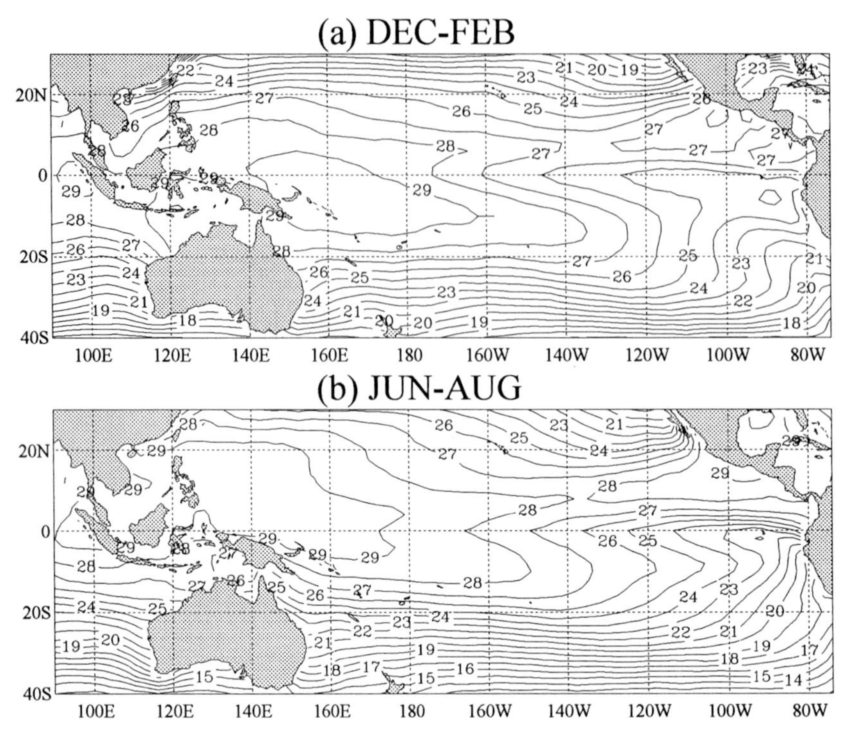

FIG. 3B.1. Sea surface temperature in °C for (a) December–February and (b) June–August, averaged for the years 1985–90.

precipitation is 230 W m^{-2}. In Fig. 3B.2, values less than this threshold have been shaded. This convective activity is associated with several important features. In DJF, for example, the axis of OLR minima stretches eastward, along 5°–10°S from the eastern Indian Ocean to the date line. It then extends southeastward toward higher latitudes, to about 40°S, 130°W. West of the date line, the near-zonal band of convection is commensurate with the ITCZ, which protrudes into northern Australia and reflects the summer monsoon season there. An excellent discussion of features associated with the Australia monsoon circulation can be found in section 3A of this chapter. East of the date line, the diagonal cloud band is referred to as the SPCZ. In DJF, it is most intense and extends farther poleward than in other seasons.

In JJA, the ITCZ cloud band has shifted into the NH; the axis of minimum values of OLR stretches from southeast Asia to 10°N, near 140°E, and then generally eastward across the entire Pacific Ocean. The convection across the Pacific, particularly that over the central and eastern Pacific, is strongest in JJA and early fall, but exists throughout the year (e.g., note the weaker convective values in DJF). An axis of minimum values also extends from New Guinea to the vicinity of 10°S and the date line in JJA. This feature marks the semipermanent "wet zone" of the western South Pacific, which stretches from the Soloman Islands to the Samoan Islands.

A comparison between the SST distributions in Fig. 3B.1 and the OLR fields in Fig. 3B.2 reveals that the NH and SH ITCZs generally lie along the axes of maximum SST. In contrast, the SPCZ cloud band is located in a region of strong SST gradients. This baroclinic structure is a well-known feature of the SPCZ (see review article by Vincent 1994, for specific references). More will be said later concerning this subject.

Another indicator of the large-scale convective activity that occurs in the ITCZ and SPCZ is midtropospheric upward motion. Maps of vertical velocity at 400 hPa (level of maximum values) are shown in Fig. 3B.3 for all four seasons. Values were obtained for the years 1985–90, directly from the WCRP TOGA archive II ECMWF uninitialized analyses. The figure shows that the major bands of rising motion occur along the ITCZ and SPCZ in all four seasons. The changing strength and seasonal meandering of these two convective bands is also evident. In DJF, the SH ITCZ and SPCZ exhibit the strongest rising motions, with an axis that stretches from the maritime continent eastward, then southeastward, to approximately 30°S, 130°W. In the other three seasons, the rising motion in the SPCZ is less intense and not as expansive, particularly in its subtropical portion. In addition, the rising motion associated with the ITCZ over the western portion of the map shifts well to the north in JJA and September–November (SON), and coincides well with

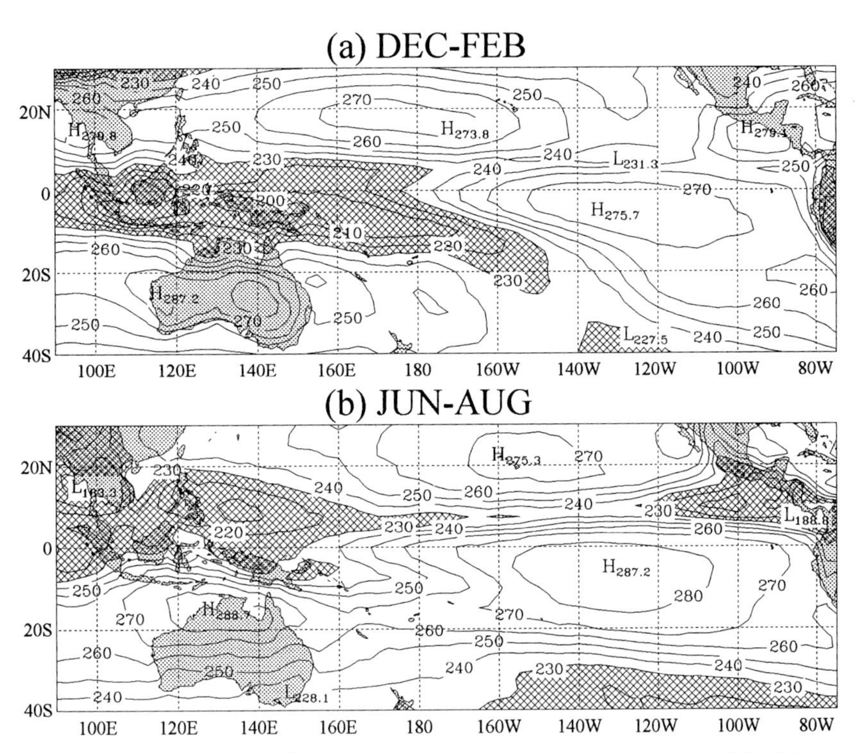

FIG. 3B.2. OLR in W m^{-2} for (a) DJF and (b) JJA, averaged for the years 1985–90. Values ≤ 230 W m^{-2} are shaded.

the NH monsoon circulation. In addition, a west–east band of rising motion between 5° and 10°N extends across the central and eastern Pacific Ocean in all seasons. This band, which is most intense in JJA and SON, corresponds to the location of the ITCZ. The patterns of rising motion, therefore, are in good agreement with the previous two variables discussed, and support the conclusion that the SPCZ circulation and convective activity is strongest and most expansive during the austral summer period. Furthermore, the SPCZ extends much farther poleward in DJF, inferring that tropical/extratropical interactions are greatest in that season. Another prominent feature in Fig. 3B.3 is the sinking motion that occurs over the semipermanent area of high pressure in the subtropical eastern South Pacific during all four seasons. More will be said about this feature later.

A convenient summary of some of the characteristics of the SPCZ and ITCZ, with the appropriate caveats, is given in Figs. 3B.4 and 3B.5. Figure 3B.4 illustrates a schematic view of the *annual* mean surface conditions associated with these two convergence zones. The figure, which was extracted from Trenberth (1991), shows that the SPCZ lies in a region of low-level moisture convergence between the predominantly northeasterly flow east of the eastern Pacific subtropical high and the cooler, predominantly southeasterly flow from higher latitudes. The caveat to be emphasized with regard to the figure is that it represents a schematic diagram of the annually averaged conditions, which undergo large seasonal, as well as interannual, variations. More will be said about these variations later. Figure 3B.5 shows maps of precipitation, SST, and surface wind convergence for January, extracted from Kiladis et al. (1989). They note that "the surface convergence maximum associated with the SPCZ lies to the south of the axis of maximum precipitation, typical of tropical convergence zones" [e.g., the Global Atmospheric Research Program Atlantic Tropical Experiment (GATE) A/B-scale ship array]. They further note, as Fig. 3B.5 shows, that "maximum precipitation lies to the south of the axis of maximum SST"—that is, in the region where SST gradients become increasingly important as one proceeds southeastward. The caveat concerning this figure is that the axes of maximum convergence and rainfall fail to show the observed spatial continuity that exists in summer between the monsoonal trough and ITCZ, over northern Australia, and the SPCZ, east of about 170°E.

The discussion now switches to the kinematics of the large-scale flow features over the tropical South Pacific. Figure 3B.6 shows the MSLP patterns for each of the four midseason months, based on the same 1985–90 ECMWF analyses that were used to describe variables in Figs. 3B.1–3B.3. In January, the most prominent feature is the trough of low pressure that extends eastward from the monsoonal low centered

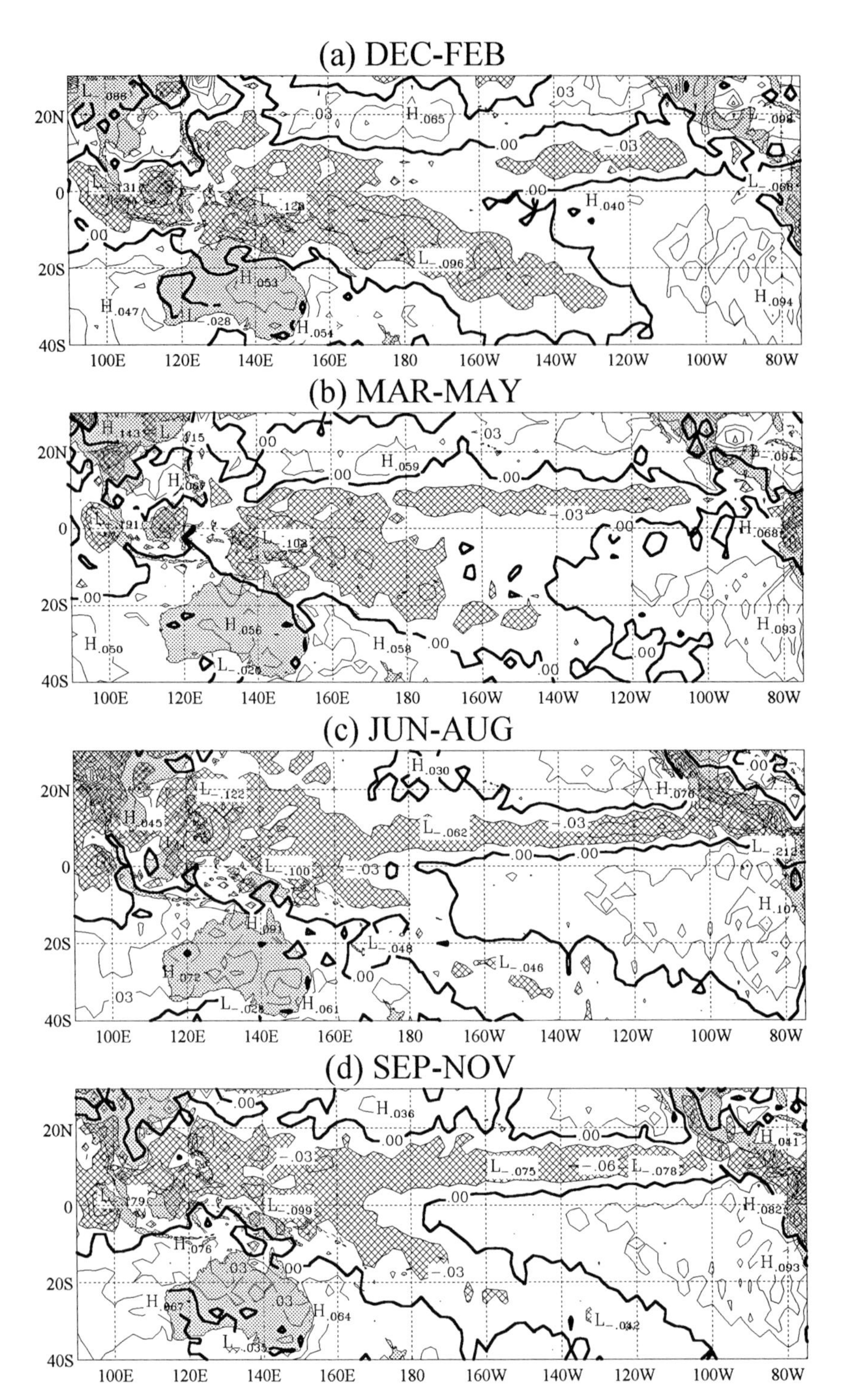

FIG. 3B.3. Vertical p-velocity (ω) at 400 hPa in Pa s^{-1} for (a) DJF, (b) March–May (MAM), (c) JJA, and (d) September–November (SON), averaged for the years 1985–90. Values ≤ -0.03 are shaded.

over northern Australia to near the date line, where it splits into two parts. One trough continues to be associated with the SH and NH ITCZs, while the other trough is associated with the SPCZ. Also shown is a sharp trough over eastern Australia. In the eastern Pacific, the well-known semipermanent area of high pressure is located near 30°S, 100°W. In April, Fig. 3B.6 shows that the equatorial trough has moved

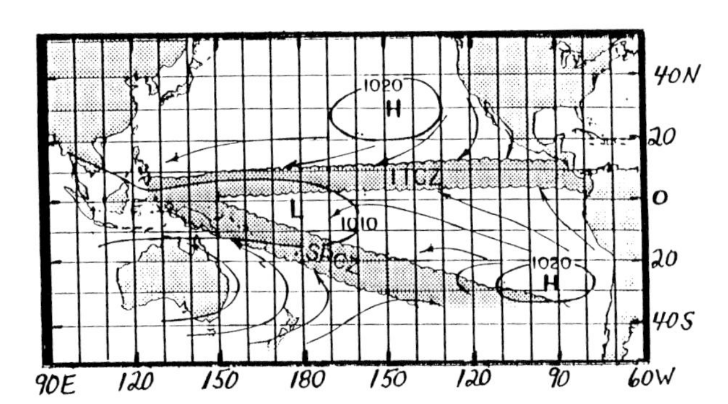

FIG. 3B.4. Schematic view of the main convergence zones, the ITCZ and SPCZ, along with the annual MSLP contours and surface wind streamlines (extracted from Trenberth 1991).

northward across most of the Pacific. The trough over eastern Australia remains, and the SPCZ trough is barely evident and has moved eastward. The eastern Pacific high has the same approximate location, but has slightly lower pressure than its value of 1022 hPa in January. By July, the pattern has changed substantially. The low-latitude trough is located well north of the Equator over the western Pacific, in conjunction with the Asian summer monsoon and the NH ITCZ. A mid-Pacific trough also occurs in the SH. In contrast to January, Australia is dominated by high pressure in July, and the SPCZ trough is essentially absent. The eastern Pacific high has shifted toward South America and has approximately the same strength as it had in April. Finally, in October, the low-latitude trough has moved slightly south of its July position over most of the Pacific. Also, the low pressure center over northern Australia, and its accompanying trough along the eastern part of the continent, have returned. The SPCZ

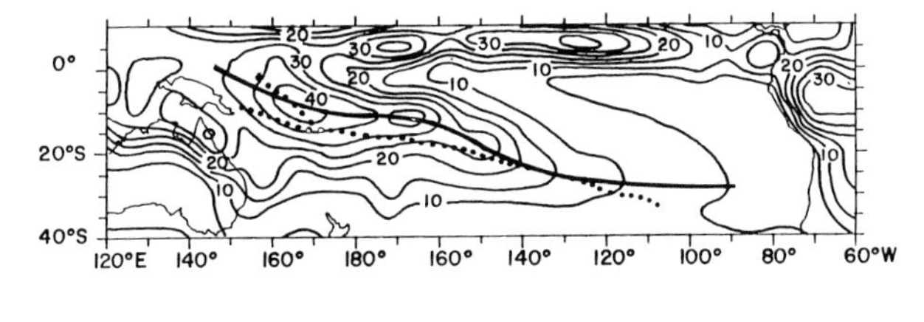

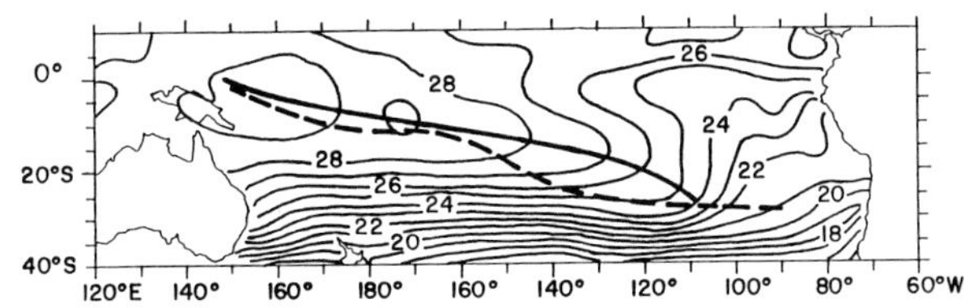

FIG. 3B.5. (a) Mean January precipitation (cm month^{-1}) for the period 1950–79. Also shown is the axis of highest precipitation (solid) and the axis of maximum surface wind convergence for the period 1961–80 (dash). (b) Mean January SST (°C) for the period 1950–79. Solid line depicts the axis of maximum SST, and the dashed line the axis of highest precipitation, taken from (a). (Extracted from Kiladis et al. 1989.)

trough, however, that was evident in January and April and is a well-known feature of the summertime circulation, has not yet returned. According to the mean monthly maps presented by Vincent and Schrage (1994), it reappears by December. The subtropical high pressure center over the eastern Pacific remains near 30°S, 90°W and strengthens by more than 3 hPa. A comparison between the MSLP patterns in Figs. 3B.6 and 3B.4 reveals that Trenberth's (1991) schematic is a reasonable blend of the four seasons.

The near-surface flow fields are what actually define the convergence zones (ITCZ and SPCZ), referred to earlier. In this context, streamlines and isotachs are provided at the 10-m level for January and July in Fig. 3B.7. In January, the streamline pattern shows that northeast trade winds extend across the entire subtropical and tropical North Pacific. Over the eastern Pacific and maritime continent region, the flow crosses the Equator and meets the southeast trade winds along the position occupied by the ITCZ. This confluence line (actually convergence) occurs near 20°S across northern Australia, where it is commensurate with the monsoon trough; then it swings north over the western Pacific, to approximately 10°–15°S. From the date line eastward, the confluence line is not as robust, but it is oriented southeast along the SPCZ. There is also some confluence along 5°N over the central and eastern Pacific, where the NH ITCZ, albeit weak, is located.

In July, southeast trades prevail over the tropical South Pacific, and little evidence exists of any confluence in the vicinity of the SPCZ. It is well known that the SPCZ is much weaker in July than in January. At all longitudes, the low-level flow is cross-equatorial from the SH. Once in the NH, the air over the Pacific meets the northeasterly trade wind flow, and results in the ITCZ between 5°–10°N. This feature is prominent throughout most of the year, but is strongest in northern summer and fall. Strong confluence also occurs in July (Fig. 3B.7c) over the summer monsoon region of southeast Asia and the adjacent western Pacific. Finally, in both January and July, over the eastern South Pacific, as well as in July over Australia, anticyclones are present. They show good agreement with the high pressure centers seen in Fig. 3B.6. In addition, the pressure and wind fields shown in Figs. 3B.6 and 3B.7 are in good agreement with those given by Sadler et al. (1987). The latter is an excellent reference of the long-term mean monthly distributions of MSLP, surface wind, and wind stress, and SSTs in the Tropics and subtropics. For example, Sadler's maps show that over the western Pacific, during the transition seasons [March–May (MAM) and SON], a double-trough structure exists in the tropical low-level flow. This frequently results in the formation of twin cyclones straddling the Equator for several days (e.g., Palmer 1952; Keen 1982; Revell 1986, 1987).

The upper-level flow in January and July is depicted in Fig. 3B.8, which shows streamlines and isotachs at

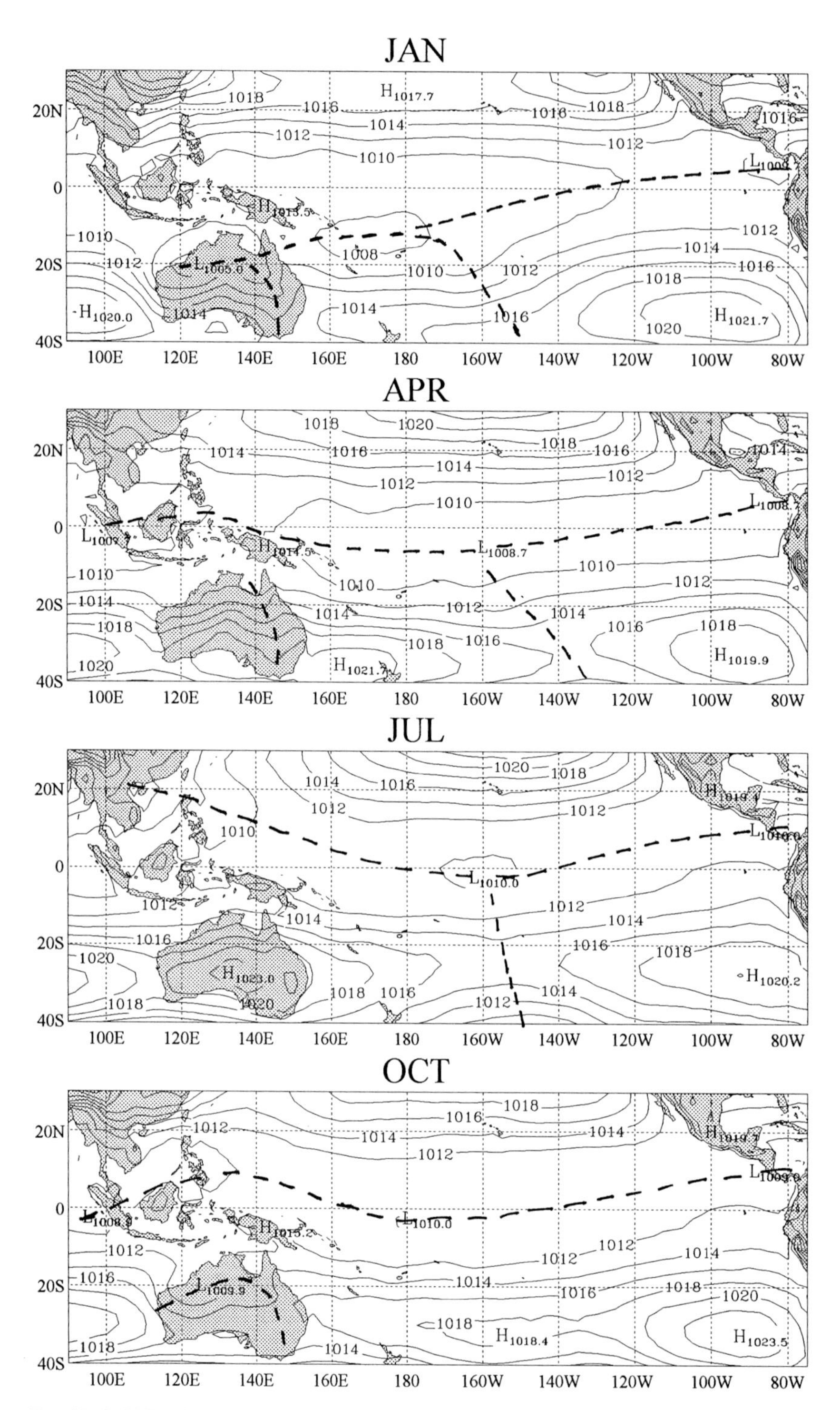

FIG. 3B.6. MSLP in hPa for (a) January, (b) April, (c) July, and (d) October, averaged for the years 1985–90. Major trough axes are dashed.

200 hPa. In January, a pair of anticyclones straddle the Equator near 160°E, with anticyclonic flow stretching westward from them to the edge of the analysis region. This feature, which has been cited in other climatological studies (e.g., Newell et al. 1972; Sadler 1975; Sadler et al. 1987), helps to provide the upper-level outflow associated with the ITCZs in the NH and SH, including the Australian monsoonal flow. An upper-level ridge (axis of anticyclonic flow) also extends from the SH anticyclone toward the southeast. This

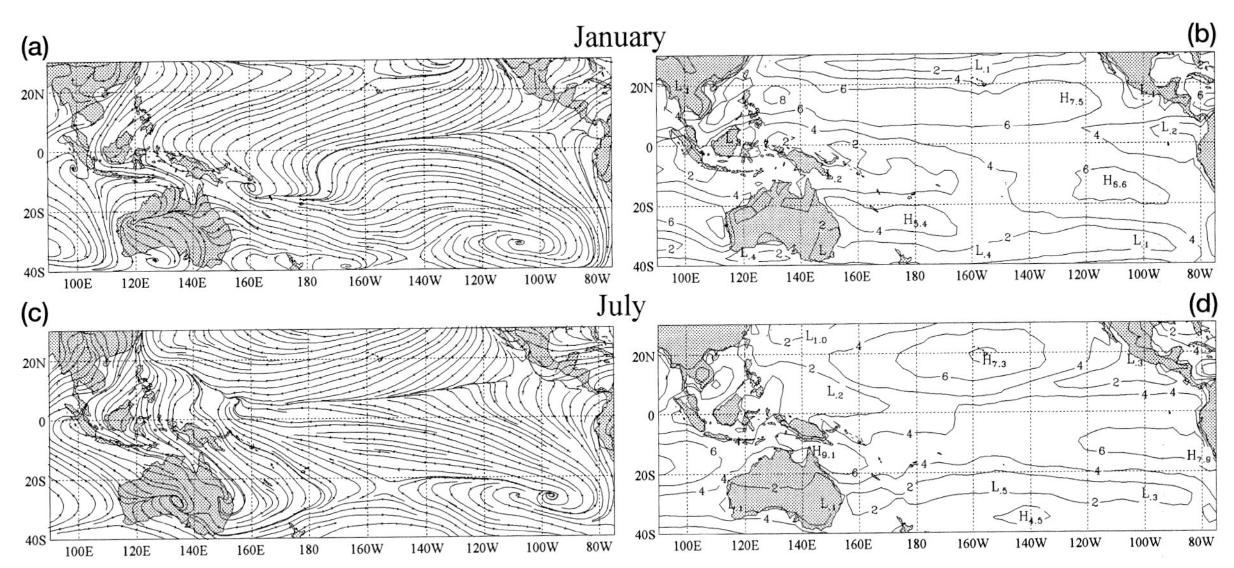

FIG. 3B.7. Streamlines and isotachs in m s^{-1} at the surface (10 m) for (a,b) January and (c,d) July, averaged for the years 1985–90.

feature, which has been noted by others (e.g., Sadler 1972, 1975; Vincent 1982), lies almost above the axis of low-level cyclonic flow (Fig. 3B.7) and is aligned along the approximate location of the SPCZ. Another important but not easily recognized feature of the flow is the axis of westerly wind maxima that extends across Australia and the western Pacific along 30°S. More is said about this feature in the next paragraph. By July, the flow pattern in Fig. 3B.8 has changed considerably. A pair of upper-level anticyclones still straddles the Equator, but both are considerably east of their January locations, especially the one in the NH. As for January, cross-equatorial flow from the summer to the winter hemisphere takes place across much of the analysis region. The subtropical jet across Australia and the western Pacific has strengthened considerably and, as pointed out by numerous investigators in the past, is a major feature of the wintertime circulation.

Although the midsummer subtropical jet over Australia and the western Pacific is much weaker than its midwinter counterpart, it is pursued further because of its uniqueness and its importance to other circulation features discussed later. Studies by Hurrell and Vincent (1990, 1991, 1992), Ko and Vincent (1995, 1996), and Vincent et al. (1997), as well as others, have shown that divergent outflow from the upper tropical troposphere, under the influence of the Corio-

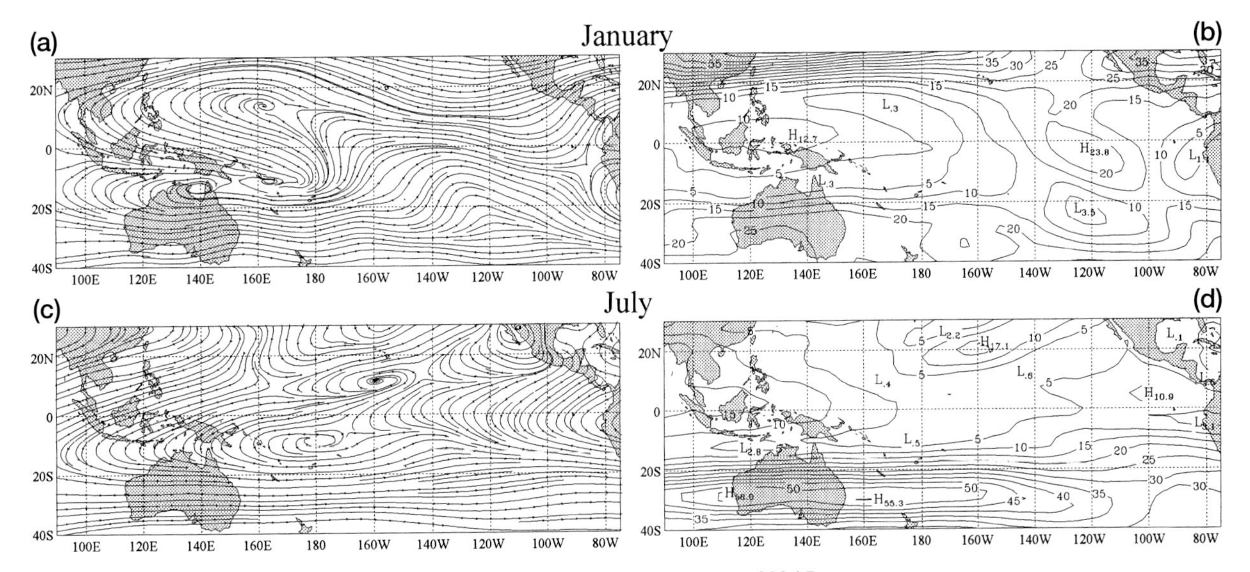

FIG. 3B.8. As in Fig. 3B.7 except at 200 hPa.

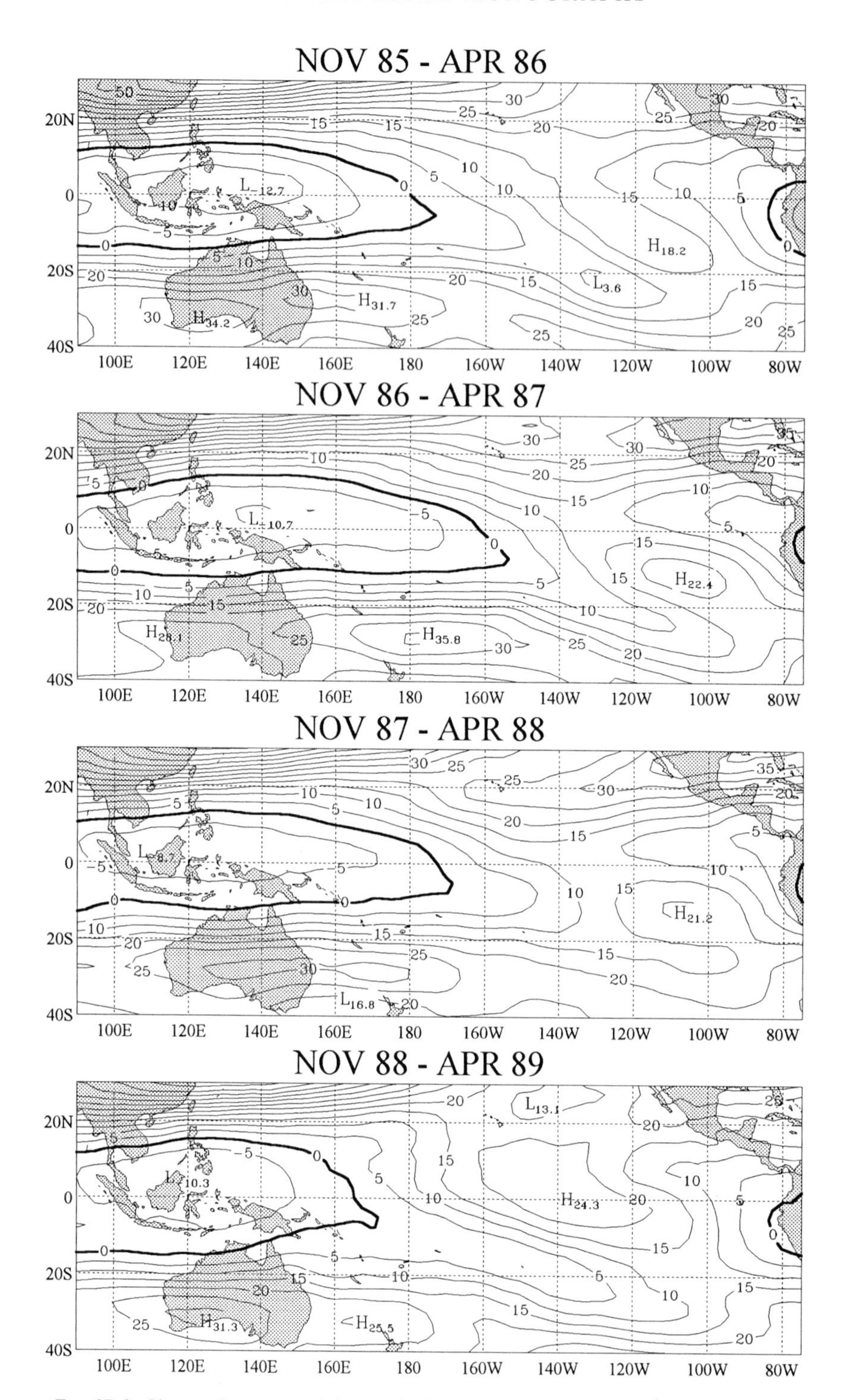

FIG. 3B.9. Six-month average of the zonal wind (u) at 200 hPa in m s^{-1} for (a) November 1985–April 1986, (b) November 1986–April 1987, (c) November 1987–April 1988, and (d) November 1988–April 1989.

lis force, makes an important contribution to the existence of the time mean jet, as well as to the eastward-propagating jet streaks that are prominent across the region in summer. Before discussing this topic further, however, it is appropriate to document some of the characteristics of the jet. Figure 3B.9 shows the 6-month mean maps of 200-hPa zonal wind for each of the four long summer seasons from 1985 to

1989. Note that the core of jet remains near 30°S in each year but meanders considerably with longitude from one year to the next and changes in intensity. For example, in 1986–87, it is located farther east and is stronger than in the other three years. This period corresponds to an El Niño event, and it is well known that a strong convective signal propagates eastward to the central Pacific during an El Niño (e.g., Rasmusson 1985; Philander 1989; also see chapter 8). As stated above, divergent outflow from a tropical heat source makes an important contribution to the existence of this subtropical jet. Furthermore, it is well documented that the SPCZ cloud band undergoes an eastward (and northward) movement during an El Niño event (e.g., Meehl 1987). In his review article, Vincent (1994) cites work by several authors that shows that a subtropical jet frequently accompanies, and is an integral part of, the SPCZ circulation. Thus, it appears that convection within the SPCZ might be related to the subtropical jet, although the relationship between these two phenomena appears to be uncertain.

The role that midlatitude eddy processes play in maintaining the summertime subtropical jet is also an unresolved issue. For example, Hurrell and Vincent (1991, 1992) used the localized EP flux equations given by Trenberth (1986) and, for a case study of the midsummer (January 1979) subtropical jet, showed that both midlatitude wave activity and divergent outflow from a tropical heat source were important in maintaining it. In their study, the diabatically driven divergent meridional wind was found to accelerate the jet in its entrance region, whereas the midlatitude forcing by divergent eddy transports of momentum and sensible heat acted to decelerate it. In the exit region, both terms acted to accelerate the jet. The net effect was an eastward progression of the jet over a 10–15-day period. In this regard, Ko and Vincent (1995, 1996) and Vincent et al. (1997) found that individual jet streaks formed over eastern Australia and moved eastward near the base of an upper-air trough at phase speeds between 5 and 10 m s^{-1}, with a periodicity of 1–2 weeks during southern summer. In particular, Vincent et al. (1997) composited 40 cases of eastward-propagating jet streaks and deduced that, in about two-thirds of the cases, tropical forcing was the primary initiation and maintenance mechanism for local jet accelerations.

Attention now is focused on the behavior of several kinematic variables in the warm pool region of the equatorial western Pacific. Some of the results discussed here were extracted from an atlas by Vincent and Schrage (1994) and a journal article by Vincent et al. (1995). Time series of monthly means of the zonal wind, meridional wind, horizontal divergence, and vertical p-velocity (ω) are presented at selected levels for the grid box centered at 2.5°S, 155°E. This point lies within the intensive flux array (IFA) of TOGA COARE (see Webster and Lukas 1992 for a detailed description) and is located within the highest SST isotherm (Fig. 3B.1). All time series have been constructed from the aforementioned ECMWF analyses and are for the same 6-yr period as used in earlier diagrams. Figure 3B.10 shows the time series of zonal wind at 200 hPa and the surface (10 m), as well as the meridional wind at 200 and 850 hPa. One of the most prominent features in the 200-hPa zonal wind is its semiannual cycle, with minimum (maximum) easterlies (westerlies) occurring in April–May and October–November and maximum easterlies occurring in January–February and July–August (Fig. 3B.10a). This feature, which has been discussed by Hendon et al. (1989) and others, is due to circulation changes that occur because of the seasonal passage of the sun across the Equator twice each year. Peak easterlies coincide with maximum convective activity associated with the NH and SH summer monsoons. Also seen in Fig. 3B.10a are the signals associated with the 1986–87 El Niño and 1988–89 La Niña events. The winds are more easterly during El Niño and more westerly, or less easterly, during La Niña. These changes are compatible with the eastward (westward) shift in the major area of convective activity and accompanying upper-level difluent flow that occur

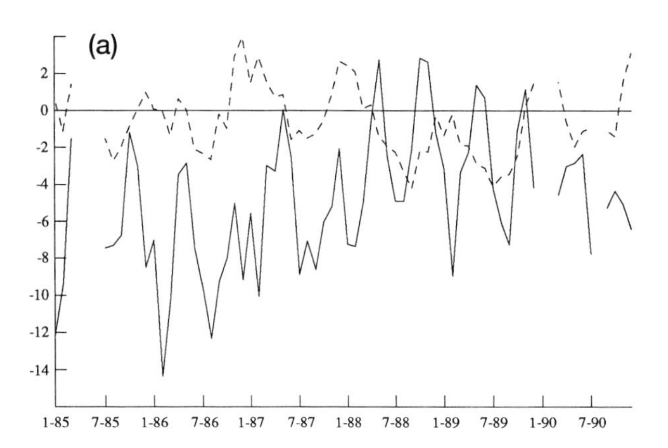

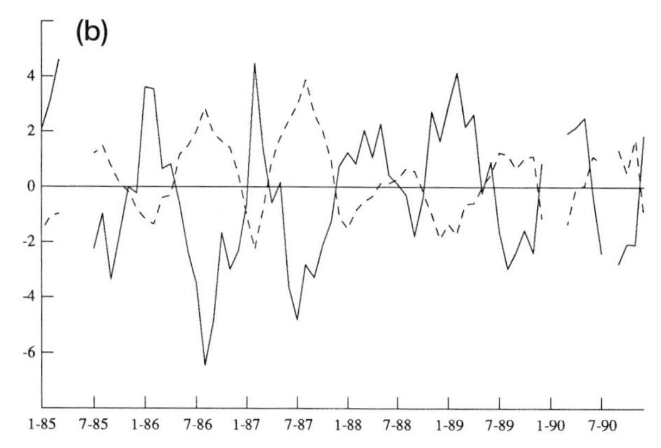

FIG. 3B.10. Time series of monthly and area-averaged (a) zonal wind (u) at 200 hPa (solid) and surface (dash), and (b) meridional wind (v) at 200 hPa (solid) and 850 hPa (dash) in m s^{-1} for the period 1985–90, for the grid box centered at 2.5°S, 155°E.

during an El Niño (La Niña). In contrast to the zonal winds aloft, the surface flow is dominated by an annual cycle. The latter are characterized by westerly maxima in January and easterly maxima in July. This feature has also been discussed elsewhere (e.g., Meehl 1987; Hendon et al. 1989). Furthermore, winds tend to become more westerly (or less easterly and, therefore, anomalous westerly) during the 1986–87 El Niño. Anomalous westerly wind bursts at the surface are a well-known component of El Niño (e.g., Philander 1989; Kiladis et al. 1993). As stated earlier, detailed descriptions of the characteristics of El Niño and La Niña are given in chapter 8 and thus will not be discussed further here.

Time series of the meridional wind at 200 and 850 hPa are illustrated in Fig. 3B.10b. At 200 hPa, the figure reveals that, unlike the zonal wind, which exhibited a semiannual cycle, the v component shows an annual cycle, with southerly (northerly) flow being a maximum approximately in February–March (August–September). The series at 850 hPa also shows an annual cycle that, as expected, is out of phase with that in the upper troposphere. At 850 hPa, peak values of northerly (southerly) flow occur in February–March (August–September). These trends in the meridional wind are indicative of the seasonal shift in the Hadley-type circulation that generally exists in the TOGA COARE region.

Time series of horizontal divergence at 200 and 850 hPa, averaged over the grid box centered at 2.5°S, 155°E, are shown in Fig. 3B.11, together with vertical motion at 400 hPa. Since these variables are somewhat "noisier" than previous variables, three-month running means are presented. At 200 hPa, divergence exists throughout the time series and an annual cycle is partially recognizable, with minimum divergence from about July to September and maximum divergence approximately six months later (Fig. 3B.11a). A gradual trend of increasing divergence prior to and during the first half of the 1986–87 El Niño event is seen to take place, followed by a steady decrease in 1987 and 1988 as the region transforms from El Niño to La Niña conditions. During the latter half of the 1988–89 La Niña, upper-level divergence increases again. At 850 hPa, the figure shows that, except for late in 1988, convergence dominates. Values show a strong annual cycle, with maximum convergence in the early months of each calendar year and minimum convergence or weak divergence about six months later. The results at 850 hPa, taken together with those at 200 hPa, generally show an out-of-phase relationship, in which divergence aloft occurs in conjunction with convergence in the lower troposphere. Furthermore, the period of most intense circulation (i.e., when maximum low-level convergence is aligned with maximum upper-level divergence) usually takes place in the first few months of each year. Lastly, unlike the time series at 200 hPa, the one at 850 hPa shows no identifiable trend with

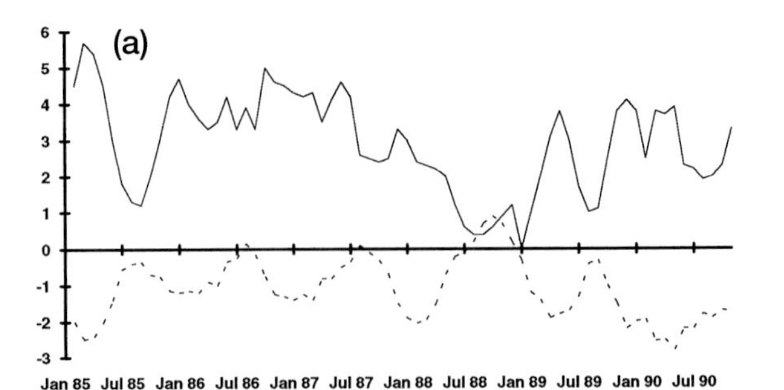

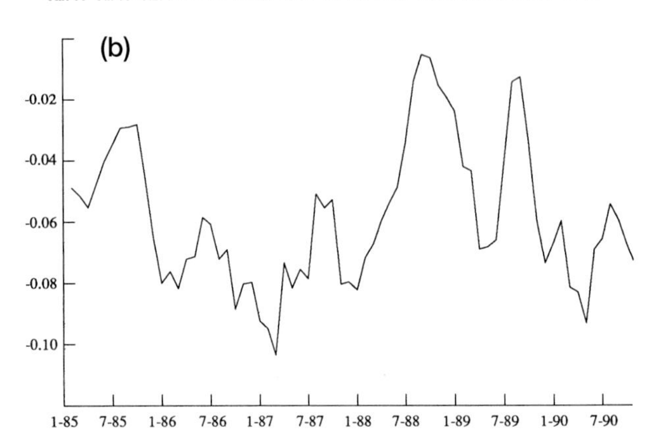

FIG. 3B.11. As in Fig. 3B.10, except (a) for the horizontal divergence at 200 hPa (solid) and 850 hPa (dash) in 10^{-6} s^{-1}, and (b) for vertical p-velocity (ω) at 400 hPa in Pa s^{-1}. Additionally, three-month running means were used in this figure.

regard to either El Niño or La Niña. The time series of vertical velocity at 400 hPa, which is near the level of maximum rising motion, is illustrated in Fig. 3B.11b; it is dominated by upward motion. A tendency toward an annual cycle exists, with the strongest upward motion centered on the months January–April and the weakest rising air occurring about six months later. A tendency also exists for stronger rising motion to occur during El Niño and weaker rising motion during La Niña.

The IFA of TOGA COARE was embedded within two larger arrays, the most extensive being the large-scale array (LSA). The boundaries of the LSA were 10°N, the date line, 10°S, and 140°E, thus encompassing much of the warm pool region. Cross sections of the zonal wind and vertical p-velocity are presented in the y–z plane between 30°N and 30°S. Figures 3B.12 and 3B.13, respectively, illustrate these two variables, which have been zonally averaged from 140°E to the date line (i.e., across the LSA longitudes) for each of the three-month seasons. The zonal-wind cross sections (Fig. 3B.12) clearly depict the seasonal variations in the location and strength of the upper-tropospheric subtropical westerly jet streams. The NH jet never advances far enough toward the Equator to be included within the latitudinal limits of the cross section. Most likely, it is nearest the Equator in DJF and appears to approach a maximum speed of about

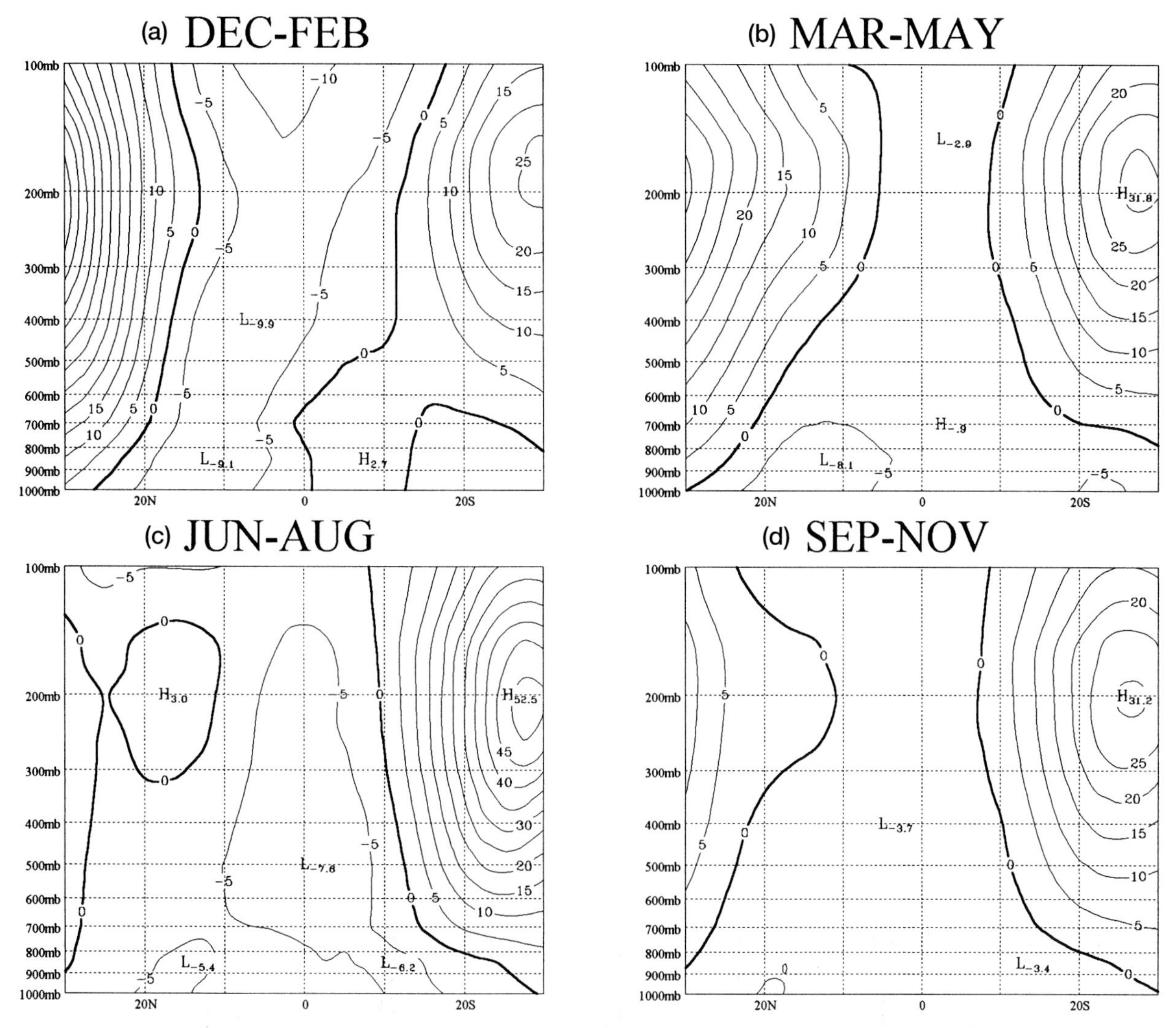

FIG. 3B.12. Meridional cross sections of zonal wind (u) in m s^{-1}, averaged between 140°E and the date line for the years 1985–90, for (a) DJF, (b) MAM, (c) JJA, and (d) SON.

60 m s^{-1} at 200 hPa. The SH jet, as expected, is a maximum in JJA, with a peak value of 52 m s^{-1} at 200 hPa between 25° and 30°S. In all four seasons, the jet in the SH is closer to the Equator than its counterpart in the NH. A comparison between the characteristics of the jets averaged from 140°E to the date line and those obtained for the general circulation (i.e., based on a global zonal average; e.g., Newell et al. 1972, p. 36) reveals that the westerly jets in both hemispheres are considerably stronger over the western Pacific sector than they are for their respective global averages. For example, Fig. 3B.12a shows a maximum of 60 m s^{-1}, whereas the global value for the same season is 35–40 m s^{-1}. It is well known that the global maximum of upper-tropospheric westerly winds occurs near the east coast of Asia; thus it is not surprising that the western Pacific value is relatively much larger. With regard to the SH westerly jet maxima, the western Pacific values for the four seasons are 30, 32, 52, and 31 m s^{-1}, while those for the globe are 23, 23, 31, and 27 m s^{-1} (Newell et al. 1972). In addition, the location of the global maxima are 5°–10° south of the location of each of the seasonal jets over the western Pacific. This is due to the zonal averaging process and occurs because, at some SH longitudes, a polar jet exists, in addition to the subtropical jet (e.g., van Loon et al. 1972, pp. 90–93).

At low latitudes, east winds dominate. Maximum low-level easterlies are centered in the winter hemisphere near 10° latitude. The largest magnitudes of these winds occur in DJF, when values are nearly −10 m s^{-1} throughout the troposphere. The weakest easterlies occur in the equinox seasons when the sun is above the Equator. Except for DJF, the latitudinal extent of the easterlies decreases with altitude, and the

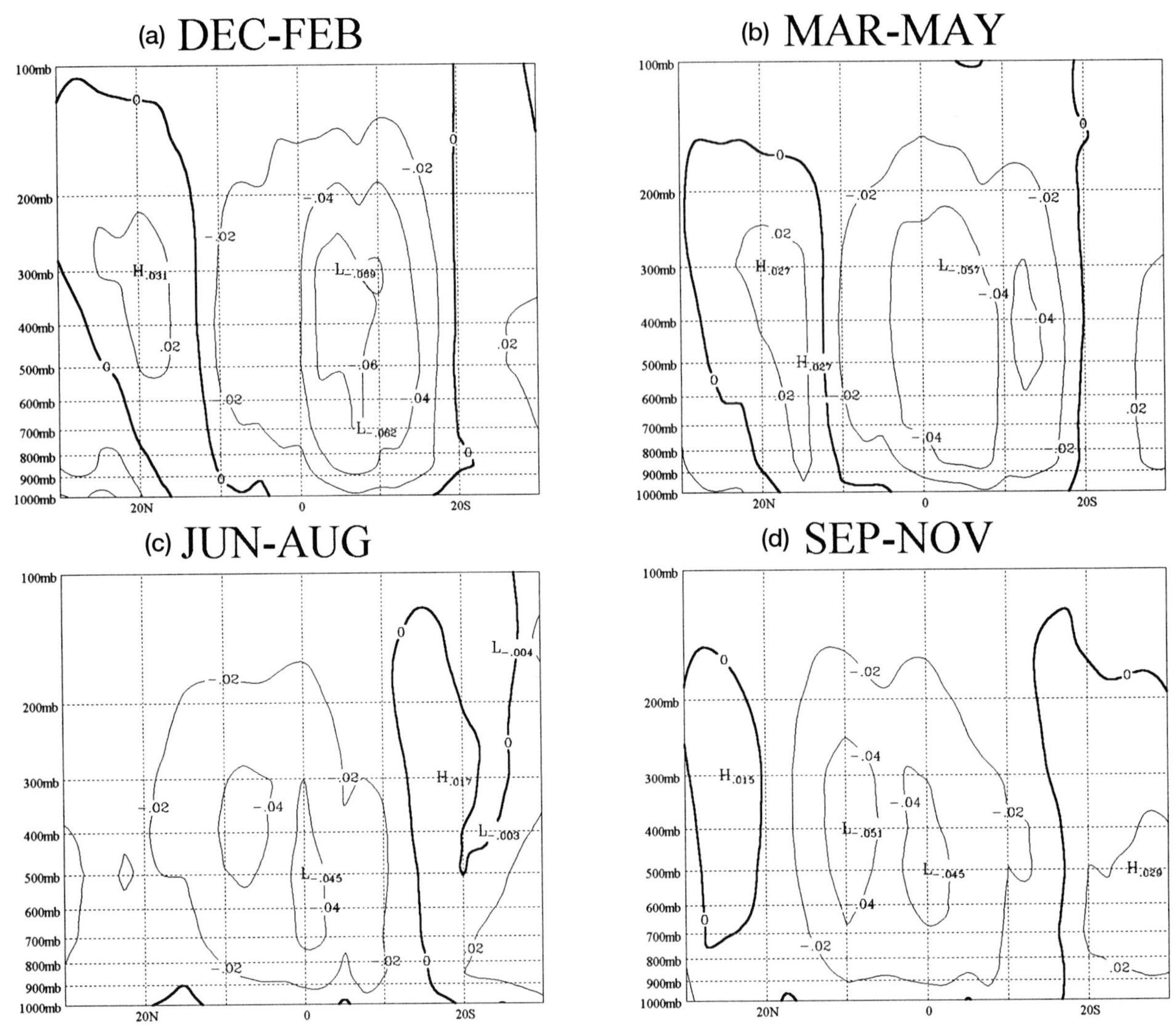

FIG. 3B.13. As in Fig. 3B.12 except for vertical p-velocity (ω) in Pa s^{-1}.

narrowest width of upper-tropospheric easterlies occurs in MAM. A comparison of the tropical easterlies over the western Pacific to those based on global averages (e.g., Newell et al. 1972) shows that the latitudinal extent and strength of the easterly winds are greater over the western Pacific. This occurs because, at some longitudes, either westerly winds or weak easterlies exist at low latitudes (e.g., Newell et al. 1972, pp. 34–35).

Cross sections of vertical motion are shown in Fig. 3B.13 for each of the four seasons. In DJF, rising motion occurs from 10°N to 20°S, with a maximum value at 5°S near 300 hPa. Downward motion is a maximum near 20°N, and it appears that a secondary maximum may occur at or near 30°S. Thus, Hadley-type cells most likely exist in both hemispheres, with their rising branch located in the LSA. This pattern persists into MAM, but maximum upward motion is less than it was in the previous season. Also, peak values of both rising and sinking motion are at lower elevations in MAM than in DJF. By JJA, the location of maximum upward motion has shifted northward, following the sun. Peak upward values occur at the Equator and between 5° and 10°N and, as for the previous season, the level of maxima is near 400 hPa. There also appears to be another region of maximum rising air near 30°N. There is no sinking motion in the NH and the sign of ω in the SH is variable. Thus, no identifiable Hadley circulation is present in this cross section. In SON, the latitudes, levels, and magnitudes of maxima in rising motion are the same as for JJA. The main change in the pattern between JJA and SON is the reappearance of centers of sinking motion in the latter season. Maxima of downward vertical velocity occur at 25°N and 25°S, suggesting that a local Hadley-type circulation has reappeared in SON.

The remainder of this section is devoted to an examination of the vertical distributions of convective

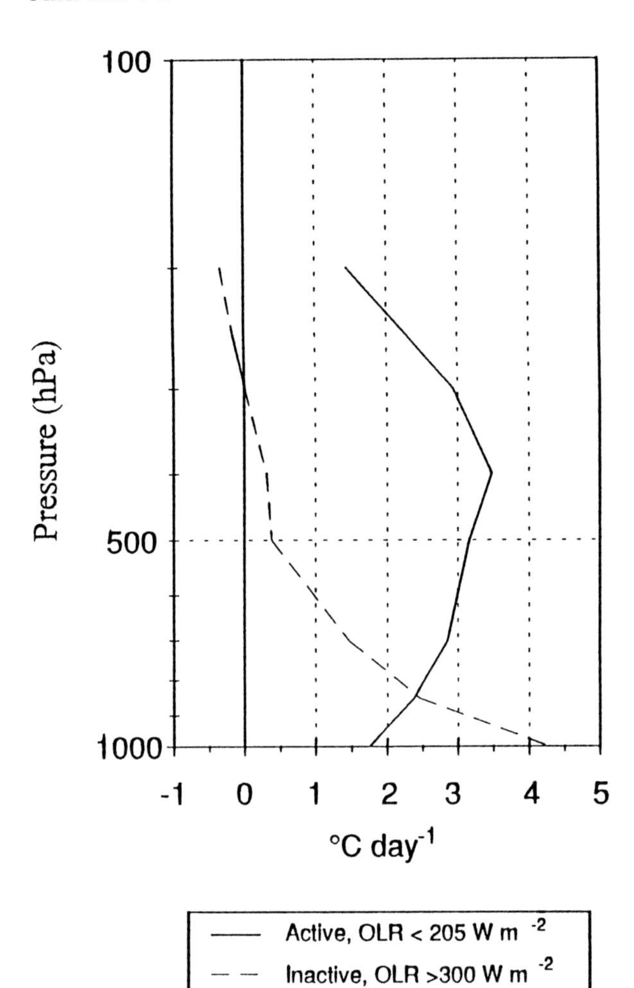

FIG. 3B.14. Vertical profiles of total convective heating in °C day^{-1} for July 1986, averaged over the region bounded by 15°N, 120°W, 15°S, and 150°E, for grid points containing active convection where OLR < 205 W m^{-2} (solid) and inactive convection where OLR > 300 W m^{-2} (dash).

heating over the tropical Pacific Ocean. It has frequently been suggested that the shape of the vertical profile of convective heating is an important contributor to the propagation speed of low-frequency tropical convective disturbances (e.g., Sui and Lau 1989). Profiles of convective heating were obtained by computing $Q_1 - Q_R$, often referred to as the total convective heating (e.g., Reed and Recker 1971; Nitta 1972; Thompson et al. 1979; McBride 1981; Miller and Vincent 1987), where Q_1 is the apparent heat source (Yanai et al. 1973) and Q_R is the net radiation. In the present results, Q_1 was calculated from the same ECMWF analyses that were used to compile several previous figures. Values of Q_R were derived according to the method described by Ramsey and Vincent (1995).

Figure 3B.14 shows an example of the differences in profiles of total convective heating between active and inactive convection events over the tropical Pacific area bounded by 15°N–15°S, 150°E–120°W for July 1986. To partition the profiles into these two categories, values of OLR (same as those used to obtain Fig. 3B.2) were applied as follows: OLR < 205 W m^2 was considered convectively "active" and OLR > 300 W m^2 was considered convectively "inactive." These criteria were chosen to select the highest and lowest 10% of the 26 784 profiles compiled for the month. Not surprisingly, Fig. 3B.14 shows that the active events yielded maximum heating at 400 hPa, while the inactive events had maximum heating near the earth's surface. From this analysis, the conventional view of convective heating maxima above the 500-hPa level, over the equatorial regions of the Pacific, is well verified and in good agreement with previous results (e.g., Reed and Recker 1971; Nitta 1972; Yanai et al. 1973; Miller and Vincent 1987; and Pedigo and Vincent 1990). In their atlas, Vincent and Schrage (1995) show numerous seasonal profiles of convective heating for a variety of regions across the tropical Pacific.

To illustrate the variability of $Q_1 - Q_R$ on shorter-than-monthly timescales, 5-day averages were compiled for a period from July 1986 to February 1987. Vertical cross sections of total convective heating were constructed, and a sample of the results is presented in Fig. 3B.16. The results can be compared to patterns of OLR anomalies, which are shown in Fig. 3B.15 in the form of Hovmöller diagrams. The OLR values were averaged over the latitude band, 10°N–10°S. Anomalies were based on 5-day averages to be compatible with the heating results. They were derived by subtracting the 6-yr (1985–90) average monthly mean from the 5-day period for the month in which the 5-day average occurred. Figure 3B.15a shows, as is well known, that deep convection is most prominent over the warmer water regions of the Indian and Pacific Oceans. The most notable feature in the diagram is the slow eastward progression of the area of maximum convection from the western to the central Pacific. This is associated with the El Niño signal of 1986–87; similar diagrams can be found in the *Climate Diagnostic Bulletin* (e.g., Weickmann 1986). The reader is referred to chapter 8, where a detailed discussion of El Niño is given.

Of particular interest here, however, are the quasi-periodic episodes of deep convection that are observed over the western Pacific (140°E–180°) between the latter part of July and the early-to-mid part of October 1986. Three distinct episodes of convection, with breaks in between, occurred on intraseasonal, but less than MJO, timescales. Each episode appears to contain one or more shorter periods of intense convection. To examine these convective periods more closely, the portion of the Hovmöller diagram in Fig. 3B.15a, between longitudes 120°E and 90°W and dates 27 July and 10 October, is enlarged. The resulting diagram,

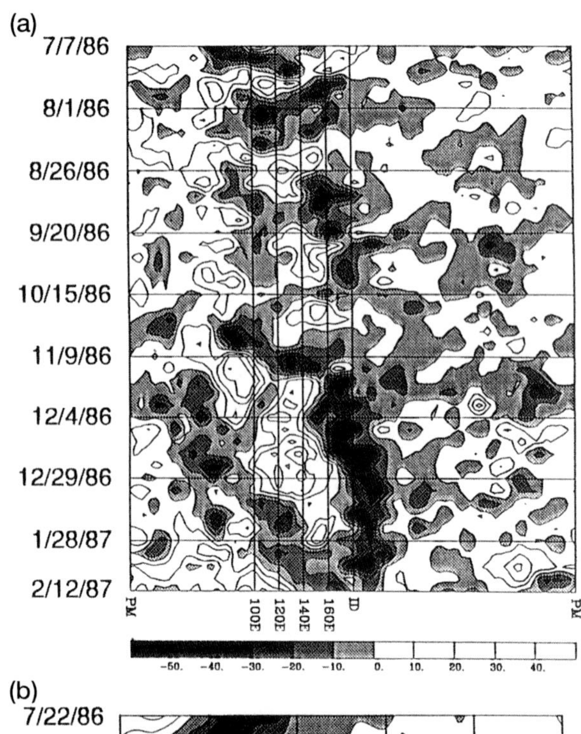

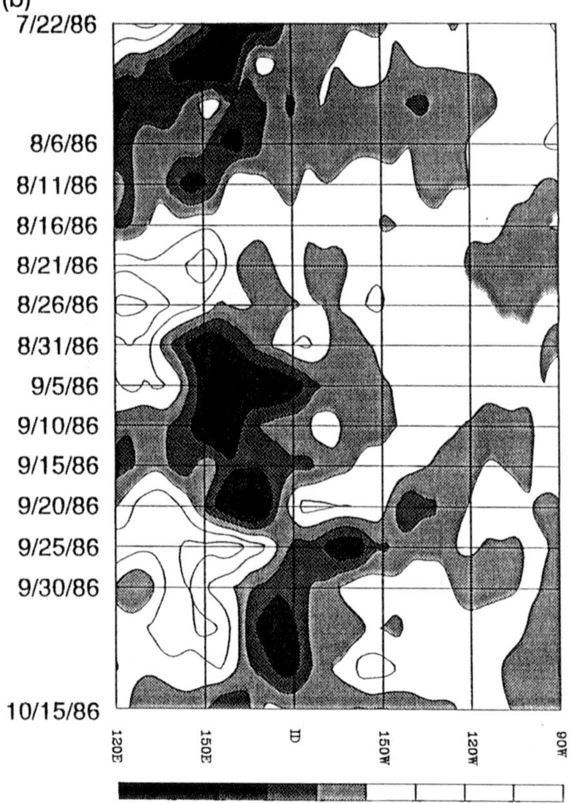

FIG. 3B.15. Hovmöller diagrams of 5-day averages of OLR anomalies in W m^{-2}: (a) 7 July 1986 to 12 February 1987, averaged between 10°N and 10°S; and (b) an enlarged portion of (a), from 22 July to 15 October 1986 and from 120°E to 90°W.

shown in Fig. 3B.15b, suggests that the most intense convective anomalies propagated slowly eastward from about 150°E to near the date line during the two-and-a-half-month period. It also appears that convective anomalies strengthened about every 10–20 days; however, the use of 5-day averages may preclude any conclusions regarding periodic behavior on this temporal scale. Nevertheless, Schrage and Vincent (1996) found numerous episodes of convective anomalies in the same region, with a period range of 7–21 days.

Figure 3B.16 shows six vertical cross sections taken along the Equator, based on 5-day averages of $Q_I - Q_R$ during the period beginning 9–13 August and ending 23–27 September 1986. To conserve space, the 5-day periods shown are not consecutive; instead, they were selected to highlight significant features in the heating fields. Each cross section was obtained by averaging data from 5°N–5°S. The first panel (Fig. 3B.16a) is for the period 9–13 August and reveals that the highest heating rates occurred near 400 hPa, just west of 150°E. This is in agreement with Fig. 3B.15b, which shows minimum values of OLR anomalies at that location. Proceeding eastward, there is good agreement between $Q_I - Q_R$ and the OLR anomaly pattern. The negative peak at 500 hPa, near 155°W, may imply that vertical fluxes are removing dry static energy from the middle and upper troposphere.

The next cross section shown is for the period 29 August–2 September 1986. During this pentad, heating rates over the western Pacific become enhanced, as also indicated by the OLR anomalies. Maxima of convective heating are near 500 hPa. East of the date line, heating is confined to the lower troposphere, and Fig. 3B.15b indicates a corresponding lack of convection. In the four remaining pentads (Figs. 3B.16c–f), excellent agreement between high values of convective heating and the location of maximum convection occurs, as implied by the OLR anomaly patterns. The third and fourth panels show that maximum heating rates have increased greatly over the western Pacific. This compares very favorably with the large OLR negative anomalies seen in Fig. 3B.15b. As the convection intensifies, the level of maximum heating moves upward from 500 to 400 hPa. In the next pentad shown (18–22 September), the largest heating rate during the entire study occurred (Fig. 3B.16e). The location of maximum heating is slightly east of previous maxima, in agreement with the shift in convection indicated by the OLR anomalies. The level of maximum heating remains at 400 hPa. A weak secondary maximum of heating located just east of 150°W also occurs, which agrees reasonably well with the smaller values of negative OLR anomalies seen there. Finally, the last pentad (Fig. 3B.16f) and the corresponding OLR anomaly pattern reveal that a drastic change has occurred across the Pacific. Figure 3B.15b implies that, during this period, maximum convection has

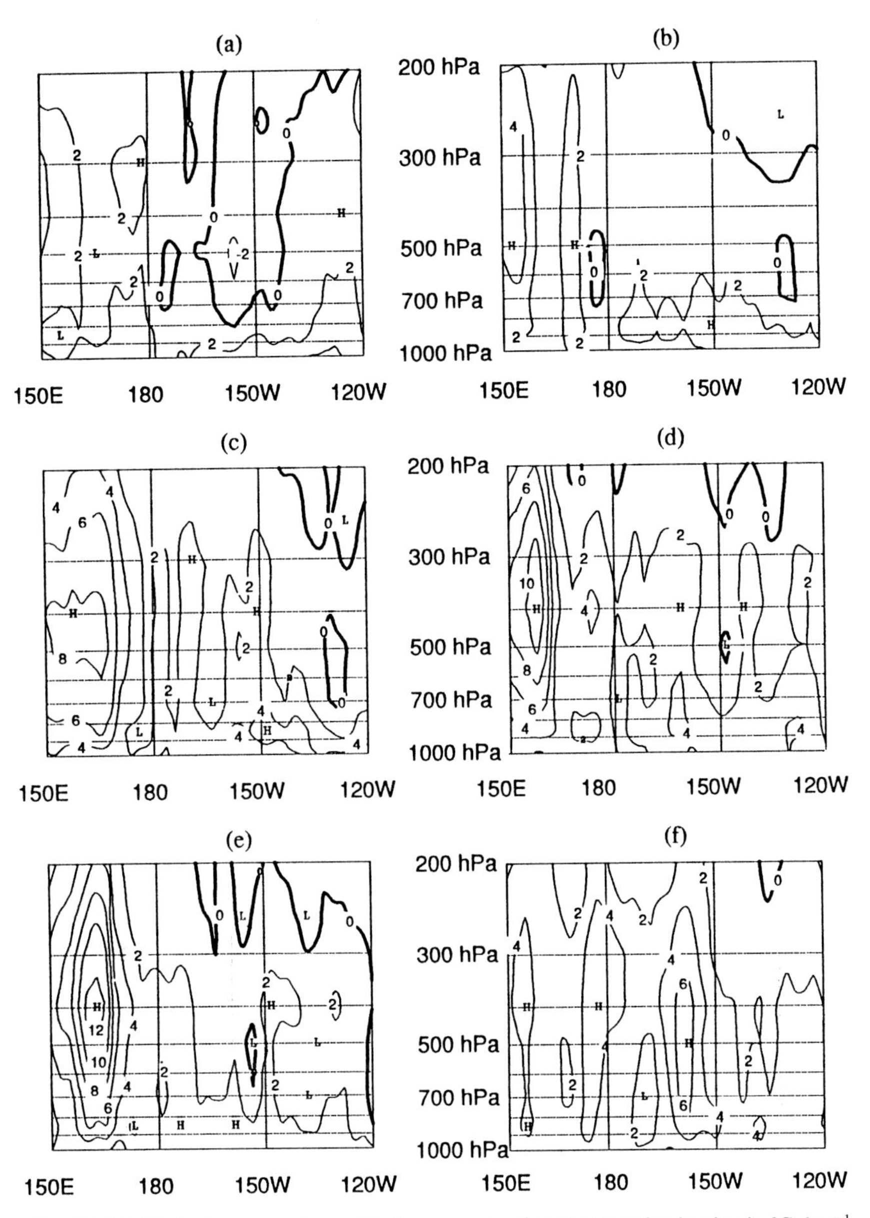

FIG. 3B.16. Vertical cross sections of 5-day averages of total convective heating in °C day^{-1}, averaged between 5°N and 5°S for (a) 9–13 August, (b) 29 August–2 September, (c) 3–7 September, (d) 8–12 September, (e) 18–22 September, and (f) 23–27 September 1986. Intervals are in 2°C day^{-1}.

shifted to near 160°W, while the western Pacific appears to be free of any strong convective activity. The distribution of convective heating rates also shows maximum values near 160°W. The level of this maximum, as well as that of the secondary maximum near the western edge of the convective activity (approximately 175°E), is again at a lower level (i.e., 500 hPa).

In conclusion, the tropical South Pacific contains many interesting circulation features. Among the more robust features, and the ones discussed in this section of chapter 3, are the ITCZ, SPCZ, western Pacific warm pool, subtropical jet stream and jet streaks, and large-scale organized convective disturbances. The South Pacific also is a region that experiences circulation changes on all temporal scales, ranging from diurnal to interannual. The focus of this section has been twofold: 1) seasonal changes, with regard to the ITCZ, SPCZ, and western Pacific warm pool; and 2) subseasonal changes, with regard to subtropical jet streaks and large-scale convective disturbances. To summarize, the ITCZ was found to be most intense in the JJA season and, throughout most of the year, was located near the axis of highest SSTs. The SPCZ, on the other hand, was most intense in DJF and was located in a region of strong SST gradient, indicating that baroclinic processes are important to its maintenance. The LSA region of the TOGA COARE field program was used to examine the climatology of the warm pool. One feature worth highlighting is that a Hadley-type circulation, with air rising over the warm pool and sinking at subtropical latitudes, was present in all seasons except JJA. The subseasonal changes associated with westerly subtropical jet streaks at 200 hPa represented a relatively new result. During the summer half of the year, a detailed study revealed that a cyclic pattern occurred that consisted of jet streaks forming over eastern Australia and subsequently moving eastward toward the date line. The period of this cycle was 1–2 weeks. Furthermore, the overall pattern just described was found to shift eastward about 20° longitude during an El Niño event. Finally, results were shown that suggested that a 10–20-day cycle exists in the large-scale convective activity along the equatorial western and central Pacific (5°N–5°S). Vertical profiles of total convective heating were computed from 5-day averages of $Q_1 - Q_R$, where Q_1 is the dry static energy or "apparent" heat source, and Q_R is the net radiative cooling. Reasonable distributions were observed to occur during the life cycles of the convective disturbances, leading to the tentative conclusion that the Q_1 budget approach is a viable one to use to obtain heating profiles as long as some temporal (and spatial) averaging is applied.

REFERENCES

Finch, C. J., 1994: *TOGA CD-ROM User's Guide.* Jet Propulsion Laboratory Physical Oceanography Distributed Active Archive Center, California Institute of Technology, 126 pp.

Hendon, H. H., N. E. Davidson, and G. Bunn, 1989: Australian summer monsoon onset during AMEX 1987. *Mon. Wea. Rev.,* **117,** 370–390.

Hennon, C., and D. G. Vincent, 1996; *Climatology during the TOGA COARE Intensive Observing Period (November 1992–February 1993) Compared to a Longer-Term Recent Climatology (1985–1900).* Purdue University, 95 pp. [Available from Department of Earth and Atmospheric Sciences, Purdue University, West Lafayette, IN 47907.]

Hurrell, J. W., and D. G. Vincent, 1990: Relationships between tropical heating and subtropical westerly maxima in the Southern Hemisphere during SOP-1, FGGE. *J. Climate,* **3,** 751–768.

———, and ———, 1991: On the maintenance of short-term subtropical wind maxima in the Southern Hemisphere during SOP-1, FGGE. *J. Climate,* **4,** 1009–1022.

———, and ———, 1992: A GCM case study on the maintenance of short-term subtropical wind maxima in the summer hemisphere during SOP-1, FGGE. *Quart. J. Roy. Meteor. Soc.,* **118,** 51–70.

Keen, R. A., 1982: The role of cross-equatorial cyclone pairs in the Southern Oscillation. *Mon. Wea. Rev.,* **110,** 1405–1416.

Kiladis, G. N., H. von Storch, and H. van Loon, 1989: Origin of the South Pacific convergence zone. *J. Climate,* **2,** 1185–1195.

———, G. A. Meehl, and K. M. Weickmann, 1993: Westerly windburst events in the TOGA COARE region. *EOS,* **74,** 133.

Ko, K.-C., and D. G. Vincent, 1995: A composite study of the quasi-periodic subtropical wind maxima over the South Pacific during November 1984–April 1985. *J. Climate,* **8,** 579–588.

———, and ———, 1996: Behavior of one- to two-week summertime subtropical wind maxima over the South Pacific during an ENSO cycle. *J. Climate,* **9,** 5–16.

Liebmann, G., and C. A. Smith, 1996: Description of a complete (interpolated) outgoing longwave radiation dataset. *Bull. Amer. Meteor. Soc.,* **77,** 1275–1277.

Madden, R. A., and P. R. Julian, 1971: Detection of a 40–50-day oscillation in the zonal wind in the tropical Pacific. *J. Atmos. Sci.,* **28,** 702–708.

———, and ———, 1972: Description of global-scale circulation cells in the Tropics with a 40–50-day period. *J. Atmos. Sci.,* **29,** 1109–1123.

McBride, J. L., 1981: An analysis of diagnostic cloud mass flux models. *J. Atmos. Sci.,* **38,** 1977–1990.

Meehl, G. A., 1987: The annual cycle and interannual variability in the tropical Pacific and Indian Ocean regions. *Mon. Wea. Rev.,* **115,** 27–50.

Miller, B. L., and D. G. Vincent, 1987: Convective heating and precipitation estimates for the tropical South Pacific during FGGE, 10–18 January 1979. *Quart. J. Roy. Meteor. Soc.,* **113,** 189–212.

Newell, R. E., J. W. Kidson, D. G. Vincent, and G. J. Boer, 1972: *The General Circulation of the Tropical Atmosphere and Interactions with Extratropical Latitudes.* Vol. 1, MIT Press, 258 pp.

Nitta, T., 1972: Energy budget of wave disturbances over the Marshall Islands during the years 1956 and 1958. *J. Meteor. Soc. Japan,* **50,** 71–84.

Palmer, C. E., 1952: Tropical meteorology. *Quart. J. Roy. Meteor. Soc.,* **78,** 126–164.

Pedigo, C. B., and D. G. Vincent, 1990: Tropical precipitation rates during SOP-1, FGGE, estimated from heat and moisture budgets. *Mon. Wea. Rev.,* **118,** 542–557.

Philander, G., 1989: El Niño and La Niña. *Amer. Sci.,* **77,** 451–459.

Ramsey, P. G., and D. G. Vincent, 1995: Computation of vertical profiles of longwave radiative cooling over the equatorial Pacific. *J. Atmos. Sci.,* **52,** 1555–1572.

Rasmusson, E. M., 1985: El Niño and variations in climate. *Amer. Sci.,* **73,** 168–177.

Reed, R. J., and E. E. Recker, 1971: Structure and properties of synoptic-scale wave disturbances in the equatorial western Pacific. *J. Atmos. Sci.,* **28,** 1117–1133.

Revell, C. G., 1986: Tropical Cyclone Namu. *Wea. and Clim.,* **6,** 67–69.

———, 1987: The 1986/87 hurricane season in the South Pacific. *Wea. and Clim.,* **9,** 587–607.

Reynolds, R. W., 1988: A real-time global sea surface temperature analysis. *J. Climate,* **1,** 75–86.

Sadler, J. C., 1972: The mean winds of the upper troposphere over the central and eastern Pacific. Tech. Rep. UHMET-72-04, 35 pp. [Available from Dept. of Meteorology, University of Hawaii, 2525 Correa Rd., HIG 331, Honolulu, HI 96822].

———, 1975: The upper tropospheric circulation over the global tropics. Tech. Rep. UHMET-75-05, 35 pp. [Available from Dept. of Meteorology, University of Hawaii, 2525 Correa Rd., HIG 331, Honolulu, HI 96822].

———, M. A. Lander, A. M. Hori, and L. K. Oda, 1987: Tropical marine climatic atlas, Vol. II: Pacific Ocean. Tech. Rep. OHMET-87-02, 27 pp. [Available from Dept. of Meteorology, University of Hawaii, 2525 Correa Rd., HIG 331, Honolulu, HI 96822].

Schrage, J. M., and D. G. Vincent, 1996: Tropical convection on 7–21-day timescales over the western Pacific. *J. Climate,* **9,** 587–607.

Sui, C.-H., and K.-M. Lau, 1989: Origin of low frequency (intraseasonal) oscillations in the tropical atmosphere. Part II: Structure and propagation of mobile wave–CISK modes and their modification by lower boundary forcings. *J. Atmos. Sci.,* **46,** 37–56.

Thompson, R. M., Jr., J. W. Payne, E. E. Recker, and R. J. Reed, 1979: Structures and properties of synoptic-scale disturbances in the intertropical convergence zone of the eastern Atlantic. *J. Atmos. Sci.,* **36,** 53–72.

Trenberth, K. E., 1986: An assessment of the impact of transient eddies on the zonal flow during a blocking episode using localized Eliassen–Palm flux diagnostics. *J. Atmos. Sci.,* **43,** 2070–2087.

———, 1991: General characteristics of El Niño–Southern Oscillation. *Teleconnections Linking Worldwide Climate Anomalies.* M. Glantz, R. W. Katz, and N. Nicholls, Eds., Cambridge University Press, 13–41.

———, 1992: Global analyses from ECMWF and atlas of 1000 to 10 mb circulation statistics. NCAR Tech. Note NCAR/TN-373+STR, 191 pp. [Available from National Center for Atmospheric Research, P.O. Box 3000, Boulder, CO 80307.]

van Loon, J., J. J. Taljaard, T. Sasamori, J. London, D. V. Hoyt, K. Labitzke, and C. W. Newton, 1972: *Meteorology of the Southern Hemisphere. Meteor. Monogr.,* No. 35, C. W. Newton, Ed., Amer. Meteor. Soc., 263 pp.

Vincent, D. G., 1982: Circulation features over the South Pacific during 10–18 January 1979. *Mon. Wea. Rev.,* **110,** 981–993.

———, 1994: The South Pacific convergence zone (SPCZ): A review. *Mon. Wea. Rev.,* **122,** 1949–1970.

———, and J. M. Schrage, 1994: *Climatology of the TOGA COARE and Adjacent Regions (1985–1990), Vol. 1: Kinematic Variables.* Purdue University, 104 pp. [Available from Dept. of Earth and Atmospheric Sciences, Purdue University, West Lafayette, IN 47907.]

———, and ———, 1995: *Climatology of the TOGA COARE and Adjacent Regions (1985–1990), Vol. 2: Thermodynamic and Moisture Variables.* Purdue University 123 pp. [Available from Dept. of Earth and Atmospheric Sciences, Purdue University, West Lafayette, IN 47907.]

———, ———, and L. D. Sliwinski, 1995: Recent climatology of kinematic variables in the TOGA COARE region. *J. Climate,* **8,** 1296–1308.

———, K.-C. Ko, and J. M. Schrage, 1997: Subtropical jet streaks over the South Pacific. *Mon. Wea. Rev.,* **125,** 438–447.

Webster, P. J., and R. Lukas, 1992: TOGA COARE: The coupled ocean–atmosphere response experiment. *Bull. Amer. Meteor. Soc.,* **73,** 1377–1416.

Weickmann, K. M., 1986: Intraseasonal (30–60 day) atmospheric oscillations. Climate Diagnostics Bulletin No. 86/13, 2–3. [Available from Climate Analysis Center, NMC/NWS/NOAA, Washington, DC.]

Yanai, M., S. Esbensen, and J. H. Chu, 1973: Determination of the bulk properties of tropical cloud clusters from large-scale heat and moisture budgets. *J. Atmos. Sci.,* **30,** 611–627.

Chapter 3C

South America

PRAKKI SATYAMURTY AND CARLOS A. NOBRE

Centro de Previsão de Tempo e Estudos Climáticos, Instituto Nacional de Pesquisas Espaciais, São Paulo, Brazil

PEDRO L. SILVA DIAS

Department de Ciencias Atmosfericas, Instituto Astronomico e Geofisico, Universidade de Sao Paulo, Sao Paulo, Brazil

3C.1. Introduction

Extending meridionally from 10°N to 60°S, South America (SA) presents features of tropical, subtropical, and extratropical weather and climate. This continent lies between the two large oceans, the Pacific and the Atlantic, and as such there is a great influence of these oceans on the meteorology of this landmass. The regional circulation characteristics over SA can only be appreciated by referring to the effects of these two oceans. An important and distinct geographical feature of the continent is the presence of a steep and narrow mountain range extending all the way from the northern tip to the southern tip along the west coast. Another important feature is the tropical Amazon jungle, occupying about 35% of the total continental area and 65% of the tropical area. With one of the world's most humid climates, this tropical forest makes the continent very unique. This region also contains some of the infamous deserts and arid regions, such as the Atacama desert in northern Chile and northeastern Brazil (NEB).

The regional atmospheric circulation over SA presents many interesting characteristics, such as the Bolivian high, the SACZ, the Chaco low in summer, and the cold surges into the equatorial region known as *friagens* in winter. The tropical eastern South Pacific and the tropical South Atlantic adjoining this continent are the only tropical oceans free of tropical storms such as hurricanes and typhoons. In the tropical Atlantic Ocean, the ITCZ presents a west-southwest to east-northeast stretching with a reasonably wide meridional seasonal migration at its southwestern end. South America is also a region of strong meridional interaction between the Tropics and the extratropics. South America's tropics and subtropics suffer a large direct influence of the El Niño (EN) and the Southern Oscillation (SO) phenomena. Early this century, Sir Gilbert Walker found a very significant correlation between SO and the Ceará (Brazil) rainfall, which has led to the investigation of the effects of global circulation anomalies in the Tropics over the climatic variability in many regions of the world. More will be said later about the impact of ENSO on SA.

3C.2. Regional circulation features

An attempt is made to give a broad description of the South American regional circulation features and their role in the meteorology and climatology of the continent. Large seasonal changes are seen in the overall characteristics of the regional circulation over South American tropics and subtropics. They are presented with the aid of the climatological (30-yr means of the period 1961–90) monthly charts of the MSLP, the 850-hPa vector winds, the OLR, and the 200-hPa vector winds for January, April, July, and October, representing the four austral seasons—summer, fall, winter, and spring—respectively. These charts are obtained from NCEP (V. Kousky 1996, personal communication).

The MSLP for the four austral seasons is shown in Fig. 3C.1. The subtropical high pressure centers in the eastern South Pacific and the South Atlantic Oceans lie on either side of SA, and relative low pressures prevail over the continent in all seasons. The MSLP over the continent is lowest in summer (January) and highest in winter (July). The Pacific high is stronger in summer and the Atlantic high is stronger in winter, and they are about the same intensity in the transition seasons. The two highs are farthest from the continent in summer, when most of the tropics and subtropics of South America have their rainy season. It can also be seen over the continent that there is a general low pressure belt around the Equator, known as the equatorial trough, which merges with the ITCZ over the oceans. The position and intensity of this trough has a strong bearing on the rainy season in the northern Amazon, Guyanas, Venezuela, and Colombia. The subtropical high in the Atlantic migrates from 15°W, 27°S in August to 5°W, 33°S in February and, in association with the southward migration of the ITCZ in February to May, determines the rainy season in the adjacent NEB region (Hastenrath 1991).

An important feature in the South Atlantic is that the subtropical high in summer is weaker than in

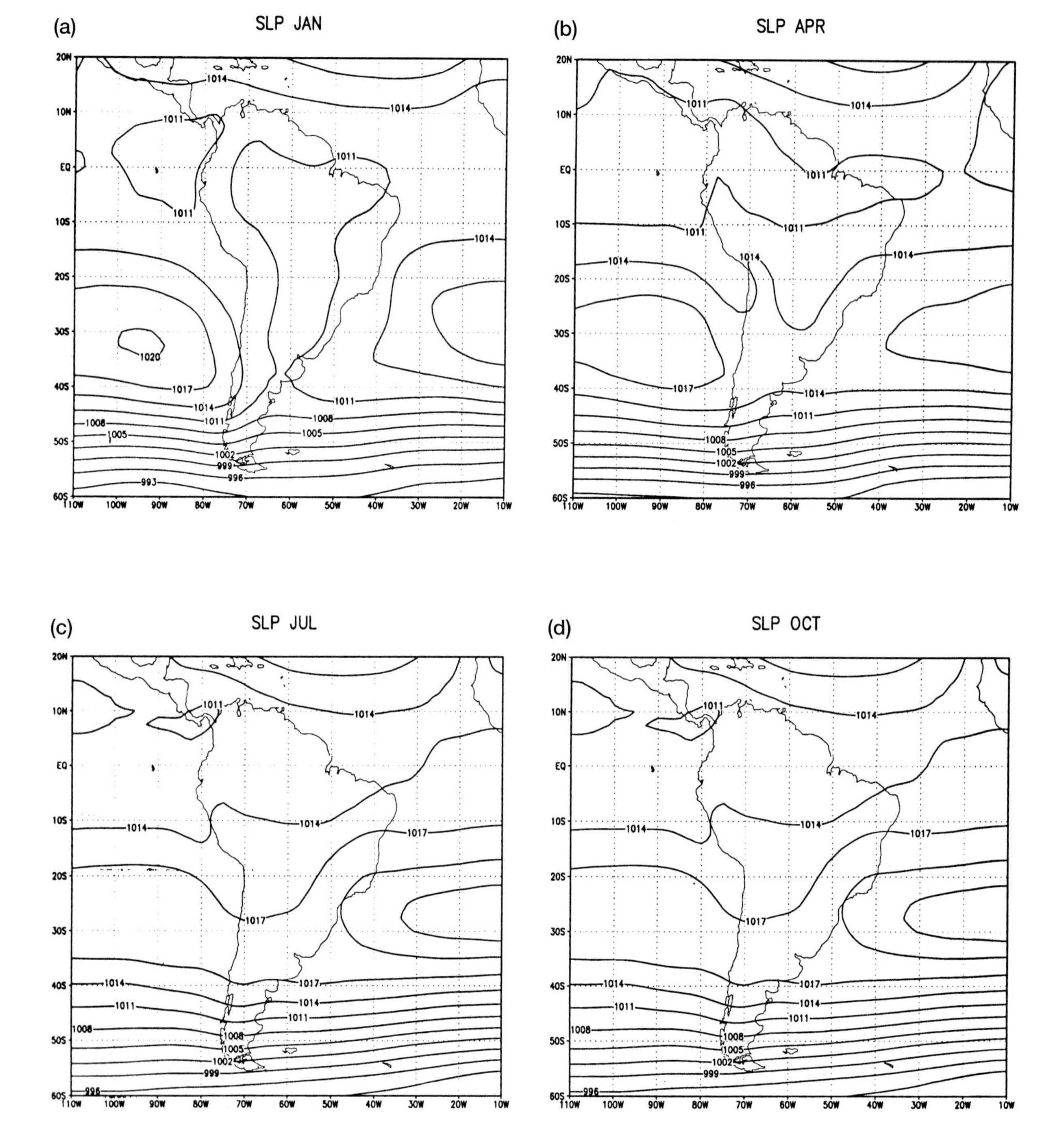

FIG. 3C.1. MSLP (hPa) distribution for January, April, July, and October.

winter, while all other subtropical highs, in general, are stronger in summer. As the subtropical high in the South Atlantic migrates southward in early summer, convection develops along the SACZ (described later in this section), intruding into the high. As a result, the subtropical high in summer is frequently split, and in the average it is weaker.

The lower tropospheric mean circulation for the four austral seasons are shown in Fig. 3C.2. In general the trade winds, coming from the Atlantic region north of the subtropical high pressure center, flow as far west as the Andes in Colombia and Peru, where they are blocked by the steep and high topography and turn gradually northerly and northwesterly east of the

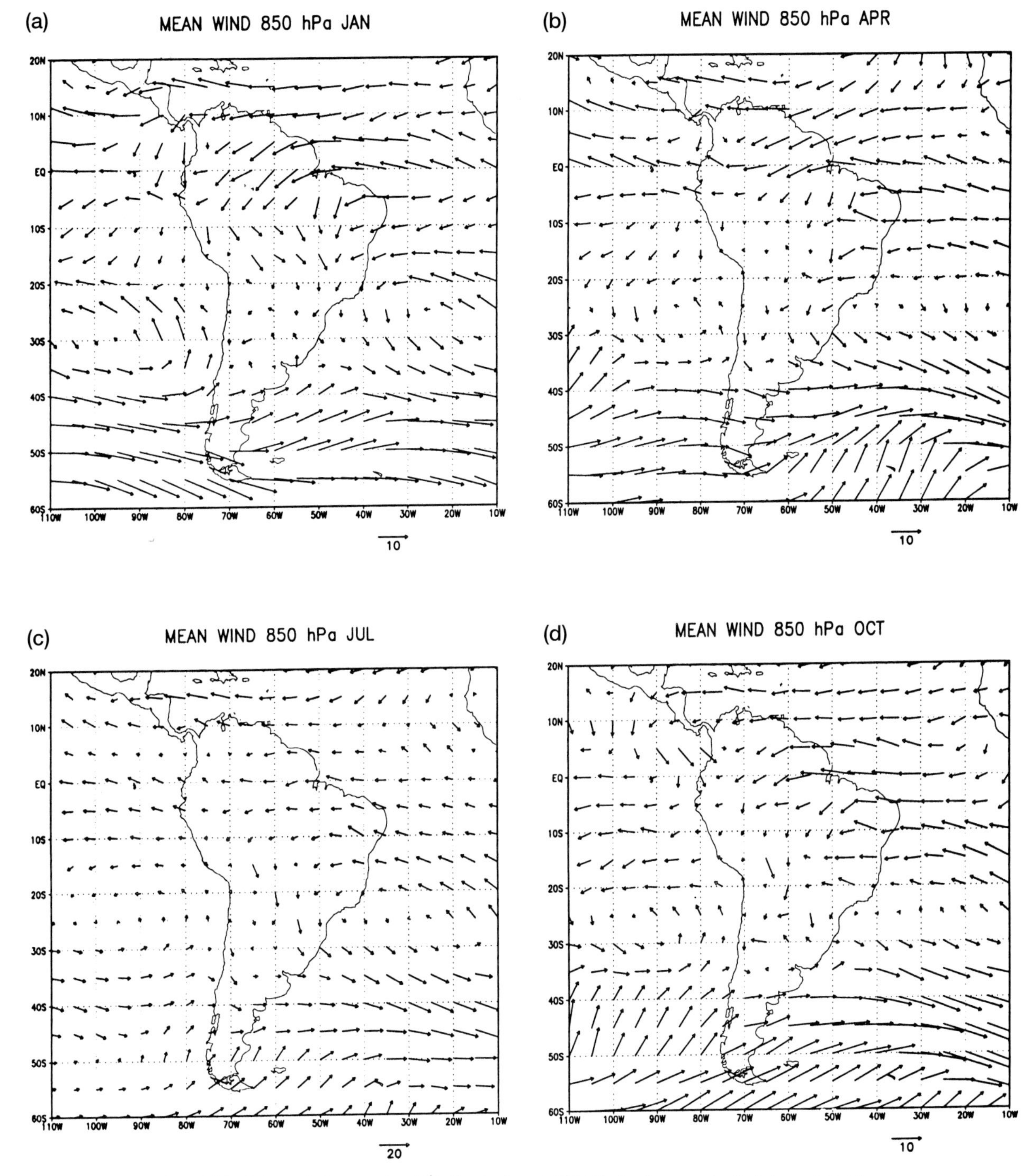

FIG. 3C.2. Mean wind vector (m s^{-1}) distribution at 850 hPa for January, April, July, and October.

mountains. Over the east coast, the trades penetrate the continent at slightly different angles and speeds in different seasons, which have a significant effect on the coastal precipitation (Kousky 1985). The col region (the region of intersection of the N–S trough and the E–W ridge) between the two subtropical anticyclones lies over the continent in the latitudinal belt between 15° and 40°S, and this region, with its strong deformation field, is frontogenetic (Satyamurty and Mattos 1989). In this region there is an indication of the presence of a low-level northerly jet (wind speed of the order of 15 m s^{-1}) below 850 hPa that is responsible for the transport of water vapor and heat to the region of Paraguay and northern Argentina from

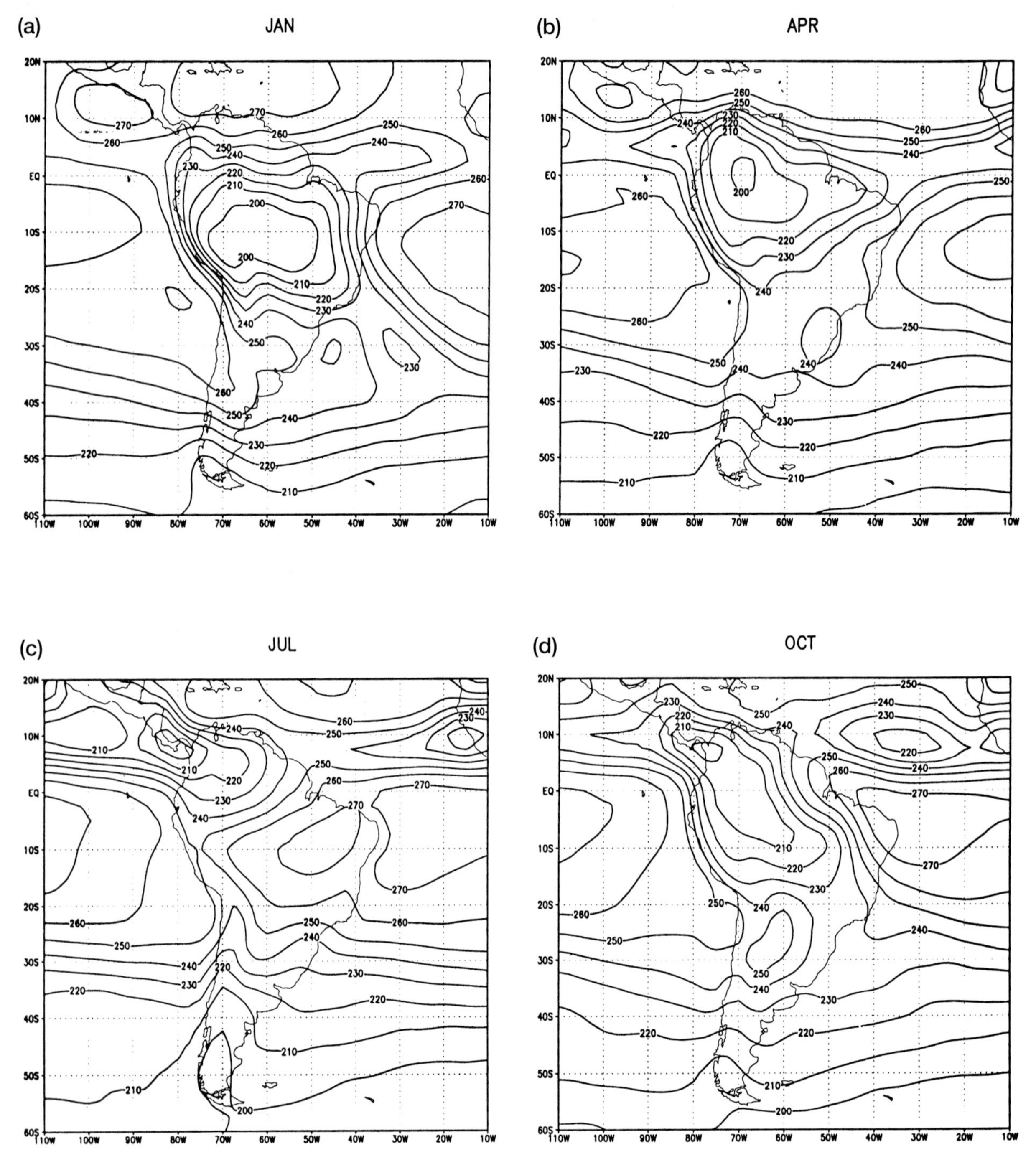

FIG. 3C.3. Mean monthly OLR (W s^{-1}) for January, April, July, and October.

the Amazon (Peagle 1987). The westerlies, south of 40°S, show a trough over southern Argentina in July and October, when the winds in the Pacific are more southerly.

In April, the trades in the Atlantic north of 10°N have a more northerly component, and the ITCZ is located at a southerly position in the Caribbean sea. The ITCZs in the western Atlantic and eastern Pacific, as viewed in terms of the axis of confluence of the trades, occupy their northernmost positions in late austral winter, around 10°N in the Atlantic and around 13°N in the Pacific. The southernmost positions are observed in late austral summer, around 3°N in the Atlantic (in February through April) and 5°N in the Pacific (in January and February) (Hastenrath 1991).

The dominance of convective activity over the tropical Americas is identified in the OLR pentad charts by considering a threshold value of 200 W m^{-2} (Horel et al. 1989). It seems reasonable that, in the monthly mean charts, tropical areas of OLR less than 230 W m^{-2} can be designated to be convective regions. Over the tropical SA, a great part of the rainfall is convective and, therefore, low values of OLR are associated with high precipitation. Figure 3C.3 shows the climatological mean monthly OLR. In austral summer (January) the convective activity over the continent is maximum, with low OLR values (< 230 W m^{-2}) occupying most of Brazil and the adjoining regions except the extreme northeastern portion. The elongated band of high convective activity emanating from the Amazon region and running southeastward into the South Atlantic Ocean is known as the SACZ (Kalnay et al. 1986; Kodama 1992; Figueroa et al. 1995). The SACZ is a dominant feature in summer and is usually absent in winter. The region of convective activity over the continent gradually diminishes and recedes northwestward from summer to winter. In July the nucleus of maximum rainfall lies over western Colombia, Panama, and adjoining seas. OLR in excess of 260 W m^{-2} is registered over the eastern parts of the continent north of 25°S, showing that most of Brazil remains dry in winter. In austral spring the convective activity starts returning to the Amazon basin from the northwest.

The upper-tropospheric circulation for the corresponding months is shown in Fig. 3C.4. The anticyclonic circulation over the continent centered around 15°S, 65°W in summer is known as the Bolivian high. The three dominant features in southern summer are the Bolivian high, first noted by Gutman and Schwerdtfeger (1965) and confirmed by Virji (1982); a cyclonic circulation near the coast of NEB studied by Kousky and Gan (1981); and a deformation region near the coast of Peru.

Gutman and Schwerdtfeger (1965), based on the heat balance of the tropospheric column in the Bolivian plateau region, have indicated that the role of sensible heating might be responsible for the formation of the Bolivian high. Other studies using simple linearized models (Silva Dias et al. 1983; DeMaria 1985) have pointed to the role of latent heating.

The basic mechanism of the formation of the Bolivian high is the regional-scale ascending motion due to intense convective activity over the Amazon. The convergence of water vapor in the low levels and the release of latent heat in the middle troposphere help sustain the rising motion and keep the region relatively warm in the upper troposphere. The anticyclonic circulation around the high is fairly strong, with relative vorticity exceeding 3.0×10^{-5} s^{-1}, especially on the southern and eastern sides.

Strong southerly winds on the eastern side of the Bolivian high soon lose their anticyclonic vorticity and turn cyclonic near the NEB coast due to conservation of absolute vorticity. Although this is one strong reason for the formation of the trough to the northeast of the Bolivian high, it is not sufficient to explain all the cyclonic vorticity observed on the daily charts of about -3.0×10^{-5} s^{-1}. The trough very often becomes a closed vortex with a relatively cold core. The center of the vortex is cloud free and its periphery (especially northern) presents convective activity (Kousky and Gan 1981), which means that the air subsides in the center and rises on the periphery of the vortex. One explanation for the cold core could be the higher vertical lapse rates of temperature in subsiding air columns and lower lapse rates in the rising columns. An air column becoming colder is equivalent to it becoming shallower, and, by the conservation of potential vorticity principle, the relative vorticity, ζ, becomes more cyclonic. The potential vorticity can be written as

$$\eta = -g(\zeta + f)(\partial\theta/\partial p),$$

where f is the Coriolis parameter (or planetary vorticity), g is the acceleration due to gravity, θ is the potential temperature, and p is the pressure. In subsiding air, the temperature profile is very nearly dry adiabatic and $|\partial\theta/\partial p|$ is small. Thus, the airstream coming from higher latitudes into the vortex with subsidence acquires more cyclonic relative vorticity.

The vortex near NEB presents a rather irregular zonal and meridional movement and, depending on its position, different regions of NEB become excessively wet or excessively dry. Using the vorticity equation, Kousky and Gan (1981) argued that the greatest convective activity is expected ahead of the moving vortex. Rao and Bonatti (1987) have conjectured that barotropic instability of the regional mean basic winds in the upper levels could be one of the causes for the formation vortices over the South Atlantic adjoining Brazil.

The features of the regional circulation, such as the Bolivian high, NEB low, and SACZ in summer, are quasi-stationary and are more impressive on daily charts than in the monthly climatological average. Figure 3C.5 shows an infrared (IR) cloud picture for a summer day and the corresponding upper-tropospheric flow, in which an anticyclone over Bolivia and a closed cyclonic vortex over northeastern Brazil are easily identified. The cloud picture shows a wide band of intense convection stretching from northeastern Peru to Rio de Janeiro in Brazil, which corresponds to the position of the SACZ on 16 February 1995. A schematic sketch of all the important synoptic features of the regional circulation over SA is presented in Fig. 3C.6.

There are not many studies about the deformation field near Peru coast, but there are some indications

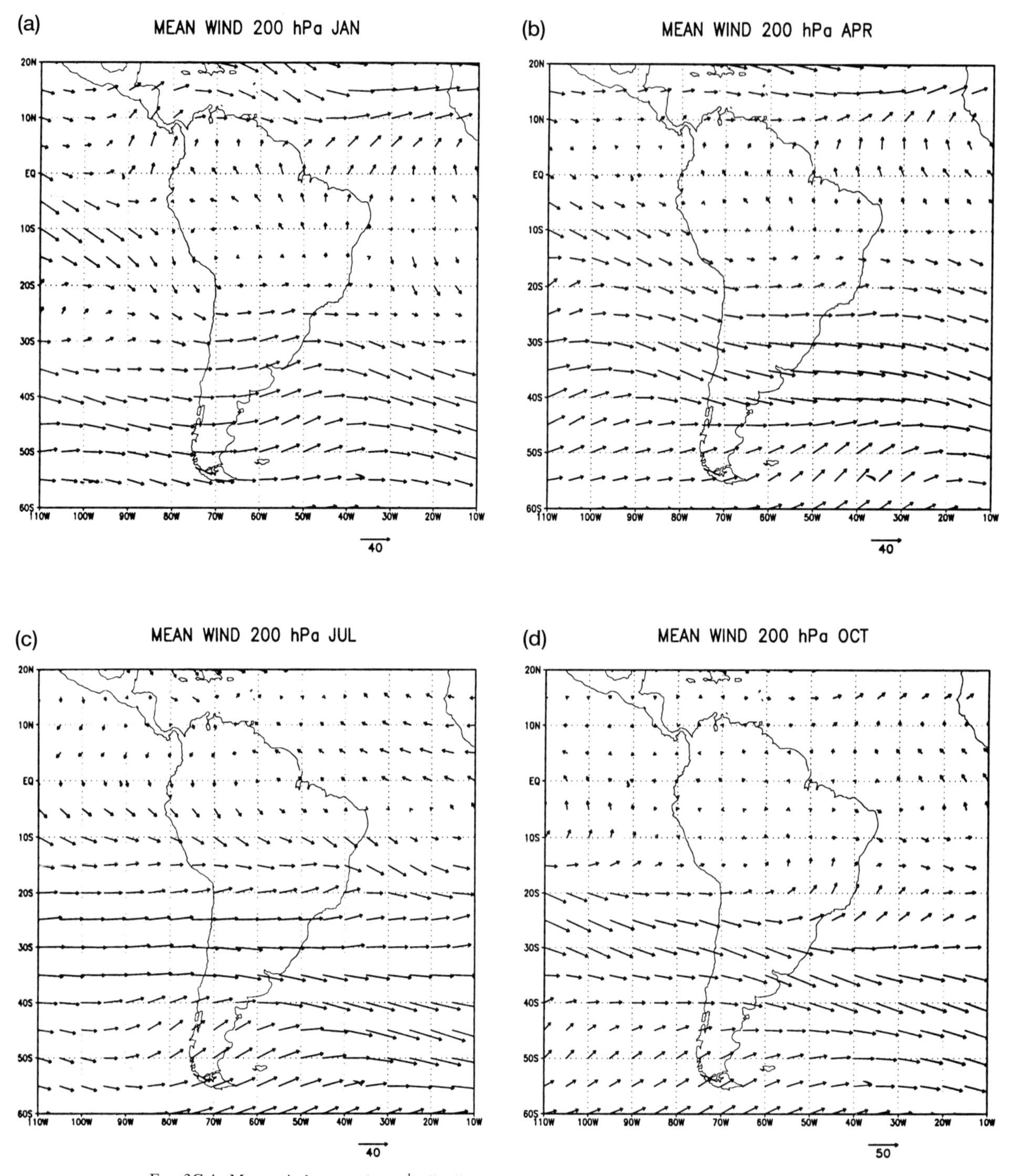

FIG. 3C.4. Mean wind vector (m s^{-1}) distribution at 200 hPa for January, April, July, and October.

that it affects the rainfall over eastern Peru and adjoining regions.

Other important characteristics observed in Fig. 3C.4 are that the northern hemispheric westerlies migrate as far south as 10°N in January in the Atlantic; the westerlies in the subtropics from 90°W westward are weak in this month; there is a clear indication of a trough in the westerlies near 45°W that is apparently associated with the SACZ; and, in July, the southern hemispheric westerlies migrate as far north as 5°S as a consequence of northward incursion of the extratropical airmasses deep into the Tropics in this season.

There have been a few attempts to simulate the regional circulation characteristics described above,

(a)

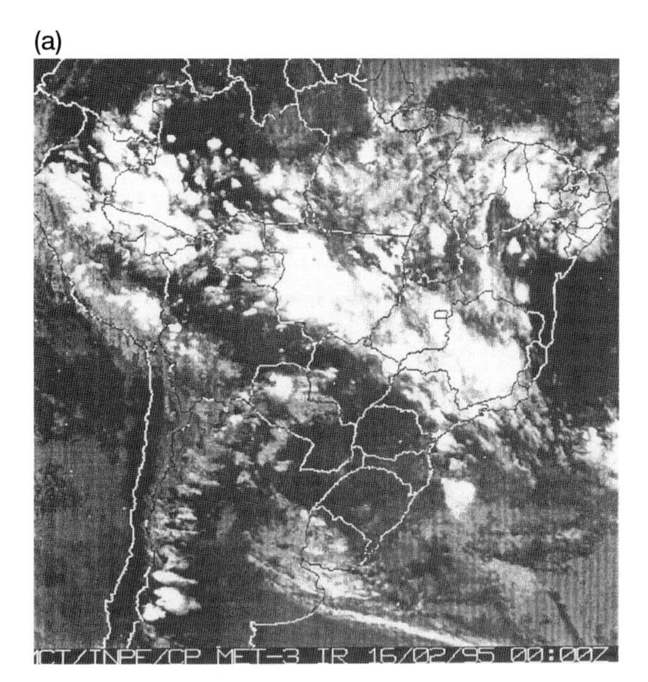

(b)

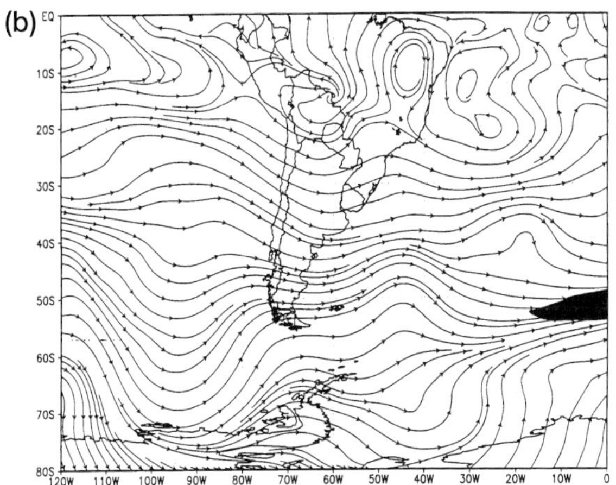

FIG. 3C.5. (a) Infrared satellite imagery at 0000 UTC on 16 February 1995 and (b) corresponding 250-hPa streamline analysis. In (b), areas with winds in excess of 50 m s^{-1} are darkened.

(a) LOWER TROPOSPHERE

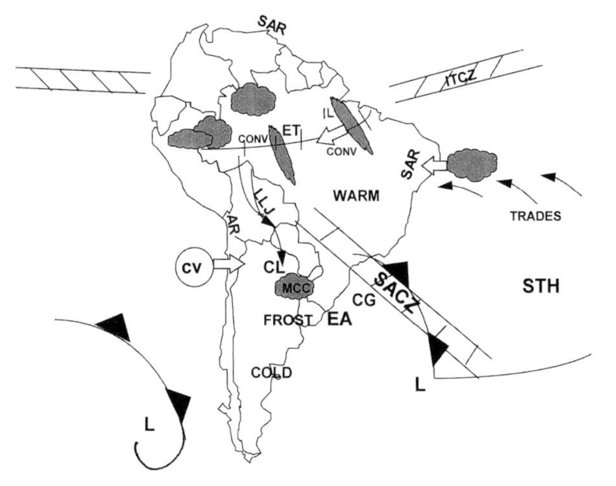

(b) UPPER TROPOSPHERE

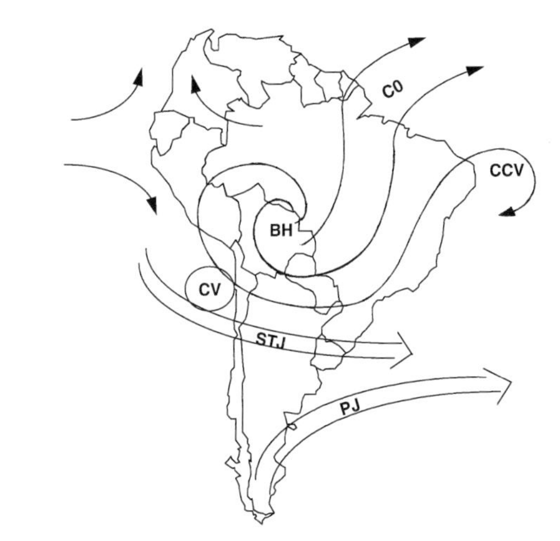

FIG. 3C.6. Schematic sketch of important atmospheric circulation features over the South American region: (a) lower troposphere and (b) upper troposphere. [IL = instability line; CL = Chaco low; LLJ = low-level jet; ET = equatorial trough; AR = arid region; SAR = semiarid region; MCC = mesoscale convective complex; CONV = convective activity; CG = cyclogenesis; STH = subtropical high; EA = extratropical anticyclone; L = low pressure center; CV = cyclonic vortex; BH = Bolivian high; CCV = cold-core vortex; CO = cirrus outflow; STJ = subtropical jet; PJ = polar jet.]

using simple regional mathematical models. Silva Dias et al. (1983) studied the transient response of the tropical atmosphere over SA to a time-varying diabatic heat source over the Amazon basin on timescales of up to a few days. They attempted to explain the large-scale features, such as the upper-level anticyclonic circulation over Bolivia, as a result of the propagation of equatorial waves. Kleeman (1989) produced a trough in the upper troposphere near the NEB coast as a result of the effect of the narrow and steep Andean topography. Gandu and Geisler (1992) conducted a study of the influence of the Andes mountains on the circulation. Although they have simulated many features, such as the Bolivian high, they did not quite produce one important summer feature, the SACZ. In a recent study, Figueroa et al. (1995), with the help of an eta-coordinate model, simulated the formation of the Bolivian high, the trough near the northeastern coast of Brazil, and an elongated convergence zone in the observed position of the SACZ. The diurnal variation of convection in the Amazon region was found to be important in those simulations.

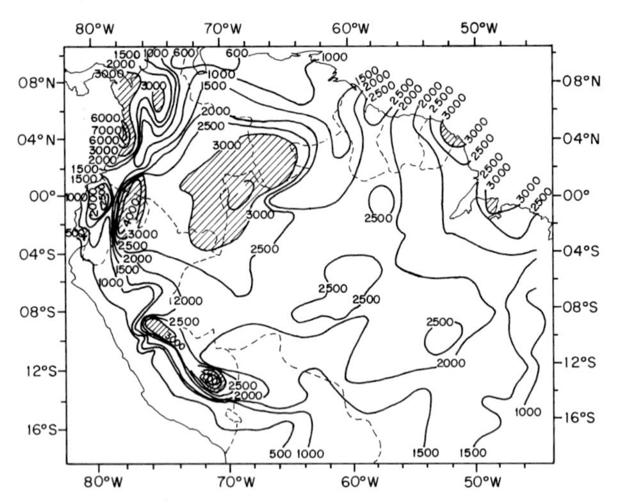

FIG. 3C.7. Total annual precipitation (mm) distribution over the Amazon basin and adjoining regions. (Adapted from Figueroa and Nobre 1990.)

3C.3. Rainfall climatology

A rainfall climatology for the tropical SA is obtained by Figueroa and Nobre (1990), based on available data in the 30-yr period 1960–89. The data used are rather heterogeneous in the sense that some stations, mostly in Brazil, have 30 years of data and some have only 7 to 8 years of data. This is the only publication, however, where one can find the annual climatology and seasonal and monthly rainfall variability over the whole of SA, north of 20°S and west of 45°W. The annual total precipitation distribution given in Fig. 3C.7 shows that the rainfall over the Amazon basin and adjoining regions varies from 1500 mm yr^{-1} in the east to 3500 mm yr^{-1} in the west and northeast (around the mouth of the Amazon River). In the central Amazon, it is about 2800 mm yr^{-1}. Highest rainfalls, exceeding 5000 mm yr^{-1}, are observed in western Colombia over a relatively small region. The rainfall is also very high (about 3000 mm yr^{-1}) in eastern Peru and adjoining regions east of the Andes mountains. Figure 3C.8 shows the seasonal rainfall distributions. The rainfall maxima are located over southern and eastern Amazon in summer (December to February). They move to northwestern and northeastern Amazon in autumn (April to June), and further northwestward in austral winter (June to August). The maximum rainfall shows a southeastward penetration into the central Amazon in spring (September to November). These observations are consistent with the OLR observations presented in Fig. 3C.3. Figure 3C.9 shows the monthly variation of the rainfall in the form of histograms for some select stations over the region. All the stations show strong seasonal variations. In the southeastern Amazon, the rainy season is around summer (November to March), and relatively dry times occur around winter (May to September). North of 5°S, November to May are relatively dry and March to September are the rainy months. One station in western Columbia has a peculiar distribution with very large rainfall. Stations west of the Andes in the rain shadow region receive very scanty rainfall, mainly in January to March.

Rao and Hada (1990) have presented charts of the rainy and dry seasons for Brazilian stations using 21 years of data (1958–78). Except for the northeast and far south, most of Brazil between 10° and 27°S has its rainy season in summer (December to February). The rainy season progresses to the north gradually, so that northernmost Brazil, north of 2°S, has its rains in April to June. Most of central Brazil from 5° to 24°S, west of 42°W, has its dry season in winter (June to August), and the dry season progresses northward to north of 2°S in austral spring (September to November). Southern Brazil, south of 28°S, has a fairly regular monthly distribution of rainfall; however, it receives relatively more rain in winter and late winter (July to September) and less rain in spring and early summer (March to June). NEB has a different rainfall distribution. The annual mean precipitation over NEB, as given by Kousky and Chu (1978), shows a large spatial variation, from more than 1000 mm in a narrow belt along the east coast and west of 44°W to less than 500 mm in the semiarid interior. The rainy season in northern NEB is from March to May, while in the semiarid interior it is from February to April, and in the eastern coastal regions it is from April to June. As a result of the convergence between the mean wind and the onshore land breeze, the NEB coastal areas experience a nocturnal maximum in the rainfall activity, whereas the regions 100 to 300 km inland experience a diurnal rainfall maximum due to the advance of the sea breeze (Kousky 1980).

3C.4. Transient disturbances

South America experiences varied types of transient disturbances, of both extratropical and tropical origin, and on all scales, from synoptic and mesoscale to organized and disorganized convection. Thus the study of regional synoptic meteorology becomes an interesting and, at the same time, very challenging subject. Some of the very violent meteorological systems found in the Tropics and subtropics (such as tropical cyclones and tornadoes), however, are either absent or are very rare in SA and the adjoining seas. One reason for the absence of tropical storms in this region could be colder SST below 27°C in the tropical South Atlantic.

Cold frontal passages are the most common transient weather events over the continent. The midlatitude cyclones coming from the Pacific cross the Andes and Argentina, south of 35°S, take a east-southeasterly course in the Atlantic, while the cold front associated with the low pressure center moves northeastward. As

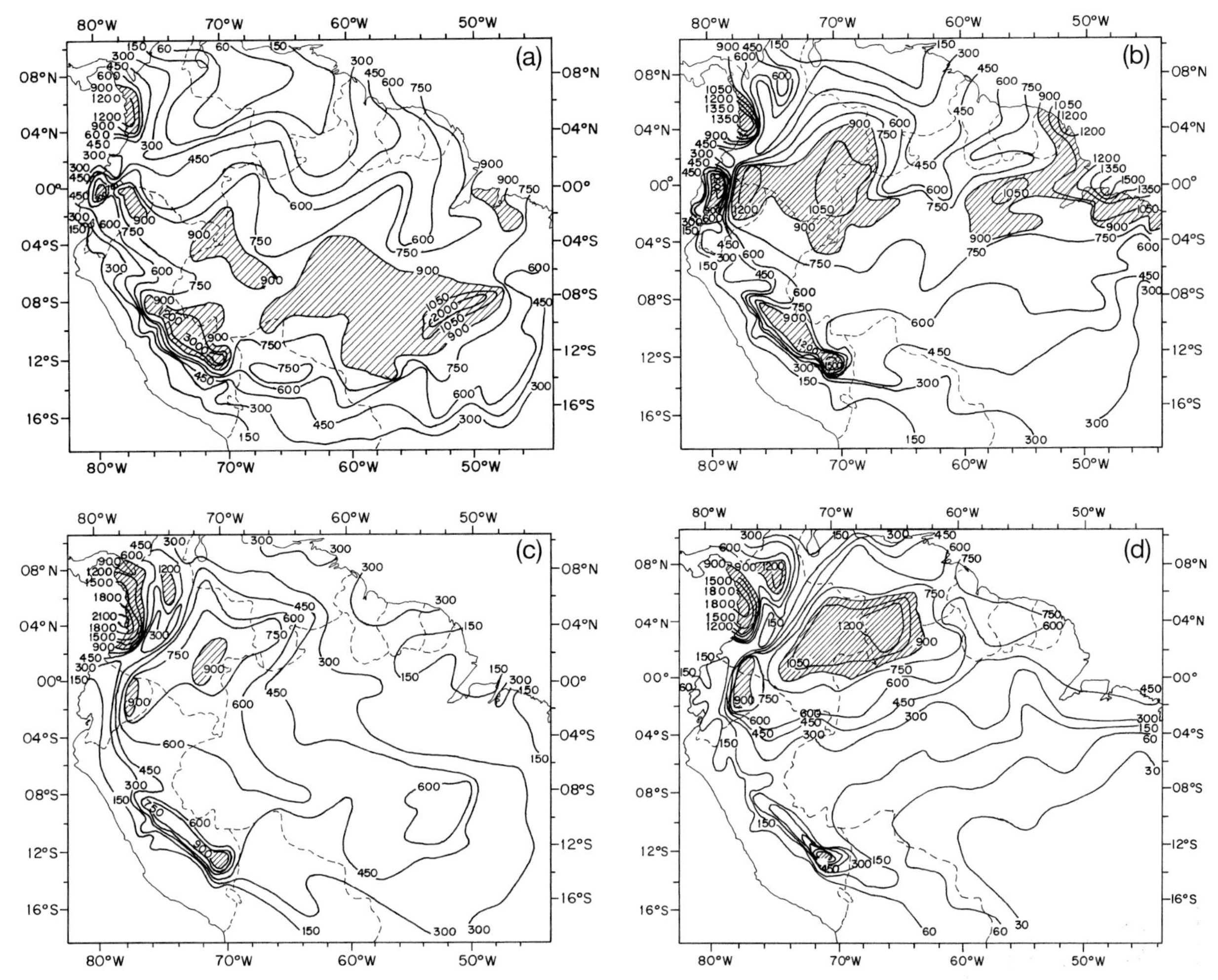

FIG. 3C.8. Seasonal precipitation (mm) distribution over the Amazon basin and adjoining regions: (a) December to February, (b) March to May, (c) June to August, and (d) September to November. (Adapted from Figueroa and Nobre 1990.)

the cold front sweeps the eastern parts of the continent, convective activity is triggered over Argentina, Bolivia, Brazil, and eastern Peru. Figure 3C.10 shows an IR satellite picture and the corresponding surface pressure distribution, in which a front lies in the Atlantic, penetrating into the continent, with many convective clusters over central parts of SA. An observational study of the frequency of frontal systems over SA by Oliveira (1986), based on 10 years of geostationary infrared imagery, from 1975 to 1984, gives useful monthly statistics of the frontal penetrations and their effect on the convective activity over subtropical and tropical SA. Figures 3C.11 and 3C.12 show the principal results of Oliveira. In general, the frontal penetrations are well spread over all seasons in all latitudinal bands. Their convective activity, however, is very low in the winter months, especially in June and July. They are more frequent in the southern latitudinal belt between 35° and 40°S (about nine per month) and less frequent in the northern belt, north of 20°S (about two per month). They are responsible for a large part of the rainfall in northern Argentina; Uruguay; Paraguay; southern, southwestern, and central western Brazil; Bolivia; and southern Peru.

The fronts quite often pass along the east coast of Brazil and enter NEB. Interior Bahia state and NEB coast owe a great part of their precipitation to fronts or their remains in the periods November–February and May–July, respectively (Kousky 1979). Over NEB the fronts do not present large thermal contrasts; however, they remain active due to large humidity contrasts.

In winter, the cold air masses with high surface pressure and low temperatures penetrate north-northeastward over the continent in the wake of the fronts, often causing intense frost south of 30°S and occasional moderate frost in the states of Rio Grande do Sul, Santa Catarina, Paraná, Mato Grosso do Sul, São Paulo, and southern Minas Gerais. They move on to the Atlantic around 20°S and gradually lose their identity and merge with the subtropical high. The occurrence and intensity of frost increase southward from 20°S and with altitude. Depending on the inten-

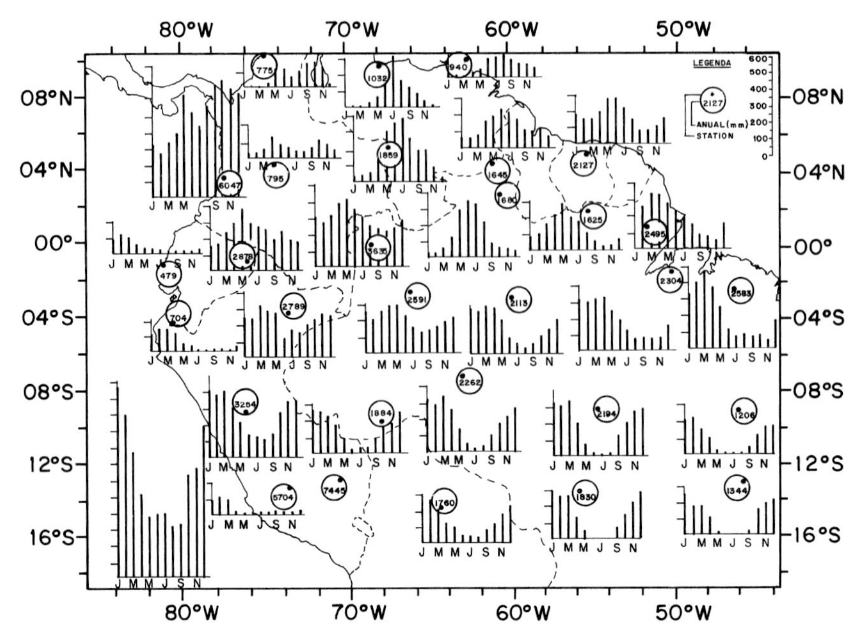

FIG. 3C.9. Monthly distribution of precipitation (mm) for several stations in tropical South America west of 45°S, in the form of histograms. (Adapted from Figueroa and Nobre 1990.)

sity, frost can cause heavy damage to agriculture, especially when it hits coffee plantations in Brazil. For example, the estimated damage caused by the June and July 1994 frosts in the coffee-growing areas of the state of São Paulo, Brazil, are recorded to be more than 50%, down to 1 890 000 *sacos* (*saco* = bag of 60 kg) from a normal production of 4 200 000 *sacos*.

Although frontal passages are common in all seasons, there are some seasonal variations, mainly in intensity and northward progression, as given in Fig. 3C.11. The most spectacular frontal invasions occur in winter, and such events, known as *friagens,* are described by Serra and Ratisbona (1942). Three such cases, one each in June, July, and August, are registered in the austral winter of 1994 (Marengo et al. 1997) and are shown in Fig. 3C.13. In the June event, for example, the dewpoint temperature dropped around 10°C from 20°C in one day, and the surface pressure rose by 10 hPa in two days at a northerly station, Rio Branco (10°S, 63°W). There are cases in which the cold front crosses the Equator in the northern Amazon region, followed by a *friagem* (singular of *friagens*) (Fortune and Kousky 1983). Prior to a deep northward penetration of the cold front north of 10°S, cyclogenesis is often observed near the east coast around 25°S. The cold-air damming effect just east of the Andes helps the cold air mass to persist for a few days and advance northward into the Amazon region affecting Bolivia, eastern and northern Peru, the Brazilian Amazon region, and southern parts of Colombia, Venezuela, and Guyanas. Figure 3C.14 shows a 10-yr (1979–89) climatology of intense anticyclone (surface pressure at its center greater than 1028 hPa) passages over the South American continent for May, June, July, and August months. It can be clearly seen that the anticyclone path is northeastward up to northern Argentina, where it bifurcates, with one branch going to the east and the other to the north. These northward-moving cold air masses cause the *friagens.*

The region of Paraguay, northern Argentina, Uruguay, and southern Brazil experience the effects of sudden development of mesoscale convective complexes (MCCs) (Maddox 1980), especially in the period November–April (Velasco and Fritsch 1987). These MCCs usually start in the early hours of the day before sunrise and have a short life cycle, less than a day. They are possibly triggered by the mountain breeze in a fairly unstable atmosphere. Near and around the intersection of the low-level northerly jet and the upper-level subtropical jet (see Fig. 3C.6), the shear (Kelvin–Helmholtz) instability is maximum, and this region is suitable for the triggering of convective activity. The low-level jet provides the necessary humidity transport for the formation of the clouds and precipitation in the complexes. The MCCs move eastward from their source region in northern Argentina and Paraguay to affect southwestern Brazil and Uruguay with intense rainfall. Silva Dias (1987) made a review study of the MCCs and discussed their relevance to short-range weather forecasting in southern Brazil and adjoining regions. The vertical profiles of the equivalent potential temperature θ_e, before and after the passage of an MCC over the city of São Paulo, show large differences in the lower middle troposphere, in the 850–650-hPa layer. Near 700 hPa,

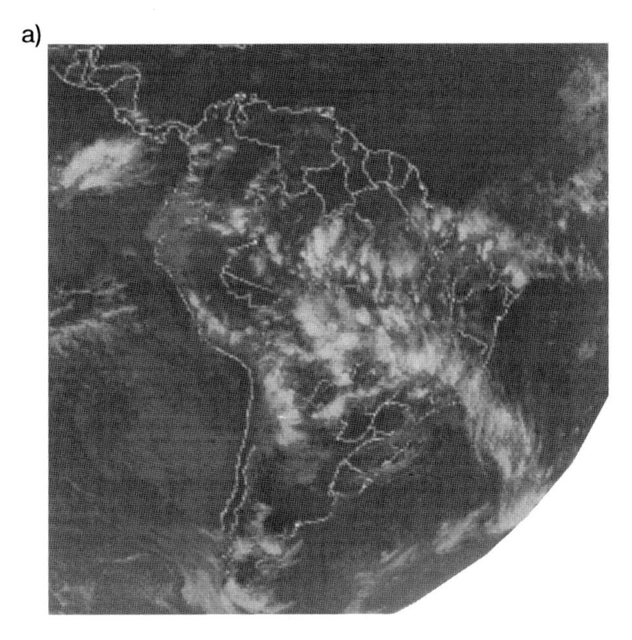

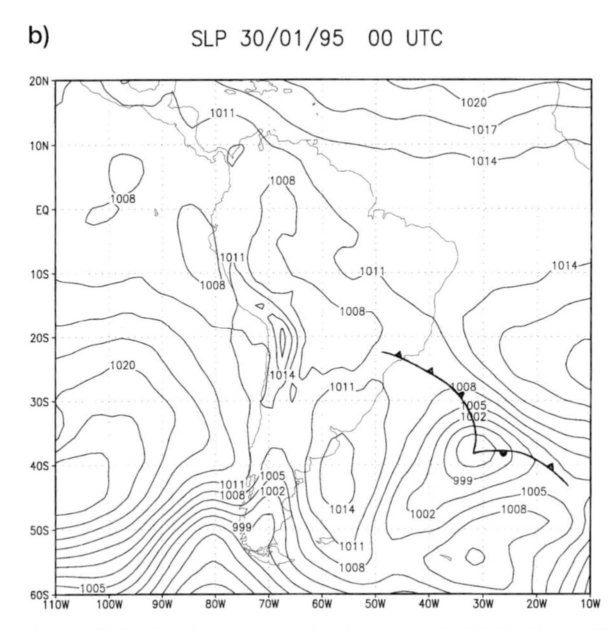

FIG. 3C.10. (a) Infrared satellite image at 0000 UTC on 30 January 1995, and (b) corresponding surface pressure (hPa) analysis showing the presence of a cold front near the east coast of Brazil.

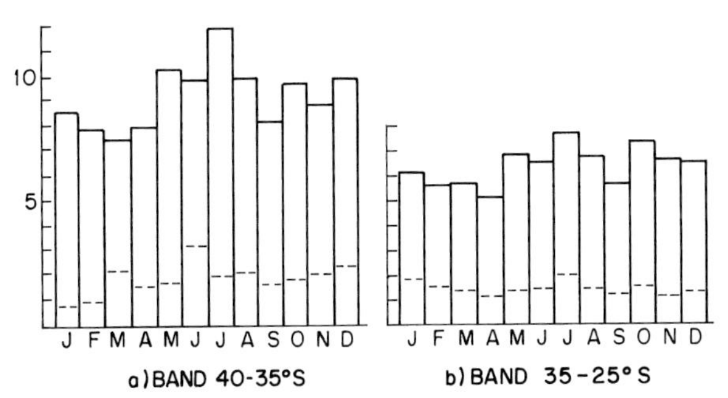

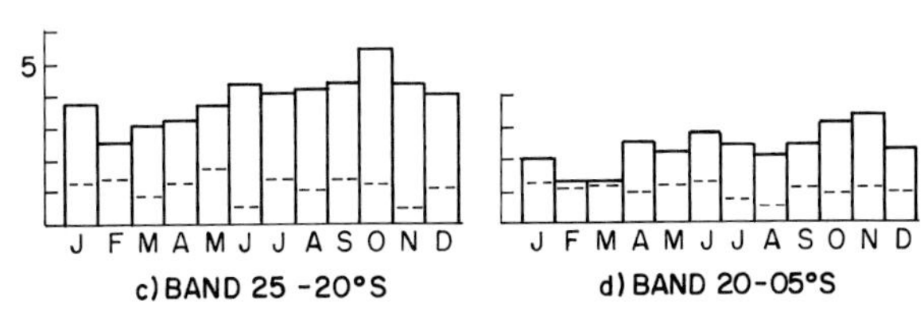

FIG. 3C.11. Mean monthly distributions of frequencies of frontal passages in the four latitudinal bands (a) 40°–35°S, (b) 35°–25°S, (c) 25°–20°S, and (d) 20°–5°S. Continuous lines are 10-yr (1975–84) means, and the broken lines are the standard deviations. (Adapted from Oliveira 1986.)

θ_e increases to 330 K after the passage of an MCC, from approximately 315 K.

The small-scale convective activity observed over the Amazon and adjoining regions is not totally disorganized. Besides having a strong diurnal variation, the activity shows intensification when a frontal system approaches southern and southeastern parts of Brazil in all seasons, except that it is less in winter (Oliveira and Nobre 1985). Bands of convective clouds form near the central and eastern parts of the north coast of SA, and some of them move west-southwestward inland over the Amazon region as lines of instability (Cohen et al. 1995). The convection develops near the boundary between the equatorial easterlies and the southwesterly land breeze, and slowly moves southwestward with the onset of sea breeze. The instability lines are more frequent in austral winter from April to August and have typical dimensions of 1400 by 170 km. They are basically driven by a low-level easterly

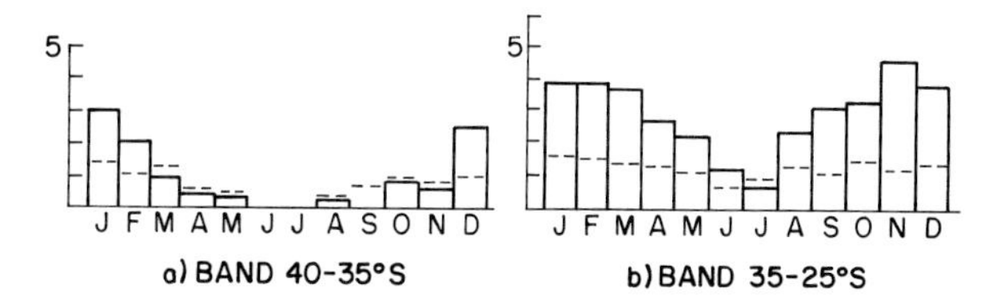

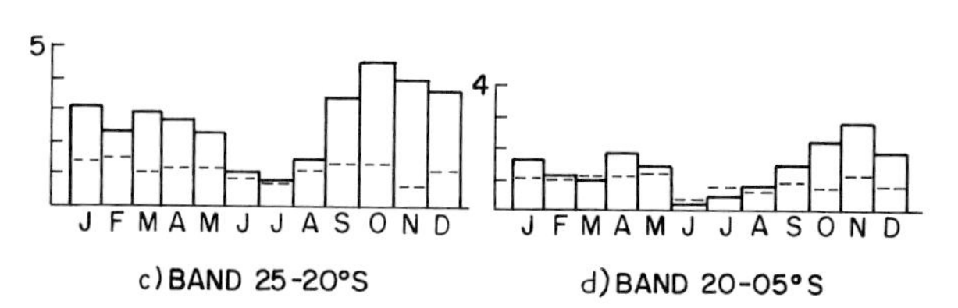

FIG. 3C.12. Same as in Fig. 3C.11 except for the frequency of convective activity associated with frontal passages. (Adapted from Oliveira 1986.)

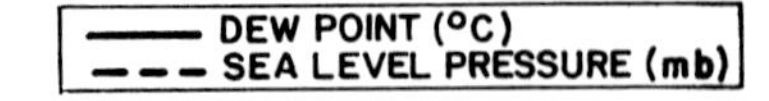

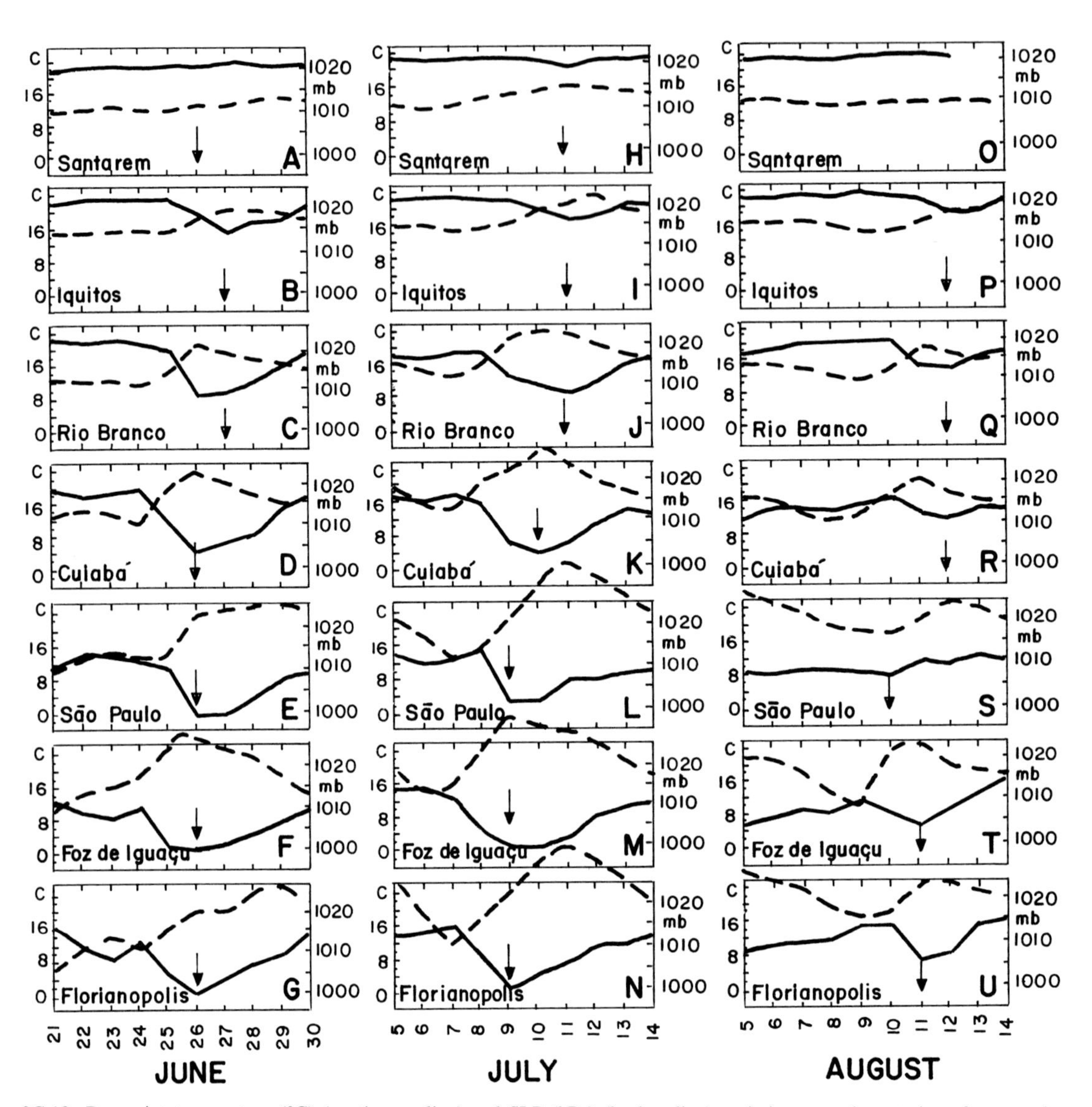

FIG. 3C.13. Dewpoint temperature (°C) (continuous line) and SLP (hPa) (broken line) variations at select stations from south to north during three episodes of cold-air invasions (*friagens*) into the tropics of SA, in the periods 21–30 June, 5–14 July, and 5–14 August 1994. The coordinates of the stations are: Florianópolis (27°40'S, 48°33'W), Foz do Iguaçu (25°31'S, 51°08'W), São Paulo (23°37'S, 46°39'W), Cuiabá (15°33'S, 56°07'W), Rio Branco (09°58'S, 67°48'W), Iquitos (03°45'S, 73°15'W), and Santarem (02°26'S, 54°43'W) (kindly given by J. Marengo).

wind maximum and can propagate as far inland as 1000 km from the coast. In general, slower-moving (12 m s^{-1}) lines (L1) penetrate less than 400 km, and the faster-moving (16 m s^{-1}) lines (L2) penetrate deeper than 400 km. The average life periods of L1 and L2 are 11 and 17 h, respectively. Vertical profiles of the humidity mixing ratio and θ_e, before and after the passage of an instability line in the interior Amazon forest area, are shown in Fig. 3C.15. The mixing ratio of 20 g kg^{-1} at the 950-hPa level, before the passage of the instability line, is reduced to less than 15 g kg^{-1} after its passage. In all levels, except at the surface, the mixing ratio is reduced by about 2 g kg^{-1}. An interesting aspect is that the humidity inversion layer near the surface and the isohygric layer between the 900- and 850-hPa levels disappear after the passage, and the profile becomes rather smooth. In the θ_e profile, before the passage, there are high temperatures in the lower troposphere below the 850-hPa level and low values above, indicating potential convective in-

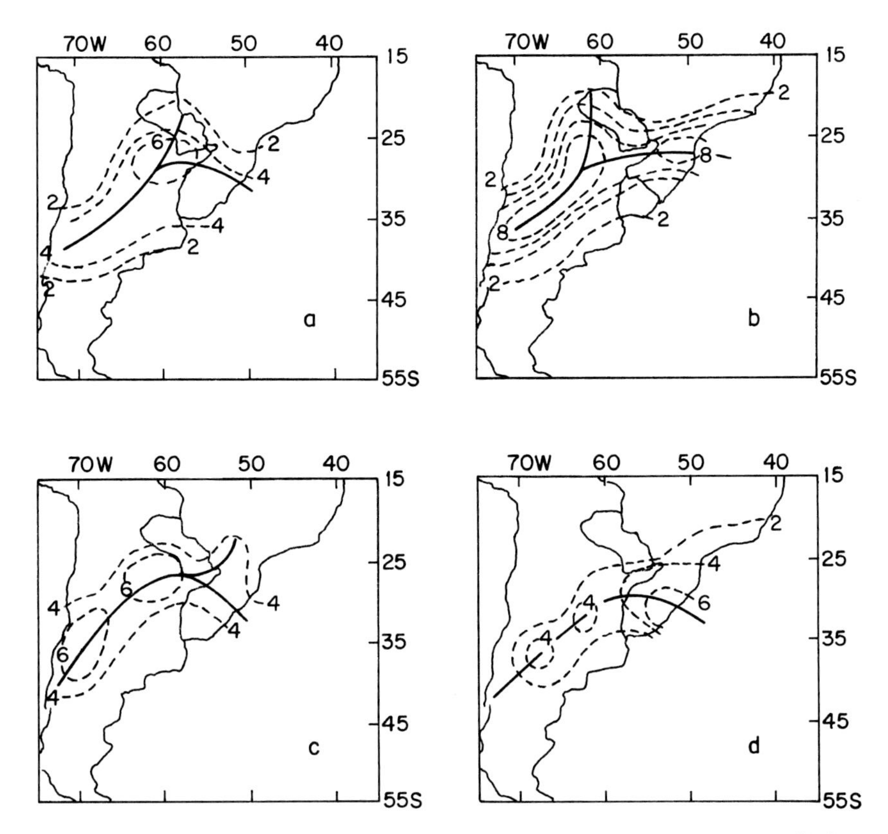

FIG. 3C.14. Tracks of intense surface anticyclones over South America in (a) May, (b) June, (c) July, and (d) August, obtained from 10-yr data (1979–89). Broken lines are isolines of mean frequency (number per month), and the continuous lines are the mean trajectories.

stability. After the passage, the lower troposphere cools and the middle troposphere turns warmer, indicating that the air column has become stable. In the case of the MCC passage in southern Brazil (Silva Dias 1987), the surface temperatures do not show appreciable cooling.

Recently, Garstang et al. (1994) and Greco et al. (1994) have studied the structure and moisture transport associated with the Amazon coastal squall lines (ACSL) using the data collected during the Amazon Boundary Layer Experiment. They noted that ACSL are discontinuous cloud clusters that propagate across the central Amazon basin at speeds of 50–60 km h^{-1}, and their life cycle is composed of six possible stages: coastal genesis, intensification, maturity, weakening, reintensification, and dissipation. They observed a deep vertical ascent in the leading edge of the convection area and a very weak vertical movement in the trailing stratiform region, with a weak unsaturated downdraft below 500 hPa in the region between. These features are shown schematically in Figs. 12 and 15 of Garstang et al. (1994).

The cloud imagery vividly shows the presence and movement of mesoscale and small-scale convective systems embedded in the trade winds approaching the northern parts of the coast of NEB from the equatorial Atlantic (Yamazaki and Rao 1977). These systems do not quite have the same characteristics as the easterly waves observed in the Caribbean; nevertheless, monitoring them is very important for weather prediction in NEB.

3C.5. Interactions with extratropics

The deep meridional penetrations of the cold fronts observed over SA are an indication of the strong interactions that take place between the Tropics and the middle latitudes and the energy propagation that occurs in the meridional direction. According to Charney (1969), there may exist critical latitudes where the time mean, zonally averaged zonal-wind component $[u]$ matches the zonal phase speed of the waves where the meridional group velocity of the Rossby waves becomes zero. Webster and Holton (1982), however, found that if the longitudinally asymmetric basic state zonal wind, u, is westerly over a region, waves smaller than the westerly "duct" may propagate from one hemisphere to the other. Arkin and Webster (1985) have, with the aid of climate datasets, tested the hypothesis that the regions of equatorial westerlies act as efficient ducts or corridors for the penetration of transient and stationary extratropical modes deep into

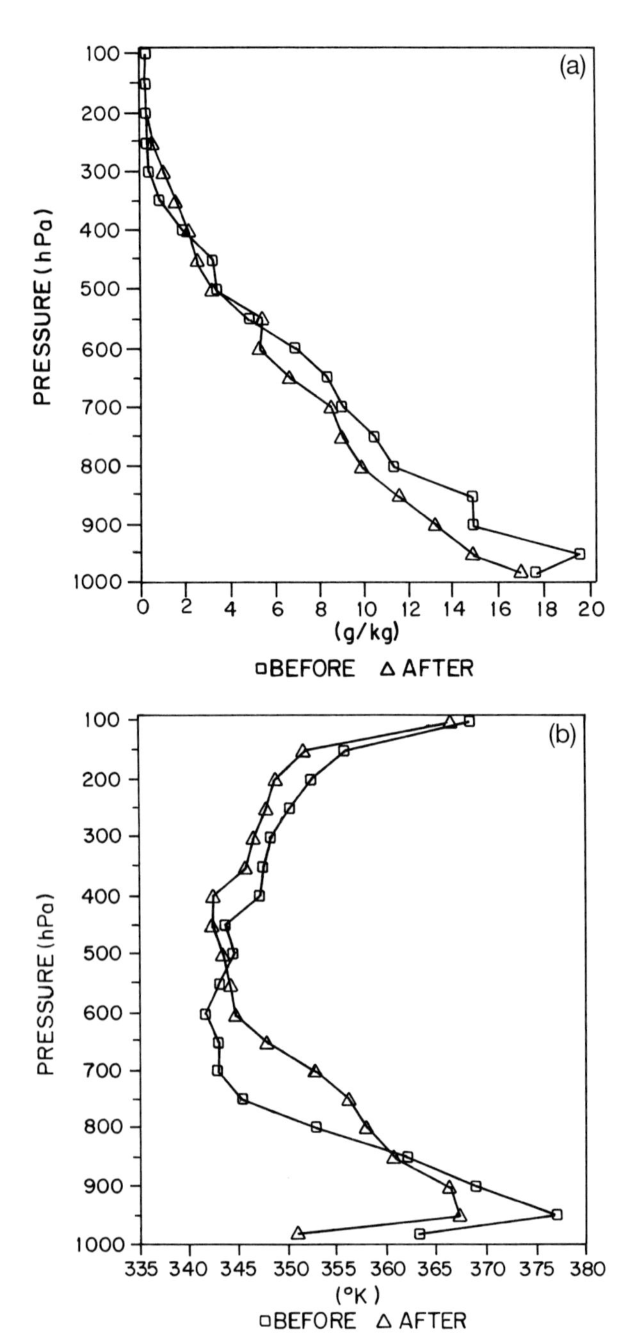

FIG. 3C.15. Vertical profiles of (a) humidity mixing ratio and (b) equivalent potential temperature at Alta Floresta station (8°20'S, 54°10'W). □—□—□ ⇒ at 1200 UTC on 6 May 1987 before the passage, and △—△—△ ⇒ 0000 UTC on 7 May 1987 after the passage of the instability line. (Adapted from Cohen et al. 1995.)

the Tropics and even cross the Equator. From Fig. 3C.4, one can see that, in winter (July), strong westerlies prevail at 200 hPa in the Tropics as far north as 5°S in the SA region, and in this season many *friagens* are experienced in the tropical SA. Recently, Tomas and Webster (1994) have conjectured that extratropical wave activity may propagate from the SH into the NH during boreal fall.

3C.6. South Atlantic convergence zone

The summer cold fronts penetrate rapidly from Argentina northeastward into the region of SACZ over southeastern Brazil, where they remain stationary for periods of about 5 to 10 days, imparting large amounts of rainfall to the regions of Rio de Janeiro, northern and eastern São Paulo, southern and western Minas Gerais, Mato Grosso do Sul, and southern and eastern Mato Grosso states of Brazil. A typical value of precipitation in an active episode of the SACZ in a period of one week is 300 mm, whereas the typical value of precipitable water is 55 mm, and there is no large change in this value from before the rainy episode to afterward. It is easy to deduce that most, if not all, of the precipitation in the region of the SACZ is caused by convergence of water vapor flux integrated in the vertical, z, and in time, t. That is,

$$P = -\int\int \nabla \cdot (\rho q \mathbf{V}) dz\, dt = \int\int \nabla \cdot (q\mathbf{V}) \frac{dp}{g}\, dt,$$

where P is the total precipitation in an episode, ρ is the density of air (considered constant in the horizontal), q is the specific humidity, $\mathbf{V}$ is the wind, g is the acceleration due to gravity, and p is the pressure. In this equation, the contribution of the evaporation from the surface is neglected, considering that the air remains largely saturated during active SACZ episodes. Assuming uniformity of conditions along the SACZ, taking Cartesian coordinates in the directions parallel to and perpendicular to the orientation of the SACZ, and designating v as the wind component perpendicular to the convergence zone, the humidity convergence simplifies to

$$P = \int\int \frac{\partial}{\partial y}(qv)\frac{dp}{g}\, dt,$$

which is approximately written as

$$P = T\int q\overline{\frac{\partial v}{\partial y}}\frac{dp}{g},$$

where an overbar indicates a time average over the interval of the episode T. Taking the average precipitation rate, P/T, to be 40 mm day^{-1} and vertically averaged specific humidity in the lowest 400-hPa layer to be 10 g kg^{-1}, the average convergence required is estimated to be $\sim 1 \times 10^{-5}$ s^{-1}. This is clearly one or two orders of magnitude larger than the commonly observed values. From this, one can see that the convergence ($\partial v/\partial y$) in the SACZ is very important for the persistence of rain.

The results of Figueroa et al. (1995) show a well-defined zone of low-level convergence in the observed mean position of the SACZ, besides other features, such as the Bolivian high and the trough to the northeast of it. Their study stresses the importance of the topography and the heat source on producing the major regional summer circulation features over SA. The observational and mechanistic experimental study of Kalnay et al. (1986) showed the existence of a relationship between the occurrence of a strong SACZ and a strong, eastward-shifted SPCZ in southern summer. Some studies link the intensification of SACZ with the approach of the 30–60-day MJO (Casarin and Kousky 1986; Grimm and Silva Dias 1995).

3C.7. Teleconnections with planetary-scale phenomena

The main teleconnections between the planetary-scale circulations of the Pacific Ocean and SA are due to ENSO. ENSO is described in detail in chapter 8 and will not be reiterated here. There have been many studies in the last two decades about the effects of ENSO on the rainfall variability over Brazil and other countries in South America. Caviedes (1973) and Markham and McLain (1977) associated the droughts in NEB directly to the El Niño event in the eastern Pacific. Kousky et al. (1984) have extended the studies to all the regions of Brazil. Aceituno (1988) has correlated the surface variables, station precipitation, SLP, and surface temperature for March–April over the whole of SA with the SO index, and his main results are shown in Fig. 3C.16. One can see that generally positive correlations are in northern SA and negative correlations are in southern SA. Positive correlation implies that, during negative (positive) SO episodes, negative (positive) rainfall anomalies prevail. In this regard, Rao et al. (1986) have presented a case study of the 1983 severe drought in NEB.

Moura and Shukla (1981) have linked the rainfall variability in NEB to SST anomalies in the Atlantic and found that warmer-than-normal waters in the South Atlantic and colder-than-normal temperatures in the North Atlantic are conducive to higher-than-normal rainfall in NEB. The influence of the Atlantic Ocean over the rainfall in NEB is further explored by Nobre and Shukla (1996), who have found that, in situations where the North Atlantic subtropical high pressure center is more intense and the South Atlantic high pressure center is less intense, conditions are favorable for the southward shift of the ITCZ and, consequently, higher-than-normal rainfall over northern parts of NEB. Negative rainfall anomalies to the south of the Equator during March–May (rainy season of NEB) are related to an early withdrawal of ITCZ over the northern tropical Atlantic. They also found that precipitation anomalies over northern and southern parts of NEB are out of phase. Drought years over northern NEB are commonly preceded by wetter years over southern NEB, and vice versa. They suggested that patterns of tropical Atlantic circulations are related to the phases of the ENSO phenomenon. Hastenrath (1991) also has described the relationship between ENSO and the rainfall over NEB.

Strong signals are also found in the rainfall variability in conjunction with the SO over northern and northeastern parts of SA from Venezuela to Ceará and over southeastern parts around Uruguay (Ropelewski and Halpert 1987, 1989). Broadly, their results show that in the years of high SO and cold SST over the central Pacific, there is wet anomaly in the northern SA and a dry anomaly in the south-central eastern SA, and vice versa.

Chile and the Andean region near the west coast of SA is another key region where interannual rainfall variability is closely related to the ENSO phenomenon. Forecasting of equatorial Pacific SST has a reasonable skill with regard to the rainfall in the northern part of central Chile (30°–35°S) for the later part of the rainy season in July and August (Aceituno and Vidal 1990). Diagnostic studies by Aceituno and Garreaud (1994) on the interannual variability of rainfall in central Chile have confirmed a well-defined tendency for abundant winter rainfall during the occurrence of EN episodes and drought during La Niña events. Using empirical methods suggested by Ropelewski and Halpert, Pisciottano et al. (1994) found that years with EN tend to have higher-than-normal rainfall from November to January in Uruguay. Years with high values of the SO index tend to have lower-than-normal rainfall from October to December. The period from March to July tends to have higher-than-normal rainfall after EN years and lower-than-normal rainfall after high SO index years.

Velasco and Fritsch (1987) observed that the number of MCCs in northern Argentina and Paraguay region in an EN year are more than double the number observed in a normal year. This is connected to the strengthening of the subtropical jet in the EN years. The increase of rainfall in southern Brazil in the EN years is due to the increased MCC activity.

It is useful to recollect the major changes observed in the circulation features over SA during the EN year 1993, which was an extension of the 1991–92 El Niño. Very strong convective activity persisted off the coast of northern Peru and Equador in the eastern tropical Pacific. As a result, the annual rainfall over that region, which is normally about 300 mm, exceeded 1000 mm. Very dry conditions persisted over the whole of tropical Brazil, especially over NEB, where many regions received less than 100 mm and some places in the interior remained completely dry, indicating a large and strong subsiding branch over SA, east of 65°W and north of 25°S. Both the Walker circulation in the east–west and the Hadley circulation in the north–south planes adversely affected this region. The subtropical jet was situated near 30°S and was stronger than normal. Excess convective activity over Paraguay, northeastern Argentina, Uruguay, and the adjoining regions in Brazil resulted in excess rainfall

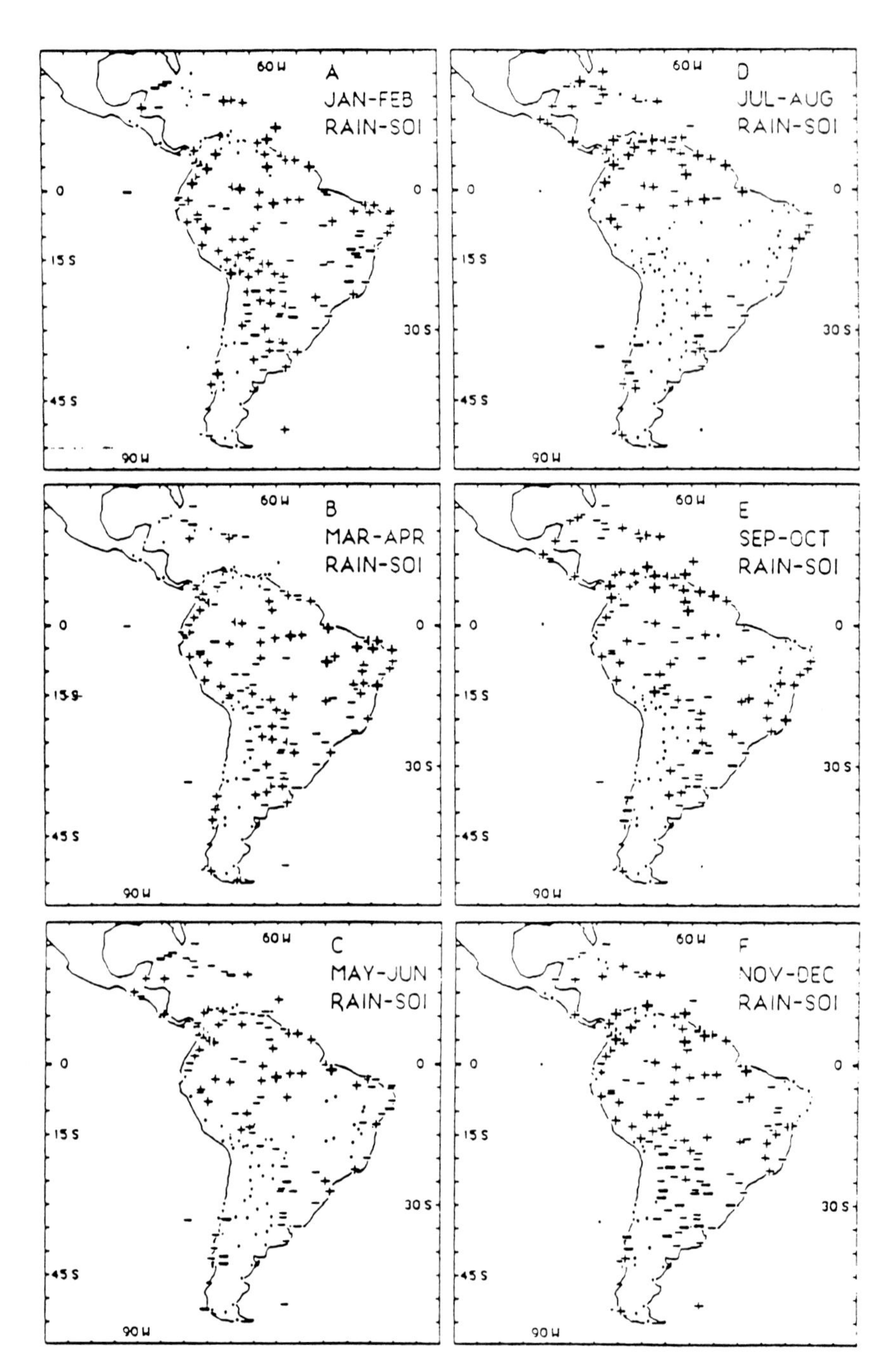

FIG. 3C.16. Bimonthly charts of correlation between SO index and station precipitation: (a) Jan–Feb, (b) Mar–Apr, (c) May–Jun, (d) Jul–Aug, (e) Sep–Oct, and (f) Nov–Dec. The symbols + and − indicate values higher than normal and lower than normal, respectively, in a low SO index year. Bold symbols indicate high correlation. Dots indicate stations with less than 100 mm rain for the bimonthly interval (Source: Aceituno 1988).

there. The cold fronts could not penetrate beyond 30°S, due to the persistence of general blocking conditions near the coast of southwest SA and the presence of the strong subtropical jet, and, as a result, the temperatures over most of Brazil were a few degrees Celsius warmer, especially in winter.

3C.8. Other evidence of interannual and intraseasonal variability

Synoptic analyses of interannual variations of the tropical summer circulation were conducted by Tanaka and Nishizawa (1983). They observed that in summers with strong Hadley-type circulations, the rainfall is above normal in coastal NEB and southern Brazil. During weak Hadley circulation periods, the rainfall is above normal in the southern interior of NEB and below normal in southern Brazil. The Hadley-type circulation over SA was found to be strong when the 850-hPa cyclonic circulation near 28°S, 65°W was well defined, and was weak when both the 850-hPa cyclonic circulation and the 150-hPa anticyclonic circulation were weak and displaced to the north toward the Amazon River basin. In these summers, cold-air invasions from the south were more frequent. The synoptic situations of relatively dry and relatively wet periods during the months of February, March, and April 1981 over NEB were presented by Alcantara et al. (1984). They found that the penetration of the South Atlantic subtropical high, in conjunction with high zonal index in the SH and low zonal index in the NH, was responsible for the dry period (20 February to 7 March). A low zonal index in the SH and a high zonal index in the NH permitted the penetration of a cold front, coming from the south, into the NEB region, which interacted with a traveling disturbance in the equatorial trough and produced intense rainfall for the region during the wet period (10 to 31 March).

3C.9. Intraseasonal variability

The 40- to 50-day oscillation, first noted by Madden and Julian (1971) in the tropical Pacific, was later observed in all tropical regions around the globe, with variable intensity, and is usually referred to as a 30–60-day oscillation. A thorough review of the literature on this topic is presented recently by Madden and Julian (1994), and considerable attention is devoted to this topic in chapter 8. With regard to its influence on SA, the MJO takes about six weeks to reach SA after it reaches the western Pacific. Its intensity diminishes and usually disappears over the eastern Pacific, but its effect can be felt when it occasionally reappears in eastern SA (Kousky and Cavalcanti 1988). It is felt that this oscillation affects, though weakly, the intensity of convection over the Amazon basin and southeastern Brazil, especially in southern summer. Kayano and Kousky (1992) have verified the MJO in the OLR variability over the SA sector through extended empirical orthogonal function analysis. There is a suggestion that the oscillation tends to have a higher frequency during the warmer episodes of ENSO phenomenon (Kuhnel 1989). A principal mode analysis of the combined OLR and 250-hPa circulation fields by Kousky and Kayano (1994) indicate that an anomalous westerly (easterly) flow over the SA tropics is associated with a positive (negative) OLR anomaly over northeastern SA and over the eastern equatorial Pacific, and these features are associated with the MJO.

A very interesting aspect of intraseasonal variability in the region south of about 15°S, east of the Andes, called *veranico* (a Portuguese word meaning "little summer"), occurs in the middle or latter half of winter, when the temperatures rise well above their seasonal mean values. This phenomenon is observed in almost all the winters. The surface temperatures cool down gradually from above 33°C in April to less than 20°C in July, in small steps of steep falls and smaller rises associated with cold frontal passages. Sometime around August, the temperatures rise to near 30°C and remain relatively high during a period of about 3 to 4 weeks before they fall again. The warm temperatures are apparently related to regional-scale subsidence over tropical and subtropical SA east of 65°W that inhibits the advance of cold air masses to north of 30°S. During *veranico,* most of SA north of 30°S remains dry, and the winter crops in this region are largely destroyed. In the Spanish-speaking subtropical countries of Uruguay, Paraguay, and northern Argentina, the term *veranillo* (also meaning "little summer") is used to refer to a warm spell within the winter season. The opposite event of a relative long cool spell in the middle of summer is also observed south of 20°S in many years. These intraseasonal variations are similar to the so-called "breaks" in the monsoon activity known in other parts of global tropics. As yet, there are no systematic studies on these phenomena, and it is difficult to say if and how they are related to the blocking phenomena in the Pacific (Lejenas 1984; Kayano and Kousky 1990) and the MJO. During blocking episodes in the eastern Pacific, rain-producing transient disturbances steer to the south downstream of the region and avoid advancing to the north of 40°S over SA.

Actually, the term *veranico* has different meanings in different regions of SA. In the northeastern and northern regions of Brazil, north of about 18°S, the term is popularly used to designate a common occurrence of a dry spell within the rainy season (Assad et al. 1993). The dry spell lasts anywhere between one and three weeks and always has a deleterious effect on agriculture, primarily when it happens in the beginning of the rainy season. It is common in the SA tropics to refer to the rainy season as "winter" due to cooler temperatures and to the dry season as "summer" due to warmer temperatures, regardless of astronomical or meteorological seasons. Therefore, a dry spell, which is warm, during the rainy season, is termed *veranico* in those regions.

In some years the dry conditions in tropical and subtropical Brazil extend to larger areas and to longer periods, as has happened recently in the 1994 winter. Figure 3C.17 shows the rainfall and its anomaly distributions in Brazil for the month of August 1994. A large area (more than 2×10^6 km^2) in southeastern Brazil received almost no rain (less than 1 mm rainfall) in August. The anomalies are negative in the

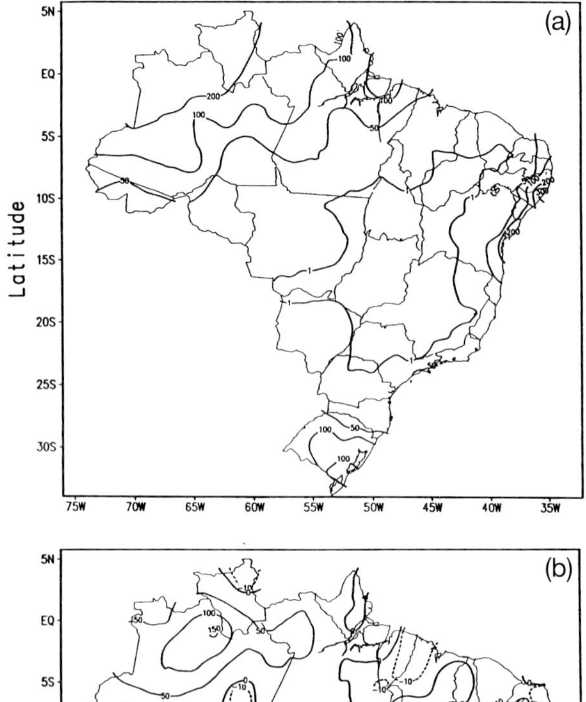

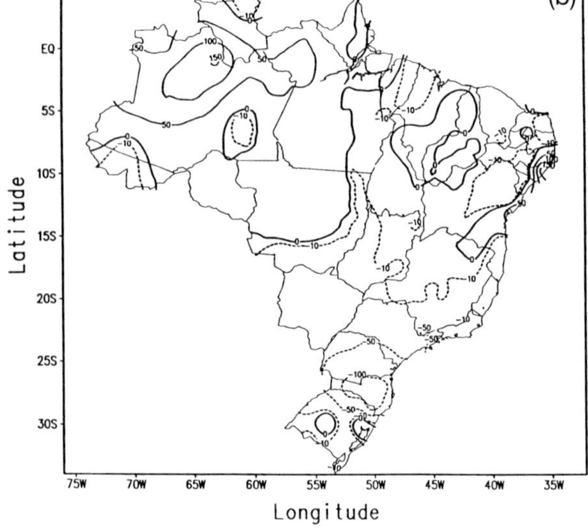

FIG. 3C.17. (a) Rainfall (mm) and (b) its anomaly (mm) over Brazil for August 1994. In (b), dashed lines indicate negative values (from Satyamurty and Pezzi 1997).

whole country, except for the northern region of the state of Amazonas, and are larger than the rainfall amounts in many parts of south-central Brazil. Very low relative humidity values (less than 25%) were reported on many days in central Brazil. The dry conditions, known locally as *estiagem,* persisted up to October, and thousands of head of cattle have died of dehydration. In this period, not only did the winter crops fail, but there was also a substantial increase in the forest fires in central Brazil. In a recent study by Satyamurty and Pezzi (1997), it was found that the planetary-scale Walker circulation associated with excessively strong convective activity over the region of Indonesia, as well as over the adjoining southwest monsoon region (southern Asia), maintained a large and persistent subsidence branch over most of the tropical and subtropical SA, east of 60°W. This caused the formation of a stationary high, which appeared somewhat like an extension of the South Atlantic subtropical high into the continent.

3C.10. Climatic impacts of Amazon deforestation

The regional and possible global effect of deforestation of the Amazon basin is a current question of major importance and scientific interest. Models used to study the climatological response to tropical deforestation contain both atmosphere and land–biosphere components (Dickinson 1989) and take a radical scenario of complete removal of the tropical forest and replacement by pasture. The model atmosphere senses the difference between the tropical forest and the pastureland mostly through the specified albedo and roughness length values. Notwithstanding the differences in the physical parameterizations and integration details, the numerical simulation experiments of Amazon deforestation conducted in the recent past (Nobre et al. 1991; Dickinson and Kennedy 1992; Dirmeyer and Shukla 1994; Manzi 1993) have produced many alarming common features for the post-deforestation climate. All the experiments have indicated an increase of temperature, varying from 0.6°C to about 2.0°C, and most of them have indicated a decrease of precipitation up to about 1.8 mm day^{-1} over the deforested Amazon pasture.

An increase in albedo over the pastureland implies less shortwave radiation absorbed by the surface, which reduces available energy for evapotranspiration and sensible heating of the air. On the other hand, a reduction in roughness over the grassland in comparison with the forest causes a decrease in the efficiency of removing heat and water vapor off the surface by means of turbulent eddies.

At the outset, it looks like the net effect is to increase the temperature and reduce the precipitation, but there are other important complex feedback mechanisms such as cloud–radiation–temperature and temperature–surface pressure–convergence of water vapor to be considered. An increase of albedo leads to a reduction of absorbed shortwave radiation, which in turn leads to a decrease in evapotranspiration. Less water vapor in the boundary layer may imply reduced cloudiness and, during daytime, leads to more shortwave radiation reaching the surface. At night, however, the thermal loss by the surface is larger. The net effect, according to Dickinson and Kennedy (1992), is still reduction of net radiation at the surface. The effect of increased temperature at the surface is to reduce the surface pressure, which increases the convergence of moisture into the basin, which, in turn, leads to increased cloudiness and precipitation. Zeng et al. (1996), using a simple model, discussed the feedback mechanisms related to water vapor convergence in terms of alterations in the Walker circulation and SST gradients. There is clearly a need for further development of the model parameterization schemes and other aspects of the simulations in order to firmly establish a probable scenario for the hydrologic effects of Amazon deforestation. These include multiyear runs to resolve natural variability in the model; more complete validation of the surface–biosphere parameterizations from field observations; better resolution of topography; and the incorporation of ocean–atmosphere coupling.

The importance of the Amazon tropical rain forest for the balance of the regional ecosystem and its implications for the global climates and changes are studied through multinational field experiments in the last two decades. The articles contained in the recent publication edited by Gash et al. (1996) give a good overview of experiments like the Amazon Boundary Layer Experiment and the Anglo Brazilian Amazonian Climate Observation Study and their results.

3C.11. Areas needing further research

Many aspects of the regional circulation over South America are not yet well understood. Among them, the most notable is the formation of SACZ in summer. Although a few authors (Figueroa et al. 1995) have attempted, through numerical simulations of simple baroclinic regional models, to explain the formation in terms of the heat source over the Amazon basin and the mountain-generated trough on the east coast of the continent, there has not been a theoretical explanation for the elongated convergence zone and its location and orientation. The relationship between the SACZ and the SPCZ and its explanation have to be further explored. Another important aspect of the summer circulation is the formation of a cold-core vortex in the upper troposphere near the NEB coast, studied by Kousky and Gan (1981), and its slow migrations onto the continent and off to the ocean. The Bolivian high also meanders quite a bit from 50° to 80°W around 15°S. The mechanisms for their meanders are not known.

One important question is why the MJO is not very prominent over SA. A frequently occurring phenomenon deserving the attention of the meteorological research community in the region of SA is the *veranico* in winter and the opposite phenomenon in summer. Why are there long breaks in precipitation in the middle of summer and winter seasons over SA? How are they related to the blockings in the Pacific? For that matter, the mechanisms of the formation of blocking highs itself need further observational and theoretical investigations. A lot more attention has to be paid to the global circulation changes to understand the occurrence of large-scale hazards such as the *estiagens*.

It is known that, on average, the advection of humidity into the Amazon basin from the surrounding seas accounts for only about 50% of the precipitation, which means that the evapotranspiration in the tropical forest is very important for the rainfall over the region, and the forest and the climate are in good cooperation by efficiently recycling the local humidity. One of the major problems in this region is to know beforehand the degradation of the regional and global climate in view of the present rate of deforestation and other adverse human activities such as the burning of biomass.

Some other important outstanding experimental and research investigations in this region are i) the determination of the intensity, position, and transport capacity of the low-level northerly jet east of the Andes; ii) understanding the triggering mechanism of the convective activity over the tropical SA by the presence of even a feeble cold front in the subtropics of the South Atlantic; and iii) simulation of the development and propagation of the instability lines in the equatorial regions of SA.

REFERENCES

Aceituno, P., 1988: On the functioning of the Southern Oscillation in the South American sector. Part I: Surface climate. *Mon. Wea. Rev.*, **116**, 505–524.

———, and F. Vidal, 1990: Variabilidad interanual en el caudal de ríos andinos en Chile central en relación con la temperatura de la superficie del mar en el Pacifico central (Interannual variability of stream flow in Andean rivers in central Chile and its relation with sea surface temperature in the central Pacific). *Rev. Soc. de Ingienería Hidráulica* (Chile), **5**, 7–19.

———, and R. Garreaud, 1994: Impacto de los fenómenos El Niño y La Niña en regímenes fluiométricos andinos (Impacts of El Niño and La Niña in Andean fluvial regimes). *Proc. (Vol. 3): XVI Congresso Latinoamericamno de Hidráulica,* Universidad Central de Chile, Santiago, Chile, 15–25.

Alcantara, F., A. S. Valença, and I. A. Vieira, 1984: Intensive rainfall for March 1981 over the northeast region of Brazil. Abstracts, *Second WMO Symp. on Meteorological Aspects of Tropical Droughts,* Fortaleza, Brazil, World Meteorological Organization, 45–48.

Arkin, P. A., and P. J. Webster, 1985: Annual and interannual variability of tropical–extratropical interaction: An empirical study. *Mon. Wea. Rev.*, **113**, 1510–1523.

Assad, E. D., E. E. Sano, R. Masutomo, L. H. R. De Castro, and F. A. M. Da Silva, 1993: Veranicos na região dos cerrados Brasileiros: Frequencia e probabilidade de ocorrencia (Dry spells in the Brazilian "cerrados" region: Frequency and probability of occurrence). *Psquisa Agropecuária Brasil.*, **28**, 993–1003.

Casarin, D. P., and V. E. Kousky, 1996: Anomalias de precipitação no sul do Brasil e variações na circulação atmosférica. (Precipitation anomalies in south Brazil and variations in the atmospheric circulation.) *Rev. Brasil. Meteorol.*, **1**, 83–90.

Cavalcanti, I. F. A., 1983: *Casos de intensa precipitação nas regiões sul e sudeste do Brasil no período de inverno de 1979 a 1983.* INPE-3743-RPE/498, Instituto Nacional de Pesquisas Espaciais (INPE), S. J. Campos, Brazil, 40 pp.

Caviedes, C. N., 1973: Secas and El Niño: Two simultaneous climatological hazards in South America. *Proc. Assoc. Amer. Geogr.*, **5**, 44–49.

Charney, J. G., 1969: A further note on large-scale motions in the Tropics. *J. Atmos. Sci.*, **26**, 182–185.

Cohen, C. P. C., M. A. F. Silva Dias, and C. A. Nobre, 1995: Environmental conditions associated with Amazonian squall lines: A case study. *Mon. Wea. Rev.*, **123,** 3163–3174.
DeMaria, M., 1985: Linear response of a stratified tropical atmosphere to convective forcing. *J. Atmos. Sci.*, **42,** 1944–1959.
Dickinson, R. E., 1989: Implications of tropical deforestation for climate: A comparison of model and observational descriptions of surface energy and hydrological balance. *Philos. Trans. R. Soc. London,* Ser. B, **324,** 423–431.
———, and P. Kennedy, 1992: Impacts on regional climate of Amazon deforestation. *Geophys. Res. Lett.*, **19,** 1947–1950.
Dirmeyer, P. A., and J. Shukla, 1994: Albedo as a modulator of climate response to tropical deforestation. *J. Geophys. Res.*, **99,** 20 863–20 877.
Figueroa, S. N., and C. A. Nobre, 1990: Precipitation distribution over central and western tropical South America. *Climanálise,* **5,** 36–45.
———, P. Satyamurty, and P. L. Silva Dias, 1995: Simulations of the summer circulation over the South American region with an eta coordinate model. *J. Atmos. Sci.*, **52,** 1573–1584.
Fortune, M., and V. E. Kousky, 1983: Two severe freezes in Brazil: Precursors and synoptic evolution. *Mon. Wea. Rev.*, **111,** 181–196.
Gandu, A. W., and J. E. Geisler, 1992: A primitive equations model study of the effect of topography on the summer circulation over tropical South America. *J. Atmos. Sci.*, **48,** 1822–1836.
Garstang, M., H. L. Massie Jr., J. Halverson, S. Greco, and J. Scala, 1994: Amazon coastal squall lines. Part I: Structure and kinematics. *Mon. Wea. Rev.*, **122,** 608–622.
Greco, S., J. Scala, J. Halverson, H. L. Massie Jr., W.-K. Tao, and M. Garstang, 1994: Amazon coastal squall lines. Part II: Heat and moisture transports. *Mon. Wea. Rev.*, **122,** 624–635.
Grimm, A. M., and P. L. Silva Dias, 1995: Analysis of tropical–extratropical interactions with influence functions of a barotropic model. *J. Atmos. Sci.*, **52,** 3538–3555.
Gutman, G. J., and W. Schwerdtfeger, 1965: The role of latent and sensible heat for development of high pressure system over the subtropical Andes in the summer. *Meteorol. Rundschau,* **18,** 69–75.
Hastenrath, S., 1991: *Climate Dynamics of the Tropics.* Kluwer Academic Publishers, 488 pp.
Horel, J. D., A. N. Hahmann, and J. E. Geisler, 1989: An investigation of the annual cycle of convective activity over the tropical Americas. *J. Climate,* **2,** 1388–1403.
Kalnay, E., K. C. Mo, and J. Peagle, 1986: Large-amplitude, short-scale stationary Rossby waves in the Southern Hemisphere: Observations and mechanistic experiments to determine their origin. *J. Atmos. Sci.*, **43,** 252–275.
Kayano, M. T., and V. E. Kousky, 1990: Southern Hemisphere blocking: A comparison between two indices. *Meteor. Atmos. Phys.*, **42,** 165–170.
———, and ———, 1992: Sobre o monitoramento das oscilações intrasazonais (On the monitoring of the intraseasonal oscillations). *Rev. Brasil. Meteorol.*, **7,** 593–602.
Kleeman, R., 1989: A modeling study of the effect of the Andes on the summertime circulation of tropical South America. *J. Atmos. Sci.*, **46,** 3344–3362.
Kodama, Y.-M., 1992: Large scale common features of subtropical precipitation zones (the Baiu frontal zone, the SPCZ and the SACZ). Part I: Characteristics of subtropical frontal zones. *J. Meteor. Soc. Japan,* **70,** 813–836.
Kousky, V. E., 1979: Frontal influences on northeast Brazil. *Mon. Wea. Rev.*, **107,** 1140–1153.
———, 1980: Diurnal rainfall variation in northeast Brazil. *Mon. Wea. Rev.*, **108,** 488–498.
———, 1985: Atmospheric circulation changes associated with rainfall anomalies over tropical Brazil. *Mon. Wea. Rev.*, **113,** 1951–1957.
———, and P. S. Chu, 1978: Fluctuation in annual rainfall for northeast Brazil. *J. Meteor. Soc. Japan,* **57,** 457–465.
———, and M. A. Gan, 1981: Upper tropospheric cyclonic vortex in the tropical South Atlantic. *Tellus,* **33,** 538–551.
———, and I. F. A. Cavalcanti, 1988: Precipitation and atmospheric circulation anomaly patterns in the South American sector. *Rev. Brasil. Meteorol.*, **3,** 199–206.
———, and M. T. Kayano, 1994: Principal modes of outgoing longwave radiation and 250-mb circulation for the South American sector. *J. Climate,* **7,** 1131–1142.
———, M. T. Kagano, and I. F. A. Cavalcanti, 1984: The Southern Oscillation: Oceanic–atmospheric circulation changes and related rainfall anomalies. *Tellus,* **36A,** 490–504.
Kuhnel, I., 1989: Spatial and temporal variation in Australia–Indonesian region cloudiness. *Int. J. Climatol.*, **9,** 395–405.
Lejenas, H., 1984: Characteristics of Southern Hemisphere blocking as determined from a time series of observational data. *Quart. J. Roy. Meteor. Soc.*, **110,** 967–979.
Madden, R. A., and P. R. Julian, 1971: Description of a 40–50-day oscillation in the zonal wind in the tropical Pacific. *J. Atmos. Sci.*, **28,** 702–708.
———, and ———, 1994: Observations of the 40–50-day tropical oscillation—A review. *Mon. Wea. Rev.*, **122,** 814–837.
Maddox, R. A., 1980: Mesoscale convective complexes. *Bull. Amer. Meteor. Soc.*, **61,** 1374–1387.
Manzi, A. O., 1993: Introduction d'un schema de transferts sol–vegetation–atmosphere dans um modele de circulation generale et application la simulation de la deforestation amazonienne (Introduction to a soil–vegetation–atmosphere transfer scheme in a general circulation model and an application to Amazon deforestation simulation). Ph.D. thesis, University Paul Sabatier, 229 pp. (Available from the library of the Université Paul Sabatier, Toulouse, France.)
Marengo, J., A. Cornejo, P. Satyamurty, C. Nobre, and W. Sea, 1997: Cold surges in tropical and extratropical South America: The strong event in June 1994. *Mon. Wea. Rev.*, **125,** 2759–2786.
Markham, C. G., and D. R. McLain, 1977: Sea surface temperature related to rain in Ceará, northeastern Brazil. *Nature,* **265,** 320–323.
Moura, A. D., and J. Shukla, 1981: On the dynamics of droughts in northeast Brazil: Observations, theory, and numerical experiments with a general circulation model. *J. Atmos. Sci.*, **38,** 2653–2675.
Nobre, C. A., P. J. Sellers, and J. Shukla, 1991: Amazonian deforestation and regional climate change. *J. Climate,* **4,** 957–988.
Nobre, P., and J. Shukla, 1996: Variations of sea surface temperature, wind shear, and rainfall over the tropical Atlantic and South America. *J. Climate,* **9,** 2464–2479.
Oliveira, A. S., 1986: Interações entre sistemas frontais na América do Sul e a convecção da Amazônia (Interactions between the South American frontal systems and the Amazonian convection). INPE-4008-TDL/239, Instituto Nacional de Pesquisas Espaciais, 115 pp. (Available from the library of the Instituto Nacional de Pesquisas Espaciais, São José dos Campos, SP, Brazil.)
———, and C. A. Nobre, 1985: Meridional penetration of frontal systems in South America and its relation to organized convection in the Amazon. INPE-3407-PRE/676, Instituto Nacional de Pesquisas Espaciais, 4 pp. (Available from the library of the Instituto Nacional de Pesquisas Espaciais, São José dos Campos, SP, Brazil.)
Paegle, J., 1987: Interactions between convective and large-scale motions over Amazonia. *Geophysiology of Amazonia: Vegetation and Climate Interactions,* R. Dickinson, Ed., Wiley, 347–390.
Pisciottano, G., A. Díaz, G. Gazes, and C. R. Mechoso, 1994: Rainfall anomalies in Uruguay associated with the extreme phases of the El Niño/Southern Oscillation phenomenon. *J. Climate,* **7,** 1286–1302.
Rao, V. B., and J. P. Bonatti, 1987: On the origin of upper tropospheric cyclonic vortices in the South Atlantic Ocean and

adjoining Brazil during the summer. *Meteor. Amos. Phys.*, **37,** 11–16.

———, and K. Hada, 1990: Characteristics of rainfall over Brazil: Annual variations and connections with the Southern Oscillation. *Theor. Appl. Climatol.*, **42,** 81–91.

———, P. Satyamurty, and J. I. B. de Brito, 1986: On the 1983 drought in northeast Brazil. *J. Climatol.*, **6,** 43–51.

Ropelewski, C. F., and M. S. Halpert, 1987: Global and regional-scale precipitation patterns associated with the El Niño/Southern Oscillation. *Mon. Wea. Rev.*, **115,** 1606–1626.

———, and ———, 1989: Precipitation patterns associated with the high index phase of the Southern Oscillation. *J. Climate,* **2,** 268–284.

Satyamurty, P., and L. F. Mattos, 1989: Climatological lower tropospheric frontogenesis in the midlatitudes due to horizontal deformation and divergence. *Mon. Wea. Rev.*, **117,** 1355–1364.

———, and L. P. Pezzi, 1997: On some large scale aspects of the generalized drought conditions in Brazil in winter and spring 1994. Preprints, *Fifth Intl. Conf. on Southern Hemisphere Meteorology and Oceanography,* Pretoria, S. Africa, Amer. Meteor. Soc., 168–169.

Serra, A., and L. Ratisbona, 1942: *As massas de ar da America do Sul.* (*The Air Masses of South America.*) Ministério da Agricultura (Brazil), Serviço de Meteorologia, 66 pp.

Silva Dias, M. A. F., 1987: Sistemas de mesoescala e previsão de tempo a curto prazo. (Mesoscale systems and short range weather forecasting.) *Rev. Brasil. Meteorol.*, **2,** 133–150.

Silva Dias, P. L., W. H. Schubert, and M. DeMaria, 1983: Large-scale response of the tropical atmosphere to transient convection. *J. Atmos. Sci.*, **40,** 2689–2707.

Tanaka, M., and T. Nishizawa, 1983: Interannual fluctuations of the tropospheric circulation and the rainfall in South America. *Latin American Studies,* **6,** 45–61.

Tomas, R. A., and P. J. Webster, 1994: Horizontal and vertical structure of cross-equatorial wave propagation. *J. Atmos. Sci.*, **51,** 1417–1430.

Velasco, I., and J. M. Fritsch, 1987: Mesoscale convective complexes in the Americas. *J. Geophys. Res.*, **92,** 9591–9613.

Virji, H., 1982: An estimate of the summertime tropospheric vorticity budget over South America. *Mon. Wea. Rev.*, **110,** 217–224.

Webster, P. J., and J. R. Holton, 1982: Cross-equatorial response to middle-latitude forcing in a zonally varying basic state. *J. Atmos. Sci.*, **39,** 722–733.

Yamazaki, Y., and V. B. Rao, 1977: Tropical cloudiness over the South Atlantic Ocean. *J. Meteor. Soc. Japan,* **55,** 205–207.

Zeng, N., R. E. Dickinson, X. Zeng, 1996: Climate impact of Amazon deforestation: A mechanistic model study. *J. Climate,* **9,** 859–883.

Chapter 3D

Africa and Surrounding Waters

JOHN VAN HEERDEN

Department of Civil Engineering, University of Pretoria, Pretoria, South Africa

J. J. TALJAARD

South African Weather Bureau (Retired), Pretoria, South Africa

3D.1. Introduction

Although the main theme of the chapter is the meteorology and basic climatology of the Tropics, presumably bounded in the south by the 25°S latitude circle, or the Tropic of Capricorn at 23° 30′S, there is no natural boundary between the weather systems to the north and south across the subcontinent or the oceans on either side. In addition, the weather along the subtropical belt has not been allocated for separate discussion; therefore, South African conditions, being intimately related with events farther north, are included in the discussion.

The area to be discussed is bounded by the Equator and 35°S, from 15°W to 75°E, but for convenience the boundaries of all relevant charts are extended another 5°. The political boundaries of the territories (as of 1996) and the locations of the islands in the area are given in Fig. 3D.1. Topography plays a major role in the circulation and weather systems of landmasses, and therefore the height contours of the subcontinent are given in Fig. 3D.2. Small patches of the land rise to 2000 m and even 3000 m above sea level, such as in the Drakensberg Escarpment of Lesotho, western Kenya, and astride the Rift Valley between Lakes Tanganyika and Mobutu. Broad river valleys, like those of the Congo (Zaïre), Zambezi, and Limpopo Rivers, form inlets along which the humid lower layers of maritime air are channeled inland, whereas escarpments and high watersheds in proximity of the coasts form barriers against the inland penetration of the maritime air. Regional features and effects of this kind will be discussed in subsequent sections.

The material presented is designed to follow a logical sequence, starting with the significant airstreams, particularly at low levels, which affect the weather on the land; the properties, stability, and modification of the air masses involved; the rainfall; the perturbations (synoptic systems) responsible for the rainfall; and, briefly, the climatic variability.

The terms "southern Africa" and "the subcontinent" will be used as alternatives for "Africa south of the Equator." "South Africa" is equivalent to the "Republic of South Africa (RSA)," but is sometimes loosely used to include also the southern parts of Botswana and Namibia south of 25°S. "East Africa" is used for the three territories Kenya, Uganda, and Tanzania.

A dearth of references to research on the meteorology of tropical southern Africa since about 1970 will be noticed. This can be attributed to the decline of meteorological observations, particularly upper-air soundings, in the region since that time and, consequently, significant research.

3D.2. The predominant atmospheric circulation systems

Charts of MSLP over the sea only are given in Figs. 3D.3a,b, for DJF and JJA, as determined from the ECMWF daily gridded data for 1985–92. The isobars for sea level are omitted over the land because SLP is fictitious and highly misleading for the elevated continent. Flow lines indicate the mean surface wind directions, those over the sea being based on the marine climatic atlases of the British, U.S., Dutch, and German meteorological services, which have utilized the observations of ships since the last century. The flow lines over the land surface are based on the data and treatises of Thompson (1965), Schulze (1965), South African Weather Bureau (1975), Torrance (1981), Leroux (1983), and others. It will be seen that, in general, the surface wind at sea is directed about 20° across the isobars, toward lower pressure, poleward of 15°S, but, closer to the Equator, the flow is more directly toward lower pressure, especially along the ITCZ.

The main features of the surface circulation in DJF (Fig. 3D.3a) are:

1) the subtropical anticyclones at 30°S, 5°W in the Atlantic Ocean and at 35°S, 85°E in the Indian

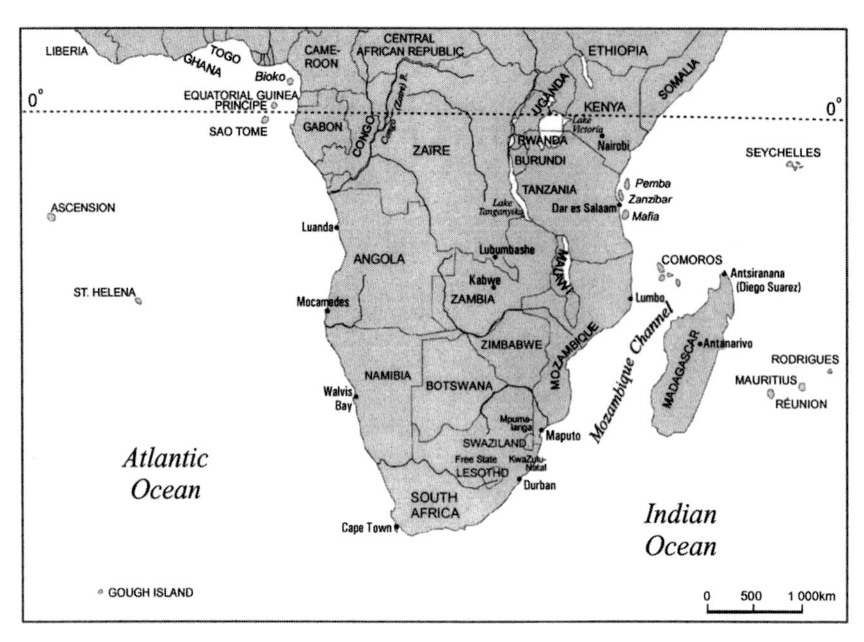

FIG. 3D.1. Political boundaries of territories in southern Africa and location of the main islands around the subcontinent.

Ocean are linked by a ridge that passes a few degrees south of the land;
2) the broad band of easterly trade winds in the Indian Ocean between the subtropical ridge and the ITCZ, which is positioned from 20°S in the Mozambique Channel to about 5°S at the eastern boundary of the diagram;
3) the northeast monsoon current, which crosses the Equator and spreads into East Africa and southward to the ITCZ;
4) the southerly to southeasterly trades between the Atlantic anticyclone and the continent, which recurve to become the southwest monsoon that enters land between 12°S and 5°N;
5) the confluence stretching along about 15°S from eastern Zambia to southern Angola [formerly called the Congo Air Boundary and more lately the Zaïre Air Boundary, but indicated in the diagram as the inter-ocean convergence zone (IOCZ)];
6) the westerlies of the Congo (Zaïre) basin, which extend eastward to the ITCZ in the north and southward to the IOCZ in the south;
7) the easterly trades over Zimbabwe and the northern provinces of the RSA, which recurve anticyclonically southward over Botswana and Namibia; and
8) the westerly current over the extreme southwestern plateau of South Africa, which splits to become either southerly or westerly over the plateau.

A quick glance creates the impression that there is little difference in the pressure fields and flow patterns of summer and winter (JJA, Fig. 3D.3b) in the Atlantic Ocean sector. The core of the subtropical anticyclone is displaced only 3° or 4° northward and about 5° westward, to 10°W. With the almost 10-hPa increase in MSLP along the South African and Namibian coasts, however, as opposed to only 5 hPa at the core of the anticyclone, the zonal pressure gradient weakens substantially so that the southerly flow diminishes markedly. In fact, the characteristic fresh southeasterlies over the southwest coast in summer (the "Cape doctor") revert to predominant northwesterlies. Farther north, both the wind directions and the speeds change little over the sea. The flow onto the land attains an increased southerly component north of 10°S in winter.

In the Indian Ocean, however, the change of the pressure and wind fields from summer to winter is dramatic. The core of the subtropical anticyclone shifts westward from 85° to 60°E, while the northward displacement is 5°. The pressure on the South African east coast rises 10 hPa. The ITCZ disappears, so that the northeast monsoon is replaced by a southerly monsoon, consisting of a broad southeasterly trade wind between 10° and 30°S, which changes direction to become the well known southwest monsoon as it flows across the Equator. Over the land the change is also dramatic, with the ITCZ being displaced to North Africa. The vestigial remains of the IOCZ linger over the Zaïre Basin, where the Atlantic maritime air penetrates to 30°E along the Equator. The core of the predominantly anticyclonic surface flow over South Africa remains almost in the same position over the east coast as in summer, but with the strong pressure rise of about 10 hPa, the circulation over two thirds of the southern African plateau reverts to a highly constant southeasterly to easterly trade-wind flow.

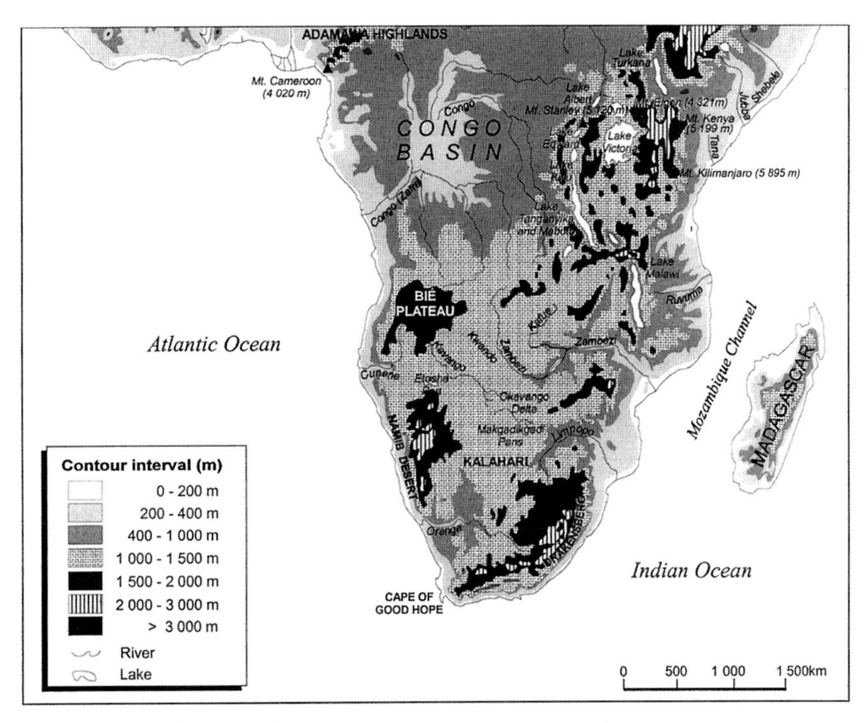

FIG. 3D.2. Topography of southern Africa and Madagascar.

The contour patterns at 850, 700, 500, and 200 hPa are shown for DJF and JJA in Figs. 3D.4 and 3D.5. Wind vectors from ECMWF analyses are also shown. The sea level circulation in summer (Fig. 3D.3) and that at 850 hPa (Fig. 3D.4) over the sea are in fair agreement, which may be interpreted to mean that the subinversion maritime air masses are, on the whole, deeper than 1500 gpm; otherwise, there ought to have been significant differences between the two circulations in some areas. A region where dissimilarity exists is over the tropical west coast and the adjacent ocean, where southerly to southwesterly flow dominates at sea level, but at 850 hPa the flow is indicated to be either light variable or easterly. This raises the point mentioned by Trewartha (1961) as to whether or not the southwest monsoon layer of moist maritime air here is deeper than 1500 m near and over the Equator. The pressure gradient at 850 hPa indicated for that area in Figure 3D.4 suggests that there should actually be westerly flow at this level in the equatorial zone.

A striking change occurs between the pressure patterns and gradients at 850 and 700 hPa in summer. The col over South Africa at 850 hPa in summer is replaced by a ridge along 25°S at 700 hPa, and both the Atlantic and Indian Ocean anticyclones are relatively much weaker at the latter level. The ITCZ trough northeast of Madagascar is still evident. A problem exists about the validity of the ECMWF analysis at 700 hPa in summer, which shows only a broad and weak trough extending southward from Zaïre over western Zambia to Botswana in this season. A large number of pilot balloon soundings in northern Rhodesia (now Zambia) (Rhodesia Meteorological Service 1958) and even a radiosonde sounding station at Elizabethville (now Lubumbashe) in the late 1950s indicated predominant moderate to fresh northerly to northwesterly flow east of a trough, or low, close to 12°S, 25°E, whereas no soundings have been available for the ECMWF analyses in the region.

The circulation pattern at 500 hPa is very similar to that at 700 hPa in summer, but the Atlantic anticyclone has weakened further compared to those existing at 850 and 700 hPa. The subtropical ridge axis at 500 hPa is located between 20° and 25°S, which is 10° equatorward of the axis at sea level. Proceeding to 200 hPa, a prominent anticyclone appears over the subcontinent between 15° and 20°S (the highest elevation of the isobaric surface is actually along 18°S). Taljaard (1995) plotted the positions of anticyclone centers at 200 hPa (often elongated ridges extending E–W) as analyzed on the daily charts of ECMWF for the 10-yr period 1981–90 (Fig. 3D.6). Most of these systems cluster astride latitude 20°S. Another notable phenomenon at 200 hPa is the northerly component of the wind over the central tropical Atlantic Ocean (this is actually the eastern flank of the pronounced trough of the upper troposphere east of Brazil). There is no evidence for a mean subtropical jet stream at 200 hPa in the South African region in summer.

As in summer, the circulations at sea level and 850 hPa are also very similar in winter (Figs. 3D.3 and 3D.5). The cores of the subtropical anticyclones at 850

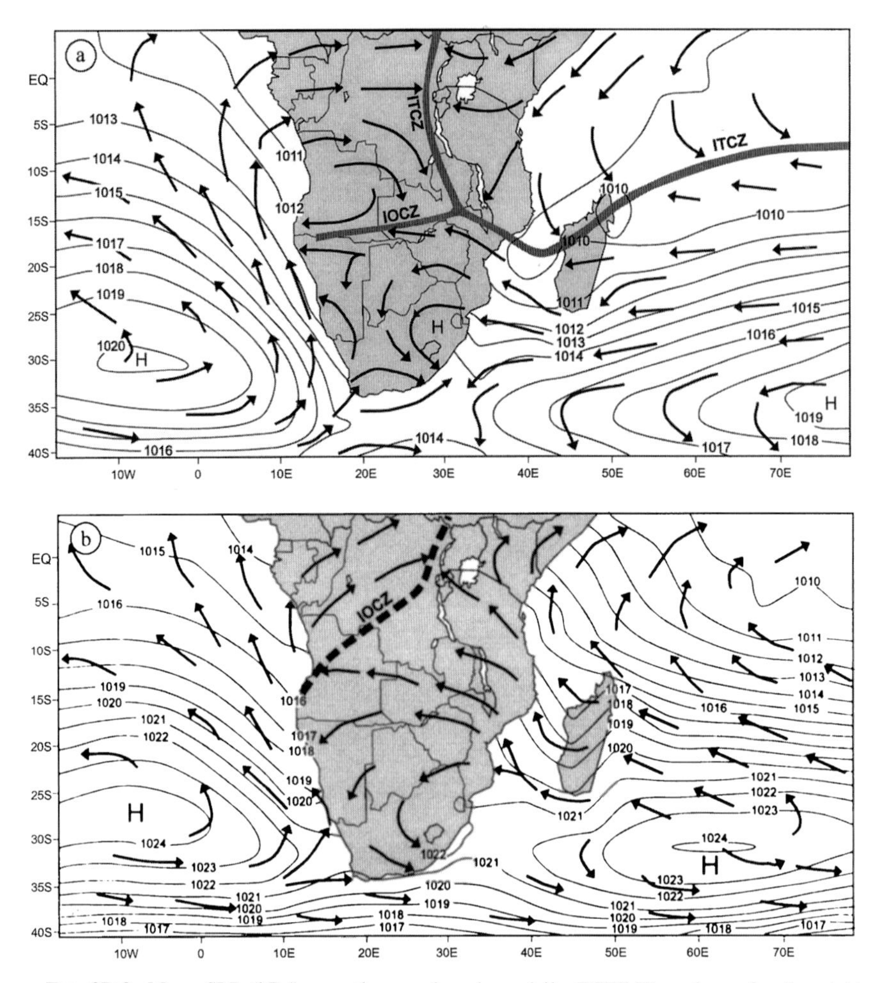

FIG. 3D.3. Mean SLP (hPa) over the sea, based on daily ECWMF analyses for the eight years 1985–1992 for (a) summer (DJF) and (b) winter (JJA). Mean flow lines on sea and land based on sources detailed in text. Thick solid lines in summer represent the ITCZ and IOCZ. Broken thick line in winter is the mean boundary between the dry continental southeast trade winds and the moist southwest monsoon air.

hPa are located a few degrees west of and only about one degree north of their sea level counterparts. As in summer, however, the easterly flow at 850 hPa over and seaward of the equatorial west coast is evidently not in accord with the indicated pressure gradient. The coexistence of the anticyclone over eastern South Africa and low pressure over the Equator at 20°–25°E produces the extensive region of easterly continental trade winds between South and East Africa, where cross-contour flow is indicated to vary between 30° and 70°. The remarkable easterly to southerly flow, reverting to the southwesterly monsoon north of the Equator in the Indian Ocean, is striking.

Findlater (1969, 1971) used the observations of the large number of pilot balloon stations and a few rawinsondes that existed at that time to study the low-level jet stream over and off the East African and northern Madagascar coasts in winter. He also used aircraft observations. He found that a core of strong southeasterly flow bypasses the northern tip of Madagascar on the average, then crosses the Equator as a southerly flow just inland of the Kenya coast, and finally recurves as a southwesterly wind maximum over Somali, just seaward of the mountain massif of Ethiopia (Fig. 3D.7a). Findlater found that the jet core wandered appreciably between southerly and easterly directions, south of the Equator, and that split cores existed at times with speeds as high as 100 kt on a few occasions. The vertical cross section along the Equator (Fig. 3D.7b) reveals pronounced lateral shear between the coast and the high ground of Kenya.

Proceeding upward from 850 to 200 hPa, the flow patterns in winter are remarkably different in the Atlantic and Indian Oceans. In the Atlantic Ocean, the regime of the westerlies extends northward to only 5° south of the Equator, while mean speeds of 30–35 m s^{-1} remain practically constant over the latitude span from 20° to 40°S. In the Indian Ocean, a ridge

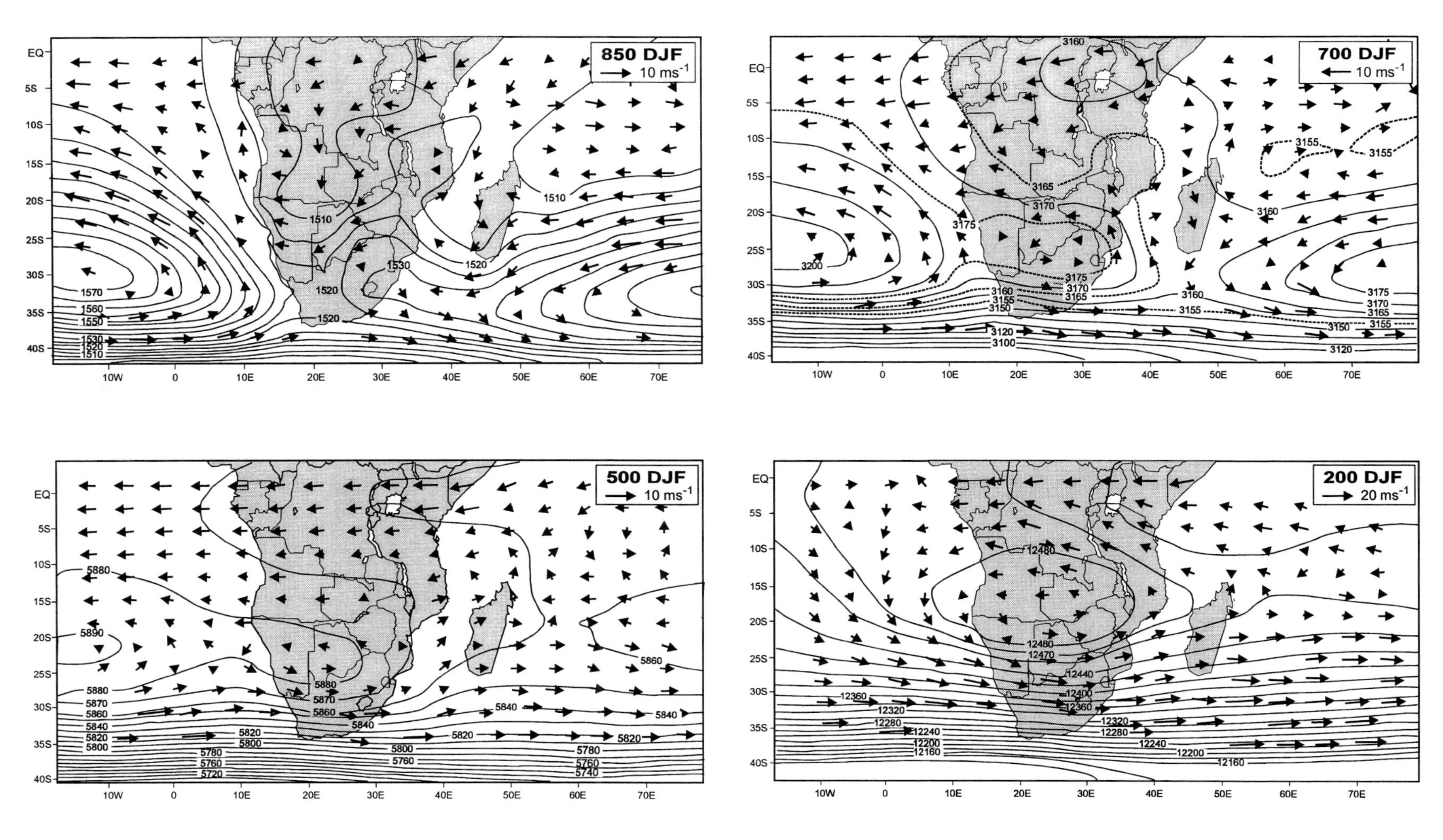

FIG. 3D.4. Mean contours of geopotential height (gpm), derived from daily ECWMF analyses (1985–1992) for the 850-, 700-, 500-, and 200-hPa levels for summer (DJF).

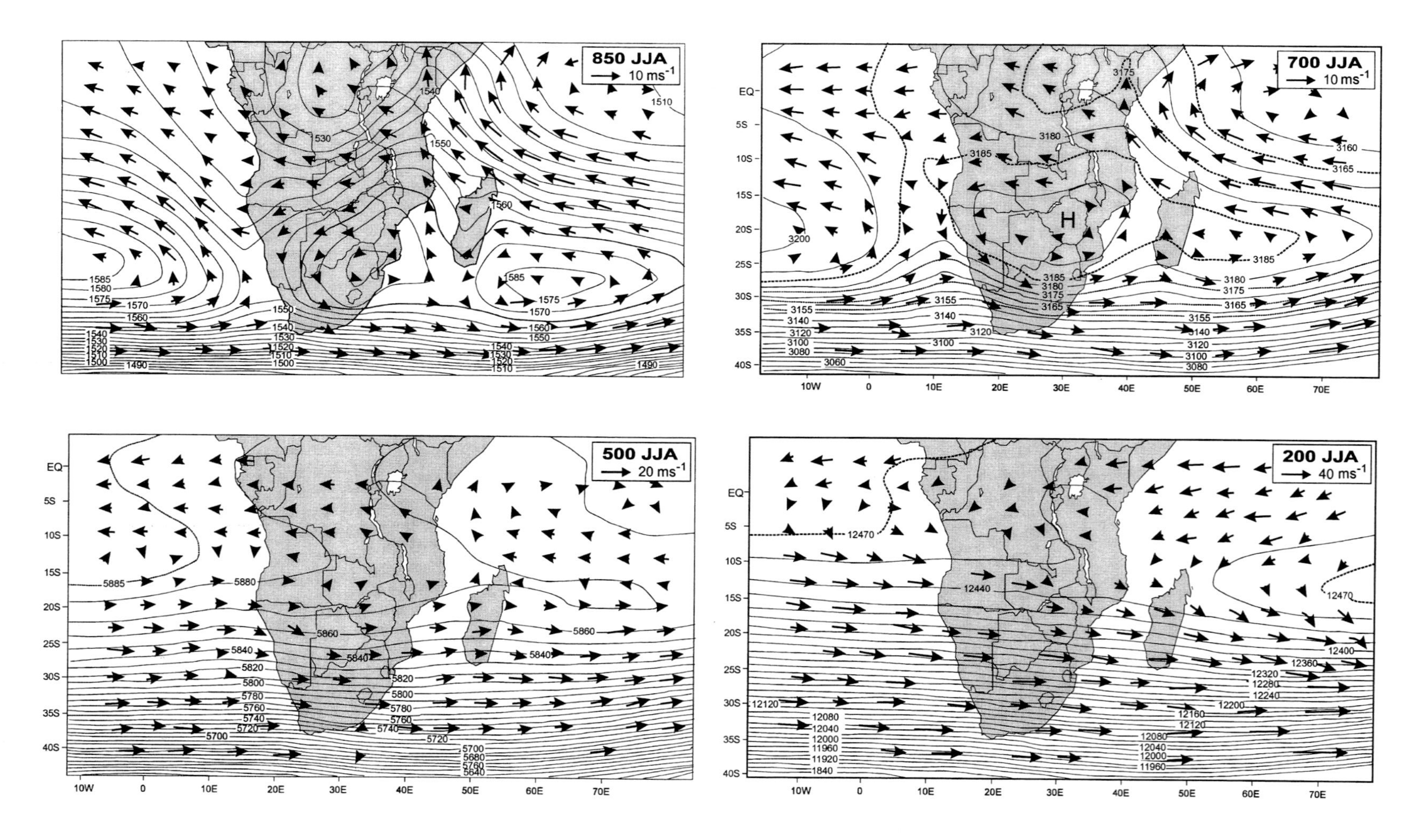

FIG. 3D.5. As for Fig. 3D.4 but for winter (JJA).

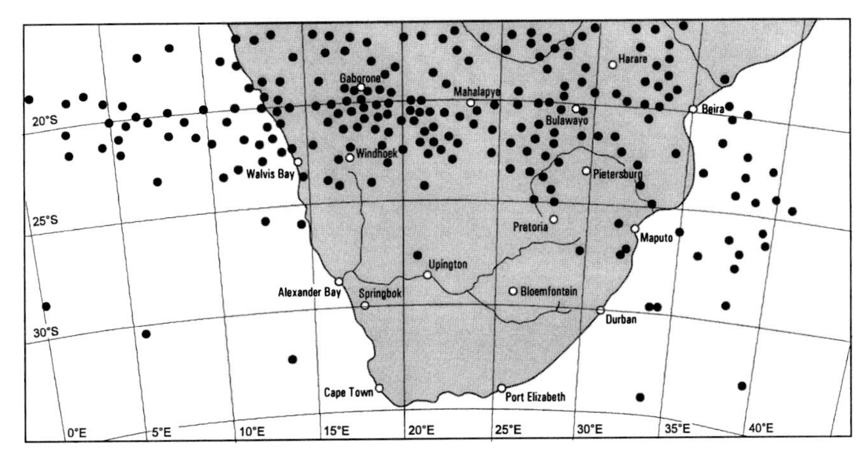

FIG. 3D.6. Positions of anticyclone centers at 200 hPa on the daily January 1200 UTC charts of ECWMF for the period 1981–1990 (after Taljaard 1995).

axis is prominent at 200 hPa along 15°S, giving rise to easterlies with a northerly component in the Tropics and westerlies of 40–45 m s^{-1} at 30° to 40°S. The northward flow at low levels and the southward component in the upper troposphere in the tropical western Indian Ocean in winter is a clear example of a Hadley circulation cell, while the increase of the pressure gradient and the westerlies in the 25°–30°S zone indicate the initiation of the pronounced subtropical jet stream of Australasia.

3D.3. Temperature

The surface temperature distributions are presented here; remarks concerning temperatures aloft are discussed later, in the form of lapse rates. The distribution of the mean land surface temperature (half the sum of maximum and minimum temperatures) in January and July (Fig. 3D.8a,b) clearly illustrates the predominant role played by land elevation on this element. Of course, other factors such as cloudiness, rainfall, and distance from the sea have effects, too. The lowest temperatures are found on the southeastern Drakensberg Escarpment of South Africa, the mountainous areas east and west of Lake Victoria, and the interior of Madagascar, whereas high temperatures prevail over the low grounds of the Zaïre River basin, Mozambique, the Kalahari (especially in summer), and the valleys of the Zambezi, Limpopo, and Orange Rivers. The narrow coastal zone of the Namib desert is an exception to this, a result of the upwelling cold water from the Benguela current.

The mean surface air temperature over the sea is approximately the same as the SST, and, therefore, the distributions of the surface air temperature around southern Africa shown in Fig. 3D.8a,b can also be taken to represent the SSTs in January and July. The most striking feature is the higher temperatures along the east coast than along the west coast. Near the Equator, the temperature is only about 2°C higher along the east coast than along the west coast. Pro-

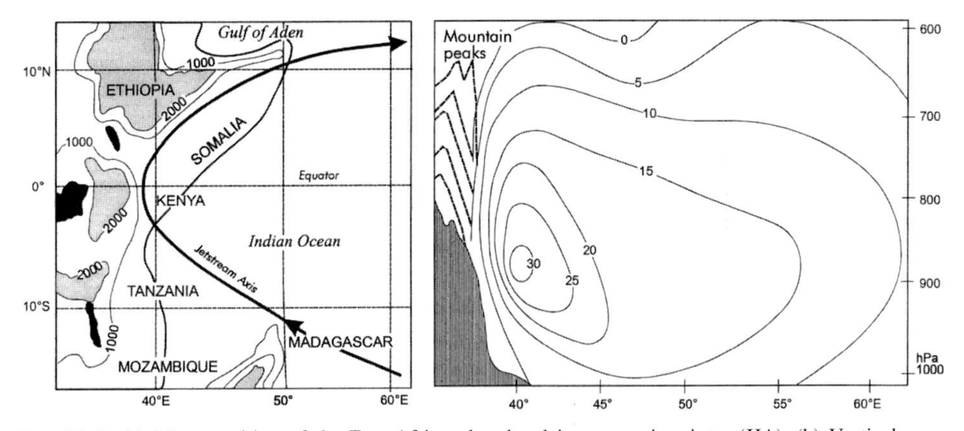

FIG. 3D.7. (a) Mean position of the East African low-level jet stream in winter (JJA). (b) Vertical cross section of this jet stream along the Equator. Isotachs in kt (adapted from Findlater 1969, 1971).

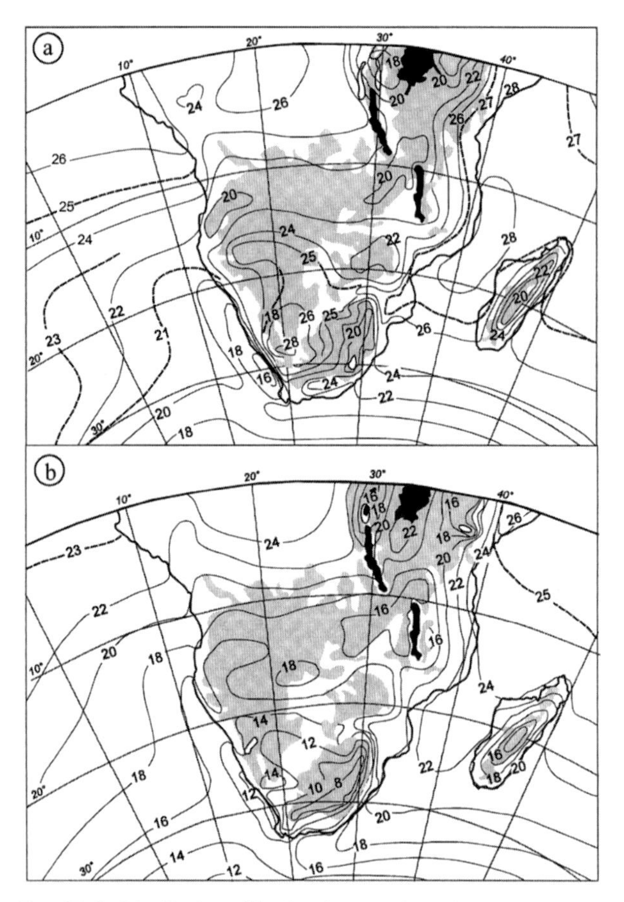

FIG. 3D.8. Distribution of land and sea surface air temperatures in (a) January and (b) July (compiled from various marine and continental climate atlases).

ceeding southward along the coast, this difference increases until, at 30°S, it is 8°C in both summer and winter. East of the land, the temperature changes very little with increasing distance from the shore, but off the west coast the temperature increases gradually. The cold upwelling zone along the west coast and the warm south equatorial, Mozambique, and Agulhas currents along the east coast are responsible for these contrasts south of 15°S.

As discussed more fully in the next section, the profound difference of surface air temperature along the west and east coasts is not maintained with increasing elevation. Taljaard (1996a) has shown that the low air temperatures along the west coast exist only in a shallow layer, usually less than 500 m deep, overlain by warm, dry subsiding air, whereas over the east coast (e.g., at Durban) a much deeper layer of fairly warm maritime air is present, particularly in summer.

3D.4. Temperature lapse rates, humidity, and stability

The distribution of the mean temperature on isobaric surfaces or, alternatively, that of the thicknesses of the layers between successive isobaric surfaces, ought to explain the changing pressure and wind patterns with increasing height. Instead of showing some of these charts, however, examples of the vertical temperature and humidity structures and the consequent stabilities will be presented. The discussion of the parameters in this section is partitioned into three regions: the west coast–Atlantic, the continent, and the east coast–Indian Ocean.

a. The west coast and the adjacent Atlantic Ocean

The persistence of a remarkable temperature inversion over the Atlantic Ocean south of the Equator was discovered during cruises of the German Atlantic *Meteor* expedition of 1925–27. The observations were first described by Von Ficker (1936), and later by Riehl (1979), Hastenrath (1991), and Asnani (1993). Briefly, the *Meteor* expedition found the base of the inversion to be lowest (less than 500 m above sea level) along the coast from Walvis Bay to a position near the Equator, increasing westward to more than 1500 m near South America and northward to over 2000 m along the ITCZ at about 5°N. The temperature increases from the bottom to the top of the inversion by over 8°C near Walvis Bay, but less than 1°C along the Equator. At the same time, the relative humidity decreases from the bottom to the top of the inversion by more than 70% in the south to less than 20% in the north. These values were obtained during individual cruises across the Atlantic Ocean; thus, they do not reflect the fluctuations or seasonal variations at individual sites.

Examples of the typical fluctuations occurring at Walvis Bay (Rooikop) in summer are shown in Fig. 3D.9a–d, for four successive days. The soundings on the individual days are compared with the mean soundings of two Januarys. On 4 January 1991, the air above the usual inversion at about 400 m above sea level was influenced by subsidence, with the inversion strength being about 12°C, but on 5 January, slightly moister air was advected from southern Angola. This continued on 6 January, when the upper air became exceptionally moist and the inversion height lifted to about 1000 m. On the fourth day, the moist air from the continent was replaced by cooler drier air from the Atlantic Ocean. These fluctuations are typical of the southern Angolan and northern and central Namibian coasts, but at Saint Helena Island (Fig. 3D.9e) conditions are more monotonous. In summer, the base of the inversion fluctuates between the 900- and 850-hPa levels, above which extremely dry air with humidity mixing ratios of only 1 to 2 g kg^{-1} prevail, interspersed with infrequent occasions when wedges of moister air are present at 600–700 hPa due to fresh easterlies from the distant continent. The latter situation is illustrated by the sounding on 29 December 1984 (Fig 3D.9e). In winter, the inversion height also

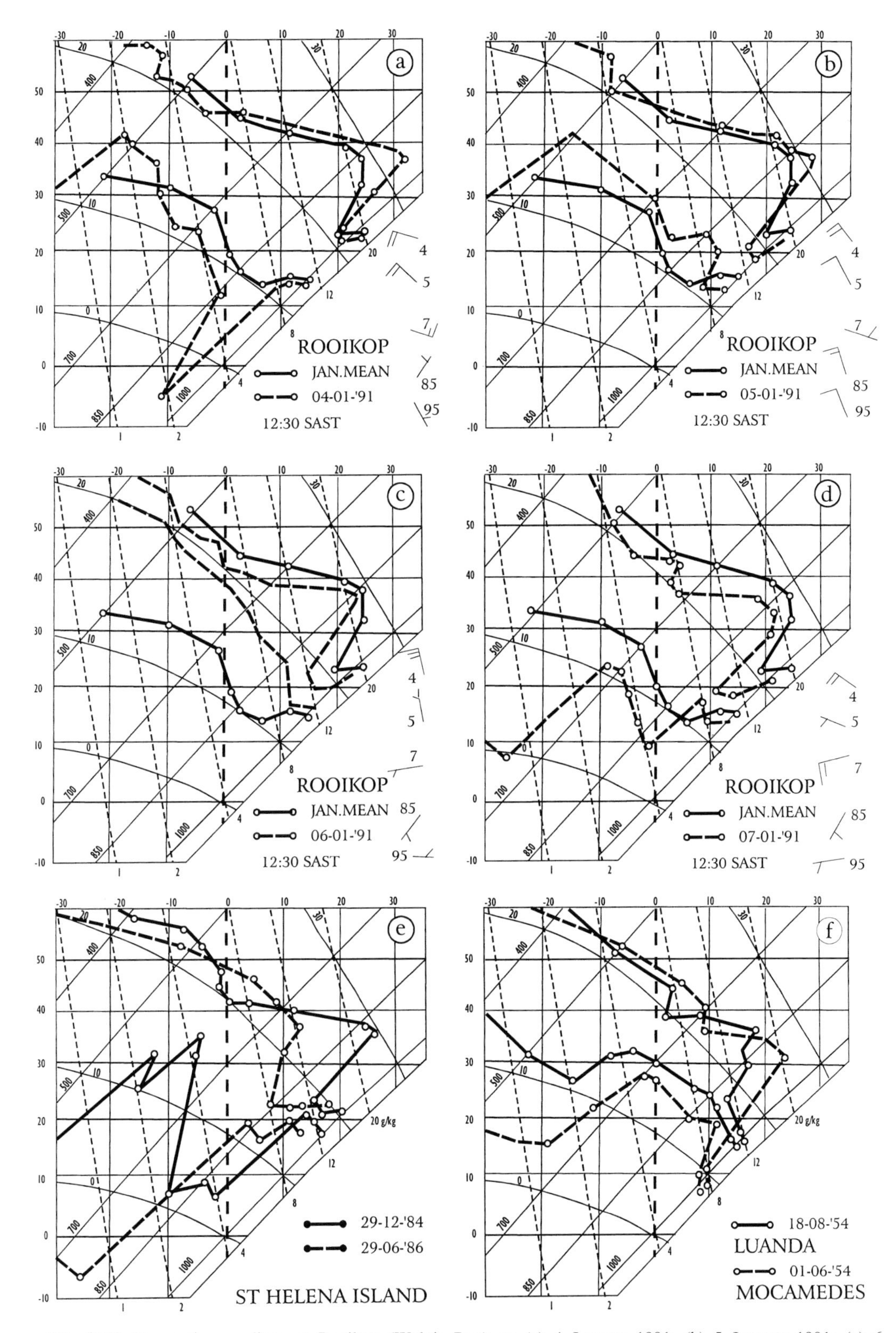

FIG. 3D.9. Upper-air soundings at Rooikop (Walvis Bay) on (a) 4 January 1991; (b) 5 January 1991; (c) 6 January 1991; and (d) 7 January 1991. Times are 1230 South African Standard Time (UTC + 2). (e) At St Helena Island on 29 December 1984 and 29 June 1996. (f) At Luanda on 18 August 1954 and at Mocamedes on 1 June 1954.

fluctuates between 900 and 850 hPa, but on rare occasions it lifts to 800 hPa or even higher when large-amplitude westerly troughs or cut-off lows extend exceptionally far northward. Farther south, it can be safely assumed that fluctuations that are characteristic of the subtropical and middle-latitude westerly belts occur with increasing frequency and intensity, such as those that regularly occur at Gough Island (40°S); however, this area falls outside the scope of the present review.

As already found by the *Meteor* expedition, the regime of the inversions over the tropical South Atlantic Ocean extends northward along the west coast, almost to the Equator. Examples of soundings at Mocamedes (15°S) and Luanda (8°S) are shown in Fig. 3D.9f. Both examples are from winter, when the continental southeasterly trade-wind current prevails on the plateau and spreads out overhead beyond the west coast. The existence of double inversions is evident at both stations, the lower one being the top of the moist maritime layer and the elevated one the top of the overrunning continental current. At Mocamedes, the lower inversion was 14°C on this occasion, which exceeds the mean inversion strength at Walvis Bay. The continental air is characterized by an almost dry-adiabatic lapse rate and a modest humidity. There are, of course, substantial interdiurnal fluctuations in the heights and magnitudes of the inversions. In summer, there is little distinction between the surface maritime layer and the superposed humid current (not necessarily of easterly origin) at Luanda, but conditions at Mocamedes remain closely similar to those at Walvis Bay.

b. The continent

The study of Taljaard (1955) drew attention to the marked stability of the atmosphere over most parts of southern Africa and the surrounding oceans in both summer and winter. Trewartha (1961) concluded that the comparatively low rainfall over much of southern Africa is due to this persistent or highly frequent condition. The Namib desert, which stretches from about 32° to 15°S, was initially ascribed solely to the coldness of the west coast upwelling regime (Benguela current) until the discovery of the phenomenal temperature inversion at a few hundred meters above the surface, which is naturally strengthened by the coldness of the sea. The situation is remarkably different over the east coast, however, but not sufficiently so to be conductive to abundant precipitation. Temperature inversions over the east coast, north of 25°S, and in the Madagascar region are commonly found at 2 to 4 km above sea level (as will be shown in the next subsection), and the air below these levels is not always saturated; hence, the air advecting onto the land in the easterlies cannot be described as humid throughout the year.

Nairobi, located at 2°S, is affected by two warm and moist trade-wind currents: the northeast monsoon in southern summer and the southeast monsoon in southern winter. The modification of these currents, however, is not sufficient to produce persistent instability and high rainfall to extensive parts of Kenya and Tanzania in the midsummer and midwinter months. Typical upper-air soundings at Nairobi during the southern midseasons (Fig 3D.10a) indicate moist air in the lowest 2–3 km, capped by inversions or stable layers and dry air aloft. The given soundings reflect early morning conditions, whereas higher surface temperatures in the afternoon promote convective instability and showery rainfall, particularly on high ground. Soundings at Dar es Salaam (on the coast at 7.5°S, not shown) support the contention that the air over the East African coast is mostly humid maritime tropical air, up to the stable discontinuity (at 2–4 km), capped by dry upper air. This structure is normally potentially unstable (wet-bulb potential temperature decreasing with height), and, therefore, when synoptic disturbances associated with low-level convergence affect the region, the soundings revert to deep saturation and a wet-adiabatic lapse rate. This occurs mainly during the rainy spells in the transitional seasons (southern spring and autumn).

Upper-air conditions at Kabwe (formerly Broken Hill), at 14°S, 28°E, (Fig. 3D.10b) can be taken as representative of those existing in summer over continental tropical southern Africa including Zaïre, Angola (except the coastal belt), Zambia, the southern part of Tanzania, and the northern part of Mozambique. The air over these regions is humid and potentially unstable with hardly any inversions or stable layers in summer. These are the regions of convergence along and astride the ITCZ and the IOCZ. The typical summer sounding at Kabwe is plotted in Fig. 3D.10b, along with another typical sounding at the station in winter. Winter is the dry season of tropical southern Africa, South Africa (excluding the south coast), Botswana, and Namibia, when marked temperature inversions are characteristic at 750 to 650 hPa. Early morning radiation inversions at the surface are surmounted by relatively warm dry air with a dry-adiabatic lapse rate up to the convective limit, above which extremely dry subsided air prevails. The humidity mixing ratio of the continental air is 3–5 g kg^{-1}, increasing gradually downwind to 7–10 g kg^{-1} over Angola and Zaïre.

Examples of upper-air soundings at Bloemfontein (29°S, 26°E) are relevant to explaining conditions pertaining to the South African plateau in summer and winter. Figure 3D.10c illustrates conditions on a wet and a dry day in summer when the flow in a deep layer over the station is northerly and southwesterly, respectively. On the wet day, the air mass is unstable and moist, with no inversion at any level. The dry day is characterized by a dry layer extending upward to nearly 500 hPa, in which the lapse rate is approximately dry adiabatic. These conditions alternate every couple of days, with many intermediate fluctuations naturally coming into play. The moist air masses

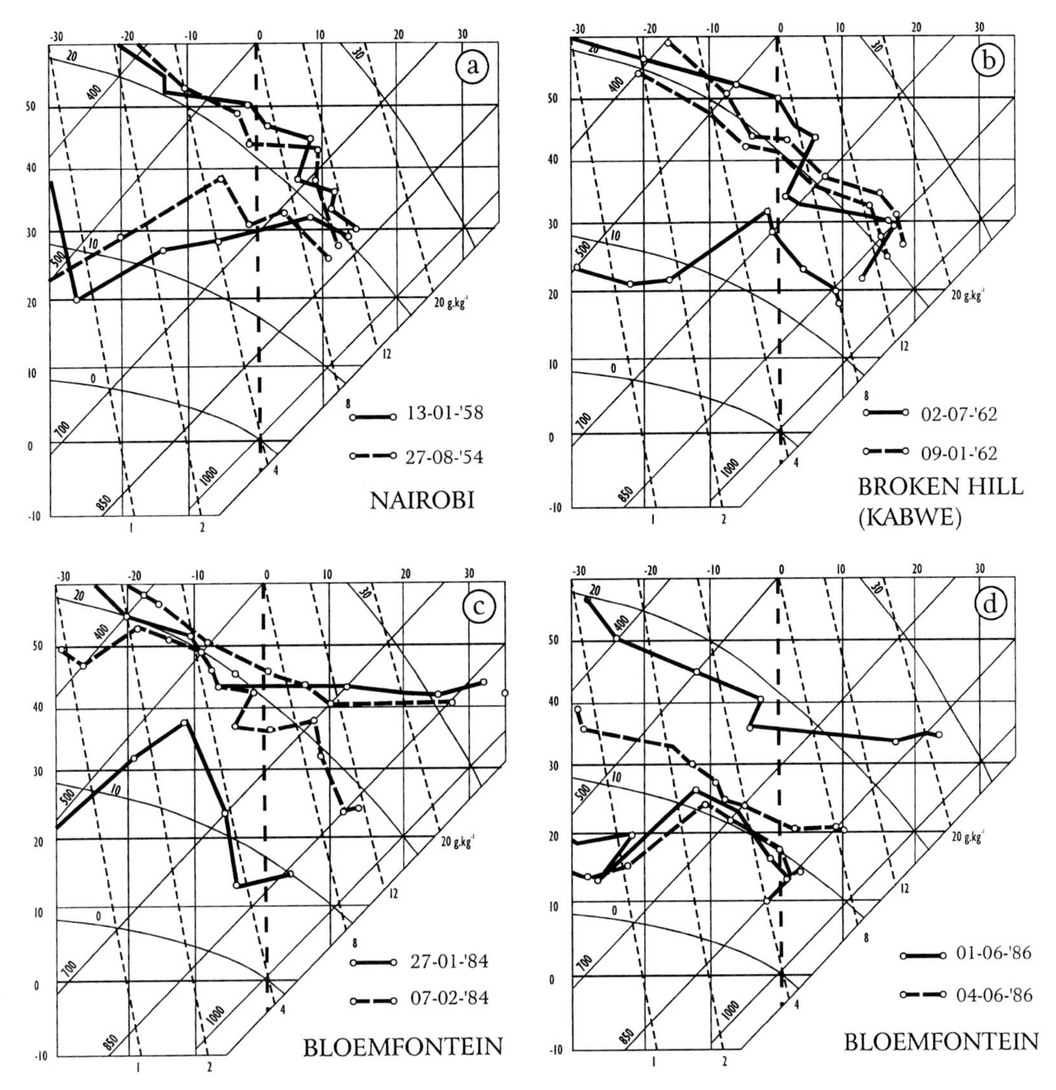

FIG. 3D.10. Soundings at (a) Nairobi, on 13 January 1958 and 27 August 1954 (early morning soundings); (b) Kabwe (formerly Broken Hill), on 2 July 1962 (early morning soundings); (c) Bloemfontein, on 27 January 1984, 7 February 1984, (d) 1 June 1986, and 4 June 1986 (midday soundings).

originate in the Indian Ocean or they advect from the humid tropical air circulating over the neighboring territories to the north. The dry air arrives from the Atlantic Ocean. In winter (Fig. 3D.10d), strong temperature fluctuations occur when relatively warm continental air flowing from north or northwest is replaced by cold postfrontal subtropical or sub-Antarctic air. Both the warm and the cold air masses are usually capped by inversions at 850 to 600 hPa, but the incidence of sharp westerly troughs or cut-off lows induce deeper instability, as well as during the rare occurrences of appreciable rain or snow.

c. The east coast and adjacent Indian Ocean

The considerably higher temperatures of the Indian Ocean, and the predominantly easterly onshore component of the flow of the air masses between the Equator and 30°S, create vastly more favorable conditions for precipitation in this region than over the west coast and the Atlantic Ocean. Not only is the depth of the maritime air below the common inversion or stable transition to the upper dry air more elevated than in the west, but the absolute humidity is also much higher (except for that small portion of the tropical west coast north of about 12°S). Representative examples of radiosonde soundings [aircraft soundings at Diego Suarez (now Antsiranana) at the northern tip of Madagascar and Tulear on the southwest coast of the island], are shown in Fig. 3D.11a,b. Antanarivo (Fig. 3D.11a) is located almost 1200 m above sea level, and, thus, the very moist surface layer of the easterly trades is not reflected in the soundings. In winter, the

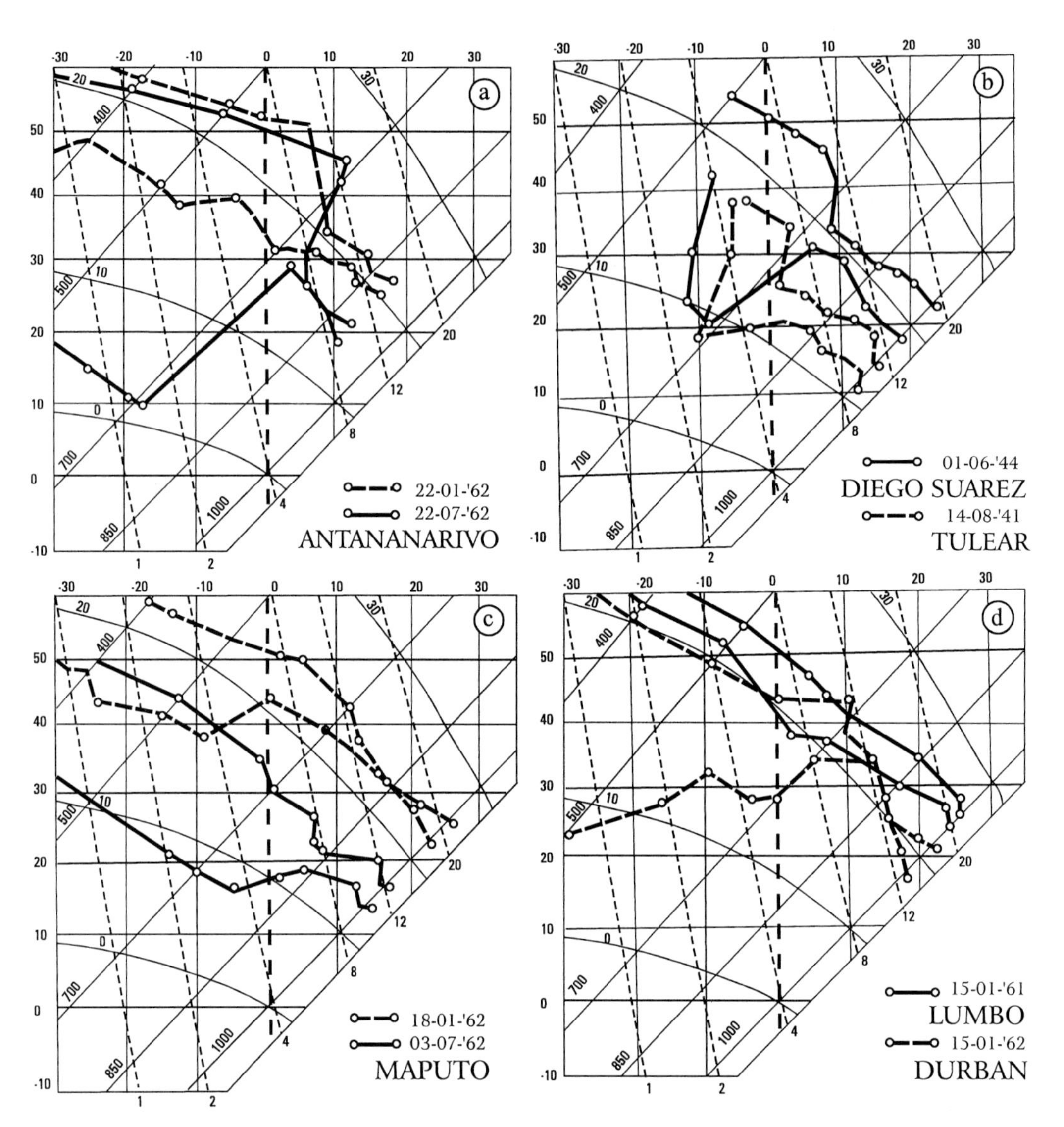

FIG. 3D.11. Soundings at (a) Antanarivo on 22 July 1962 and 22 January 1962; (b) Diego Suarez on 1 June 1944 and Tulear on 14 August 1941 (the latter two are aircraft soundings); (c) Maputo on 18 January 1962 and 3 July 1962; (d) Lumbo on 15 January 1961 and Durban on 15 January 1962.

air below the usually pronounced inversion at 3 km above sea level cannot be labeled humid, but the moisture is sufficient to cause ample precipitation in the onshore flow over the east coast of the island. Soundings at Mauritius are similar to those over Madagascar. In summer, the soundings fluctuate between being stable and unstable when the air becomes saturated and the lapse-rate approximates to the wet adiabatic (e.g., when tropical cyclones occur). The aircraft soundings at Diego Suarez and Tulear (Fig. 3D.11b) are in accord with those at Antanarivo, except for the fictitiously high dewpoints due to the freezing of the wet-bulb thermometers above about 650 hPa.

A summer and a winter sounding at Maputo (formerly Lourenco Marques) shows fluctuations between summer and winter conditions at 25°S on the continental east coast (Fig. 3D.11c). Fluctuations at this subtropical latitude are pronounced due to the westerly troughs and cold fronts that affect the RSA and southernmost Mozambique. The air becomes humid and unstable in summer preceding upper troughs, which occasionally proceed from west to east across South Africa. Lumbo is located at 15°S on the coast of Mozambique, very close to the position where the ITCZ enters the land. The few soundings available for this station in summer all reveal a practically saturated column of potentially unstable air, as shown in Fig. 3D.11d. Winter soundings are not available for Lumbo, but it is practically certain that the moist unstable layer extends up to the stable boundary, with dry air above 2 to 4 km. Durban, at 30°S on the South African east coast, shows a conditionally unstable

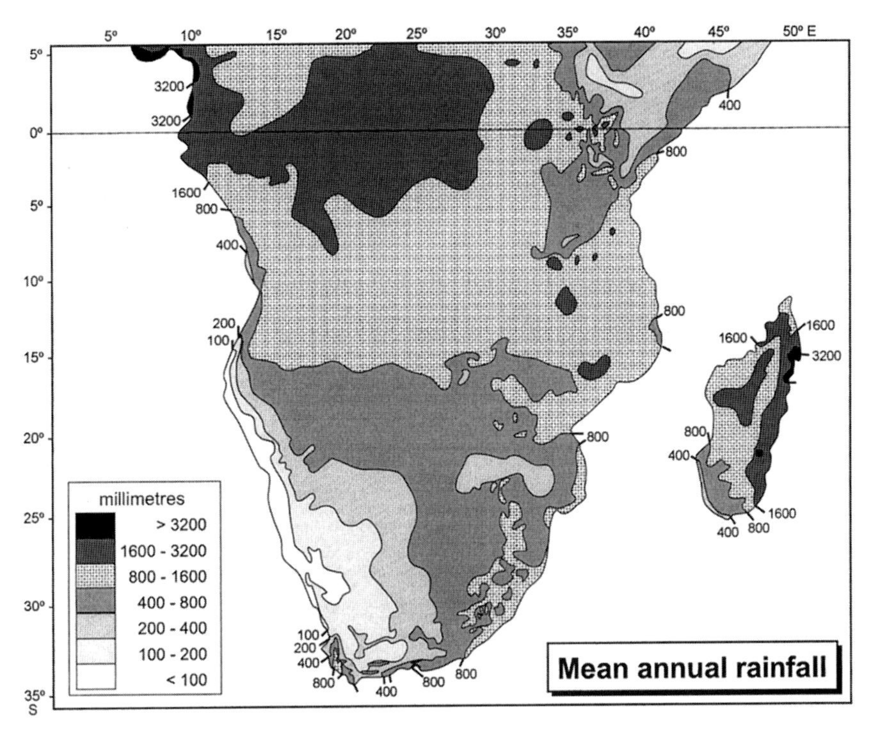

FIG. 3D.12. Annual mean rainfall over tropical and southern Africa (adapted from Thompson 1965).

layer up to 900 hPa, a saturated layer up to 800 hPa, capped by an inversion extending to 700 hPa.

3D.5. Rainfall climatology

The distribution of mean annual and midseasonal (January, April, July, and October) rainfall for southern Africa according to Thompson (1965) are given in Figs. 3D.12 and 3D.13. Several regional rainfall climatologies and studies on the many aspects of rainfall exist for the region. Among these are Jackson (1961), Schulze (1965), Landsberg (1972), Van Rooy (1972, 1980), Tyson et al. (1975), Torrance (1981), Leroux (1983), Tyson (1986), and Nicholson et al. (1988). Some of these have a bearing on climate variability and will be referred to again in section 3D.7. For the present purpose, Thompson's less-detailed charts will suffice.

The prominent features of the annual rainfall (Fig. 3D.12) are 1) the large area of heavy precipitation, in excess of 1600 mm over the central Zaïre Basin, linked to heavy rain over Gabon and Congo; 2) similar heavy rain over the east coast and the high ground of northern Madagascar, 3) the contrasting low rainfall (200–800 mm) over equatorial East Africa, except over patches of high ground; 4) the striking east–west gradient of rainfall across South Africa from more than 800 mm in the east to less than 100 mm in the Namib desert, which stretches for 2000 km along the west coast from about 32°S to almost 10°S; and 5) the relatively dry Limpopo River valley between South Africa and Zimbabwe.

The seasonal march of the rainfall is discussed in great detail in Nicholson et al. (1988) for all of southern Africa; Schulze (1965) and Van Rooy (1972) for South Africa; Torrance (1981) for Zimbabwe; and various other regional publications. The following brief remarks concern only the rainfall in the four midseason months January, April, July, and October (Fig. 3D.13).

Figure 3D.13a shows that in January the rainfall abruptly decreases northward in the zone between the Equator and 5°N, but northernmost Kenya and the adjacent Somaliland also receive little rain. The zone with average rainfall exceeding 200 mm stretches across the continent between latitudes 10° and 15°S (excluding coastal Angola) and probably continues right across the Mozambique Channel to Madagascar, where the northwestern and southeastern littorals receive over 400 mm. The extensive arid-to-semiarid region of Namibia, Botswana, the western RSA, and the Limpopo River valley is striking.

Figure 3D.13b shows the marked withdrawal of the summer rains from the territories south of about 12°S in autumn. This decrease in the rainfall sets in during March and is most rapid in April/May. It is notable that the rainfall peaks in most parts of the equatorial belt between about 10°S and 5°N in April; this is ascribed to intermittent converging air currents from the two hemispheres as the ITCZ shifts northward. Rainfall is modest over the coastal belts and the eastern high ground of the RSA. The east coast of Madagascar still receives frequent heavy rain.

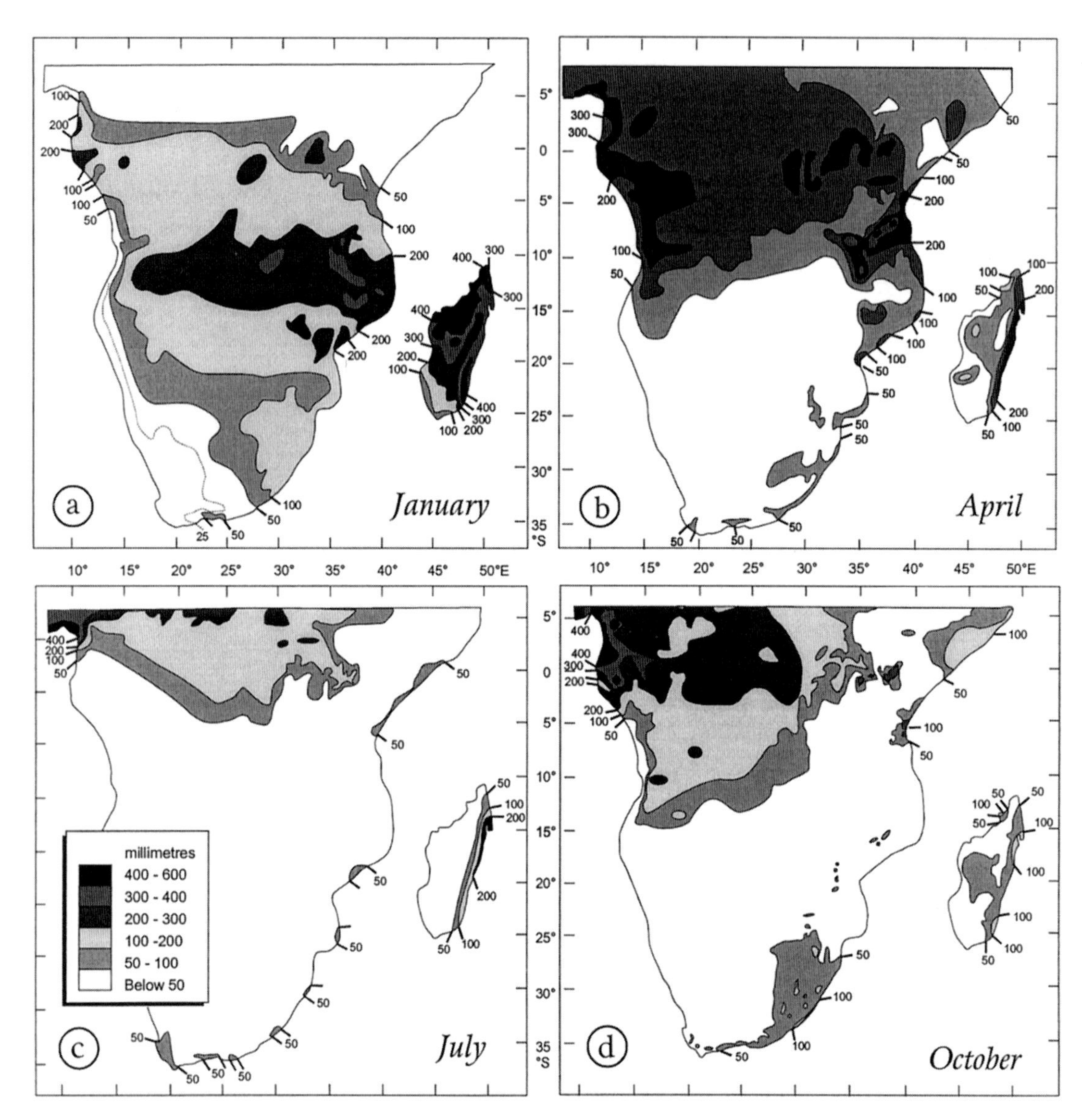

FIG. 3D.13. Mean rainfall over southern Africa during the midseason months January, April, July, and October (adapted from Thompson 1965).

In July (Fig. 3D.13c), significant rainfall is limited to the northern fringes of Zaïre, the inland mountains of Kenya and Uganda, the east coast of Madagascar, and the coastal belts of the southern and southwestern RSA. October sees the return of moderate to heavy rain (Fig. 3D.13d) over the northern parts of Angola and Zaïre, but not yet to East Africa. Whereas generally light rains also return to the eastern provinces of the RSA, Madagascar rains are at a low ebb.

3D.6. Circulation systems over tropical and subtropical southern Africa

a. Tropical southern Africa

Meteorologists of the British school in East Africa and the Rhodesias (now Zimbabwe and Zambia), and the French school in equatorial western Africa, attempted during 1940–70 (which might be called the golden period of African tropical meteorology) to identify the circulation systems favoring rainfall in the region. A lucid account of their findings up to 1960 is given in Trewartha (1961). Further information is contained in Bargman (1962), Thompson (1957, 1965), Taljaard (1972), Torrance (1981), and Leroux (1983), with the latter being a comprehensive treatise on the weather and climate of tropical Africa. Torrance (1981) compiled a detailed description of the climate of Zimbabwe, but this publication also deals briefly with the synoptic systems and processes responsible for the weather of the region. It includes a bibliography of the publications of the Rhodesian Meteorological Service before 1981—thus also of the

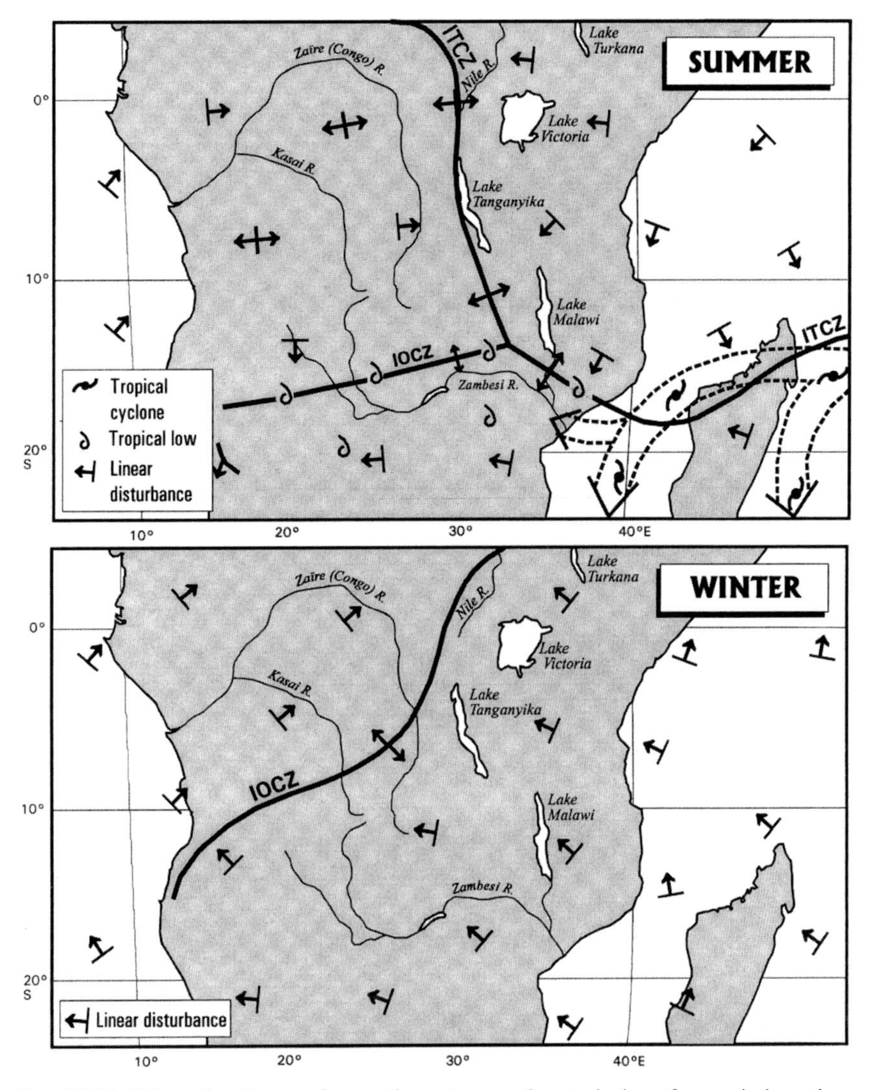

FIG. 3D.14. Schematic picture of synoptic systems and perturbations for tropical southern Africa and surrounding oceans for (a) summer and (b) winter.

rain- and drought-producing systems. Remarks on the weather systems of the African territories whose climates are described in Landsberg (1972) add to a comprehension of the meteorology of the subcontinent.

Figure 3D.14 is a simplified schematic of the known and, in some cases, surmised large-scale conditions for tropical southern Africa and surrounding oceans in summer and winter. In summer, the indicated positions of the ITCZ and the IOCZ troughs are the mean locations of the cores of the normally broad zones where convergence and active weather dominate. The zones are displaced laterally from time to time and are even severed occasionally when particularly strong surges occur from one side to the other. East African meteorologists all agree that surges form the west transport deep humid air from the Zaïre basin across the continental divide, mainly over Tanzania and southern Kenya, causing particularly wet spells. This occurs when tropical cyclones penetrate to the northern part of the Mozambique Channel or when they proceed westward as tropical lows or "easterly waves," across Mozambique, Malawi, Zimbabwe, and Zambia. Occasionally, the ITCZ/IOCZ is displaced southward by the northerly flow east of the tropical lows, which normally travel westward and from which troughs extend southward across Zimbabwe and Botswana toward South Africa. An example of the latter will be discussed later.

Three types of low pressure systems (tropical cyclones, tropical lows, and heat lows) generally dominate at low latitudes. The tropical cyclones that enter the Mozambique Channel mostly follow tracks southward well off the coast, but a few approach close enough, or even enter the land, to cause strong wind and heavy rain. Once every year or two, a cyclone will continue tracking inland, but then it rapidly weakens to become a tropical low on the high ground of Zimbabwe, Malawi, and Zambia. One such system was described by Kreft (1953). Once they have advanced to Zimbabwe, Zambia, or Botswana, they cannot be distinguished from the tropical systems that develop on the land along the IOCZ/ITCZ and can be recognized at the surface as lows with 850-hPa geopotential contours 20–30 gpm lower than at 200–300 km around their centers.

Tropical lows can be clearly recognized on loops of half-hourly geostationary satellite photos, with the low-level circulation (up to about 600 hPa) being cyclonic and that in the upper troposphere anticyclonic, thus implying that the systems are cool core in the lowest 3–4 km and warm core higher up. The majority of them slowly move westward but seldom reach the west coast of Africa, while a very few have been found to move eastward. Weak to moderate tropical lows occur much more frequently in the Mozambique Channel and to the northeast of Madagascar along the ITCZ.

The third type of low is the heat low, which frequents the meeting ground of the northeast and southwest monsoons along the mountain chains and blocks between Zaïre and East Africa. They generally are shallow systems caused by relatively high temperatures in the lowest few kilometers.

Trewartha (1961) used the views of previous authors to describe the weather systems of the Zaïre basin and East Africa in summer. There is little doubt that the humid air masses of Zaïre and Angola are mainly derived from the Atlantic Ocean via the southwest monsoon. Once this air passes over the warm sea in the Gulf of Guinea and then proceeds eastward along or immediately north of the Equator, it is transformed to a humid unstable mass in which convection and thorough mixing is a daily event. It is natural that the air arriving overhead at 800–600 hPa from East Africa (including Ethiopia) in the prevailing upper easterly to northerly flow adds to the total moisture over Zaïre. A further important process that is mostly ignored is humidification through evapotranspiration (i.e., evaporation from falling rain, wet soil, open water, and transpiration from vegetation). This process might be called recycling of water vapor that initially arrives from the sea and is particularly important in regions where low-level convergence dominates. A debated point has been the depth of the southwest monsoon current, which is generally considered as being mostly less than 1500 m. The radiosonde soundings at Luanda in summer, however, have more often than not shown the landward flux of humid air from the Gulf of Guinea to be up to at least the 700-hPa level.

As already stated, Fig. 3D.14 is an attempt to indicate the recognized and, in some cases, presumed directions of propagation of disturbances associated with the large-scale circulation. These might be in the form of line squalls, shear lines, or elongated zones with thicker (greater vertical extent) cloud than normal and with associated precipitation. They might originate from dissipating cold fronts that penetrate across the subtropical high pressure belt or simply from pressure and wind pulses caused by the waxing and waning of subtropical anticyclones. Changing middle and upper-tropospheric circulation systems are also likely to induce convergence or divergence at low levels, thereby affecting the elevation of the almost omnipresent inversions in the trade-wind/monsoon systems. Where the inversion is raised sufficiently, instability and precipitation might ensue, such that line phenomena and vertical circulations are created. The areas of the disturbances need not always be perpendicular to the wind direction, but they ought to propagate mainly downwind. The latter remarks apply particularly to the tropical Indian Ocean, where the inversion is characteristically 2 to 4 km above sea level and the subinversion layer is warm and humid.

The following description and remarks with regard to tropical systems are based mainly on the expositions of Forsdyke (1949), Taljaard (1955), Thompson (1957, 1965), Trewartha (1961), and Johnson and Mörth (1962). Trewartha describes the accepted views, as well as some opposing beliefs, of meteorologists who practiced in Zaïre and East Africa before 1960. One group of disturbances definitely approach from the Atlantic Ocean, and they propagate eastward overland between about 10°S and 5°N. The radiosonde soundings at Saint Helena show oscillations in the height and intensity of the trade-wind inversion, as well as in the temperature and humidity of the maritime layer. All meteorologists agree on the existence of these pulses from the Atlantic Ocean, but they disagree, for instance, with Bultot (1952), who maintained that most of the rain over Zaïre is associated with eastward-propagating perturbations. Rainteau (1954) and Jeandidier (1956) favor the role of westward-propagating line squalls or zones of enhanced rainfall. They consider the predominating easterly to northerly winds at 700 hPa and higher to act as steering currents for the deep cumulonimbi, particularly those that develop far inland or along the high continental divide. Interaction between the surface westerlies and the upper northeasterlies result in the formation of line squalls or north–south elongated rain zones. Although the air mass over Zaïre is humid and unstable on practically all summer days, random convection and thundershowers are considered to contribute less than half of the total precipitation.

Considerable support has been gained for the concepts of Johnson and Mörth (1962) that certain pressure and wind patterns in the lower half of the troposphere over the equatorial zone are favorable or unfavorable for rain. These are the duct, bridge, and drift patterns. The duct is defined as a zone of low pressure and easterly flow along the Equator, flanked by anticyclones to both sides. It is deduced that the area of confluence (and convergence) at the entrance of the easterly current is favorable for rain. The bridge is a zone of relatively high pressure and westerly flow along the Equator, flanked by low pressure systems. The western entrance to this type of flow is most favorable for rain. The drift pattern develops when pressure is relatively high on one side of the Equator and low on the other. The flow is easterly on the high pressure side and then crosses the Equator to become westerly on the other side, where convergence, lifting, and rainfall ensues. These arguments are theoretically based, and practical experience seems to support them, but the actual pressure and flow patterns are, as a rule, more complicated.

Johnson and Mörth (1962) also studied the daily rainfall distributions over East Africa, and they concluded that there is little or no evidence for elongated narrow perturbations that move predominantly in any particular direction over the region. The rain is decidedly associated with elongated but broad rain zones, however, which are displaced slowly in some, but not always the same, direction. According to the meager upper-air information, these large patches of enhanced precipitation coincide with "humidity pools," in which the common temperature inversion is nonexistent or weak and present at greater heights than normal. The associated pressure and wind fields aloft, however, could not be determined with confidence. These authors and Thompson (1957, 1965) are adamant that random convective storms contribute only a small fraction of East African rainfall. Nevertheless, it is the high ground that receives rain on 50% to 80% of the days of the rainiest months (Thompson 1965; Leroux 1983).

In winter, the large-scale circulation pattern that controls the synoptic systems is much less complicated. Half of the plateau is under the influence of stable divergent easterly trade winds, in which there naturally are pulses in speed and direction, but they are inconsequential in regard to rain. These pulses emanate from the subtropical anticyclones that propagate eastward along about 30°S across South Africa, causing the injection of shallow cool to cold maritime air along the eastern seaboard. Meager orographic precipitation occurs over the high ground of Zimbabwe, northern Mozambique, and Tanzania. Evidence exists (Forsdyke 1949) that these surges affect East Africa, where they produce a modest amount of rain along the coast of Kenya. They are reinforced by disturbances in the southeast monsoon that bring appreciable rain to the east coast of Madagascar. Since the southeast monsoon has a long trajectory over warm waters, it is warm and moist below the normal subsidence inversion at 2–4 km, such that the portion of it that enters Kenya contributes largely to the heavy rain over the high ground of northwestern Kenya and Uganda along and north of the Equator. The IOCZ boundary, which crosses northern Angola and central Zaïre in winter, is not active weatherwise, because the southwest monsoon to the north of it is a shallow moist current except where it approaches and crosses the Equator, where marked convergence and precipitation sets in.

Further evidence for westward-moving disturbances, other than tropical cyclones, in the 10°–20°S latitude band of the southwestern Indian Ocean during summer is contained in Jury et al. (1990), who used satellite cloud mosaics for the period 1972–84 to determine the daily displacement of deep convective cloud masses. They also employed OLR anomaly patterns for identifying and verifying the existence and movement of the areas of deep convective activity. Other authors who described the westward propagation of disturbances in this oceanic region are Gichuiya (1974) and Okoola (1989).

Midway points on either side of January and July, the months of April and October can be taken as the months of most rapid transition between summer and winter conditions in southern Africa. The IOCZ/ITCZ trough system becomes diffuse in April and is rapidly displaced northward, whereas the northeast monsoon still lingers on, though weakening. A broad zone of predominantly easterly flow comes into being, and this humid easterly flow proceeds inland, causing the heaviest rainfall of the year over Tanzania and the tropical east coast. Rainfall also increases sharply over Kenya and Uganda, however, along the Equator. The far southward extension of heavy rain over Angola in April is also notable, the cause of which has not been described. Similarly, the moderate rainfall over inland Angola in October, in contrast to the lack of appreciable rain over inland East Africa, is striking yet unexplained. Thompson (1965) suggests that the rapid southward spread of the rain over Angola in spring may be related to the strong heating and trough development that occurs over Zambia, Zimbabwe, and Botswana in October.

b. Subtropical southern Africa

For discussing the significant synoptic circulation systems and perturbations that influence the subtropics, the subcontinent and adjacent seas are divided into tropical and subtropical regions, taking the 15°–20°S belt as the dividing zone. Considering the circulation systems in the subtropical zone, extended wet and dry months or periods are described first. Finally, rain-producing synoptic-scale systems that are linked to tropical systems or have their moisture source in the Tropics are described. The considerable range of circulation patterns responsible for rain and drought in South Africa, some of which are described here, have been extensively described by Taljaard (1981, 1985, 1987, 1989, 1996b), Taljaard and Steyn (1988), Triegaardt and Kits (1963), Triegaardt and Landman (1995), Tyson

(1981, 1984, 1986), Harrison (1984; 1986, unpublished M.S. thesis), Schulze (1984), and others.

1) CIRCULATION PATTERNS DURING WET SUMMER MONTHS

For periods of a month or more, wet conditions over the central interior of subtropical southern Africa are associated with positive surface pressure anomalies to the southwest, south, or southeast of the country. The SLP is usually highest along 40°S but the anomalies are strongest along approximately 45°S. At the same time, upper-air geopotential heights are below normal over the western or central RSA. The presence of slow-moving anticyclones along 40°S leads to the influx of moist maritime air from the Indian Ocean to the eastern parts of the region. At 850 hPa, a deep trough or low, linked to the IOCZ, extends southward over the central plateau. On the eastern side of this system, moist tropical air, sometimes originating from as far north as the Zaïre basin or Tanzania, is advected southward to 30°S. In the upper air, generally above the 500-hPa level, strong synoptic-scale divergence is present east of a slow-moving or quasi-stationary upper trough. The simultaneous occurrence of these features, their respective phases, and the influence of the topography, as well as their duration, results in above-normal rainfall mostly over the central and eastern parts of the region. Figure 3D.15a,b (from Taljaard 1987) illustrates the surface pressure anomalies and the 500-hPa geopotential anomalies for December 1975. During that month, large areas of subtropical Africa received more than double the long-term average rainfall. Large parts of the southern Kalahari recorded more than five times the long-term average rainfall for December. Note the large positive MSLP and 500-hPa anomalies near and south of 40°S. At 500 hPa, large negative anomalies over the region indicate the frequent occurrence of troughs or cut-off low pressure systems. Van Heerden et al. (1988) linked the occurrence of this type of circulation system to the cold phase of ENSO (La Niña). Harrison (1983), Lindesay (1988, unpublished Ph.D. thesis), and others also described the relationship between phases of ENSO and the rainfall over southern Africa on a monthly to seasonal scale.

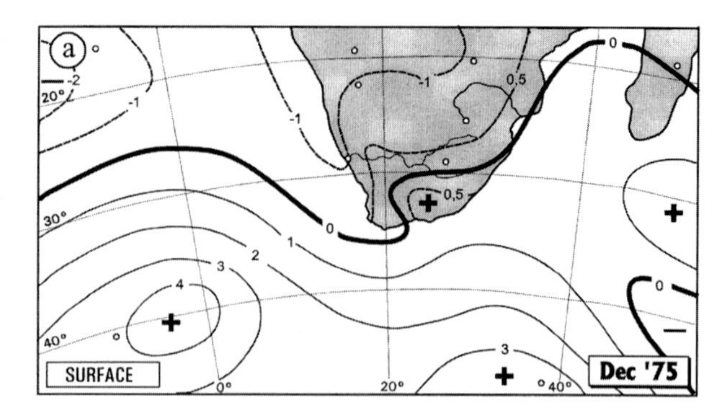

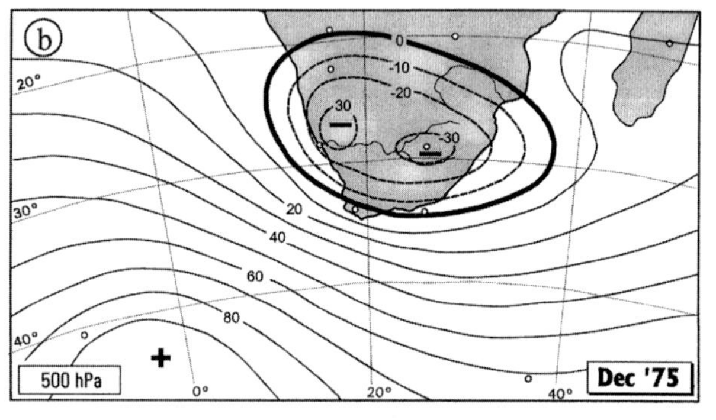

FIG. 3D.15. Anomalies of (a) surface pressure (hPa) and (b) 500-hPa geopotential height (gpm) during the very wet December 1975 (after Taljaard 1987).

2) CIRCULATION PATTERNS DURING DRY SUMMER MONTHS

A notable feature of the mean monthly circulation during dry summer months is the northward displacement of pressure systems. An example is shown for February 1983 (Fig. 3D.16). The surface westerlies increase and extend northward to the southern and southwestern coasts of South Africa. The midlatitude upper-tropospheric jet stream is displaced to a position north of 40°S, while a closed upper-tropospheric anticyclone often predominates over the subtropical regions for several weeks. The upper anticyclonic circulation often extends down to below 700 hPa, while a shallow heat low appears over the central and southern interior. Large-scale subsidence prevails over most of the subcontinent and the presence of dry subsiding air over the region is the primary cause for the lack of rainfall. High surface temperatures, low humidity, and cloudless conditions accompany the subsidence. At the surface, the Atlantic anticyclone, displaced well to the north of 30°S, tends to invade the country from the west and, in conjunction with fast-moving cyclones off the south coast, causes dry westerlies to prevail over the region south of 30°S. In general, this circulation pattern is typical of winter. Cyclonic systems affect the southwestern coastal regions, bringing anomalous summer rainfall to this Mediterranean climatic region.

Taljaard (1989) described the circulation patterns during the dry summer of 1982/83, which coincided with the severe warm phase of ENSO. Note the SLP anomalies for February 1983 (Fig. 3D.16), indicating the Atlantic anticyclone that invaded the land from the west. Also shown is the 300-hPa height anomalies, which illustrate the ridge extending eastward from the Atlantic Ocean across the land north of 30°S. The rainfall map for February 1983 (Fig. 3D.17) shows the severe deficiency of rainfall over most of the subtropical continent. At the same time, several unseasonal cyclonic (frontal) systems reached the extreme southwest-

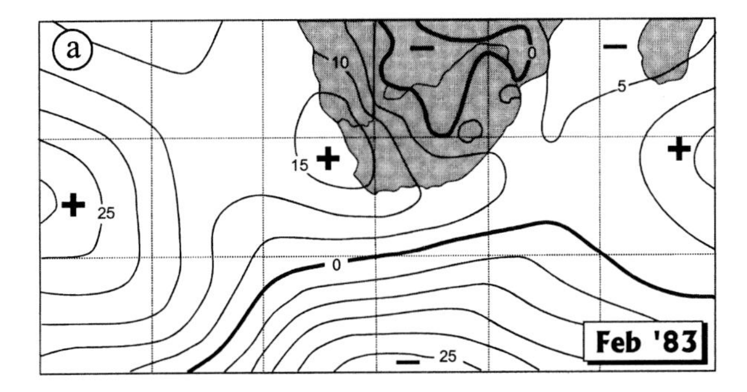

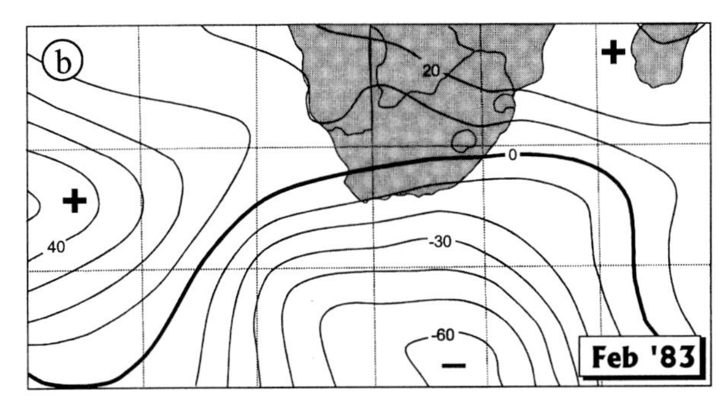

FIG. 3D.16. Anomalies of (a) 1000-hPa and (b) 300-hPa geopotential contours (gpm) during the very dry February 1983 (after Taljaard 1989).

ern coastal belt, resulting in rainfall of more than 300% of the long-term mean. Lindesay (1988, unpublished Ph.D. thesis), Harrison (1983), and Van Heerden et al. (1988) described the association of the anomalous circulation patterns prevailing during wet and dry months and the phase of the SO. The reversal of anomalous geopotential heights follows the phase change of the SO, with the dry circulation patterns described here being common during the warm phase of ENSO.

3) CIRCULATION PATTERNS DURING WET AND DRY SPELLS LASTING ONE TO SEVERAL WEEKS

Triegaardt and Landman (1995), expanding on earlier studies by South African researchers, described the circulation prevailing during wet and dry periods lasting more than a week. Their results, which are based on ECMWF products, identify these patterns in the analyses of the general circulation model (GCM) output. Rainfall predictions from climate models and GCMs are generally not good for periods as short as one week. Identification of persistent anomalous circulation patterns in the GCM output may be far more useful in the analysis of climate scenarios and medium-term rainfall prediction. Jordaan (1996, unpublished Ph.D. thesis) identified the presence of these waves in 30-day ensemble predictions from a GCM and was able to determine the most probable longitudinal position, as well as the probability of occurrence.

For the central and eastern plateau of South Africa, Triegaardt and Landman (1995) identified the following features of wet and dry periods. 1) Wet periods

- A long-wave upper ridge (zonal wavenumber 3 or 4) is present southwest, south, or southeast of South Africa;
- the strength of the upper westerlies decreases markedly at 30°–40°S;
- positive surface pressure anomalies are present southwest, south, or southeast of South Africa;
- a cut-off low over South Africa often accompanies these features; and
- when a large amplitude upper trough is present over the west coast, divergent upper circulation downwind of this trough causes large-scale upward motion and rainfall, provided the surface circulation is favorable for the advection of moisture.

2) Dry periods

- A long-wave upper trough is present southwest, south, or southeast of South Africa;
- stronger-than-normal upper westerlies and a northward shift of the jet stream occurs over South Africa;
- a well-developed surface and/or upper trough is present south of South Africa;
- a closed upper-air high associated with subsiding air over southern Africa suppresses rainfall in the face of other favorable circulation patterns or sufficient moisture; and
- a deep tropical low or tropical cyclone is present over the Mozambique Channel.

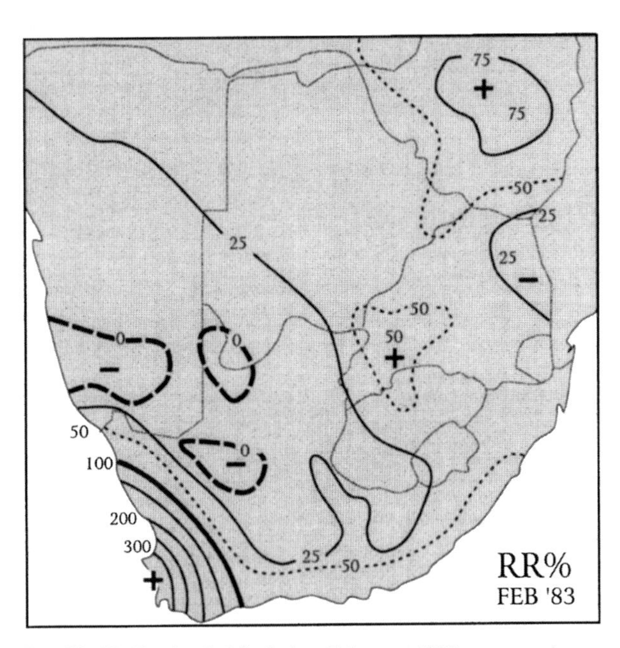

FIG. 3D.17. Total rainfall during February 1983, expressed as a percentage of the monthly mean (after Taljaard 1989).

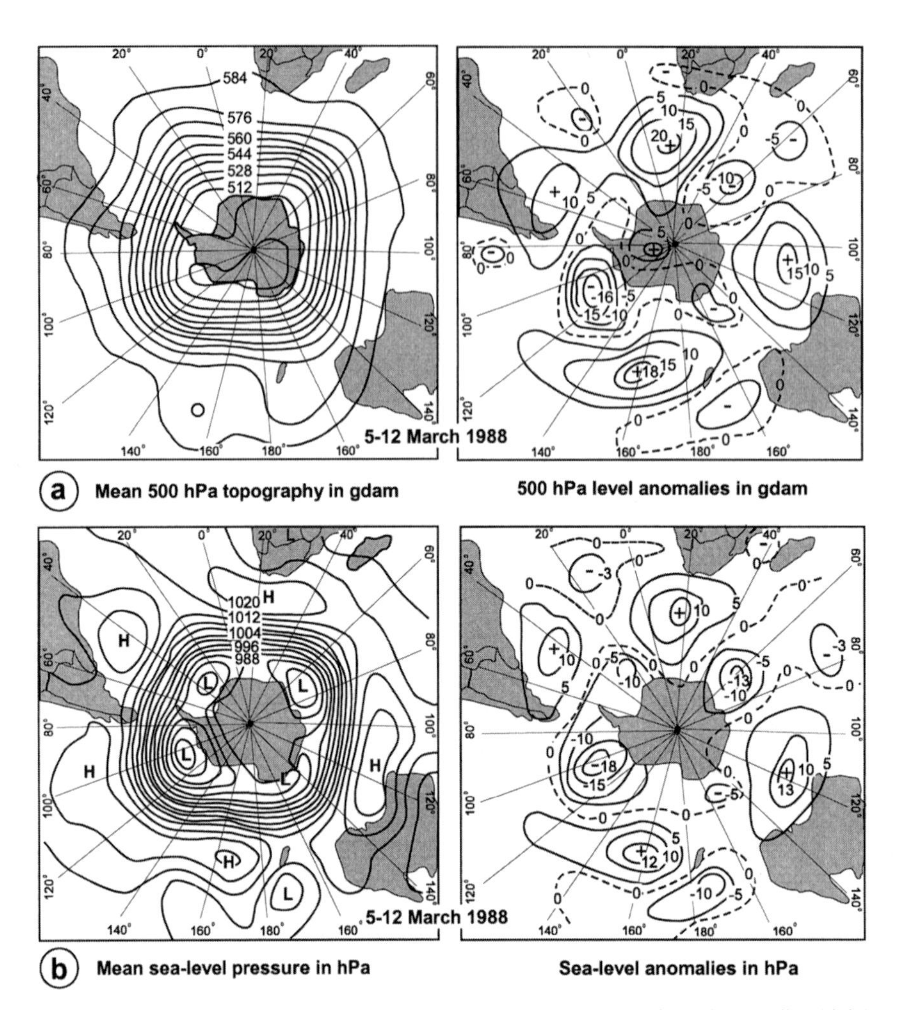

FIG. 3D.18. (a) Mean hemispheric 500-hPa geopotential contours (left) and anomalies (right) in geopotential dam for the wet period 5–12 March 1988; and (b) mean hemispheric sea level pressure (left) and anomalies (right) in hPa for the same period (after Triegaardt and Landman 1995).

Figure 3D.18a,b (from Triegaardt and Landman 1995) shows the atmospheric circulation and its anomalies at 500 hPa and at the surface during the wet spell of 5–12 March 1988. During this period, rainfall varied between 63 and 112 mm per station per province, and several stations reported total falls in excess of 200 mm. At the 500-hPa level, circulation over the higher latitudes was dominated by a four-wave pattern. An upper trough was present over the west coast, and the 500-hPa anomalies clearly indicate the presence of a strong upper ridge south of the land. The SLP anomalies also indicate the southward extension of the anomalously high pressure. In contrast, during 24–28 January 1983, more than 80% of South African rainfall stations west of the escarpment and 61% of the stations over Kwazulu-Natal reported no rain. Those stations reporting rainfall generally had well below 7 mm in total for this period. Figure 3D.19a,b depicts the 500-hPa and SLP fields and anomalies during this dry spell. A four-wave pattern clearly dominated the upper circulation, with the anomaly fields indicating large negative anomalies south of the land. A closed 500-hPa high was present over South Africa. At sea level, a pronounced trough was situated south of the land, pressure being 10 hPa below the January long-term average at 50°S. Barclay (1992, unpublished M.S. thesis) studied the difference between westerly troughs producing wet and dry conditions over South Africa and concluded that wet troughs must have large amplitudes and extend well into the subtropics. Tyson (1986) also showed that the average longitudinal position of large amplitude westerly troughs over southern Africa can be linked to extended wet or dry spells over southern Africa.

Taljaard and Steyn (1991) amended and expanded the statement by Tyson (1986): "The development of above normal pressures over the Atlantic Ocean to the southwest of Cape Town in the vicinity of Gough Island, over time scales of months to decades, produces conditions that are favourable for increased

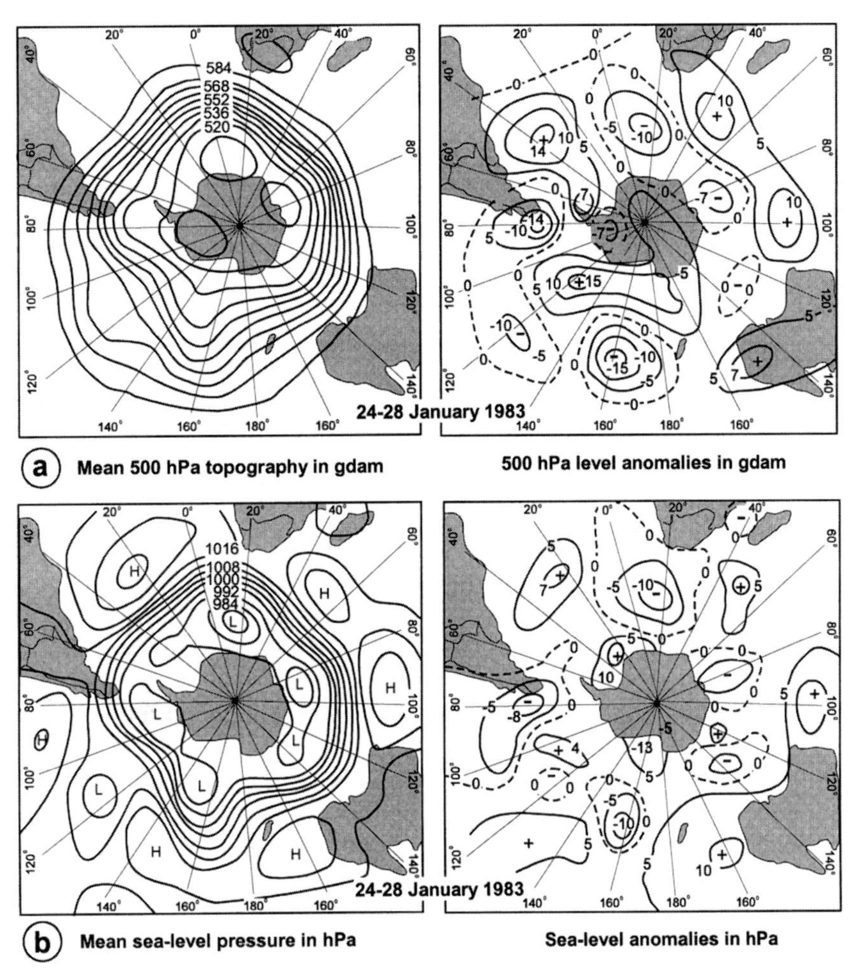

FIG. 3D.19. Same as Fig. 3D.18 but for the dry period 24–28 January 1983 (after Triegaardt and Landman 1995).

rainfall over the summer rainfall regions of southern Africa." They found that, for wet spells of a few days (as well as for anomalously wet months, seasons, and even years), the subtropical high pressure belt is displaced southward through several degrees of latitude, but with the cores of greatest positive anomaly south instead of southwest of the land. This situation is achieved by a high frequency of ridging high pressure systems from the South Atlantic Ocean propagating well south of the normal axis of the subtropical high pressure belt. At the same time, upper troughs or cut-off low pressure systems form over the western parts of southern Africa, extending down to the 850-hPa level.

4) RAIN-PRODUCING CIRCULATION SYSTEMS AFFECTING SUBTROPICAL AFRICA

Several classification schemes have been developed for systems that produce rain over South Africa (e.g., Harrison 1986, unpublished M.S. thesis); Tyson 1986; Diab et al. 1991). More recently, Taljaard (1996b) proposed a comprehensive classification for both summer and winter based on surface pressure systems and upper-tropospheric circulation patterns. Included in this scheme, which is closely similar to that of Tyson (1986), are several typical circulation patterns favoring significant rainfall. He recognized five patterns for summer, two for winter, and two year-round. An abbreviated list is given below, followed by examples for the most effective rain-producing systems.

1) Summer
   - the westerly trough (tropical/temperate trough)
   - the southward-extending tropical easterly trough or low
   - the ridging high
   - the blocking high
   - the tropical cyclone
2) Winter
   - the northward-displaced wave cyclone
   - the postfrontal cold low

3) All seasons
   - the cold-front trough (more pronounced in winter)
   - the cut-off low.

All of the rain-producing weather systems described below have strong links with the Tropics. Two of these systems, the westerly trough and the cut-off low, are also linked to midlatitude waves (i.e., illustrating tropical and extratropical interaction). The V- or U-shaped tropical trough generally remains stationary, while the effect of tropical cyclones is restricted to the east coastal areas.

(i) Example 1: The westerly trough in summer

After satellite images became available on a regular basis, the existence of cloud bands over southern Africa, often with linear western boundaries, gained prominent recognition, notably by Harrison (1983; 1984; 1986, unpublished M.S. thesis) and Tyson (1986). The cloud bands of subtropical southern Africa are confined mainly to the land and eastern fringes of the Atlantic Ocean. They develop some hundreds of kilometers ahead (east) of the westerly troughs (tropical/temperate troughs), in moist air advecting from the Indian Ocean and tropical southern Africa. The displacement of the upper anticyclones of the subtropical zone are controlled by the eastward passage of deep middle-latitude cyclones, as well as mobile disturbances in the easterlies north of 20°S. The westward and southward flux of the moisture required for cloud-band formation is a complex result of the locations of the tropical lows along the IOCZ/ITCZ, the subtropical upper anticyclones along 20°–25°S, and the cold frontal troughs farther south. Harrison (1984; 1986, unpublished M.S. thesis) and Tyson (1986) described the cloud bands as ducts, channels, or conduits along which the southward flux of high-energy air over southern Africa takes place; in this way, it contributes to the north–south exchange of energy required to maintain the earth's energy balance. Tropical/temperate troughs are very important summer rain-producing systems in subtropical southern Africa.

Taljaard (1996b) describes the formation of these cloud bands. They form in the broad northerly airstreams west of the subtropical upper anticyclones (700–200 hPa) and east of approaching troughs in the upper westerlies over the Atlantic Ocean. Quasi-geostrophic theory, adapted for the Southern Hemisphere (Holton 1992), requires upward motion and upper divergence ahead of an approaching trough in order to ensure that the vorticity changes remain geostrophic in the presence of negative vorticity being advected downwind. Considerations of continuity then require low-level convergence and, if abundant moist tropical air is available, extensive cloud formation along the duct. Harrison described these bands as direct links between the tropical lows of the IOCZ/ITCZ trough and midlatitude systems in the south. More numerous are the instances when these bands appear along the western boundaries of the upper anticyclones that flank approaching westerly (tropical/temperate) troughs. When a tropical low develops along the IOCZ/ITCZ, latent heat release in convective cloud towers in the middle and upper troposphere results in the development of upper-tropospheric anticyclonic circulation, which also produces upper southward flux and cloud-band formation.

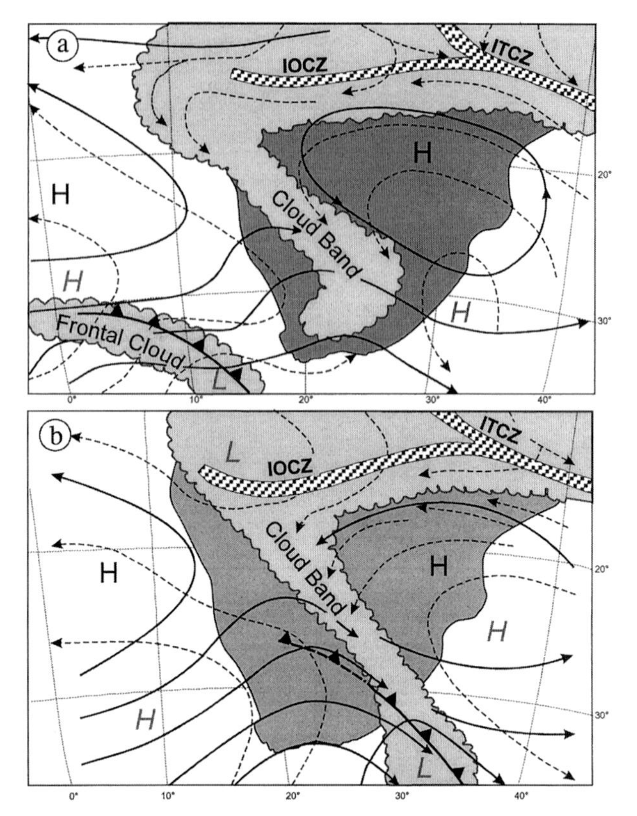

FIG. 3D.20. Schematic representation of cloud bands associated with subtropical and cold-front troughs on two successive days. The thin broken lines indicate the 850-hPa circulation, and the thin solid lines the 700–500-hPa circulation (after Taljaard 1996b).

Figure 3D.20a,b is a schematic representation (Taljaard 1996b) of the cloud bands associated with the westerly trough. The flow lines at 850 hPa illustrate the relationship between the flux of easterly tropical air and the cloud band forming along the western flank of the upper anticyclone ahead of the subtropical trough. Figure 3D.21a,b shows the 250 and 850 hPa contours and the *Meteosat* cloud images on 7 and 8 January 1991. The upper-trough axis is some distance west of the cloud band on both days. The cold front over the southern coastal regions and its associated cloud band are distinctly separate phenomena from the subtropical trough and cloud band, merging only over and south of the land on 8 January. Indications are that the approaching cold front in-

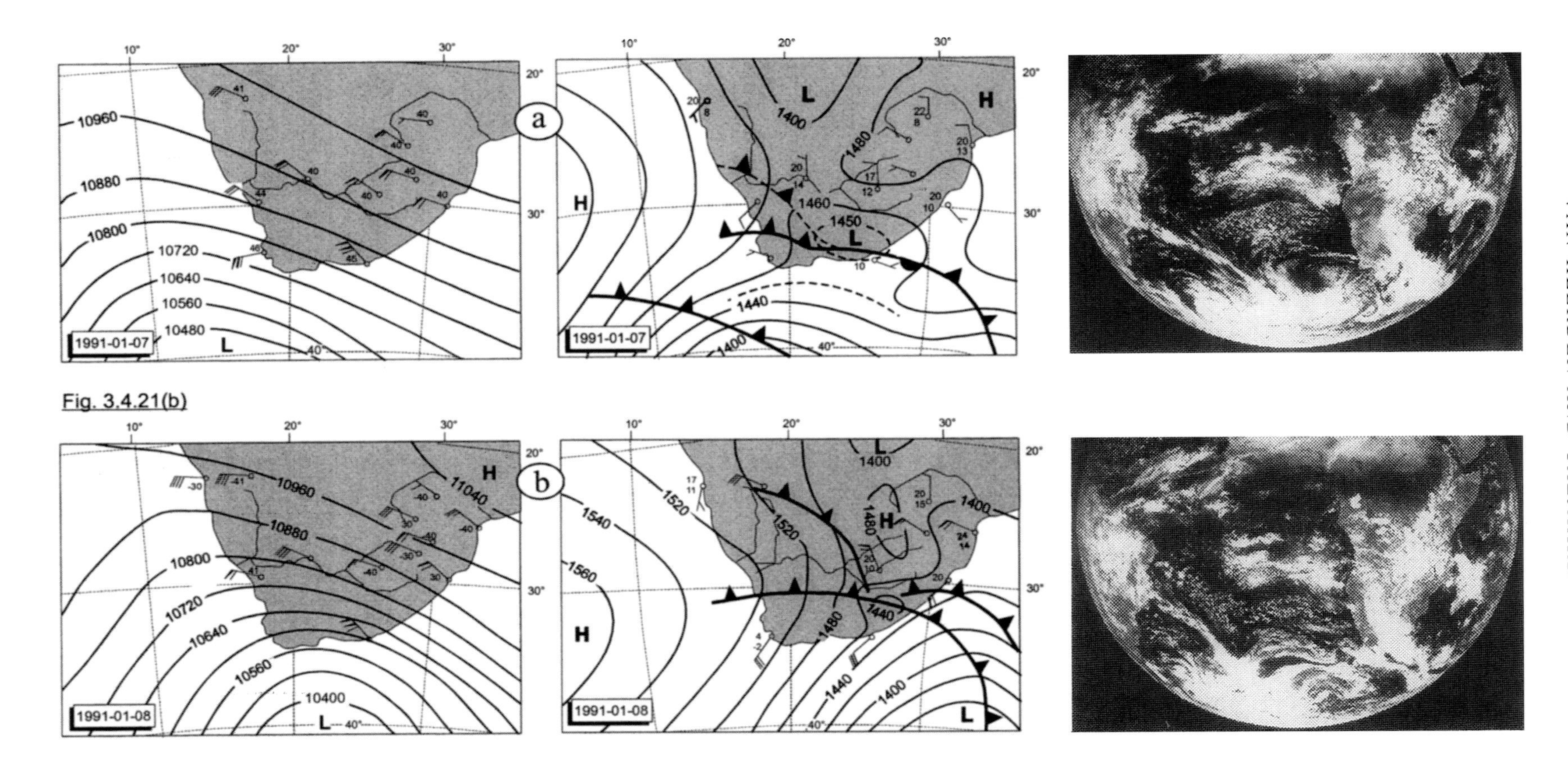

FIG. 3D.21. The 850- and 250-hPa height and the Meteosat cloud distributions on 7 and 8 January 1991 (after Taljaard 1996b).

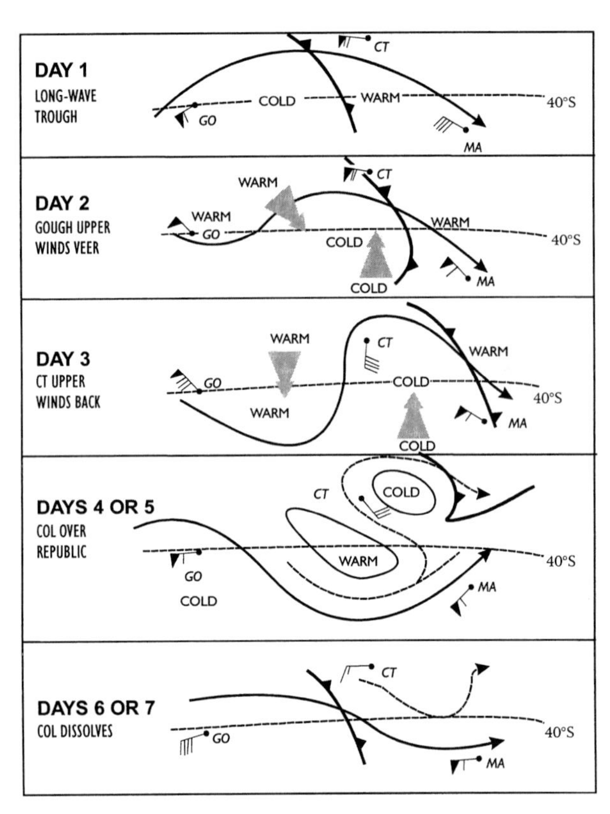

FIG. 3D.22. Schematic representation of the development of a cut-off low in the upper troposphere in the South African region (GO = Gough Island; CT = Cape Town; MA = Marion Island). Adapted from Taljaard (1985).

duced the subtropical trough to intensify and accelerate eastward.

(ii) Example 2: The all-season cut-off low

This type of large-amplitude synoptic-scale disturbance over South Africa was first described by Le Roux and Taljaard (1952) and identified as a separate upper cold pool. Subsequently, Taljaard (1952) called this feature a cut-off low pressure system. Triegaardt and Kraus (1957) described the formation of a cold upper low over the western interior and pointed out the similarities with "anticyclonic wave disruption." The cut-off low has strong links with midlatitude waves and is discussed here because it affects rainfall as far north as 15°S.

Figure 3D.22 depicts the stages of development and decay of cut-off systems (Taljaard 1985) in the region, but naturally there are many deviations from the mean picture. On day 1, a broad cold-front trough (one of the many normal middle and upper-tropospheric wave troughs of the region) is present between Gough Island and Cape Town. The first clue to probable cut-off development is the rapid veering of the upper wind to northwest and appreciable warming of the air column at Gough Island on day 2. Further veering of the wind and warming at the island, accompanied by backing of the wind at Cape Town on day 3, is indicative of an amplifying wave. The cutting-off process is complete on days 4 and 5, with the end product being a slow-moving warm anticyclone southwest or south of Cape Town and a cold low over or close to the South African west coast. The advection of cold air south of the upper trough and warm surface air south of the upper ridge results in the rapid deepening (increasing amplitude) of the trough–ridge system (Holton 1992). The anticyclonic curvature of the isohypses at the southeastern sector of the upper ridge soon becomes too large for the swift-moving northwesterly air parcels to follow. The flow becomes ageostrophic, slipping across the isohypses to link with the upper westerlies south of the Indian Ocean ridge. In this way, an upper cold pool becomes isolated from the westerlies.

The surface high pressure system, at the stage of the upper-air development just described, often appears banana shaped in the way it is curved around the south coast. Along its eastern flank, this high advects cool maritime air northward over the Mozambique Channel, often disrupting and modifying the flow along the ITCZ. At the 850-hPa level, a well-defined trough or low is the near-surface reflection of the cut-off upper low. On the eastern flank of the 850-hPa low, tropical air originating from as far north as 15°S is advected southward, often reaching the south coast. This continental tropical airstream normally advances southward west of the Drakensberg Escarpment and provides the moisture required for widespread rain over the central interior. At the mature stage of the cut-off low system, the sea level high along the south and east coasts causes the moist maritime air of the Indian Ocean to flow inland across the east and south coasts and escarpments. Strong surface winds are maintained by the tight pressure gradient between the high along the coast and the interior low. The southward advection of the continental tropical and Indian Ocean air masses takes place below the upper-level divergence ahead of the cut-off low. Further enhanced by orographical lifting and undercutting by the cool maritime air, widespread rains (and often flooding) result over the eastern parts of South Africa. Once the surface high has extended south of the country, it starts to cut off the advection of cold surface air below the upper cut-off low. The deepening of the cut-off low then ceases. In a day or so, the upper cut-off low departs from the land and the circulation gradually returns to predominantly zonal.

Triegaardt et al. (1988) described the devastating floods of 26–29 September 1987 in Kwazulu–Natal that resulted from the development of a deep cut-off low. Figure 3D.23 shows the SLP and 850-hPa geopotential distribution over land on 27 September 1987, while Fig. 3D.24 shows the 200-hPa geopotential

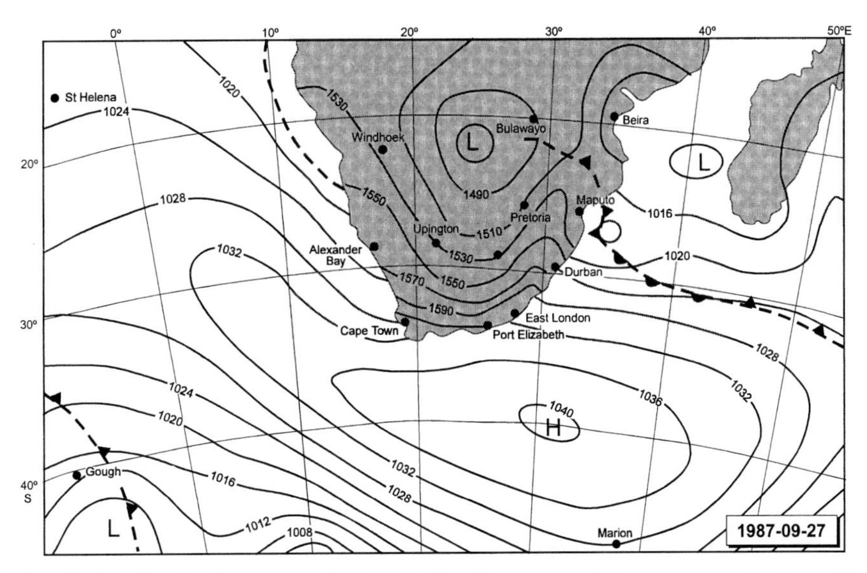

FIG. 3D.23. Sea level pressure (hPa) and 850-hPa geopotential contours on land on 27 September 1987. Adapted from Triegaardt et al. (1988).

height field on this day. During the period 26–29 September 1987, large areas of the Kwazulu–Natal province received more than 600 mm of rain, while some northern coastal areas recorded amounts exceeding 900 mm. These floods were responsible for the death of more than 300 people and caused approximately one billion rand of damage to the infrastructure of Kwazulu–Natal.

Tennant and Van Heerden (1994) used a version of the U.K. Meteorological Office GCM to model this cut-off low. They concluded that the topography of the subcontinent is vital for the typical evolution and development of cut-off lows. Without the high South African plateau, the very strong upper-level divergence, the resulting surface convergence, and the strong ascent did not develop in their model simulations. In addition, the surface pressure systems were much weaker, and the cut-off low moved eastward rapidly. These authors also concluded that ascent at the first range of mountains was responsible for much of the rain during this event.

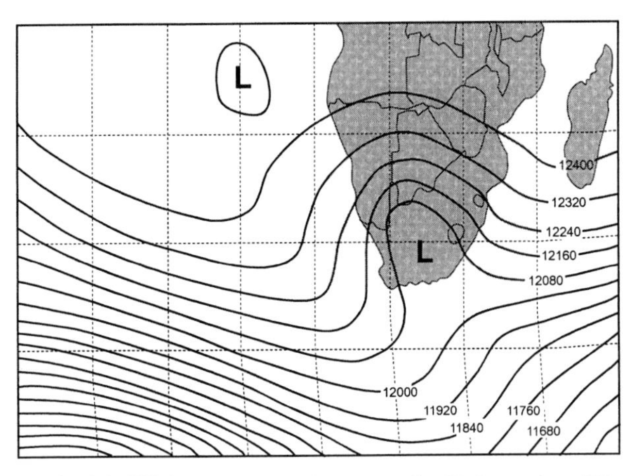

FIG. 3D.24. 200-hPa geopotential contours for 27 September 1987. Adapted from Triegaardt et al. (1988).

(iii) Example 3: The southward-extending tropical easterly trough or low (V-shaped trough or Botswana Low)

Taljaard (1996b) describes the way in which a southward-extending tropical trough normally develops when an anticyclone invades the eastern parts of South Africa, while a ridge extending from the Atlantic Ocean prevails over the western parts of the land. The two ridges are separated by a V- or U-shaped trough. As often as not, a closed tropical low can be drawn in the 850-hPa geopotential height field and, since it is mostly located over Botswana, it has become known as the Botswana low. An important feature of this system is the high humidity of the tropical air flowing southward or southwestward along the eastern flank of the low. Dewpoints of 18° to 20°C at the surface are surmounted by air in which the lapse rate is approximately wet adiabatic above 700 hPa. This type of system differs from the cut-off low in that it is warm cored above about 600 hPa, and the circulation gradually reverts to anticyclonic above 500 hPa. This situation is maintained by latent heat release in the extensive convective towers developing over the eastern flank of the trough/low.

During 1988, a system best described as a warm-cored tropical low brought devastating floods to the southern Free State province of South Africa, when more than 300 mm rain fell during the five-day period

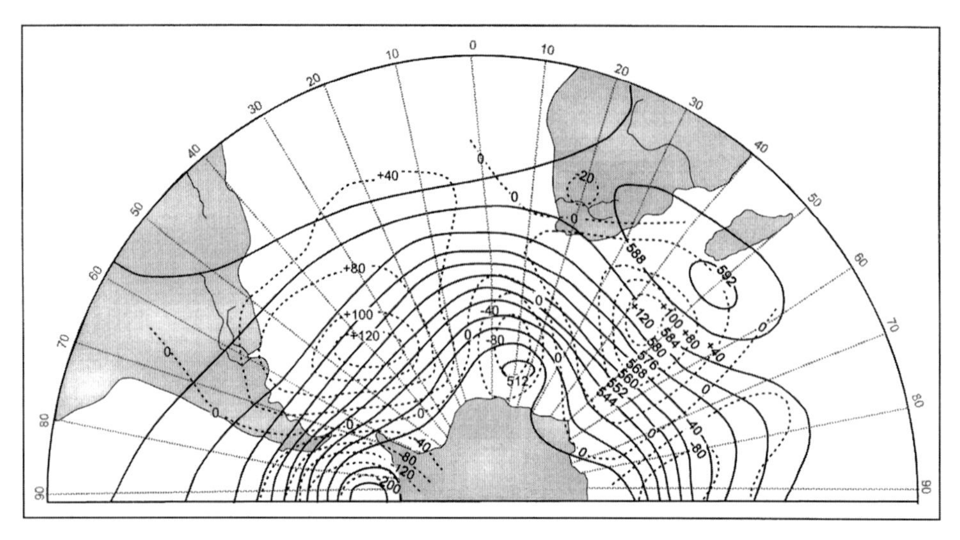

FIG. 3D.25. Mean 500-hPa contour pattern (solid lines, gp dam) for period 19–23 February 1988. Stippled lines are departures from normal (after Triegaardt et al. 1991).

of 19–23 February. Rainfall was restricted to a narrow band (250 km wide) across the southern Free State. This region is a gentle westward-sloping plateau (1400 m), and this contributed to the severity of the floods that disrupted all north–south land communications for several days. Triegaardt et al. (1991) investigated the system in some detail using ECMWF initialized analyses. They concluded that the hemispheric upper-tropospheric circulation was dominated by a large-amplitude, wavenumber 4 circulation system in the middle-latitude westerlies. Figure 3D.25 shows the 500-hPa geopotential height field averaged from 19 to 23 February. The warm-core nature of this system is clearly illustrated by the mean 850- and 200-hPa height fields (Fig. 3D.26a,b) for the period. The horizontal mass divergence field in units of tenths of kilograms at 200 hPa (Fig. 3D.26b) was located directly above the area of maximum vertically integrated mean horizontal flux of water vapor (not shown).

Triegaardt et al. (1991) concluded that a short-wave upper trough, moving through the long-wave upper trough on 19 and 20 February, deepened and formed a large-amplitude trough over the western interior. This maintained large upper-tropospheric horizontal velocity divergence above the 850-hPa Botswana low. At the same time, a strong influx of surface moisture from the east over Zimbabwe, curving southward over Botswana and southeastward over the southern Free State, was maintained by the high east of the land, as shown by the 850-hPa height field (Fig. 3D.26a). Conditions were favorable for large-scale deep convection and the release of latent heat in the upper troposphere. This maintained the upper-tropospheric anticyclone and horizontal velocity divergence required to strengthen the 850-hPa low. It is concluded, therefore, that the intensification of the continental tropical low (Botswana low) resulted from interaction between large-amplitude disturbances in the circumpolar westerlies and the advection of water vapor westward and southward over the central interior by a subtropical surface anticyclone situated east of the land.

(iv) Example 4: The tropical cyclone

Tropical cyclones do not occur in the South Atlantic Ocean. Over the southwestern Indian Ocean, their average frequency is 11 per year (Dunn 1985). Over the southwestern Indian Ocean, tropical cyclone frequencies peak during the midsummer months, December to March (Crutcher and Quayle 1974). Because the tropical cyclone Domoina devastated large parts of Mozambique, Swaziland, and northern Kwazulu–Natal in January 1984, renewed interest was kindled in the southwest Indian Ocean tropical cyclones. This resulted in studies by Poolman and Terblanche (1984), Dunn (1985), Padya (1989), de Coning (1990), Jury et al. (1990), Jury (1993), and others. Dunn (1985) lists 430 tropical cyclones east of longitude 45°E, as opposed to 159 west of 45°E, during the years 1912 to 1984. J. Olivier, in a contribution to Alexander (1993), referring to earlier work by Jury and Pathack (1991), determined the genesis points, tracks, and average speeds of tropical cyclones (Fig. 3D.27) over the southwestern Indian Ocean during the period 1939–63. Considering that the direct impact of a tropical cyclone in terms of rainfall and wind is restricted to a few hundred kilometers from the center of the system, it can be assumed that only those cyclones within or near the Mozambique Channel affect the mainland. Table 1 (after Darlow 1990, personal communication) lists the tropical cyclones in the Mozambique Channel during the period 1950–88. This table provides the

(a)

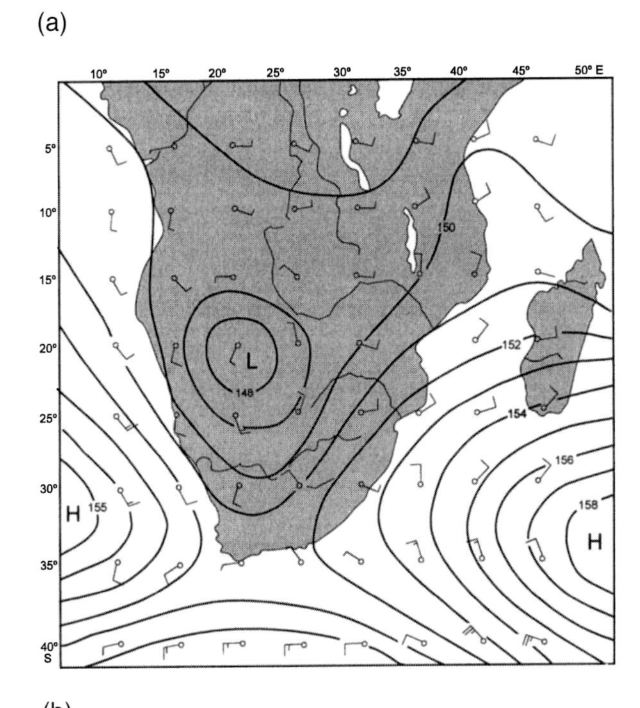

(b)

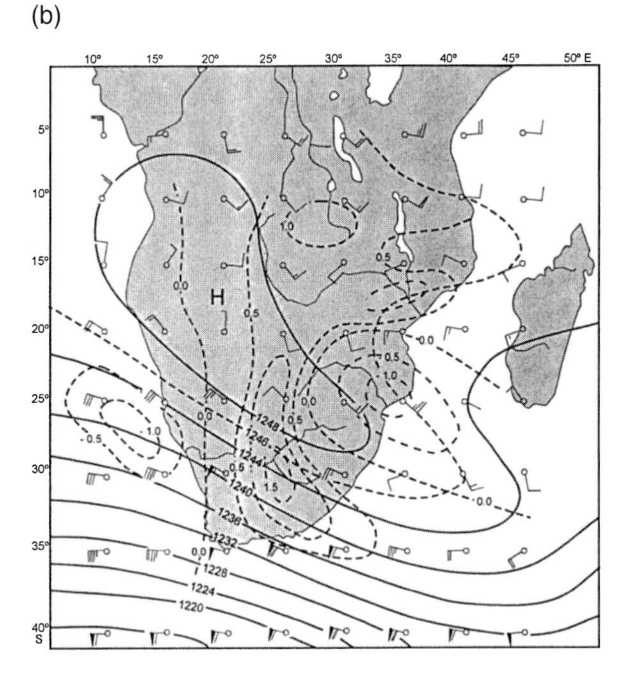

FIG. 3D.26. (a) Mean 850-hPa contour and wind patterns (geopotential dam and kt) for period 19–23 February 1988; (b) mean 200-hPa contour and wind patterns for same period. The stippled lines indicate the mean divergence in intervals of 0.5×10^{-5} s^{-1} (after Triegaardt et al. 1991).

area and estimated average rainfall depth of the cyclones affecting southern Africa.

Olivier (contribution to Alexander 1993) compared the average monthly circulation during January 1975 and January 1984 and concluded that the following

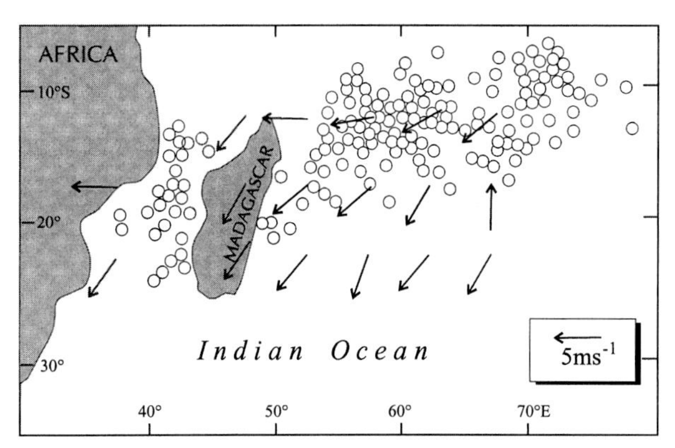

FIG. 3D.27. Genesis points and mean directions of tracks of tropical cyclones over the southwestern Indian Ocean during 1939–63 (after Jury and Pathack 1991).

conditions are conducive to the movement of tropical cyclones onto the African mainland:

- above-average SLPs in and south of the Mozambique Channel;
- a trough axis in the easterlies situated alongside the South African east coast; and
- strong easterlies in the channel extending south of Madagascar.

Figure 3D.28 shows the SLP anomalies during the January of 1976 and 1984. The factors noted by Olivier are clearly apparent. The SLP anomalies over the southeastern Atlantic Ocean differed considerably during the two months and do not seem to have played significant roles in the movement of the cyclones. Darlow (1990, personal communication), however, stated that tropical cyclones reaching Madagascar are all steered by ridging anticyclones from the South Atlantic and are influenced by the intensities of anticyclones south of the Mozambique Channel. He sug-

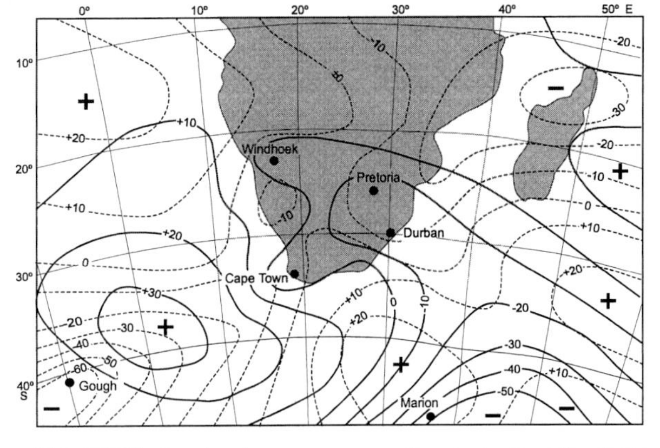

FIG. 3D.28. Sea level pressure anomalies in tenths hPa. Solid lines are for January 1975 (Blandine and Deborah); dashed lines are for January 1983 (Domoina) (after Olivier's contribution to Alexander 1993).

TABLE 3D.1. Cyclones entering the Mozambique Channel, 1950–88 (Source: Darlow 1990, personal communication).

Affecting South African rainfall				Not affecting South African rainfall	
Name	Date	Area (km^2)	Depth (mm)	Name	Date
A	11/02/56–16/02/56	65 000	155	B	19/01/56–31/01/56
Astrid	24/12/57–02/01/58	82 000	230	Kate	10/03/62–15/03/62
Brigette	30/01/60–01/02/60	49 000	155	Irma	21/02/67–24/02/67
Claude	30/12/65–05/01/66	24 000	175	Georgette	09/01/68–31/01/68
Caroline	04/02/72–13/02/72	10 200	170	Blandine	06/01/75–10/01/75
Eugenie	05/02/72–22/02/72	101 000	150	Deborah	24/01/75–04/02/75
Danae	21/01/76–30/01/76	66 000	175	Doaza	24/01/88–01/02/88
Emilie	01/02/77–06/02/77	53 000	180		
Domoina	16/01/84–31/01/84	107 000	370		
Imboa	10/02/84–21/02/84	26 000	155		

gests that the timing of the push from the South Atlantic is important in determining whether the channel cyclones will move onto the mainland coast or whether they will recurve southeastward from Madagascar. If the cyclone is still in the central parts of the channel when the Atlantic high extends south of the continent, the probability for landfall increases. If a cyclone has already reached the southern part of the channel, however, it will start recurving into the Indian Ocean. Olivier (in Alexander 1993) also considers the presence of an abnormally strong Indian Ocean high in the vicinity of Marion Island to be important in preventing the cyclone from departing from the channel. The eastern escarpment, rising to altitudes of 1500 to 2000 m in the north and 3000 m in Lesotho, seems to be an effective barrier to the westward movement of tropical cyclones. Cyclones A, Astrid, Eugenie, Danae, and Emilie all reached longitudes west of the main escarpment but remained north of the Limpopo River, and they all weakened to tropical low status soon after crossing the coast.

Jury (1993), investigating synoptic features favoring the development of more or fewer tropical cyclones in the southwestern Indian Ocean during midsummer, identified the following features:

1) More tropical cyclones
   - enhanced upper easterlies 0°–10°S, accompanied by enhanced upper divergence;
   - enhanced lower-tropospheric equatorial (5°–10°S) westerlies, 1000–700-hPa level;
   - enhanced cyclonic vorticity in the zone 10° to 15°S;
   - area with SSTs of 28°C over the southwest Indian Ocean increases considerably;
   - SSTs decrease along the East African coast; and
   - southern Hadley cell becomes more intense.
2) Fewer tropical cyclones
   - enhanced upper westerlies to 15°S, pronounced westerly wind shear;
   - lower-tropospheric anticyclone (1000–700 hPa) at 25°S, 55°E, resulting in enhanced northward advection of modified maritime polar air;
   - reduced lower-tropospheric equatorial westerlies;
   - reduced lower-tropospheric easterly trades across the subtropics (15°–30°S);
   - reduced cyclonic vorticity along the ITCZ; and
   - weaker northern Hadley cell with Indian monsoon outflow directed to East Africa.

In a modeling study by Van Heerden et al. (1995), using the CSIRO-9 GCM, a +2°C SST anomaly was forced over the tropical southwestern Indian Ocean northeast of Madagascar. The resultant circulation anomalies that developed in the model fields indicated enhanced low-level westerlies over equatorial and tropical Africa extending eastward to 60°E. Upper-air easterlies were considerably enhanced over the area north of Madagascar, extending westward to central Africa. At the same time, the surface flow over the Mozambique Channel became southerly, advecting moist maritime air northward over eastern Africa. Model results also indicated enhanced subsidence of upper Atlantic air over western and central southern Africa. In general, the CSIRO-9 GCM results support the analysis by Jury (1993) and explain why central southern Africa experiences drought and heat when a tropical cyclone is in the Mozambique Channel.

Poolman and Terblanche (1984) described the havoc caused by Domoina on 28 to 31 January 1984 when it crossed the Mozambique coastline, moved westward to the Lowveld of the Mpumalanga province, and then advanced southward across Swaziland to northern Kwazulu–Natal. Figure 3D.29 shows the track Domoina followed after identification on 19 January 1994 east of Madagascar. More than 800 mm of rain fell in the four days over Swaziland and northern Kwazulu–Natal. These amounts have been exceeded only by those that caused the devastating floods caused by the cut-off low of September 1987. Floodwaters from Domoina washed away most bridges or bridge-approach embankments over northern Kwazulu–Natal, but fortunately very little loss of life occurred.

3D.7. Climate variability

Climate variability can be defined as the fluctuation of climate over periods of a few months to a few

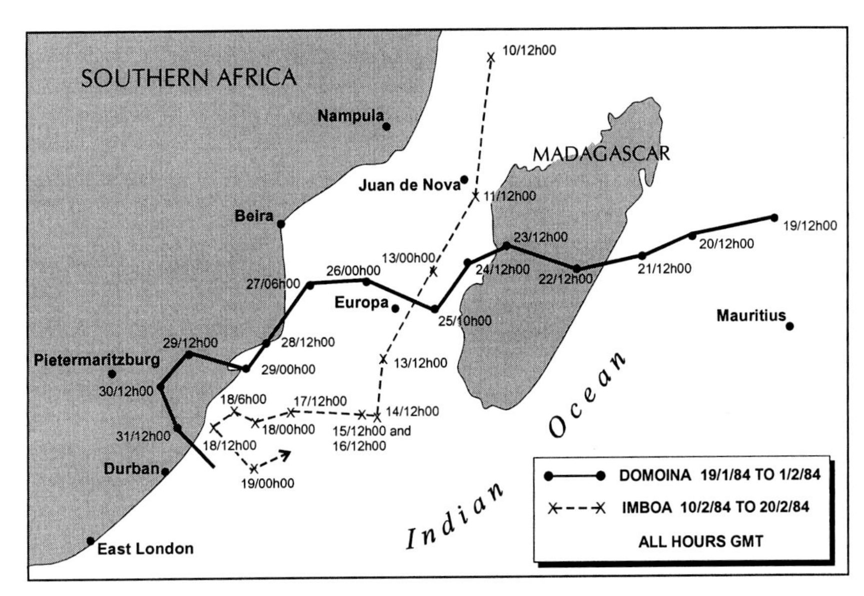

FIG. 3D.29. Track of cyclone Domoina from 19 January to 1 February 1984 (after Poolman and Terblanche 1984).

decades. This definition excludes the very short-term fluctuations caused by synoptic weather systems, as well as climatic change that implies the variation of mean conditions that persist over several decades, centuries, and longer. The short-term conditions associated with synoptic circulation systems and the mean conditions that existed over southern Africa during the seven years 1985–92 are described in the preceding sections—the latter admittedly being too short a period to qualify as long term.

The attention of meteorologists and climatologists has turned increasingly during the past three decades to studies of climatic change and variability rather than to the very short-term fluctuations of synoptic systems. This was partly sparked by Bjerknes' (1969) revival of interest in the Southern Oscillation and El Niño phenomena of the tropical Pacific Ocean, but also by the capabilities introduced by satellites to determine SSTs and other climate elements quite accurately worldwide, as well as by the rapid advance of computer technology. It is well taken into account that the values of parameters used to determine climate anomalies are normally monthly means, and that these monthly means are often strongly affected by one or two extreme events that last only a few days. The frequency, intensity, and duration of synoptic disturbances have quite profound effects on the calculated mean conditions.

Examples of comprehensive thorough treatises of climate are those of Schulze (1965) for South Africa and Torrance (1981) for Zimbabwe. Less-detailed discussions for subregions of southern Africa are contained in Landsberg (1972) and in the publication series of individual territorial meteorological services. Schulze and Torrance discussed the rainfall, wind, and temperature characteristics of their respective regions in considerable detail, including the occurrence and duration of long wet and dry periods. The associated atmospheric circulation systems were treated only briefly, however, pressure not being a climate element. Sea temperature was recognized as an important control of climate, but the role played by ENSO had not yet been accorded its present-day status. In addition, knowledge of the SST fluctuations in the South Atlantic and Indian Oceans did not exist.

Tyson et al. (1975) revealed the existence of a quasi–18-yr oscillation in the summer rainfall of South Africa. The mean rainfall during the successive 9-yr phases of the oscillation deviate by about 10% from the long-term mean, but it is notable that 2–3 yr of each 9-yr spell show deviations of opposite sign. Other less-prominent oscillations of 10 and 2–3 yr also exist in South Africa. Van Rooy (1980) provided independent verification of the 18-yr oscillation in a review of 1–5-yr wet and dry periods in South Africa. Surprisingly, these periodicities do not reflect the oscillations of ENSO.

Taljaard (1981, 1987, 1989) described the atmospheric circulations, rainfall, and temperature distributions that lasted for three- to six-month periods over the subcontinent south of 23°S. The summers of early 1974 and 1975–1976 were extremely wet, whereas the summer of 1982–1983 was extremely dry. The wet seasons were cool and the dry season was hot, this negative correlation of rain and temperature being characteristic; SLP was anomalously high south of the land during the wet periods and anomalously low during the prolonged drought. Upper winds showed northerly and westerly anomalies during the wet and dry periods, respectively, over South Africa. Sharp westerly troughs and cut-off lows were frequent when it was wet and rare and weak during the drought, pointing to a high meridional index during the rainy periods and a strong zonal circulation during the drought. Harrison (1983; 1986, unpublished M.S. thesis) found from satellite mosaics that cloud bands stretching from NNW to SSE frequent the western and central parts of subtropical southern Africa during wet periods, but the bands are displaced toward the eastern parts of this region and to Madagascar during drought. This is in accord with the dominant northerly and westerly circulations described in the previous paragraph.

Tyson (1986) synthesized the knowledge gained on climatic variability that existed at that time and developed a model of the dominant zonal and meridional wind components in vertical tropospheric cross sections along 10° and 30°E during wet and dry spells. These spells are the 9-yr wet and dry spells of the quasi–18-yr oscillation first described by Tyson et al. (1975). Because of the dearth of upper-air observations over southern Africa, the wind components shown in these cross sections are subject to a degree of uncertainty and should be verified by using the data of sophisticated analysis models such as the ECMWF model, in conjunction with all the upper-air soundings to date.

Nicholson (1986) found that rainfall anomalies south of the Equator are spatially coherent throughout the subcontinent, but that anomaly patterns indicate a strong and consistent opposition between equatorial and subtropical latitudes (i.e., 0°–15°S vs. 15°–32°S). Nicholson et al. (1988) produced a comprehensive atlas and statistics on African rainfall in which a large variety of the aspects of rainfall are discussed. Attention is given to the characteristics of rainfall variability, but this centers mainly on short-term temporal and spatial distributions and on the longer-term fluctuations that have been defined here as climatic change.

Lindesay (1988, unpublished Ph.D. thesis) found significant positive correlation between the Southern Oscillation Index (SOI) of the Pacific Ocean and the late summer-month rainfalls of South Africa, as well as with pressure and wind in the South African region. Van Heerden et al. (1988) improved the rainfall correlation by rearrangement of the months. Matarira (1989) also found significant positive correlations between the rainfall over southeastern central Africa (Zambia, Zimbabwe, Mozambique) and the SOI, as well as corresponding pressure and temperature anomalies associated with anomalous rainfalls, as those described by Taljaard (1981, 1987, 1989).

Since about 1990, the attention focused on correlations between SST and rainfall in southern Africa, and attempts have been made to explain the significant correlations found in terms of certain atmospheric circulation patterns. Jury (1995) published a review of the research done prior to 1995. Walker (1989, unpublished Ph.D. thesis) pioneered this research, by correlating the South African rainfall in non–El Niño years with SST in various ocean areas around South Africa. She found correlation coefficients of +0.5 and higher with the SSTs of the eastern Agulhas current region and that enhanced easterly flow existed east of the land in wet years. Mason (1992, unpublished Ph.D. thesis) extended and verified these results and found improved correlations when the phase of the quasi-biennal circulation is taken into account, the west phase being most favorable. Pathack (1993, unpublished Ph.D. thesis) established that the equatorial Indian Ocean is a key determinant of summer rainfall variability over the interior plateau of South Africa. The SST deviations of the central Indian Ocean are well correlated with the ENSO fluctuations of the tropical Pacific Ocean. It now seems to be well established that above-normal SST over the central Indian Ocean is unfavorable for rain over southern Africa. The reverse is true for wet summers, as suggested by the study of Jury and Lutjeharms (1993). They presented a diagnostic model of the anomalous horizontal and vertical circulations over southern Africa and the central western Indian Ocean (Fig. 3D.30) during dry summers over subtropical southern Africa. The essential features are increased upper westerlies over and west of the subcontinent, associated with enhanced subsidence, and increased low-level cyclonic circulation over and east of Madagascar, accompanied by above-normal SST. Increased vertical circulation comes into being in the latter region, resulting in anticyclonic outflow in the upper troposphere, leading in turn to anomalous easterly flow over tropical south-

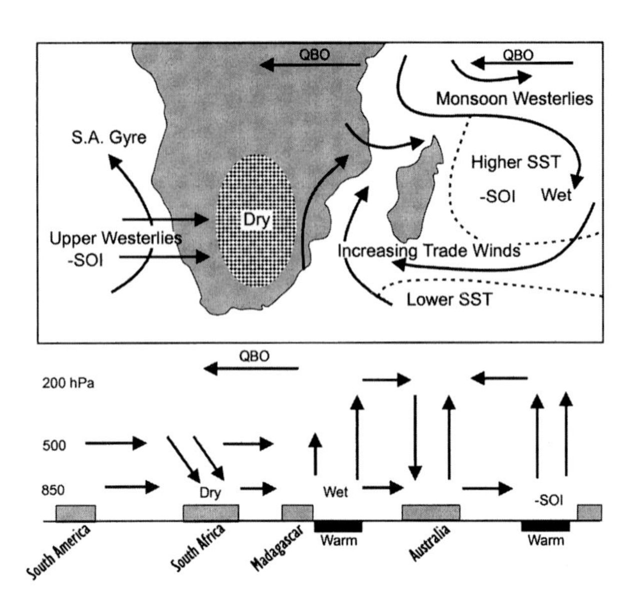

FIG. 3D.30. Results of a diagnostic model showing the circulation, SST, and convective anomalies during droughts over southern Africa. The lower panel depicts the anomalous vertical and horizontal relative motion (after Jury and Lutjeharms 1993).

ern Africa. Over the tropical Pacific Ocean the SOI is then characteristically below normal. Van Heerden et al. (1995), in an independent modeling study using the CSIRO-9 CGM, found that a +2°C SST anomaly over the southwestern tropical Indian Ocean results in circulation anomalies similar to those described by Jury and Lutjeharms (1993).

Recent research by Landman (1996, unpublished M.S. thesis) building on earlier work by Walker (1989, unpublished Ph.D. thesis), Mason (1992, unpublished Ph.D. thesis), Jury (1995), and Hastenrath et al. (1995), among others, used multivariate analysis to determine the statistical relationships between anomalous Indian and Atlantic Ocean SSTs and rainfall over southern Africa. Landman's statistical model links regional rainfall anomalies with SST anomalies over specific parts of the tropical Indian and Atlantic Oceans.

Long-range rainfall predictions over subtropical southern Africa based on ENSO and SST anomalies have shown considerable promise, but only time will decide the success of these methods.

3D.8. Summary

1) The surface and near-surface (850-hPa) pressure and airflow patterns of the eastern South Atlantic Ocean show relatively little change from summer (DJF) to winter (JJA). The subtropical anticyclone is displaced only 3° and 5° northward and westward, respectively, in the cold season, allowing the southern westerlies to increase at 35°S and thereby introducing the winter rains of the extreme southwestern coastal belt. The southerly to southeasterly trade winds of this ocean sector decrease modestly in winter as a result of a 10-hPa pressure rise over South Africa, compared to 5 hPa at the midocean anticyclone. In contrast, the circulation changes dramatically in the western Indian Ocean, where the core of the subtropical anticyclone is displaced from 35°S, 60°E to 30°S, 85°E, while the pressure also rises 10 hPa over the South African east coast. The broad belt of easterly trade winds of summer revert to an extensive fresh southeasterly current that crosses the Equator to become the southwestern monsoon of the northern Indian Ocean. The northeastern monsoon of tropical East Africa in summer is thereby fully reversed. Over the land, the summertime surface trough at 15°S along the ITCZ and its westward extending branch (here called the IOCZ) is replaced by a stable dry southeasterly to easterly continental trade-wind current encompassing most of the subcontinent north of the South African anticyclone.
2) The normal equatorward increase of mean temperature causes the sea level high pressure belt to shift northward with increasing height in both summer and winter. In summer, the oceanic surface anticyclone disappears aloft, while a marked zonally elongated anticyclone is established at 200 hPa across the land along about 15°S, which is almost vertically above the low-level ITCZ/IOCZ trough. This superimposition can be explained as the result of maximum latent heat release where the rainfall is heaviest. A meridionally aligned trough is indicated below 700 hPa along 20°–25°E in summer, but the mean winds do not clearly reflect this feature, no doubt due to the lack of upper-air sounding stations during the brief period (1985–92) employed for the construction of the charts. The upper-tropospheric westerlies expand remarkably far north, to about 5°S in the Atlantic Ocean in winter, compared to the Indian Ocean, where a ridge develops along 15°S, creating easterlies on its equatorward side and enhanced westerlies to its south (these westerlies actually lead to the pronounced subtropical jetstream at 28°S over the eastern Indian Ocean and Australasia). The reversal of the low-level and upper-tropospheric winds off tropical East Africa in winter is a remarkable example of inter-hemispheric air-mass exchange.
3) Only surface mean temperature distributions for January and July are displayed (the upper-air mean temperature distributions being reflected by the vertical changes in the isobaric-contour patterns). Over the sea, the influence of the upwelling cold water of the Benguela current along the west coast and the southward transport of warm water by the Mozambique and Agulhus current along the east coast is reflected by coastal temperature differences of 6°–8°C at similar latitudes. Over the land, the effects of height above sea level and the seasonal positions of the sun predominate. The most pronounced features are the cold surface over eastern South Africa in winter and the hot spot over western South Africa in summer. Attention is focused on the significant vertical temperature and humidity structures over the two ocean sectors and the land. Over the eastern South Atlantic Ocean and the adjacent desert coastal belt south of about 10°S, the structure is extremely stable throughout the year, with intense inversions at 0.5 to less than 1.5 km separating cool, moist maritime air from warm, dry subsided air above. Over and off the east coast, inversions or stable layers at 750 to 650 hPa are also frequent in both seasons but more so in winter. In summer, the air over the Congo (Zaire) basin, eastern Angola, Zambia, and astride the ITCZ over central Mozambique and northern Madagascar becomes predominantly humid and unstable, but perturbations produce great variability over all continental and east coast regions, including South Africa, Botswana, and the eastern half of Namibia. The extensive southeasterly trade-wind current over the land from about 20° to 5°S in winter is stable and dry above 700 hPa.

4) Charts of annual mean rainfall and for the midseason months January, April, July, and October are displayed. Many local problems require explanation, but stability and topographical effects play major roles.
5) The short- and medium-term synoptic patterns of the region are discussed separately for tropical and subtropical southern Africa, with more attention being given to those favorable for rainfall. The most clearly defined systems of the tropical zone in summer are the tropical cyclones at 10° to 20°S that frequent Madagascar and the Mozambique Channel and the tropical lows that move at variable speeds, mostly westward, along the ITCZ/IOCZ trough at 15°S. They possess characteristic cyclonic circulations below about 500 hPa but revert to anticyclones in the upper troposphere. Although the perturbations of the equatorial zone north of 10°S have been studied, mostly prior to 1965, their three-dimensional structures are poorly known, mainly due to a dearth of upper-air soundings. Slow-moving weak troughs associated with convergence and deeper-than-normal cloud masses evidently produce most of the rain. These systems seem to propagate mostly downwind and occur both over the sea and the land. Those that occur over East Africa do not obviously originate over the sea, but eastward-advancing disturbances are common over the equatorial west coast. Many disturbances over the Congo (Zaire) basin, however, travel slowly westward from the continental divide. Zonally aligned ridges and troughs over and on either side of the Equator are theoretically associated with rain and dry weather. The subject of rain and drought over the equatorial zone of Africa requires considerable clarification.
6) The rain- and drought-producing circulation systems of subtropical southern Africa are well known and are described in fair detail in this paper, with abbreviated case studies being included. The short-term rain-producing systems in summer are tropical/temperate troughs approaching from the west, southward-extending tropical easterly troughs, cut-off lows, and ridging or blocking anticyclones south of the land. A few tropical cyclones over or in proximity of the east coast are prone to produce floods over the adjacent land. In winter, the cold-front troughs of the westerly waves, postfrontal cold troughs, and cut-off lows favor rainfall over the southwestern coastal belt. When the ridging highs or blocking anticyclones are slow moving or recur in quick succession along the 35°–45°S zone, they become known as long-wave ridges and give rise to protracted wet spells. Dry weather over the South African summer rainfall region is associated with wave troughs in the westerlies, and the longer they persist the more severe the drought becomes. The jet streams associated with the planetary ridges and troughs occur anomalously far south and north, respectively.
7) A large number of studies on climate variability, particularly on correlations between rainfall in southern Africa and SST fluctuations around the Southern Hemisphere, are briefly described and discussed, but because these studies are still being continued and are not conclusive yet, the results and comments will not be summarized here. Many research papers linking SST anomalies and rainfall over Africa have been written. Few of these pay any attention to the variability of synoptic-scale weather systems during ENSO events.

REFERENCES

Alexander, W. J. R., 1993: *Flood Risk Reduction Measures.* Dept. of Civil Engineering, University of Pretoria, 220 pp.

Asnani, G. C., 1983: *Tropical Meteorology.* Noble Printers, 1202 pp.

Bargman, D. J., 1962: Tropical meteorology in Africa. *Proc. Munitalp Symp.,* Nairobi, Kenya, WMO and the Munitalp Foundation, 445 pp.

Bjerknes, J., 1969: Atmospheric teleconnections from the equatorial Pacific. *Mon. Wea. Rev.,* **97,** 163–172.

Bultot, F., 1952: On the related nature of the rain in the Belgian Congo. Publ. l'Inst. Natl. l'Étude du Congo Belge, Bureau Climatologique, No. 6, 16 pp.

Crutcher, H. L., and R. G. Quayle, 1973: *Mariner's Worldwide Climatic Guide to Tropical Storms at Sea.* Naval Weather Service, Washington.

de Coning, E., 1990: Predicrability of tropical cyclones in the Southwestern Indian Ocean. South African Weather Bureau Newsletter No. 492, 1–5.

Diab, R. D., R. A. Preston-Whyte, and R. Washington, 1991: Distribution of rainfall by synoptic type over Natal. *Int. J. Climatol.,* **11,** 877–888.

Dunn, P., 1985: An investigation into tropical cyclones in the Southwest Indian Ocean. Flood Studies Tech. Note No. 1, 24 pp. [Available from Dept. of Water Affairs and Forestry, Private Bag X313, Pretoria 0001, South Africa.]

Findlater, J., 1969: A major low-level air current near the Indian Ocean during northern summer. *Quart. J. Roy. Meteor. Soc.,* **95,** 362–380.

———, 1971: Mean monthly air flow at low levels over the western Indian Ocean. Geophysical Memoirs No. 115, Her Majesty's Stationery Office, London, 53 pp.

Forsdyke, A. J., 1949: Weather forecasting in tropical regions. *Geophys. Mem.,* No. 82, British Meteor. Office, 82 pp.

Gichuiya, S. N., 1974: Easterly disturbances in the southeast monsoon. East African Meteorological Division Tech. Memo No. 21, 30 pp. [Available from the Kenya Meteorological Office, Nairobi.]

Harrison, M. S. J., 1983: The Southern Oscillation, zonal equatorial circulation cells and South African rainfall. Preprints, *First Intl. Conf. Southern Hemisphere Meteorology,* Sao Jose dos Campos, Brazil, Amer. Meteor. Soc., 302–305.

———, 1984: The annual rainfall cycle over the central interior of South Africa *S. African Geograph. J.,* **66,** 47–64.

Hastenrath, S., 1991: *Climate Dynamics of the Tropics.* Kluwer Academic Publishing, 488 pp.

———, L. Greischar, and J. Van Heerden, 1995: Prediction of the summer rainfall over South Africa. *J. Climate,* **8,** 1511–1518.

Holton, J. R., 1992: *An Introduction to Dynamic Meteorology.* Second ed., Academic Press, 391 pp.

Jackson, S. P., 1961: *Climatological Atlas of Africa.* CCTA/CSA, Government Printer (Pretoria), 55 plates.

Jeandidier, G., 1956: Storms and monsoons of the Congo Basin. Publication of the Meteorological Service No. 3, French Equatorial Africa, Brazzaville.

Johnson, D. H., and H. T. Mörth, 1962: Forecasting research in East Africa. *Proc. Munitalp Symp.,* Nairobi, Kenya, WMO and the Munitalp Foundation, 56–137.

Jury, M. R., 1993: A preliminary study of climatological associations and characteristics of tropical cyclones in the southwestern Indian Ocean. *Meteor. Atmos. Phys.,* **51,** 101–115.

———, 1995: A review of research on ocean–atmospheric interactions and South African climate variability. *S. African J. Sci.,* **91,** 289–294.

———, and B. Pathack, 1991: A study of climate and weather variability over the tropical southwestern Indian Ocean. *Meteor. Atmos. Phys.,* **47,** 37–48.

———, and J. R. E. Lutjeharms, 1993: The structure and probable driving forces for the 1991–1992 drought in southern Africa (in Afrikaans). *J. Science and Technology* (S. Africa), **12,** 8–16.

———, B. Pathack, C. Macarthur, and S. Mason, 1990: A climatological study of synoptic weather systems over the tropical southwest Indian Ocean. *Meteorology and Atmospheric Physics,* 46 pp.

———, ———, G. Campbell, B. Wang, and W. Landman, 1991: Transient convective waves in the tropical southwestern Indian Ocean. *Meteor. Atmos. Phys.,* **47,** 27–36.

Kreft, J., 1953: Descriptive account of a cyclone of February 1950. *Notos 3,* South African Weather Bureau, 149–157.

Landsberg, H. E., 1972: *World Survey of Climatology, Vol. 10: Climate of Africa.* Elsevier Publishing Company, 604 pp.

Le Roux, J. J., and J. J. Taljaard, 1952: An example of unusual winter weather over South Africa during July 1952. *Notos 1,* South African Weather Bureau, 70–78.

Leroux, M., 1983: *The Climate of Tropical Africa.* Champion, 250 pp.

Matarira, C. H., 1989: Drought over Zimbabwe in a regional and global context. Preprint, *Int. Conf. Southern Hemisphere Meteorology,* Buenos Aires, Argentina, Amer. Meteor. Soc., 218–219.

Nicholson, S. E., 1986: The nature of rainfall variability in Africa south of the Equator. *J. Climate,* **6,** 515–530.

———, J. Kim, and J. Hoopingarner, 1988: *Atlas of African Rainfall and Its Interannual Variability.* Florida State University, 237 pp.

Okoola, R. E., 1989: Westward moving disturbances in the Southwest Indian Ocean. Kenya Meteorological Department, 30 pp.

Padya, B. M., 1989: *Weather and Climate of Mauritius.* MGI, Mauritius.

Poolman, E., and D. Terblanche, 1984: Tropical cyclones Domoina and Imboa. South African Weather Bureau Newsletter No. 420, 37–46.

Rainteau, P., 1954: Study of weather change over the Congo Basin. Publication of the Meteorological Service No. 3, French Equatorial Africa, Brazzaville.

Rhodesia Meteorological Service, 1958: Mean upper winds over the Federation of Rhodesia and Nyasaland. *Rhodesia Meteorological Notes,* Series A, **14,** 18 pp.

Riehl, H., 1979: *Climate and Weather of the Tropics.* Academic Press, 611 pp.

Schulze, B. R., 1965: *Climate of South Africa. Part 8: General Survey.* South African Weather Bureau, 330 pp.

Schulze, G. C., 1984: A survey and discussion of pressure variability and synoptic patterns in the Southern Hemisphere during the period October 1982 to February 1983. *S. African J. Sci.* **80,** 94–97.

South African Weather Bureau, 1975: *Climate of South Africa. Part 12: Surface Winds.* South African Weather Bureau, 79 pp.

Taljaard, J. J., 1952: Heavy rains along the South Coast. South African Weather Bureau Newsletter No. 38.

———, 1955: Stable stratification in the atmosphere over southern Africa. *Notos 4,* South African Weather Bureau, 217–230.

———, 1972: Synoptic meteorology of the Southern Hemisphere. *Meteorology of the Southern Hemisphere, Meteor. Monogr.,* No. 35, Amer. Meteor. Soc., 139–213.

———, 1981: The anomalous climate and weather systems of January to March 1974. South African Weather Bureau Tech. Paper No. 9, 92 pp. [Available from the South African Weather Bureau, Private Bag X097, Pretoria 0001, South Africa.]

———, 1985: Cut-off lows in the South African region. South African Weather Bureau Tech. Paper No. 14, 153 pp. [Available from the South African Weather Bureau, Private Bag X097, Pretoria 0001, South Africa.]

———, 1987: The anomalous climate and weather systems over South Africa during summer 1975–1976. South African Weather Bureau Tech. Paper No. 16, 80 pp. [Available from the South African Weather Bureau, Private Bag X097, Pretoria 0001, South Africa.]

———, 1989: Climate and circulation anomalies in the South African region during the dry summer of 1982/83. South African Weather Bureau Tech. Paper No. 21, 45 pp. [Available from the South African Weather Bureau, Private Bag X097, Pretoria 0001, South Africa.]

———, 1995: Atmospheric circulation systems, synoptic climatology, and weather phenomena of South Africa. Part 2: Atmospheric circulation systems in the South African region. South African Weather Bureau Tech. Paper No. 28, 65 pp. [Available from the South African Weather Bureau, Private Bag X097, Pretoria 0001, South Africa.]

———, 1996a: Atmospheric circulation systems, synoptic climatology and weather systems of South Africa. Part 5: Temperature phenomena in South Africa. South African Weather Bureau Tech. Paper No. 31, 42 pp. [Available from the South African Weather Bureau, Private Bag X097, Pretoria 0001, South Africa.]

———, 1996b: Atmospheric circulation systems, synoptic climatology weather systems of South Africa, Part 6: Rainfall in South Africa. South African Weather Bureau Tech. Paper No. 32, 100 pp. [Available from the South African Weather Bureau, Private Bag X097, Pretoria 0001, South Africa.]

———, and P. C. L. Steyn, 1988: Relationships between rainfall and weather systems in the catchment of the Vaal Dam. South African Weather Bureau Tech. Paper No. 20, 57 pp. [Available from the South African Weather Bureau, Private Bag X097, Pretoria 0001, South Africa.]

———, and ———, 1991: Relationships between atmospheric circulation and rainfall in the South African region. South African Weather Bureau Tech. Paper No. 24, 62 pp. [Available from the South African Weather Bureau, Private Bag X097, Pretoria 0001, South Africa.]

Tennant, W. J., and J. Van Heerden, 1994: The influence of orography and local sea-surface temperature anomalies on the development of the 1987 Natal floods: A general circulation model study. *S. African J. Sci.,* **90,** 45–49.

Thompson, B. W., 1957: Some reflections on equatorial and tropical forecasting. East African Meteorological Department, Tech. Memo No. 7, 14 pp. [Available from East African Meteorological Department, Nairobi, Kenya.]

———, 1965: *The Climate of Africa.* Oxford University Press, 15 pp. plus 132 charts.

Torrance, J. D., 1981: *Climate Handbook of Zimbabwe.* Department of Meteorological Services (Zimbabwe), 222 pp.

Trewartha, G. T., 1961: *The Earth's Problem Climates.* The University of Wisconsin Press, 334 pp.

Triegaardt, D. O., and G. M. Kraus, 1957: An example of wave disruption over South Africa. *Notos 6,* South African Weather Bureau, 6–12.

———, and A. Kits, 1963: The pressure distribution at several levels over southern Africa and adjacent oceans during 5-day rainy and dry spells in southern Transvaal and northern Free State during the summer of 1960–61. South African Weather Bureau Newsletter No. 168, 37–43. [Available from the South African Weather Bureau, Private Bag X097, Pretoria 0001, South Africa.]

———, and W. A. Landman, 1995: The influence of atmospheric long-waves on summer rainfall in the Transvaal, Orange Free State and Natal. South African Weather Bureau Tech. Paper No. 26, 30 pp. [Available from the South African Weather Bureau, Private Bag X097, Pretoria 0001, South Africa.]

———, D. E. Terblanche, J. Van Heerden, and M. V. Laing, 1988: The Natal flood of September 1987. South African Weather Bureau Tech. Paper No. 19, 62 pp. [Available from the South African Weather Bureau, Private Bag X097, Pretoria 0001, South Africa.]

———, J. Van Heerden, and P. C. L. Steyn, 1991: Anomalous precipitation and floods during February 1988. South African Weather Bureau Tech. Paper No. 23, 25 pp. [Available from the South African Weather Bureau, Private Bag X097, Pretoria 0001, South Africa.]

Tyson, P. D., 1981: Atmospheric circulation variations and the occurrence of extended wet and dry spells over southern Africa. *Int. J. Climatol.,* **1,** 115–130.

———, 1984: The atmospheric modulation of extended wet and dry spells over South Africa, 1958–1978. *Int. J. Climatol.,* **4,** 621–635.

———, 1986: *Climatic Change and Variability in Southern Africa.* Oxford University Press, 220 pp.

———, T. G. J. Dyer, and M. N. Mametse, 1975: Secular changes in South African rainfall. *Quart. J. Roy. Meteor. Soc.,* **101,** 817–833.

Van Heerden, J., D. E. Terblanche, and G. C. Schulze, 1988: The Southern Oscillation and South African summer rainfall. *Int. J. Climatol.* **8,** 577–597.

———, C. J. de W. Rautenbach, and M. Truter, 1995: Techniques for seasonal and long-term rainfall prediction in South Africa. Water Research Commission, South Africa, Rep. No. 306/1/95, 68 pp.

Van Rooy, M. P., 1972: *District Rainfall for South Africa and the Annual March of Rainfall over Southern Africa.* South African Weather Bureau, 60 pp.

———, 1980: Extreme rainfall anomalies over extensive parts of South Africa during periods of 1 to 5 successive summer years. South African Weather Bureau Tech. Paper No. 8, 32 pp.

Von Ficker, H., 1936: *Die Passatinversion.* Veroffenlichungen, Meteorol. Institut. Universität Berlin, Vol. 4, 33 pp.

Chapter 4

Meteorology of the Antarctic

DAVID H. BROMWICH

Polar Meteorology Group, Byrd Polar Research Center, The Ohio State University

THOMAS R. PARISH

Department of Atmospheric Science, University of Wyoming

4.1. Basic geographical characteristics

The great antarctic ice sheets and the surrounding Southern Ocean environs are large heat sinks in the global energy budget. Their geographical position limits the amount of solar insolation incident at the surface at such latitudes, and the high reflectivity of the ice fields of the Antarctic continent reduces the effective heating (see Carleton 1992). There are pronounced differences between the north and south polar regions. The Northern Hemisphere polar region consists essentially of an ice-covered ocean surrounded by continental landmasses, while the Southern Hemisphere features a continental landmass about the pole surrounded by an ocean. The most significant consequence of the vastly differing polar geographies is that the meridional temperature gradients become enhanced in the SH, resulting in a semipermanent baroclinic zone surrounding Antarctica and in effect thermally isolating the Antarctic continent to a degree unparalleled in the NH. As a result, Antarctica experiences the coldest and most harsh climate on earth. The intensified temperature contrast also supports a large west-to-east thermal wind component. Upper-tropospheric westerlies, which circumscribe the Antarctic continent, are considerably stronger than their NH counterparts (Schwerdtfeger 1984, pp. 223–226).

A second factor critical for the pronounced hemispheric polar contrast is the continental orography of Antarctica. With a mean elevation of nearly 2400 m, it is by far the highest continent on earth. There are dynamical constraints brought about by the continental orography that shape the meteorology of the high southern latitudes. Among these is the establishment of a mean meridional circulation between the continent and more northerly latitudes. The lower branch of this consists primarily of the katabatic wind regime, the ubiquitous feature of the lower atmosphere over the continent and an important mechanism for the transport of mass and cold air northward. The elevated antarctic terrain also serves as a barrier to the inland progression of extratropical cyclones, which frequent the continental periphery. There is evidence that the antarctic orography serves as an anchor for the upper-level tropospheric and lower-stratospheric circulations, thereby reducing the wandering of the center of the circumpolar vortex.

Figure 4.1 serves as a reference to the antarctic orography and some of the geographical features and stations. A complete listing of the historical antarctic stations and dates of operation can be found in Schwerdtfeger (1970); selected climatic records can also be found in Schwerdtfeger (1984). The Antarctic continent is essentially entirely covered with a permanent ice sheet that, in places over East Antarctica, exceeds 4000 m in thickness. Inspection of the antarctic ice topography reveals marked differences in the terrain slope over the continent. Both the gently sloping wide expanse of the Antarctic interior and the steep near-coastal ice slopes are meteorologically significant and control the near-surface wind and temperature regimes throughout the year. Depicted also in Fig. 4.1 is the mean seasonal extent of the pack ice in the high southern latitudes, from Gloersen et al. (1992) for the years 1978–87 based on satellite observations. The sea ice surrounding the continent undergoes vast seasonal variations, receding from a maximum area of approximately 2×10^7 km in early spring to 4×10^6 by autumn (Zwally et al. 1983). Such pronounced variations result in strong seasonal surface heat budget changes over the Southern Ocean. Carleton (1981, 1983) notes that the zonal frequency of new cyclones in high southern latitudes may be related to the advance and retreat of the sea ice edge.

4.2. The temperature and radiation regimes

Given its geographical position surrounding the South Pole and elevated ice surface, it is not surprising that the radiative budget of Antarctica is strongly negative for most of the year (Radok 1981). The snow

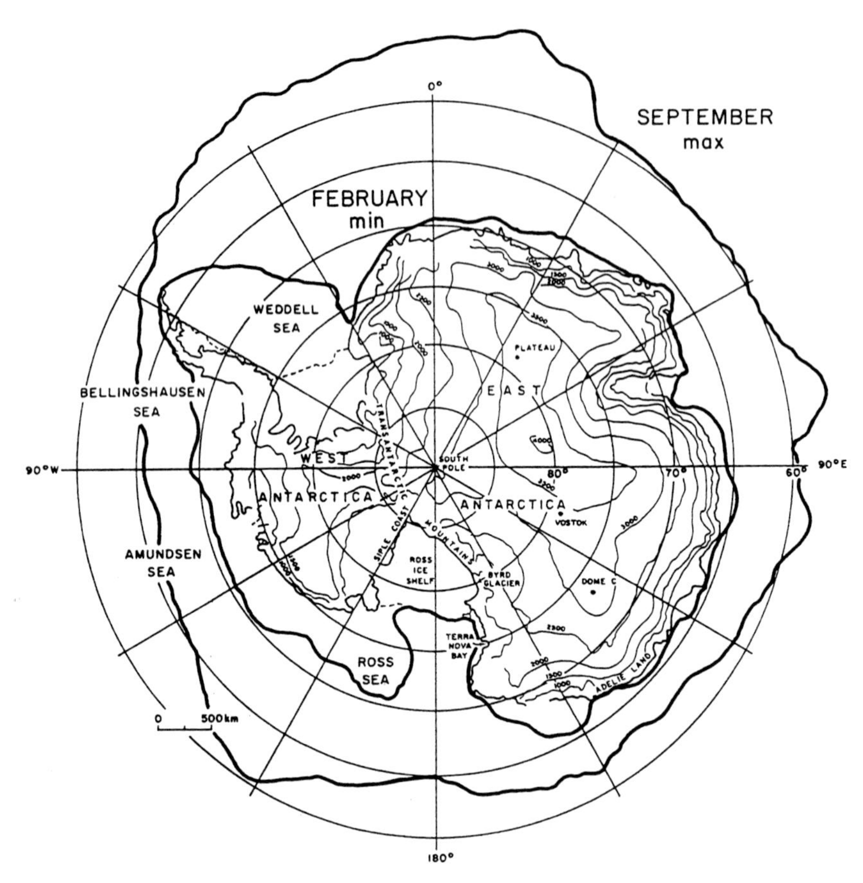

FIG. 4.1. The Antarctic continent and selected geographical references. Thin lines are height contours in meters. Thick lines are maximum and minimum sea-ice extents for the years 1978–87, after Gloersen et al. (1992).

and ice surface acts nearly as a black body; emissivity values in excess of 0.90 have been measured over the antarctic surface (Carroll 1982). Continental albedo values reported by a number of authors (e.g., Kuhn et al. 1977; Carroll and Fitch 1981; Wendler et al. 1988) are generally around 0.80 or higher, implying that only about 20% of the already weak solar insolation is available to warm the continental ice slopes. Back radiation from the warmer atmosphere above the inversion limits the surface cooling somewhat (Kuhn and Weickmann 1969; King 1996) but, except under dense stratus conditions, is far less than the surface radiative flux lost to space. Climatologically speaking, the extremely low water vapor content above the continent and the prevalence of clear sky conditions (Schwerdtfeger 1970; Yamanouchi and Kawaguchi 1992) restrict the downwelling atmospheric component.

A number of studies have examined the heat budget over the continental environs to show the importance of radiational forcing (Kuhn et al. 1977; Bintanja and van den Broeke 1995; King et al. 1995; King 1996). King et al. (1995) conducted a surface energy budget study over an ice shelf during winter and found that turbulent transports and heat conductivity through the ice are small compared to the radiative fluxes. Modeling work (Parish and Waight 1987) suggests that turbulent heat fluxes may form a significant component of the surface energy budget during strong katabatic wind episodes near the coast but are secondary to radiative fluxes over the gently sloping terrain of the vast Antarctic interior (see also King 1990, 1993a).

The result of the strong radiative deficit is the establishment of the great antarctic surface temperature inversion, which persists over nearly the entire continent during all but the brief midsummer period. Such inversions are generally contained within a depth of only a few hundred meters, with the most extreme vertical temperature changes in the lowest 100 m. Depictions of the pronounced inversions over the interior plateau of East Antarctica can be found in Kuhn et al. (1977) and are summarized in Schwerdtfeger (1984). Estimates of the mean wintertime inversion strength can be seen in the initial work of Phillpot and Zillman (1970) and the slightly modified version shown in Schwerdtfeger (1970). An illustration of the wintertime inversion strength (warmest tropospheric temperature minus the near-surface value) is shown as

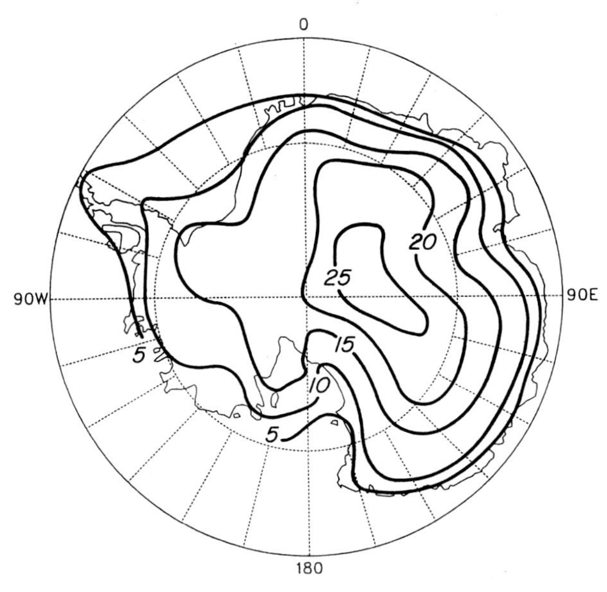

FIG. 4.2. Mean wintertime (June–September) inversion strength (°C) over Antarctica (after Phillpot and Zillman 1970; Schwerdtfeger 1970).

Fig. 4.2. The isopleths of inversion strength in some sense mirror the underlying terrain, with the strongest inversions in excess of 25 K over the highest portions of the continent. Daily measurements from the high interior stations such as Plateau and Vostok indicate that inversions approaching 35 K are not uncommon. The terrain slope is probably as important as elevation in the development or destruction of the great temperature inversion over the continent. The steeply sloping surfaces encourage a vigorous vertical momentum exchange in the lowest portion of the atmosphere via the katabatic wind regime, which limits the strength of the near-coastal inversions. Ample observational evidence exists to support mechanical mixing as an important process in destroying the inversion over the interior (Kuhn et al. 1977). In addition, Carroll (1994) has shown that a strong correspondence is seen between warming events at the South Pole and the longwave radiation budget. The influx of thick cloud shields over the continent, which are associated with cyclonic activity near the coast, can also disrupt the radiative budget over the interior and lead to a breakdown of the inversion.

Antarctica experiences extreme continentality, owing to its geographical location and physical characteristics of the ice surface. Large seasonal temperature changes are evident on the high plateau of Antarctica in response to the solar and longwave radiational forcing. Automatic weather station (AWS) data (Keller et al. 1994, for example) for the period 1980–91 indicate that Dome C, like other sites situated in the Antarctic interior, undergoes a pronounced annual temperature cycle, with the coldest mean monthly temperatures 35°C colder than those found during the summer months (see Fig. 4.3). The outstanding features of this temperature regime are the rapid seasonal transitions and the peculiar flat midwinter temperature trend. The summer period is very short, only six weeks or so in duration, centered around 1 January. Individual three hourly records from the Dome C AWS indicate that autumn temperatures become obvious by early February and the onset of winter is rapid. Mean monthly surface temperatures over the high interior decrease by nearly 25°C in the two-month period from late January to late March. The winter temperature regime over the Antarctic interior becomes established by April, and only minor variations in temperature are seen for the next five months. As suggested in Fig. 4.3, the "coreless winter" is pronounced at all interior stations. Temperature records for stations situated near the coast suggest that the coreless winter phenomenon is still present, although for a shorter period of time—from approximately May through August (see Bromwich et al. 1993, for example).

The coreless winter has been discussed by many authors since first described in the early part of this century. Schwerdtfeger (1970, 1984) and Wendler and Kodama (1993) have summarized early studies. The coreless winter is a consequence of the radiatively forced heat budget near the surface over the Antarctic interior. As the solar insolation of the ice slopes decreases, a pronounced net cooling of the surface takes place. The outgoing longwave radiation from the snow surface, which radiates nearly as a black body, exceeds the sum of the incoming radiation and downward sensible heat flux and other smaller heat fluxes toward the surface, such as heat conduction from the ground and sublimation. As the surface cools, the emitted longwave radiation from the surface decreases according to the Stefan–Boltzman law. Eventually, a near balance is reached between the upward radiation from the surface and the combined effects of primarily

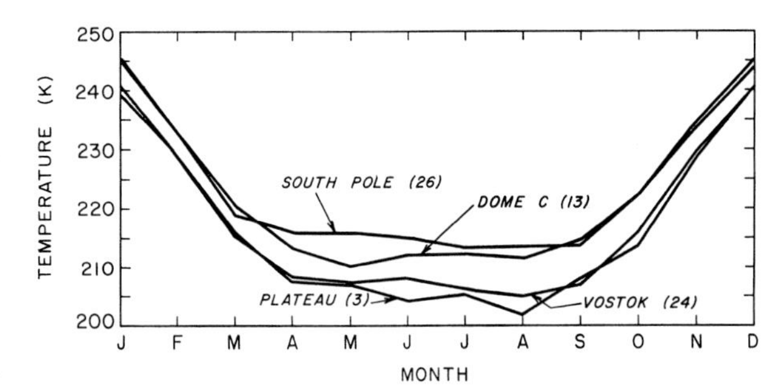

FIG. 4.3. Mean monthly temperatures at interior stations South Pole, Dome C, Vostok, and Plateau, based on multi-year records. Number in parentheses indicates the number of years in each record.

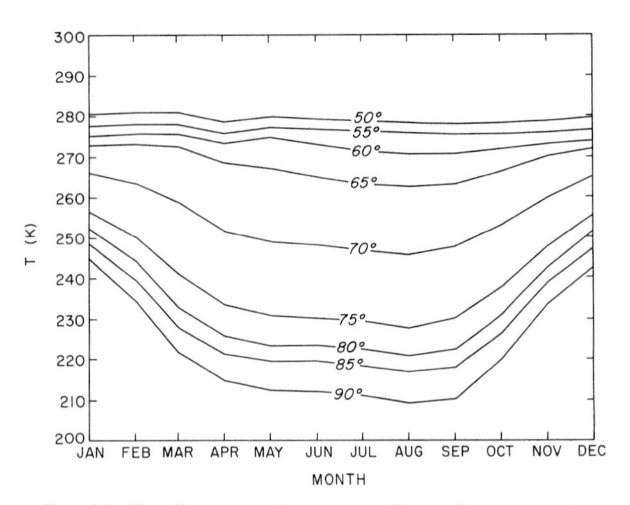

FIG. 4.4. Zonally averaged mean monthly surface temperatures for latitudes 50°–90°S, based on ECMWF analyses, averaged for 1985–94, in K.

the downwelling radiation and sensible heat exchange from the now much warmer atmosphere just above the surface. It may also be the case that as the surface cools past −80°C or so, the efficiency of the cooling process decreases. At such low temperatures the maximum longwave radiation emitted from the surface shifts toward longer wavelengths. The atmosphere becomes increasingly opaque to longwave radiation beyond a wavelength of 15 μm or so, due to absorption by CO_2. This implies that a natural buffer exists to prevent the surface temperatures from decreasing past a certain point in an absolute sense as well. The lowest surface temperature ever recorded on the surface of the earth was −89.2°C at Vostok in 1983.

Surface temperatures over the high southern latitudes are greatly influenced by the antarctic ice sheets. As an illustration, Fig. 4.4 represents the zonally averaged mean monthly surface temperatures from the ECMWF (see Trenberth 1992) model for a 10-yr period, 1985–94. Recent work has shown that the ECMWF model output reveals close agreement with available surface and upper-air observations using a variety of statistical measures for this time period (e.g., Cullather et al. 1997); in addition, climatological analyses of snow accumulation and surface temperature match time-averaged ECMWF depictions (Genthon and Braun 1995). Continent–ocean contrasts are evident in the annual course of surface temperature over high southern latitudes. South of approximately 70°, the range of the annual cycle of surface temperature exceeds 20 K and approaches 30 K south of 80°S. The meridional surface temperature gradients between Antarctica and the subpolar latitudes also show an annual cycle. The latitudinal surface temperature differences increase dramatically during austral autumn and remain large until the return of the sun at antarctic latitudes in the springtime, at which time the gradients relax.

4.3. The surface wind regime

The strong radiative deficit over the continental ice slopes of Antarctica has important dynamical constraints on the near-surface windfield. The often intense radiative cooling of the sloping ice surface implies that a density contrast becomes established with the coldest and hence most dense air near the surface. A horizontal pressure-gradient force thus becomes established in the lowest portion of the atmosphere over the sloping terrain, directed in a downslope sense. Lettau and Schwerdtfeger (1967) coined the phrase "sloped-inversion" pressure-gradient force to underscore the fundamental roles of both the sloping terrain and temperature inversion near the surface in producing this force. Radok (1973, 1981) has also discussed the forcing of the katabatic wind regime and the potential role of the drainage circulation in the energetics of the antarctic atmosphere. The intensity of the katabatic wind is proportional to the steepness of the underlying terrain, and consequently the most intense katabatic winds are found near the steep coastal stretch of the continent. As is seen from Fig. 4.1, the coastal stretch of East Antarctica contains especially steep ice slopes and is the site of the most intense surface wind conditions found on earth. Parish (1982a, 1988) describes some of the fundamental characteristics of the katabatic wind and has listed multi-year resultant wind statistics for a number of the historical stations. Such records underscore the importance of the terrain in forcing the surface wind over Antarctica.

A number of other factors have been suggested as potentially significant in forcing the continental drainage flows as well. Radok (1973) and Wendler and Kodama (1984) have noted that along-slope temperature gradient from the coast of Adélie Land to Dome C situated on the high plateau of East Antarctica is generally superadiabatic. This implies that as the air moves downslope, it is colder than the air it replaces and hence is subject to acceleration. Others have commented on the potential role of drift snow in enhancing the drainage flows (e.g. Radok 1973; Loewe 1974; Kodama et al. 1985; Gosink 1989). It is thought that the sublimational cooling of embedded snow particles may enhance the downslope forcing of the airstream.

There is now a vast assortment of surface wind data from the Antarctic continent and some of the main features of these drainage winds are now well described and understood. Since 1980, the number of wind observations has been enhanced considerably with the development of AWS (e.g., Stearns and Savage 1981; Stearns 1982; Allison and Morrissy 1983; Stearns et al. 1993). Currently, more than 50

AWS units are operating, and much of the data collected has been made widely available (Keller et al. 1994, for example). Some of the deployments have enabled detailed assessments of the katabatic wind regime, such as the arrays of stations in Adélie Land (Wendler et al. 1993; Périard and Pettré 1993) and inland of Terra Nova Bay (Bromwich et al. 1993).

The katabatic wind system is among the most persistent surface flow regimes in the world, rivaling even the trade-wind regime (Radok 1973). The directional constancy, a ratio of the vector-average wind speed to the scalar-average wind speed, is generally 0.90 or greater in analyses of annual wind statistics for antarctic stations (Parish 1982a, 1988; Wendler and Kodama 1984, 1985), indicating the unidirectional nature of the katabatic winds. In nearly all cases, the resultant wind direction is oriented to the left of the fall line some 20°–50°, consistent with Coriolis deflection of a pure gravity-driven flow. It is significant that the wind directions at antarctic stations are constrained to follow topographic pathways. This implies that the topographically induced forcing is the predominant forcing mechanism. Observations of manned and automatic stations along the katabatic-prone stretch of coastal East Antarctica show that approximately 90% of the wind observations occur in a narrow directional sector of 30° or so (see Madigan 1929; Loewe 1974; Keller et al. 1994). This is especially impressive given the frequency and intensity of cyclonic disturbances in the highly baroclinic zone along the periphery of the Antarctic continent.

Much of the early literature (Ball 1956, 1957, 1960; Streten 1963; Mather and Miller 1966, 1967a; Lettau and Schwerdtfeger 1967; Schwerdtfeger and Mahrt 1968; Weller 1969; Radok 1973) explored the dynamical relationship between the inversion strength, terrain slope, and friction to explain the observed katabatic wind characteristics. Subsequent observational (e.g., King 1989, 1993b) and modeling efforts (Parish 1984; Parish and Waight 1987; Parish and Wendler 1991; Gallée and Schayes 1992, 1994; Gallée et al. 1996; Hines et al. 1995) have also examined the terrain-induced forcing. It became clear that, to a first approximation, the near-steady flows in the interior of the Antarctic continent could be envisaged as a result of the balance between the terrain-induced horizontal pressure gradient force, the Coriolis force, and the friction force, as modified by the horizontal pressure gradient force in the free atmosphere. Recent work by Neff (1992, 1994) and Carroll (1994) based on measurements from acoustic sounders (Neff 1981) near the South Pole suggests that transitory forcing from synoptic influences presents a complicated boundary-layer structure and that the upper-level forcing is important in the lower atmospheric flow regime. Coastal flows can become highly non-uniform, and ample evidence (Tauber 1960; Streten 1968, 1990; Wendler and Kodama 1985; Sorbjan et al. 1986; Parish et al. 1993a,b; Yasunari and Kodama 1993; van den Broeke and Bintanja 1995) exists to suggest that the synoptic environment acts to modulate their intensity. In particular, Parish et al. (1993a) and Yasunari and Kodama (1993) show that the strongest katabatic winds are associated with favorable synoptic-scale pressure gradients, which generally imply relatively weak upper-level westerly winds that accompany the strong lower-tropospheric easterlies. Modeling work by Murphy and Simmonds (1993) has also shown the importance of synoptic forcing in producing strong katabatic wind conditions at Casey near the coast of East Antarctica.

Observations also show significant seasonal modulation is present in the continental-scale katabatic wind regime, with marked winter maxima at exposed coastal stations (Mather and Miller 1967b). Wendler et al. (1988) and Kodama et al. (1989) have conducted detailed field studies during summer periods over the near-coastal region of Adélie Land and shown that wind speeds undergo a modest diurnal variation due to solar radiation, with the maximum wind speeds seen during the early morning hours and minimum speeds early in the afternoon (see also Périard and Pettré 1993). One unexpected finding was that the directional constancy remained high throughout the observing period, despite the relatively strong solar insolation, due to the pressure gradient associated with the land–ocean thermal contrast.

Parish and Bromwich (1987) have shown that a qualitatively accurate diagnosis of the time-averaged winter pattern of the surface airflow over the Antarctic continent is possible given reasonable estimates of the sloped-inversion pressure gradient force and some representative value of friction. Such an analysis exploits the strong control of the underlying terrain on the surface windfield and the attendant high directional constancy of the wind. Figure 4.5 is adapted from that work and illustrates the preferred cold-air drainage channels over the face of the continental ice sheets. Available evidence, including both surface observations and surveys of sastrugi orientation (Parish and Bromwich 1987; Parish 1988), offers support to the simulated broad-scale streamline pattern. As seen in Fig. 4.5, the drainage pattern is not uniform. Rather, katabatic wind streamlines become concentrated in a finite number of channels over the continent. The convergence of drainage streamlines implies that cold, negatively buoyant air from a broad horizontal area becomes concentrated into a restricted pathway. Such confluence zones represent areas of enhanced cold-air supplies available to feed katabatic winds downstream. The accumulation of negatively buoyant reserves allows the downstream katabatic wind regime to become anomalously intense and persistent.

Confluence regions upslope from the very windy areas in Adélie Land (Parish 1981, 1984; Parish and Wendler 1991; Wendler et al. 1993; Wendler et al.

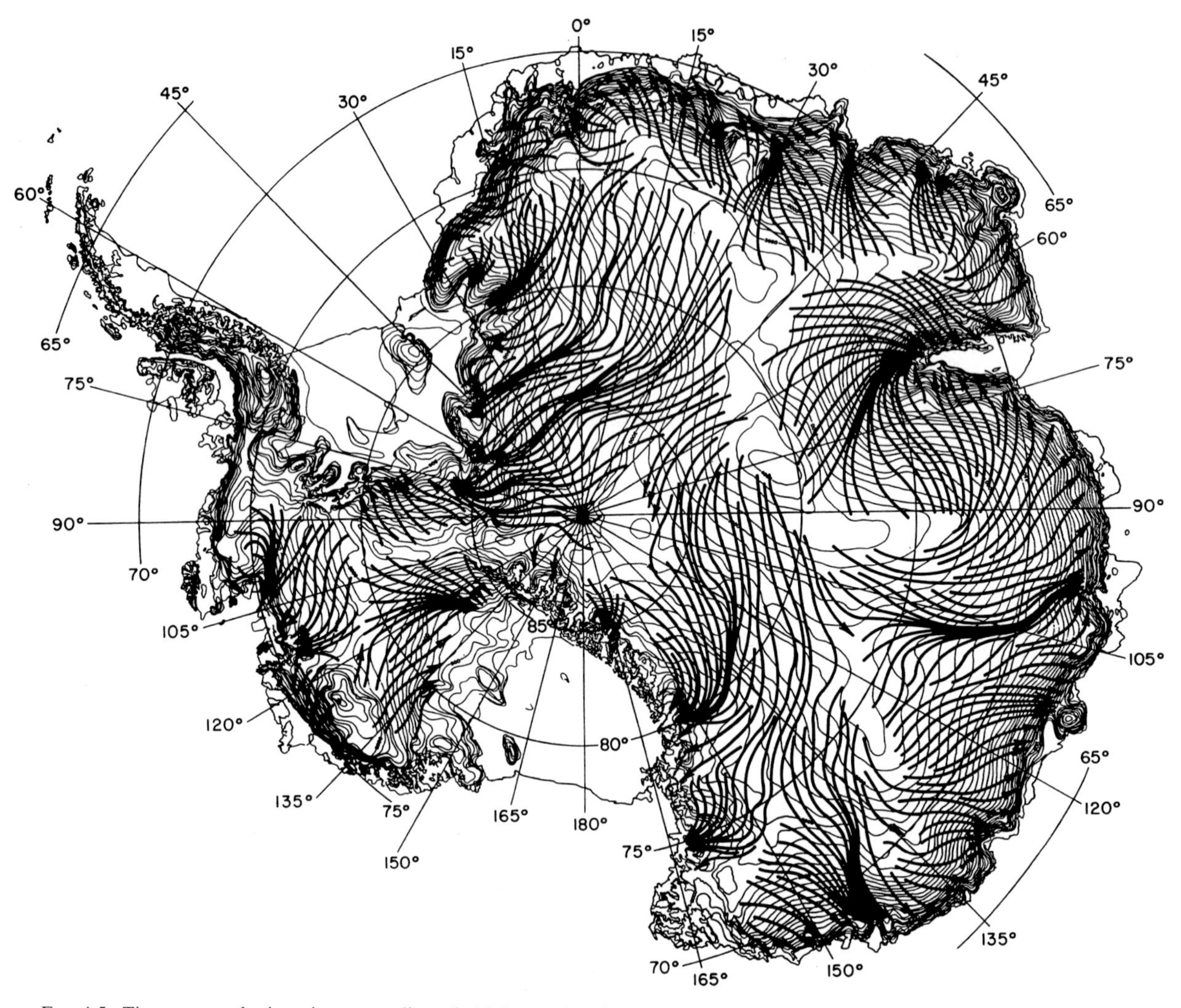

FIG. 4.5. Time-averaged wintertime streamlines (bold lines) of surface airflow over Antarctica, after Parish and Bromwich (1987). Thin lines are elevation contours in increments of 100 m.

1997) and at Terra Nova Bay (Bromwich and Kurtz 1984; Bromwich 1989a,b; Parish and Bromwich 1989; Bromwich et al. 1993) have previously been discussed. The katabatic wind at Adélie Land is especially impressive. Mawson's 1911–14 expedition set up the base camp of Cape Denison, at which katabatic winds were incessant for nearly two years. The wind statistics are impressive (Mather and Miller 1967b; Loewe 1972, 1974). The mean wind speed during the two-year period of observations was 19.8 m s^{-1}. The highest mean monthly wind speed was 24.9 m s^{-1} during July 1913; the windiest day was recorded on 16 August 1913, during which time the mean wind speed was 36.0 m s^{-1}. Three-hour reports of wind speeds in excess of 20 m s^{-1} were recorded on approximately 50% of all days. Even the lowest mean monthly wind speed during the two-year period (11.7 m s^{-1} during February 1912) is exceptional, since it is comparable to the typical winter monthly mean wind speed at katabatic-prone coastal stations.

Local intensification of katabatic airstreams is also suggested in the simulation along the Siple Coast stretch of West Antarctica, upstream from Byrd Glacier in the Transantarctic Mountains and the western side of the Amery Ice Shelf, with numerous smaller confluence zones along the coastal stretch of East Antarctica from 0° to 60°E. Strong evidence (Bromwich 1986, 1989a; Bromwich et al. 1992; Bromwich and Liu 1996; Breckenridge et al. 1993; Carrasco and Bromwich 1993) has been presented that supports the confluence zones at both the Siple Coast and Byrd Glacier sites. The drainage through the Siple Coast region can be enhanced by the prevailing synoptic situation. Bromwich et al. (1992, 1994) note that cyclone centers frequently pass north of the Ross Ice Shelf and over the Amundsen Sea and act to intensify

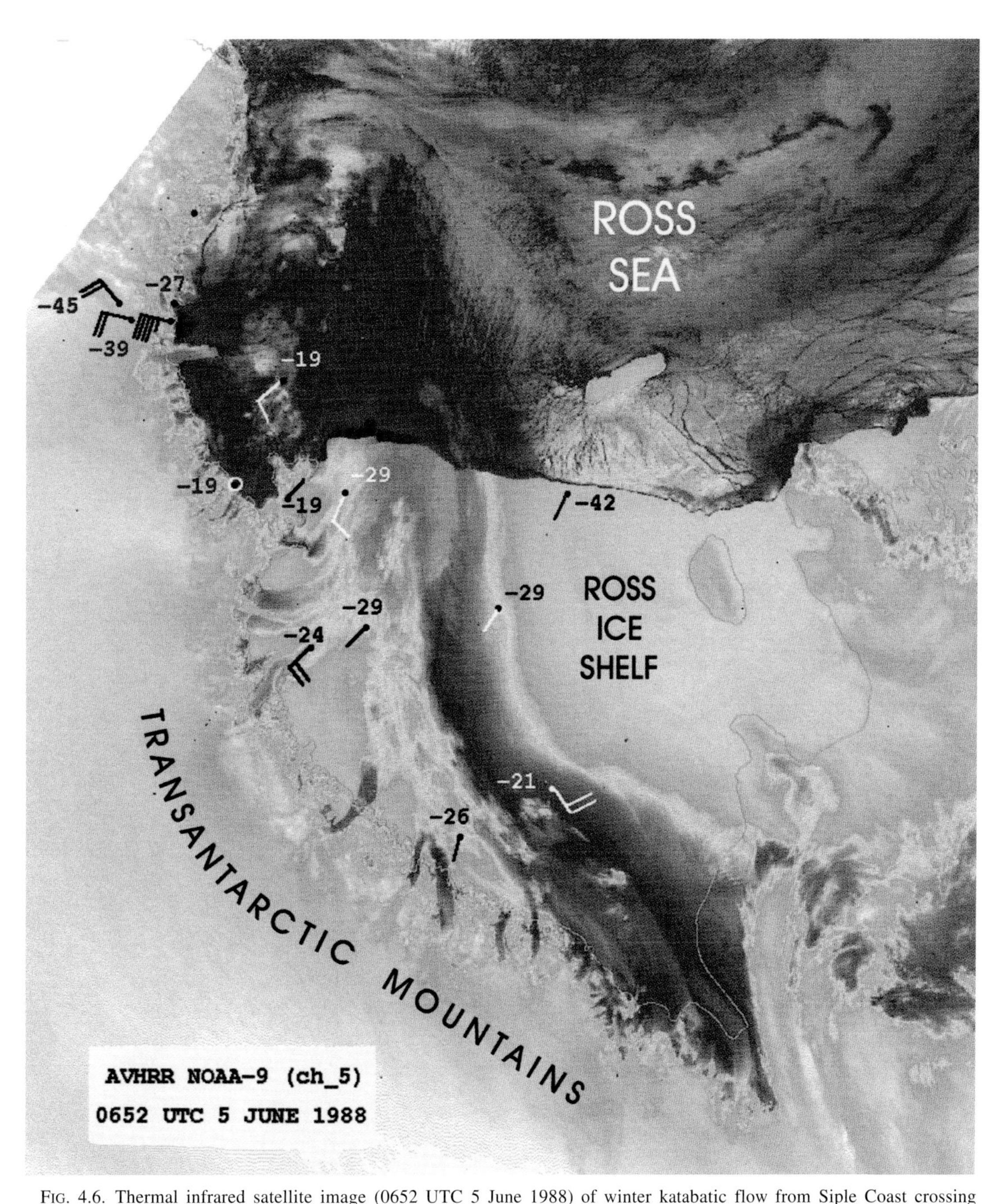

FIG. 4.6. Thermal infrared satellite image (0652 UTC 5 June 1988) of winter katabatic flow from Siple Coast crossing the Ross Ice Shelf (dark and warm signature) to the Ross Sea. Simultaneous AWS observations of air temperature in °C and near-surface winds using conventional notation are plotted. Brightness temperatures near 0°C over the Ross Sea are shown in red to resolve areas of open water and/or thin ice.

the drainage flow through and beyond the confluence zone (Fig. 4.6).

A similar favorable synoptic pressure field over the Ross Ice Shelf also supports the flow through the various glaciers in the Transantarctic Mountains (Bromwich 1989a,b), as can be seen in Fig. 4.6. Breckenridge et al. (1993) note that thermal infrared imagery reveals evidence of katabatic flows through glacier valleys on over 50% of the days on which satellite coverage is obtainable, and they suggest that such drainage conditions are found on nearly 100% of the cloud-free days. The presence of a katabatic wind signature on thermal infrared images was first proposed by Swithinbank (1973), who noticed a number of dark (warm) streaks issuing from the major glacier valleys off the high plateau of East Antarctica. The paradox of dark (i.e., warm) signatures representing what should be predominantly a negatively buoyant katabatic airstream has been discussed by a number of authors including D'Aguanno (1986), Bromwich (1989a,b), Bromwich et al. (1992), Breckenridge et al. (1993), and Carrasco and Bromwich (1993). The extreme horizontal extent of the thermal signatures supports the notion of a negatively buoyant drainage current. In addition, the aircraft observations by Parish and Bromwich (1989) near Terra Nova Bay, in which the airstream was potentially colder than the environment by approximately 5°C and yet appeared dark on the thermal infrared imagery, supports the concept of negatively buoyant airstreams. The dark signature may reflect the strong mixing process associated with katabatic winds, which may result in warmer skin temperatures as compared to the more tranquil surrounding areas (Bromwich 1989a,b), a result that can also be inferred from numerical modeling of surface winds and temperatures (Parish and Waight 1987).

The persistence of the katabatic wind regime over the face of the continent is also a bit puzzling since the coastal periphery about Antarctica is the home of frequent and intense cyclonic disturbances. The horizontal pressure gradients associated with these Southern Ocean cyclones are often large and extend over wide geographical areas. The cyclones, however, tend to follow the strong baroclinic zone, which generally follows the continental perimeter with limited penetration into the interior of the continent (e.g., Mechoso 1980; Parish 1982a). On occasion, when cyclones move onto the ice continent, the antarctic orography acts to reduce the circulation. This can be understood in terms of the conservation of potential vorticity. As a cyclone impinges upon the continent, it experiences a sharp increase in the height of the ice surface. The depth of the vortex column, associated with the circulation of the cyclone, must therefore decrease. The process is much the same as the well-observed vertical shrinking and consequent circulation decrease of extratropical cyclones in North America as the Rocky Mountain barrier is reached. The fact that the ice sheet serves as a natural orographic buffer is one reason in explaining the well-documented persistence of the continental katabatic wind circulation.

To provide some insight as to the natural variability of the continental katabatic wind streamlines, analyses from the ECMWF model for a period of rapid pressure change over the Antarctic continent and high southern latitudes in the winter of 1988 were examined. The analyses are based on twice-daily (0000 and 1200 UTC) reports gridded to a 2.5° latitude/longitude scale. Figure 4.7a illustrates the streamlines of the surface wind for 0000 UTC 28 June 1988. The resolution of the antarctic orography is relatively coarse as compared to the actual terrain, as illustrated in Fig. 4.5, and certain details of the interior ice topography have been smoothed considerably. The broad-scale katabatic wind pattern in Fig. 4.7a is, however, similar to that in Fig. 4.5, with radially diffluent drainage off the high interior of East and West Antarctica. A number of intense cyclonic disturbances can be seen in the streamline depiction of Fig. 4.7a. In each case it is evident that the horizontal pressure gradients associated with the cyclonic circulation draw in cold continental air and thereby appear to enhance the katabatic wind regime on the western flank of the cyclone. At the same time, there is little evidence that even intense cyclones result in upslope motion on the eastern side

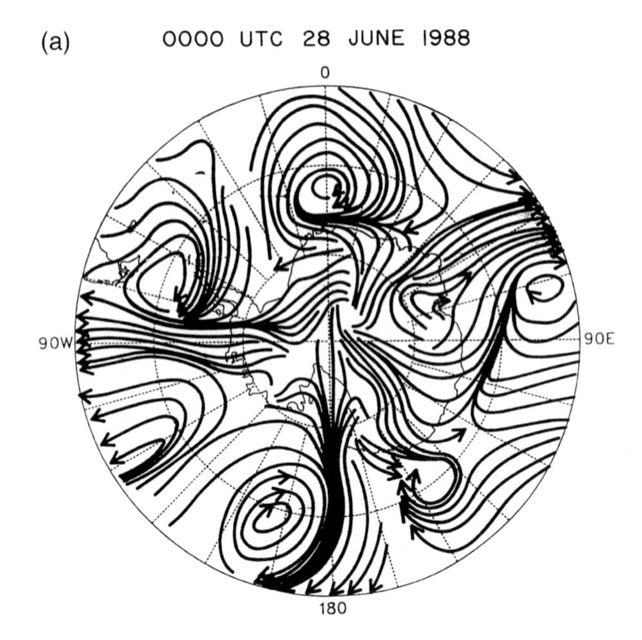

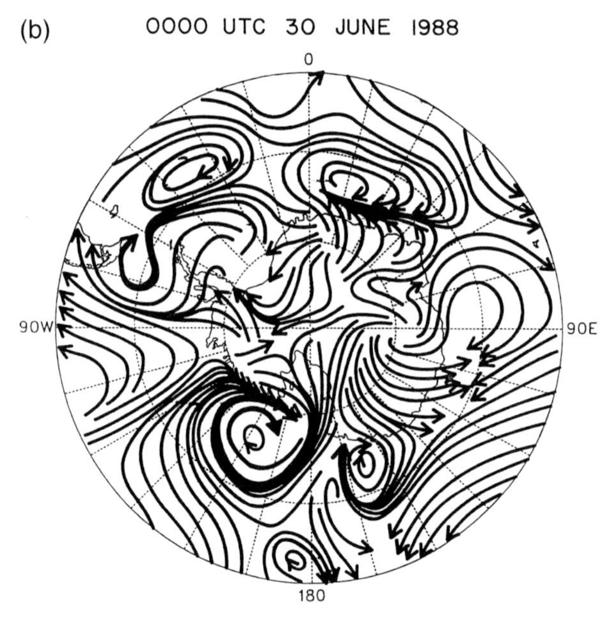

FIG. 4.7. Surface-wind streamline pattern over the high southern latitudes, based on ECMWF analyses for (a) 0000 UTC 28 June 1988 and (b) 0000 UTC 30 June 1988.

of the circulation pattern. The continental katabatic wind regime in the ECMWF maintains the downslope tendency over the steep coastal slopes of East Antarctica. Figure 4.7b illustrates the surface wind streamlines on 0000 UTC 30 June 1988. Inspection of the katabatic wind streamlines in the vicinity of the cyclonic vortices along the coastal rim north of the continent suggests synoptic interaction. Again, there is little evidence that the ambient horizontal pressure gradients are able to force upslope motion on the eastern half of the cyclone. Note that pronounced lines of convergence appear over the ocean to the east of the cyclone centers, with streamlines over the continent reflecting climatological drainage patterns. There is a suggestion that drainage paths become altered over the large interior ice dome in East Antarctica in response to the position and intensity of the cyclonic activity. The katabatic drainage through the Ross Sea corridor on 0000 UTC 30 June taps cold-air reserves over a much larger area of the Antarctic interior than that seen on 28 June. Wind shifts of up to 90° occur over a section of the gently sloping East Antarctic plateau, although the broad-scale drainage off the continent is not appreciably different. Model evidence supports the plethora of observations, which suggest that the continental-scale katabatic windfield over the continent is remarkably robust.

As the katabatic winds move northward across the coastline of Antarctica, the horizontal pressure gradient force decreases abruptly. The ageostrophic offshore propagation distance of katabatic airstreams varies considerably, from typical dissipation in hydraulic jumps close to the East Antarctic coastline (Pettré and André 1991; Schwerdtfeger 1970; Gallée et al. 1996) to frequent propagation for tens to hundreds of kilometers offshore from Terra Nova Bay (Bromwich 1989a; Parish and Bromwich 1989). Gallée (1997) explains the latter behavior as being due to offshore spreading of the laterally confined katabatic airstream blowing from Terra Nova Bay. The adjustment of these airflows to the offshore circulation is responsible for a mass increase to the north of the continent. This piling up of mass offshore from the continental coastline is responsible for the development of a northward-directed horizontal pressure gradient force in the lowest levels of the atmosphere. This pressure field supports a low-level easterly current that nearly encircles the continent. Numerical simulations of the katabatic wind regime, as shown in Parish et al. (1997), clearly depict this band of easterlies, which extend several hundred kilometers northward from the coast. These circumpolar easterlies are prominent climatological features of the low-level wind regime about the Antarctic continent. It has been proposed (Parish et al. 1997) that such a band of sea level easterlies derive dynamical support from the mass flux associated with the time-averaged katabatic wind regime. Such an idea is not new. Previously, Kidson (1947; see also Court 1951) echoed the same idea, noting that the cold-air accumulation just offshore from the Antarctic continent produces a pressure surplus of approximately 6 hPa and that such effects extend some 300 km from the coast. Schwerdtfeger (1984, p. 108) has also proposed that the katabatic winds act "to reinforce and maintain the circumpolar easterlies near sea level."

An alternate way of viewing the circumpolar easterlies is in terms of simple inertial turning of the cold katabatic airstream as it outruns its continental forcing. The horizontal scale and intensity of the easterlies roughly match this idealized picture. The damming of cold air up against the steep continental escarpment provides the northward-directed horizontal pressure gradient force. In this sense the easterlies are an example of a "barrier" wind (Bromwich 1988). Schwerdtfeger (1975, 1979) was the first to describe this motion. The initial antarctic application was along the Antarctic peninsula. He reasoned that as cold, stable air is advected by east winds across the Weddell Sea and reaches the mountain chain, the air resists moving over the barrier and becomes dammed up against the windward side. The hydrostatic increase in pressure supports a low-level southerly airflow directed parallel to the mountains. Numerical experiments (Parish 1983) suggest that the barrier winds may extend some 300 km or more away from the foot of the mountains and that the damming effect may result in a pressure surplus of up to 8 hPa. Similar barrier wind phenomena have been proposed along the Transantarctic Mountains on the western Ross Ice Shelf (Schwerdtfeger 1984; Slotten and Stearns 1987; O'Connor and Bromwich 1988; Bromwich et al. 1992; O'Connor et al. 1994) and have been documented along mountain ranges in middle latitudes as well (Parish 1982b).

4.4. Variations in atmospheric pressure

There is no doubt that the circumpolar belt surrounding the Antarctic continent and extending into the Southern Ocean is one of the most active cyclonic regions on earth (e.g., Streten and Troup 1973). As noted by Schwerdtfeger (1984), the band of low pressure surrounding the continent is reflective of the highest frequency of cyclones. Synoptic analyses have long been hampered by the relative lack of reporting data. In addition, construction of sea level pressure fields over the antarctic orography is difficult, owing to inherent problems of reduction to sea level. Extrapolation of pressure through several kilometers of ice typically results in excessively high sea level values. Thus, sea level synoptic analyses over the high plateau regions are subject to considerable uncertainty. Observations from manned coastal stations, as well as reports from numerous AWS units, clearly show considerable synoptic variability in the surface pressures. Parish (1982a) has noted that significant differences in the variation of surface pressure exist between the Antarctic coastal regions and high continental plateau,

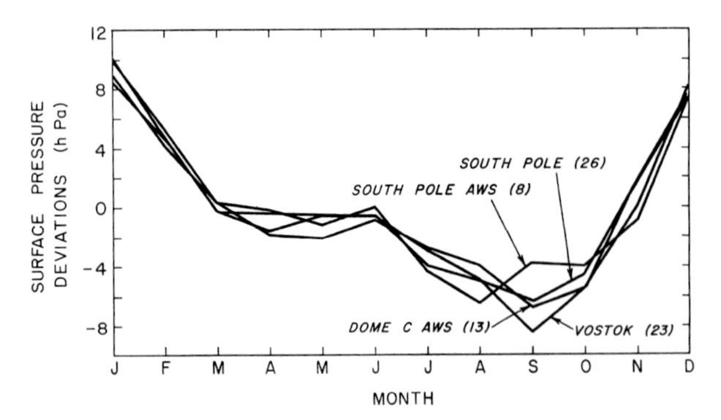

FIG. 4.8. Mean monthly surface-pressure deviations from mean annual surface pressure at interior stations South Pole, Dome C, Vostok, and South Pole AWS. Number in parentheses indicates the number of years in each record.

which can be attributed to the diminishing influence of cyclones on the interior sites. It is instructive to examine the relative importance of synoptic events in contributing to the total variance in surface pressure. The results of power spectrum analyses from the year 1958 for numerous coastal and interior manned stations show that the percentage variance of surface pressure explained over the synoptic time periods of 2.5 to 5 days for coastal stations is nearly three times greater than at the interior sites. Recent work using AWS records near Terra Nova Bay (Parish 1992a; Bromwich et al. 1993), as well as output from the ECMWF 10-yr averages, validates the above conception and again indicates the role of the antarctic orography in buffering the interior plateau from the influence of transient cyclones.

Antarctic orography is also responsible for a large-amplitude seasonal pressure cycle over the high southern latitudes. As an example, Fig. 4.8 illustrates the multi-year averages of mean monthly pressure deviations from the mean annual surface pressure over the interior of the continent for the South Pole and Vostok manned stations, as well as the Dome C and South Pole AWS. Data from the South Pole and Vostok records were taken from Schwerdtfeger (1984) and are for 26 and 23 yr, respectively. The Dome C data were obtained from the AWS records from 1980–93, and the South Pole AWS data record spans 1986–93 (e.g., Keller et al. 1994). Surface pressures at each site display rapid and pronounced changes, which are especially evident during the transitional periods September–December and January–April. Surface pressures over the high plateau undergo a decrease of approximately 12 hPa from January to April and a rise of 16 hPa from September to December. Such changes in the atmospheric mass loading over Antarctica reflect the large-scale thermodynamic and dynamic adjustment near the surface of the continent associated with the rapid changes in the intensity of solar insolation (Radok et al. 1996; Parish and Bromwich 1997). Of note also is the secondary maximum during the midwinter period. A second harmonic is apparent in the annual course of surface pressures over Antarctica, with maxima at the solstices and minima at the equinox periods. The SAO in surface pressure has been discussed extensively (e.g., Schwerdtfeger and Prohaska 1956; Schwerdtfeger 1960, 1967; van Loon 1967, 1972a; van Loon and Rogers 1984; Meehl 1991; Hurrell and van Loon 1994). Schwerdtfeger and Prohaska (1956) suggest that the SAO is based on the differential solar heating of different latitude belts and point out that the meridional difference in solar insolation between 50° and 80°S reaches maxima in March and September. Van Loon (1967) and, more recently, Meehl (1991) note that the SAO mechanism involves the different annual temperature cycles between the Antarctic continent and the surrounding oceans. A discussion of the SAO and modeling considerations are given by Meehl in chapter 10 of this monograph.

Figure 4.9 illustrates the mean monthly zonally averaged surface pressure deviations from the January mean for the years 1985–94 from the ECMWF analyses. It can be seen that the phase of the SAO reverses between approximately 55° and 60°S. The meridional pressure differences between, for example, 70° and 50°S reach maxima during the equinox periods. This phase reversal is responsible for a semiannual oscillation in the geostrophic winds in the lower troposphere surrounding Antarctica (e.g., Schwerdtfeger 1970). The second harmonic in ECMWF surface pressure output from the 1985–94 period explains nearly 40% of the total variance at latitudes 65°–70°S and approximately 25% over the antarctic latitudes 75°–80°S.

Surface pressure changes are especially pronounced during the austral springtime (September–December) and autumn (January–April) transition periods. The largest changes during both periods appear to be maximized over the high interior of Antarctica and

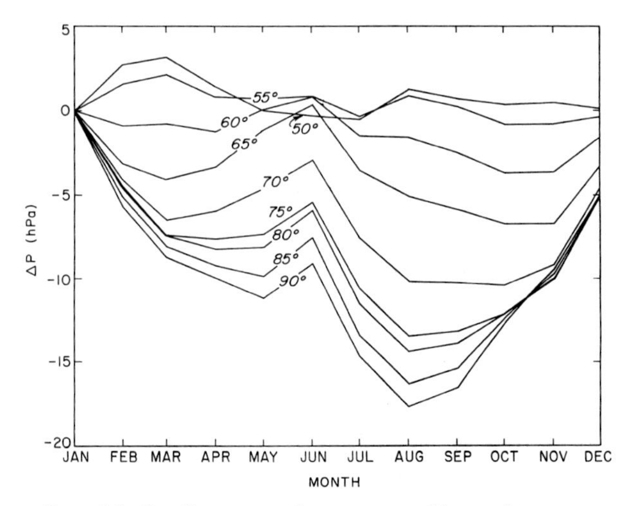

FIG. 4.9. Zonally averaged mean monthly surface-pressure deviations from January mean for latitudes 50°–90°S, based on ECMWF analyses, averaged for 1985–94, in hPa.

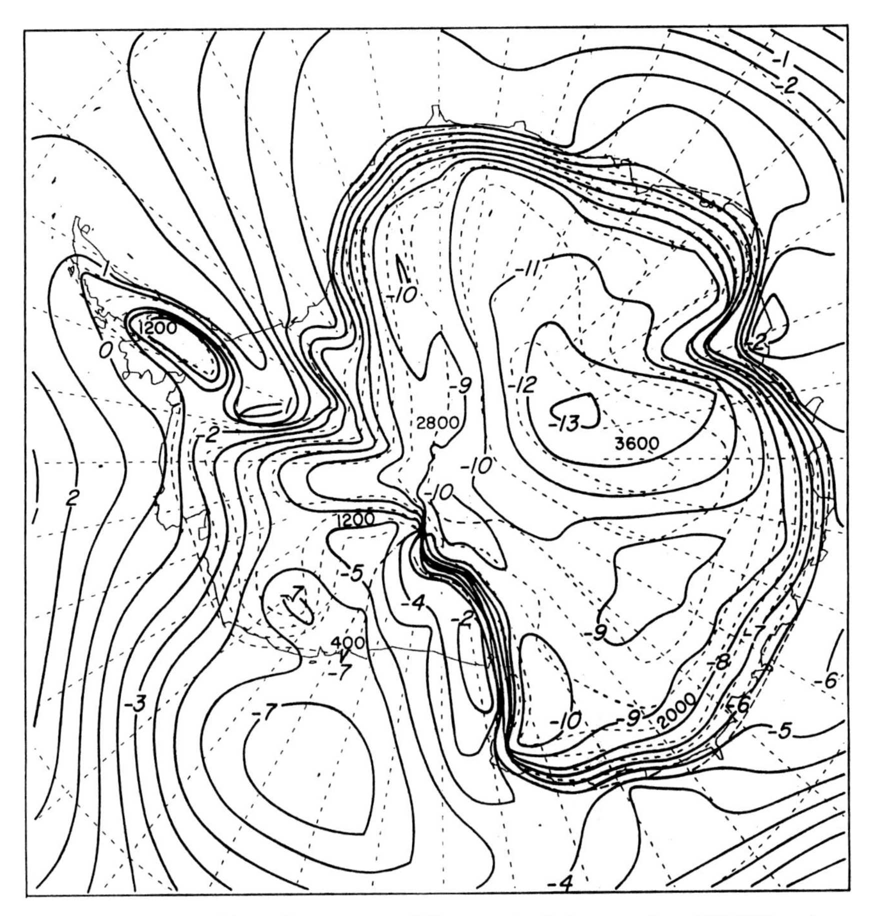

FIG. 4.10. Mean monthly surface-pressure difference, April–January, from ECMWF analyses, averaged for 1985–94, in hPa. Dashed lines are elevation contours in 400-m increments.

decrease rapidly northward from the continental periphery, such that the phase of the mass loading cycle becomes reversed between 50° and 60°S. As an example, Fig. 4.10 illustrates the mean monthly changes in surface pressure from January to April based on the ECMWF analyses for the years 1985–94. Such a depiction is consistent with surface observations, as well as data collected by AWS atop the Antarctic interior (e.g., Keller et al. 1994). Parish et al. (1997) and Parish and Bromwich (1997) have shown that similar patterns of surface pressure increase occur during the austral springtime period, which match the changes discussed in Schwerdtfeger (1984, see his Fig. 6.8).

Observed surface pressure changes in Fig. 4.10 appear to be intimately linked to the underlying antarctic orography. The pattern of pressure differences appears to follow the general orientation of the topography. This is especially obvious along the steeply sloping coastal region, which in Fig. 4.10 is the site of a strong pressure-difference gradient. Such a seasonal pressure-difference pattern is the result of the rapid diabatic cooling of the sloping terrain and lower levels of the atmosphere in response to the diminishing intensity of solar insolation over the continent. Hydrostatic considerations suggest that the pressure difference between vertical levels is dependent on the mean temperature of the atmospheric column. As the mean temperature in a particular atmospheric column decreases, the pressure difference between the surface and some vertical level increases. Assuming sea level pressure does not change appreciably, cooling of the lower layers will appear to result in an apparent pressure decrease at a fixed height. Such is the case over Antarctica during the austral autumn transition period. Figure 4.11 depicts a vertical cross section of the zonally averaged temperature changes during the austral autumn months from January to April, 1985–94. The largest changes are seen near the surface and in the stratosphere.

Hydrostatic considerations alone, however, cannot explain the seasonal surface pressure changes. The decrease in pressure observed over the interior of the Antarctic continent implies a net northward mass flux away from Antarctica. Parish et al. (1997) propose that the mean meridional circulation between Antarctica and the subpolar latitudes becomes modulated at low levels during the austral autumn period. It is the

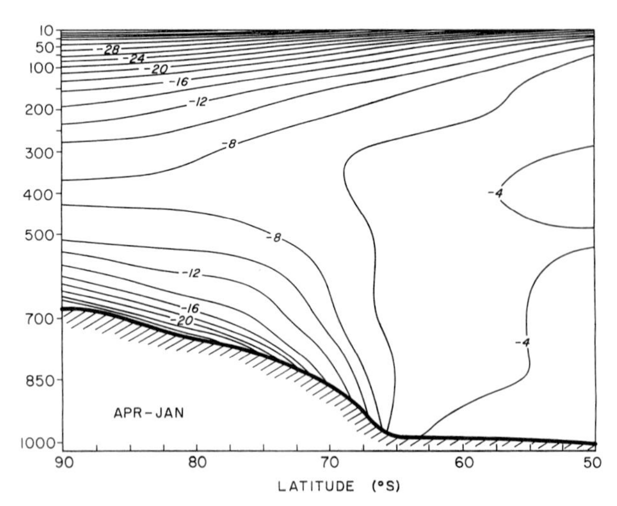

FIG. 4.11. Vertical cross section of mean April minus January zonally averaged isobaric temperatures for latitudes 50°–90°S, based on ECMWF analyses, averaged for 1985–94, in K.

katabatic wind regime that undergoes the most significant change and is the fundamental agent of the northward mass transport during autumn. Midtropospheric level transports toward Antarctica do not undergo the same magnitude of change as the near surface flows, and, hence, a net imbalance in mass transport occurs. During the austral springtime, the sequence of events reverses. The rapidly increasing solar insolation tends to reduce the northward transport via the katabatic wind regime. As a result, the net inflow toward the pole in the middle to upper troposphere, and in the stratosphere, exceeds the outflow in the lower levels, and atmospheric mass loading over Antarctica occurs (Wendler and Pook 1992; Radok et al. 1996). In each seasonal example, the low-level changes in wind and temperature are mutually consistent, and the changes in surface pressure can be viewed as a consequence of the large-scale adjustment process over Antarctica.

4.5. Precipitation characteristics

Precipitation over the grounded antarctic ice sheets is a significant aspect of sea level variations because it represents a removal of water from the global ocean amounting to 7 mm yr^{-1}. For steady-state conditions, an equal amount of water substance is returned to the sea by ice flow and calving and melting at the margin. An increase in precipitation following from circulation anomalies happens essentially instantaneously, but the increase in ice flow to restore steady-state balance takes much longer, resulting in a sea level decrease that persists for hundreds to thousands of years. Understanding precipitation controls is also important to interpreting the climatic records from cores drilled from the ice, as this fell as snow in the distant past.

Direct determination of precipitation amounts in Antarctica is frustrated by a number of factors particularly relating to wind effects, which in cold climates are a notorious obstacle to accurate gauge measurements of precipitation (e.g., Groisman et al. 1994). An additional complication arises for the interior of the continent, where snowfall amounts are very small and are invariably less than the minimum resolution of snow gauges (Bromwich 1988). Accumulation of snow provides a convenient approach for depicting the time-averaged spatial distribution of annual precipitation (Fig. 4.12, after Giovinetto and Bentley 1985). The effects of sublimation to the atmosphere, melting, and drift snow transport are thought to be small on the continental scale in relation to precipitation, and thus, to first order, precipitation equals accumulation. There is growing evidence, however, that the neglect of sublimation, which becomes largest in the summer half-year, may be an invalid assumption (e.g., Stearns and Weidner 1993). Figure 4.12 shows that precipitation decreases by an order of magnitude from the coast to the interior of East Antarctica in conjunction with the dramatic increase in surface elevation. The precipitation rate over West Antarctica is much higher than East Antarctica. This follows from the much greater impact of synoptic forcing and from the time-averaged flow of moisture into this area in conjunction with the circumpolar vortex (Lettau 1969; Bromwich et al. 1995).

Many aspects of antarctic precipitation characteristics are accessible via indirect methods. A promising approach uses the atmospheric moisture budget derived from numerical analyses. The difference between precipitation and evaporation/sublimation ($P - E$) for an area is calculated from the convergence of the moisture transport into the overlying atmospheric volume plus the impact of moisture storage changes (Yamazaki 1992; Budd et al. 1995; Bromwich et al. 1995); note that $P - E$ or net precipitation closely approximates the accumulation rate. This approach is appropriate for averages over large areas and for time periods of seasons and longer. The ECMWF analyses have been compared against antarctic rawinsonde moisture transports and found to closely reproduce these input observations (Bromwich et al. 1995). To begin, Fig. 4.13 shows the derived $P - E$ for the Antarctic continent from the atmospheric moisture budget. The convergence values have been derived at the individual gridbox level and then adjusted to achieve dry-air mass balance (Trenberth 1991). The ECMWF data are sampled at 2.5° resolution from the original spectral data and, as a result, aliasing effects occur that are particularly noticeable in the convergence field (Trenberth 1992, see p. 80). Here, zonal averaging on a length scale of 250 km is used to obtain numerical stability for the estimates. In polar regions, this method is appropriate because the spectral data describe information on progressively smaller

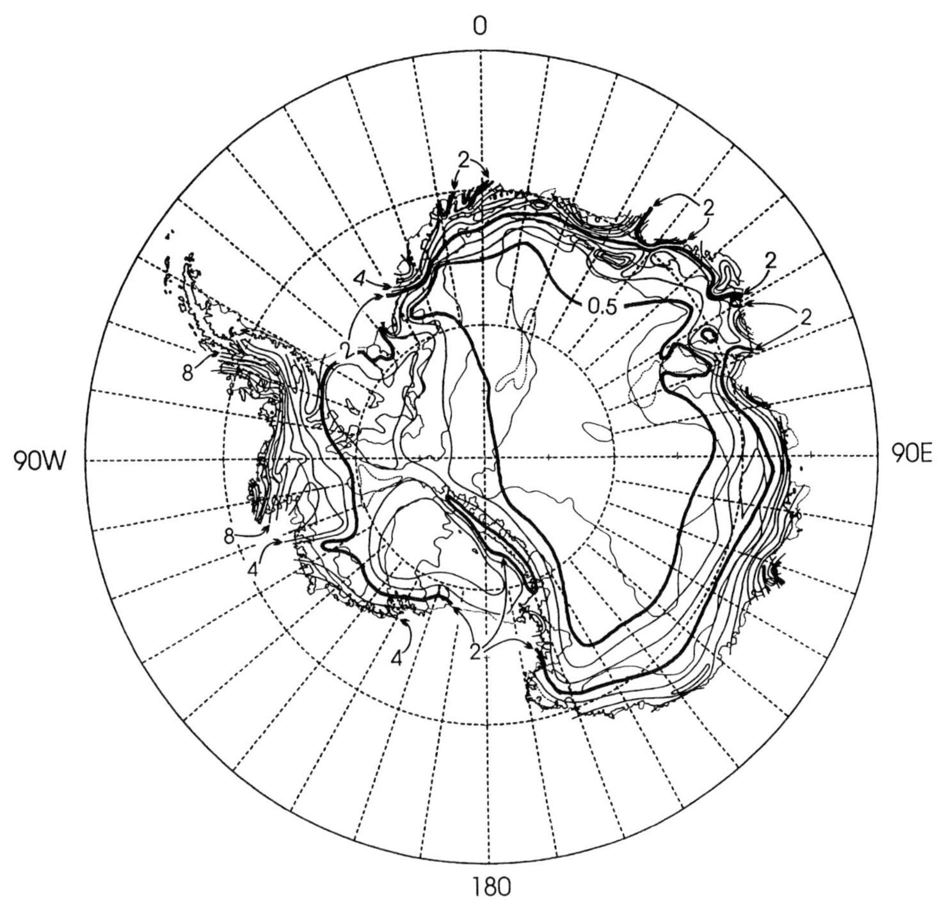

FIG. 4.12. Annual accumulation for Antarctica, adapted from Giovinetto and Bentley (1985). Solid lines are accumulation isopleths in 100 kg m^{-2} yr^{-1} (or, equivalently, 100 mm yr^{-1}); 0.5 and 2.0 contours are bolded. Light solid lines are elevation contours in km.

and more unrealistic spatial scales with increasing latitude. The loss of precision using gridpoint data must be considered negligible relative to the expected accuracy in data-sparse polar regions.

As in the observations, large values are found along the coast of East Antarctica in Fig. 4.13, and an extensive area with values less than 5 cm yr^{-1} (or 0.5 of the units used in Fig. 4.12) is located over the East Antarctic plateau. The moisture budget calculation resolves the marked accumulation gradient across West Antarctica and the large values along the Bellingshausen Sea coast and the west side of the Antarctic peninsula. Significant problems are evident in the Ross Sea–Ross Ice Shelf area; however, this sector is characterized by large decadal variability of moisture convergence (Cullather et al. 1996) that may invalidate the comparison between the 1985–94 moisture budget calculation and the long-term accumulation depiction. The negative accumulation values ($E > P$) in East Antarctica are close to and not much larger in area than net ablation zones resolved by Giovinetto and Bentley (1985). To summarize, there is a good overall fit in terms of features and magnitudes between the observed accumulation pattern (Fig. 4.12) and that calculated from the atmospheric moisture budget (Fig. 4.13).

Figure 4.14 shows the annual cycle of $P - E$ for the Antarctic continent, derived from the ECMWF analyses. There is a broad winter maximum and a summer minimum, consistent with earlier evaluations (Bromwich 1988). In addition, there is a suggestion of semiannual maxima in the fall and spring. Precipitation and accumulation records from the Weddell Sea and Antarctic peninsula areas show that this semiannual signal dominates in that region (e.g., Eicken et al. 1994), due to the semiannual cycle in the intensity and latitudinal location of the circumpolar trough (Turner et al. 1997). Similar annual precipitation cycles apparently characterize the ocean areas just to the north of Antarctica (van Loon 1972b). Also included in Fig. 4.14 is the derived annual cycle for areas above 1500 and 2500 m elevation contours, following Budd et al.

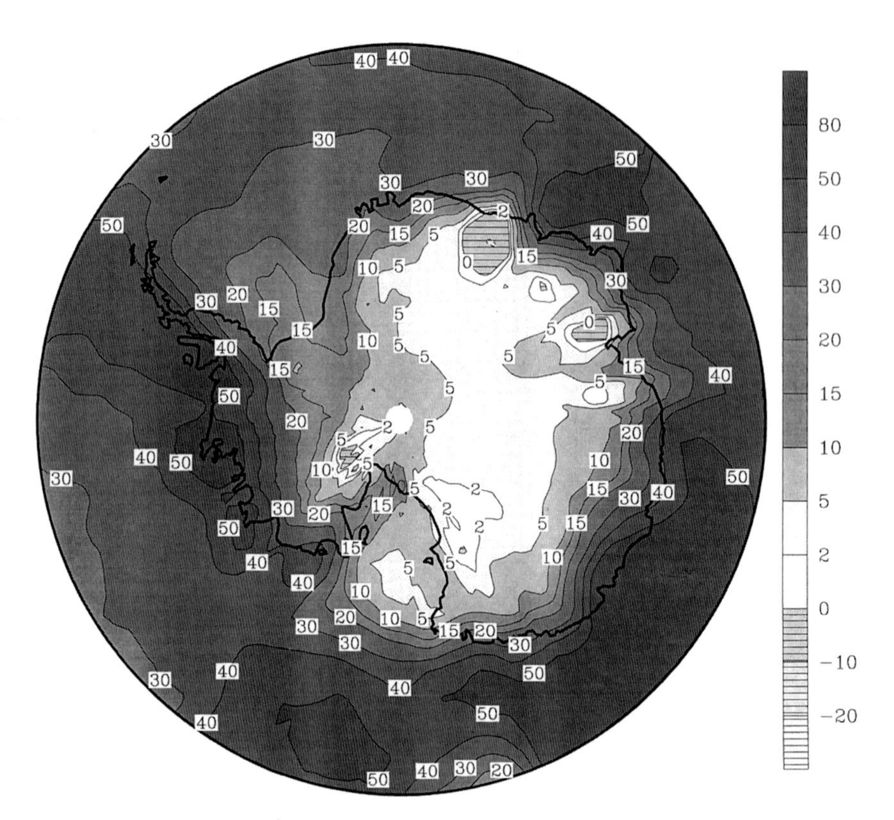

FIG. 4.13. Average accumulation distribution for Antarctica, computed from the ECMWF atmospheric moisture budget for 1985–94, in cm yr^{-1}.

(1995). Values are nearly constant from March to October, with little evidence of a semiannual variation; this is contrary to the results of Budd et al. (1995), but consistent with the limited surface observations (Bromwich 1988). The southward dissipation of the semiannual precipitation cycle from the Southern Ocean to the East Antarctic plateau is consistent with the parallel decrease of the importance in the semiannual pressure oscillation (section 4.4).

Episodic synoptic processes dominate precipitation formation in the coastal margins (e.g., Turner et al. 1995). As an illustration of this, Fig. 4.15 shows the annual mean eddy moisture transport vectors in relation to the sea level pressure field. The eddy transports (with respect to monthly means) are larger just to the east of the quasi-stationary lows in the circumpolar trough, as can be seen more clearly in rawinsonde-derived transports (Bromwich 1988). The eddy transports are almost everywhere directed poleward. For the continent as a whole, about 90% of the moisture convergence (or equivalently $P - E$) is due to the eddy component. The impact of synoptic forcing decreases with elevation in East Antarctica (compare Jonsson 1995) such that the semicontinuous fall of ice crystals from apparently clear skies becomes the dominant precipitation formation mechanism (Radok and Lile 1977; Schwerdtfeger 1984; Bromwich 1988).

In at least some coastal areas, mesoscale cyclones are found to contribute significantly to the total precipitation, as suggested by some general circulation model studies (e.g., Tzeng et al. 1993). To the north of McMurdo Station on Ross Island lies the very active mesoscale cyclogenesis area near Terra Nova Bay (Bromwich 1991; Carrasco and Bromwich 1994, 1996), which is associated with that area's intense katabatic winds. Rockey and Braaten (1995) estimate that about 38% of the McMurdo Station precipitation is associated with mesoscale cyclones. Modeling studies by Gallée (1996) reveal how this phenomenon occurs and resolve a persistent, moist return flow, which could contribute to the precipitation over the Antarctic interior.

Temporal variability in antarctic precipitation is also accessible via the atmospheric moisture budget. Cullather et al. (1996) examined the ECMWF analyses from 1980 to 1994 and found that the greatest variability in moisture convergence took place in the South Pacific sector of the continent. As Fig. 4.16 illustrates, the variability for part of this area (120°W–180°, 75°–90°S) was highly correlated and in phase with the SOI from 1980 to 1990 but became anticorrelated after 1990, coinciding with the start of the extended series of ENSO events in the early 1990s (Trenberth and Hoar 1996). The ENSO variability is associated with

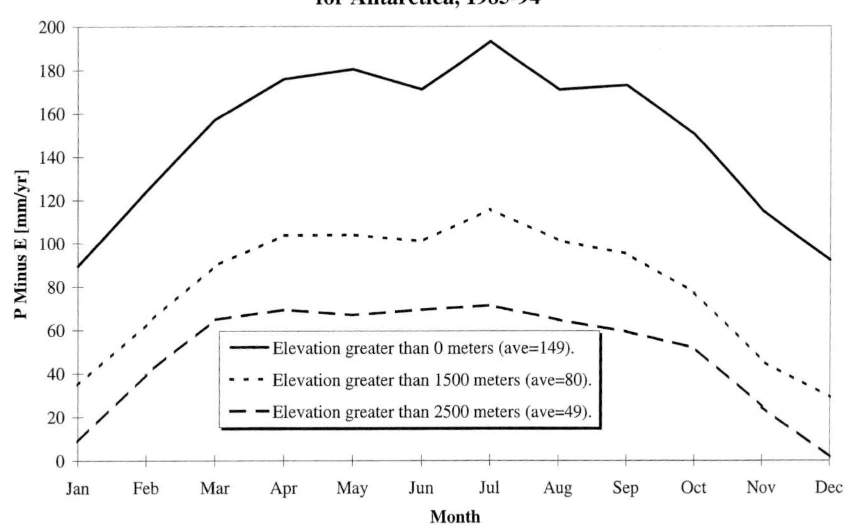

FIG. 4.14. The annual variation in accumulation rate derived from the ECMWF atmospheric moisture budget for various elevation domains, averaged for 1985–94, in cm yr^{-1}.

migrations in the Amundsen low. As shown in Fig. 4.17 during normal precipitation periods in this West Antarctic sector, here characterized by 1980 (left), the Amundsen Sea low is located near its long-term average position to the northeast of the Ross Sea, and moisture enters the West Antarctic sector from almost due north. During reduced precipitation periods (1982, right), the Amundsen Sea low is 1500 km to the east of its position in 1980, and the moisture wraps around the low to enter the sector mostly from the east. Marked blocking was found to persist over the Wilkes Land part of East Antarctica, starting in 1991 and extending to at least 1994; this coincided with the establishment of the anticorrelation between the SOI index and West Antarctic sector precipitation.

The moisture budget analyses also provide some idea of the temporal trends in antarctic precipitation. Figure 4.18 depicts net precipitation for Antarctica and its adjacent permanent ice shelves from 1980 to 1994. A statistically significant upward trend of 3 mm yr^{-1} (or 2% yr^{-1}) is found, consistent with an earlier synthesis of mostly ground-based accumulation observations from the 1960s and 1970s (Bromwich and Robasky 1993). The confidence in this trend is tempered by the lower reliability of the WMO version of the ECMWF analyses (Trenberth 1995) than the WCRP archive (Cullather et al. 1996) and the major impact of the 1982/83 El Niño event. In glaciological studies, mention is often made of the high spatial correlation between long-term averages of accumulation and surface air temperature (e.g., Robin 1977). To test whether this idea applies to temporal variations, the 500-hPa temperatures (surface temperatures were not available for the entire period but gave similar results for 1985–94) are compared in Fig. 4.18 with the derived antarctic accumulation amounts. On the annual timescale, accumulation and temperature have little association ($r = 0.37$), but there is an upward trend over the 15 years that is as significant as that for accumulation. The latter suggests that, over the decadal timescale, accumulation and temperature are positively correlated; dividing the temporal regression slopes gives a sensitivity of about 20% precipitation increase per °C temperature rise, consistent with that quoted in some glaciological studies (compare Morgan et al. 1991).

4.6. Far-field impacts of high southern latitude processes

Given the persistence of the radially diffluent katabatic wind regime over the continental surface, it is not surprising that the low-level drainage elicits a large-scale tropospheric response throughout the high southern latitudes. Continuity requirements dictate that the low-level divergence associated with the katabatic drainage must be compensated by an inflow of relatively warm air at middle to upper levels of the troposphere over Antarctica, which converges and subsides over the continent. It is through this process that the katabatic wind regime is resupplied. The katabatic wind regime thus forms a significant component of the lower branch of a time-averaged thermally direct meridional circulation. As an example of the time-averaged conditions, Fig. 4.19 depicts the zonally averaged meridional and vertical velocity components for the month of July, based on analyses from the ECMWF 12-hourly output for the 10-yr period 1985–

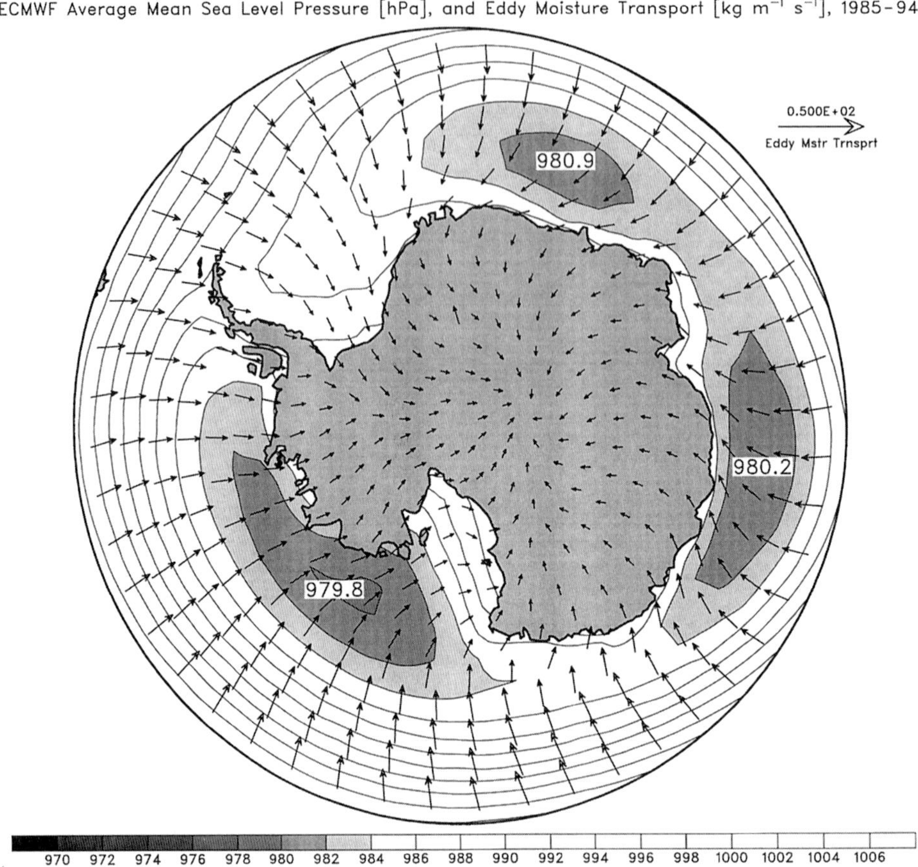

FIG. 4.15. Average MSLP field for 1985–94, contoured every 2 hPa, with average vertically integrated eddy moisture transport (arrows), in kg m^{-1} s^{-1}.

94. The meridional circulation is well defined and consists of broad subsidence over the continent south of approximately 70°S, which feeds the lower-level circulation including the katabatic wind regime. The return flow toward Antarctica in the middle and upper troposphere is deep and relatively weak. The lower-level branch is by far the most intense segment of the meridional circulation and primarily represents the katabatic component. Both ECMWF and the Community Climate Model Version 1 of the National Center for Atmospheric Research (NCAR CCM1; Parish et al. 1994) analyses suggest that this mean meridional circulation extends throughout most of the troposphere. The rising branch situated between 65° and 55°S appears to be associated with cyclone activity to the north of the Antarctic continent. There may, however, be some forcing of this rising motion from the katabatic outflow. As the drainage flows move away from the continental terminal slopes, deceleration occurs as the airstream outruns the dynamic support provided by the sloping ice surfaces. The attendant convergence of the katabatic flow just offshore from Antarctica supports the rising motion branch adjacent to the continent.

The existence of this time-averaged circulation has been discussed by a number of authors, including Radok (1973) and Mechoso (1980, 1981). Egger (1985) and James (1988, 1989) have noted that such a mean meridional circulation has important dynamic constraints on the circulation in the free atmosphere. The middle- to upper-tropospheric convergence over the Antarctic continent implies that cyclonic vorticity is continually being generated. James (1989) suggests that the drainage flow off the Antarctic continent may help center the polar vortex over Antarctica. This "anchoring" of the upper-tropospheric vortex over Antarctica has also been suggested by Parish (1992a). The mean July 300-hPa circumpolar vortex from the 1985–94 ECMWF record (Fig. 4.20) shows a zonal flow pattern tied to the Antarctic continent. There is little evidence of the time-averaged standing waves seen in the NH. In addition, the center of the vortex is directly over the Antarctic continent, in contrast to the more diffuse position of the NH vortex center. Two-dimensional numerical simulations by Egger (1985), James (1989), and Parish (1992a,b) show that an upper-level vortex forms in response to the meridional

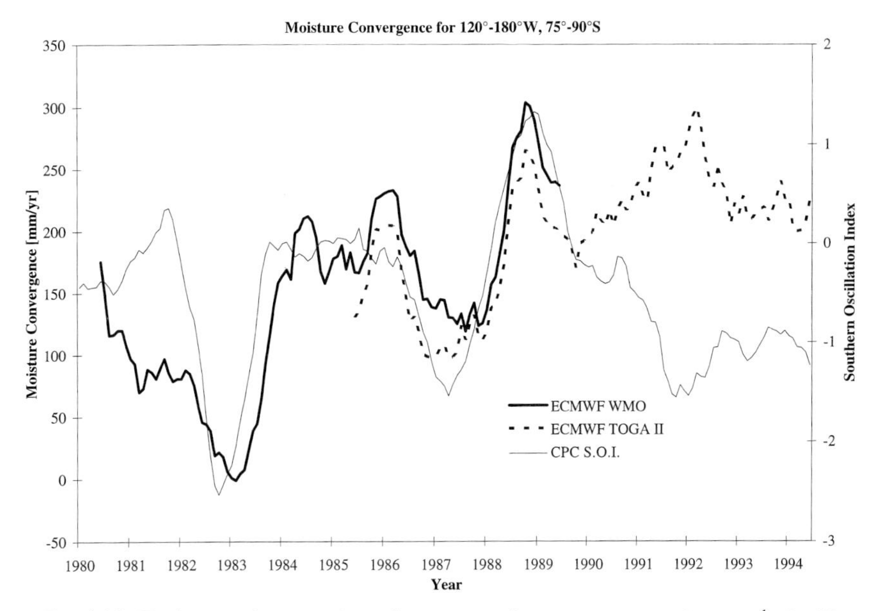

FIG. 4.16. Twelve-month centered running mean moisture convergence (mm yr^{-1}) for West Antarctic sector bounded by 75°S and 120°W–180° in relation to the SOI (hPa). Adapted from Cullather et al. (1996).

convergence above the Antarctic continent (see also Simmonds and Law 1995). Horizontal pressure gradients develop in the upper troposphere, which oppose the katabatic wind regime and, on the timescale of a week or so, reach anomalously large values such that the drainage flow spuriously decays. Egger (1985) and James (1989) note, in contrast to the depictions by the above axisymmetric experiments, that the real atmosphere over Antarctica must rid itself of cyclonic vorticity to prevent the upper-tropospheric vortex from reaching such an intensity as to completely shut down the katabatic circulation. Egger (1985, 1991, 1992) and James (1989) suggest that synoptic eddies are responsible for the transport of angular momentum northward so as to relieve the high southern latitudes of this vorticity buildup. Recently, Juckes et al. (1994) have evaluated the momentum budget over the Antarctic continent using results from a GCM simulation. Their results support the idea that the eddy mass flux associated with transient cyclones transports angular momentum northward from the high southern latitudes.

The mean meridional circulation between Antarctica and subpolar latitudes is also responsible for the seasonal cycle of mass loading onto the continent during the austral springtime and net transport away during the austral autumn. As indicated earlier, such mass transports are the result of large-scale thermodynamic and dynamic adjustments in response to the solar insolation cycle. The magnitude of the surface pressure changes and the areal extent, as illustrated in Fig. 4.10, suggests that significant mass fluxes take place across the Antarctic coastline, and a response in the more northerly latitudes of the SH must occur. Figure 4.21a depicts the zonally averaged surface pressure changes during the transition periods January–April and September–December, based on the ECMWF 12-h analyses for the years 1985–94. The antarctic surface pressure changes are by far the largest, yet there is a compensating surface pressure change that extends to the subtropics. Assuming that most of the pressure change signal is the result of SH processes and little mass transport from the NH is seen, it appears as though the seasonal surface pressure changes over Antarctica influence mass redistribution over nearly the entire hemisphere. An approximate mass budget can be obtained by weighting the surface pressure changes by the cosine of the latitude and then integrating northward from the South Pole. The results for the ECMWF 1985–94 period are illustrated in Fig. 4.21b and show that the surface pressure changes over the Antarctic continent require a compensating atmospheric mass adjustment that reaches the subtropics. Implications of this seasonal transport of atmospheric mass across the Antarctic are not known, although there appears to be a teleconnection with surface pressures over the continental landmasses of Australia, southern Africa, and South America, which show pressure trends of opposite sign during the transition periods (not shown).

The surrounding oceanic region to the north of Antarctica is the site of numerous transient cyclonic disturbances, which significantly influence the day-to-day wind and pressure fields along the coast and the time-averaged fields north of the continent. As noted by Schwerdtfeger (1984), the circumpolar sea level low pressure belt is a reflection of the maximum in

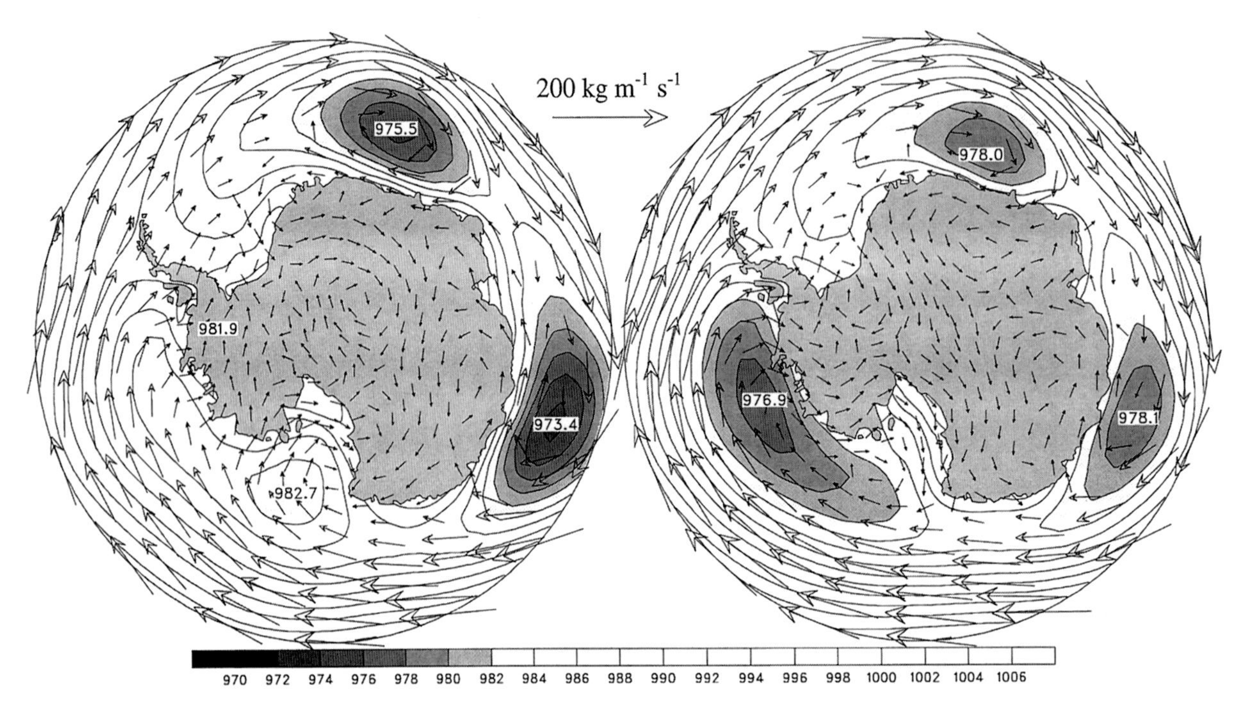

FIG. 4.17. Typical annual MSLP patterns for normal (left, 1980) and reduced (right, 1982) West Antarctic precipitation events. Vectors represent vertically integrated moisture flux (plotted maximum vector is 200 kg m^{-1} s^{-1}). MSLP field is contoured every 2 hPa. Adapted from Cullather et al. (1996).

cyclonic activity. Figure 4.22 is a depiction of the mean sea level pressure field for July based on the individual 0000 and 1200 UTC ECMWF analyses for the years 1985–94. The position of the circumpolar trough is seen to be situated between 58° and 66°S, some 400–700 km north of the coastline. Schwerdtfeger (1984) estimates the mean position of the sea level trough to be about 62°S adjacent to the East Antarctic coastline and 65° elsewhere. The position of the individual low pressure centers in Fig. 4.22 fits well with the mean climatology, as indicated in Tzeng et al. (1993). It is probably more than coincidence that the low centers at 150°W and 100°E seem to be situated on the western side of noted confluence features. Such a synoptic configuration maximizes the cold-air transport from the continent by tapping into the katabatic confluence zones. Bromwich et al. (1992) have noted that strong katabatic surges are common over the Ross Sea when synoptic-scale low pressure disturbances are present over the western Amundsen Sea. It may be the case that the topographically constrained drainage features play a role in establishing preferred locations of cyclonic disturbances about the continent.

There is increasing evidence that teleconnections exist between Antarctica and the subtropics. In particular, evidence of the ENSO signal in antarctic data has been documented. Such a relationship between ENSO events and antarctic conditions was first reported by Savage et al. (1988), who observed that the 1983 air temperature at the South Pole was the coldest since observations began in 1957. They noted that this was accompanied by an increase in the surface winds over a wide area of the continent. These changes followed the strongest ENSO event this century during the 1982/83 austral summer (National Research Council 1983). Smith and Stearns (1993a,b) continued this work by examining the surface pressure and temperature records from Antarctica and found distinct pattern changes from the year before to the year after the minimum in the SOI index (closely associated with El Niño events in the SSTs of the tropical Pacific Ocean) in composites of the six "warm events" between 1957 and 1984. No physical basis for the observed teleconnections was provided by these studies. Karoly (1989), Kitoh (1994), and Chen et al. (1996) note that the behavior of the split jet stream over the South Pacific Ocean is associated with the SOI variations. As shown by Fig. 4.23, adapted from Chen et al. (1996), the subtropical jet near 30°S strengthens in conjunction with SOI minima, whereas the polar front jet near 60°S is strongest during periods when the SOI is near zero or is positive. Using the 1986–89 time period as a case study, when a moderate warm phase was immediately followed by a pronounced cold phase, Chen et al. showed that the subtropical jet variations were a direct consequence of the tropical heating anomalies associated with the Pacific sea surface temperature changes. The acceleration of the polar front jet in the La Niña event of 1988/89 (SOI > 0) was

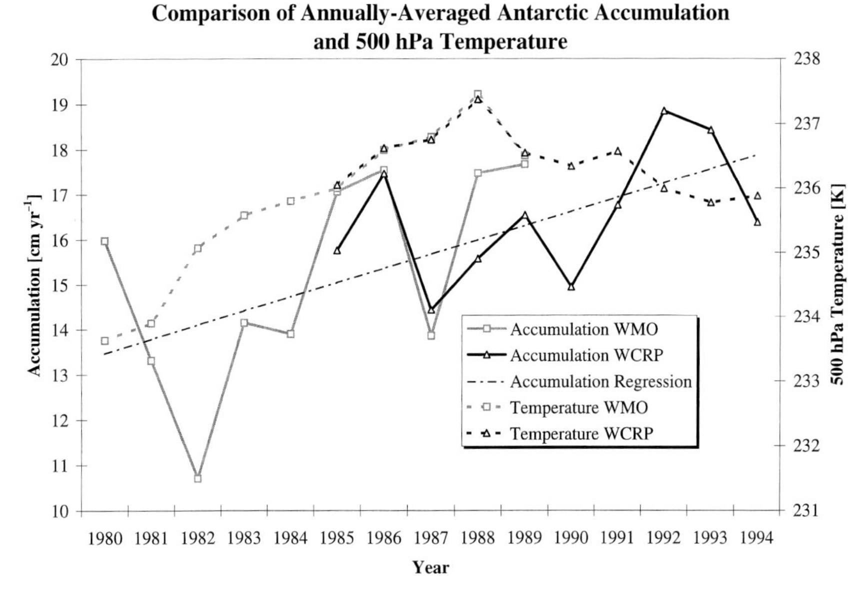

FIG. 4.18. Comparison of annually averaged antarctic accumulation (cm yr^{-1}), computed from the ECMWF atmospheric moisture budget, and spatially averaged 500-hPa temperature (K), for 1980–94.

found to result from the poleward propagation of cyclonic disturbances from the SPCZ and the equatorward propagation of disturbances from Antarctica.

This ENSO teleconnection is responsible for the zonal migrations in the Amundsen Sea low discussed earlier and the strong modulation of the precipitation rate over West Antarctica. Particularly numerous are reports of an ENSO impact on antarctic sea-ice extent (e.g., Chiu 1983; Carleton 1989; Xie et al. 1994; Simmonds and Jacka 1995; Gloersen 1995). Xie et al. (1994) present results showing strong correlations between sea-ice extent and equatorial SST anomalies and refer to the relation between ENSO events and antarctic sea ice as the Southern Oceanic Oscillation. Because of the broad variety of observations including sea-ice extent and meteorological records, the SOI phenomenon in Antarctica is perhaps the most thor-

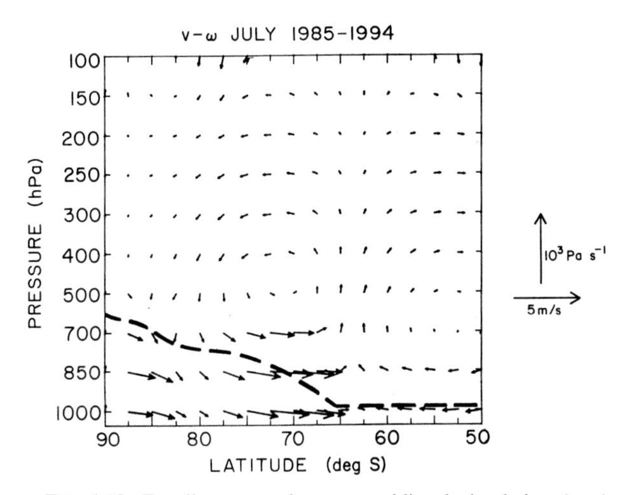

FIG. 4.19. Zonally averaged mean meridional circulation (v–w) interpolated to standard isobaric levels for July, based on ECMWF analyses for the 10-yr period 1985–94. Thick dashed line represents zonal-mean orography of Antarctica.

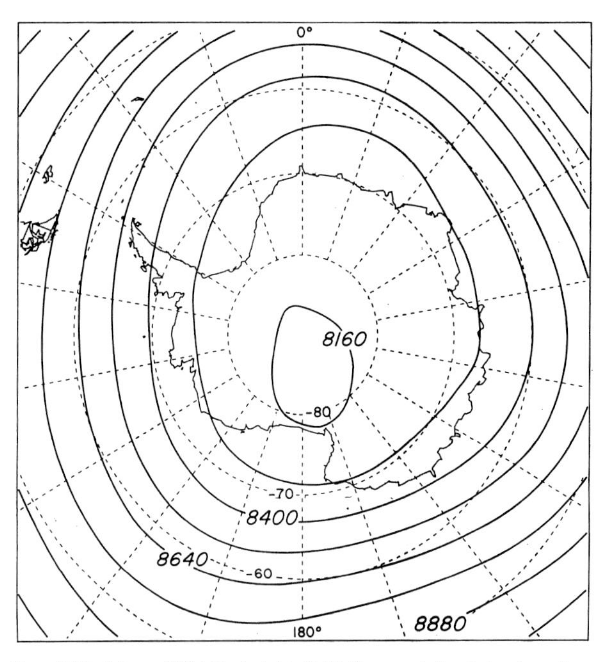

FIG. 4.20. Mean 300-hPa height field (in gpm) for July, based on ECMWF analyses for the 10-yr period 1985–94.

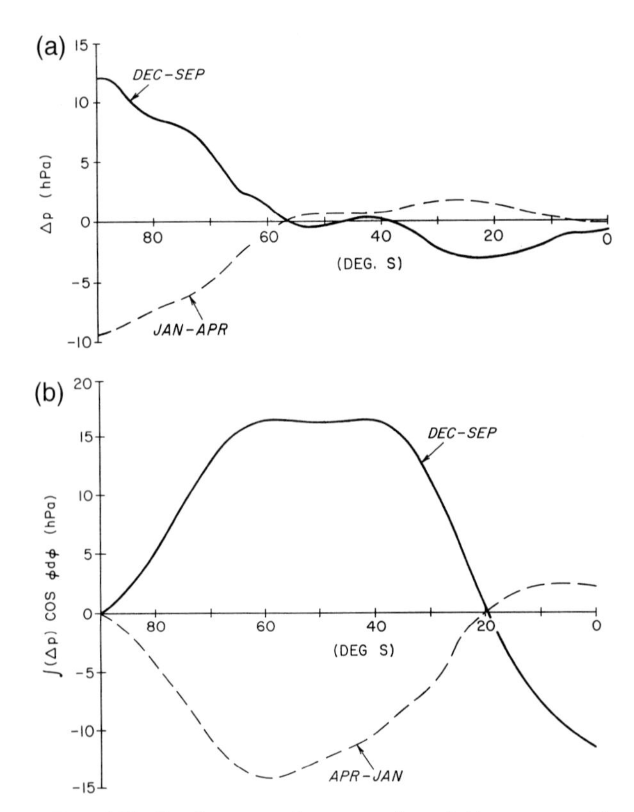

FIG. 4.21. Zonally averaged representation of (a) mean monthly latitudinal variation of surface-pressure differences December minus September and April minus January; and (b) mean latitudinal area-weighted mass budget integrated northward from South Pole for transition seasons April–January and December–September, based on ECMWF analyses for the 10-yr period 1985–94.

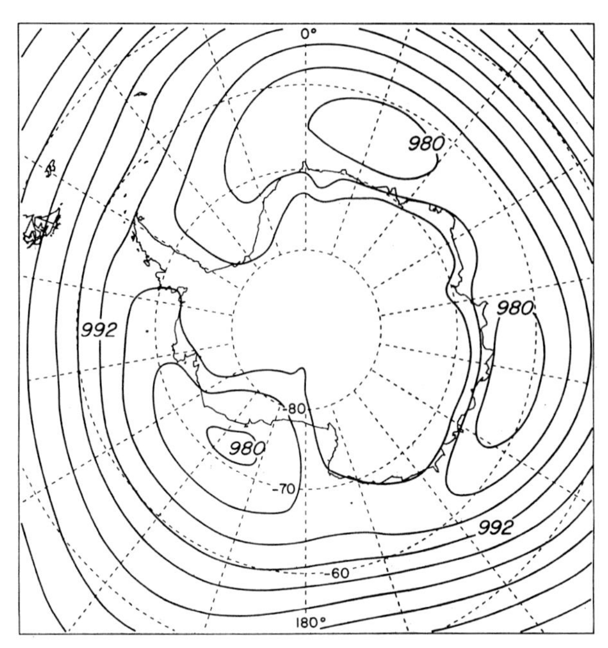

FIG. 4.22. MSLP pattern for July, based on ECMWF analyses for 10-yr period 1985–94.

oughly documented interannual variability in high southern latitudes.

4.7. Discussion

The vast antarctic ice sheet plays fundamental roles in its perpetuation (e.g., Oglesby 1989). The height and reflectivity of the surface guarantee that minimal melting of the ice sheet will be confined to the margins during the short antarctic summer. The blocking effect of the ice sheet limits the poleward propagation of cyclones, with most precipitation falling on the flanks of the ice sheet and leaving little moisture to precipitate in the high interior. Thus, the ice sheet exerts fundamental controls on the atmosphere that help to maintain its present configuration, where snow accumulation is nearly balanced by ice flow to the ocean.

The antarctic katabatic wind regime, which is a consequence of the strong radiative cooling adjacent to the elevated and sloping ice surface, is found to exert influences far beyond the shallow boundary layer. The need to supply the near-surface air blowing to the ocean couples the entire troposphere to the boundary layer. The coupling between the troposphere and the stratosphere is a relatively unexplored topic that may have major consequences for the low winter stratospheric temperatures that are a prerequisite for the recently enhanced springtime depletion of ozone, known as the antarctic ozone hole (see chapter 6). Mass imbalances that arise in the katabatic wind circulation during seasonal transitions result in large mass exchanges between high and middle/low latitudes. The consequences of variations in this exchange's timing and magnitude offer the potential of a major role for Antarctica in Southern Hemisphere climate variability.

The broader-scale importance of antarctic meteorology discussed above, in conjunction with the acknowledged role of the antarctic hydrologic cycle on eustatic change, strongly argues for increased monitoring of the antarctic climate. At present, however, the total amount of manned station observations, particularly the number of upper-air reports, is in decline as a result of former Soviet base closures (Cullather et al. 1997). Data scarcity places additional importance on the remaining observational resources, particularly automatic weather station and satellite data. For example, the 500-hPa geopotential height over the Antarctic interior can be reasonably estimated from AWS surface pressure and temperature readings (Phillpot 1991; Radok and Brown 1996). In the future, however, the maintenance of an adequate observational network will have to become a priority if a complete understanding of the antarctic climate system and its remote impacts is to be realized.

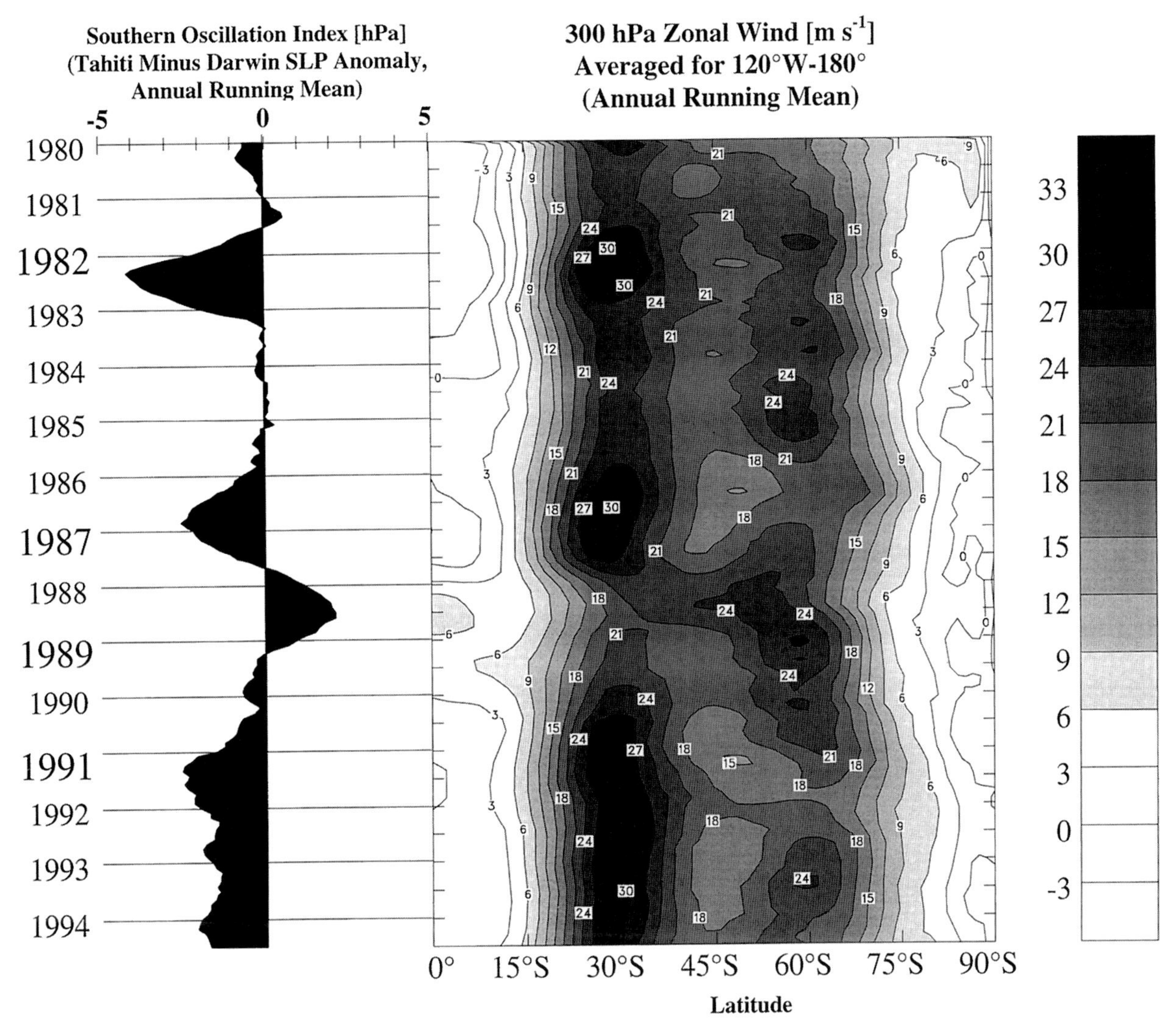

FIG. 4.23. Hovmöller diagram of annual running mean 300-hPa zonal wind, averaged over 120°W–180°, from ECMWF analyses, in m s^{-1}. On the left is the annual running mean SOI, in hPa. Adapted from Chen et al. (1996).

The increased importance of utilizing all available observations in a comprehensive data assimilation has stimulated an international collaboration to examine analyses and forecasts produced by the operational weather forecasting centers. Under the FROST program (Antarctic First Regional Observing Study of the Troposphere; Turner et al. 1996), an assessment is being made of the capability to utilize meteorological data over and around Antarctica and improve the derivation of atmospheric products from satellite data for the region. As can be seen from the present results based on ECMWF output, the availability of a complete and dynamically consistent dataset allows many aspects of antarctic atmospheric behavior to be explored in a more comprehensive fashion than previously.

Numerical modeling is increasingly important in attempts to understand the antarctic atmospheric environment and provides a physically consistent approach to interpolation between the sparse observation points. Substantial improvement in the performance of general circulation models in high southern latitudes has been achieved over the last decade, such that many aspects of antarctic climate (surface temperature, inversion strength, accumulation, surface winds, etc.) can now be captured with some fidelity (Schlesinger 1984; Simmonds 1990; Tzeng et al. 1994; Connolley and Cattle 1994; Krinner et al. 1997). Improvements in climate model simulations of the Southern Hemisphere extratropical circulation have been related to higher resolution, which is necessary for resolving the steep antarctic coastal slopes (chapter 10). Mesoscale models are being used to study katabatic winds, as extensively discussed earlier; airflow around complex terrain (Seefeldt 1996); mesoscale cyclone formation (e.g., Engels and Heinemann 1996); and coastal polynya formation (Gallée 1997), among other things. Initial efforts have been made to couple GCM and mesoscale models for antarctic latitudes (Hines et al. 1997a). A prerequisite for almost all such investigations is adaptation of the planetary boundary-layer parameterization to treat the stable antarctic conditions (e.g., Hines et al. 1995; Krinner et al. 1997). In addition, major problems can occur in the simulation of the atmospheric hydrologic cycle, such as the gross oversimulation by some models of the cloud amount and thickness over the East Antarctic plateau during winter (Hines et al. 1997b) and anomalous moisture diffusion over the coastal slopes (Connolley and King 1996).

For long-term investigations, there is a vital need to obtain a more complete picture of the spatial and temporal variability of antarctic atmospheric behavior. A primary goal of the International Trans-Antarctic Scientific Expedition is the collection and annual interpretation of environmental parameters (accumulation rate, temperature, etc.) on a continent-wide basis for the last 200 years through the use of shallow ice cores (Mayewski 1996). Should the full potential of this project be realized, Antarctica will be transformed from the continent with the most poorly documented climatic variations to one of the better known.

In summary, antarctic meteorology has evolved considerably in recent years due to the increased effort in numerical modeling and greater availability of satellite and automatic weather station data. These advances augment what is otherwise the most data-sparse land area on earth. Resulting analyses have revealed a complex atmospheric circulation that has large seasonal and interannual variations and *interacts* with lower latitudes on a variety of timescales.

Acknowledgments. The authors are grateful for the sustained support from the Office of Polar Programs of the National Science Foundation and, more recently, from the National Aeronautics and Space Administration. The authors appreciate the assistance of Richard Cullather in the preparation of this manuscript. This chapter is contribution 1111 of Byrd Polar Research Center.

REFERENCES

Allison, I., and J. V. Morrissy, 1983: Automatic weather stations in the Antarctic. *Aust. Meteor. Mag.,* **31,** 71–76.

Ball, F. K., 1956: The theory of strong katabatic winds. *Aust. J. Phys.,* **9,** 373–386.

———, 1957: The katabatic winds of Adélie Land and King George V Land. *Tellus,* **9,** 201–208.

———, 1960: Winds on the ice slopes of Antarctica. *Antarctic Meteorology, Proc. Symp. in Melbourne 1959,* Melbourne, Australia, Pergamon, 9–16.

Bintanja, R., and M. van den Broeke, 1995: The surface energy balance of antarctic blue ice and snow. *J. Appl. Meteor.,* **34,** 902–926.

Breckenridge, C. J., U. Radok, C. R. Stearns, and D. H. Bromwich, 1993: Katabatic winds along the Transantarctic Mountains. *Antarctic Meteorology and Climatology: Studies Based on Automatic Weather Stations,* Antarctic Research Series, Vol. 61, D. H. Bromwich and C. R. Stearns, Eds., Amer. Geophys. Union, 69–92.

Bromwich, D. H., 1986: Surface winds in West Antarctica. *Antarct. J. U.S.,* **21**(5), 235–237.

———, 1988: Snowfall in high southern latitudes. *Rev. Geophys.,* **26,** 149–168.

———, 1989a: Satellite analyses of antarctic katabatic wind behavior. *Bull. Amer. Meteor. Soc.,* **70,** 738–749.

———, 1989b: An extraordinary katabatic wind regime at Terra Nova Bay, Antarctica. *Mon. Wea. Rev.,* **117,** 688–695.

———, 1991: Mesoscale cyclogenesis over the southwestern Ross Sea linked to strong katabatic winds. *Mon. Wea. Rev.,* **119,** 1736–1752.

———, and D. D. Kurtz, 1984: Katabatic wind forcing of the Terra Nova Bay polynya. *J. Geophys. Res.,* **89,** 3561–3572.

———, and F. M. Robasky, 1993: Recent precipitation trends over the polar ice sheets. *Meteor. Atmos. Phys.,* **51,** 259–274.

———, and Z. Liu, 1996: An observational study of the katabatic wind confluence zone near Siple Coast, West Antarctica. *Mon. Wea. Rev.,* **124,** 462–477.

———, J. F. Carrasco, and C. R. Stearns, 1992: Satellite observations of katabatic wind propagation for great distances across the Ross Ice Shelf. *Mon. Wea. Rev.,* **120,** 1940–1949.

———, T. R. Parish, A. Pellegrini, C. R. Stearns, and G. A. Weidner, 1993: Spatial and temporal characteristics of the intense katabatic winds at Terra Nova Bay, Antarctica. *Antarctic Meteorology and Climatology: Studies Based on Automatic Weather Stations,* Antarctic Research Series, Vol. 61, D. H. Bromwich and C. R. Stearns, Eds., Amer. Geophys. Union, 47–68.

———, Y. Du, and T. R. Parish, 1994: Numerical simulation of winter katabatic winds from West Antarctica crossing Siple Coast and the Ross Ice Shelf. *Mon. Wea. Rev.,* **122,** 1417–1435.

———, F. M. Robasky, R. I. Cullather, and M. L. Van Woert, 1995: The atmospheric hydrologic cycle over the Southern Ocean and Antarctica from operational numerical analyses. *Mon. Wea. Rev.,* **123,** 3518–3538.

Budd, W. F., P. A. Reid, and L. J. Minty, 1995: Antarctic moisture flux and net accumulation from global atmospheric analyses. *Ann. Glaciol.,* **21,** 149–156.

Carleton, A. M., 1981: Monthly variability of satellite-derived cyclone activity for the Southern Hemisphere winter. *Int. J. Climatol.,* **1,** 21–38.

———, 1983: Variations in antarctic sea ice conditions and relationships with Southern Hemisphere cyclonic activity, winters 1973–77. *Arch. Meteor. Geophys. Bioklimatol.,* **B32,** 1–22.

———, 1989: Antarctic sea–ice relationships with indices of the atmospheric circulation of the Southern Hemisphere. *Climate Dyn.,* **3,** 207–220.

———, 1992: Synoptic interactions between Antarctica and lower latitudes. *Aust. Meteor. Mag.,* **40,** 129–14.

Carrasco, J. F., and D. H. Bromwich, 1993: Satellite and automatic weather station analyses of katabatic surges across the Ross Ice Shelf. *Antarctic Meteorology and Climatology: Studies Based on Automatic Weather Stations,* Antarctic Research Series, Vol. 61, D. H. Bromwich and C. R. Stearns, Eds., Amer. Geophys. Union, 93–108.

———, and ———, 1994: Climatological aspects of mesoscale cyclogenesis over the Ross Sea and Ross Ice Shelf regions of Antarctica. *Mon. Wea. Rev.,* **122,** 2405–2425.

———, and ———, 1996: Mesoscale cyclone activity near Terra Nova Bay and Byrd Glacier, Antarctica, during 1991. *Global Atmos. Ocean Sys.,* **5,** 43–72.

Carroll, J. J., 1982: Long-term means and short-time variability of the surface energy balance components at the South Pole. *J. Geophys. Res.,* **87,** 4277–4286.

———, 1994: Observations and model studies of episodic events over the south polar plateau. *Antarct. J. U.S.,* **29**(5), 322–323.
———, and B. W. Fitch, 1981: Effects of solar elevation and cloudiness on snow albedo at the South Pole. *J. Geophys. Res.,* **86,** 5271–5276.
Chen, B., S. R. Smith, and D. H. Bromwich, 1996: Evolution of the tropospheric split jet over the South Pacific Ocean during the 1986–1989 ENSO cycle. *Mon. Wea. Rev.,* **124,** 1711–1731.
Chiu, L. S., 1983: Antarctic sea ice variations 1973–1980. *Variations in the Global Water Budget,* A. Street-Perrott, M. Beran, and R. Ratcliffe, Eds., Reidel, 301–311.
Connolley, W. M., and H. Cattle, 1994: The antarctic climate of the UKMO unified model. *Antarct. Sci.,* **6,** 115–122.
———, and J. C. King, 1996: A modeling and observational study of East Antarctic surface mass balance. *J. Geophys. Res.,* **101,** 1335–1343.
Court, A., 1951: Antarctic atmospheric circulation. *Compendium of Meteorology,* T. F. Malone, Ed., Amer. Meteor. Soc., 917–941.
Cullather, R. I., D. H. Bromwich, and M. L. Van Woert, 1996: Interannual variations in antarctic precipitation related to El Niño–Southern Oscillation. *J. Geophys. Res.,* **101,** 19 109–19 118.
———, ———, and R. W. Grumbine, 1997: Validation of operational numerical analyses in antarctic latitudes. *J. Geophys. Res.,* **102,** 13 761–13 784.
D'Aguanno, J., 1986: Use of AVHRR data for studying katabatic winds in Antarctica. *Int. J. Remote Sens.,* **7,** 703–713.
Egger, J., 1985: Slope winds and the axisymmetric circulation over Antarctica. *J. Atmos. Sci.,* **42,** 1859–1867.
———, 1991: On the mean atmospheric circulation over Antarctica. *Geophys. Astrophys. Fluid Dyn.,* **58,** 75–90.
———, 1992: Topographic wave modification and the angular momentum balance of the Antarctic troposphere. *J. Atmos. Sci.,* **49,** 327–334.
Eicken, H., M. A. Lange, H.-W. Hubberten, and P. Wadhams, 1994: Characteristics and distribution of snow and meteoric ice in the Weddell Sea and their contribution to the mass balance of sea ice. *Ann. Geophys.,* **12,** 80–93.
Engels, R., and G. Heinemann, 1996: Three dimensional structures of summertime antarctic mesoscale cyclones, Part II: Numerical simulations with a limited area model. *Global Atmos.–Ocean Sys.,* **4,** 181–208.
Gallée, H., 1996: Mesoscale atmospheric circulations over the southwestern Ross Sea sector, Antarctica. *J. Appl. Meteor.,* **35,** 1129–1141.
———, 1997: Air–sea interactions over Terra Nova Bay during winter: Simulations with a coupled atmosphere–polynya model. *J. Geophys. Res.,* **102,** 13 835–13 849.
———, and G. Schayes, 1992: Dynamical aspects of katabatic wind evolution in the antarctic coastal zone. *Bound.-Layer Meteor.,* **59,** 141–161.
———, and ———, 1994: Development of a three-dimensional meso-γ primitive equations model, katabatic winds simulation in the area of Terra Nova Bay, Antarctica. *Mon. Wea. Rev.,* **122,** 671–685.
———, P. Pettré, and G. Schayes, 1996: Sudden cessation of katabatic wind in Adélie Land, Antarctica. *J. Appl. Meteor.,* **35,** 1142–1152.
Genthon, C., and A. Braun, 1995: ECMWF analyses and predictions of the surface climate of Greenland and Antarctica. *J. Climate,* **8,** 2324–2332.
Giovinetto, M. B., and C. R. Bentley, 1985: Surface balance in ice drainage systems of Antarctica. *Antarct. J. U.S.,* **20**(4), 6–13.
Gloersen, P., 1995: Modulation of hemispheric sea-ice cover by ENSO events. *Nature,* **373,** 503–506.
———, W. J. Campbell, D. J. Cavalieri, J. C. Comiso, C. L. Parkinson, and H. J. Zwally, 1992: *Arctic and Antarctic Sea Ice, 1978–1987: Satellite Passive–Microwave Observations and Analysis.* National Aeronautics and Space Administration, 290 pp.
Gosink, J. P., 1989: The extension of a density current model of katabatic winds to include the effects of blowing snow and sublimation. *Bound.-Layer Meteor.,* **49,** 367–394.
Groisman, P. Y., R. G. Quayle, and D. R. Easterling, 1994: Reducing biases in estimates of precipitation over the United States. Preprints, *6th Conf. on Climate Variations,* Nashville, TN, Amer. Meteor. Soc., 165–169.
Hines, K. M., D. H. Bromwich, and T. R. Parish, 1995: A mesoscale modeling study of the atmospheric circulation of high southern latitudes. *Mon. Wea. Rev.,* **123,** 1146–1165.
———, ———, and Z. Liu, 1997a: Combined global climate model and mesoscale model simulations of antarctic climate. *J. Geophys. Res.,* **102,** 13 747–13 760.
———, ———, and R. I. Cullather, 1997b: Evaluating moist physics for antarctic mesoscale simulations. *Annals Glaciol.,* **25,** 282–286.
Hurrell, J. W., and H. van Loon, 1994: A modulation of the atmospheric annual cycle in the Southern Hemisphere. *Tellus,* **46A,** 325–338.
James, I. N., 1988: On the forcing of planetary-scale Rossby waves by Antarctica. *Quart. J. Roy. Meteor. Soc.,* **114,** 619–637.
———, 1989: The Antarctic drainage flow: Implications for hemispheric flow on the Southern Hemisphere. *Antarct. Sci.,* **1,** 279–290.
Jonsson, S., 1995: Synoptic forcing of wind and temperature in a large cirque 300 km from the coast of East Antarctica. *Antarct. Sci.,* **7,** 409–420.
Juckes, M. N., I. N. James, and M. Blackburn, 1994: The influence of Antarctica on the momentum budget of the southern extratropics. *Quart. J. Roy. Meteor. Soc.,* **120,** 1017–1044.
Karoly, D. J., 1989: Southern Hemisphere circulation features associated with El Niño–Southern Oscillation events. *J. Climate* **2,** 1239–1252.
Keller, L. M., G. A. Weidner, and C. R. Stearns, 1994: *Antarctic Automatic Weather Station Data for the Calendar Year 1992.* Department of Atmospheric and Oceanic Sciences, University of Wisconsin–Madison, 356 pp. [Available from the Department of Atmospheric and Oceanic Sciences, University of Wisconsin–Madison, Madison, WI 53706.]
Kidson, E., 1947: Daily weather charts extending from Australia and New Zealand to the Antarctic continent. *Austr. Antarct. Expedition 1911–1914,* Sci. Rep. Ser. B, Vol. 7, Government Printer (Sydney), 31 pp.
King, J. C., 1989: Low level wind profiles at an antarctic coastal station. *Antarct. Sci.,* **1,** 169–178.
———, 1990: Some measurements of turbulence over an antarctic ice shelf. *Quart. J. Roy. Meteor. Soc.,* **116,** 379–400.
———, 1993a: Contrasts between the antarctic stable boundary layer and the midlatitude nocturnal boundary layer. *Waves and Turbulence in Stratified Flows,* S. D. Mobbs and J. C. King, Eds., Clarendon Press, 105–120.
———, 1993b: Control of near-surface wind over an antarctic ice shelf. *J. Geophys. Res.,* **98,** 12 949–12 953.
———, 1996: Longwave atmospheric radiation over Antarctica. *Antarct. Sci.,* **8,** 105–109.
———, P. S. Anderson, M. C. Smith, and S. D. Mobbs, 1995: Surface energy and water balance over an antarctic ice shelf in winter. Preprint, *4th Conf. on Polar Meteorology and Oceanography,* Dallas, TX, Amer. Meteor. Soc., 79–81.
Kitoh, A., 1994: Tropical influence on the South Pacific double jet variability. *Proc. NIPR Symp. Polar Meteor. Glaciol.,* **8,** 34–45.
Kodama, Y., G. Wendler, and J. Gosink, 1985: The effect of blowing snow on katabatic wind in Antarctica. *Ann. Glaciol.,* **6,** 59–62.
———, ———, and N. Ishikawa, 1989: The diurnal variation of the boundary layer in summer in Adélie Land, eastern Antarctica. *J. Appl. Meteor.,* **28,** 16–24.

Krinner, G., C. Genthon, Z.-X. Li, and P. Le Van, 1997: Studies of the antarctic climate with a stretched-grid general circulation model. *J. Geophys. Res.,* **102,** 13 731–13 745.

Kuhn, M., and H. K. Weickmann, 1969: High altitude radiometric measurements of cirrus. *J. Appl. Meteor.,* **8,** 147–154.

———, L. S. Kundla, and L. A. Stroschein, 1977: The radiation budget at Plateau Station, Antarctica, 1966–67. *Meteorological Studies at Plateau Station,* Antarct. Res. Ser., Vol. 25, J. A. Businger, Ed., Amer. Geophys. Union, 41–73.

Lettau, B., 1969: The transport of moisture into the Antarctic interior. *Tellus,* **21,** 331–340.

Lettau, H. H., and W. Schwerdtfeger, 1967: Dynamics of the surface-wind regime over the interior of Antarctica. *Antarct. J. U.S.,* **2**(5), 155–158.

Loewe, F., 1972: The land of storms. *Weather,* **27**(3), 110–121.

———, 1974: Considerations concerning the winds of Adélie Land. *Gletscherk. Glazialgeol.,* **10,** 189–197.

Madigan, C. T., 1929: Tabulated and reduced records of the Cape Denison station, Adélie Land. *Austr. Antarct. Expedition 1911–1914,* Sci. Rep. Ser. B, Vol. 4, Government Printer (Sydney), 286 pp.

Mather, K. B., and G. S. Miller, 1966: Wind drainage off the high plateau of eastern Antarctica. *Nature,* **209,** 281–284.

———, and ———, 1967a: The problem of the katabatic wind on the coast of Terre Adélie. *Polar Rec.,* **13,** 425–432.

———, and ———, 1967b: Notes on topographic factors affecting the surface wind in Antarctica with special reference to katabatic winds and bibliography. Tech. Rep. UAG-R-189, Geophysical Institute, University of Alaska–Fairbanks, 125 pp.

Mayewski, P. A., Ed., 1996: Science and implementation plan for the U.S. contribution to the International Trans Antarctic Scientific Expedition (ITASE). Research Rep., University of New Hampshire, 55 pp. [Available from Climate Change Research Center, EOS-Morse Hall, University of New Hampshire, Durham, NH 03824-3525.]

Mechoso, C. R., 1980: The atmospheric circulation around Antarctica: Linear stability and finite-amplitude interactions with migrating cyclones. *J. Atmos. Sci.,* **37,** 2209–2233.

———, 1981: Topographic influences on the general circulation of the Southern Hemisphere. *Mon. Wea. Rev.,* **109,** 2131–2139.

Meehl, G. A., 1991: A reexamination of the mechanism of the semiannual oscillation in the Southern Hemisphere. *J. Climate,* **4,** 911–926.

Morgan, V. I., I. D. Goodwin, D. M. Etheridge, and C. W. Wookey, 1991: Evidence from antarctic ice cores for recent increases in snow accumulation. *Nature,* **354,** 58–60.

Murphy, B. F., and I. Simmonds, 1993: An analysis of strong wind events simulated in a GCM near Casey in the Antarctic. *Mon. Wea. Rev.,* **121,** 522–534.

National Research Council, 1983: *El Niño and the Southern Oscillation: A Scientific Plan.* National Academy Press, 63 pp.

Neff, W. D., 1981: An observational and numerical study of the atmospheric boundary layer overlying the East Antarctic ice sheet. NOAA Tech. Memo. ERL WPL-67, 272 pp.

———, 1992: Synoptic influence on inversion winds at the South Pole. Preprints, *3rd Conf. on Polar Meteorology and Oceanography,* Portland, OR, Amer. Meteor. Soc., (J2) 24–28.

———, 1994: Studies of variability in the troposphere and atmospheric boundary layer over the South Pole: 1993 experimental design and preliminary results. *Antarct. J. U.S.,* **29**(5), 302–304.

O'Connor, W. P., and D. H. Bromwich, 1988: Surface airflow around Windless Bight, Ross Island, Antarctica. *Quart. J. Roy. Meteor. Soc.,* **114,** 917–938.

———, ———, and J. F. Carrasco, 1994: Cyclonically forced barrier winds along the Transantarctic Mountains near Ross Island. *Mon. Wea. Rev.,* **122,** 137–150.

Oglesby, R. J., 1989: A GCM study of antarctic glaciation. *Climate Dyn.,* **3,** 135–156.

Parish, T. R., 1981: The katabatic winds of Cape Denison and Port Martin. *Polar Rec.,* **20,** 525–532.

———, 1982a: Surface airflow over East Antarctica. *Mon. Wea. Rev.,* **110,** 84–90.

———, 1982b: Barrier winds along the Sierra Nevada Mountains. *J. Appl. Meteor.,* **21,** 925–930.

———, 1983: The influence of the Antarctic peninsula on the windfield over the western Weddell Sea. *J. Geophys. Res.,* **88,** 2684–2692.

———, 1984: A numerical study of strong katabatic winds over Antarctica. *Mon. Wea. Rev.,* **112,** 545–554.

———, 1988: Surface winds over the Antarctic continent: A review. *Rev. Geophys.,* **26,** 169–180.

———, 1992a: On the interaction between antarctic katabatic winds and tropospheric motions in the high southern latitudes. *Aust. Meteor. Mag.,* **40,** 149–167.

———, 1992b: On the role of antarctic katabatic winds in forcing large-scale tropospheric motions. *J. Atmos. Sci.,* **49,** 1374–1385.

———, and D. H. Bromwich, 1987: The surface windfield over the antarctic ice sheets. *Nature,* **328,** 51–54.

———, and K. T. Waight, 1987: The forcing of antarctic katabatic winds. *Mon. Wea. Rev.,* **115,** 2214–2226.

———, and D. H. Bromwich, 1989: Instrumented aircraft observations of the katabatic wind regime near Terra Nova Bay. *Mon. Wea. Rev.,* **117,** 1570–1585.

———, and G. Wendler, 1991: The katabatic wind regime at Adélie Land. *Int. J. Climatol.,* **11,** 97–107.

———, and D. H. Bromwich, 1997: On the forcing of seasonal changes in surface pressure over Antarctica. *J. Geophys. Res.,* **102,** 13 785–13 792.

———, P. Pettré, and G. Wendler, 1993a: The influence of large scale forcing on the katabatic wind regime of Adélie Land, Antarctica. *Meteor. Atmos. Phys.,* **51,** 165–176.

———, ———, and ———, 1993b: A numerical study of the diurnal variation of the Adélie Land katabatic wind regime. *J. Geophys. Res.,* **98,** 12 933–12 947.

———, D. H. Bromwich, and R.-Y. Tzeng, 1994: On the role of the Antarctic continent in forcing large-scale circulations in the high southern latitudes. *J. Atmos. Sci.,* **51,** 3566–3579.

———, Y. Wang, and D. H. Bromwich, 1997: On the forcing of seasonal pressure changes over the Antarctic continent. *J. Atmos. Sci.,* **54,** 1410–1422.

Périard, C., and P. Pettré, 1993: Some aspects of the climatology of Dumont d'Urville. *Int. J. Climatol.,* **13,** 313–327.

Pettré, P., and J. C. André, 1991: On the surface pressure change through Loewe's phenomenon and the katabatic flow jumps: Study of two cases in Adélie Land. *J. Atmos. Sci.,* **48,** 557–571.

Phillpot, H. R., 1991: The derivation of the 500 hPa height from automatic weather station surface observations in the Antarctic continental interior. *Aust. Meteor. Mag.,* **39,** 79–86.

———, and J. W. Zillman, 1970: The surface temperature inversion over the Antarctic continent. *J. Geophys. Res.,* **75,** 4161–4169.

Radok, U., 1973: On the energetics of surface winds of the antarctic ice cap. Energy fluxes over polar surfaces. WMO Tech. Note 129, 69–100.

———, 1981: The lower atmosphere of the polar regions. *Geol. Rundsch.,* **70,** 703–724.

———, and R. C. Lile, 1977: A year of snow accumulation at Plateau Station. *Meteorological Studies at Plateau Station,* Antarct. Res. Ser., Vol. 25, J. A. Businger, Ed., Amer. Geophys. Union, 17–26.

———, and T. J. Brown, 1996: Antarctic 500 hPa heights and surface temperatures. *Aust. Meteor. Mag.,* **45,** 55–58.

———, I. Allison, and G. Wendler, 1996: Atmospheric surface pressure over the interior of Antarctica. *Antarct. Sci.,* **8,** 209–217.

Robin, G. de Q., 1977: Ice cores and climatic change. *Phil. Trans. Roy. Soc. London,* **B.280,** 143–168.

Rockey, C. C., and D. A. Braaten, 1995: Characterization of polar cyclonic activity and relationship to observed snowfall events at McMurdo Station, Antarctica. Preprints, *4th Conf. on Polar Meteorology and Oceanography,* Dallas, TX, Amer. Meteor. Soc., 244–245.

Savage, M. L., C. R. Stearns, and G. A. Weidner, 1988: The Southern Oscillation signal in Antarctica. Preprints, *2nd Conf. on Polar Meteorology and Oceanography,* Madison, WI, Amer. Meteor. Soc., 141–144.

Schlesinger, M. E., 1984: Atmospheric general circulation model simulations of the modern antarctic climate. *Environment of West Antarctica: Potential CO_2-Induced Changes,* National Academy Press, 155–196.

Schwerdtfeger, W., 1960: The seasonal variation of the strength of the southern circumpolar vortex. *Mon. Wea. Rev.,* **88,** 203–208.

———, 1967: Annual and semi-annual changes of atmospheric mass over Antarctica. *J. Geophys. Res.,* **72,** 3543–3547.

———, 1970: The climate of the Antarctic. *World Survey of Climatology,* Vol. XIV, S. Orvig, Ed., Elsevier, 253–355.

———, 1975: The effect of the Antarctic peninsula on the temperature regime of the Weddell Sea. *Mon. Wea. Rev.,* **103,** 41–51.

———, 1979: Meteorological aspects of the drift of ice from the Weddell Sea toward the middle-latitude westerlies. *J. Geophys. Res.,* **84,** 6321–6327.

———, 1984: *Weather and Climate of the Antarctic.* Elsevier, 261 pp.

———, and F. Prohaska, 1956: The semiannual pressure oscillation, its causes, and effects. *J. Meteor.,* **13,** 217–218.

———, and L. Mahrt, 1968: The relation between terrain features, thermal wind and surface wind over Antarctica. *Antarct. J. U.S.,* **3**(5), 190–191.

Seefeldt, M. W., 1996: Wind flow in the Ross Island region, Antarctica, based on the UW–NMS with a comparison to automatic weather station data. M.S. thesis, Department of Atmospheric and Oceanic Sciences, University of Wisconsin—Madison, 96 pp. [Available from Department of Atmospheric and Oceanic Sciences, University of Wisconsin—Madison, Madison, WI 53706.]

Simmonds, I., 1990: Improvements in general circulation model performance in simulating antarctic climate. *Antarct. Sci.,* **2,** 287–300.

———, and T. H. Jacka, 1995: Relationships between interannual variability of antarctic sea ice and the Southern Oscillation. *J. Climate,* **8,** 637–647.

———, and R. Law, 1995: Associations between antarctic katabatic flow and the upper level winter vortex. *Int. J. Climatol.,* **15,** 403–422.

Slotten, H. R., and C. R. Stearns, 1987: Observations of the dynamics and kinematics of the atmospheric surface layer of the Ross Ice Shelf, Antarctica. *J. Clim. Appl. Meteor.,* **26,** 1731–1743.

Smith, S. R., and C. R. Stearns, 1993a: Antarctic pressure and temperature anomalies surrounding the minimum in the Southern Oscillation index. *J. Geophys. Res.,* **98,** 13 071–13 083.

———, and ———, 1993b: Antarctic climate anomalies surrounding the minimum in the Southern Oscillation index. *Antarctic Meteorology and Climatology: Studies Based on Automatic Weather Stations,* Antarctic Research Series, Vol. 61, D. H. Bromwich and C. R. Stearns, Eds., Amer. Geophys. Union, 149–174.

Sorbjan, A., Y. Kodama, and G. Wendler, 1986: Observational study of the atmospheric boundary layer over Antarctica. *J. Clim. Appl. Meteor.,* **25,** 641–651.

Stearns, C. R., 1982: Automatic weather stations on the Ross Ice Shelf, Antarctica. *Antarct. J. U.S.,* **17**(5), 217–219.

———, and M. L. Savage, 1981: Automatic weather stations, 1980–81. *Antarct. J. U.S.,* **16**(5), 190–191.

———, and G. A. Weidner, 1993: Sensible and latent heat flux estimates in Antarctica. *Antarctic Meteorology and Climatology: Studies Based on Automatic Weather Stations,* Antarctic Research Series, Vol. 61, D. H. Bromwich and C. R. Stearns, Eds., Amer. Geophys. Union, 109–138.

———, L. M. Keller, G. A. Weidner, and M. Sievers, 1993: Monthly mean climatic data for antarctic automatic weather stations. *Antarctic Meteorology and Climatology: Studies Based on Automatic Weather Stations,* Antarctic Research Series, Vol. 61, D. H. Bromwich and C. R. Stearns, Eds., Amer. Geophys. Union, 1–21.

Streten, N. A., 1963: Some observations of antarctic katabatic winds. *Aust. Meteor. Mag.,* **42,** 1–23.

———, 1968: Some characteristics of strong wind periods in coastal east Antarctica. *J. Appl. Meteor.,* **7,** 46–52.

———, 1990: A review of the climate of Mawson—A representative strong wind site in East Antarctica. *Antarct. Sci.,* **2,** 79–89.

———, and A. J. Troup, 1973: A synoptic climatology of satellite-observed cloud vortices over the Southern Hemisphere. *Quart. J. Roy. Meteor. Soc.,* **99,** 56–72.

Swithinbank, C., 1973: Higher resolution satellite pictures. *Polar Rec.,* **16,** 739–751.

Tauber, G. M., 1960: Characteristics of antarctic katabatic winds. *Antarctic Meteorology, Proc. Symposium in Melbourne 1959,* Melbourne, Australia, Pergamon, 52–64.

Trenberth, K. E., 1991: Climate diagnostics from global analyses: Conservation of mass in ECMWF analyses. *J. Climate,* **4,** 707–722.

———, 1992: Global analyses from ECMWF and atlas of 1000 to 10 mb circulation statistics. NCAR Tech. Note NCAR/TN-373+STR, 191 pp., plus 24 fiche.

———, 1995: Atmospheric circulation climate changes. *Clim. Change,* **31,** 427–453.

———, and T. J. Hoar, 1996: The 1990–1995 El Niño–Southern Oscillation event: Longest on record. *Geophys. Res. Lett.,* **23,** 57–60.

Turner, J., T. A. Lachlan-Cope, J. P. Thomas, and S. R. Colwell, 1995: The synoptic origins of precipitation over the Antarctic peninsula. *Antarct. Sci.,* **7,** 327–337.

———, and Coauthors, 1996: The Antarctic First Regional Observing Study of the Troposphere (FROST) project. *Bull. Amer. Meteor. Soc.,* **77,** 2007–2032.

———, S. R. Colwell, and S. Harangozo, 1997: Variability of precipitation over the coastal western Antarctic peninsula from synoptic observations. *J. Geophys. Res.,* **102,** 13 999–14 007.

Tzeng, R.-Y., D. H. Bromwich, and T. R. Parish, 1993: Present-day antarctic climatology of the NCAR community climate model version 1. *J. Climate* **6,** 205–226.

———, ———, ———, and B. Chen, 1994: NCAR CCM2 simulation of the modern antarctic climate. *J. Geophys. Res.,* **99,** 23 131–23 148.

van den Broeke, M. R., and R. Bintanja, 1995: Summertime atmospheric circulation in the vicinity of a blue ice area in Queen Maud Land, Antarctica. *Bound.-Layer Meteor.,* **72,** 411–438.

van Loon, H., 1967: The half-yearly oscillations in middle and high southern latitudes and the coreless winter. *J. Atmos. Sci.,* **24,** 472–486.

———, 1972a: Pressure in the Southern Hemisphere. *Meteorology of the Southern Hemisphere, Meteor. Monogr.,* No. 35, Amer. Meteor. Soc., 59–86.

———, 1972b: Cloudiness and precipitation in the Southern Hemisphere. *Meteorology of the Southern Hemisphere, Meteor. Monogr.,* No. 35, Amer. Meteor. Soc., 101–111.

———, and J. C. Rogers, 1984: Interannual variations in the half-yearly cycle of pressure gradients and zonal wind at sea level on the Southern Hemisphere. *Tellus,* **36A,** 76–86.

Weller, G. E., 1969: A meridional surface wind speed profile in MacRobertson Land, Antarctica. *Pure Appl. Geophys.,* **77,** 193–200.

Wendler, G., and Y. Kodama, 1984: On the climate of Dome C, Antarctica, in relation to its geographical setting. *Int. J. Climatol.,* **4,** 495–508.

———, and ———, 1985: Some results of the climate of Adélie Land, eastern Antarctica. *Z. Gletscherk. Glazialgeol.,* **21,** 319–327.

———, and M. Pook, 1992: On the half-yearly pressure oscillation in eastern Antarctica. *Antarc. J. U.S.,* **27**(5), 284–285.

———, and Y. Kodama, 1993: The kernlose winter in Adélie Land. *Antarctic Meteorology and Climatology: Studies Based on Automatic Weather Stations,* Antarctic Research Series, Vol. 61, D. H. Bromwich and C. R. Stearns, Eds., Amer. Geophys. Union, 139–147.

———, N. Ishikawa, and Y. Kodama, 1988: On the heat budget of an icy slope of Adélie Land, eastern Antarctica. *J. Appl. Meteor.,* **27,** 52–65.

———, J. C. André, P. Pettré, J. Gosink, and T. Parish, 1993: Katabatic winds in Adélie Land. *Antarctic Meteorology and Climatology: Studies Based on Automatic Weather Stations,* Antarctic Research Series, Vol. 61, D. H. Bromwich and C. R. Stearns, Eds., Amer. Geophys. Union, 23–46.

———, C. Stearns, G. Weidner, G. Dargaud, and T. Parish, 1997: On the extraordinary katabatic winds of Adélie Land. *J. Geophys. Res.,* **102,** 4463–4474.

Xie, S., C. Bao, Z. Xue, L. Zhang, and C. Hao, 1994: Interaction between antarctic sea ice and ENSO events. *Proc. NIPR Symp. Polar Meteor. Glaciol.,* **8,** 95–110.

Yamanouchi, T., and S. Kawaguchi, 1992: Cloud distribution in the Antarctic from AVHRR data and radiation measurements at the surface. *Int. J. Remote Sens.,* **13,** 111–127.

Yamazaki, K., 1992: Moisture budget in the antarctic atmosphere. *Proc. NIPR Symp. Polar Meteor. Glaciol.,* **6,** 36–45.

Yasunari, T., and S. Kodama, 1993: Intraseasonal variability of katabatic wind over East Antarctica and planetary flow regime in the Southern Hemisphere. *J. Geophys. Res.,* **98,** 13 063–13 070.

Zwally, H. J., J. C. Comiso, C. L. Parkinson, W. J. Campbell, F. D. Carsey, and P. Gloersen, 1983: *Antarctic Sea Ice, 1973–1976: Satellite Passive–Microwave Observations.* National Aeronautics and Space Administration, 206 pp.

Chapter 5

Mesoscale Meteorology

MICHAEL J. REEDER
Centre for Dynamical Meteorology and Oceanography, Monash University, Victoria, Australia

ROGER K. SMITH
Meteorological Institute, University of Munich, Munich, Germany

---

5.1. Introduction

In the present book's predecessor, *Meteorology of the Southern Hemisphere*, mesoscale phenomena hardly rated a mention. Such a comment is certainly no criticism of that work, for their omission simply represented the state of the science: mesoscale meteorology was still in its infancy. Since then our understanding of the mesoscale aspects of weather in both hemispheres has progressed in leaps and bounds. The progress made over the past two decades in understanding mesoscale phenomena has been stimulated in large part by the practical benefits of such research, since mesoscale phenomena account for most of our significant "weather." Nonetheless, it should be acknowledged that we still have a long way to go in understanding mesoscale weather systems.

The aim of this review is to paint a broad-brush picture of Southern Hemisphere mesoscale meteorology, focusing on the refereed English-language literature. Throughout the review, we adopt a very liberal interpretation of the word *mesoscale*. Although the fundamental dynamics of mesoscale weather systems are, of course, hemisphere independent, regional topography often plays a critical and sometimes subtle role in shaping the evolution and structure of the systems. For this reason, Southern Hemisphere mesoscale systems often possess characteristics that distinguish them from their Northern Hemisphere counterparts. We try to emphasize these distinguishing characteristics in this review. It is important to bear in mind that most of the theoretical and modeling work on mesoscale phenomena has been stimulated by observations of phenomena in the Northern Hemisphere. It is not yet clear to what extent these theories and modeling results carry over the Southern Hemisphere.

The most serious obstacle to documenting the morphology of mesoscale weather systems in the Southern Hemisphere and understanding their dynamics is the difficulty of obtaining the necessary observations. The routine observational network over virtually the entire Southern Hemisphere is totally inadequate for a proper study of mesoscale phenomena; in fact it is barely adequate for the detailed analysis of synoptic-scale features. For this reason, the substantial progress made over the last two decades has been due in large measure to satellite observations and a number of major, specially designed observational programs. Satellite data, for instance, have been the principal means by which the life cycle, morphology, and climatology of Southern Hemisphere midlatitude and high-latitude cyclones have been deduced. Although not exhaustive, the specially designed observational programs include the Amazon Boundary-Layer Experiments (ABLE 2A and 2B); the Australian Monsoon Experiment (AMEX); the Central Australian Fronts Experiment (CAFE); the Cold Fronts Research Programme (CFRP); the Equatorial Mesoscale Experiment (EMEX); the Island Thunderstorm Experiment (ITEX); the Southerly Buster Observational Programme (SUBOP); the Southerly Change Experiment (SOUCHEX); the Stable Antarctic Boundary Layer Experiment (STABLE); the Tropical Rainfall Measuring Mission; and numerous "morning glory" field experiments. The majority of these experiments have been conducted in the Australasian region. Much, but by no means all, of the present chapter is devoted to discussing their findings,[1] together with what has been learned from satellite observations.

Before proceeding further, a few words on what we mean by *mesoscale* are in order. It is often convenient, both from a conceptual and mathematical point of view, to think of the atmosphere as comprising a number of more-or-less dynamically distinct scales of motion, the mesoscale being one such scale. Those phenomena possessing fundamentally similar force

[1] The results of very recent field experiments, such as the Maritime Continent Thunderstorm Experiment, TOGA COARE, the Southern Alps Experiment, and the 1996 Central Australian Fronts Experiment, are not reviewed, as results from these are only just appearing in the literature.

and thermodynamic balances may be usefully grouped together. On the other hand, attaching labels like *mesoscale* can be decidedly unhelpful if it means introducing divisions where none exist naturally, which is the case more often than not. Moreover, one must always bear in mind that the assumption that such scales remain distinct is not always justifiable because the governing equations are highly nonlinear. In this review we have adopted a highly flexible, eclectic, and ultimately subjective approach. Mesoscale phenomena are simply characterized by two dynamical properties: their length scales are not larger than the locally defined Rossby radius of deformation, and their locally defined Rossby numbers are not small compared with unity. In other words, rotation may be important but not dominant. Generally, the advective nonlinearity plays an important part in the evolution of mesoscale weather systems. Although unquestionably important, we have decided that the following topics fall outside the scope of this review: studies concerned primarily with the properties of the boundary layer; cloud microphysics; and mesoscale numerical weather prediction.

5.2. Extratropical and high-latitude vortices

Synoptic and mesoscale analyses have been, and continue to be, hamstrung by the lack of data over the vast tropical and southern oceans of the Southern Hemisphere. The data problem notwithstanding, the arrival of the satellite era a quarter of a century ago provided a major step forward: for the first time, identification and classification of cyclonic developments over the ocean became possible. Important early examples in the extratropical and polar regions include the satellite-based studies by Troup and Streten (1972), Zillman and Price (1972), and Streten and Troup (1973). Nevertheless, satellite images are not always easy to interpret, and, in many instances, dynamically based theories for observed developments have still to be constructed. As valuable as satellite-based studies are, they provide only limited information on the vertical structure of cyclones and the underlying dynamical processes.

There have been very few case studies of extratropical or high-latitude cyclogenesis in the Southern Hemisphere and virtually none incorporating in situ measurements. Those studies that have been undertaken have been based almost entirely on objective analyses of routine data and on satellite imagery. For synoptic-scale extratropical cyclones, see: Ryan and Medows 1979; Hill 1980; Bonatti and Rao 1987; Jury and Laing 1990; Veldon and Mills 1990; Mills 1991; and Mills and Russell 1992. For mesoscale midlatitude cyclones, see: Holland et al. 1987; Lynch 1987; Pascoe et al. 1990; and Revell and Ridley 1995. For high-latitude cyclones, see: Carrasco and Bromwich 1993; Turner et al. 1993b; and Heinemann 1996. One exception is the observation study by Griffiths et al. (1998). These authors documented the structure and evolution of an intense cut-off low in detail unprecedented for the Australian region by combining high time-resolution radiosonde profiles with collocated wind profiler observations and routine Australian Bureau of Meteorology analyses. The cut-off low developed over the ocean south of Australia and subsequently interacted with a subtropical frontal system over the central part of the continent. Throughout the period of observation, the strongest temperature gradients lay through the subtropics. Moreover, the subtropical temperature gradients strengthened, while the midlatitide temperature gradients associated with the cut-off low weakened. The system documented was similar in structure to the instant occlusion investigated by Browning and Hill (1985), the major difference being that the Australian cut-off low developed poleward of a subtropical front. The cut-off low was cold-cored and produced a very deep tropopause fold. Over Adelaide (34°37′S, 138°29′E), the tropopause descended to an altitude of about 5 km, and the temperature at 500 hPa fell to −30°C. The western edge of the low-level cold dome was marked by a strong secondary warm front, which passed over Adelaide about 8 h after the tropopause height minimum. The cut-off low produced very high rainfalls, resulting in flash flooding and much damage in the Adelaide Hills area. The most severe convection took place in the region behind the lowered tropopause around the passage of the warm front. The cut-off low has also been examined by Mills and Wu (1995) using Australian Bureau of Meteorology analyses.

There have been few theoretical studies of the dynamics of Southern Hemisphere cyclones. One example is the study by Bonatti and Rao (1987), who compared the structure of observed extratropical cyclones over South America with the structure predicted by linear, baroclinic instability theory. A second example is that by Reeder et al. (1991). These authors investigated the structure of a nonlinear, three-dimensional, baroclinic wave and its associated cold front, and showed that, despite the relative simplicity of the model, the solution showed surprisingly close agreement with the typical summertime developments off southern Australia. The studies by Orlanski and Katzfey (1991) and Orlanski et al. (1991) examined development in the eastern Pacific using ECMWF analyses and limited-area numerical modeling. The key results of Orlanski and Katzfey are that decaying cyclones redistribute energy downstream by an ageostrophic geopotential flux and that this energy source plays an important, and often overlooked, part in initiating new cyclones. Finally, using ECMWF analyses and a limited-area numerical model, Innocentini and Caetano Neto (1996) examined a particularly intense storm that developed in the lee of the Andes, although these authors focused on the generation of ocean waves by the storm.

The continents cover a far smaller fraction of the Southern Hemisphere than the Northern Hemisphere; 81% of the Southern Hemisphere is covered by water, compared with 61% of the Northern Hemisphere. This difference in land coverage is particularly striking between about 45° and 65°S, where the Southern Ocean encircles the hemisphere, obstructed only by the southern part of South America. In contrast, only a small fraction of the area poleward of about 66°S is ocean. Despite the high degree of zonal symmetry in the Southern Hemisphere extratropics and polar latitudes, cyclones develop in localized, zonally asymmetric regions known as storm tracks, as in the Northern Hemisphere. Southern Hemisphere storm tracks are covered in greater depth by Hurrell et al. in chapter 1 of this monograph, and the summary below is meant simply to set the scene for later discussion.

Storm track climatologies based on satellite data may be found in the papers by van Loon (1965), Taljaard (1967), Streten and Troup (1973), Carleton (1979, 1981a), and Physick (1981), and those based on routine analyses are found in Trenberth (1991), Jones and Simmonds (1993), and Sinclair (1994, 1995a,b, 1996a), among others. These studies have shown that, by and large, cyclones are born in the midlatitudes, spend their lives spiraling eastward and poleward, and eventually die south of 60°S in the antarctic circumpolar trough. The preferred regions for development are near the principal upper-tropospheric jet streams, namely those over the southern Atlantic Ocean, the southern-central Pacific Ocean, the Indian Ocean, and south of Australia. Sinclair (1995a,b) found that the eastern coasts of Australia and South America and the lee side of the Andes are also important sites for cyclogenesis. According to Gan and Rao (1994), cyclones and fronts weaken as they approach the Andes and, after crossing the mountain range, strengthen on the lee side. Over the oceans, wintertime cyclogenesis is closely linked to the regions of strongest SST gradient. Rapid cyclogenesis is most common east of South America, southeast of Africa, south of Australia, and near New Zealand (Sinclair 1995a). A climatology of cyclones forming on the eastern Australian seaboard has been constructed by Hopkins and Holland (1997a). Although the frequency of cyclonic developments peaks during the winter months, storm tracks in the Southern Hemisphere exhibit little seasonal variation compared with those in the Northern Hemisphere. Kidson and Sinclair (1995) have examined the interannual variability of Southern Hemisphere storm tracks. A recent review of Southern Hemisphere vortices can be found in the article by Carleton (1992). An important finding from these studies is that, compared with the Northern Hemisphere, synoptic-scale cyclones (also known as transient eddies) play a more central role in the eddy transports of heat and momentum. This suggests that understanding the dynamics of these cyclones may be a prerequisite to understanding in detail the mechanisms driving the general circulation of the Southern Hemisphere.

We end this section with a note of caution: cyclone climatologies are sensitive to the method used to identify and track the cyclones. For further discussion of this point with particular relevance to South America, see Satyamurty et al. (1990), Gan and Rao (1991), Sinclair (1994, 1995a, 1996b), and Rao and Gan (1996).

a. Satellite-based classifications of synoptic-scale cyclones

Guided by satellite imagery over the Southern Ocean, Zillman and Price (1972) suggested that cyclonic developments fall into one of three basic types: *comma developments*, *instant occlusions*, and *frontal waves*, as are illustrated in Fig. 5.1. Unfortunately, there is no uniformity in the terminology used, and comma developments also go by the names *polar lows*, *polar troughs*, and *cold-air vortices*, the last name being preferred here. In the Southern Hemisphere they are sometimes called *inverted comma developments*. Zillman and Price identified two distinct features of cyclogenesis in the satellite imagery: 1) a mobile, short-wave, upper-level disturbance that appears in the imagery as a convective cloud cluster, and 2) a low-level preexisting baroclinic zone that is marked as a cold front in Fig. 5.1 and is seen in the imagery as an extensive (and usually) stratiform cloud band. Each type of development culminates in the comma-shaped cloud pattern characteristic of mature extratropical cyclones. Browning (1990) and Evans et al. (1994) reviewed Zillman and Price's paradigm for cyclogenesis from a Northern Hemisphere perspective. In particular, Evans et al. concluded that all satellite-derived classification schemes are, in effect, variants on that originally devised by Zillman and Price.

In line with modern theories and observations of cyclogenesis [see, for example, the reviews by Hoskins (1990), Reed (1990), and Uccellini (1990)], Zillman and Price's (1972) classification scheme explicitly recognizes that cyclogenesis arises from the interaction between finite-amplitude disturbances, such as an upper-level short-wave trough and a preexisting baroclinic zone or front. Nowadays, such disturbances are identified with cyclonic potential vorticity (PV) anomalies. For example, Hoskins et al. (1985) discuss the sequence of events that follow from the advection of an upper-level cyclonic PV anomaly over a low-level cyclonic PV anomaly, as in Zillman and Price's frontal-wave case. Since the low-level temperature gradients are normally weak and appear to play only minor roles in the development, cut-off lows fall presumably into either Zillman and Price's cold-air vortex or instant occlusion categories.

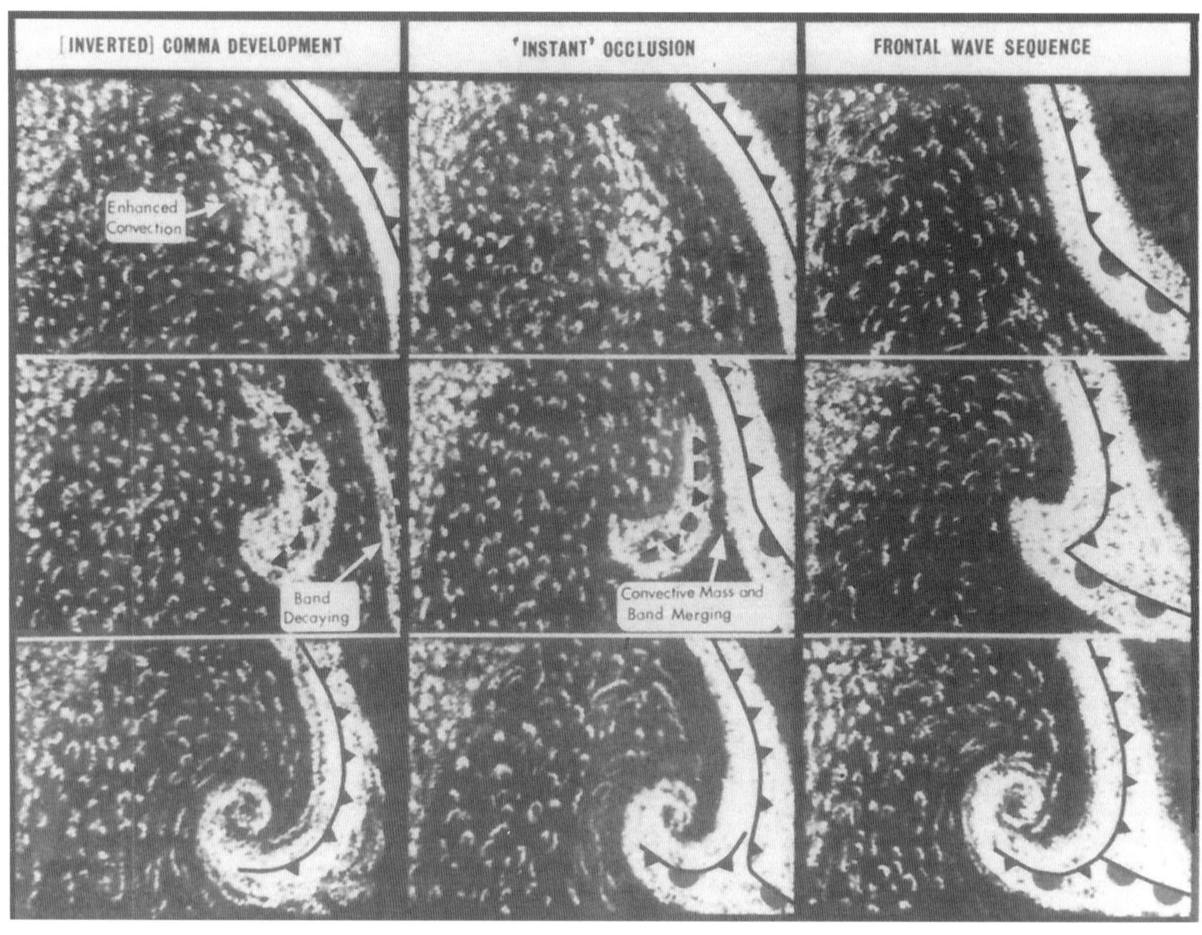

FIG. 5.1. Schematic depiction of three basic sequences of vortex development evident in satellite imagery. (a) Development of a simple comma cloud, or cold-air vortex, within the cold air. (b) Development of an instant occlusion. (c) Development of a frontal wave. (From Zillman and Price 1972.)

Although the theory has yet to be fully developed, satellite observations suggest that the mutual evolution of the short-wave trough and baroclinic zone depends crucially on the horizontal distance separating them, which is presumably a measure of how strongly the two PV anomalies interact. In the frontal wave case, cyclogenesis occurs when the trough is advected over the baroclinic zone. It is this type of development that most closely resembles the classical Norwegian model, and it appears to be most frequent between 30° and 50°S (Carleton 1979). If the short-wave trough and baroclinic zone are separated by large distances during their interaction, as in the cold-air vortex (or comma development), they retain their separate identities. This case is discussed further in section 5.2b. An intermediate possibility is that a limited interaction occurs between the short-wave trough and baroclinic zone and the cloud bands associated with these two features merge so that, in the satellite imagery at least, the system resembles a mature occluded cyclone. In a study of five Southern Hemisphere winters, Carleton (1980) found that instant occlusions formed most commonly in the longitudes of Australasia and the eastern Indian Ocean between about 30° and 55°S. An example of such a development is documented by Griffiths et al. (1998).

In a recent study of major extratropical cyclogenesis over a region encompassing eastern Australia and the western Tasman Sea, Feren (1995) found that the formation of a distinctive delta-shaped cloud, clearly discernible in the GMS imagery, often precedes explosive development by up to 24 h. Generally, the cloud has a banded structure thought to be associated with radiating internal gravity waves. Feren referred to these banded, delta-shaped clouds as *striated deltas*. Striated deltas are about 250 to 500 km wide at their base and roughly twice that in length. In general, the cloud tops are high, being comprised mostly of cirrus. Striated deltas have a life span of only about 3–6 h before taking on a more conventional comma shape. Like most major extratropical developments, they form beneath the poleward-exit region of an upper-level jet streak and diffluent midlevel trough. Amplification of the upper trough and a shortening of the

FIG. 5.2. GMS infrared satellite image for 1130 UTC 29 August 1992, showing an instant occlusion and striated delta cloud. Note particularly the transverse banding thought to be caused by internal gravity waves. (From Mills and Wu 1995.)

distance between it and the downstream ridge generally precede the formation of a striated delta. An example of a striated delta cloud from Mills and Wu (1995) is reproduced in Fig. 5.2. This particular cloud accompanied the extratropical development south of the Australian continent studied by Griffiths et al. (1998).

b. Cold-air vortices

We turn our attention now to cold-air vortices, which Zillman and Price (1972) called comma developments. Such developments are especially frequent in the winter months and generally take place over the ocean, poleward of the principal polar jet stream, although a significant number also form near the Antarctic coastline. According to Sinclair and Cong (1992), "it is difficult to find a satellite image of the Southern Hemisphere that does *not* include some postfrontal cold air development," while Streten and Troup (1973) note that cold-air vortices account for up to half of all Southern Hemisphere cyclones. Further details on the satellite-derived climatological aspects of these vortices can be found in the papers by Streten and Troup (1973), Carleton (1979, 1980, 1981b), and Carleton and Carpenter (1989, 1990). A climatology of cold-air vortices based on ECMWF analyses has been carried out by Sinclair and Cong (1992).

Unfortunately, there is little detailed information on the structure of cold-air vortices, and their dynamical origin is not well understood. A case study of one such development off Cape Town was reported by Jury and Laing (1990), but this, like virtually all other investigations of cold-air vortices in the Southern Hemisphere, was based mostly on routine surface and upper-air observations and satellite imagery. The study documented the development of a cold-air vortex west of a larger-scale trough. As is almost always the case, the development occurred beneath the diffluent exit region of an upper-level jet streak. One of the most interesting aspects of this storm was that it rapidly intensified as it passed over the Aghulas current. On the poleward side of the current, the SST gradients were almost 10 K (10 km)$^{-1}$. Jury and Laing presented evidence to show that the development was enhanced by moderately strong surface heat fluxes in the warm sector of the storm as it moved over the warm side of the sea surface gradient, and their case study underscores the importance of low-level baroclinic zones in extratropical cyclogenesis. In southern Africa, there exists a semipermanent coastal front due to the SST gradients across the Aghulas current. This front was investigated by Jury (1993a), who analyzed aircraft observations taken over a four-day midwinter period during prefrontal flow, finding a sharp coastal front confined to the marine boundary layer over the northern edge of the Aghulas current. The boundary layer was 400 m higher on the warm side of the front. Furthermore, the boundary layer on the warm side was convectively unstable, and the surface heat fluxes and surface wind stress were five times larger there. A pronounced low-level jet was also observed. (See also the discussion in chapter 3 by McBride et al. concerning cut-off lows in the southern Africa region.)

Small-scale cold-air vortices with length scales on the order of a few hundred kilometers to about 1000 km are also observed (e.g., Bromwich 1987, 1989c; Carrasco and Bromwich 1993; Turner et al. 1993a; Turner et al. 1993b). These vortices go by a variety of names including *polar low*; here we use the term *mesoscale cyclone*. Mesoscale vortices develop predominantly near the almost permanent, strong baroclinic zone (or front) surrounding Antarctica, separating cold continental air from relatively warm maritime air. It appears that mesoscale cyclogenesis may be modulated by the convergence of katabatic flows off Antarctica. Although mesoscale cyclones are mostly comma shaped (Turner et al. 1993b), they are sometimes spiral shaped and occasionally possess a hurricane-like eye (e.g., Carleton 1981b; Turner and Row 1989; Carleton and Carpenter 1990; Heinemann 1990; Bromwich 1991; Fitch and Carleton 1992; Carleton and Fitch 1993; Carrasco and Bromwich 1994). According to Turner et al. (1993b), antarctic mesoscale vortices generally lack the deep convection that characterizes their Northern Hemisphere counterparts (i.e., polar lows).

Sensible and latent heat fluxes provided by the strong ice-sea temperature gradients presumably play important roles, although to date there have been few quantitative assessments made. In this connection, Fantini and Buzzi (1993) have used a highly idealized numerical model to show that the turbulent transfer of sensible heat and water vapor between the relatively warm ocean and the overlying cold continental air plays an important role in development of mesoscale vortices.

Mesoscale cyclones are not confined to the high latitudes. *East-coast cyclones* are a class of infrequent (one or two per year on average) but very intense mesoscale cyclones that form during winter in subtropical latitudes near the east coast of Australia. Like tropical cyclones, east-coast cyclones generally possess eyes. Although likely, we are unaware of any studies to date that indicate whether or not similar storms develop on the eastern seaboards of other Southern Hemisphere continents. The size of east-coast cyclones range typically between 100 and 500 km. Heavy rainfall and strong winds, often of hurricane force, are the norm for such storms.

A necessary condition for the development of east-coast cyclones appears to be a trough in the easterly flow over eastern Australia. Such a configuration is known in Australia as an *easterly dip*. The synoptic configuration appears to have elements in common with the so-called *reversed shear flow*, in which polar lows to the west and north of Norway often develop (Duncan 1978). Cyclogenesis generally takes place poleward of a pronounced subtropical jet stream, beneath the right exit of an upper-level jet streak embedded within a diffluent trough. Central ingredients for the formation of east-coast cyclones are the tongue of relatively warm water that extends down the eastern Australian coastline and the warm-cored eddies shed from its southernmost part. Such eddies are transient features of the East Australian current. According to Holland et al. (1987), the land–sea temperature gradient is most pronounced at night after the continent has cooled, explaining the preference for nocturnal development by east-coast cyclones. These authors find that east-coast cyclones are warm cored (at least at low levels) and highly convective, often possessing spiral cloud bands and an eye. Apparently, the environmental conditions favorable for east-coast cyclogenesis are similar to those conducive for explosive cyclogenesis off the North American seaboard. Figure 5.3 shows an MSLP chart and GMS satellite image showing the development of an east-coast cyclone on 27 July 1984. Holland et al. group east-coast cyclones into three types, according to essentially two criteria: their size and where they form relative to the trough axis. A more comprehensive account of the structure and evolution of east-coast cyclones may be found in Holland et al. and in Lynch (1987).

Modeling studies to date include those by Golding and Leslie (1993), Leslie et al. (1987), Hess (1990), McInnes and Hess (1992), and Revell and Ridley (1995). The latter authors have focused particularly on the evolution of the low-level PV. Their chief conclusion is that, once the cyclone is well established, the evolution of the low-level PV is well described by Snyder and Lindzen's (1991) theory of baroclinic wave–CISK (conditional instability of the second kind). In other words, Revell and Ridley find that the low-level PV anomaly is produced chiefly by the vertical gradient of latent heating in a preexisting frontal cloud band. In contrast to Holland et al. (1987), Revell and Ridley deemphasize the role played by

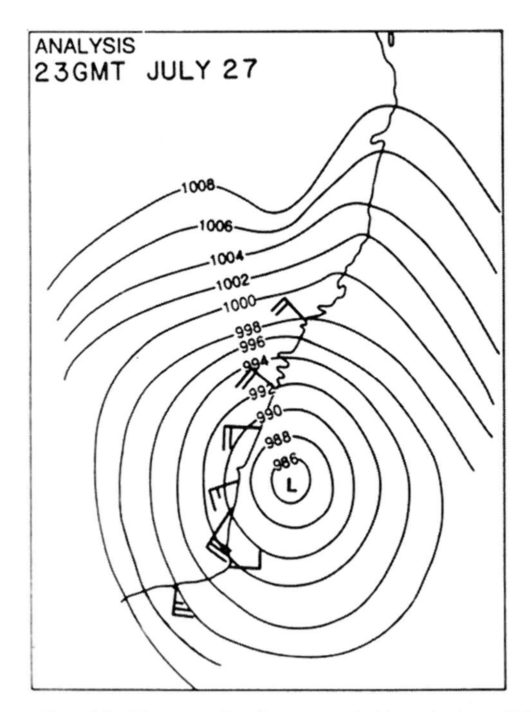

FIG. 5.3. Mean sea level pressure (mb) analysis at 2300 UTC 27 July 1984 and GMS satellite image at 0000 UTC 28 July 1984 showing an east-coast cyclone. The arrow on the satellite image points to the surface position of the cyclone. (From Holland et al. 1987.)

low-level temperature gradients and surface fluxes of heat and moisture, except insofar as they precondition the atmosphere for cyclogenesis (see also Revell 1997). It should be noted, however, that theories pertaining to wave-CISK have been challenged by a number of authors (e.g., Ooyama 1982; Raymond 1994; Hopkins and Holland 1997b).

5.3. Cold fronts

Much of the significant weather experienced in the midlatitude and subtropical regions of all Southern Hemisphere continents is associated with frontal passages. To date, the systematic study of the mesoscale structure and dynamics of cold fronts in the Southern Hemisphere has been confined to the Australasian region. In other regions, the emphasis of research has been on the statistical aspects of rainfall associated with frontal systems (e.g., Kousky 1979; Smith 1985; Barclay et al. 1993) and the interaction of fronts with topographic barriers (see section 5.5c). Frontal activity in Amazonia is discussed by Paegle (1987).

a. Frontogenetic regions

One measure of the frontogenetic tendency in a flow is the *frontogenesis function*, $D|\nabla_h\theta|/Dt$, first introduced by Petterssen (1936). The frontogenesis function represents the material rate of change of horizontal potential temperature gradient, $|\nabla_h\theta|$. In the special case of horizontal flow,

$$\frac{D}{DT}|\nabla_h\theta| = -\frac{1}{2}|\nabla_h\theta|(D - E\cos 2\beta),$$

where D is the divergence, $(\partial u/\partial x + \partial v/\partial y)$, E is the total deformation, $[(\partial u/\partial x - \partial v/\partial y)^2 + (\partial v/\partial x + \partial u/\partial y)^2]^{1/2}$, and β is the angle between the axis of dilatation and the isentropes (see, for example, Keyser et al. 1988).

Figure 5.4, taken from Satyamurty and De Mattos (1989), shows the climatological fields of total deformation and frontogenesis at 850 mb in the latitude band from 45°S to 45°N. These fields were constructed from the U.S. National Meteorological Center 5° × 5° analyses and averaged over the period from 1975 to 1981. The climatological pattern of frontogenesis in the Southern Hemisphere is very different from that in the Northern Hemisphere. In the Southern Hemisphere, the frontogenesis function and total deformation are maximized in four northwest–southeast sloping bands located over the mid-Pacific, the western South Atlantic, the western Indian Ocean, and western Australia. The most pronounced region of frontogenesis is found over South America, while the bands of strong frontogenesis located over the mid-Pacific and the western Atlantic correspond to the SPCZ and SACZ, respectively (see chapter 3 by

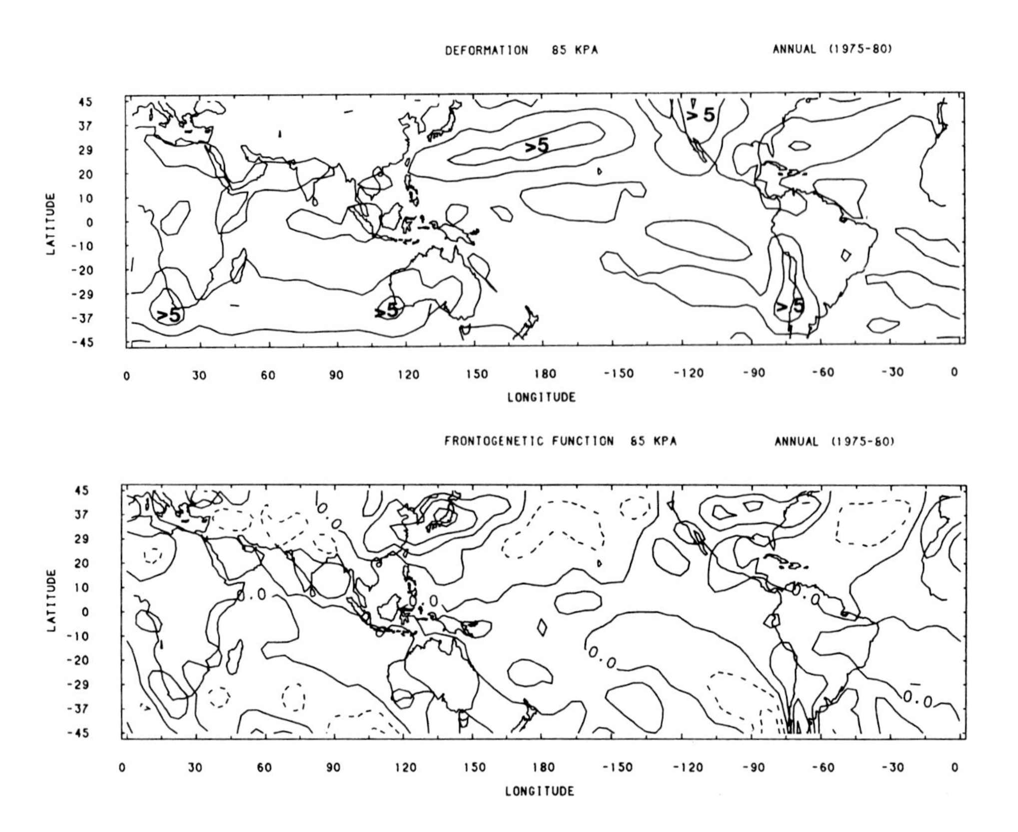

FIG. 5.4. Seven-year mean wind fields of 850-mb total deformation and frontogenesis function between 45°N and 45°S. (a) Total deformation; contour interval 5×10^{-6} s^{-1}. (b) Frontogenesis function; contour interval 2.5×10^{-12} K m^{-1} s^{-1}. (From Satyamurty and De Mattos 1989.)

McBride et al.). Figure 5.4 shows that, on the whole, frontogenesis is weaker in the Southern Hemisphere than in the Northern Hemisphere.

The frontogenesis maxima in the southwestern parts of Australia, Africa, and South America appears at first glance to have no more than a tenuous connection with extratropical cyclogenesis, since the storm tracks are generally far to the south of these continents. At the latitudes of the total deformation maxima, anticyclones over the oceans and troughs over the continents feature prominently in the average mean sea level isobars shown in chapter 3. During the winter, these anticyclones move 5°–10° equatorward from their mean summertime position, an anticyclone forms over the Australian continent, and the anticyclone in the Indian Ocean moves around 30° westward during winter. Frontogenesis over the continent generally takes place in the trough region between two subtropical anticyclones, although the fronts extend southward to a parent depression at much higher latitudes. As a cold front approaches the southwestern coast of Australia, Africa, or South America, the wind fields associated with the parent depression to the south, the two anticyclones, and the geographically fixed trough over the continent to the north combine to produce a pattern of synoptic-scale deformation over and to the south of the continent. Thus, a deformation field acts to strengthen the temperature contrast between the coast and the continental trough.

b. Frontal structure

The present section focuses on the mesoscale structure and dynamics of Australian cold fronts, although it seems likely that conclusions drawn about the behavior of Southern Ocean cold fronts that sweep across the Australian continent would be applicable also to their counterparts that cross southern Africa and South America.

The summertime cold fronts that affect the southern part of Australia are commonly referred to as *cool changes*. At times, cool changes are accompanied by large temperature falls, typically 10°C and in extreme cases 20°C, in half an hour. A recent review of these fronts is given by Reeder and Smith (1992). In the late spring and early summer of 1981, 1982, and 1984, a series of field experiments was conducted in southeastern Australia as part of CFRP to investigate the structure and behavior of the cool change (Smith et al. 1982a; Ryan 1982, 1983; Ryan et al. 1985), and the data obtained were synthesized to produce a conceptual model for a mature cool change. The conceptual

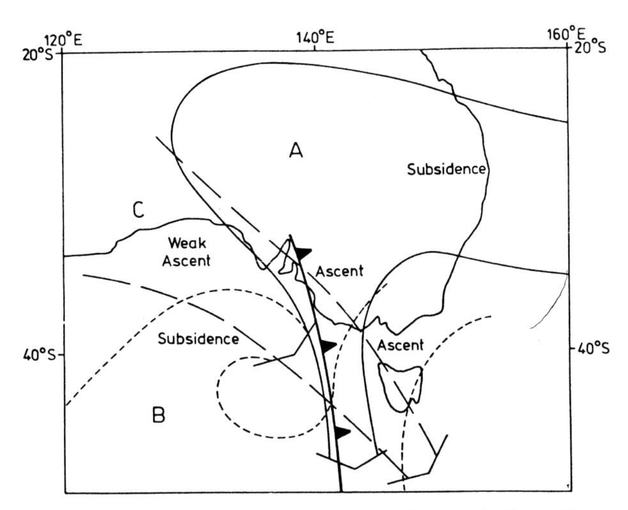

FIG. 5.5. Schematic representation of the isentropic flow relative to an observer moving with a cold front. Flow is shown on (B) 295 K, (A) 305 K, and (C) 315 K surface. (From Wilson and Stern 1985.)

model is described in detail by Wilson and Stern (1985) and Ryan and Wilson (1985), who suggested that the kinematics of the cool change may be understood simply in terms of three distinct airstreams. A schematic representation of these airstreams relative to an observer moving with the cool change is shown in Fig. 5.5. Each airstream is depicted on an isentropic surface. Displaying the airflow on an isentropic surface is a useful technique because, insofar as the flow is adiabatic, an air parcel is constrained by conservation of potential temperature to remain on the same isentropic surface.

The principal airstream on the 305-K isentropic surface is labeled "A" in Fig. 5.5. Air parcels follow a trajectory that originates over the Tasman Sea before subsiding and turning anticyclonically southward. This hot northerly airflow ascends from the continental boundary layer and is analogous to the *warm conveyor belt* identified as being a prominent feature of Northern Hemisphere fronts (e.g., Browning 1990). Unlike its Northern Hemisphere counterpart, however, it is relatively dry, so the prefrontal cloud base is generally high, typically about 700 mb. The prefrontal part of the airstream on the 295-K surface, labeled "B" in Fig. 5.5, originates ahead of the surface front but, in contrast to the warm conveyor belt, has a more easterly maritime trajectory. Consequently, this airstream, termed the cold conveyor belt, is moister and cooler than the warm conveyor belt. Parcels follow a trajectory that ascends while moving southwestward before crossing the surface front beneath the warm conveyor belt. The postfrontal part of the airstream on this isentropic surface turns cyclonically behind the surface front while subsiding. In general, this postfrontal cold air has been drawn from two distinct regions. It may have originated well ahead of the surface front and been part of the cold conveyor belt, or it may have subsided from upper levels while turning cyclonically around the upper trough. This is perhaps the feature of the conceptual model that shows the greatest variability from case to case. At upper levels on the 315-K surface, a northwesterly flow (relative to the moving front) is evident. This flow is very dry, has relatively low potential temperature, and generally caps the prefrontal middle-level cloud. The confluence of the

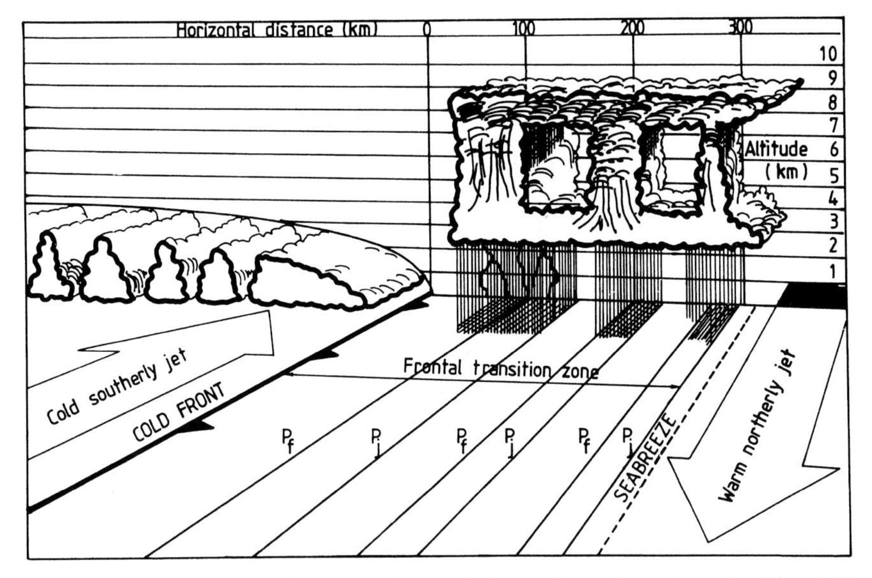

FIG. 5.6. Major features of a summertime cool change in southeast Australia. PJ and PF stand for *pressure jump* and *pressure fall*, respectively. The sea breeze may or may not be present. When present, the sea breeze may produce the most significant change. (From Ryan et al. 1988.)

descending dry middle-level airstream behind the front and the ascending warm conveyor belt airstream commonly produces midlevel convection, particularly during late spring and early summer.

Figure 5.6, taken from Ryan et al. (1988), shows an idealized section through a mature summertime cool change. The region between the prefrontal northerly jet and the cold southerly jet behind the front (airstreams A and B in Fig. 5.6) delineates the frontal transition zone, in which there may occur several change lines at the surface; the first of these is often a sea breeze front, and the remainder are generally associated with the passage of squall lines. A case study of such a squall is reported by Physick et al. (1985). During summer, the most significant change is often associated with the sea breeze front. The analyzed cold front marks the place at which the pressure begins its steady rise and corresponds to the arrival of the cold air mass. Another noteworthy feature is the shallowness of the cold air. A more comprehensive account of the mesoscale structure of the cool change can be found in Garratt and Physick (1983), Garratt et al. (1985), Garratt (1986), and May et al. (1990).

Simple dynamical models of the cool change have been formulated and explored by Reeder (1986), Garratt and Physick (1986, 1987), Reeder and Smith (1987, 1988a,b), Physick (1988), and Reeder et al. (1991). One of the simplest models of frontogenesis in the Australian region involves the formation of a cold front in a growing baroclinic wave. As the baroclinic waves amplify and deepen, they develop patterns of horizontal deformation capable of rapidly intensifying temperature gradients, thereby developing strong fronts. At first sight it may not be obvious that the classical baroclinic instability mechanism is responsible for frontogenesis over southern Australia, particularly as frontogenesis tends to occur in the trough between two broad anticyclones with parent depression well to the south of the continent. Reeder and Smith (1986), however, integrated a two-dimensional nonlinear anelastic numerical model with a simple turbulence parameterization using Eady's (1949) linear normal mode solution for an unstable baroclinic wave as an initial condition. The nonlinear baroclinic wave develops a ridge-trough structure and surface front that replicates remarkably well the important features of the cool change and the environment in which it evolves. This calculation was extended by Reeder et al. (1991), who investigated the structure of a mature three-dimensional baroclinic wave and the cold front that forms within it. McInnes et al. (1994) examined forecasting aspects of the cool change using a three-dimensional numerical weather prediction model.

In contrast to the nonlinear normal mode calculations summarized above, Reeder and Smith (1987) numerically integrated an initial value problem in which the initial conditions are characteristic of the atmospheric state prior to the Ash Wednesday bushfire front of 16 February 1983. A low-level prefrontal northerly airstream in thermal wind balance is prescribed at the initial time and is embedded within a zonal shear flow as in the Eady problem. While nonlinear normal mode calculations tell us what kinds of general structures to expect, the initial value problem is required to complete the picture and is the basis for forecasting frontogenesis. The solutions to the initial value problem develop the same front-relative airstreams as in the conceptual model based on a synthesis of the CFRP observations. The initial value problem was extended by Reeder (1986) to include a simple parameterization heating over the land.

The foregoing conceptual model with its multiple change lines is called a Type 1 cool change by Garratt (1988) and corresponds to the complex changes described in the earlier studies. Notwithstanding the relative dryness of cool changes in comparison with cold fronts studied in the Northern Hemisphere, moist processes can play a significant role in springtime fronts. Stratiform precipitation occurs frequently in both the warm and cold air, while rainbands and squall lines may be found in the warm prefrontal air. Although precipitation is relatively common in the prefrontal air, much of it often evaporates in the dry subcloud-layer air, producing virga. Such evaporative cooling below cloud base frequently leads to the formation of shallow prefrontal change lines in the frontal transition zone, which may obscure the final synoptic change. Indeed, Ryan et al. (1989) show that, at times, evaporative cooling may be the dominant process in the development of the change. They analyzed a case (18 November 1984) where the synoptic front is greatly weakened by evaporative cooling and a new region of enhanced temperature gradient is established several hundred kilometers ahead of it. Ryan et al. (1990) used a numerical model to investigate the dynamics of warm conveyor-belt convection, which, as noted above, is relatively common in the spring and early summer. They find that evaporatively driven downdrafts play a central role in organizing the convective circulation and that destabilization of the cloud tops by radiative cooling is important in maintaining the convection.

In some cases, moist processes appear to play only a minor role, especially in mid- and late-summertime fronts, which normally exhibit a single well-defined change line without a complex transition zone. Strong differential heating across the southern Australian coastline appears to be a crucial element in the dynamics of this kind of front, which is referred to as Type II by Garratt (1988) (see also Reeder 1986 and Physick 1988). Sometimes a particular front may be characterized by a single dry change inland while possessing a more complex multiple-line structure near the coast or nearer the parent cyclone center. This characteristic is presumably related to the variation of moisture and surface heating along the front. Type II fronts have characteristics of the predominantly dry fronts that penetrate deeply over the Australian continent at other times of the year; these are discussed in section 5.3c.

Finally, we note that while wintertime fronts are probably not fundamentally different to those observed during the CFRP, they have received little attention to date (see Mills 1989; May et al. 1990; Abbs and Jensen 1993).

c. Subtropical fronts

Cold fronts often cross the Southern Hemisphere continents to quite low latitudes even though they are not always analyzed as fronts at these latitudes. For example, frontal systems, or their remnants, commonly penetrate the tropical and subtropical latitudes of Brazil at all times of the year (Kousky 1979; Paegle 1987) and are responsible for a significant fraction of the rainfall in that region. Likewise, frontal passages through southern Africa account for much of Zimbabwe's late spring and summer rainfall maximum (Smith 1985; see also chapter 3 by McBride et al.).

In South America, the most spectacular cold frontal passages occur during the winter months. Wintertime cold-air surges that reach the upper Amazon basin are known as *friagens* in Brazil (Myers 1964). On occasion, these cold fronts cross the Equator, penetrating the Northern Hemisphere Tropics. A number of cases have been documented, including those by Myers (1964), Parmenter (1976), Hamilton and Tarifa (1978), and Fortune and Kousky (1983). Winter fronts are accompanied by strong postfrontal ridging, and severe frosts are common in the cold postfrontal air. The fronts generally move equatorward, while the parent extratropical cyclone moves eastward or southeastward. On occasion, these fronts can initiate severe convection in the subtropics, such as in the case reported by Myers. Cold fronts that reach the Tropics, however, generally tend to suppress the convection in the ITCZ.

In Australia, wintertime frontal passages often penetrate deep into the subtropics also and appear to have much in common with South American friagens. Indeed, they are common through central Queensland during the late dry season (September to November), with cooler air sometimes reaching the Gulf of Carpentaria. In the first half of this period, they are normally without precipitation and often without cloud, but they may produce dust storms. As the wet season approaches, however, they may trigger deep convection as they progress toward the northeast coast of Queensland. Typically, the fronts are shallow (~1 km deep) and move into a deep, convectively well-mixed boundary layer (~3–4 km deep), which, during the night, overlies a strong but shallow radiation inversion.

The synoptic environment of these fronts is similar to that of the summertime cool-change of southeastern Australia described above in section 5.3b, with fronts forming in the trough region between the two subtropical anticyclones, relatively far from the center of the parent cyclone. A unique feature of the region is the presence of a heat trough over northeastern western Australia and northwestern Queensland, with which the frontal trough eventually merges. Generally, the frontal passage is followed by strong ridging from the west.

Much insight into the structure and dynamics of subtropical cold fronts has been gained from CAFE, which took place in 1991 (Smith et al. 1995) and built on earlier studies (Smith et al. 1986; Smith and Ridley 1990). The data obtained during the CAFE experiment highlight the large diurnal variation of frontal structure associated with diabatic processes. The fronts are often difficult to locate during the late morning and afternoon when convective mixing is at its peak, but develop strong surface signatures in the evening as the convection subsides and a surface-based radiation inversion develops. Moreover, there appears to be a ubiquitous tendency in the early morning for the formation of a nonlinear wave-like structure at the leading edge of the frontal zone as the inversion strengthens. This aspect of their dynamics is discussed further in section 5.6c. Subtropical fronts pose significant analysis and forecasting problems in the Southern Hemisphere.

5.4. Topographically forced circulations

We turn our attention now to mesoscale circulations forced by variations in the local topography. These include thermally driven circulations due to land-water temperature contrasts and circulations forced by variations in the orographic slope.

a. Sea breezes

Sea breezes feature prominently in the coastal circulation of most tropical and subtropical regions of the globe. Inquiry into the nature of sea breezes dates back to antiquity, as discussed by Neumann (1973). Today, their theoretical basis is reasonably well understood: in essence, sea breezes behave dynamically like evolving gravity currents, their leading edge often developing strong frontal gradients in wind and temperature (e.g., Kraus et al. 1990). Nonetheless, sea breezes still attract the interest of research meteorologists, since the details of their structure and evolution, frequently essential for determining the dispersal of pollutants, for instance, can be quite complex, depending as much on the local topography as on the prevailing meteorological conditions. Regional sea-breeze studies in the Southern Hemisphere include those by Physick (1976, 1982), Physick and Byron-Scott (1977), Kousky (1980), Thompson and Neale (1981), Sturman and Tyson (1981), Dexter et al. (1985), Abbs (1986),

McKendry et al. (1986, 1987, 1988), McKendry (1989), Lloyd (1990), Physick and Abbs (1991, 1992), Skinner and Tapper (1994), and Finkele et al. (1995). An up-to-date review with an emphasis on the Southern Hemisphere is given by Abbs and Physick (1992).

In Australia, three aspects of sea-breeze dynamics have attracted particular attention: their remarkable ability at low latitudes to propagate several hundred kilometers inland; their role in generating nonlinear wave disturbances; and their ability to organize convection. It is the first aspect that will be addressed now; the other aspects are covered in sections 5.6b and 5.7, respectively.

Clarke (1955) was probably the first to make the point that sea-breeze circulations need not be confined to the coastal margin but could be long lived, propagating several hundred kilometers inland after sunset. Since then, many studies, both observational and theoretical, have confirmed Clarke's ideas (e.g., Clarke 1961, 1983a,b, 1989a; Reid 1957; Garratt 1985; Garratt and Physick 1985; Physick and Smith 1985). For example, during the Koorin boundary-layer experiment in northern Australia, the sea breeze generally disrupted the evolution of the nocturnal boundary layer and the nocturnal low-level jet at Daly Waters, located some 300 km inland (Garratt 1985). The sea-breeze circulation strengthens and the sea-breeze front accelerates as the turbulent daytime boundary layer changes to stable nighttime conditions. Physick and Smith also observed that sea breezes in northern Australia continue to propagate far inland well after sunset, and, in a companion numerical modeling study, they showed that the nocturnal structure of the sea breeze is strikingly dissimilar to the daytime structure. The sea breeze undergoes a dramatic transformation once the advection of cooler sea air is cut off: after about midnight a vortex develops at the leading edge of the sea breeze. As the sea breeze evolves further, it decelerates and the circulation weakens until no closed circulation remains. A similar pattern of demise was observed for sea breezes in southeastern Australia by Clarke (1972).

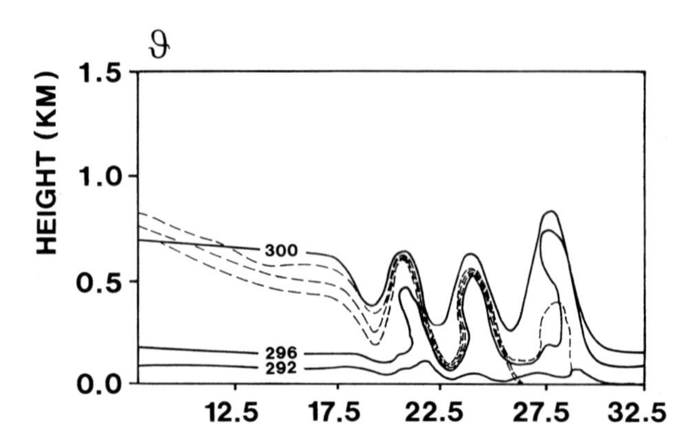

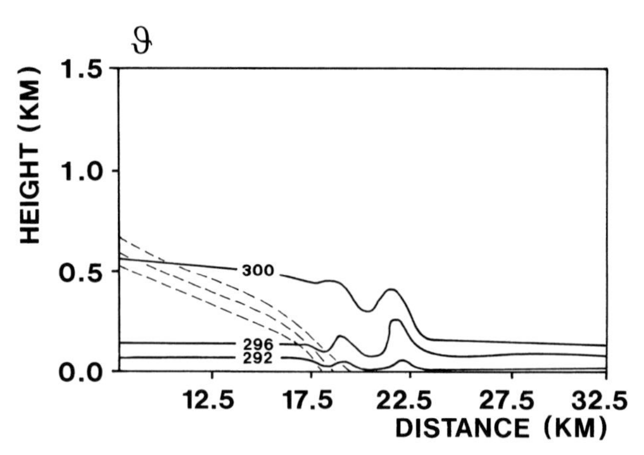

FIG. 5.7. Isentropes θ (solid lines) and 0.1, 0.2, and 0.3 tracer isopleths (dashed lines) for two numerically simulated gravity currents with different inflow Froude numbers, Fr, moving from left to right into a surface-based stable layer 150 m deep with a potential temperature gradient of 6 K (100 m)$^{-1}$ (a) Fr = 1.13, (b) Fr = 0.56. The tracer has the value zero in the environment, and unity for the cold air, as it enters the left boundary of the domain with a mean speed $\bar{u}$. In (a), the gravity current head becomes wavelike, but the waves contain cold air. In (b), a wavelike disturbance moves ahead of the cold air. (From Haase and Smith 1989.)

A comparable transformation has been observed in laboratory experiments (Wood and Simpson 1984; Rottman and Simpson 1989) and in related numerical experiments (Haase and Smith 1989), in which a gravity current moves into an environment that is neutrally stratified above a shallow, surface-based stable layer. Like the nocturnal sea breeze, the gravity-current head in these experiments breaks away from the feeder flow that supplies it with cold air and advances as a horizontal roll vortex. In cases where the stable layer is sufficiently deep and strong, the roll vortex evolves as a solitary-wave–type disturbance or as a bore wave that propagates on the stable layer. Examples of such evolution taken from Haase and Smith are illustrated in Fig. 5.7. This figure shows the isentropes and isopleths of a passive tracer for two numerically simulated gravity currents with different inflow Froude numbers. Here the Froude number is defined as Fr = $\bar{u}/c^*$, where $c^* = (g'h_i)^{1/2}$, $g' = g\Delta\theta/\Theta$, $\overline{\Delta\theta}$ is the mean potential temperature difference between the cold air and its environment, h_i is the depth of the cold-air inflow, Θ is the (constant) ambient potential temperature above the stable layer, and g is the acceleration due to gravity. The gravity current moves from left to right into a surface-based stable layer 150 m deep with a potential temperature gradient of 6 K (100 m)$^{-1}$. The passive tracer has the value zero in the environment and unity for the cold air as it enters the left boundary of the domain with a mean speed $\bar{u}$. In the case where Fr = 1.13, the gravity current head becomes wavelike, but the waves contain cold air. In contrast, when Fr = 0.56, a wavelike disturbance moves ahead of the cold air. As we shall see, the formation of a roll vortex or the evolution into a propagating nonlinear wave disturbance are common paradigms that arise in connection with buoyancy-driven mesoscale flows such as sea breezes and cold fronts.

Before closing this section on sea breezes, we note that Tapper (1988a, 1991) and Physick and Tapper (1990) have examined the sea breeze–like circulation that develops over the large salt lakes of central Australia. Further, McGowan et al. (1995) have examined the structure and evolution of an alpine lake breeze in the southern Alps of New Zealand. Finally, Holland and McBride (1989) discuss observations of the sea-breeze head based on a fortuitous radiosonde sounding as the sonde fell through the head on its descent back to the ground.

b. Katabatic flow

Katabatic flows are common in all mountainous regions. In the Southern Hemisphere, katabatic flows have received most attention in Antarctica. The dynamics of Southern Hemisphere high-latitude meteorology is dominated by the massive antarctic ice dome. Strong radiative cooling at the surface produces persistent katabatic winds over most of the continent, although the fastest winds are generally recorded on the steep coastal escarpments where the flow is directed down the orographic gradient. The katabatic winds are particularly strong along the very steep slopes of East Antarctica. Parish et al. (1993) investigated the role of synoptic pressure patterns on the development of strong katabatic flows and found that such flows are strongest when they are directed down the synoptic pressure gradient. Antarctic katabatic flows are so persistent and large scale that the earth's rotation plays a critical role, deflecting the flow to the left and establishing a mean anticyclonic circulation around Antarctica. The intensity and persistence of these flows appear to be linked to the supply of cold air in the interior of the continent. A more comprehensive account of the meteorology of high southern latitudes is provided by Bromwich and Parish in chapter 4 of this monograph and in the review article by Parish (1988).

The first theoretical treatment of antarctic katabatic flows is that by Ball (1956, 1957, 1960). Ball's model assumes a balance between the pressure gradient associated with the continental slope, advection, the Coriolis force, and friction. In the last of these papers, Ball showed that the pressure gradient associated with the continental slope is responsible for driving the flow, and that the flow is proportional to both the steepness of the slope and the strength of the inversion. Ball's hydraulic approach was extended by Manins and Sawford (1979), who allowed for mixing between the cold katabatic flow and the environmental air above, although they neglected the effects of rotation. Nevertheless, they found that entrainment of potentially warmer air from the environment into the katabatic flow could not be neglected. Indeed, entrainment proved to be far more important than surface friction in retarding the flow. Manins and Sawford compared their theory with observations of katabatic flow taken in southern Australia.

Using Ball's (1960) model, Parish (1982) and Parish and Bromwich (1986, 1987) calculated the steady-state katabatic wind field over the Antarctic continent and compared their theoretical estimate with time averages of the available data. Although the data are sparse, their time averages appear to be reasonably well predicted by Ball's model. Perhaps the most striking aspect of the katabatic flow close to the Antarctic coastal escarpments is the way in which the orography channels the flow. Pronounced regions of confluence and diffluence there were identified by Parish and Bromwich (1987). According to these authors, "the coastal sections of Adelie Land experience the strongest and most persistent surface winds on the face of the earth" because of orographic focusing of the katabatic flow. Observations at nearby Cape Denison show that the average wind speed is about 19.5 m s^{-1}. Similarly, Bromwich (1989a) reports that the wintertime katabatic flow at Inexpressible Island in Terra Nova Bay averages 17 m s^{-1}, but may reach 30 m s^{-1}. At times, the katabatic flows develop hydraulic jumps (e.g., Pettré and André 1991; Gallee et al. 1996). For example, André et al. (1993) investigated four strong katabatic flows and associated hydraulic jumps observed over Adelie Land. One of the principal conclusions reached by these authors was that entrainment from the environment into the cold katabatic flow could not be neglected in accordance with the findings of Manins and Sawford (1979). There is evidence that, at other times, katabatic flows propagate hundreds of kilometers without being significantly deflected by the Coriolis force, although the dynamics are unclear at this stage (Bromwich 1989b; Bromwich et al. 1992). So far, mesoscale modeling studies of antarctic katabatic flow have focused principally on the question of how well the models reproduce the observed mean conditions. Such studies include those by Parish (1984), Parish and Waight (1987), Parish and Bromwich (1991), Hines et al. (1991), Bromwich et al. (1994), and Gallee and Schayes (1994).

5.5. Topographically modified fronts, troughs, and ridges

We review now how fronts, troughs, and ridges are modified by topography.

a. Coastal lows and ridges

Coastal lows play a significant part in the mesoscale meteorology of southern Africa (e.g., Taljaard 1972;

FIG. 5.8. Surface synoptic pressure maps at 1400 UTC daily, obtained from the South African Weather Bureau. The map borders for each panel are 20°–45°S, 5°–45°E. In each figure, H refers to the ridging anticyclone, the bold line to the trailing midlatitude frontal system, and the arrow to the coastal low. Case 1 extends over 6–11 February 1981; case 2 over 15–20 September 1985; and case 3 over 18–23 April 1980. Pressures over the sea are given as the difference in mb from 1000 mb. Isobars are contoured at 2-mb intervals. Over the continent, the isolines refer to the 850-mb surface contoured at 10-geopotential-meter intervals. (From Reason and Jury 1990.)

Torrence 1995). They are shallow disturbances that are observed to hug the coastline, propagating eastward every six days or so throughout the year (Preston-Whyte and Tyson 1973). As noted by Reason and Jury (1990), coastal lows are always preceded by a migratory anticyclone and are followed by a Southern Ocean cold front.

Coastal lows are linked inextricably to the topography. In southern Africa they are bounded near the coast by the steep escarpment marking the edge of the Great Central Plateau. Figure 5.8, taken from Reason and Jury (1990), shows a sequence of MSLP analyses for three typical events, denoted as cases 1, 2, and 3, that where observed in February, April, and September, respectively. Each sequence begins with a prominent migratory anticyclone to the southeast of the continent directing an easterly flow across the mountains. One of the striking features of the analyses is the relatively small seasonal differences, the chief difference being the poleward shift in the anticyclone during summer. As the ridge builds to the east, a small-scale low develops on the west coast in the lee of the mountains. As noted in section 5.3a, this is one of the statistically preferred regions for frontogenesis in the Southern Hemisphere (see Fig. 5.4). Subsequently, the low propagates around Cape Agulhas, all the while sandwiched in the trough region between the original anticyclone and the next one in the sequence. An extratropical cyclone is analyzed farther south, and in each case a cold front is analyzed extending equatorward in the trough. Note that the synoptic pattern and frontal sequence is very similar to the summertime cool change of southeastern Australia described in section 5.3b. In each of the three situations shown, the extratropical cyclone overtakes the coastal low, although this is not always the case. Sometimes the coastal low dissipates over the northern coast of Natal.

It has been suggested that coastal lows are a type of forced, coastally trapped, internal Kelvin wave. In particular, Gill (1977), Nguyen and Gill (1981), and Bannon (1981) have examined the interaction between idealized synoptic disturbances and the southern African escarpment and have demonstrated that Kelvin waves are generated in the process; see also the recent theoretical treatments by Reason (1994) and Rogerson and Samelson (1995). Theory predicts a lateral scale for the coastal low on the order of a deformation radius from the coast. In this case, the deformation radius would be defined as $L_R = (g'H/f)^{1/2}$, where H is the depth of the marine inversion, g' is the reduced gravity, based on the potential temperature contrast between the marine air and the overlying air mass, and f is the Coriolis parameter. Like coastally trapped Kelvin waves, the disturbance amplitude is observed to be greatest near the coast and to decay seaward (Reason and Jury 1990). Bannon, who developed his theory on a beta plane, concluded that Rossby waves were not excited for realistic forcing frequencies.

The offshore environment through which coastal lows propagate has a two-layer structure, comprising a cool maritime boundary layer capped at about 900 mb by a strong subsidence inversion below the height of the escarpment with weak winds aloft. It has been suggested that the inversion traps the wave energy in the marine boundary layer (Gill 1977). Reason and

Jury (1990) suggest that the coastal lows decelerate, weaken, and may eventually dissipate on the east coast because of the weaker inversion. The weaker inversion is probably related to the surface heat flux provided by the Aguhlas current, which flows along the coast from about Durban to Port Elizabeth. They note that the widening of the coastal plain on the east coast may play a role also.

With the observed structure of the marine environment in mind, all theoretical studies to date have used one- or two-layer shallow-water-wave models to examine the dynamics of coastal lows (Gill 1977; Nyugen and Gill 1981; Bannon 1981; Reason and Jury 1990; Rogerson and Samelson 1995). Such treatments cannot address the effectiveness of the waveguide in ducting the disturbance. In the light of the recent observations of the morning glory environment (discussed later in section 5.6b), it is possible that the waveguide is quite leaky.

According to Reason and Jury (1990), the passage of coastal lows is generally accompanied by a depression of the inversion, a marked wind shift, and a temperature fall. In line with their observations, Gill (1977) showed that nonlinear wave steepening causes the disturbance to evolve a frontal-like discontinuity at its leading edge. Moreover, close to the escarpment, rotation exerts little control over the alongshore wind, since the cross-shore flow vanishes there. This allows air parcels to accelerate down the pressure gradient, unaffected by the Coriolis force, and the disturbance can develop the character of a solitary Kelvin wave or a trapped gravity current. The conditions under which each type of development is favored is not known, but it is possibly related to the length scale of the initial disturbance; localized large-amplitude disturbances may give rise to solitary Kelvin waves, whereas broader-scale intrusions of cold air may steepen to form trapped gravity-current–like disturbances.

Although far less common than coastal troughs, *coastal ridges* are observed off southern Africa and elsewhere, most notably southern Australia (Holland and Leslie 1986) and the west coast of North America (e.g., Mass and Albright 1987). It is those disturbances affecting southern Africa, however, that were studied first and have received most attention. Reason and Jury (1990) examine the preference for coastal lows over coastal highs along the South African coastline and conclude that the offshore flow is responsible, rather than wave steepening as first proposed by Gill (1977). The offshore flow advects warm, well-mixed air to the lee of the continent, producing prefrontal foehnic heating, depressing the marine boundary layer, and consequently producing a lee trough. The dynamics of mesoscale coastal ridging is discussed by Reason and Steyn (1990, 1992).

b. Topographically generated troughs

For relatively narrow mountain ranges such as the New Zealand Southern Alps, the lee trough is confined relatively close to the mountains, but for continental-scale orography, such as the Andes, the influence is felt much farther downstream. In the latter case, the lee trough is a significant feature at upper levels (Satyamurty et al. 1980) and is consequently a preferred site for cyclogenesis (e.g., Gan and Rao 1994; Sinclair 1996b).

The trough in the lee of the New Zealand Southern Alps is especially pronounced in northwesterly flows. Such conditions typically precede the passage of a cold front and often produce strong föhn winds on the eastern side of the Southern Alps, known locally as a *nor'wester*. [The föhn winds of South Africa and South America are known locally as the *berg wind* and the *Zonda*, respectively (Brinkmann 1971).] Disorganized wave clouds are common under these conditions. Upstream of the Southern Alps, the northwesterly flow commonly splits and is channeled around the ends of the South Island through Cook and Foveaux Straits. The airstream through Cook Strait subsequently recurves at low levels into the surface lee trough, producing a low-level northeasterly (Sturman et al. 1984, 1985; McKendry 1983; McKendry et al. 1986, 1987; McGowan and Sturman 1996). The interaction between the nor'wester and the northeasterly flow over the Canterbury plains has been studied by McGowan and Sturman (1996). The low-level flow on the east coast of the South Island is complicated further by sea breezes embedded within this onshore flow (Sturman and Tyson 1981; McKendry et al. 1988). At times, a low-level barrier jet forms on the western side of the Alps (McGowan and Sturman 1996).

During the warmer months, typically September through to April, surface troughs are observed to form regularly in central Western Australia and to the west of the great Dividing Range along the northeast coast of Queensland. Generally, these surface troughs are associated with heat lows in the northwest of Western Australia and northwestern Queensland. According to Adams (1986), the trough almost always marks a dry line separating dry continental air from moist maritime air, and during summer the difference in dewpoint across the dry line is commonly as much as 20°C.

The dynamics of such surface troughs is examined using a two-layer quasigeostrophic model by Fandry and Leslie (1984), who conclude that the Western Australian troughs are substantially due to surface heating, whereas in the case of the eastern Australian troughs, both surface heating and orography are important. Kepert and Smith (1992) show that one can account for many features of the two heat troughs using a single-layer shallow-water model in which the effects of heating and cooling are represented by a distribution of mass sources and sinks. Adams (1986, 1993) carries out a more sophisticated analysis of the effects of heating and orography on continuously stratified easterly and westerly airstreams over northern Australia using linear perturbation theory. Leslie and Skinner (1994) examine the formation of the Western Australian heat trough with the aid of a high-resolution regional numerical model [see also the studies by Leslie (1980), Skinner and Leslie (1982), and Watson (1984)]. A basic study of the dynamics of heat lows is contained in a recent paper by Rácz and Smith (1998).

With the approach of a cold front from over the Southern Ocean, the Western Australian trough is observed to translate eastward. Under these circumstances, the trough is commonly referred to as the *prefrontal trough*, a climatology of which can be found in Hanstrum et al. (1990a). At times, rapid frontogenesis takes place within the preexisting trough at the expense of the original Southern Ocean front, which subsequently weakens. A case study illustrating this process was presented by Hanstrum et al. (1990b). The analysis and prediction of such events represents a major forecasting problem.

Finally, we note that McGregor and Kimura (1989) and McKendry and Revell (1992) have modeled cyclonic eddies shed by the orography in the Melbourne and Auckland areas, respectively. The eddies tend to form under stable, light wind conditions.

c. *Pampero Secco, southerly buster, and southerly change*

Pampero Secco is the name given to an intense form of cold front that occurs along the eastern side of the Andes. A similar phenomenon occurs along the east coast of New Zealand's South Island, where it is commonly referred to as a *southerly change* (Ridley 1990; Sturman et al. 1990; Smith et al. 1991), and along the southeast coast of Australia, where it is called a *southerly buster*.[2] Southerly busters, which often herald the *cool change* for Sydney, occur also mainly in the spring and summer months. The name describes the arrival of the front, which is usually marked by the sudden onset of strong southerly wind squalls of cool maritime origin that typically replace hot northwesterlies originating over the continent.

The intensity of these coastal fronts and their special characteristics appear to be closely linked with topographic features. As summertime cold fronts cross the southern part of Chile or the east coast of Australia or advance toward the south coast of New Zealand, they are deformed by the mountain ranges, suffering retardation on the inland side and being accelerated along the coast. Such fronts show a characteristic S-shape when analyzed on synoptic charts. An example of this kind of frontal deformation is shown in Fig. 5.9. In an early theoretical study of the Australian southerly buster, Baines (1980) suggested that the

(a)

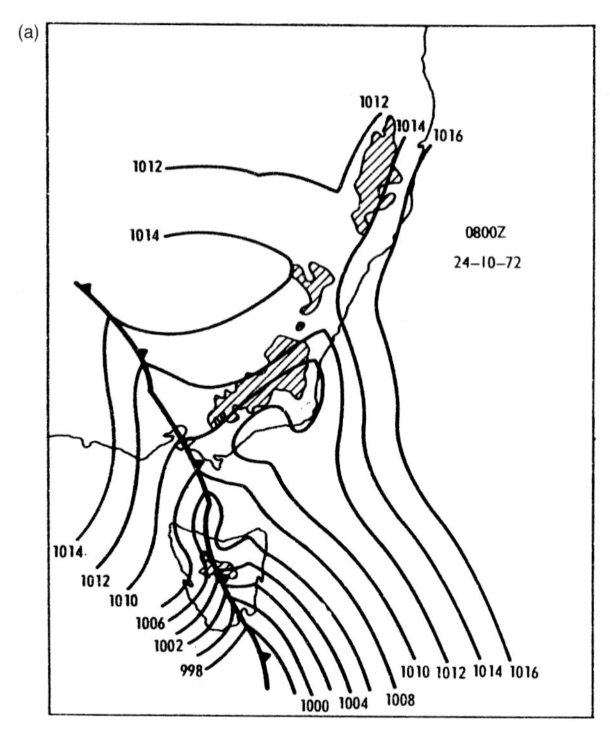

(b)

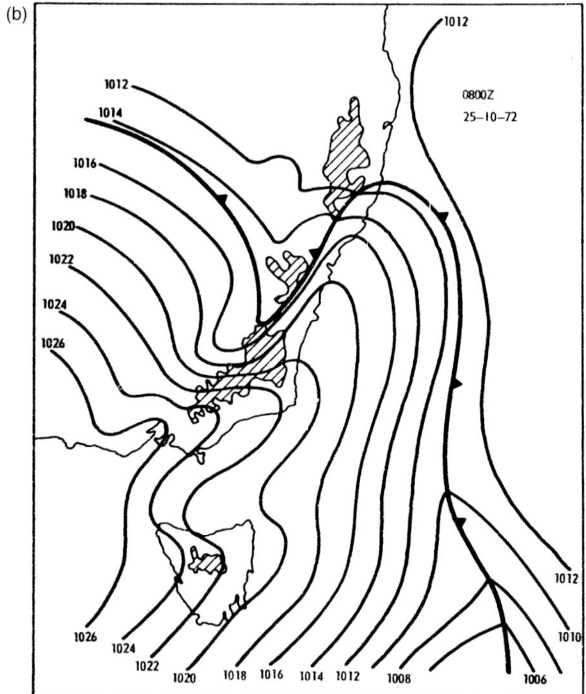

FIG. 5.9. Mean sea level pressure charts for 0800 UTC 24 October and 0800 UTC 25 October 1972, showing the development of a southerly buster on a cold front. (From Baines 1980.)

[2] Southerly busters are occasionally called *southerly bursters*, especially in the older literature.

coastal acceleration is caused by the blocking of postfrontal cold air by the mountains. He hypothesized that this leads to the formation of a coastally trapped gravity current in which the increase in pressure due to an increase in the cold-air depth toward the mountain range is balanced by the Coriolis force associated with the component of flow parallel with the range, much as in an internal Kelvin wave as discussed earlier in section 5.5a. In a related study, Garratt et al. (1989) examined the role of coastal topography in modifying the motion of an idealized cold front.

In the late spring of 1982, a field experiment was organized to investigate the structure of the southerly buster in more detail and, in particular, to test Baines' hypothesis. The experiment (SUBOP) documented two well-developed fronts during a three-week period using an instrumented research aircraft to obtain measurements of frontal structure over the sea. The finding of this experiment, along with those of other investigations using conventional data, are reported by Colquhoun et al. (1985) and Coulman et al. (1985). In the two SUBOP fronts, the depth of the cold air was relatively shallow, not exceeding the average height of the mountain barrier (about 1.5 km), but it was not found to decay away from the barrier on the scale of the deformation radius as predicted by Baines' theory. The propagation speed of the front agreed surprisingly well with the speed formula for a gravity current, at least during some 10 h and over 500 km of the front's evolution along the coast (Coulman et al. 1985). In itself, this does not provide a sensitive test of the gravity current hypothesis for reasons given by Smith and Reeder (1988, section 3). While both of the SUBOP fronts had a pronounced roll-type circulation extending some 50 km behind the surface front, the cold air farther behind was advancing more slowly than the surface front. In this sense, the flow was fundamentally different in structure from steady gravity-current flows in laboratory tank experiments, which are continuously being supplied with colder fluid in a low-level *feeder flow* that extends back to the cold source (Simpson 1982). It was found that the cold-air mass adjusts to the sea-surface conditions by the formation of a neutrally stratified convective marine boundary layer. In contrast, the warm prefrontal air is modified rapidly as it adjusts to the cooler sea, forming a shallow, strongly stable boundary layer, no more than 100–200 m deep. This may be an important factor in the evolution of the roll vortex structure. As discussed in section 5.4a, similar features have been observed in laboratory experiments (Wood and Simpson 1984; Rottman and Simpson 1989) and related numerical experiments (Haase and Smith 1989). On occasion, morning-glory–like roll clouds are observed to accompany or precede the cold surge, with pressure signatures strikingly similar to those recorded during the passage of morning glories over northeastern Australia (Christie 1992; see section 5.6b).

Despite these small-scale features of frontal behavior that are intrinsically unsteady, it appears that the frontal motion remains controlled to a large degree by the synoptic-scale flow patterns and their evolution. This is evidenced by the ability of mesoscale numerical weather prediction models to forecast frontal movement with reasonable accuracy, even though their resolution is far too coarse to capture the small-scale dynamical aspects described above (e.g., Gauntlett et al. 1984; Howells and Kuo 1988; McBride and McInnes 1993; McInnes 1993; McInnes and McBride 1993). Indeed, analyses of numerical simulations of one of the SUBOP fronts by Howells and Kuo and by McInnes and McBride have helped to identify the important factors in the development of the southerly buster. According to Howells and Kuo, these include the blocking and channeling of cool maritime air around the southeastern part of the mountain ranges; blocking of the flow on the western side of the mountains; adiabatic subsidence (or föhnic) heating of the prefrontal westerly air flow, which increases the temperature contrast between the warm and cold air masses; sensible heating of the prefrontal continental air mass; reduced frictional coupling of the warm prefrontal westerlies over the cooler ocean; and stronger frictional retardation over the land to slow the progress there of the front.

Using a mesoscale model, Howells and Kuo (1988) explored the relative importance of these factors in a series of sensitivity experiments in which one or more of the processes were suppressed. They presented one calculation without orography in which the "southerly jet" was "almost as strong as that of the control experiment." They interpreted this as indicating that "the Great Dividing Range is not essential for developing a strong coastal jet," although they went on to show that the range does enhance the intensity of the jet by the first three mechanisms listed above. A more recent numerical study of the same front is reported by McInnes and McBride (1993). While the model simulations of the SUBOP front by McInnes and McBride and by Howells and Kuo are, for the most part, in close agreement, the conclusions reached in each of these studies differ markedly. McInnes and McBride show that orography is the major factor enhancing the movement of the front along the coast. Like Howells and Kuo, McInnes and McBride investigate the roles of orography, heating, and friction through a series of sensitivity experiments. The effect of each process is determined by analyzing thc frontal position in a particular model run relative to its position in a model run excluding topography, heating, and friction. In brief, topography alone is shown to produce a surge of cold air along the coast; friction alone retarded the front on both sides of the mountain ridge, with the effect being less pronounced on the coastal side; and heating alone is shown to accelerate the front on the inland side of the ridge while acting to produce only a very slight surge on the coastal side. McInnes and McBride go on to show that doubling the height of the orography leads to an even more pronounced and faster-moving coastal jet, a result with implications for southerly changes in New Zealand. Finally, we note that models such as those used by Howells and Kuo and McInnes and McBride are capable of simulating the mesoscale flow in which the southerly buster develops but cannot resolve the surface front itself. Clarification of the dynamics of the surface front, including the processes leading to intense cross-frontal wind and temperature gradients, awaits further work.

Southerly changes on the east coast of New Zealand appear in many cases to have very similar characteristics to the southerly buster. In four summertime cases documented by Sturman et al. (1990), Smith et al. (1991), and Sturman (1992), the postfrontal southerly flow remained shallow for several hours after the surface wind change, well below the mean ridge height of the Southern Alps. In three of these cases, the prefrontal low-level flow was dominated by a warm northwesterly föhn. In one case of an extremely intense front, the adverse pressure gradient along the northeast coast of the South Island, associated with the strong westerly flow through Cook Strait, appeared sufficient to halt the northeastward progression of the front to the North Island. Frontal movement along the west coast of the South Island appears to be impeded by northeasterly flow parallel with the Southern Alps caused by blocking. Moreover, the enhanced frontal temperature contrast provided along the east coast by prefrontal föhnic heating is absent, so that the front is also less intense along the west coast. One of the key differences between intense southerly changes and Australian southerly busters is that the latter are normally preceded by hot continental northerlies. Recently, Katzfey (1995a,b) modeled the orographic precipitation associated with the passage of fronts across the South Island.

5.6. Mesoscale wave generation and propagation

Internal gravity waves play an important role in coupling the troposphere to the middle atmosphere by redistributing momentum and energy. In particular, upward-radiating gravity waves have an associated vertical flux of horizontal momentum. When such waves are transient or suffer attenuation by whatever means, momentum may be transferred to the mean flow. In this way, gravity waves can exert a stress on the atmosphere far from their source region and can profoundly affect the atmospheric circulation at scales much larger than their own. Recent interest in the generation of gravity waves, as well as their propagation characteristics and eventual demise, has been fueled by the central position they hold in determining the large-scale circulation of the middle atmosphere. The weight of observational evidence suggests that gravity waves are predominantly generated in the troposphere by flow over orography, by convection, and by jet/front systems.

a. Orographically generated waves

Lee waves are commonly observed over the east coast of the South Island of New Zealand during nor'westers. Cherry (1972) has compiled the statistics of lee waves observed at Canterbury during the course of one year. Sturman (1980) examined a lee-wave train that developed in the southwesterly flow over Bank's Peninsula near Christchurch; see also Farkas (1958), Revell (1983), and Auer (1992). Mitchell et al. (1990) observed a train of waves in satellite imagery over Macquarie Island (54°S, 159°E) and analyzed their structure using linear wave theory. In Australia, the emphasis has been on trapped lee waves and strong downslope flows associated with wave breaking aloft (Pitts and Lyons 1987, 1989, 1990; Blockley and Lyons 1994; Grace 1991; Grace and Holton 1990; Sha et al. 1996).

Orographically generated gravity waves have also been studied in Antarctica. The British Antarctic Survey maintains an array of meteorological instruments at Halley Station, Antarctica, and has conducted a series of experiments designed to investigate turbulence and internal gravity waves in the strongly stable boundary layer. Only the gravity-wave aspects will be discussed here. The experimental program is known as STABLE. Two intensive observational phases were carried out, the first in 1986 and the second in 1991. A summary of the two experiments is contained in articles by King et al. (1989) and Rees et al. (1994). In strongly stable conditions such as those encountered during STABLE, turbulence is suppressed, and the boundary-layer flow is dominated by internal gravity waves. During both phases of STABLE, high-frequency, small-amplitude gravity-wave activity was almost always observed, the dominant horizontal wavelengths lying between 500 and 2000 m. The regional topography plays the central role in generating these waves (Rees and Mobbs 1988; Rees et al. 1994). Halley Station is located on the Brunt Ice Shelf about 10 km from the sea-ice boundary. Forty kilometers to the southeast, the continental shelf rises to a height of about 3 km. A series of irregular undulations run parallel to the *Hinge Zone*, a narrow region connecting the ice shelf to the continent. The ridges and troughs have a maximum amplitude of about 5 m, decreasing toward the sea, and a wavelength of approximately 1 km. Rees and Mobbs argue that the small-amplitude waves that are almost always present in stable conditions at Halley are generated by katabatic airflow over the undulations in the Hinge Zone. Culf and McIlveen (1993) report that undulations commonly seen in their SODAR measurements, also taken at Halley station, have phase lines that are vertical. The amplitude of these disturbances increases with height within the strong surface-based inversion, above which it decays. Such features are characteristic of trapped internal gravity waves.

Occasionally, the occurrence of small-amplitude waves is punctuated by brief bursts of large-amplitude wave activity. The large-amplitude waves are thought to be radiating Kelvin–Helmholtz instabilities, a conjecture that is supported both by linear instability calculations (King et al. 1987) and by the results of a fully nonlinear numerical model (Rees 1987). King (1993) reports that, on occasion, isolated large-amplitude waves are detected also. Such waves tend to be sharply peaked with flat troughs and are reminiscent of conoidal waves. *Cats-eyes*, the hallmark of large-amplitude Kelvin–Helmholtz instabilities, have been observed by Culf and McIlveen (1993) in their SODAR observations.

b. Morning glories

Perhaps the mesoscale phenomenon most closely associated with the Southern Hemisphere is the *morning glory* of northeastern Australia, first studied by Clarke (1972). Morning glory is the name given to a distinctive cloud formation, characterized by a spectacular low-level roll cloud or series of roll clouds, that may be seen in the southern part of the Gulf of Carpentaria region of northwestern Queensland. The clouds are parallel to one another, often stretching for several hundred kilometers. The leading roll is usually smooth, whereas successive rolls are more ragged in texture, reflecting a higher degree of turbulence. As the name suggests, these disturbances form in the early hours of the morning and normally decay within a few hours after sunrise. The associated disturbance is by far the best documented and most widely known example of an atmospheric undular bore. Although similar disturbances have been observed in many parts of the globe and apparently even on Mars (Pickersgill 1984), a noteworthy feature of northeasterly morning glories is the remarkable regularity with which they are generated in a particular geographical location. Reviews on the morning glory are given by Smith (1988) and Christie (1992); see also Clarke (1983c), Neal et al. (1977), and Reeder et al. (1995).

Two photographs of morning-glory roll clouds are shown in Fig. 5.10. The first is an aerial photograph showing the almost two-dimensional nature of the disturbance, as is often the case. The second is a photograph of a more three-dimensionally structured disturbance. An interesting and unusual feature of this disturbance is the series of instabilities along the roll cloud. Christie (1992) shows other photographs of this particular event and reports that the unstable perturbations subsequently broke, leading to the vortex cloud in the photograph. Christie suggests that the localized instability is the result of an interaction between the large visible roll cloud and a similar, but smaller-amplitude, clear-air disturbance.

Although they are generated sporadically at all times of the year, morning glories are most frequent during the late dry season, September to October. The disturbances are accompanied by pressure fluctuations of several millibars, wind gusts up to about 10 m s^{-1}, and usually propagate with a speed of about 10–15 m s^{-1} relative to the ground. The most common direction of movement is from the northeast, but cloud lines oriented east–west and moving from the south are also observed. Southerly morning glories, which are often associated with the passage of cold fronts across central Australia, are discussed further in section 5.6c. On occasion, northeasterly and southerly morning glories are generated on the same evening, and the interaction between two such disturbances is not uncommon; one such case is documented by Reeder et al. (1995), and a photograph from the ground showing such an interaction is published in Smith et al. (1986). Recently, Christie (1992) reports a few cases of morning glories moving from the southeast, but the origin of these is unknown.

Over the past two decades, northeasterly morning glories have been studied in some detail. The early investigations were based on the meager routine data from the region (e.g., Clarke 1972; Neal et al. 1977), while later ones were based on special field experiments (e.g., Clarke et al. 1981; Smith et al. 1982b; Christie and Muirhead 1983a,b; Smith and Morton 1984; Smith and Page 1985; Clarke 1989b; Menhofer et al. 1997a,b). Northeasterly disturbances are generated during the late evening over the western side of Cape York peninsula as the circulation associated with the east-coast sea breeze develops a marked line of convergence (Clarke 1983b, 1984, 1985; Clarke et al. 1981; Noonan and Smith 1986, 1987; Smith and Noonan 1998). The precise details of the generation mechanism are still uncertain, but it appears that the lifting of stably stratified air from near the surface along the convergence line is an important component. The formation of this convergence line so far inland is aided by an easterly geostrophic wind across the peninsula, which, at the same time, restricts the inland penetration of the west-coast sea breeze. The deep inland penetration is aided by the smallness of the Coriolis parameter at these latitudes; the longevity of tropical sea breezes has been discussed earlier in section 5.4a. Except possibly in the south of the peninsula, the east-coast sea-breeze surge encounters the decaying west-coast sea-breeze circulation on the western side of the peninsula, typically some 350 km inland from the east coast, and moves into a strengthening nocturnal inversion. In this way, a propagating mesoscale convergence line and associated pressure jump are generated in the lower atmosphere. The disturbance is observed to propagate southwestward on the low-level stable layer produced by the onshore

(a)

(b)

FIG. 5.10. Northeasterly morning-glory roll-cloud formation near Burketown in northwestern Queensland, Australia. (a) An aerial photograph showing the common quasi-two-dimensional nature of the disturbance. (b) A more structured cloud line photographed at about 0730 LST on 27 October 1983. An unusual feature of this disturbance is the series of elevated cloud peaks possibly resulting from a breaking instability near the crest of the wave. These localized peaks are believed to be a manifestation of the interaction of the larger-amplitude, visible bore wave with a smaller-amplitude, clear-air solitary wave disturbance. (Photograph by R. K. Smith.)

branch of the sea-breeze circulations along the west coast of the peninsula and the southern coast of the Gulf of Carpentaria. The depth of the cool air behind the east-coast sea breeze is typically 2 km, whereas that behind the west-coast sea breeze is only 500–700 m deep (Physick and Smith 1985). As the disturbance evolves, an amplitude-ordered family of stable solitary waves develop at its leading edge. It is these waves that are responsible for the characteristic morning-glory roll clouds. The roll clouds are formed as moist low-level air is lifted and cooled adiabatically in the crest of the approaching wave, and are eroded as the air descends and is adiabatically warmed in the wave trough.

The recent model calculations by Smith and Noonan (1998) suggest that the sea-breeze collision is not a prerequisite for bore formation; indeed, in the extreme south of the peninsula, the east sea-breeze front sharpens up to form a marked convergence line prior to its collision, and, further south still, collision may never occur. Bore formation may still occur as the nocturnal radiation inversion strengthens, as happens over central Australia in association with a cold front—see section 5.6c.

Theoretical descriptions of morning-glory propagation have been developed within the framework of the weakly nonlinear wave theory first developed by Benjamin (1967), Davis and Acrivos (1967), and Ono (1975). These theories consider long internal gravity waves propagating on a shallow stable layer of fluid with a deep layer of neutrally stable fluid above and assume the wave amplitude is small compared with the depth of the stable layer. The resulting equation for the wave amplitude is sometimes known as the *BDO equation*. The case considered by Benjamin was extended by Clarke et al. (1981) to include shear and applied to the morning glory. Most calculations pertaining to the morning glory have been of this type (e.g., Christie 1989; Christie and Muirhead 1983a,b). One exception is the study by Egger (1984), which takes the Korteweg–de Vries equation as a starting point; this equation describes weakly nonlinear waves in a single shallow layer of homogeneous fluid. Noonan and Smith (1985) present a detailed comparison between the linear theory, the weakly nonlinear theory, and morning-glory observations. The linear long-wave speed gives good agreement with observed propagation speeds, but some uncertainties arise because of an arbitrariness in the choice of the total layer depth in the calculations, upon which the calculated speed depends. Surprisingly, the speed predictions of weakly nonlinear theory are marginally worse, and there is a significant disagreement between the observed and predicted half-widths of the bore waves. Time-dependent integrations of the BDO equation by Christie (1989) and Christie and Muirhead (1983a,b) have illustrated the asymptotic development of amplitude-ordered solitary wave families from smooth initial conditions. Detailed reviews of weakly nonlinear theories applied as to the morning glory can be found in two articles by Christie (1989, 1992).

As mentioned in section 5.4a, Clarke (1984), Crook and Miller (1985), and Haase and Smith (1989) have used two-dimensional numerical models to investigate the generation and propagation of undular bores and solitary waves as a gravity current moves into a surface-based stable layer. If the stable layer is sufficiently deep and/or strong so that the phase speed of waves on the layer is sufficiently large compared to the speed of the gravity current, a nonlinear wave disturbance forms and propagates ahead of the gravity current. If the stable layer is too weak and/or shallow for this to be the case, the presence of the stable layer has a profound effect on the gravity current head, which develops one or more large-amplitude waves enveloping the colder air (see section 5.4a and especially Fig. 5.7). In this case, the large-amplitude waves advect isolated regions of intruding cold air ahead of the main body of the gravity current. Haase and Smith show that the number of waves is related to the relative depth of the stable layer. As the stable-layer depth increases, the number of waves increases also, although their amplitudes decrease. Crook and Miller find that when the effects of condensation are included in their model, the wavelength increases and amplitude decreases. This behavior can be explained in terms of a reduction in static stability associated with the condensation of water vapor.

Questions regarding the effectiveness of the morning-glory wave guide were first raised by Clarke et al. (1981). These authors made the point that theory predicted attenuation rates of morning-glory waves that were far too large to account adequately for their observed longevity. Their predictions did not take into account the vertical wind structure of the morning-glory environment, however, which was largely unknown at the time except in the lowest 2–3 km. Crook (1986, 1988) reexamined the problem, exploring a range of mechanisms that could plausibly lead to the trapping of wave energy in the lower atmosphere and assessing the relevance of these mechanisms by conducting a series of numerical modeling experiments. His calculations were guided by radiosonde and other sounding data that were available at that time. All the mechanisms explored by Crook (1988) involve reductions of the Scorer parameter above the stable layer that lead to wave trapping; these include opposing winds in the middle and upper troposphere, reflection of wave energy from a higher-level inversion, and curvature in the lower wind profile that opposes the motion of the wave. The last mechanism appears to be the one of most relevance to the morning glory.

Recently, the issue of trapping has been reexamined by Smith and Noonan (1998) and Menhofer et al. (1997a). The latter authors compare radiosonde soundings taken at Burketown, located near the Gulf of Carpentaria coastline, on days on which morning-glory disturbances were known to have been generated over the Cape York peninsula but failed to reach Burketown, with days on which morning glories were observed there. Their chief conclusion is that the leakiness of the wave guide is *not* the main factor in determining whether or not morning glories are able to propagate large distances from their origin. This finding is consistent with the recent numerical modeling study by Smith and Noonan (1998), who conclude that energy losses associated with the leakiness of the wave guide in morning-glory wave disturbances may be at least partially offset by energy gains associated with the evolving mesoscale patterns generated by sea-breeze circulations. This hypothesis cannot be tested using Smith and Noonan's model: first, the model resolution is far too coarse to represent the individual waves, and second, the model is hydrostatic, an assumption that would preclude the formation of solitary waves and undular bores in deep fluids (e.g., Christie 1992).

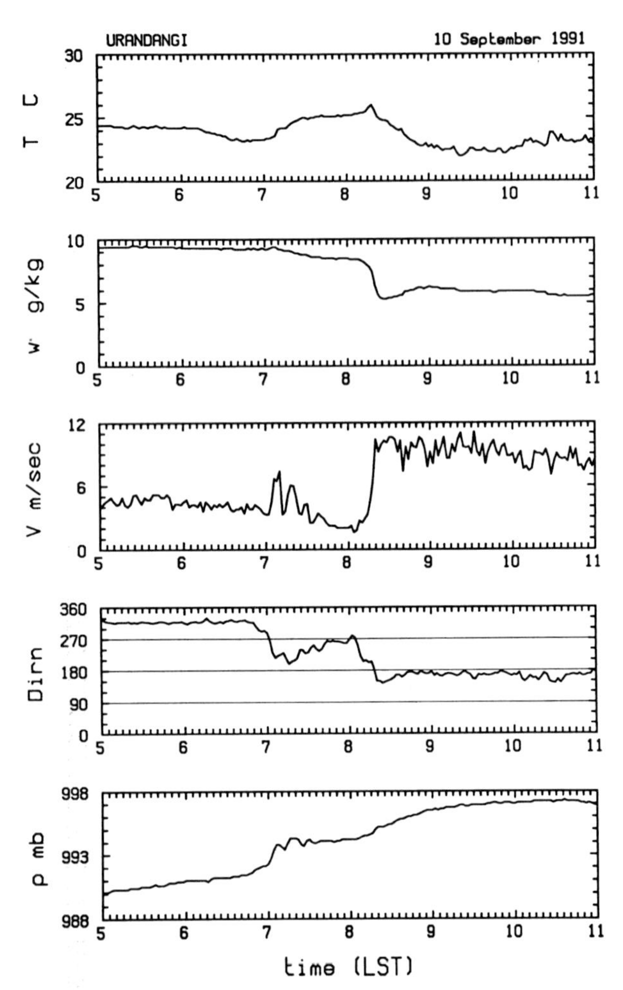

FIG. 5.11. Surface station data at Urandangi (approximately 160 km southwest of Mount Isa in northern Queensland, Australia) surrounding the passage there on 10 September 1991 of the first front during CAFE. (a) Temperature, (b) mixing ratio, (c) wind speed, (d) wind direction, and (e) pressure. (From Smith et al. 1995.)

Although most attention has been focused on the morning glory of northeastern Australia, especially the northeasterly disturbances, similar disturbances have been reported elsewhere in the Southern Hemisphere (e.g., Christie et al. 1978, 1979; Christie et al. 1981; Robin 1978; Drake 1984, 1985; Clarke 1986; Physick 1986; Smith 1986; and Manasseh and Middleton 1995). Physick and Tapper (1990) have suggested that the sea-breeze–like circulations that develop over large central Australian salt lakes may also generate solitary wave disturbances.

c. Large-amplitude waves generated by cold fronts

As noted in the previous section, it appears to be common for subtropical cold fronts over the central part of Australia to generate large-amplitude waves during the late evening or early morning before sunrise (Smith et al. 1995). As the waves develop, they propagate ahead of the air-mass change on the preexisting inversion, giving the front the character of a double change. This behavior is exemplified by the surface data at Urandangi (approximately 160 km southwest of Mount Isa) for the first CAFE event. There the front showed a notable double structure (Fig. 5.11). The leading change at 0700 LST was characterized by a pressure jump (1.8 mb) and a sharp change in wind direction, followed by a series of pressure waves that coincided with fluctuations in the wind speed and direction. The observed wind and pressure signatures are typical of those associated with an amplitude-ordered family of solitary waves or undular bore. As a moderate northwesterly wind was blowing before the disturbance arrived, the temperature rise accompanying these waves was relatively small ($< 1°C$). Sometimes large rises in temperature (up to 7°C) accompany the frontal passage at the surface because of the destruction of a shallow but strong radiation inversion by turbulent mixing at the front. The second change at Urandangi, at 0815 LST, marked the air-mass boundary and was heralded by a strong surge in the wind, a further wind direction change, a sharp decrease in dewpoint temperature and mixing ratio, and a further jump in pressure, albeit smaller than with the leading change.

In a second event observed during CAFE, the nonlinear waves generated a pair of spectacular morning-glory roll clouds as they approached moister air near the Gulf of Carpentaria. Visible satellite imagery shows that the roll clouds were over 300 km long and separated by a distance of about 20 km (Fig. 5.12). The observations made during the second event are unique in providing the first clear evidence of the formation of a southerly morning glory in the gulf region from a cold front in the south. Robin (1978), Clarke (1986), and Physick (1986) have documented several pressure jump lines and roll clouds that may have been generated in a similar way by cold fronts over southeastern Australia.

The precise formation mechanism for such nonlinear waves remains uncertain and is the subject of continued research. One possibility is that the disturbances are generated by the enhanced low-level convergence within the frontal trough, or possibly the inland heat trough, that accompanies the formation of a nocturnal jet. Whatever the case, there is evidence that the mechanism of formation of the bore waves themselves may not depend crucially on the details of the forcing (Christie 1989).

d. Gravity-wave activity and cold fronts

Using VHF wind-profiler observations, Eckermann and Vincent (1993) examined in detail the characteristics of wind oscillations that accompanied the passage

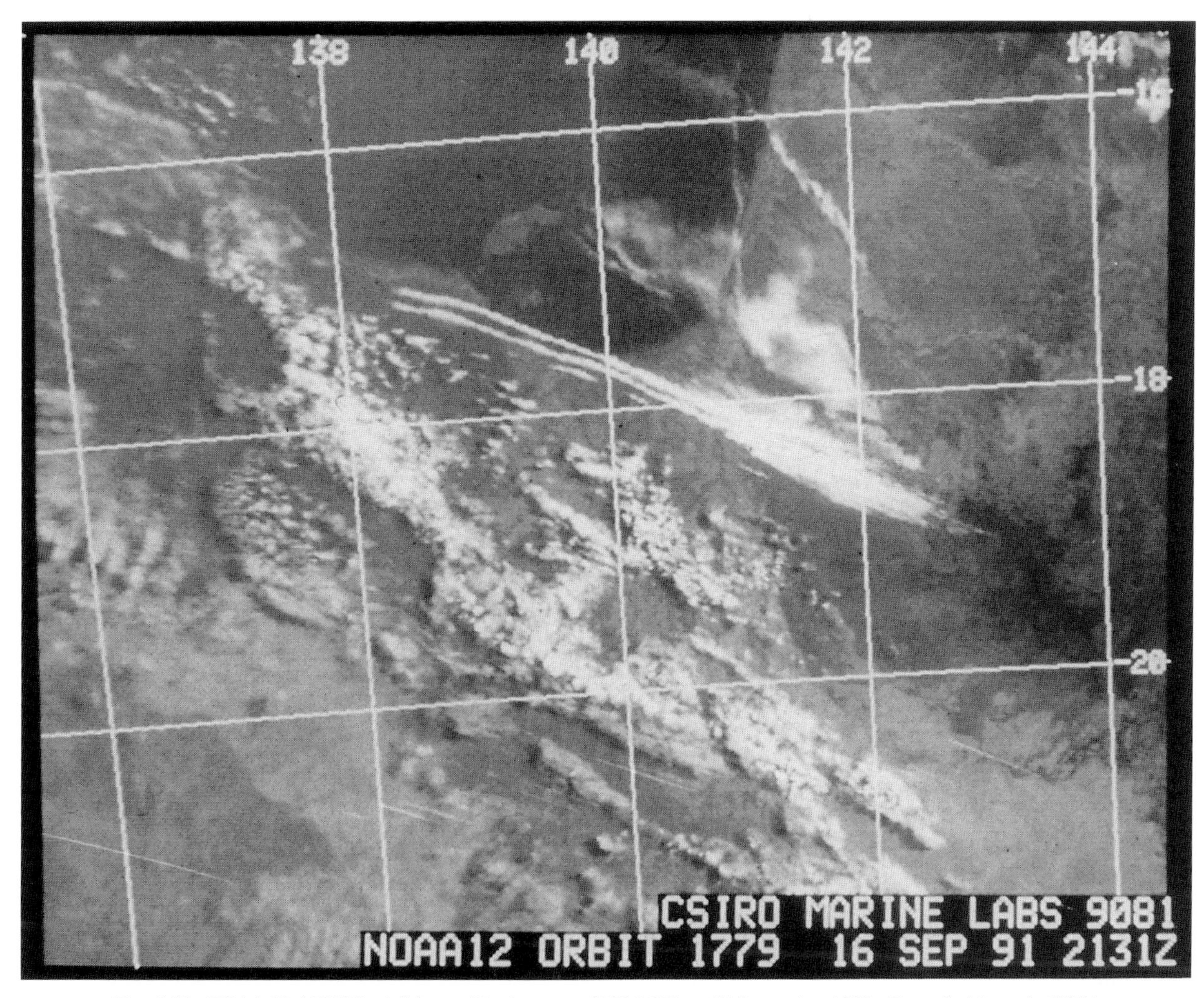

FIG. 5.12. *NOAA-12* AVHRR visible satellite image at 0730 LST on 17 September 1991. (From Smith et al. 1995.)

of six cold fronts across southern Australia during the second half of 1988. They found that with the passage of a front, the variance of the horizontal wind rose by about an order of magnitude along a tongue that was roughly coincident with the frontal boundary. Analysis showed that these fluctuations were consistent with low-frequency, long-wavelength inertia-gravity waves with large, ground-based phase speeds. Moreover, these waves were close to their predicted saturation amplitudes and therefore were presumably exerting a stress on the mean flow. Vertical-velocity variances, the hallmark of high-frequency waves, were found to be largest in the troposphere, both ahead of and behind the cold fronts in regions where strong cross-frontal circulations were detected. Less-intense oscillations were observed also, far ahead and far behind the fronts. In addition, ducted gravity waves were identified.

5.7. Tropical convection

While the problem of tropical convection may seem to be truly hemisphere independent, there have been a number of major studies devoted to convective systems in the Southern Hemisphere, and some of these systems have been found to have special regional characteristics. For example, convection over the Indonesian archipelago, often referred to as the *maritime continent* (Ramage 1968), is the deepest over the globe and is an integral part of the ascending region of the Hadley and Walker circulations. This region includes the equatorial western Pacific and a part of northern Australia, where the surrounding seas are among the warmest on earth. Holland and Keenan (1980) have suggested that deep convection in this region, and by implication a significant fraction of precipitation and latent heat release, is associated with diurnal island thunderstorms (see also Keenan et al. 1989a). Accordingly, it seems appropriate here to include a review of studies in the part of the maritime continent south of the Equator, especially because the region has been the location of many important field programs to investigate tropical convection. Other significant locations for deep convective activity lie in the tropical continental region of South America (Newell et al. 1972; Paegle 1987) and equatorial southern

Africa, and we review studies that have been carried out in these regions as well. Since convection is influenced by, and in turn influences, many scales of motion, we have been especially liberal in our interpretation of *mesoscale* in the present section and include climatological studies of convection. Tropical weather systems that are strongly influenced by convection—namely, tropical depressions and tropical cyclones—are discussed later in section 5.8a, and tropical upper-tropospheric vortices are discussed in section 5.8b. (See also chapter 3 in this volume by McBride et al., which is devoted to the larger-scale aspects of the Southern Hemisphere Tropics.)

a. Southern Africa

We begin with southern Africa, where mesoscale studies of convection appear to be few. Garstang et al. (1987) carried out a climatological study of the generation of convective storms over the escarpment of northeastern South Africa in terms of *synoptic*, *meso*, and *local-scale* forcing. They found that severe storms there, such as those that produce large hail, are associated with midtropospheric troughs in the westerlies, which interact with a major topographic boundary. They also found that a mesoscale capping inversion exists in the lee of the escarpment on storm days. The subsidence inversion, which is anchored to the escarpment, is deepened during the day by the advection of the daytime mixed layer off an elevated heat source. The development of severe storms occurs selectively where easterly low-level moist inflow can penetrate the capping inversion (see also Held 1977).

Lyons (1991) investigated the convective variability over Africa during three austral summers, based on OLR satellite data. He found that a major source of OLR/convective variability over the rainy region of equatorial southern Africa in this season is associated with the interaction between midlatitude wave disturbances embedded in the westerlies and the quasi-stationary tropical upper-tropospheric trough in the vicinity of the southwest equatorial Indian Ocean.

b. South America

One of the early studies of convection over South America was by Kousky (1980), who carried out a study of diurnal rainfall variations in northeastern Brazil for the period 1961–70. Most coastal areas were found to experience a nocturnal maximum in rainfall activity, which, it was suggested, is probably due to convergence between the mean onshore flow and the offshore land breeze. Areas 100–300 km inland were found to experience a daytime rainfall maximum associated with the development and inland advance of convective systems triggered by sea-breeze convergence. The diurnal rainfall variability at most interior locations appeared to be associated with mountain-valley breezes. Seasonal rainfall variations were also investigated.

Interactions between large-scale and convective motions within the Tropics in general, and over Amazonia in particular, are discussed in an article by Paegle (1987). Horel et al. (1989) used OLR data to describe the annual cycle of convection that resides over the Amazon basin during the austral summer and over Central America and the adjacent waters of the Pacific during the austral winter. They determined, inter alia, the preferred locations of convective activity during the wet season in the respective hemispheres, as well as the beginning and end of these seasons. Velasco and Fritsch (1987) used enhanced infrared satellite imagery and conventional surface and sounding data to document the existence and climatological characteristics of MCCs over midlatitude South America and in the tropical region between North and South America. Prominent features of the satellite images taken over Amazonia are the rapidly expanding stratiform anvils. The clouds in such images are commonly called *popcorn cumulonimbus* and are thought to be responsible for a large fraction of the region's precipitation (Wallace 1987).

Modeling studies of the effects of transient convective heating and diurnally varying heating on the tropical circulation over South America have been carried out by Silva Dias et al. (1983, 1987), respectively. Horel et al. (1994) describe numerical simulations of the atmospheric circulation over South America using NCAR's MM4 regional model. Modeling by Paegle (1987) suggests that low-level nocturnal jets over Amazonia may play an important part in initiating convection in that region.

Narrow bands of convective cloudiness over the northern coast of South America, from Guiana to the state of Maranhao in Brazil, are often seen in satellite imagery (Kousky 1980; Sun and Orlanski 1981a,b; Cohen et al. 1995). Some of the cloud bands move inland as squall lines, while others dissipate near the coast. Sun and Orlanski pointed out that well-organized banded cloud clusters and squall lines often form in the early hours of the morning in this area. The cloud bands are generally separated by a few hundred kilometers and are frequently aligned parallel to the Brazilian coastline. Kousky maintains that both propagating and nonpropagating lines are initiated by sea-breeze circulations. Sun and Orlanski offered an explanation of the observations of a banded structure in terms of parametrically excited inertia-gravity waves. Theoretical studies by Orlanski (1973) and Sun and Orlanski (1981a,b) have demonstrated that inertia-gravity waves associated with parametric instability (also known as trapeze instability) can be initiated by the sea-breeze circulation. The diurnal variation of the boundary-layer Brunt-Väisälä frequency may parametrically excite inertia-gravity waves and may be responsible for the observed organization, although the growth rates are small. The waves are excited near the coastline and propagate inland. Little wave activity is predicted (or observed) out to sea. Although the period depends on the latitude, inertia-gravity waves with periods of one to two days are generated, and, like those observed, the most unstable modes have wavelengths of a few hundred kilometers. Sun and Orlanski further support the relevance of their parametric instability theory by pointing to observations taken during the Global Atmospheric Research Program (GARP) Atlantic Tropical Experiment (GATE) that show a pronounced two-day wave (Orlanski and Polinsky 1977). Parametric instability, however, is sensitive to the latitude, being stronger at low latitudes, and the frictional stress and background wind shear (Rotunno 1976). Moreover, nonlinear calculations suggest that the instability may be realized only under conditions of strong diurnal heating or latent heat release (Sun and Orlanski 1981b). On balance, the relevance of this mechanism in organizing convection in the Tropics is uncertain.

Studies of convective systems over the central Amazon basin have resulted from ABLE 2A and 2B, which were carried out in the periods July–August 1985 and April–May 1987, respectively (Garstang et al. 1990). A primary goal of the ABLE experiments was to determine the sources, sinks, concentrations, and transport of trace gases and aerosols originating from the tropical rainforest soils, wetlands, and vegetation. Convective systems, however, were also of major interest because of the relevance of convective mass transport in interpreting the chemical measurements. The findings of ABLE 2A are the subject of a series of papers published in the *Journal of Geophysical Research* in February 1988. The studies of convective systems that were observed during ABLE 2B are described in section 5.7c. In another part of the experiment, De Oliveira and Fitzgarrald (1993) carried out an observational study of the diurnal evolution of the planetary boundary layer over the Amazon rain forest near the confluence of the Solinoes and Negro Rivers. They note the existence there of a diurnal rotation of the low-level wind and the frequent presence of low-level wind maxima. They suggest that this behavior can be plausibly attributed to a river and land-breeze circulation induced by the thermal contrast between the rivers and the adjacent forest.

c. *Amazon coastal squall lines*

Greco et al. (1990) classified the organized convective systems observed during the ABLE 2B experiment into three main types: *coastally occurring systems*, *basin-occurring systems*, and *locally occurring systems*. The coastally occurring systems are large mesoscale to synoptic-scale systems of generally linear organization that form along the northern coast of Brazil during the afternoon and travel across the

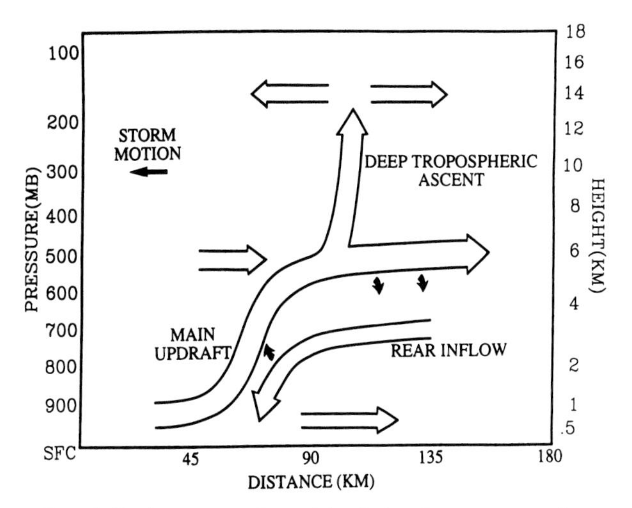

FIG. 5.13. A conceptual model of the flow structure for a mature ACSL, constructed from a combination of vertical velocity and storm-relative flow calculations and two-dimensional cloud model simulations. (From Garstang et al. 1994.)

central Amazon basin as squall lines at speeds of 50–60 km h^{-1}. Garstang et al. (1994), who refer to these systems as Amazon coastal squall lines (ACSLs), studied in detail three of the eight events that were documented during the ABLE 2B experiment. The eight storms contributed to a large proportion (nearly 90%) of the total rainfall recorded during the experiment. Garstang et al. identified six possible stages in the life cycle of ACSLs, including coastal genesis, intensification, maturity, weakening, reintensification, and dissipation. The lines were composed of three different cloud components: a prestorm region that often contained towering cumuli; leading-edge convection with deep vertical ascent; and multiple, precipitating cloud layers in a trailing stratiform region. There was substantial shear in the low-level inflow in all three cases studied. ACSLs have lifetimes between 24 and 48 h, and there is evidence that the shear in the low-level inflow plays a role in their longevity. The storms travel 1000 km or more during their lifetime. Vertical profiles of the equivalent potential temperature θ_e taken from the prestorm, leading-edge convection and trailing stratiform regions indicate that ACSLs stabilize the troposphere in their wake and remove the tropospheric minimum of θ_e. Figure 5.13 shows a conceptual model of the flow structure for a mature ACSL. The heat and moisture budgets of ACSLs are discussed by Greco et al. (1994), while Cohen et al. (1995) examine the environmental conditions associated with squall lines during ABLE 2B. The latter authors find that squall-line days have in common a stronger and deeper low-level jet when compared with days without squall lines. They suggest that the intensification of the jet may be associated in the tropical Atlantic with propagating easterly waves that eventually reach Manaus. A further contributory factor is the localized heat sources over the western Amazon basin.

Silva Dias and Ferreira (1992) describe a linear nonhydrostatic spectral model that was initialized with the vertical profiles of temperature and wind observed prior to convective development along the northern coast of South America during ABLE 2B. The model produced unstable modes with mesoscale wavelengths and propagation speeds comparable with those observed in Amazonian squall lines. The role of convection in trace gas transport and ozone production has been addressed in papers by Scala et al. (1990) and Pickering et al. (1991, 1992).

d. Australia and the Pacific

An early detailed case study of a convective system in the Australian region is that of Drosdowsky (1984), who described a major tropical squall line that affected large areas of northern Australia in December 1981. Keenan and Brody (1988) investigated the synoptic-scale modulation of convection during the Australian summer monsoon, based on GMS (Japanese geostationary meteorology satellite) infrared imagery in conjunction with wind fields obtained from the Australian Numerical Meteorology Research Centre[3] (ANMRC) tropical analysis scheme. They found that the organization of convection into synoptic-scale bands was associated with the 200-mb Southern Hemisphere flow. The troughs and associated subtropical jet streaks had amplified from the south, interacting with and enhancing the monsoon convection. Areas of enhanced convective banding were found east of upper-level tropospheric troughs. West of the troughs, in the subsiding air, convection was suppressed.

In recent years there have been several observational studies of convective systems in the northern Australian region, especially those associated with the Australian monsoon but also systems occurring during the dry and transition seasons. A recent review of research on the Australian monsoon by Manton and McBride (1992) includes a summary of investigations of convective systems that are an integral part of the monsoon and highlights some of the special regional aspects. Two major field experiments carried out in northern Australia were AMEX and EMEX. The former, described by Holland et al. (1986), was carried out in two phases. The first phase, held in October 1986, was designed to investigate the so-called North Australian cloud line, a phenomenon that is reviewed in section 5.7e. The second and much larger phase was carried out simultaneously with EMEX early in 1987 and was designed to study the broad-scale features of the Australian summer monsoon. The EMEX experiment itself was designed to investigate maritime convective systems, which are an integral feature of the monsoon. The results of these experiments are reviewed in section 5.7f.

Observational studies have also been carried out in the Australian region to investigate the special behavior of tropical island thunderstorms; these studies include ITEX and the recent Maritime Continent Thunderstorm Experiment (MCTEX). The results of the former and those of a related modeling study are reviewed in section 5.7g.

Following the two AMEX experiments, a number of mesoscale studies of tropical convection have been conducted in the Darwin region of northern Australia (Keenan et al. 1988; Rutledge et al. 1992). Keenan and Carbone (1992) carried out a morphological study of convective systems during active and break conditions of the summer monsoon and identified several types. These were compared with other systems worldwide. The formation of convective lines perpendicular to the low-level shear and the apparent balance between advancing shallow cold pools and the shear are ubiquitous features. Significant differences were found, however, between systems that develop during active and break conditions. Monsoon convection is generally weak, but *can* have large CAPE. Squall-line structures are generated in the monsoon, often within large stratiform decks. During break and transition periods, the dynamics of storms are dominated by the 700-mb wind maximum and the underlying shear, the shear being slightly larger than for both severe and nonsevere storms documented in the United States by Bluestein et al. (1987). In both break and monsoon periods, storms showed a tendency for the evolution of squall-line structures oriented perpendicular to the low-level wind vector, even when the convection was initiated parallel to the low-level shear. The types of storms observed in the study were compared also with those observed during GATE. Keenan and Rutledge (1993) present a case study of the evolution in the Darwin region of a monsoonal rainband in an environment of low CAPE and weak low-level wind shear, based on an analysis of dual-Doppler-radar data. Cifelli and Rutledge (1994) used data from a wind profiler located in Darwin to observe the vertical motion in four mesoscale convective systems, while May and Rajopadhyaya (1996) carried out a similar study of a tropical squall line using the profiler data to infer aspects of the precipitation microphysics. Keenan and Anderson (1987) present observations of surface wind direction at 40-km resolution over large areas surrounding both developing and nondeveloping tropical cloud clusters, obtained using the Australian Jindalee skywave radar. Preliminary results of an experiment designed at improving short-term forecasts of tropical convection using the Doppler radar facility at Darwin are described by Keenan et al. (1991).

A climatological study of MCCs over the western Pacific region using satellite imagery from the GMS

[3] Now the Bureau of Meteorology Research Centre.

FIG. 5.14. Aerial photograph of an NACL. (Photograph by R. K. Smith.)

satellite was carried out by Miller and Fritsch (1991). They found that MCCs are common in the region, both in the Northern and Southern Hemispheres, and that these systems display many of the same characteristics as those found in the Americas. The systems are nocturnal and tend to form over or in the immediate vicinity of land. In the Southern Hemisphere, they typically move to the left of the climatological mean 700–500-mb flow, in the direction of the source region of highest equivalent potential temperature air. A few MCCs that moved from land to water formed tropical storms, while a few tropical systems that moved over land formed MCCs.

Finally, Miller and Vincent (1987) used data from the tropical South Pacific for a nine-day period during FGGE to estimate convective heating and precipitation in an area containing the SPCZ (see chapter 3 of this monograph by McBride et al.).

e. North Australian cloud lines

The North Australian cloud line (NACL) is a westward-moving line of convective cloud, sometimes extending many hundreds of kilometers across the entire gulf (Drosdowsky and Holland 1987; Drosdowsky et al. 1989). It occurs principally during the dry season and during break (i.e., inactive) periods of the summer monsoon. An aerial photograph of a typical dry-season line is shown in Fig. 5.14. Drosdowsky and Holland (1987) carried out a climatological study of these cloud lines based on satellite imagery and identified three basic types. These include:

- long thin cloud lines, either straight or in the form of mesoscale arcs, generally composed of small cumuli, although isolated large cumuli or cumulonimbi and precipitation may also occur;
- extensive areas of broken cloud, either stratiform or convective, up to 400 km across with a sharp leading edge; and
- overcast deep convective systems, such as tropical squall lines, consisting of a sharp leading edge of deep convective cells and a trailing mesoscale stratiform anvil.

The first two types are observed in all seasons, with a maximum frequency of occurrence in the premonsoon months of September and October.

One theory is that the NACL is initiated by a convergence line that forms in response to the diurnal circulations over Cape York peninsula when the prevailing gradient-level flow is easterly (Noonan and Smith 1986, 1987; Smith and Noonan 1998). The phenomenon was the subject of a modest field experiment carried out in October 1986, the results of which are reported by Drosdowsky et al. (1989). During the AMEX Phase I experiment, most of the cloud lines developed on the western side of deep convection cells that were initiated along the sea-breeze front. The subsequent evolution of the lines appeared to be similar to the African squall-line developments described by Bolton (1984), in which the presence of an easterly jet seems to play an important role. As in the African situation, the air over Cape York peninsula is relatively dry aloft, having undergone dry convective mixing. Thus, evaporatively driven downdrafts may be expected to be especially vigorous and, presumably, assist the propagation of the lines over the gulf. Recent numerical calculations by Smith and Noonan (1998) show that even in dry models, eastward-moving con-

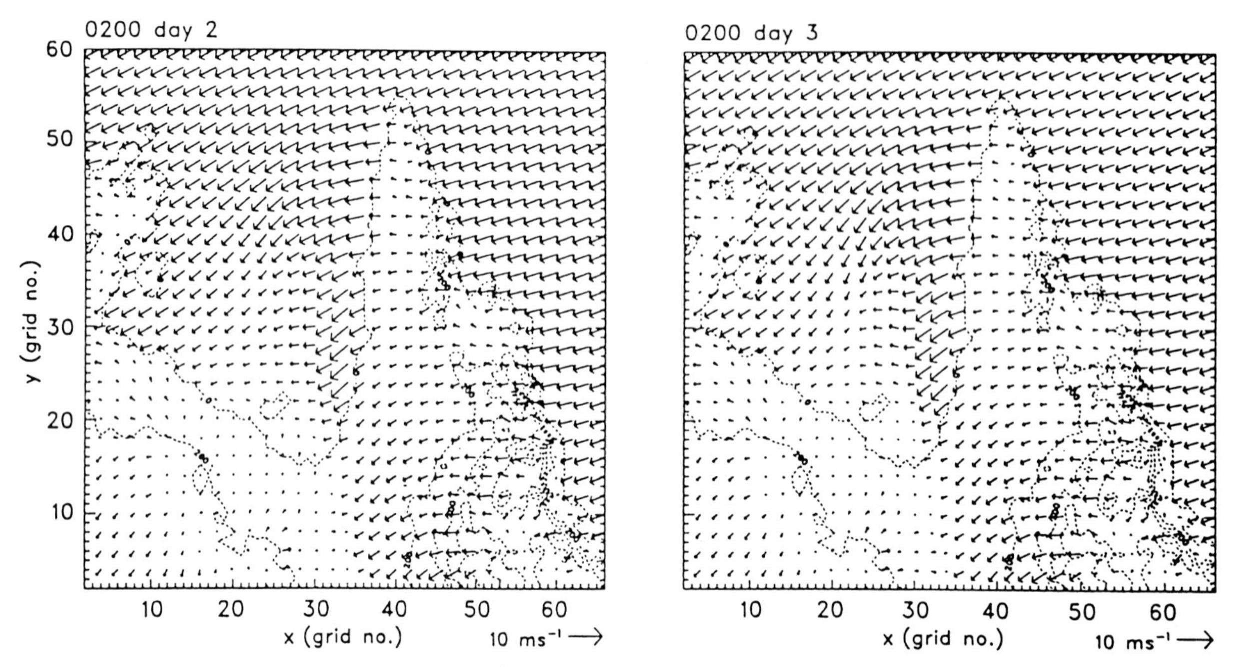

FIG. 5.15. Vector plots of the surface wind at 0200 LST: (a) on day 2 and (b) on day 3, in the calculation of Smith and Noonan (1998), in which there is a uniform northeasterly geostrophic flow over northeastern Australia. Note the surface convergence line over the eastern part of the Gulf of Carpentaria on each day.

vergence lines are pronounced features of the mesoscale circulations over the gulf when the broad-scale zonal component of low-level flow is easterly. Figure 5.15 illustrates the structure of these lines in the surface flow patterns over the gulf region at 0200 LST on consecutive mornings in a numerical-model simulation where a uniform, large-scale 8 m s^{-1} northeasterly geostrophic flow was present aloft. Subsequently, the line moves westward across the gulf as observed. The relative contribution of dry and moist processes to the propagation of NACLs has yet to be determined.

Finally, we note that Kingwell (1984) observed a long-lived semicircular squall line in a satellite image taken off northwestern Australia. The squall line appeared to be organized by a gust front from convection over the land and had many similarities to the NACL.

f. The AMEX and EMEX experiments

While the AMEX experiment focused on the synoptic-scale environment of the Australian summer monsoon, the EMEX experiment was based on airborne measurements of mesoscale convective systems embedded within the monsoon (Webster and Houze 1991). The former ran for a period of 36 days from 10 January 1987, and the latter for a slightly shorter period. A meteorological overview of the pre-AMEX and AMEX periods over the Australian region was given by Love and Garden (1988), and some early results were reported by McBride and Holland (1989). Data from the experiment were used by Davidson and Puri (1992) to test and verify short-term forecasts from the Australian Bureau of Meteorology Research Centre (BMRC) tropical prediction scheme, with special attention paid to the prediction of monsoon events (onset, active, and break phases) and to tropical-cyclone behavior (motion and strength).

A primary aim of AMEX was to determine the vertical profiles of convective heating and moistening associated with deep convection and compare these with profiles obtained in other parts of the world (e.g., in GATE, which was held over the Atlantic off the west coast of Africa in 1974). The findings concerning these aspects were reported by McBride et al. (1989) and Frank and McBride (1989). While the gross structure of cloud clusters from GATE and AMEX have similar life cycles, the peak rainfall in AMEX was about 25% larger, apparently as a result of the warmer SST and moister environment during AMEX. The heating profiles are different, however: the AMEX profiles have a maximum in the middle troposphere, the level of which varies little with time, while in GATE the maximum heating is lower but shifts upward with time as the systems mature. McBride and Frank attribute these differences to the more rapid formation of stratiform rain in the AMEX systems.

Six-hourly analyses based on the special AMEX dataset were used by Davidson (1995a) to investigate the large-scale vorticity budget of a range of tropical weather situations. The study pointed, inter alia, to two important characteristics of convective systems in

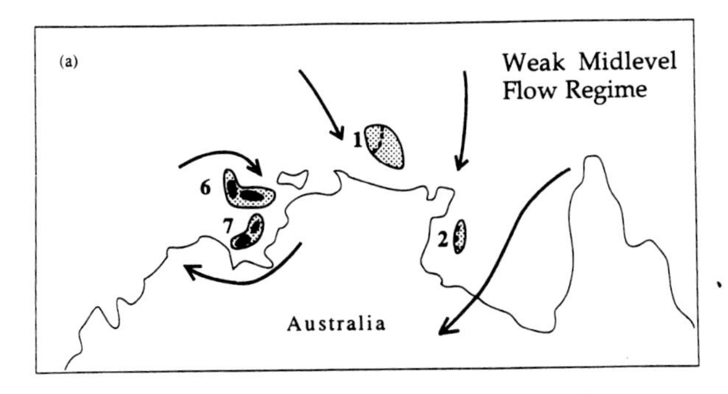

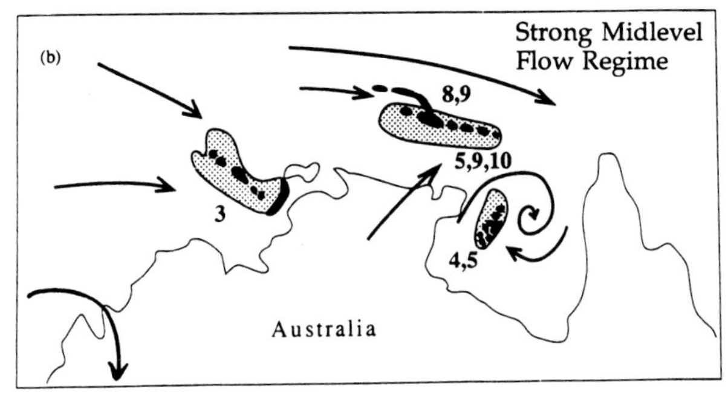

FIG. 5.16. Schematic guide to the MCS's documented during EMEX in relation to midlevel flow streamlines. Light outlines represent radar-echo boundaries; convective areas are black, while stratiform precipitation areas are stippled. In panel (a), a weak-wind regime is depicted, while panel (b) represents the strong-wind regime. (From Mapes and Houze 1992.)

the Tropics: a midlevel convergence maximum for cloud clusters embedded in a weak vorticity environment; and significant apparent vorticity sources associated with both convective and stratiform regions. Davidson pointed out that failure to represent these apparent sources in numerical models may lead to significant forecast errors. In a companion paper, Davidson (1995b) developed a procedure to include these sources in tropical numerical weather prediction models and showed examples of typical integrations using the scheme.

Keenan et al. (1989a) used the AMEX dataset, including satellite and radar data, to study diurnal variations in tropical cloudiness and tropospheric winds. They found the diurnal variation in cloudiness to be a consistent feature of both disturbed and undisturbed conditions. Averaged over the entire AMEX period, the cloudiness peaked during the morning over the ocean and during the afternoon over the land, the amplitude being much larger over the land (see also Hendon and Woodberry 1993 and Holland and Keenan 1980). The AMEX dataset has also been used to extensively test and verify short-term forecasts from the BMRC tropical prediction scheme, including its ability to predict tropical-cyclone motion (Davidson and Puri 1992).

Figure 5.16 shows a schematic guide to the mesoscale convective systems (MCS's) that were documented during the EMEX experiment, detailed studies of which have been carried out by Alexander and Young (1992), Mapes and Houze (1992, 1993a,b, 1995), and Ryan et al. (1992). Alexander and Young found, inter alia, that convective lines are typically oriented nearly normal to the low-level shear vector (950–750 mb) when the magnitude of this vector exceeded 5 m s^{-1}; otherwise, they are oriented along the middle-level shear vector (800–400 mb). Line speeds are best correlated with the maximum rear-to-front flow below 300 mb. The analyses of Mapes and Houze (1992) suggest that several positive feedback mechanisms act to keep thermodynamic conditions favorable for convection and that monsoon convection, once initiated, persisted as self-exciting *superclusters*, composed at any one time of many distinct MCS's. Indeed, the analyses suggest that convection is modulated by low-level processes rather than by large-scale ascent. Among these low-level processes are

- evaporation enhancement by convectively induced surface winds;
- humidification of the dry middle troposphere by convection;
- the formation of boundary-layer cold pools, which trigger additional convection through intense low-level lifting at their edges;
- the removal of these cold pools by surface fluxes on the time scale of half a day; and
- frictional convergence toward regions of low surface pressure, which, inter alia, increases the depth of the moist surface layer and lowers the level of free convection, thereby reducing the amount of convective inhibition.

In the same paper, Mapes and Houze noted that, at least during AMEX, "monsoon convection, once triggered, lasted until tropical cyclones formed and eventually died of the usual causes (landfall or poleward drift)." As a result, they argue that "active phases of the monsoon, once initiated, can be regarded as superclusters of overlapping, interacting deep convection, which at any given instant are composed of distinct individual mesoscale convective systems." Of course, the foregoing processes will differ in importance for convection over land, in comparison with maritime convection, the study of which was the main thrust of EMEX. Mapes and Houze considered also the vorticity dynamics of the monsoon circulation and showed that vortex stretching accounts qualitatively for the time-mean circulation during AMEX.

Lucas et al. (1994) studied the collective properties of convective cores (updrafts and downdrafts) in oceanic convection, based on aircraft data gathered during EMEX. Data on the surface energy balance at representative land stations during both phases of AMEX are presented by Tapper (1988b).

FIG. 5.17. Photograph of a hector taken from the Bureau of Meteorology Regional Forecast Centre in Darwin. (Photograph by R. K. Smith.)

Based on analyses of the EMEX data, Mapes (1993) shows that a heat source with a vertical profile like that of observed tropical MCS's causes, through inviscid gravity-wave dynamics, upward displacements at low levels in a mesoscale region surrounding the heating. Since such displacements are destabilizing (i.e., they reduce the convective inhibition to convection), Mapes argues that conditions near an MCS, but beyond the area of low-level outflow therefrom, become more favorable for the development of additional convection. Accordingly, cloud clusters should be *gregarious*. Observational evidence is provided in support of this contention.

Several tropical cyclones formed during AMEX, and the additional data provided unprecedented documentation of these events. These events will be discussed later in section 5.8a.

g. Tropical island thunderstorms

The ITEX experiment conducted in December 1991 was a modest experiment designed to study tropical thunderstorms that form regularly over the Tiwi Islands (Bathhurst and Melville Island) just north of Darwin during the months leading up to the summer monsoon and during monsoon breaks (Keenan et al. 1989b). The storms are spectacular features of the northern skyline from the Darwin area (Fig. 5.17) and are known locally as *hectors*. A climatology of these storms and a description of their evolution is presented by Keenan et al. (1990). Keenan et al. (1994) describe the structure and evolution of a particular storm during ITEX based on Doppler radar data. After the initial development of convection cells along a line of sea-breeze convergence, individual storms were observed to merge and form squall-like structures that move off the islands. Keenan et al. (1994) found that the vertical velocity and radar structure has more similarity with continental rather than oceanic convection. In particular, maximum updraft velocities are nearly an order of magnitude larger than those typical of oceanic convection. Simpson et al. (1993) describe the rainfall and merging characteristics of these storms on nine days in 1988, one of which developed under a cirrus overcast sky. They found, inter alia, that 90% of the rainfall originates from large merged systems. Skinner and Tapper (1994) presented a climatology of the sea breezes that formed over the Tiwi Islands during the ITEX experiment. Golding (1993) used a version of the United Kingdom Meteorological Office mesoscale weather prediction model to simulate two cases of deep tropical convection from the ITEX experiment. He found good agreement between the simulations and the observed early shower development, in which convection is first initiated along the sea-breeze front and subsequently reinforced by downdraft outflows. The later movement of the storms, however, is less well represented. Cloud merger is reproduced and in the model appears to be associated with colliding downdraft outflows. Most recently, a further, more ambitious field experiment, MCTEX, has been carried out to investigate the life cycle and physical processes involved in island thunderstorms, again concentrating on the hectors that form over the Tiwi Islands north of Darwin.

5.8. Tropical vortices

Many studies of tropical cyclones in the Southern Hemisphere have been concerned with the climatological or synoptic-scale aspects of genesis or motion, rather than mesoscale aspects. No doubt this reflects the paucity of available mesoscale observations. Nevertheless, for completeness we review briefly the literature on all these aspects (see also chapter 3 by McBride et al.).

a. Tropical cyclones and monsoon depressions

Cliatological studies of tropical cyclones and tropical cyclogenesis in the Australian region have been conducted by Holland (1984a,b,c), McBride and Keenan (1982), and Foley and Hanstrum (1994). Annual reports of tropical cyclones in the South Pacific and southeast Indian Ocean in recent years have been provided by Drosdowsky and Woodcock (1991), Ready and Woodcock (1992), Bannister and Smith (1993), and Gill (1994). Some aspects of tropical cyclones in the south Indian Ocean are discussed by De Angelis (1977). A climatology of tropical cyclones over the southwest Indian Ocean can be found in Jury (1993b). The relationship between South Pacific tropical cyclones and the Southern Oscillation has been investigated by Revell and Goulter (1986). Vincent (1985) investigated a cyclone development in the SPCZ during FGGE. Keenan and Templeton (1983) compared radiosonde soundings obtained within 180 km of tropical cyclones in the Australian region with similar sets of hurricane and typhoon data.

Studies of tropical cyclogenesis include those of Foster and Lyons (1984, 1988) for developments in the Western Australian region. Four tropical cyclones, Connie, Irma, Damien, and Jason, developed over northern Australia during AMEX. Two of these, Irma and Jason, developed entirely within the AMEX observational network and are analyzed in detail by Davidson et al. (1990). Numerical simulations of Irma and Connie were carried out by Davidson and Kumar (1990) and Kurihara et al. (1996), respectively. The possible role of cross-equatorial interactions during tropical cyclogenesis in the eastern hemisphere is examined by Love (1985). Keenan and Brody (1988) found that outbreaks of monsoon convection, at least, are virtually unrelated to Northern Hemisphere cold surges. Robertson et al. (1989) used FGGE-data to examine the two-dimensional structure of the region containing the SPCZ during a period when two cyclones formed and propagated along the zone into midlatitudes. Pascoe et al. (1990) and Sinclair (1993a) have examined tropical cyclones that have been transformed into midlatitude storms, while Foley and Hanstrum (1994) discuss tropical cyclones on the western coast of Australia that interact with midlatitude fronts.

The only two studies of the mesoscale aspects of tropical cyclones carried out in the Australian region are the detailed study of the boundary layer of Tropical Cyclone Kerry (1979) by Black and Holland (1995) and the study by Ryan et al. (1992) of a wide rainband in developing Tropical Cyclone Irma that occurred during the AMEX experiment. Both studies were made possible by the availability of data from research aircraft.

Holland et al. (1984) present observations of a very large upper-tropospheric mesoscale temperature perturbation arising from the interaction between dissipating Tropical Cyclone Kerry and a middle-latitude trough in the westerlies. Holland and Merrill (1984) investigate possible physical mechanisms leading to tropical-cyclone structural changes, using a diagnostic model in conjunction with observations in the Australian/southwest Pacific region. Observations of record cold cloud-top temperatures in Tropical Cyclone Hilda (1990) over the Coral Sea off eastern Australia are reported by Ebert and Holland (1992).

On the forecasting side, Keenan (1982) presents an analysis of operational tropical cyclone position and intensity forecasts in the Australian region during the period 1973 to 1980, while Lazic (1990, 1993a,b) investigates forecasts of AMEX tropical cyclones using a numerical model with step mountains. Le Marshall and Mills (1995) have demonstrated that the inclusion of TOVS data can improve tropical-cyclone forecasts in the Australian region. Sinclair (1993a) carried out a synoptic case study of extratropical transition of southwest Pacific Tropical Cyclone Patsy (1986), based on ECMWF analyses. The cyclone underwent a rapid demise as a result of the shearing effect of an approaching middle-latitude trough from the west. It was regenerated by the approach of the next trough, however. Sinclair emphasizes certain differences in behavior compared with extratropical transitions in the Northern Hemisphere, where land–sea contrasts may play a more prominent role. Sinclair (1993b) examines the case of Tropical Cyclone Bola, which, after it became extratropical, caused extremely heavy rainfall as it crossed the North Island of New Zealand. Keenan and Anderson (1987) have shown the capability of the Australian Jindalee skywave radar (see section 5.7d) to locate the rotation center of tropical cyclones and present an example where a developing vortex was located before it was classified as a tropical cyclone.

Lajoie and Butterworth (1984) investigate oscillations of high-level cirrus cloud and heavy precipitation around 11 tropical cyclones in the Australian region, the maximum cirrus being at ± 3 h of 0300 Local Mean Solar Time (LMST) and minimum at ± 3 h of 1800 LMST. In the mean, the ratio of maximum to minimum area is 1.7:1. The diurnal oscillation is 12 h out of phase with that over the North Atlantic. Heavy precipitation associated with tropical cyclones measured on the only atoll in the southwest Pacific with long rainfall records has a similar diurnal variation to that of the area of cirrus canopy.

At best, monsoon depressions lie at the upper end of what might be considered to be mesoscale phenomena, but they are certainly important components of the Australian summer monsoon and are responsible for a significant fraction of tropical rainfall during the monsoon season. Despite this, they have received relatively little study, and there exists no detailed climatology of these disturbances. A brief description of their main characteristics is given by Davidson and Holland (1987), who analyze two particular cases in some detail. These authors note that there are about five such disturbances per year, of which one or two will be the dominant Australian weather feature over a period of a few days. Like tropical cyclones, monsoon depressions are warm-cored systems in the mid- to upper troposphere, with the maximum tangential circulation in the lower troposphere. Indeed, the basic structure of a monsoon depression is similar to that of the outer region of a tropical cyclone, some originating as a tropical cyclone that moves over land (McBride and Keenan 1982). The tangential circulation can be as strong as that of a minimal tropical cyclone, with surface winds exceeding 20 m s^{-1}, and depressions can develop into tropical cyclones if they drift over the warm ocean (Foster and Lyons 1984; McBride and Keenan 1982).

Love and Garden (1984) studied aspects of the mean monsoon circulation over Australia in the austral summer of 1973/74, a year in which the SOI showed its largest positive excursion since records began in 1932. The broadscale monsoon circulation was found to be cold cored from the surface to 700 mb and warm cored between 700 and 150 mb. Love and Garden also carried out a detailed wind analysis on three moist isentropic surfaces (335, 350, and 365 K) on one day of the period and found that the anomalously strong southeast trade winds associated with a blocking pattern over the Tasman Sea were an important feature in the maintenance of the quasi-stationary low pressure system that was centered over the continent. They argue that descent in the trade-wind flow feeding into the low on its western side undercuts the moist northwesterly airstream entering to the north and west, providing an upslope mechanism to sustain the observed quasi-steady nimbostratus cloud mass.

The diagnostic analyses of the two monsoon depressions studied by Davidson and Holland (1987) are based on objective analyses using the (then operational) ANMRC tropical analysis scheme. Both depressions formed over the continent (a relatively data-rich area) in the monsoon trough in the vicinity of the climatological northwest Australian heat low. They subsequently moved southeastward, eventually losing their identity over southeastern Australia over a week later. This movement is in contrast to Indian depressions, which generally move westward. Davidson and Holland analyzed the heat and angular momentum budgets of the two systems and described in detail their evolution and structure. A similar diagnostic analysis was carried out by Hell and Smith (1998), who examine the structure and evolution of an intense monsoon depression that developed near the coast of northwestern Australia in February 1994. The rapid in situ development of the depression was preceded about three days earlier by a steady strengthening of the monsoon trough, but it was not obviously influenced by neighboring weather systems, as appears to have been the situation in other documented cases in the Australian region. Of special interest is the structure and evolution of the potential vorticity field as the storm intensified, which was strongly influenced by the effects of cumulus convection. This is evidenced by the sharp increase in amplitude of the material rate of change of potential vorticity concurrently with the rapid increase in the amplitudes of the apparent heat source and moisture sink. Idealized numerical model calculations relevant to formation of the monsoon depression are described by Dengler and Smith (1998).

Mills and Zhao (1991) carried out numerical predictability experiments in relation to a monsoon depression that brought record rainfall over Australia. They found, inter alia, that the forecast intensification of the low as it crossed the continent was significantly aided by feedback from latent heat release. Interestingly, they found that this feedback contributed to the forecast intensification of the subtropical jet well downstream of the depression. In a companion paper, Zhao and Mills (1991) described the subsequent synoptic evolution of this depression, noting that it did not transform into a cold-cored baroclinic system, but rather the baroclinic low that crossed the continent developed as a separate entity downstream of it. The monsoon depression itself weakened after it had crossed the coast.

Davidson and Hendon (1989) presented evidence of a downstream development process that operates across the entire longitudinal span of the Australasian monsoon, with cyclonic and anticyclonic vortices developing and moving eastward in the monsoon trough at speeds on the order of 5 m s^{-1}.

The formation of deep depressions at low latitudes over the southern African continent appears to be a rare occurrence, but one such development is described by Kumar (1977). Similarly, tropical storms are rarely observed over the tropical oceans on either side of South America. Presumably, the lack of tropical storms over southern Africa and South America is related to the relatively cold waters off these continents.

b. Tropical upper-tropospheric vortices

In a study of summertime tropospheric circulation patterns over South America, Virji (1981) drew attention to the occurrence of upper-tropospheric closed cyclonic vortices that form just off the east coast of Brazil. In general, these track westward over Brazil at a speed of 4°–6° longitude per day. Virji found that as the vortices move westward, the convective activity over the Amazon basin gradually becomes suppressed and the 200-mb Bolivian anticyclone weakens. Kousky and Gan (1981) carried out a further analysis of such vortices over the tropical South Atlantic and confirmed that they are, indeed, primarily a summertime phenomenon. The vortices, which generally form near the axis of the mid-oceanic troughs, are cold cored with a direct thermal circulation, the colder air sinking and warmer air on the periphery rising. The cloud patterns associated with the cyclones are found to depend on the direct thermal circulation, the vortex location, and its direction of movement. Kousky and Gan consider mechanisms for the formation of vortices and discuss their effects on Brazilian weather.

5.9. Conclusion

We have tried to present a broad outline of the progress made over the past two decades in understanding the mesoscale meteorology of the Southern Hemisphere. Southern Hemisphere mesoscale meteorology suffers from a lack of high resolution in situ measurements. Because of this, a great deal of the research in this area has been of a statistical nature or has been based upon specially designed field experiments. In comparison to the number of observational studies, there have been few theoretical or numerical modeling studies of Southern Hemisphere mesoscale phenomena. In our opinion, major progress in further understanding the dynamics of Southern Hemisphere mesoscale weather systems is most likely to be achieved through a greater emphasis on specially designed field experiments, in conjunction with the judicious use of mesoscale numerical models.

Acknowledgments. We would like to thank Ligia Bernardet, Steve Eckermann, Jeff Kepert, Diane MinFa, Linda Hopkins, Bill Physick, and Andy Sturman for their helpful comments.

REFERENCES

Abbs, D. J., 1986: Sea-breeze interactions along a concave coastline in southern Australia: Observations and numerical modeling study. *Mon. Wea. Rev.,* **114,** 831–848.

———, and W. L. Physick, 1992: Sea-breeze observations and modelling: A review. *Aust. Meteor. Mag.,* **41,** 7–19.

———, and J. B. Jensen, 1993: Numerical modeling of orographically forced postfrontal rain. *Mon. Wea. Rev.,* **121,** 189–206.

Adams, M., 1986: A theoretical study of the inland trough of north-eastern Australia. *Aust. Meteor. Mag.,* **34,** 85–92.

———, 1993: A linear study of the effects of heating and orography on easterly airstreams with particular reference to northern Australia. *Aust. Meteor. Mag.,* **42,** 69–80.

Alexander, G. D., and G. S. Young, 1992: The relationship between EMEX mesoscale precipitation feature properties and their environmental characteristics. *Mon. Wea. Rev.,* **120,** 554–564.

André, J.-C., P. Pettré, G. Wendler, and M. Zephoris, 1993: Vertical structure and downslope evolution of antarctic katabatic flows. *Waves and Turbulence in Stably Stratified Flows,* S. D. Mobbs and J. C. King, Eds., Oxford University Press, 91–104.

Auer, A. H., 1992: Wave cloud formations in the lee of southern New Zealand. *Weather,* **47,** 103–105.

Baines, P. G., 1980: The dynamics of the southerly buster. *Aust. Meteor. Mag.,* **28,** 175–200.

Ball, F. K., 1956: The theory of strong katabatic winds. *Aust. J. Phys.,* **9,** 373–386.

———, 1957: The katabatic winds of Adelie Land and King George V Land. *Tellus,* **9,** 201–208.

———, 1960: Winds on the ice slopes of Antarctica. *Antarctica Meteorology,* Permagon Press, 9–16.

Bannister, A. J., and K. J. Smith, 1993: The South Pacific and south-east Indian Ocean tropical cyclone season 1990–1991. *Aust. Meteor. Mag.,* **42,** 175–182.

Bannon, P. R., 1981: Synoptic-scale forcing of coastal lows: Forced double Kelvin waves in the atmosphere. *Quart. J. Roy. Meteor. Soc.,* **107,** 313–327.

Barclay, J. J., M. R. Jury, and W. Landman, 1993: Climatological and structural differences between wet and dry troughs over southern Africa in the early summer. *Meteor. Atmos. Phys.,* **51,** 41–54.

Benjamin, T. B., 1967: Internal waves of permanent form in fluids of great depth. *J. Fluid Mech.,* **29,** 559–592.

Black, P. G., and G. J. Holland, 1995: The boundary layer of Tropical Cyclone Kerry (1979). *Mon. Wea. Rev.,* **123,** 2007–2028.

Blockley, J. A., and T. J. Lyons, 1994: Airflow over a two-dimensional escarpment. III: Nonhydrostatic flow. *Quart. J. Roy. Meteor. Soc.,* **120,** 79–110.

Bluestein, H. B., G. T. Marks, and M. H. Jain, 1987: Formation of mesoscale lines of precipitation: Nonsevere squall lines in Oklahoma during the spring. *Mon. Wea. Rev.,* **115,** 2719–2727.

Bolton, D., 1984: Generation and propagation of African squall lines. *Quart. J. Roy. Meteor. Soc.,* **110,** 695–722.

Bonatti, J. P., and V. B. Rao, 1987: Moist baroclinic instability in the development of North Pacific and South American intermediate-scale disturbances. *J. Atmos. Sci.,* **44,** 2657–2667.

Brinkmann, W. A. R., 1971: What is foehn? *Weather,* **26,** 230–239.

Bromwich, D. H., 1987: A case study of mesoscale cyclogenesis over the south-western Ross Sea. *Antarctic J. U.S.,* **22,** 254–256.

———, 1989a: An extraordinary katabatic wind regime at Terra Nova Bay, Antarctica. *Mon. Wea. Rev.,* **117,** 688–695.

———, 1989b: Satellite analyses of antarctic katabatic wind behavior. *Bull. Amer. Meteor. Soc.,* **70,** 738–349.

———, 1989c: Subsynoptic-scale cyclone development in the Ross Sea sector of the Antarctic. *Polar and Arctic Lows,* P. F. Twitchell, E. A. Rasmussen, and K. L. Davidson, Eds., A. Deepak Publishing, 331–345.

———, 1991: Mesoscale cyclogenesis over the southwestern Ross Sea linked to strong katabatic winds. *Mon. Wea. Rev.,* **119,** 1736–1752.

———, J. F. Carrasco, and C. R. Stearns, 1992: Satellite observations of katabatic-wind propagation great distances across the Ross ice shelf. *Mon. Wea. Rev.,* **120,** 1940–1949.

———, Y. Du, and T. R. Parish, 1994: Numerical simulation of winter katabatic winds from West Antarctica crossing Siple Coast and the Ross Ice Shelf. *Mon. Wea. Rev.,* **122,** 1417–1435.

Browning, K. A., 1990: Organization of clouds and precipitation in extratropical cyclones. *Extratropical Cyclones: The Erik Palmén Memorial Volume,* C. W. Newton and E. O. Holopainen, Eds., Amer. Meteor. Soc., 129–165.

———, and F. F. Hill, 1985: Mesoscale analysis of a polar trough interacting with a polar front. *Quart. J. Roy. Meteor. Soc.,* **111,** 445–462.

Carleton, A. M., 1979: A synoptic climatology of satellite-observed extratropical cyclone activity for the Southern Hemisphere winter. *Arch. Meteor. Geophys. Bioklim.,* **27B,** 265.

———, 1980: Climatology of the "instant occlusion" phenomenon for the Southern Hemisphere winter. *Mon. Wea. Rev.,* **109,** 177–181.

———, 1981a: Monthly variability of satellite derived cyclonic activity for the Southern Hemisphere winter. *J. Climatol.,* **1,** 21–38.

———, 1981b: Ice–ocean–atmosphere interactions at high southern latitudes in winter from satellite observations. *Aust. Meteor. Mag.,* **29,** 183–195.

———, 1992: Synoptic interactions between Antarctica and lower latitudes. *Aust. Meteor. Mag.,* **40,** 129–127.
———, and D. A. Carpenter, 1989: Satellite climatology of "polar air" vortices for the Southern Hemisphere winter. *Polar and Arctic Lows,* P. F. Twitchell, E. A. Rasmussen, and K. L. Davidson, Eds., A. Deepak Publishing, 331–345.
———, and ———, 1990: Satellite climatology of "polar lows" and broadscale climate associations for the Southern Hemisphere. *Int. J. Climatol.,* **10,** 219–246.
———, and M. Fitch, 1993: Synoptic aspects of antarctic mesocyclones. *J. Geophys. Res.,* **98,** 12 997–12 018.
Carrasco, J. F., and D. H. Bromwich, 1993: Mesoscale cyclogenesis dynamics over the south-western Ross Sea, Antarctica. *J. Geophys. Res.,* **98,** 12 973–12 995.
———, and ———, 1994: Climatological aspects of mesoscale cyclogenesis over the Ross Sea and Ross Ice Shelf regions of Antarctica. *Mon. Wea. Rev.,* **122,** 2405–2425.
Cherry, N. J., 1972: Winds and lee waves over Canterbury, New Zealand. *N.Z. J. Sci.,* **15,** 597–600.
Christie, D. R., 1989: Long nonlinear waves in the lower atmosphere. *J. Atmos. Sci.,* **46,** 1462–1491.
———, 1992: The morning glory of the Gulf of Carpentaria: A paradigm for nonlinear waves in the lower atmosphere. *Aust. Meteor. Mag.,* **41,** 21–60.
———, and K. J. Muirhead, 1983a: Solitary waves: A hazard to aircraft operating at low altitudes. *Aust. Meteor. Mag.,* **31,** 97–109.
———, and ———, 1983b: Solitary waves: A low-level wind shear hazard to aviation. *Int. J. Aviation Safety,* **1,** 169–190.
———, ———, and A. L. Hales, 1978: On solitary waves in the atmosphere. *J. Atmos. Sci.,* **35,** 805–825.
———, ———, and ———, 1979: Intrusive density flows in the lower troposphere: A source of atmospheric solutions. *J. Geophys. Res.,* **84,** 4959–4970.
———, ———, and R. H. Clarke, 1981: Solitary waves in the lower atmosphere. *Nature,* **293,** 46–49.
Cifelli, R., and S. A. Rutledge, 1994: Vertical motion structure in maritime continent mesoscale convective systems: Results from a 50-MHz profiler. *J. Atmos. Sci.,* **51,** 2631–2652.
Clarke, R. H., 1955: Some observations and comments on the sea breeze. *Aust. Meteor. Mag.,* **11,** 47–68.
———, 1961: Mesostructure of dry cold fronts over featureless terrain. *J. Meteor.,* **12,** 715–735.
———, 1972: The morning glory: An atmospheric hydraulic jump. *J. Appl. Meteor.,* **11,** 304–311.
———, 1983a: Fair weather nocturnal wind surges and atmospheric bores. Part I: Nocturnal wind surges. *Aust. Meteor. Mag.,* **31,** 133–145.
———, 1983b: Fair weather nocturnal wind surges and atmospheric bores. Part II: Internal atmospheric bores in northern Australia. *Aust. Meteor. Mag.,* **31,** 147–160.
———, 1983c: The "morning glory" of northern Australia. *Weatherwise,* **36,** 134–137.
———, 1984: Colliding sea-breezes and the creation of internal atmospheric bore waves: Two-dimensional numerical studies. *Aust. Meteor. Mag.,* **32,** 207–226.
———, 1985: Geostrophic wind over Cape York peninsula and pressure jumps around the Gulf of Carpentaria. *Aust. Meteor. Mag.,* **33,** 7–10.
———, 1986: Several atmospheric bores and a cold front over southern Australia. *Aust. Meteor. Mag.,* **34,** 65–76.
———, 1989a: Sea-breezes and waves: The "Kalgoorlie sea-breeze" and the "Goondiwindi breeze." *Aust. Meteor. Mag.,* **37,** 99–107.
———, 1989b: On the mortality of morning glories. *Aust. Meteor. Mag.,* **37,** 167–172.
———, R. K. Smith, and D. G. Reid, 1981: The morning glory of the Gulf of Carpentaria: An atmospheric undular bore. *Mon. Wea. Rev.,* **109,** 1726–1750.
Cohen, J. C. P., M. A. F. Silva Dias, and C. A. Nobre, 1995: Environmental conditions associated with Amazonian squall lines: A case study. *Mon. Wea. Rev.,* **123,** 3163–3174.
Colquhoun, J. R., D. J. Shepherd, C. E. Coulman, R. K. Smith, and K. L. McInnes, 1985: The southerly buster of southeastern Australia: An orographically forced cold front. *Mon. Wea. Rev.,* **113,** 2090–2107.
Coulman, C. E., J. R. Colquhoun, R. K. Smith, and K. L. McInnes, 1985: Orographically forced cold fronts: Mean structure and motion. *Bound.-Layer Meteor.,* **32,** 57–82.
Crook, N. A., 1986: The effect of ambient stratification and moisture on the motion of atmospheric undular bores. *J. Atmos. Sci.,* **43,** 171–181.
———, 1988: Trapping of low-level internal gravity waves. *J. Atmos. Sci.,* **45,** 1533–1541.
———, and M. J. Miller, 1985: A numerical and analytical study of atmospheric undular bores. *Quart. J. Roy. Meteor. Soc.,* **111,** 225–242.
Culf, A. D., and J. F. R. McIlveen, 1993: Acoustic observations of the peripheral Antarctica boundary layer. *Waves and Turbulence in Stably Stratified Flows,* S. D. Mobbs and J. C. King, Eds., Oxford University Press, 139–154.
Davidson, N. E., 1995a: Vorticity budget for AMEX. Part I: Diagnostics. *Mon. Wea. Rev.,* **123,** 1620–1635.
———, 1995b: Vorticity budget for AMEX. Part II: Simulations of monsoon onset, midtropospheric lows, and tropical cyclone behavior. *Mon. Wea. Rev.,* **123,** 1636–1659.
———, and G. J. Holland, 1987: A diagnostic analysis of two intense monsoon depressions over Australia. *Mon. Wea. Rev.,* **115,** 380–392.
———, and H. H. Hendon, 1989: Downstream development in the Southern Hemisphere monsoon during FGGE/WMONEX. *Mon. Wea. Rev.,* **117,** 1458–1470.
———, and A. Kumar, 1990: Numerical simulation of the development of AMEX Tropical Cyclone Irma. *Mon. Wea. Rev.,* **118,** 2001–2019.
———, and K. Puri, 1992: Limited area tropical prediction for AMEX. *Aust. Meteor. Mag.,* **40,** 179–190.
———, G. J. Holland, J. L. McBride, and T. D. Keenan, 1990: On the formation of AMEX cyclones Irma and Jason. *Mon. Wea. Rev.,* **118,** 1981–2000.
Davis, R. E., and A. Acrivos, 1967: Solitary internal waves in deep water. *J. Fluid Mech.,* **29,** 593–607.
De Angelis, D., 1977: World of tropical cyclones: South Indian Ocean. *Mariners Wea. Log,* **21,** 240–244.
Dengler, K., and R. K. Smith, 1998: A monsoon depression over northwestern Australia, part II: A numerical study. *Aust. Meteor. Mag.,* **47,** 135–144.
De Oliveira, A. P., and D. R. Fitzgarrald, 1993: The Amazon River breeze and the local boundary layer. Part I: Observations. *Bound.-Layer Meteor.,* **63,** 141–162.
Dexter, P. E., M. L. Heron, and B. T. McGann, 1985: A tropical winter sea-breeze observed with an HF groundwave radar. *Aust. Meteor. Mag.,* **33,** 117–128.
Drake, V. A., 1984: A solitary wave disturbance of the marine boundary layer over Spencer Gulf revealed by radar observations of migrating insects. *Aust. Meteor. Mag.,* **32,** 131–135.
———, 1985: Solitary wave disturbances of the nocturnal boundary layer revealed by radar observations of migrating insects. *Bound.-Layer Meteor.,* **31,** 269–286.
Drosdowsky, W., 1984: Structure of a northern Australian squall line system. *Aust. Meteor. Mag.,* **32,** 177–183.
———, and G. J. Holland, 1987: North Australian cloud lines. *Mon. Wea. Rev.,* **115,** 2645–2659.
———, and F. Woodcock, 1991: The South Pacific and south-east Indian Ocean tropical cyclone season 1988–1989. *Aust. Meteor. Mag.,* **39,** 113–129.
———, G. J. Holland, and R. K. Smith, 1989: Structure and evolution of North Australian cloud lines observed during AMEX Phase I. *Mon. Wea. Rev.,* **117,** 1181–1192.

Duncan, C. N., 1978: Baroclinic instability in a reversed shear flow. *Meteor. Mag.*, **107,** 17–23.

Eady, E. T., 1949: Long waves and cyclones. *Tellus,* **1,** 33–52.

Ebert, E. E., and G. J. Holland, 1992: Observations of record cloud-top temperatures in Tropical Cyclone Hilda (1990). *Mon. Wea. Rev.*, **120,** 2240–2251.

Eckermann, S. D., and R. A. Vincent, 1993: VHF radar observations of gravity-wave production by cold fronts over southern Australia. *J. Atmos. Sci.,* **50,** 785–806.

Egger, J., 1984: On the theory of the morning glory. *Beitr. Phys. Atmos.,* **57,** 123–134.

Evans, M. S., D. Keyser, L. F. Bosart, and G. M. Lackman, 1994: A satellite-derived clarification scheme for rapid maritime cyclogenesis. *Mon. Wea. Rev.,* **122,** 1381–1416.

Fandry, C. B., and L. M. Leslie, 1984: A two-layer quasigeostrophic model of summer trough formation in the Australian subtropical easterlies. *J. Atmos. Sci.,* **41,** 807–818.

Fantini, M., and A. Buzzi, 1993: Numerical experiments on a possible mechanism of cyclogenesis in the Antarctic region. *Tellus,* **45A,** 99–113.

Farkas, E., 1958: Mountain waves over Bank's Peninsula, New Zealand. *N.Z. J. Geol. Geophys.,* **1,** 677–683.

Feren, G., 1995: The "striated delta" cloud system—A satellite imagery precursor to major cyclogenesis in the eastern Australian–western Tasman Sea region. *Wea. Forecasting.,* **10,** 286–309.

Finkele, K., J. M. Hacker, H. Kraus, and R. A. D. Byron-Scott, 1995: A complete sea-breeze circulation cell derived from aircraft observations. *Bound.-Layer Meteor.,* **73,** 299–317.

Fitch, M., and A. M. Carleton, 1992: Antarctic mesocyclone regimes from satellite and conventional data. *Tellus,* **44A,** 180–196.

Foley, G. R., and B. N. Hanstrum, 1994: The capture of tropical cyclones by cold fronts off the west coast of Australia. *Wea. Forecasting,* **9,** 577–572.

Fortune, M. A., and V. E. Kousky, 1983: Two severe freezes in Brazil: Precursors and synoptic evolution. *Mon. Wea. Rev.,* **111,** 181–196.

Foster, I. J., and T. J. Lyons, 1984: Tropical cyclogenesis: A comparative study of two depressions in the north-west of Australia. *Quart. J. Roy. Meteor. Soc.,* **110,** 105–119.

———, and ———, 1988: The development of tropical cyclones in the north-west of Australia. *Quart. J. Roy. Meteor. Soc.,* **114,** 1187–1199.

Frank, W. M., and J. L. McBride, 1989: The vertical distribution of heating in AMEX and GATE cloud clusters. *J. Atmos. Sci.,* **46,** 3464–3478.

Gallee, H., and G. Schayes, 1994: Development of a three-dimensional meso-γ primitive equation model: Katabatic winds simulation in the area of Terra Nova Bay, Antarctica. *Mon. Wea. Rev.,* **122,** 671–685.

———, P. Pettré, and G. Schayes, 1996: Sudden cessation of katabatic wind in Adelelie Land, Antarctica. *J. Appl. Meteor.,* **35,** 1142–1152.

Gan, M. A., and V. B. Rao, 1991: Surface cyclogenesis over South America. *Mon. Wea. Rev.,* **119,** 1293–1302.

———, and ———, 1994: The influence of the Andes Cordillera on transient disturbances. *Mon. Wea. Rev.,* **122,** 1141–1157.

Garratt, J. R., 1985: The inland boundary layer at low latitudes. I. The nocturnal jet. *Bound.-Layer Meteor.,* **32,** 307–327.

———, 1986: Boundary-layer effects on cold fronts at a coastline. *Bound.-Layer Meteor.,* **36,** 101–105.

———, 1988: Summertime cold fronts in southeast Australia—Behavior and low-level structure of main frontal types. *Mon. Wea. Rev.,* **116,** 636–649.

———, and W. L. Physick, 1983: Low-level wind response to mesoscale pressure systems. *Bound.-Layer Meteor.,* **27,** 69–88.

———, and ———, 1985: The inland boundary layer at low latitudes. II: Sea breeze influences. *Bound.-Layer Meteor.,* **33,** 209–231.

———, and ———, 1986: Numerical study of atmospheric gravity currents. I: Simulations and observations of cold fronts. *Beitr. Phys. Atmos.,* **59,** 282–300.

———, and ———, 1987: Numerical study of atmospheric gravity currents. II: Evolution and external influences. *Beitr. Phys. Atmos.,* **60,** 88–102.

———, ———, R. K. Smith, and A. J. Troup, 1985: The Australian summertime cool change. Part II: Mesoscale aspects. *Mon. Wea. Rev.,* **113,** 202–223.

———, P. A. C. Howells, and E. Kowalczyk, 1989: The behavior of dry cold fronts traveling along a coastline. *Mon. Wea. Rev.,* **117,** 1208–1220.

Garstang, M. S., B. E. Kelbe, G. D. Emmitt, and W. B. London, 1987: Generation of convective storms over the escarpment of northeastern South Africa. *Mon. Wea. Rev.,* **115,** 429–443.

———, and Coauthors, 1990: The Amazon Boundary Layer Experiment (ABLE2B): A meteorological perspective. *Bull. Amer. Meteor. Soc.,* **71,** 19–32.

———, H. L. Massie Jr., J. Halverson, S. Greco, and J. Scada, 1994: Amazon coastal squall lines. Part I: Structure and kinematics. *Mon. Wea. Rev.,* **122,** 608–622.

Gauntlett, J., L. M. Leslie, and L. W. Logan, 1984: Numerical experiments in mesoscale prediction over southeast Australia. *Mon. Wea. Rev.,* **112,** 1170–1182.

Gill, A. E., 1977: Coastally trapped waves in the atmosphere. *Quart. J. Roy. Meteor. Soc.,* **103,** 431–440.

Gill, J. P., 1994: The South Pacific and south-east Indian Ocean tropical cyclone season 1991–1992. *Aust. Meteor. Mag.,* **43,** 181–192.

Golding, B. W., 1993: A numerical investigation of tropical island thunderstorms. *Mon. Wea. Rev.,* **121,** 1417–1433.

———, and L. M. Leslie, 1993: The impact of resolution and formation on model simulations of an east coast low. *Aust. Meteor. Mag.,* **42,** 105–116.

Grace, W., 1991: Hydraulic jump in a fog bank. *Aust. Meteor. Mag.,* **39,** 205–209.

———, and I. Holton, 1990: Hydraulic jump signatures associated with Adelaide downslope winds. *Aust. Meteor. Mag.,* **38,** 43–52.

Greco, S. J., R. Swap, M. Garstang, S. Ulanski, M. Shipham, R. C. Harriss, R. Talbot, M. O. Andrae, and P. Artaxo, 1990: Rainfall and surface kinematic conditions over central Amazonia during ABLE2B. *J. Geophys. Res.,* **93,** 17 001–17 014.

———, J. Scada, J. Halverson, H. L. Massie Jr., W.-K. Tao, and M. Garstang, 1994: Amazon coastal squall lines. Part II: Heat and moisture transports. *Mon. Wea. Rev.,* **122,** 623–635.

Griffiths, M., M. J. Reeder, D. J. Low, and R. A. Vincent, 1998: Observations of a cut-off low over Southern Australia. *Quart. J. Roy. Meteor. Soc.,* **124,** 1109–1132.

Haase, S. P., and R. K. Smith, 1989: The numerical simulation of atmospheric gravity currents. Part II: Environments with stable layers. *Geophys. Astrophys. Fluid Dyn.,* **46,** 35–51.

Hamilton, M. G., and J. R. Tarifa, 1978: Synoptic aspects of a polar outbreak leading to frost in tropical Brazil. *Mon. Wea. Rev.,* **106,** 1545–1556.

Hanstrum, B. N., K. J. Wilson, and S. L. Barrell, 1990a: Prefrontal troughs over southern Australia. I: A climatology. *Wea. Forecasting,* **5,** 22–31.

———, ———, and ———, 1990b: Prefrontal troughs over southern Australia. II: A case study of frontogenesis. *Wea. Forecasting,* **5,** 32–46.

Heinemann, G., 1990: Mesoscale vortices in the Weddell Sea region (Antarctica). *Mon. Wea. Rev.,* **118,** 779–793.

———, 1996: A wintertime polar low over the eastern Weddell Sea (Antarctica): A study with AVHRR, TOVS, SSM/I and conventional data. *Meteor. Atmos. Phys.,* **58,** 83–102.

Held, G., 1977: Description of the unusual behavior of a prefrontal squall line in South Africa. *J. Appl. Meteor.,* **16,** 651–653.

Hell, R. M., and R. K. Smith, 1998: A monsoon depression over northwestern Australia, part I: A case study. *Aust. Meteor. Mag.*, **47,** 21–40.
Hendon, H. H., and K. Woodberry, 1993: The diurnal cycle of tropical convection. *J. Geophys. Res.*, **98,** 16 623–16 637.
Hess, G. D., 1990: Numerical simulation of the August 1986 heavy rainfall event in the Sydney area. *J. Geophys. Res.*, **95,** 2073–2082.
Hill, H. W., 1980: A case of intense cyclogenesis in the Tasman Sea. *N.Z. J. Sci.*, **23,** 111–127.
Hines, K. M., D. H. Bromwich, and T. R. Parish, 1995: A mesoscale modeling study of the atmospheric circulation of high southern latitudes. *Mon. Wea. Rev.*, **123,** 1146–1165.
Holland, G. J., 1984a: On the climatology and structure of tropical cyclones in the Australian/south-west Pacific region: I. Data and tropical storms. *Aust. Meteor. Mag.*, **32,** 1–15.
———, 1984b: On the climatology and structure of tropical cyclones in the Australian/south-west Pacific region: II. Hurricanes. *Aust. Meteor. Mag.*, **32,** 17–31.
———, 1984c: On the climatology and structure of tropical cyclones in the Australian/south-west Pacific region: III. Major hurricanes. *Aust. Meteor. Mag.*, **32,** 33–46.
———, and T. D. Keenan, 1980: Diurnal variations of convection over the maritime continent. *Mon. Wea. Rev.*, **108,** 223–225.
———, and R. T. Merrill, 1984: On the dynamics of tropical cyclone structural changes. *Quart. J. Roy. Meteor. Soc.*, **110,** 723–745.
———, and L. M. Leslie, 1986: Ducted coastal ridging over southeastern Australia. *Quart. J. Roy. Meteor. Soc.*, **112,** 731–748.
———, and J. L. McBride, 1989: Quasi-trajectory analysis of a sea-breeze front. *Quart. J. Roy. Meteor. Soc.*, **115,** 571–580.
———, T. D. Keenan, and G. D. Crane, 1984: Observations of phenomenal temperature perturbations in tropical cyclone Kerry (1979). *Mon. Wea. Rev.*, **112,** 1074–1082.
———, J. L. McBride, R. K. Smith, D. Jasper, and T. D. Keenan, 1986: The BMRC Australian Monsoon Experiment: AMEX. *Bull. Amer. Meteor. Soc.*, **67,** 1466–1472.
———, A. H. Lynch, and L. M. Leslie, 1987: Australian east-coast cyclones. Part I: Synoptic overview and case study. *Mon. Wea. Rev.*, **115,** 3024–3036.
Hopkins, L. C., and G. J. Holland, 1997a: Australian heavy-rain days and associated east-coast cyclones: 1958–92. *J. Climate,* **10,** 621–635.
———, and ———, 1997b: Comments on "The origin and evolution of low-level potential vorticity anomalies during a case of Tasman Sea cyclogenesis." *Tellus,* **49A,** 401–403.
Horel, J., A. Hahman, and J. Geisler, 1989: An investigation of the annual cycle of convective activity over the tropical Americas. *J. Climate,* **2,** 1388–1403.
———, J. Pechmann, A. N. Hahman, and J. E. Geisler, 1994: Simulations of the Amazon basin circulation with a regional model. *J. Climate,* **7,** 56–71.
Hoskins, B. J., 1990: Theory of extratropical cyclones. *Extratropical Cyclones: The Erik Palmén Memorial Volume,* C. W. Newton and E. O. Holopainen, Eds., Amer. Meteor. Soc., 64–80.
———, M. E. McIntyre, and A. W. Robertson, 1985: On the use and significance of isentropic potential vorticity maps. *Quart. J. Roy. Meteor. Soc.*, **111,** 877–947.
Howells, P. A. C., and Y.-H. Kuo, 1988: A numerical study of the mesoscale environment of a southerly buster event. *Mon. Wea. Rev.*, **116,** 1771–1788.
Innocentini, V., and E. S. Caetano Neto, 1996: Case study of the 9 August 1988 South Atlantic storm: Numerical simulations of the wave activity. *Wea. Forecasting,* **11,** 78–88.
Jones, D. A., and I. Simmonds, 1993: A climatology of Southern Hemisphere extratropical cyclones. *Climate Dyn.*, **9,** 131–145.
Jury, M. R., 1993a: A thermal front within the marine atmospheric boundary layer over the Agulhas current south of Africa: Composite aircraft observations. *J. Geophys. Res.*, **99,** 3297–2204.
———, 1993b: A preliminary study of the climatological associations and characteristics of tropical cyclones in the SW Indian Ocean. *Meteor. Atmos. Phys.*, **51,** 101–115.
———, and M. Laing, 1990: A case study of marine cyclogenesis near Cape Town. *Tellus,* **42A,** 246–258.
Katzfey, J. J., 1995a: Simulation of extreme New Zealand precipitation events. Part 1: Sensitivity to orography and resolution. *Mon. Wea. Rev.*, **123,** 737–754.
———, 1995b: Simulation of extreme New Zealand precipitation events. Part II: Mechanics of precipitation development. *Mon. Wea. Rev.*, **123,** 755–775.
Keenan, T. D., 1982: A diagnostic study of tropical cyclone forecasting in Australia. *Aust. Meteor. Mag.*, **30,** 69–80.
———, and J. I. Templeton, 1983: A comparison of tropical cyclone, hurricane and typhoon mass, and moisture structure. *Mon. Wea. Rev.*, **111,** 320–327.
———, and S. J. Anderson, 1987: Some examples of surface wind field analysis based on Jindalee skywave radar data. *Aust. Meteor. Mag.*, **35,** 153–161.
———, and L. R. Brody, 1988: Synoptic-scale modulation of convection during the Australian summer monsoon. *Mon. Wea. Rev.*, **116,** 71–85.
———, and R. E. Carbone, 1992: A preliminary morphology of precipitation systems in tropical northern Australia. *Quart. J. Roy. Meteor. Soc.*, **118,** 283–326.
———, and S. A. Rutledge, 1993: Mesoscale characteristics of monsoonal convection and associated stratiform precipitation. *Mon. Wea. Rev.*, **121,** 352–374.
———, G. J. Holland, M. J. Manton, and J. Simpson, 1988: TRMM ground truth in a monsoon environment: Darwin, Australia. *Aust. Meteor. Mag.*, **36,** 81–90.
———, J. L. McBride, G. J. Holland, N. Davidson, and B. Gunn, 1989a: Diurnal variations during the Australian Monsoon Experiment (AMEX) Phase II. *Mon. Wea. Rev.*, **117,** 2535–2552.
———, B. R. Morton, M. J. Manton, and G. J. Holland, 1989b: The Island Thunderstorm Experiment (ITEX): A study of tropical maritime continent thunderstorms. *Bull. Amer. Meteor. Soc.*, **70,** 152–159.
———, ———, Y. S. Zhang, and K. Nyguen, 1990: Some characteristics of thunderstorms over Bathurst and Melville Islands near Darwin, Australia. *Quart. J. Roy. Meteor. Soc.*, **116,** 1153–1172.
———, R. J. Potts, and J. Wilson, 1991: The Darwin Area Forecasting Experiment: Description and preliminary results. *Aust. Meteor. Mag.*, **39,** 211–222.
———, B. Ferrier, and J. Simpson, 1994: Development and structure of a maritime continent thunderstorm. *Meteor. Atmos. Phys.*, **53,** 185–222.
Kepert, J., and R. K. Smith, 1992: A simple model of the Australian west coast trough. *Mon. Wea. Rev.*, **120,** 2042–2055.
Keyser, D., M. J. Reeder, and R. J. Reed, 1988: A generalization of Petterssen's frontogenesis function and its relation to vertical motion. *Mon. Wea. Rev.*, **116,** 762–780.
Kidson, J. W., and M. R. Sinclair, 1996: The influence of persistent anomalies on Southern Hemisphere storm tracks. *J. Climate,* **8,** 1938–1950.
King, J. C., 1993: Contrasts between the antarctic stable boundary layer and the midlatitude nocturnal boundary layer. *Waves and Turbulence in Stably Stratified Flows,* S. D. Mobbs and J. C. King, Eds., Oxford University Press, 105–120.
———, S. D. Mobbs, M. S. Darby, and J. M. Rees, 1987: Observations of an internal gravity wave in the lower troposphere at Halley, Antarctica. *Bound.-Layer Meteor.*, **39,** 1–13.
———, ———, J. M. Rees, P. S. Anderson, and A. D. Culf, 1989: The STABLE antarctic boundary layer experiment at Halley Station. *Weather,* **44,** 398–405.
Kingwell, J., 1984: Observations of a quasi-circular squall line off north-west Australia. *Weather,* **39,** 343–346.

Kousky, V. E., 1978: Fluctuations in annual rainfall for northeastern Brazil. *J. Meteor. Soc. Japan,* **56,** 457–465.
———, 1979: Frontal influences on northeast Brazil. *Mon. Wea. Rev.,* **107,** 1140–1153.
———, 1980: Diurnal rainfall variations in northeast Brazil. *Mon. Wea. Rev.,* **108,** 488–498.
———, and M. A. Gan, 1981: Upper tropospheric cyclonic vortices in the tropical South Atlantic. *Tellus,* **33,** 538–551.
Kraus, H., J. M. Hacker, and J. Hartmann, 1990: An observational aircraft-based study of sea-breeze frontogenesis. *Bound.-Layer Meteor.,* **53,** 223–265.
Kumar, S., 1977: An unusual case of a deep depression over southern Africa. *Weather,* **32,** 291–296.
Kurihara, K., N. Davidson, K. Puri, R. Bowen, and M. Ueno, 1996: Simulation of Tropical Cyclone Connie from the Australian Monsoon Experiment. *Aust. Meteor. Mag.,* **45,** 101–111.
Lajoie, F. A., and I. J. Butterworth, 1984: Oscillation of high-level cirrus and heavy precipitation around Australian region tropical cyclones. *Mon. Wea. Rev.,* **112,** 535–544.
Lazic, L., 1990: Forecasts of AMEX tropical cyclones with a step-mountain model. *Aust. Meteor. Mag.,* **38,** 207–215.
———, 1993a: Eta model forecasts of tropical cyclones from Australian Monsoon Experiment: Dynamical adjustment of initial conditions. *Meteor. Atmos. Phys.,* **51,** 101–111.
———, 1993b: Eta model forecasts of tropical cyclones from Australian Monsoon Experiment: Model sensitivity. *Meteor. Atmos. Phys.,* **51,** 113–127.
Le Marshall, J. F., and G. A. Mills, 1995: Tropical Cyclone Bobby—A notable example of the impact of local TOVS data. *Aust. Meteor. Mag.,* **44,** 293–297.
Leslie, L. M., 1980: Numerical modeling of the heat low over Australia. *J. Appl. Meteor.,* **19,** 381–387.
———, and T. C. L. Skinner, 1994: Real-time forecasting of the Western Australian summertime trough: Evaluations of a new regional model. *Wea. Forecasting.,* **9,** 371–383.
———, G. J. Holland, and A. H. Lynch, 1987: Australian east-coast cyclones. Part II: Numerical modeling study. *Mon. Wea. Rev.,* **115,** 3037–3053.
Lloyd, L. M., 1990: The sea-breeze at Darwin: A climatology. *Meteor. Mag.,* **119,** 105–112.
Love, G., 1985: Cross-equatorial interactions during tropical cyclogenesis. *Mon. Wea. Rev.,* **113,** 1499–1509.
———, and G. Garden, 1984: The Australian monsoon of January 1974. *Aust. Meteor. Mag.,* **32,** 185–194.
———, and ———, 1988: A meteorological overview of the pre-AMEX periods over the Australian region. *Aust. Meteor. Mag.,* **36,** 91–100.
Lucas, C., E. J. Zipser, and M. A. LeMone, 1994: Vertical velocity in oceanic convection off tropical Australia. *J. Atmos. Sci.,* **51,** 3183–3193.
Lynch, A. H., 1987: Australian east-coast cyclones. Part III: Case study of the storm of August 1986. *Aust. Meteor. Mag.,* **35,** 163–170.
Lyons, S. W., 1991: Origins of convective variability over equatorial southern Africa during summer. *J. Climate,* **4,** 23–29.
Manasseh, R., and J. H. Middleton, 1995: Boundary-layer oscillations from thunderstorms at Sydney Airport. *Mon. Wea. Rev.,* **123,** 1166–1177.
Manins, P. C., and B. L. Sawford, 1979: Katabatic winds: A field case study. *Quart. J. Roy. Meteor. Soc.,* **105,** 1011–1025.
Manton, M. J., and J. L. McBride, 1992: Recent research on the Australian monsoon. *J. Meteor. Soc. Japan,* **70,** 275–285.
Mapes, B. E., 1993: Gregarious tropical convection. *J. Atmos. Sci.,* **50,** 2026–2037.
———, and R. A. Houze Jr., 1992: An integrated view of the 1987 Australian monsoon and its mesoscale convective systems. I: Horizontal structure. *Quart. J. Roy. Meteor. Soc.,* **118,** 927–963.
———, and ———, 1993a: An integrated view of the 1987 Australian monsoon and its mesoscale convective systems. Part II: Vertical structures. *Quart. J. Roy. Meteor. Soc.,* **119,** 733–754.
———, and ———, 1993b: Cloud clusters and superclusters over the oceanic warm pool. *Mon. Wea. Rev.,* **121,** 1398–1415.
———, and ———, 1995: Diabatic divergence profiles in western Pacific mesoscale convective systems. *J. Atmos. Sci.,* **52,** 1807–1827.
Mass, C. F., and M. D. Albright, 1987: Coastal southerlies and alongshore surges of the West Coast of North America: Evidence of mesoscale topography trapped response to synoptic forcing. *Mon. Wea. Rev.,* **115,** 1707–1738.
May, P. T., and D. K. Rajopadhyaya, 1996: Wind profiler observations of vertical motion and precipitation microphysics of a tropical squall line. *Mon. Wea. Rev.,* **124,** 623–633.
———, K. J. Wilson, and B. F. Ryan, 1990: VHF radar studies of cold fronts traversing southern Australia. *Beitr. Phys. Atmos.,* **63,** 257–269.
McBride, J. L., and T. D. Keenan, 1982: Climatology of tropical cyclone genesis in the Australian region. *J. Climate,* **2,** 13–33.
———, and G. J. Holland, 1989: The Australian Monsoon Experiment (AMEX): Early results. *Aust. Meteor. Mag.,* **37,** 23–35.
———, and K. L. McInnes, 1993: Australian southerly busters. Part II: The dynamical structure of the orographically modified front. *Mon. Wea. Rev.,* **121,** 1921–1935.
———, B. W. Gunn, G. H. Holland, T. D. Keenan, N. E. Davidson, and W. M. Frank, 1989: Time series of total heating and moistening over the Gulf of Carpentaria radiosonde array during AMEX. *Mon. Wea. Rev.,* **117,** 2701–2713.
McGowan, H. A., and A. P. Sturman, 1996: Regional and local scale characteristics of foehn wind events over the South Island of New Zealand. *Meteor. Atmos. Phys.,* **58,** 151–164.
———, I. F. Owens, and A. P. Sturman, 1995: Thermal and dynamical characteristics of alpine lake breezes, Lake Tekapo, New Zealand. *Bound.-Layer Meteor.,* **76,** 3–24.
McGregor, J. L., and F. Kimura, 1989: Numerical simulation of mesoscale eddies over Melbourne. *Mon. Wea. Rev.,* **117,** 50–66.
McInnes, K. L., 1993: Australian southerly busters. Part III: The physical mechanism and synoptic conditions contributing to development. *Mon. Wea. Rev.,* **121,** 3261–3281.
———, and G. D. Hess, 1992: Modifications to the Australian limited area model and their impact on an east coast low event. *Aust. Meteor. Mag.,* **40,** 21–31.
———, and J. L. McBride, 1993: Australian southerly busters. Part I: Analysis of a numerically simulated case study. *Mon. Wea. Rev.,* **121,** 1904–1920.
———, ———, and L. M. Leslie, 1994: Cold fronts over southeastern Australia: Their representation in an operational numerical weather prediction model. *Wea. Forecasting,* **9,** 384–409.
McKendry, I. G., 1989: Numerical simulation of sea breezes over Auckland region, New Zealand—Air quality implications. *Bound.-Layer Meteor.,* **49,** 7–22.
———, and C. G. Revell, 1992: Mesoscale eddy development over South Auckland—A case study. *Wea. Forecasting,* **7,** 134–142.
———, A. P. Sturman, and I. F. Owen, 1986: A study of interacting multi-scale wind systems, Canterbury Plains, New Zealand. *Meteor. Atmos. Phys.,* **35,** 242–252.
———, ———, and ———, 1987: The Canterbury Plains northeasterly. *Wea. Climate,* **7,** 61–74.
———, ———, and ———, 1988: Numerical simulation of local thermal effects on the wind field of the Canterbury Plains, New Zealand. *N.Z. J. Geology Geophys.,* **31,** 511–524.
Menhofer, A., R. K. Smith, M. J. Reeder, and D. R. Christie, 1997a: On the structure of "morning-glory" disturbances and the environment in which they propagate. *J. Atmos. Sci.,* **54,** 1712–1725.
———, ———, ———, and ———, 1997b: The bore-like character of three morning glories observed during the Central Australian Fronts Experiment. *Aust. Meteor. Mag.,* **46,** 277–285.

Miller, B. L., and D. G. Vincent, 1987: Convective heating and precipitation estimates for tropical South Pacific during FGGE, 10–18 January 1979. *Quart. J. Roy. Meteor. Soc.,* **113,** 189–212.

Miller, D., and J. M. Fritsch, 1991: Mesoscale convective complexes in the western Pacific region. *Mon. Wea. Rev.,* **119,** 2978–2992.

Mills, G. A., 1989: Dynamics of rapid cloud band development over southeastern Australia. *Mon. Wea. Rev.,* **117,** 1402–1422.

———, 1991: The impact of increased TOVS data on the numerical forecasts of a case of cyclogenesis. *Aust. Meteor. Mag.,* **39,** 223–235.

———, and S. Zhao, 1991: A study of a monsoon depression bringing record rainfall over Australia. Part I: Numerical predictability experiments. *Mon. Wea. Rev.,* **119,** 2053–2073.

———, and I. Russell, 1992: The April 1990 floods of eastern Australia: Synoptic description and assessment of regional NWP guidance. *Wea. Forecasting,* **7,** 636–668.

———, and B.-J. Wu, 1995: The "Cudlee Creek" extratropical cyclone—An example of synoptic scale forcing of a mesoscale event. *Aust. Meteor. Mag.,* **44,** 201–218.

Mitchell, R. M., R. P. Cechet, P. J. Turner, and C. C. Elsum, 1990: Observations and interpretation of wave clouds over Macquarie Island. *Quart. J. Roy. Meteor. Soc.,* **116,** 741–752.

Myers, V. A., 1964: A cold front invasion of southern Venezuela. *Mon. Wea. Rev.,* **92,** 513–521.

Neal, A. B., I. J. Butterworth, and K. M. Murphy, 1977: The morning glory. *Weather,* **32,** 176–183.

Neumann, J., 1973: The sea and land breezes in classical Greek literature. *Bull. Amer. Meteor. Soc.,* **54,** 5–8.

Newell, R. E., J. W. Kidson, D. G. Vincent, and G. J. Boer, 1972: *The General Circulation of the Tropical Atmosphere and Interactions with Extratropical Latitudes,* Vols. 1 and 2. MIT Press, 628 pp.

Nguyen, N. A., and A. E. Gill, 1981: Generation of coastal lows by synoptic-scale waves. *Quart. J. Roy. Meteor. Soc.,* **107,** 521–530.

Noonan, J. A., and R. K. Smith, 1985: Linear and weakly nonlinear internal wave theories applied to "morning glory" waves. *Geophys. Astrophys. Fluid Dyn.,* **33,** 123–143.

———, and ———, 1986: Sea-breeze circulations over Cape York peninsula and the generation of Gulf of Carpentaria cloud-line disturbances. *J. Atmos. Sci.,* **43,** 1679–1693.

———, and ———, 1987: The generation of North Australian cloud lines and the "morning glory." *Aust. Meteor. Mag.,* **35,** 31–45.

Ono, H., 1975: Algebraic solitary waves in stratified fluids. *J. Phys. Soc. Japan,* **39,** 1082–1091.

Ooyama, K., 1982: Conceptual evolution of the theory and modelling of the tropical cyclone. *J. Meteor. Soc. Japan,* **60,** 369–379.

Orlanski, I., 1973: Trapeze instability as a source of internal gravity waves. Part I. *J. Atmos. Sci.,* **30,** 1007–1016.

———, and R. Polinsky, 1977: Spectral distribution of cloud cover over Africa. *J. Meteor. Soc. Japan,* **55,** 483–494.

———, and J. Katzfey, 1991: The life cycle of a cyclone wave in the Southern Hemisphere. Part I: Eddy energy budget. *J. Atmos. Sci.,* **48,** 1972–1998.

———, ———, C. Menendez, and M. Marino, 1991: Simulation of an extratropical cyclone in the Southern Hemisphere: Model sensitivity. *J. Atmos. Sci.,* **48,** 2293–2311.

Paegle, J., 1987: Interactions between convective and large-scale motions over Amazonia. *The Geophysiology of Amazonia: Vegetation and Climate Interactions,* R. Dickerson, Ed., Wiley-Intersciences, 347–390.

Parish, T. R., 1982: Surface winds over East Antarctica. *Mon. Wea. Rev.,* **110,** 84–90.

———, 1984: A numerical study of strong katabatic winds over Antarctica. *Mon. Wea. Rev.,* **112,** 545–554.

———, 1988: Surface winds over the Antarctic continent: A review. *Rev. Geophys.,* **26,** 169–180.

———, and D. H. Bromwich, 1986: The inversion wind pattern over West Antarctica. *Mon. Wea. Rev.,* **114,** 849–860.

———, and ———, 1987: The surface wind field over the antarctic ice sheets. *Nature,* **328,** 51–54.

———, and K. T. Waight III, 1987: The forcing of antarctic katabatic winds. *Mon. Wea. Rev.,* **115,** 2214–2226.

———, and D. H. Bromwich, 1991: Continental-scale simulation of the antarctic katabatic wind regime. *J. Climate,* **4,** 135–146.

———, P. Pettre, and G. Wendler, 1993: The influence of large-scale forcing on the katabatic wind regime at Adelie Land, Antarctica. *Meteor. Atmos. Phys.,* **51,** 165–176.

Parmenter, F. C., 1976: A Southern Hemisphere cold-front passage near the Equator. *Bull. Amer. Meteor. Soc.,* **57,** 1435–1440.

Pascoe, R. M., J. H. A. Lopdell, and M. R. Sinclair, 1990: The Canterbury/Kaikoura storm of January 19, 1988. *Wea. Climate,* **10,** 16–23.

Petterssen, S., 1936: Contributions to the theory of frontogenesis. *Geophys. Publ.,* **11**(6), 1–27.

Pettré, P., and J.-C. André, 1991: Surface-pressure change through Loewe's phenomena and katabatic flow jumps: Study of two cases in Adélie Land, Antarctica. *J. Atmos. Sci.,* **48,** 557–571.

Physick, W. L., 1976: A numerical model of the sea-breeze phenomenon over a lake or gulf. *J. Atmos. Sci.,* **33,** 2107–2135.

———, 1981: Winter depression tracks and climatological jet streams in the Southern Hemisphere during the FGGE year. *Quart. J. Roy. Meteor. Soc.,* **107,** 883–898.

———, 1982: Sea breezes in the Latrobe Valley. *Aust. Meteor. Mag.,* **30,** 255–263.

———, 1986: Observations of a solitary wave train in Melbourne, Australia. *Aust. Meteor. Mag.,* **34,** 163–172.

———, 1988: Mesoscale modeling of a cold front and its interaction with a diurnally heated land mass. *J. Atmos. Sci.,* **45,** 3169–3187.

———, and R. A. D. Byron-Scott, 1977: Observations of the sea breeze in the vicinity of a gulf. *Weather,* **32,** 373–381.

———, and R. K. Smith, 1985: Observations and dynamics of sea breezes in northern Australia. *Aust. Meteor. Mag.,* **33,** 51–63.

———, and N. J. Tapper, 1990: A numerical study of circulations induced by a dry salt lake. *Mon. Wea. Rev.,* **118,** 1029–1042.

———, and D. J. Abbs, 1991: Modeling of summertime flow and dispersion in the coastal terrain of southeastern Australia. *Mon. Wea. Rev.,* **119,** 1014–1030.

———, and ———, 1992: Flow and plume dispersion in a coastal valley. *J. Appl. Meteor.,* **31,** 64–73.

———, W. K. Downey, A. J. Troup, B. F. Ryan, and P. J. Meighen, 1985: Mesoscale observations of a prefrontal squall line. *Mon. Wea. Rev.,* **113,** 1958–1969.

Pickering, K. E., A. M. Thompson, J. R. Scala, W.-K. Tao, J. Simpson, and M. Garstang, 1991: Photochemical ozone production in tropical squall line convection during NASA/GTE/ABLE 2A. *J. Geophys. Res.,* **96,** 3099–3114.

———, ———, ———, ———, R. R. Dickerson, and J. Simpson, 1992: Free tropospheric ozone production following entrainment of urban plumes into deep convection. *J. Geophys. Res.,* **97,** 17 985–18 000.

Pickersgill, A. O., 1984: Martian bore waves of the Tharsis region: A comparison with Australian atmospheric waves of elevation. *J. Atmos. Sci.,* **41,** 1461–1473.

Pitts, R. O., and T. J. Lyons, 1987: The influence of topography on Perth radiosonde observations. *Aust. Meteor. Mag.,* **35,** 17–23.

———, and ———, 1989: Airflow over a two-dimensional escarpment. I: Observations. *Quart. J. Roy. Meteor. Soc.,* **115,** 965–981.

———, and ———, 1990: Airflow over a two-dimensional escarpment. II: Hydrostatic flow. *Quart. J. Roy. Meteor. Soc.,* **116,** 363–378.

Preston-Whyte, R. A., and P. D. Tyson, 1973: Note on pressure oscillations over South Africa. *Mon. Wea. Rev.,* **101,** 650–659.

Rácz, Z. S., and R. K. Smith, 1998: The dynamics of heat lows. *Quart. J. Roy. Meteor. Soc.*, in press.

Ramage, C. S., 1968: Role of a tropical "maritime continent" in the atmospheric circulation. *Mon. Wea. Rev.*, **96,** 365–370.

Rao, V. B., and M. A. Gan, 1996: Comments on "Climatology of cyclogenesis for the Southern Hemisphere." *Mon. Wea. Rev.*, **124,** 2614–2615.

Raymond, D., 1994: Cumulus convection and the Madden–Julian oscillation of the tropical troposphere. *Physica D,* **77,** 1–22.

Ready, S., and F. Woodcock, 1992: The South Pacific and southeast Indian Ocean tropical cyclone season 1989–1990. *Aust. Meteor. Mag.*, **40,** 111–121.

Reason, C. J. C., 1994: Orographically trapped disturbances in the lower atmosphere: Scale analysis and simple models. *Meteor. Atmos. Phys.*, **53,** 131–136.

———, and M. R. Jury, 1990: On the generation and propagation of the Southern African coastal low. *Quart. J. Roy. Meteor. Soc.*, **116,** 1133–1151.

———, and D. G. Steyn, 1990: Coastally trapped disturbances in the lower atmosphere: Dynamic commonalities and geographic diversity. *Prog. Phys. Geog.*, **14,** 178–198.

———, and ———, 1992: The dynamics of coastally trapped mesoscale ridges in the lower atmosphere. *J. Atmos. Sci.*, **18,** 1677–1692.

Reed, R. J., 1990: Advances in understanding of extratropical cyclones during the past quarter century. *Extratropical Cyclones: The Erik Palmén Memorial Volume,* C. W. Newton and E. O. Holopainen, Eds., Amer. Meteor. Soc., 27–45.

Reeder, M. J., 1986: The interaction of a surface cold front with a pre-frontal thermodynamically well-mixed boundary layer. *Aust. Meteor. Mag.*, **34,** 137–148.

———, and R. K. Smith, 1986: A comparison between frontogenesis in the two-dimensional Eady model of baroclinic instability and summertime cold fronts in the Australian region. *Quart. J. Roy. Meteor. Soc.*, **112,** 293–313.

———, and ———, 1987: A study of frontal dynamics with application to the Australian summertime "cool change." *J. Atmos. Sci.*, **44,** 687–705.

———, and ———, 1988a: On air motion trajectories in cold fronts. *J. Atmos. Sci.*, **45,** 4005–4007.

———, and ———, 1988b: On the horizontal resolution of fronts in numerical weather prediction models. *Aust. Meteor. Mag.*, **36,** 11–16.

———, and ———, 1992: Australian spring and summer cold fronts. *Aust. Meteor. Mag.*, **41,** 101–124.

———, D. Keyser, and B. D. Schmidt, 1991: Three-dimensional baroclinic instability and summertime frontogenesis in the Australian region. *Quart. J. Roy. Meteor. Soc.*, **117,** 1–28.

———, D. R. Christie, R. K. Smith, and R. Grimshaw, 1995: Interacting morning glories over northern Australia. *Bull. Amer. Meteor. Soc.*, **76,** 1165–1171.

Rees, J. M., 1987: The propagation of internal gravity waves in the stably stratified boundary layer. *Ann. Geophys.*, **5B,** 421–432.

———, and S. D. Mobbs, 1988: Studies of internal gravity waves at Halley Base, Antarctica, using wind observations. *Quart. J. Roy. Meteor. Soc.*, **114,** 939–966.

———, I. McConnell, P. S. Anderson, and J. C. King, 1994: Observations of internal gravity waves over an antarctic ice shelf using a microbarograph array. *Stably Stratified Flows: Flow and Dispersion over Topography,* I. P. Castro and N. J. Rockcliff, Eds., Oxford University Press, 61–79.

Reid, D. G., 1957: Evening wind surges in south Australia. *Aust. Meteor. Mag.*, **16,** 23–32.

Revell, C. G., 1982: High resolution satellite imagery of the New Zealand area: A view of lee waves. *Wea. Climate,* **2,** 23–29.

———, 1983: High resolution satellite imagery of the New Zealand area: A view of mesoscale cloud vortices. *Wea. Climate,* **3,** 11–19.

———, and S. W. Goulter, 1986: South Pacific tropical cyclones and the Southern Oscillation. *Mon. Wea. Rev.*, **114,** 1138–1145.

Revell, M. J., 1997: Reply to comments on "The origin and evolution of low-level potential vorticity anomalies during a case of Tasman Sea cyclogenesis." *Tellus,* **49A,** 404–405.

———, and R. N. Ridley, 1995: The origin and evolution of low-level potential vorticity anomalies during a case of Tasman Sea cyclogenesis. *Tellus,* **47A,** 779–796.

Ridley, R. N., 1990: Southerly buster events in New Zealand. *Wea. Climate,* **10,** 35–54.

Robertson, F. R., D. G. Vincent, and D. M. Kann, 1989: The role of diabatic heating in maintaining the upper-tropospheric baroclinic zone in the South Pacific. *Quart. J. Roy. Meteor. Soc.*, **115,** 1253–1271.

Robin, A. G., 1978: Roll cloud over Spencer Gulf. *Aust. Meteor. Mag.*, **26,** 125.

Rogerson, A. M., and R. M. Samelson, 1995: Synoptic forcing of coastal-trapped disturbances in the marine atmosphere boundary layer. *J. Atmos. Sci.*, **52,** 2025–2040.

Rottman, J. W., and J. E. Simpson, 1989: The formation of internal bores in the atmosphere: A laboratory model. *Quart. J. Roy. Meteor. Soc.*, **115,** 941–963.

Rotunno, R., 1976: Trapeze instability modified by a mean shear flow. *J. Atmos. Sci.*, **33,** 1663–1667.

Rutledge, S. A., E. R. Williams, and T. D. Keenan, 1992: The Down Under Doppler and Electricity Experiment (DUNDEE): Overview and preliminary results. *Bull. Amer. Meteor. Soc.*, **73,** 3–16.

Ryan, B. F., 1982: A perspective of research into cold fronts and associated mesoscale phenomena into the 1980s. *Aust. Meteor. Mag.*, **30,** 123–131.

———, 1983: The Cold Fronts Research Programme: Working together to study the "summertime cool change." *Weatherwise,* **36,** 113.

———, and J. Medows, 1979: A relative-flow isentropic analysis of a cut-off low. *Aust. Meteor. Mag.*, **27,** 43–57.

———, and K. J. Wilson, 1985: The Australian summertime cool change. Part III: Subsynoptic and mesoscale models. *Mon. Wea. Rev.*, **113,** 224–240.

———, ———, J. R. Garratt, and R. K. Smith, 1985: Cold Fronts Research Programme: Progress, future plans, and research directions. *Bull. Amer. Meteor. Soc.*, **66,** 1116–1122.

———, J. R. Garratt, and W. L. Physick, 1988: The prognosis of weather change lines over the ocean: The role of reconnaissance aircraft. *Aust. Meteor. Mag.*, **36,** 69–80.

———, K. F. Wilson, and E. J. Zipser, 1989: Modification of the thermodynamic structure of the lower troposphere by the evaporation of precipitation ahead of a cold front. *Mon. Wea. Rev.*, **117,** 138–153.

———, G. J. Tripoli, and W. R. Cotton, 1990: Convection in high stratiform cloud bands: Some numerical experiments. *Quart. J. Roy. Meteor. Soc.*, **116,** 943–964.

———, G. M. Barnes, and E. J. Zipser, 1992: A wide rainband in a developing tropical cyclone. *Mon. Wea. Rev.*, **120,** 431–447.

Satyamurty, P., and L. F. De Mattos, 1989: Climatological lower tropospheric frontogenesis in the midlatitudes due to horizontal deformation and divergence. *Mon. Wea. Rev.*, **117,** 1355–1364.

———, R. P. Santos, and M. A. M. Lemes, 1980: On the stationary trough generated by the Andes. *Mon. Wea. Rev.*, **108,** 510–519.

———, C. C. Ferreira, and M. A. Gan, 1990: Cyclonic vortices over South America. *Tellus,* **42,** 194–201.

Scala, J. R., and Coauthors, 1990: Cloud draft structure and trace gas transport. *J. Geophys. Res.*, **95,** 17 015–17 030.

Sha, W., W. Grace, and W. Physick, 1996: A numerical experiment on the Adelaide gully wind of South Australia. *Aust. Meteor. Mag.*, **45,** 19–40.

Silva Dias, M. A. F., and R. N. Ferreira, 1992: Application of a linear spectral model to the study of Amazonian squall lines during GTE/ABLE 2B. *J. Geophys. Res.*, **97,** 20 405–20 419.

Silva Dias, P. L., W. H. Schubert, and M. DeMaria, 1983: Large-scale response of the tropical atmosphere to transient convection. *J. Atmos. Sci.,* **40,** 2689–2707.

———, J. P. Bonatti, and V. Kousky, 1987: Diurnally forced tropical tropospheric circulation over South America. *Mon. Wea. Rev.,* **115,** 1465–1478.

Simpson, J. E., 1982: Gravity currents in the laboratory, atmosphere and ocean. *Ann. Rev. Fluid Mech.,* **14,** 213–234.

———, T. D. Keenan, B. Ferrier, R. H. Simpson, and G. J. Holland, 1993: Cumulus mergers in the maritime continent region. *Meteor. Atmos. Phys.,* **51,** 73–99.

Sinclair, M. R., 1993a: Synoptic-scale diagnosis of the extratropical transition of a southwest Pacific tropical cyclone. *Mon. Wea. Rev.,* **121,** 941–960.

———, 1993b: A diagnostic study of extratropical precipitation resulting from Tropical Cyclone Bola. *Mon. Wea. Rev.,* **121,** 2690–2707.

———, 1994: An objective cyclone climatology for the Southern Hemisphere. *Mon. Wea. Rev.,* **122,** 2239–2256.

———, 1995a: A climatology of cyclogenesis for the Southern Hemisphere. *Mon. Wea. Rev.,* **123,** 1601–1619.

———, 1995b: An extended climatology of extratropical cyclones over the Southern Hemisphere. *Wea. Climate,* **15,** 21–32.

———, 1996a: A climatology of anticyclones and blocking for the Southern Hemisphere. *Mon. Wea. Rev.,* **124,** 245–263.

———, 1996b: Reply to comments on "Climatology of cyclogenesis for the Southern Hemisphere." *Mon. Wea. Rev.,* **124,** 2615–2618.

———, and X. Cong, 1992: Polar airstream cyclogenesis in the Australian region: A composite study using ECMWF analyses. *Mon. Wea. Rev.,* **120,** 1950–1972.

Skinner, T., and L. M. Leslie, 1982: Predictability experiments with the Australian west coast trough. *Aust. Meteor. Mag.,* **30,** 241–249.

———, and N. J. Tapper, 1994: Preliminary sea-breeze studies over Bathurst and Melville Islands, Northern Australia, as part of the Island Thunderstorm Experiment (ITEX). *Meteor. Atmos. Phys.,* **53,** 77–94.

Smith, A. V., 1985: Studies of the effects of cold fronts during the rainy season in Zimbabwe. *Weather,* **40,** 198–202.

Smith, R. K., 1986: Evening glory wave-cloud lines in northwestern Australia. *Aust. Meteor. Mag.,* **34,** 27–33.

———, 1988: Travelling waves and bores in the lower atmosphere: The "morning glory" and related phenomena. *Earth-Sci. Rev.,* **25,** 267–290.

———, and B. R. Morton, 1984: An observational study of northeasterly "morning glory" wind surges. *Aust. Meteor. Mag.,* **32,** 155–175.

———, and M. A. Page, 1985: Morning glory wind surges and the Gulf of Carpentaria cloud line of 25–26 October 1984. *Aust. Meteor. Mag.,* **33,** 185–194.

———, and M. J. Reeder, 1988: On movement and low-level structure of cold fronts. *Mon. Wea. Rev.,* **116,** 1927–1944.

———, and R. Ridley, 1990: Subtropical continental cold fronts. *Aust. Meteor. Mag.,* **38,** 191–200.

———, and J. A. Noonan, 1998: On the generation of low-level mesoscale convergence lines over northeastern Australia. *Mon. Wea. Rev.,* **126,** 167–185.

———, B. F. Ryan, A. J. Troup, and K. J. Wilson, 1982a: Cold fronts research: The Australian summertime "cool change." *Bull. Amer. Meteor. Soc.,* **63,** 1028–1034.

———, N. A. Crook, and G. Roff, 1982b: The morning glory: An extraordinary undular bore. *Quart. J. Roy. Meteor. Soc.,* **108,** 937–956.

———, M. J. Coughlan, and J. L. Lopez, 1986: Southerly nocturnal wind surges and bores in northeastern Australia. *Mon. Wea. Rev.,* **114,** 1501–1518.

———, R. N. Ridley, M. A. Page, J. T. Steiner, and A. P. Sturman, 1991: Southerly changes on the east coast of New Zealand. *Mon. Wea. Rev.,* **119,** 1259–1282.

———, M. J. Reeder, N. J. Tapper, and D. R. Christie, 1995: Central Australian cold fronts. *Mon. Wea. Rev.,* **123,** 16–38.

Snyder, C., and R. S. Lindzen, 1991: Quasigeostrophic wave–CISK in an unbounded baroclinic shear. *J. Atmos. Sci.,* **48,** 76–86.

Streten, N. A., and A. J. Troup, 1973: A synoptic climatology of satellite observed cloud vortices over the Southern Hemisphere. *Quart. J. Roy. Meteor. Soc.,* **99,** 56–72.

Sturman, A. P., 1980: A case study of lee waves in south-westerly airflow over Bank's Peninsula, New Zealand. *Weather,* **35,** 32–39.

———, 1992: Dynamic and thermal effects on surface airflow associated with southerly changes over the South Island, New Zealand. *Meteor. Atmos. Phys.,* **47,** 229–236.

———, and D. P. Tyson, 1981: Sea breezes along the Canterbury coast in the vicinity of Christchurch, New Zealand. *J. Climatol.,* **1,** 203–219.

———, A. Trewinnard, and P. Gorman, 1984: A study of the atmospheric circulation over the South Island of New Zealand. *Wea. Climate,* **4,** 53–62.

———, S. J. Fitzsimons, and L. M. Holland, 1985: Local winds in the Southern Alps, New Zealand. *J. Climatology,* **5,** 145–160.

———, R. K. Smith, M. A. Page, R. N. Ridley, and J. T. Steiner, 1990: Mesoscale surface wind changes associated with the passage of cold fronts along the eastern side of the Southern Alps, New Zealand. *Meteor. Atmos. Phys.,* **42,** 133–143.

Sun, W.-Y., and I. Orlanski, 1981a: Large mesoscale convection and sea-breeze circulation. Part I: Linear stability analysis. *J. Atmos. Sci.,* **38,** 1675–1693.

———, and ———, 1981b: Large mesoscale convection and sea-breeze circulation. Part II: Nonlinear numerical model. *J. Atmos. Sci.,* **38,** 1694–1706.

Taljaard, J. J., 1967: Development, distribution, and movement of cyclones and anticyclones in the Southern Hemisphere during the IGY. *J. Appl. Meteor.,* **6,** 973–987.

———, 1972: Synoptic meteorology of the Southern Hemisphere. *Meteorology of the Southern Hemisphere, Meteor. Monogr.,* No. 35, Amer. Meteor. Soc., 139–213.

Tapper, N. J., 1988a: Some evidence for a mesoscale thermal circulation at The Salt Lake, New South Wales. *Aust. Meteor. Mag.,* **36,** 101–102.

———, 1988b: Surface energy balance studies in Australia's seasonally wet tropics: Results from AMEX Phase I and II. *Aust. Meteor. Mag.,* **36,** 61–68.

———, 1991: Evidence for a mesoscale thermal circulation over dry salt lakes. *Palaeogeog., Paleoclimatol., Paleaoceol.,* **84,** 259–269.

Thompson, G. H., and A. A. Neale, 1981: The Nelson sea breeze in south-westerly airstreams. *Wea. Climate,* **1,** 61–68.

Torrence, J. D., 1995: Some aspects of the South African coastal low and its rouge waves. *Weather,* **50,** 163–170.

Trenberth, K. E., 1991: Storm tracks in the Southern Hemisphere. *J. Atmos. Sci.,* **48,** 2159–2178.

Troup, A. J., and N. A. Streten, 1972: Satellite observed Southern Hemisphere cloud vortices in relation to conventional observations. *J. Appl. Meteor.,* **11,** 909–917.

Turner, J., and M. Row, 1989: Mesoscale vortices in the British Antarctic territory. *Polar and Arctic Lows,* P. F. Twitchell, E. A. Rasmussen, and K. L. Davidson, Eds., A. Deepak Publishing, 347–356.

———, T. A. Lachlan-Cope, D. E. Warren, and C. N. Duncan, 1993a: A mesoscale vortex over Halley station, Antarctica. *Mon. Wea. Rev.,* **121,** 1317–1336.

———, ———, and J. P. Thomas, 1993b: A comparison of arctic and antarctic mesoscale vortices. *J. Geophys. Res.,* **98,** 13 019–13 034.

Uccellini, L. W., 1990: Processes contributing to the rapid development of extratropical cyclones. *Extratropical Cyclones: The Erik Palmén Memorial Volume,* C. W. Newton and E. O. Holopainen, Eds., Amer. Meteor. Soc., 27–45.

van Loon, H., 1965: A climatological study of the atmospheric circulation in the Southern Hemisphere during the IGY. Part I: 1 July 1957–31 March 1958. *J. Appl. Meteor.*, **4,** 479–491.

Velasco, I., and J. M. Fritsch, 1987: Mesoscale convective complexes in the Americas. *J. Geophys. Res.*, **92,** 9591–9613.

Veldon, C. S., and G. A. Mills, 1990: Diagnostics of upper-level processes influencing an unusually intense extratropical cyclone over southeast Australia. *Wea. Forecasting,* **5,** 449–482.

Vincent, D. G., 1985: Cyclone development in the South Pacific convergence zone during FGGE, 10–17 January 1979. *Quart. J. Roy. Meteor. Soc.*, **111,** 155–172.

Virji, H., 1981: A preliminary study of summertime tropospheric circulation patterns over South America estimated from cloud winds. *Mon. Wea. Rev.*, **109,** 599–610.

Wallace, J. M., 1987: Comments on "Interactions between convective and large-scale motions over Amazonia." *The Geophysiology of Amazonia: Vegetation and Climate Interactions,* R. Dickerson, Ed., Wiley-Intersciences, 387–390.

Watson, I. D., 1984: Density change in an Australian west coast trough. *Aust. Meteor. Mag.*, **32,** 123–129.

Webster, P. J., and R. A. Houze Jr., 1991: The Equatorial Mesoscale Experiment (EMEX): An overview. *Bull. Amer. Meteor. Soc.*, **72,** 1481–1505.

Wilson, K. J., and H. Stern, 1985: The Australian summertime cool change. Part I: Synoptic and subsynoptic scale aspects. *Mon. Wea. Rev.*, **113,** 177–201.

Wood, I. R., and J. E. Simpson, 1984: Jumps in miscible fluids. *J. Fluid Mech.*, **140,** 329–342.

Zhao, S., and G. A. Mills, 1991: A study of a monsoon depression bringing record rainfall over Australia. Part II: Synoptic–diagnostic description. *Mon. Wea. Rev.*, **119,** 2047–2094.

Zillman, J. W., and P. G. Price, 1972: On the thermal structure of mature Southern Ocean cyclones. *Aust. Meteor. Mag.*, **20,** 34–48.

Chapter 6

The Stratosphere in the Southern Hemisphere

WILLIAM J. RANDEL

National Center for Atmospheric Research, Boulder, Colorado

PAUL A. NEWMAN

NASA/Goddard Space Flight Center, Greenbelt, Maryland

6.1. Introduction

The stratosphere over Antarctica is one of the most inaccessible places on the planet. During the antarctic winter, it extends from about 8 to 55 km above the surface, has temperatures colder than −90°C, and winds that are greater than 100 m s^{-1}. Yet even this terribly remote and hostile region has felt man's impact. The antarctic ozone hole is a clear example of how our industrial society can affect the atmosphere even in this remote corner of the earth. The tremendous ozone losses over Antarctica observed each spring have ultimately resulted from man-made chlorine compounds, and these ozone losses have led to increased levels of biologically harmful ultraviolet radiation at the earth's surface. Understanding the meteorology of the southern stratosphere is the key to our understanding of the antarctic ozone hole.

Knowledge and understanding of the stratosphere in the Southern Hemisphere has increased considerably since appearance of the first monograph on meteorology of the Southern Hemisphere in 1972. The analyses of Labitzke and van Loon (1972) in that publication were original and insightful in many aspects but necessarily hampered by lack of observational data (being based on a single year of monthly mean rawinsonde observations in the lower stratosphere, together with one year of satellite radiance measurements). Over the last 25 years, substantial improvements in the observational database in the SH stratosphere have occurred, due to several factors.

1) The refinement and continued availability of global satellite measurements, leading to daily meteorological analyses of stratospheric circulation.
2) The development of stratospheric data-assimilation systems, based on a blending of observations with global models.
3) Appearance of the ozone hole over Antarctica in the middle 1980s, which prompted observational aircraft campaigns and detailed studies of SH stratospheric dynamics and chemistry.
4) Analyses of stratospheric chemical constituent data from the limb infrared monitor of the stratosphere (LIMS), the total ozone mapping spectrometer (TOMS), the Stratospheric Aerosol and Gas Experiment (SAGE), and the Upper Atmosphere Research Satellite (UARS). In particular, UARS provided a wide range of simultaneous measurements of key constituents tied to stratospheric transport and ozone photochemistry.

Observations of dynamical and constituent behavior from this rich database have fueled parallel leaps in theoretical understanding of stratospheric dynamics, radiation, chemistry, and constituent transport (cf. McIntyre 1992; Holton et al. 1995).

The objective of this chapter is to give an overview of several aspects of the SH stratosphere, based primarily on the long record of observational data now available. Part of our focus is on contrasting behavior in the SH versus that in the (somewhat better-known) NH stratosphere; such contrasts often lead to better conceptual understanding of the SH itself. The general circulation of the SH stratosphere is considered in section 6.3, with focus on the seasonality of the zonal-mean fields and relationships between the mean flow, eddies, and radiative forcing. Daily variability of the stratosphere is discussed in section 6.4, including stratospheric warmings, transient and traveling planetary waves, and inferred smaller-scale features. Interannual variability of the SH stratosphere is analyzed in section 6.5, while section 6.6 focuses on global ozone characteristics and the antarctic ozone hole.

6.2. Data

A majority of the meteorological data presented here are derived from daily stratospheric analyses from NCEP (formerly called the National Meteorological Center). Daily global analyses of stratospheric geopotential height and temperature have been made since October 1978 for pressure levels covering 1000–0.4 mb (approximately 0–55 km). Lower-level data (over 1000–100 mb) are output of the Global Data Assimilation System used at NCEP. Stratospheric level data (at 70, 50, 30, 10, 5, 2, 1, and 0.4 mb) are daily operational analyses produced at the Climate Prediction Center of NCEP. The stratospheric analyses are produced using available radiosonde observations (only in the NH) and satellite-derived temperature retrievals in a modified successive correction (Cressman) analyses (Gelman and Nagatani 1977). Details of these data may be found in Gelman et al. (1986), Randel (1992), and Finger et al. (1993). Zonal-mean and eddy horizontal winds are derived from the geopotential data using balance assumptions (details discussed in Randel 1992).

We also use stratospheric meteorological analyses from the United Kingdom Meteorological Office (UKMO) stratospheric data-assimilation system (Swinbank and O'Neill 1994). This system uses a global numerical model of the atmosphere, with fields continuously adjusted toward available wind and temperature observations as the model is integrated forward in time. Output of these analyses includes temperatures and three-dimensional winds, spanning the altitude range 1000–0.3 mb (0–57 km). Data are available from November 1991 to present. Comparisons between circulation statistics from NCEP-derived products and the UKMO assimilation show reasonable agreement (e.g., Manney et al. 1996); notable differences include an underestimate of the tropical quasi-biennial oscillation (QBO) in the NCEP analyses (compared to rawinsonde observations) and a somewhat colder SH polar vortex in the UKMO data.

6.3. Observed general circulation

A simplified perspective of the stratospheric general circulation is provided by analysis of the structure and seasonal variation of the zonally averaged flow. It is straightforward to separate the governing equations (for momentum, thermodynamic energy, and mass conservation) into zonal-mean and eddy components, yielding the so-called Eulerian-mean equations (see Andrews et al. 1987, section 3.3). Interpretation of the resulting equations in terms of physically separating mean and eddy effects is often confusing, however, because the Eulerian-mean meridional circulation is not independent from the Eulerian eddy fluxes (for example, there is strong cancellation between the eddy heat-flux convergence and adiabatic cooling associated with $\bar{w}$). In fact, in the idealized case where wave transience and dissipation is neglected, the eddy and mean circulation terms exactly cancel, producing zero net forcing of the zonally averaged flow (Andrews and McIntyre 1978; Boyd 1976). An alternative and conceptually more physical separation of zonal-mean and eddy quantities is provided by the so-called transformed Eulerian-mean (TEM) budget equations [written here based on quasigeostrophic scaling assumptions, following Andrews et al. (1987, section 7.2)].

$$\frac{\partial \bar{u}}{\partial t} - 2\Omega \sin\phi \overline{v^*} = \bar{G} \text{ (zonal momentum balance)} \tag{6.1}$$

$$\frac{\partial \bar{T}}{\partial t} + \overline{w^*}\left(\frac{N^2 H}{R}\right) = \bar{Q} \text{ (thermodynamic balance)} \tag{6.2}$$

$$\frac{1}{a\cos\phi}\frac{\partial}{\partial\phi}(\overline{v^*}\cos\phi) + \frac{1}{\rho_0}\frac{\partial}{\partial z}(\rho_0\overline{w^*}) = 0 \text{ (mass continuity)} \tag{6.3}$$

$$2\Omega\sin\phi\frac{\partial \bar{u}}{\partial z} + \frac{R}{H}\frac{1}{a}\frac{\partial \bar{T}}{\partial\phi} = 0 \text{ (geostrophic thermal wind)} \tag{6.4}$$

Notation is standard and defined in the Appendix. This set of TEM equations results from defining transformed mean meridional circulation components ($\overline{v^*}$, $\overline{w^*}$) as

$$\overline{v^*} = \bar{v} - \rho_0^{-1}\frac{R}{H}\frac{\partial}{\partial z}\left(\rho_0\frac{\overline{v'T'}}{N^2}\right)$$

$$\overline{w^*} = \bar{w} + \frac{R}{H}\frac{1}{a}\frac{\partial}{\partial\phi}\left(\frac{\overline{v'T'}}{N^2}\right).$$

The terms on the right-hand side describe the mean circulation components associated with eddy heat-flux divergence. The components ($\overline{v^*}$, $\overline{w^*}$) then represent the difference between the Eulerian means ($\bar{v}$, $\bar{w}$) and these eddy-associated terms; this ($\overline{v^*}$, $\overline{w^*}$) circulation is hence termed the residual mean circulation. This residual circulation is more directly linked with diabatic processes that determine the mean mass flow in the stratosphere and provides a concise view of transport in the meridional plane.

From this TEM perspective, changes in the zonal-mean flow are driven by radiative forcing ($\bar{Q}$) of the thermodynamic field, together with wave driving of the zonal (or angular) momentum ($\bar{G}$). The thermodynamic and zonal momentum fields are in turn coupled via thermal wind balance, and the residual circulation ($\overline{v^*}$, $\overline{w^*}$) acts to redistribute the radiative and wave forcing effects nonlocally, in such a manner so as to maintain thermal wind balance. The wave driving term ($\bar{G}$) is often partitioned into components associated with large-scale planetary waves (the so-called Eliassen–Palm (or EP) flux divergence, $\rho_0^{-1}\nabla\cdot\mathbf{F}$) and those attributable to smaller or unresolved scales ($\bar{X}$) (e.g., gravity waves):

$$\bar{G} = \rho_0^{-1}\nabla \cdot \mathbf{F} + \bar{X}.$$

Components of the quasigeostrophic EP flux are given by

$$\mathbf{F} = (F_\phi, F_z)$$

$$= \rho_0 a \cos\phi\left(-\overline{u'v'},\ 2\Omega \sin\phi \frac{R}{H}\frac{\overline{v'T'}}{N^2}\right), \quad (6.5)$$

and estimates of the planetary-wave EP flux and its divergence may be made using stratospheric meteorological analyses. The structure of the stratospheric mean flow, its seasonal evolution, and the fundamental asymmetries between the NH and SH can be understood to first order by analyses of this set of equations [Eqs. (6.1)–(6.4)], using observations and/or models to derive the radiative heating and wave-driving distributions.

As a note, the full primitive equation form of the TEM equations (Andrews et al. 1987, section 3.5) includes an additional component in F_z proportional to the vertical eddy momentum flux $\overline{w'u'}$. Although it is negligible for large-scale motions, it is the dominant component for gravity-wave forcing of the mean flow. Gravity-wave forcing (which is of considerable importance for the SH stratosphere, as discussed later) may be calculated directly via this covariance term, or included (using a parameterized expression) in the $\bar{X}$ component of $\bar{G}$.

a. Zonal-mean temperature

The observed zonal-mean temperature over 0–50 km during January, April, July, and October is shown in Fig. 6.1, illustrating the overall temperature structure of the troposphere and stratosphere, together with its seasonal cycle. These and the following climatological means are based on NCEP data averaged over 1990–95. The heavy dashed line in each panel in Fig. 6.1 denotes the approximate position of the tropopause, as defined by a strong vertical gradient of the static stability parameter N^2. The stratosphere is characterized by increasing temperature with altitude above the tropopause, maximizing at the stratopause near 50 km (the upper limit of the data shown in Fig. 6.1). The temperature increase with altitude (i.e., the existence of the stratosphere) is due to the absorption of solar ultraviolet radiation by ozone, the structure of which is strongly latitudinally and seasonally dependent (following the solar declination). Radiative cooling in the stratosphere occurs mainly via infrared emission by carbon dioxide, which is principally a function of stratospheric temperature and less strongly dependent on latitude. The net (total) radiative heating is thus latitudinally and seasonally dependent, resulting in a polar stratosphere that is warm relative to low midlatitudes in SH summer (January) and relatively cold throughout the rest of the year. The cold polar stratosphere is particularly evident in midwinter (July), due to the complete absence of high-latitude heating during polar night.

b. Zonal-mean winds

Figure 6.2 shows the zonal-mean zonal wind for January, April, July, and October. There is a strong easterly jet in the SH summer (January) stratosphere over low latitudes, with winds near -50 m s^{-1}, which is somewhat stronger than the NH summer easterly jet in July (maximum near -40 m s^{-1}). Throughout the rest of the year, strong westerly winds cover middle and high latitudes of the SH stratosphere, associated (via thermal wind balance) with the cold polar stratosphere seen in Fig. 6.1. This strong wind maximum is termed the polar night jet, and the associated circulation system is termed the stratospheric polar vortex. Note that the SH winter polar night jet is much stronger than that observed in the NH winter, and the SH winter polar temperatures are much colder than those in the NH; the reason for this asymmetry is discussed below. The seasonal evolution of the SH polar night jet/polar vortex follows the pattern of appearing first in the upper stratosphere in early fall (April), strengthening and descending with time until middle winter, and then disappearing from the top down in spring. These radiative-driven changes occur first in the upper stratosphere because the radiative relaxation time there (of order 5 days) is much faster than that in the lower stratosphere (of order 50 days) (e.g., Gille and Lyjak 1987). This smooth seasonal evolution is interrupted by episodic large-scale disturbances (planetary waves) that originate in the troposphere, propagate vertically, and interact with the mean flow (these are associated with so-called stratospheric warming events). The transition from spring westerlies to summer easterlies in the stratosphere (termed breakdown of the vortex or the final warming) is often accompanied by one of these transient wave events (Newman 1986; Mechoso et al. 1988). The SH vortex breakdown is delayed until very late spring, significantly later than in the NH (as shown further below).

c. Wave driving and residual circulation

A concise diagnostic for the propagation characteristics and mean-flow forcing of large-scale planetary waves is provided by EP flux cross sections (Edmon et al. 1980; Andrews et al. 1987). Figure 6.3 shows climatological average EP flux diagrams for January, April, July, and October. In these diagrams, arrows show components of F_ϕ, F_z [Eq. (6.5)], denoting the direction and magnitude of planetary-wave propagation in the meridional plane, and contours show the EP

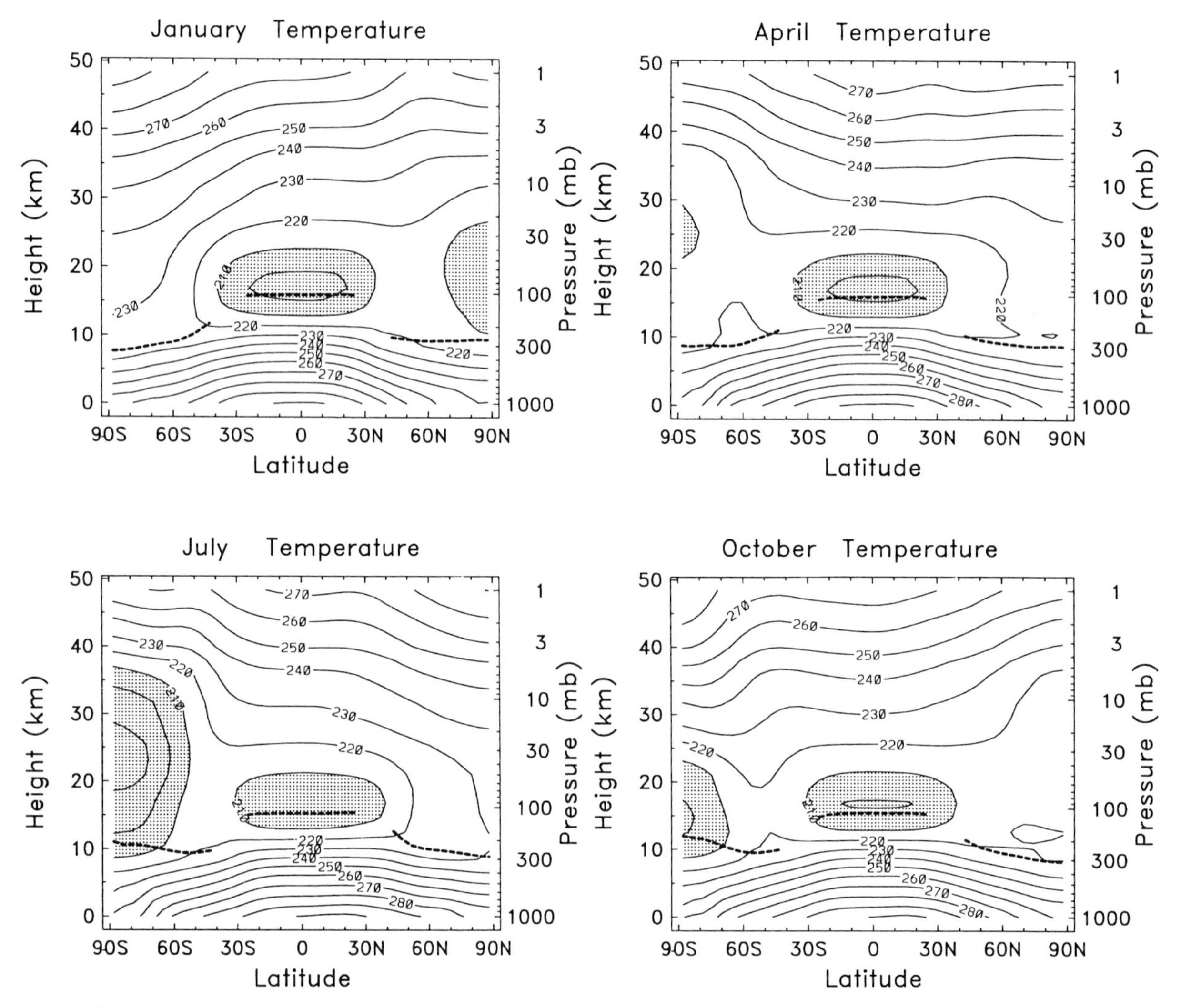

FIG. 6.1. Climatological zonal-mean temperature in January, April, July, and October. Values below 210 K are shaded. Heavy dashed lines denote approximate location of the tropopause.

flux divergence ($\rho_0^{-1}\nabla \cdot \mathbf{F}$), which is the planetary-wave component of the wave driving $\bar{G}$ in Eq. (6.1). The majority of tropospheric eddy activity (predominantly eastward-traveling, medium-scale zonal waves 4–7; Randel and Held 1991) does not propagate into the stratosphere but rather propagates equatorward or is absorbed in the middle upper troposphere (as evidenced by the maxima in wave driving in that region and strong decrease across the tropopause in Fig. 6.3). This lack of propagation of the tropospheric waves into the stratosphere is due to the background wind and thermal structures acting to restrict vertical propagation for all but the largest planetary scales (zonal waves 1–2) during winter (Charney and Drazin 1961; Dickinson 1969). Note that the tropospheric eddies do contribute a component of drag somewhat above the tropopause in Fig. 6.3, and that the residual circulation associated with tropospheric waves extends nonlocally, so that their influence is very important for behavior of the lower stratosphere (Dunkerton 1988).

The amount of wave activity in the stratosphere in Fig. 6.3 exhibits a strong seasonal cycle, with near-zero wave driving by planetary waves in the summer hemisphere (due to exclusion of vertical propagation in summer easterlies—see Fig. 6.2). The wave activity that is observed to propagate into the stratosphere is predominantly of planetary scale, although a rich spectrum of smaller scales are generated in the stratosphere by nonlinear and mixing processes (as described below). There is a strong hemispheric asymmetry in the observed planetary-scale wave driving, as EP fluxes during midwinter in the NH are much stronger than the corresponding time in the SH (compare the January and July patterns in Fig. 6.3). This is ultimately due to the stronger generation of planetary-scale waves in the NH troposphere due to orographic and

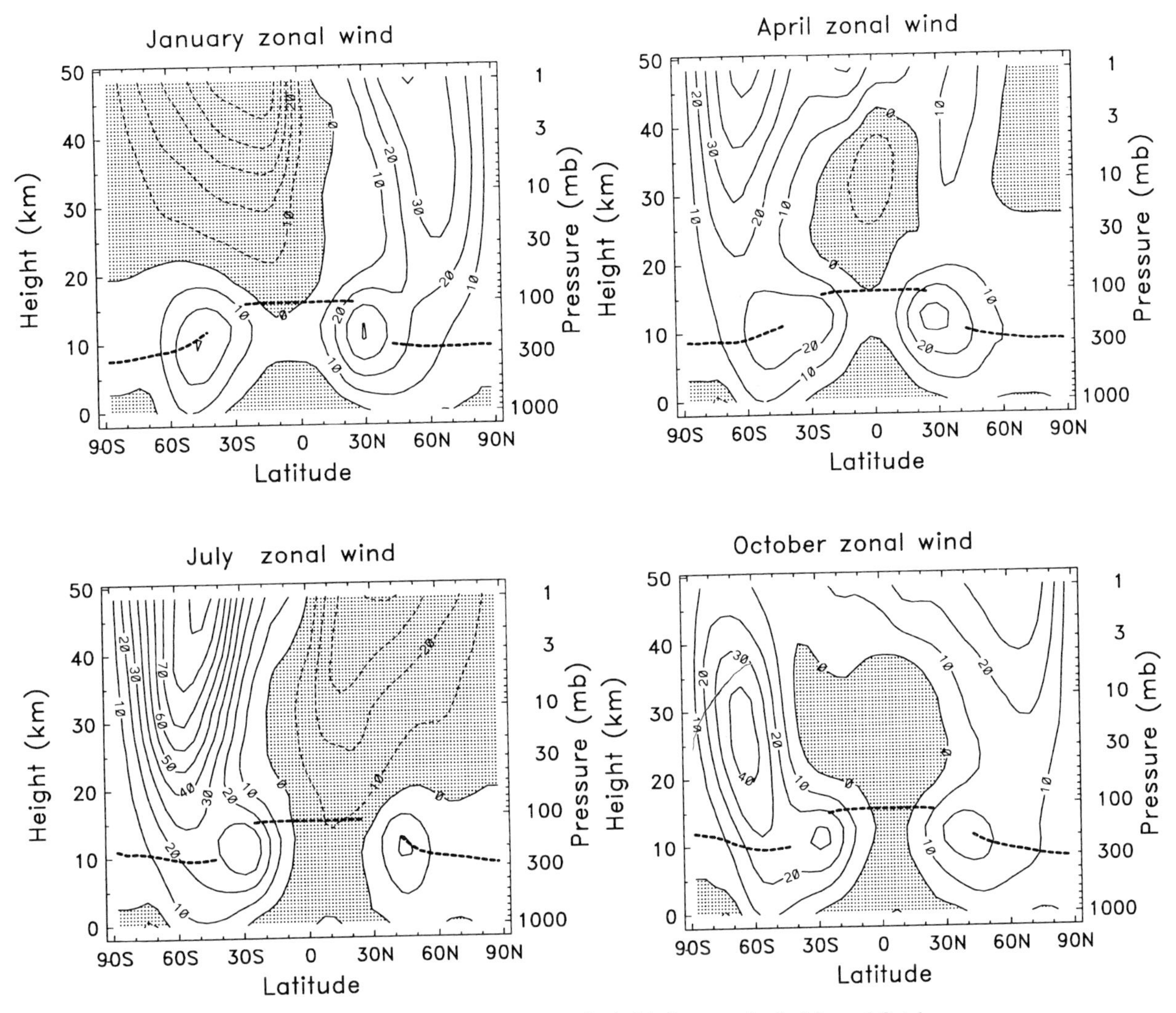

FIG. 6.2. Climatological zonal-mean zonal wind in January, April, July, and October. Contour interval is 10 m s^{-1} and negative values are shaded.

thermal forcing, which is absent in the SH. During spring, stratospheric EP fluxes are larger in the SH than in the NH, so that the seasonal cycle is markedly different between the hemispheres. This hemispheric difference in planetary-wave driving is the primary cause of the differences in wind, temperature, and polar vortex structures between the NH and SH.

The absorption of vertically propagating gravity waves [and the associated wave driving contribution to $\bar{G}$ in Eq. (6.1)] also exerts a strong influence on the circulation of the middle atmosphere. Gravity waves generated in the troposphere (by topographic forcing, convection, or frontogenesis, for example) propagate vertically and grow exponentially with height, until they reach an altitude where their horizontal phase speed equals the background winds (a critical level) or their amplitudes become so large that they become convectively unstable and "break" (leading to turbulence, small-scale mixing, and dissipation; see Prusa et al. 1996). This breaking typically occurs at mesospheric altitudes and is associated with a convergence of the waves' vertical momentum flux and acceleration of the background mean flow (tending to drag the mean flow toward the phase speed of the waves). The presence of critical lines in the troposphere and stratosphere sharply limits the phase speeds of waves that reach the mesosphere, however; such filtering effects lead to a strong gravity-wave drag in the mesosphere (of order 50–100 m s^{-1} per day) in both summer and winter (Lindzen 1981; Holton 1983). This momentum deposition is balanced by an intensified mean meridional circulation in the mesosphere (toward the winter pole), and the associated vertical circulations explain the remarkably cold summer and warm winter polar mesopause (~90 km) temperatures, which are both far from radiative equilibrium (see Andrews et al. 1987, section 7.2).

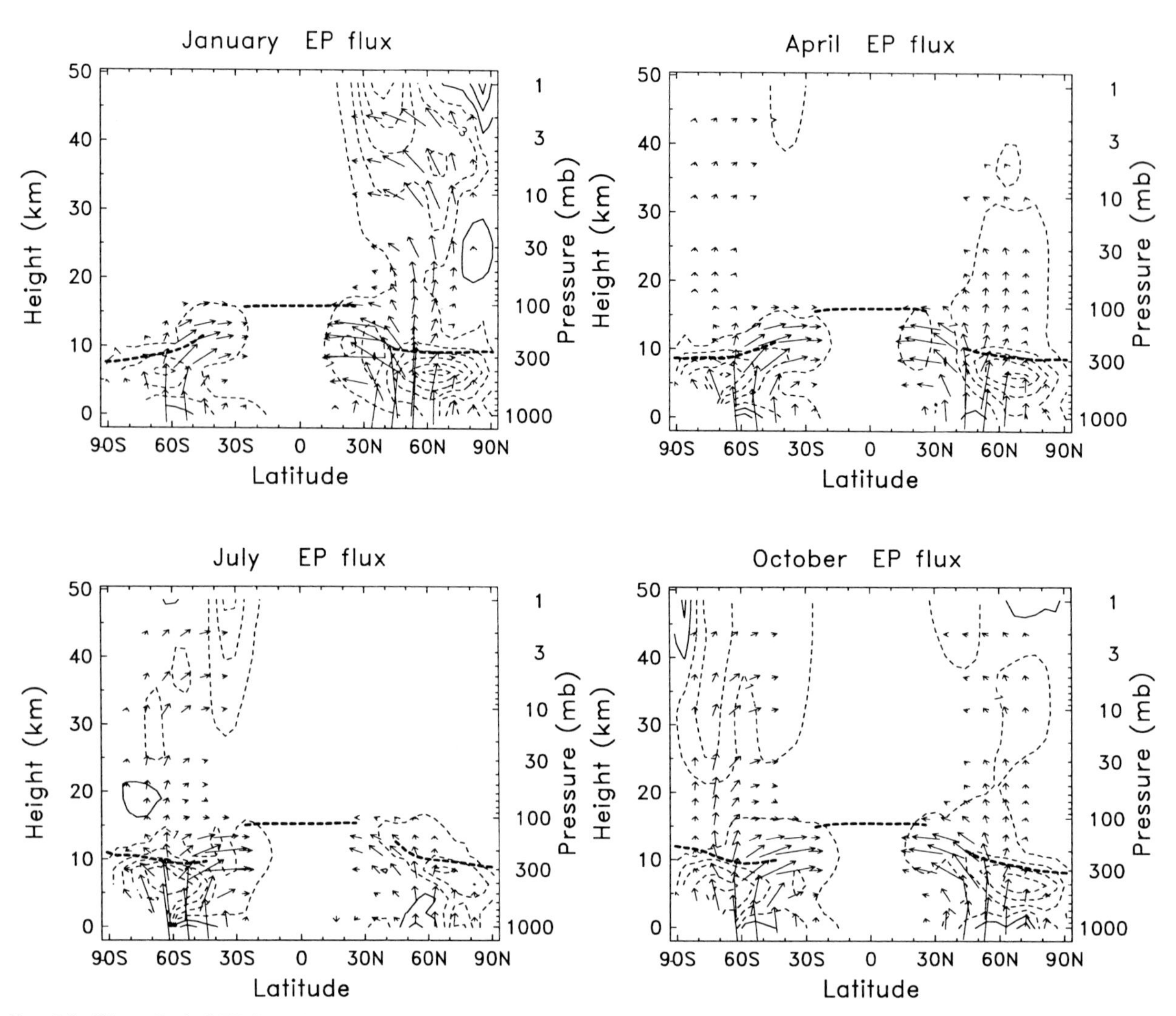

FIG. 6.3. Climatological EP-flux diagrams for January, April, July, and October. Arrows denote EP-flux vectors [components given by Eq. (5)], and contours show wave driving $\rho_o \nabla \cdot \mathbf{F}$ (contour intervals are $\pm 1, 3, 5, \ldots$ m s^{-1} per day).

Although gravity waves are primarily absorbed in the mesosphere, their influence extends well down into the stratosphere (via the nonlocal residual circulation, or "downward control"; see Haynes et al. 1991). Garcia and Boville (1994) have studied this in some detail, using a numerical model that includes parameterizations of both planetary- and gravity-wave breaking. Their results demonstrate that mesospheric gravity-wave drag can significantly affect the temperature in the polar winter stratosphere to altitudes as low as 20 km. The effect is more important when planetary-wave driving is relatively weak, as in the SH during midwinter. An important consequence of this result is that an inadequate amount of parameterized gravity-wave drag in numerical models results in a too-cold, intense polar vortex (the so-called "cold pole" problem—e.g., Mahlman and Umscheid 1987 and Boville 1995). Hitchman et al. (1989) discuss similar gravity-wave influences on variations of the stratopause in the SH.

An important component of the global stratospheric flow is the residual mean meridional circulation ($\overline{v^*}$, $\overline{w^*}$), whose structure and seasonal evolution is shown in Fig. 6.4. Components of this circulation are calculated from the coupled thermodynamic and continuity equations [Eq. (6.2)–(6.3)], using accurate radiative heating calculations to estimate net radiative forcing (Shine 1987; Rosenfield et al. 1987; Olaguer et al. 1992; Rosenlof 1995). The important effects of clouds and latent heat transfer preclude such straightforward calculations for the troposphere, and only the inferred stratospheric circulation is shown in Fig. 6.4. The overall stratospheric circulation is upward in the Tropics and downward in winter high latitudes, with strong flow from the summer to winter hemisphere in the upper stratosphere (see Dunkerton 1978; Garcia 1987).

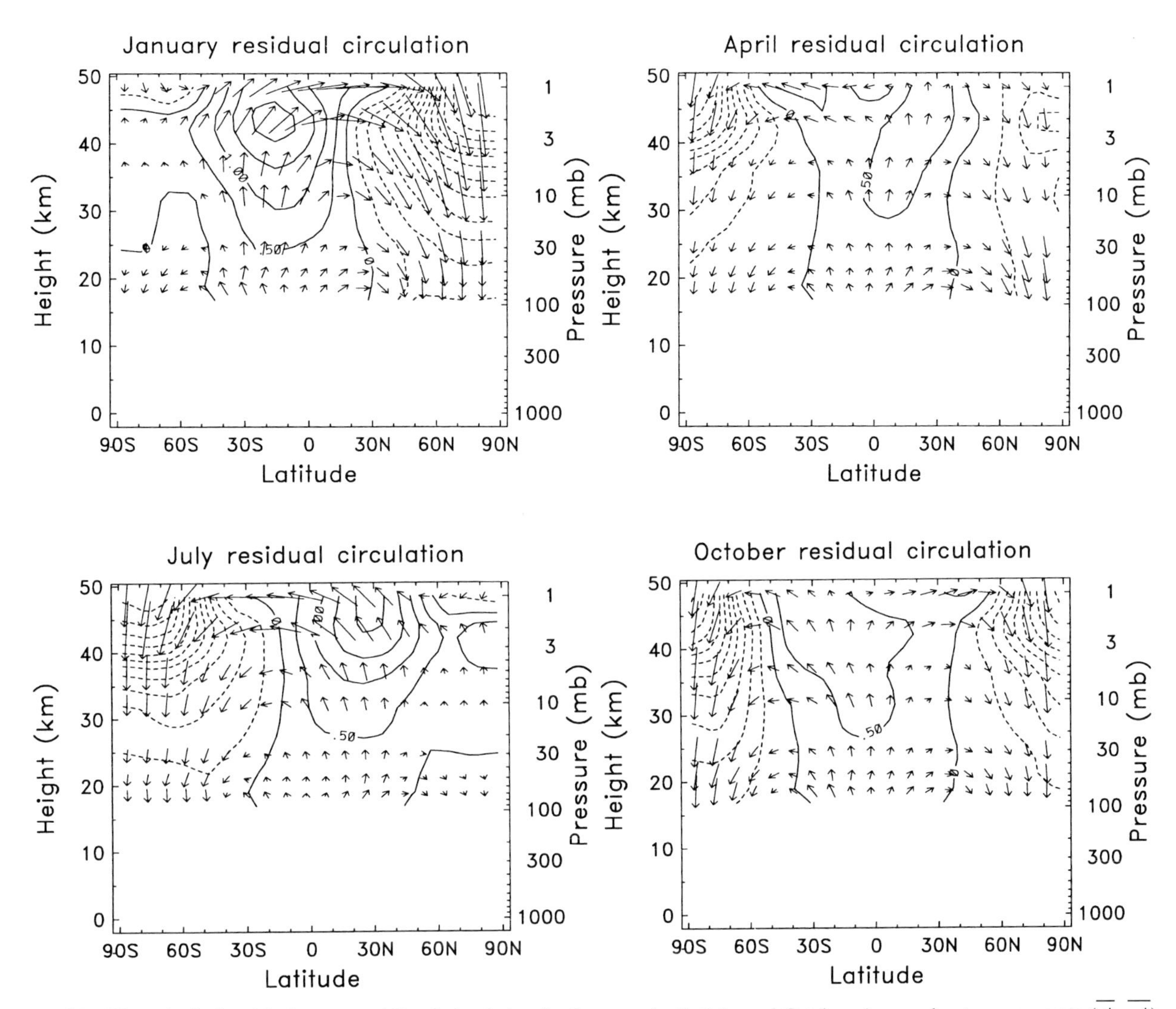

FIG. 6.4. Climatological residual mean meridional circulation for January, April, July, and October. Arrows denote components ($\overline{v^*}$, $\overline{w^*}$) with arbitary scaling, and contours show values of $\overline{w^*}$ (contour interval is 0.5 mm s^{-1}).

There is a strong seasonal cycle and hemispheric asymmetry observed in the residual circulation, with more intense poleward and downward motion observed during midwinter in the NH than in the SH winter. The upward branch of this circulation in the Tropics also varies seasonally, with upward motion centered in the respective summer subtropics (stronger during NH winter); this circulation is associated with the observed annual cycle in tropical stratospheric temperature and trace constituents (Yulaeva et al. 1994; Rosenlof 1995). Seasonality in the residual circulation is forced by seasonal variations in both radiative forcing ($\bar{Q}$) and wave driving ($\bar{G}$), but the strong circulations during winter (and spring in the SH) are attributable primarily to wave forcing (Andrews et al. 1987, section 7.2; Haynes et al. 1991; Holton et al. 1995; see also Garcia and Boville 1994). During these times, the wave driving acts to push the stratosphere away from radiative equilibrium, and the radiative forcing is largely a response to this departure from equilibrium (since Q is primarily a function of temperature). During equinox seasons, the radiative forcing is comparably more important; radiative cooling over the polar cap in autumn (when wave driving is small) leads to a buildup of the meridional pressure gradient and development of the polar jet through the lowering of mass surfaces in polar latitudes relative to lower latitudes.

The large NH–SH differences in residual circulation seen in Fig. 6.4 are due principally to the observed NH–SH differences in wave driving $\bar{G}$ (see Fig. 6.3), as seen from the steady-state version of Eq. (1), $f\overline{v^*} = \bar{G}$. Thus, the weaker midwinter residual circulation in the SH is directly related to the smaller SH wave forcing. This weaker circulation is also consistent with the colder and more intense SH polar vortex (Figs. 6.1–6.2), in that the residual circulation acts to maintain departures from radiative equilibrium (the stron-

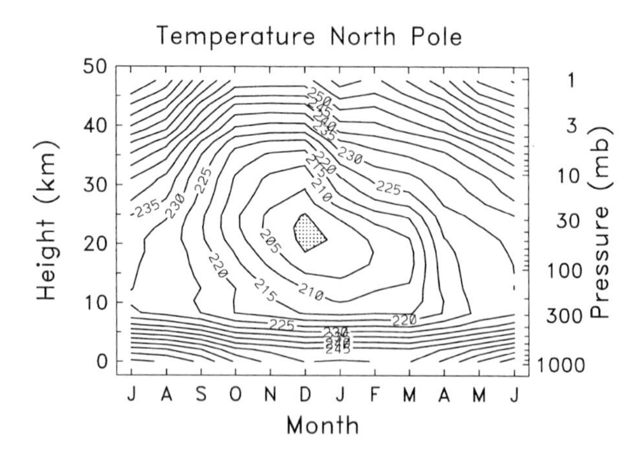

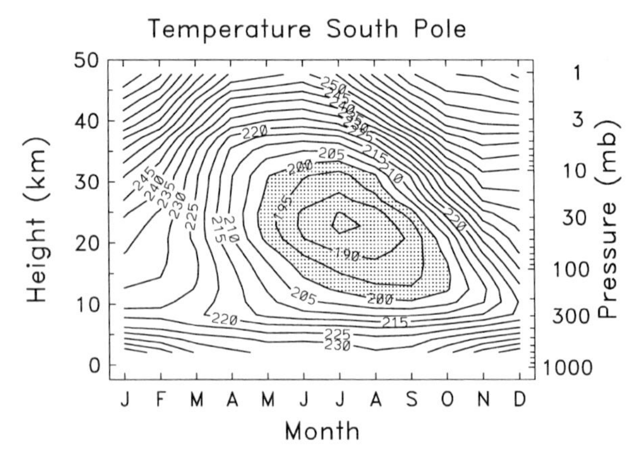

FIG. 6.5. Climatological height–time sections of zonal-mean temperature at the North Pole (top) and South Pole (bottom). Values below 200 K are shaded. Note that the time axes are shifted by six months between the NH and SH.

ger NH circulation is associated with warmer NH polar temperatures). This latter point is illustrated in the seasonal march of temperatures and residual circulation vertical velocity at the North and South Poles, respectively, shown in Figs. 6.5–6.6. Note that the time axis has been shifted by 6 months between the NH and SH in these plots, to facilitate direct comparisons. These figures clearly show that the much colder SH middle and lower polar stratosphere is associated with a weaker downward circulation in midwinter. A similar comparison of springtime conditions shows stronger downward motion in the SH, in balance with the larger SH wave driving.

The coupled seasonal variations of zonal wind in the middle stratosphere (10 mb), the large-scale wave forcing into the lower stratosphere, and residual mean vertical velocity at 50 mb is illustrated in Fig. 6.7. Here, the 50-mb heat flux is used as a measure of wave activity entering the stratosphere [this is proportional to the vertical component of the EP flux F_z—Eq. (6.5)]. [We note that heat flux is a derived quantity whose estimation is sensitive to details of the meteorological analyses used, particularly in the data-sparse SH. The 1990–95 NCEP heat fluxes shown here are approximately 20% lower in the SH lower stratosphere than corresponding UKMO estimates, but the space–time patterns are nearly identical (differences are less than 10% in the NH). NCEP-based estimates during the 1980s were significantly smaller then these more recent data, possibly due to data problems.] Figure 6.7 demonstrates that planetary waves propagate into the stratosphere only during winter–spring in the respective hemispheres, when the background zonal winds are westerly. In the SH, the seasonal maximum in wave driving is observed in spring (September–October), whereas strong values are observed throughout midwinter (November–March) in the NH. The strength of wave activity entering the stratosphere is approximately twice as big in the NH in midwinter, although SH spring values are similar to those in NH winter. The wave-driving seasonal cycles are echoed in the vertical circulation patterns, with strongest downward motion over high latitudes during NH midwinter and SH spring (these latter values are in good agreement

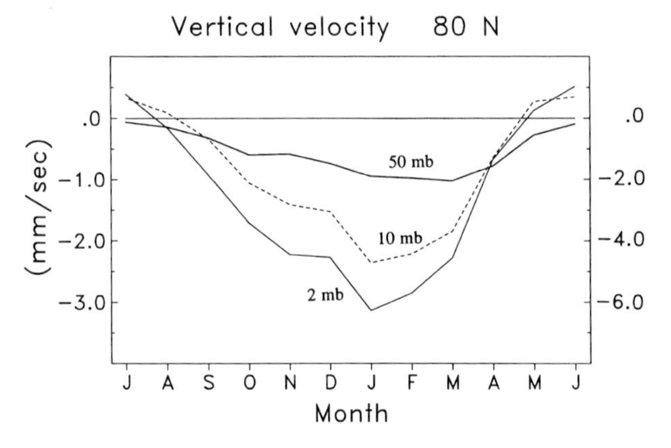

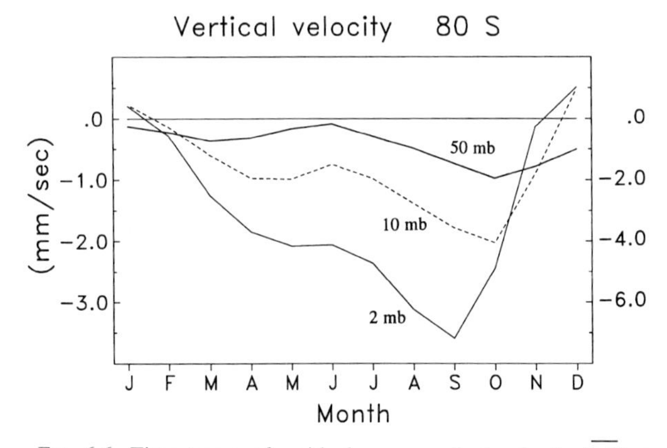

FIG. 6.6. Time traces of residual mean vertical velocity ($\overline{w^*}$) at 80°N (top) and 80°S (bottom), for the 50-, 10-, and 2-mb levels. The left axis refers to the 50- and 10-mb levels, and the right axis is for the 2-mb level. Note the time axes are shifted by six months between the NH and SH.

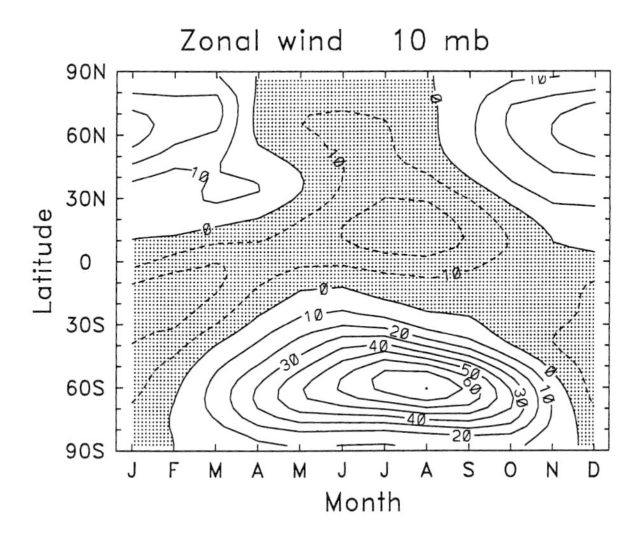

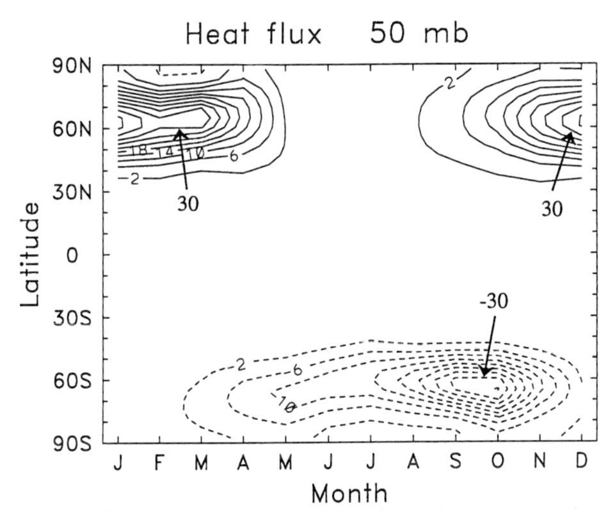

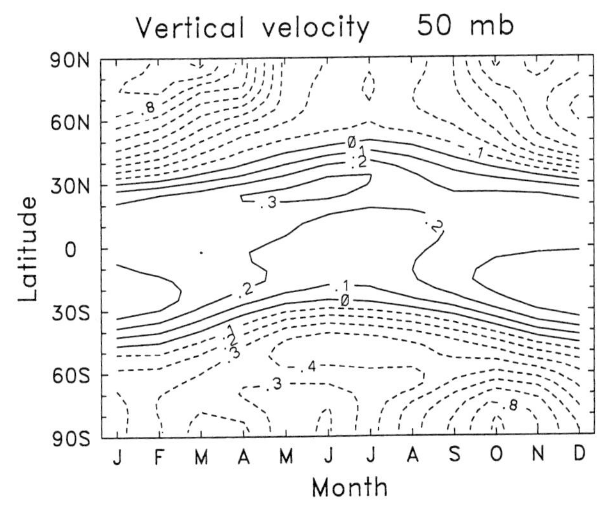

FIG. 6.7. Climatological latitude–time sections of (top) zonal-mean zonal wind at 10 mb (contour interval of 10 m s^{-1}); (middle) zonal-average eddy heat flux at 50 mb (contours of ±2, 6, 10, . . . K–m s^{-1}); and (bottom) residual mean vertical velocity at 50 mb (contour interval of 0.1 mm s^{-1}).

with downward velocities inside the SH polar vortex inferred from trace constituent observations by Schoeberl et al. 1995; see also Rosenfield et al. 1994). Note also the seasonality in the tropical upwelling in Fig. 6.7, as discussed above. These variations are consistent [in balance, via Eqs. (6.1)–(6.4)] with the colder, more intense SH midwinter polar vortex.

d. Trace constituent transport

Much has been learned about the overall mass flow in the stratosphere by analysis of quasi-conservative chemical tracers. The evolution of such tracers depends on the distribution of sources and sinks, and the relative timescales for transport versus chemical processes. The overall structure of species with long photochemical lifetimes (long compared to transport timescales of weeks to months, such as ozone in the lower stratosphere, nitrous oxide, or methane) is strongly influenced by and bears the signature of transport. Conversely, species with short lifetimes (such as ozone in the upper stratosphere) adjust quickly to the ambient background and are not sensitive to their transport history.

A useful species with a reasonably long lifetime to observe transport effects is methane (CH_4). Methane is generated in the troposphere by biotic activity, is transported into the stratosphere in the Tropics (by the general circulation), and oxidized in the middle and upper stratosphere (yielding a source of stratospheric water vapor). The photochemical lifetime of methane below 45 km is > 100 days, so that the stratospheric distribution is determined mainly by the circulation. Methane measurements from the stratospheric and mesospheric sounder have been used in a number of studies to deduce stratospheric circulation features (Jones and Pyle 1984; Solomon et al. 1986b; Holton and Choi 1988; Stanford et al. 1993). The overall characteristics of methane are similar to other constituents with tropospheric sources and stratospheric photochemical sinks, namely nitrous oxide (N_2O) and the chlorofluorocarbons.

Figure 6.8 shows meridional cross sections of stratospheric methane throughout the seasonal cycle derived from the Halogen Occulation Experiment (HALOE; Russell et al. 1993a) and cryogenic limb array etalon spectrometer (CLAES; Kumer et al. 1994) instruments on UARS. These data were organized according to the background potential vorticity fields (using so-called equivalent latitude mapping); results over most latitudes are derived from HALOE, and CLAES data are used to fill in polar regions (Randel et al. 1998). The methane contours in Fig. 6.8 bulge upward near the Equator, reflecting the mean upward circulation in the Tropics (see Fig. 6.4). The tropical peak moves latitudinally with season, due to seasonal changes in the transport circulation (compare the January and July panels in Figs. 6.4 and 6.8); there is also

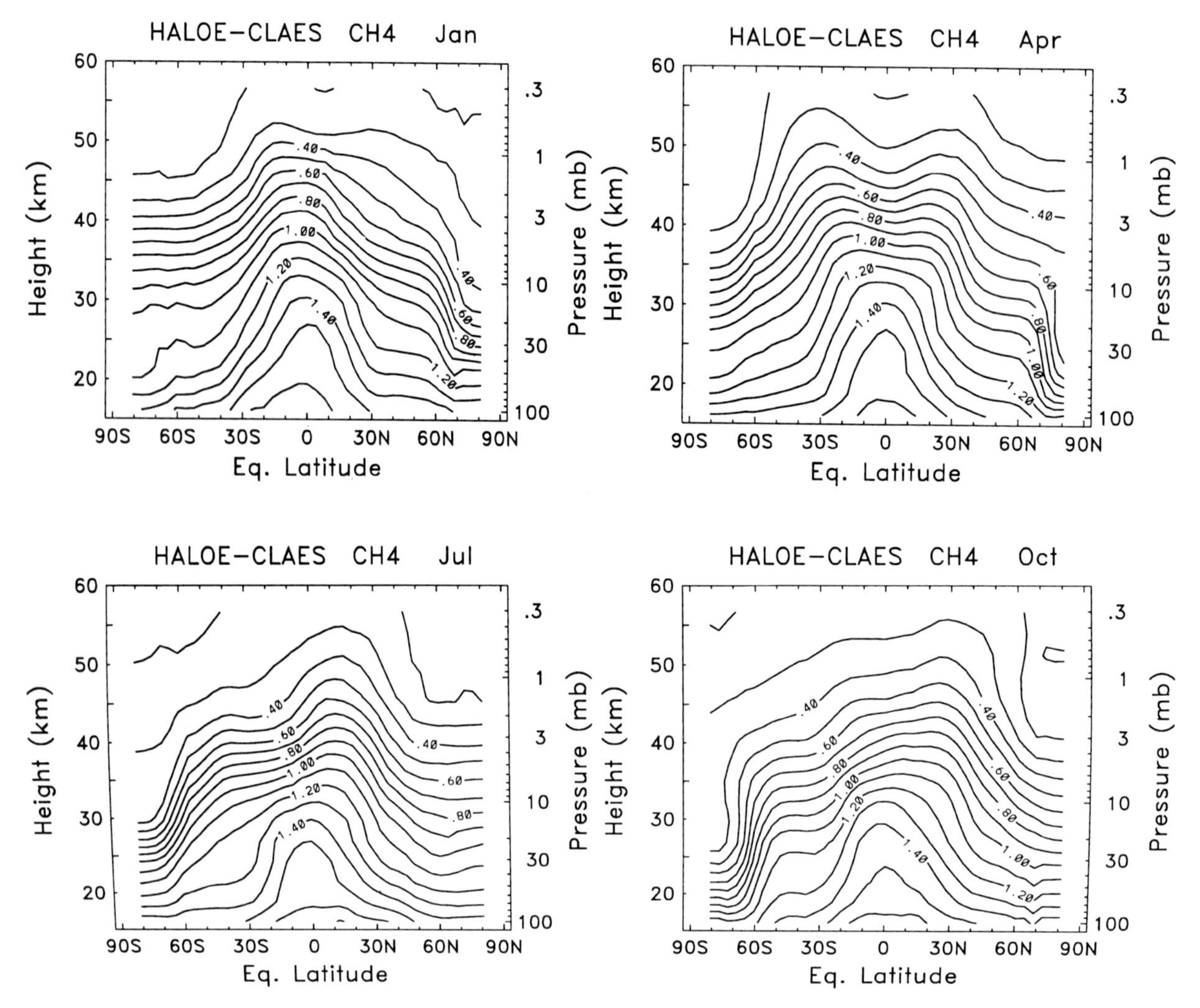

FIG. 6.8. Meridional cross sections of methane mixing ratio in January, April, July, and October, derived from HALOE and CLAES satellite data spanning 1991–96. Contour interval is 0.1 ppmv.

a double-peaked tropical pattern in April related to the tropical SAO. Enhanced meridional gradients in methane are observed in the subtropics and across the polar vortex regions; these stronger gradients denote regions of minimal horizontal mixing and are sometimes referred to as the subtropical and polar vortex "mixing barriers." The SH winter–spring polar vortex is particularly evident in these data, with remarkably strong methane gradients (near 60°S) resulting from strong downward flow inside the vortex and small horizontal transport across the vortex edge (Hartmann et al. 1989; Schoeberl et al. 1992; Fisher et al. 1993; Manney et al. 1994; Sutton 1994). In contrast, the methane isolines are relatively flat over winter midlatitudes, due to rapid horizontal mixing associated with the midlatitude "surf zone" (McIntyre and Palmer 1983, 1984; Leovy et al. 1985).

Recent simulations of the material flow and transport in the stratosphere have shown behavior very similar to these global observations. Figure 6.9 shows an example of such structure in the SH stratosphere, derived from a seasonal cycle simulation using a three-dimensional transport model (from Sutton 1994; similar results are shown in Fisher et al. 1993; Manney et al. 1994; and Eluszkiewicz et al. 1995). Figure 6.9 shows the locations of particles during the middle of SH winter (30 June), after horizontally stratified initialization three months earlier (1 April) at the locations denoted on the right in Fig. 6.9. Evolution of the particles shows upward motion in the Tropics and poleward–downward motion in the SH extratropics. Particles inside the vortex (poleward of 60°S) remain isolated, due to minimal transport across the vortex edge; conversely, strong mixing is observed in the midlatitude surf-zone region (~20°–50°S). The combination of strong descent and weak mixing in the vortex results in vortex interior air exhibiting characteristics of the mesosphere (as observed by Russell et

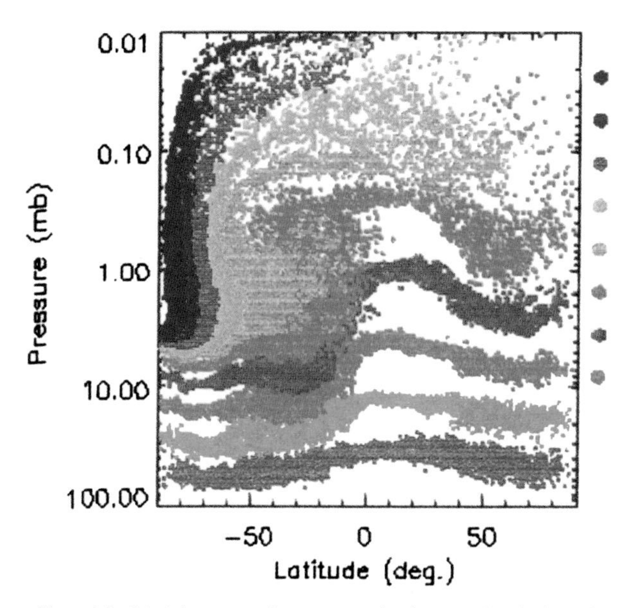

FIG. 6.9. Model-generated pattern of air parcels during the middle of SH winter (30 June), three months after horizontally stratified initialization (on 1 April) at the altitudes denoted on the right side (from Sutton 1994). Note the similarity to the SH winter methane patterns shown in Fig. 6.8.

al. 1993b). Overall, the modeled structure is remarkably similar in many aspects to that observed (i.e., the July methane patterns shown in Fig. 6.8), and such modeling allows detailed studies of the transport mechanisms that lead to the observed structure.

e. Potential vorticity perspective

In addition to chemical tracers of fluid motion, dynamical tracers are often used to study circulation and transport. Dynamical tracers consist of quasiconservative field variables derived from conventional meteorological analyses, and two important dynamical tracers are potential temperature and potential vorticity. Potential temperature θ is defined as the temperature a parcel of dry air at temperature T would acquire if expanded or compressed adiabatically to a reference pressure p_s = 1000 mb:

$$\theta = T\left(\frac{p_s}{p}\right)^{2/7}. \tag{6.6}$$

For atmospheric motions that are adiabatic, fluid parcels remain on constant θ (isentropic) surfaces. Because potential temperature is mainly a function of altitude, θ may be regarded as a vertical coordinate for fluid parcels; for adiabatic conditions, three-dimensional motion is reduced to two-dimensional flow on isentropic surfaces. In the lower and middle stratosphere, diabatic heating rates are typically less than 1 K day^{-1}, making an adiabatic approximation valid for timescales of 1–2 weeks. Thus, tracer distributions and evolution on such timescales are often analyzed on isentropic surfaces.

Potential vorticity is a derived dynamical quantity that is materially conserved for adiabatic frictionless flow. Mathematically, PV is defined as the dot product of the absolute vorticity vector and the potential temperature gradient. For the stratosphere, this can be approximated as

$$\mathrm{PV} = -g(\zeta + 2\Omega \sin \phi)\frac{\partial \theta}{\partial p}, \tag{6.7}$$

where ζ is the (local) vertical component of the relative vorticity derived from the horizontal wind field. The motivation behind PV conservation is discussed at length in McIntyre and Palmer (1983, 1984) and Hoskins et al. (1985) and references therein; PV conservation is the fluid-dynamical generalization of angular momentum conservation in Newtonian mechanics. For adiabatic flow (as noted above, a good approximation for timescales up to several weeks in the lower-middle stratosphere), contours of PV behave as material lines and allow tracing of fluid motions. These ideas are confirmed by the observed good agreement between derived PV and chemical tracer fields for transient events (e.g., Hartmann et al. 1989; Leovy et al. 1985; Randel et al. 1993; Manney et al. 1995).

A potential vorticity perspective of the time-mean SH stratospheric structure and seasonal evolution is shown in Fig. 6.10, for the 465 and 840 K isentropic levels. These isentropic levels are near 20 and 30 km, respectively, representing the lower and middle stratosphere. The winter polar vortex is associated with strong horizontal PV gradients; air inside the vortex (at polar latitudes) has PV values that are much higher than those outside, signifying distinctive air masses. The region of strong gradients is associated with the "edge" of the polar vortex and is near the locus of maximum zonal winds. The vortex edge may be objectively defined as the location of the strongest latitudinal gradient of PV (Nash et al. 1996), and the heavy contours in Fig. 6.10 denote the respective vortex edges for each month. Due to PV conservation discussed above, these strong PV gradients present a barrier to north–south mass flux across the vortex edge; the associated restoring mechanism is responsible for oscillatory behavior for large-scale vortex perturbations (planetary Rossby waves). Detailed transport calculations (Bowman 1993a,b; Chen et al. 1994; Chen 1994; Manney et al. 1994) confirm small net mass flux from inside to outside the vortex and vice-versa (discussed later in more detail). The midlatitude region outside the winter vortex exhibits relatively weaker PV gradients; this is the so-called "surf zone" (McIntyre and Palmer 1984), where north–south transport and mixing is less restricted (and readily observed; see section 6.4). Figure 6.10 also illustrates

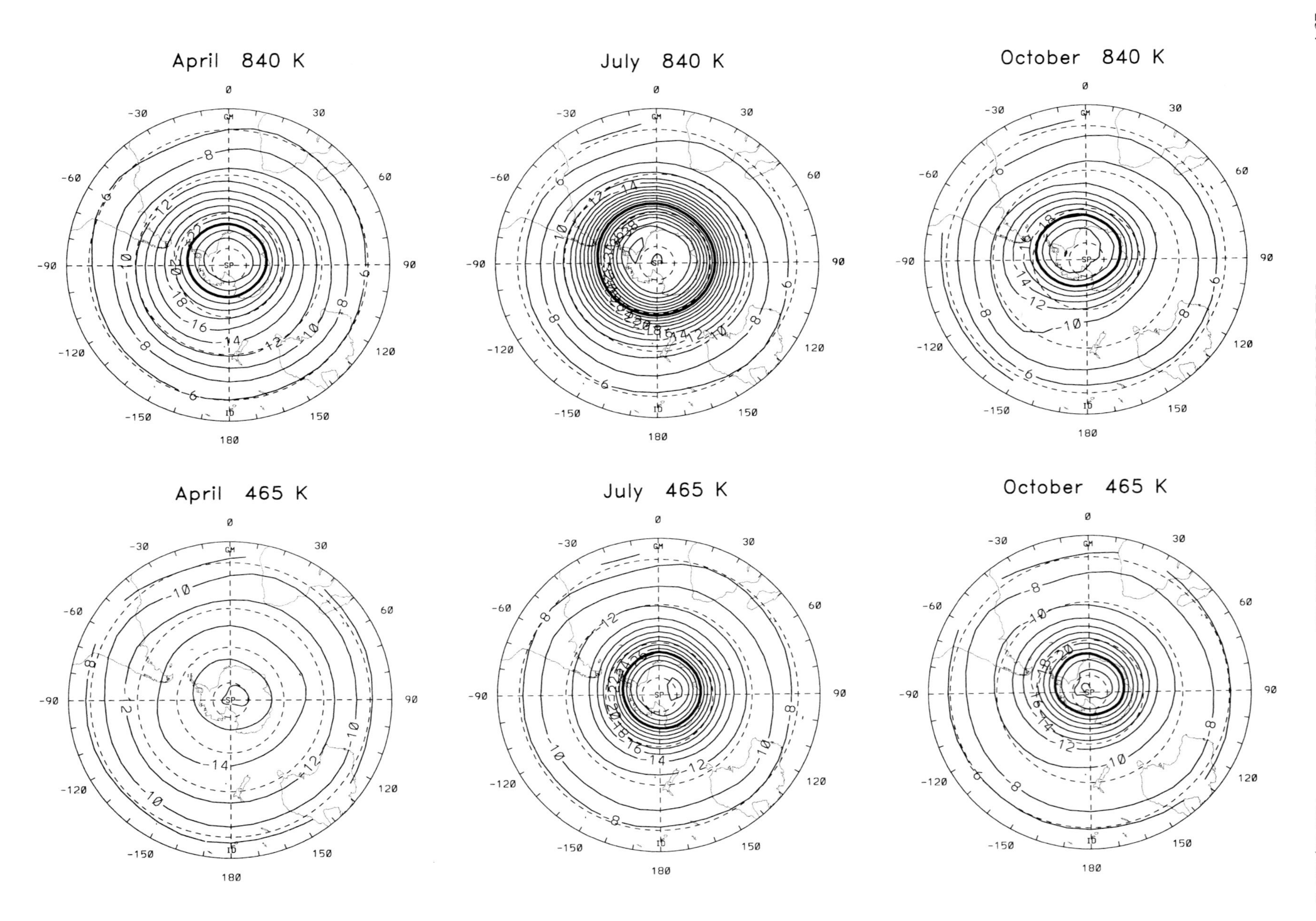
April 840 K
July 840 K
October 840 K
April 465 K
July 465 K
October 465 K

1982, or Randel and Held 1991). Statistics are calculated from data covering July–October from each year. Also included are numerical values summing the total transient and stationary components for comparison. There is a substantial amount of variability in details of the spectra from year to year, although the overall character is similar, showing eastward-traveling wave 2 and stationary wave 1 contributing the largest components. The overall heat flux is smaller in 1995, due mainly to relatively weak transient waves. The transient wave 2 spectra show the majority of covariance traveling eastward with periods of 8–25 days (corresponding to zonal phase speeds of 5–10 m s^{-1}); a relatively sharp spectral peak with period near 17 days is seen in both 1991 and 1993. This eastward-traveling wave 2 is a commonly observed feature in the SH stratosphere (Harwood 1975; Hartmann 1976; Leovy and Webster 1976; Mechoso and Hartmann 1982; Hartmann et al. 1984; Mechoso et al. 1985; Randel 1987; Shiotani and Hirota 1985; Shiotani et al. 1990; and Manney et al. 1991b). Its origin has been discussed in terms of vertical propagation from the troposphere (Randel 1987), in situ generation from instability mechanisms in the stratosphere (Mechoso and Hartmann 1982; Hartmann 1983; Manney et al. 1991b), and internal, nonlinear vortex interactions (Lahoz et al. 1996).

6.4. Synoptic variability

In this section we examine the synoptic (day-to-day) variability of the SH stratosphere during 1995, as an example year. We include comparisons with the NH winter 1994–95, to demonstrate the significant differences between hemispheres in day-to-day variability. We analyze time sections of the various quantities discussed in section 6.3, together with synoptic maps illustrating development of a planetary-wave event and the final warming. We also briefly discuss some traveling planetary waves observed in the SH stratosphere.

a. Temperatures, winds, and eddy heat fluxes

The SH stratosphere is characterized by significantly less day-to-day variability than the NH. Figure 6.14 displays the 50-hPa zonal-mean temperatures at 80°N (top) and 80°S (bottom) for the years 1994–95 and 1995, respectively (solid dark line, with the long-term mean shown as the thick white line and the envelope of 1979–95 data indicated by shading). This figure demonstrates minimal variance (on both daily and interannual timescales) in the SH in midwinter. Typically, the 80°S zonal-mean temperatures vary by only a few degrees on a day-to-day timescale during the austral summer, fall, and early winter (January through June). Only in the late winter–spring does

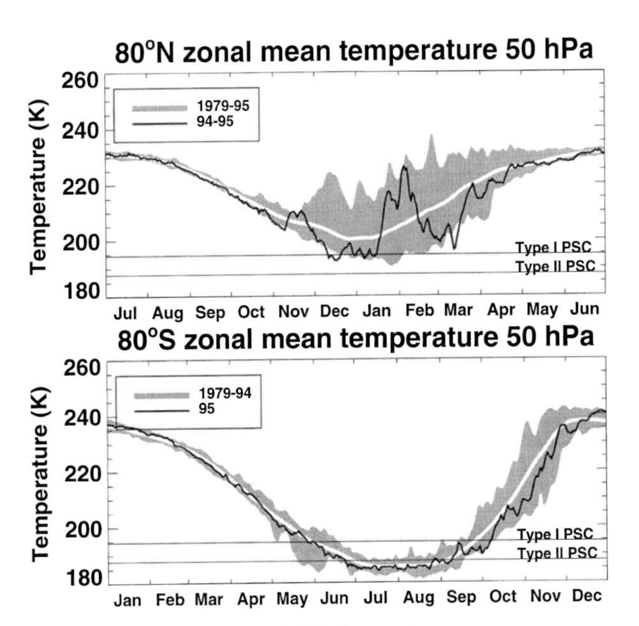

FIG. 6.14. Time series of 50-mb zonal-mean temperature at 80°N during 1994–95 (top) and 80°S during 1995 (bottom). White lines show 1979–95 means, and shaded regions denote extrema of daily values over those years.

this variability begin to increase, prior to polar vortex breakup.

In sharp contrast to the SH, the NH shows a high degree of day-to-day variability in the stratosphere during midwinter, with smaller variability in late spring (top of Fig. 6.14). One of the most striking aspects of these NH temperatures are the rapid warmings during midwinter. These rapid warmings are termed stratospheric sudden warmings and are associated with episodic planetary-wave driving ("bursts") from the troposphere (as shown below). The polar night jet is also decelerated strongly during those polar warming events. Midwinter stratospheric warmings vary in intensity, but are quite common in the NH. On the other hand, the SH winter period is almost completely devoid of significant sudden warmings.

As with the temperatures, the zonal winds in the SH exhibit much less daily variability than the NH. Figure 6.15 displays the 10-mb zonal-mean zonal wind at 60°N (top) and 60°S (bottom) for the years 1994–95 and 1995, respectively, together with long-term means and envelope of daily extremes. Rapid decelerations of the jet are observed during midwinter in the NH, associated with the warming events seen in Fig. 6.14. In contrast, the SH winter variations are weak. The fact that the episodic warming events are tied to bursts of wave activity from the troposphere is demonstrated in Fig. 6.16, which shows the respective eddy heat-flux time series (at 50 mb, 60°N–S) for these years. In the NH, there were five large bursts of wave activity that were associated with reductions of the northern jet

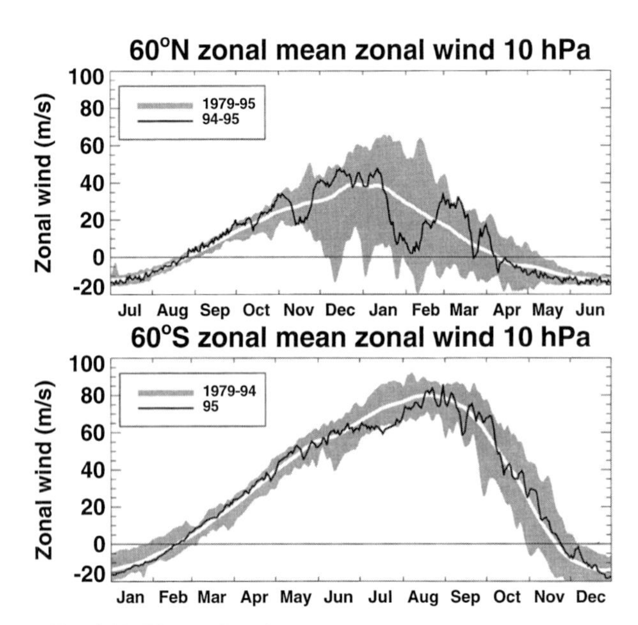

FIG. 6.15. Time series of 10-mb zonal-mean zonal wind at 60°N during 1994–95 and 60°S during 1995, along with the respective means and extrema over 1979–95.

(mid-November, late December, late January to early February, mid-March, and early April). As seen in Figs. 6.15–6.16, the first wave activity burst reduced the 10-mb zonal wind by about 15 m s^{-1}. The second burst stopped the strengthening of the jet, and the January burst (or combination of bursts) nearly reversed the zonal-mean flow. The fourth burst of wave activity in mid-March slowed the jet yet further, and

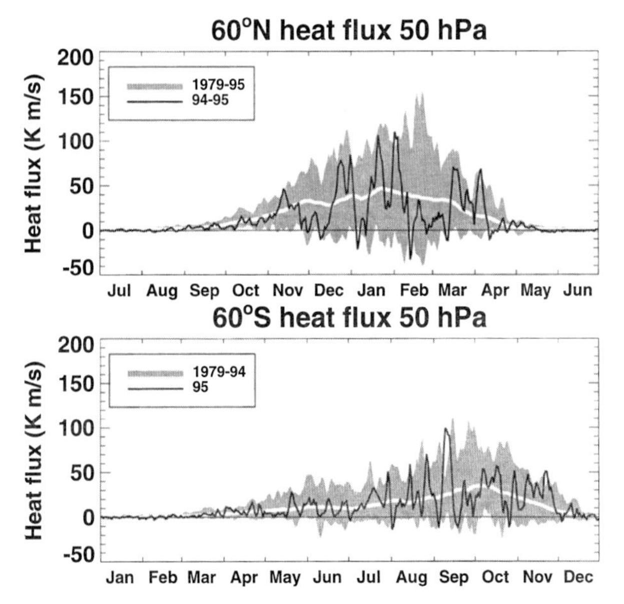

FIG. 6.16. Time series of 50-mb eddy heat flux at 60°N during 1994–95 and 60°S during 1995, along with the respective means and extrema over 1979–95.

the final burst in early April reversed the flow to the summer easterly regime. In addition to slowing the zonal wind, this wave activity warms the polar region by increasing the downward circulation in the polar region (see Fig. 6.14). These wave events are the source of the large day-to-day variability observed in the NH.

In contrast to these NH wave events, the SH shows a much weaker wave forcing in midwinter. In early September 1995, there is a burst of wave activity in the SH that decreases the zonal wind at 10 mb by over 15 m s^{-1}. In late September–October 1995, an extended burst of wave activity acts to decrease the vortex strength by a substantial amount. By late November, the 10-hPa zonal wind has reversed to easterlies. These data show that enhanced wave driving during SH spring acts to decelerate the jet, erode the vortex, and reverse the circulation to the austral summer easterly wind regime.

b. The day-to-day evolution of the polar vortex

The time series above quantify variations in zonal-mean quantities on a day-to-day basis. Evolution of the three-dimensional vortex may be viewed in terms of vortex area and by inspection of daily maps. Evolution of the area of the vortex during 1995 at 465 and 840 K is shown in Fig. 6.17, together with the 10-mb zonal wind for reference. These diagrams display the area within individual PV contours throughout the seasonal cycle, calculated in terms of "equivalent latitude" (i.e., the latitude of an equivalent PV distribution arranged symmetrically about the pole; see Buchart and Remsberg 1986). The polar vortex is identified by strong meridional PV gradients (see Fig. 6.10); note the earlier formation of the vortex at 840 K and longer persistence of the vortex at 465 K (well into spring). Figure 6.17 also shows a lack of significant high-frequency variability in vortex area throughout winter in the SH, consistent with the zonal-mean view discussed above.

O'Neill and Pope (1990) calculated PV as a function of equivalent latitude and time for the years 1979 to 1988 on the 850-K surface. In their paper, they observed that the surf zone of the Southern Hemisphere begins to develop as the eddy activity begins to increase during late winter and early spring. This is evident in Fig. 6.17 as the area of weak PV gradients at 840 K near 40°S, beginning in August and continuing through spring. A similar strong seasonality in the surf zone is not observed in the lower stratosphere (465 K).

For comparison, Fig. 6.18 shows corresponding plots for the NH winter 1994–95. Large variations in vortex area are observed in midwinter; in particular, there is a significant decrease in the vortex area throughout the stratosphere during the warming events of late January–early February. Note also

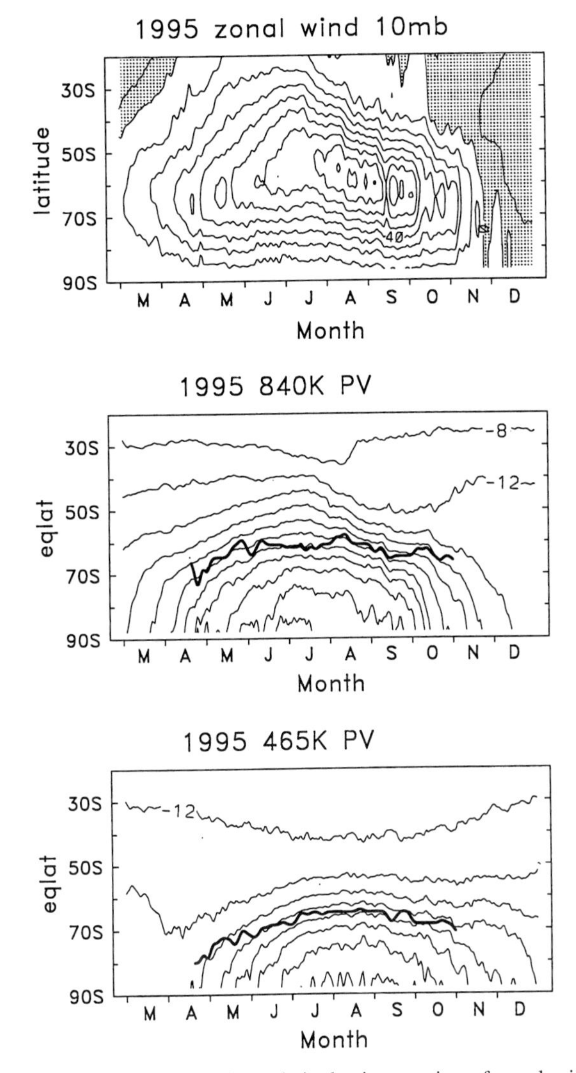

FIG. 6.17. Top panel shows latitude–time section of zonal wind at 10 mb for SH winter 1995; contours are 10 m s^{-1}. Lower panels show area evolution of individual contours of potential vorticity at 840 K (middle) and 465 K (bottom); area is expressed as equivalent latitude (see text). The PV contour interval is 4 × 10^{-4} and 4 × 10^{-5} K m^{-2} kg^{-1} s^{-1} at 840 and 465 K, respectively. Heavy lines denote the polar-vortex edge (maximum PV gradient) at the respective levels.

clear development of the surf zone (weak midlatitude PV gradients) at 840 K, beginning in January; the NH surf-zone area is significantly wider than that in the SH.

c. A case study of synoptic and planetary-wave activity

The wave-driving event of early September 1995 presents an interesting case study of vortex variability during a springtime wave event in the SH. The heat flux displayed in Fig. 6.16 shows a burst of wave activity between 9 and 12 September. Figure 6.19 displays false-color images of PV on the 460-K isentropic surface (~18 km) during this period. The flow is clockwise, approximately parallel to PV isolines; the jet core lies along the polar-vortex edge where the PV gradient is largest. The panels in Fig. 6.19 all illustrate the basic structure of the polar vortex: 1) strongly negative values of PV in the polar region (i.e., the polar vortex); 2) a strong PV gradient at about 60°S (the vortex edge); and 3) a broad region of weak PV gradients extending to the subtropics (the surf zone).

The first panel of Fig. 6.19 (6 September) shows a relatively circumpolar vortex, which begins to elongate as time progresses, extending well northward of

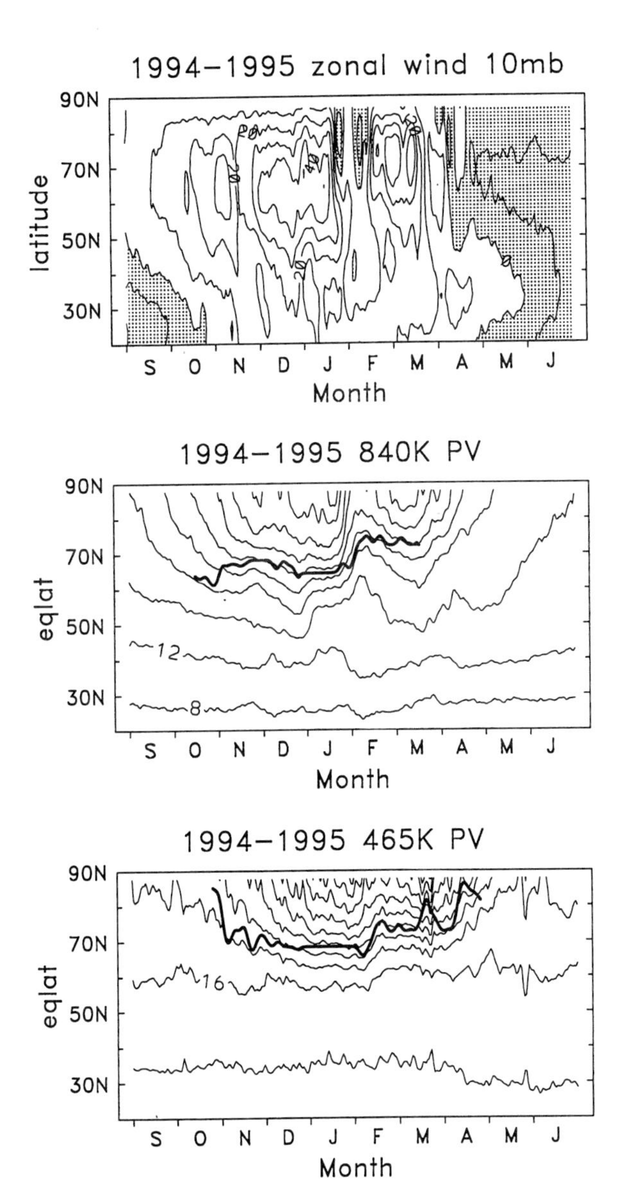

FIG. 6.18. Evolution at 10-mb zonal wind together with 840 and 465 K potential vorticity during NH winter 1994–95. Contours as in Fig. 6.17.

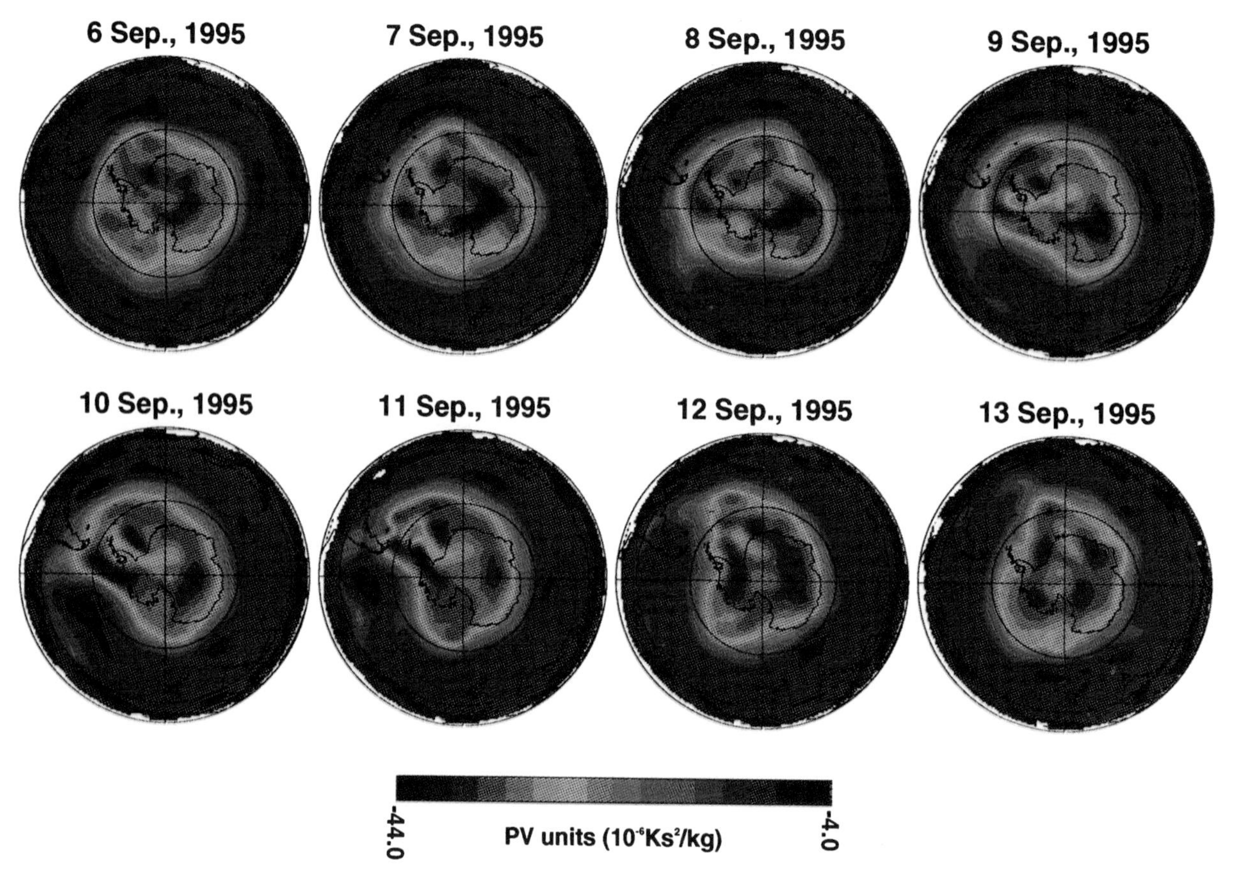

FIG. 6.19. Evolution of SH potential-vorticity fields at 465 K during 6–13 September 1995.

60°S. This "lobe" of the vortex is evident on 9 September, reaching the southern tip of South America. The lobe moves slowly eastward between 9 and 13 September and has started to decay by 13 September. On 11 September, a tongue of material extends away from the vortex, up toward South America, and then westward across South America and over the Pacific Ocean. This tongue is an example of wave breaking in the SH (McIntyre and Palmer 1983, 1984). The preferred wave breaking in the South American region is a climatological feature, related to displacement of the vortex toward this region in SH spring (associated with stationary planetary wave 1—see Fig. 6.10). In this particular case, and in most springtime wave-breaking cases, the PV is stripped out of the vortex edge region and does not involve material from deep inside the vortex. This weak transport of material across the edge of the polar vortex is consistent with a number of numerical studies showing that the southern polar vortex is relatively isolated from midlatitudes over the winter period (above 400 K) and that exchange of air across the polar-vortex boundary is relatively weak (Bowman 1993a,b; Bowman and Mangus 1993; Chen 1994; Chen et al. 1994; Manney et al. 1994; McIntyre 1995).

In the middle stratosphere, this September wave event was evidenced in deceleration of the zonal wind at 10 mb by approximately 20 m s^{-1} (Fig. 15). Figure 6.20 displays false-color images of PV on the 850-K isentropic surface (~30 km) during this period. The behavior of the PV at 850 K is qualitatively similar to that displayed at 460 K. In contrast to 460 K, however, a strong tongue of subtropical air is pulled southward around the edge of the vortex. This tongue is observed as the low PV (white colors), which is pulled out of the subtropics on 10 September over the South Atlantic Ocean, moves across the polar region, and is evident as low PV to the south of Australia on 13 September. A similar event was observed in UARS constituent observations by Randel et al. (1993) and analyzed further by Waugh (1993b).

d. A case study of the final warming

The most spectacular wave events in the SH occur during the vortex breakup in late spring (October–December). The heat flux displayed in Fig. 6.16 shows a number of bursts of wave activity in the October through December period, with diminishing intensity as the vortex weakens. Figure 6.21 displays a sequence of false-color images of PV on the 460-K isentropic

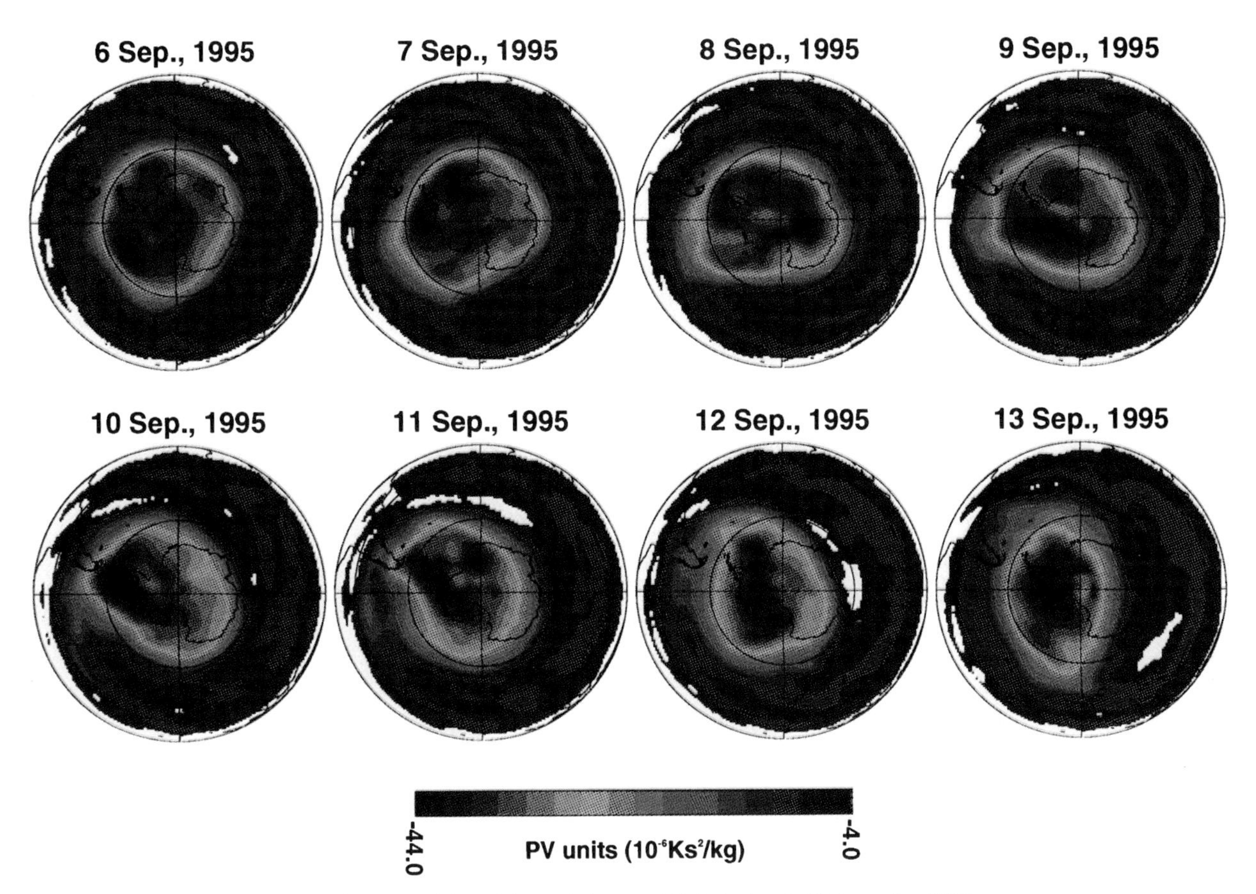

FIG. 6.20. Potential-vorticity variations at 840 K during 6–13 September 1995.

surface during this breakup phase. The vortex temperatures have risen considerably by the November–December time period (see Fig. 6.5). The first panel (6 November) shows a relatively intact circumpolar vortex, which has significantly decreased in strength since the early September period (see Fig. 6.20; note the color-scale change). This vortex continues to decrease in strength during November and early December, with substantial wave-like deformations. By 18 December, the vortex has virtually disappeared. This breakup was atypical in that the vortex persisted into mid-December at this altitude.

In the middle stratosphere, the breakup occurs significantly earlier than at lower levels. Figure 6.22 displays false-color images of PV on the 850-K isentropic surface during the breakup. The images shown in Fig. 6.22 are constructed using trajectory reverse domain filling (RDF), which produces a high-resolution representation of the PV field (see Sutton 1994; Newman and Schoeberl 1995; and Schoeberl and Newman 1995). The RDF field is represented by a dense regular grid of points; these gridded points are then advected backward in time with the trajectory model to an earlier time (in our case, 10 days prior to the date shown), and the PV observation from the analysis at this earlier time is plotted on the dense regular grid for the shown date. The evolution in Fig. 6.22 shows wave-breaking events that pull streamers of polar PV off of the vortex into long tongues of material extending well into the midlatitudes. Such tongues are ubiquitous features of large-amplitude (nonlinear) wave events in the middle stratosphere (e.g., Juckes and McIntyre 1987; Waugh 1993a; Norton 1994; Schoeberl and Newman 1995). These images demonstrate the rapid dispersion of vortex air into midlatitudes during the final warming (see similar results in Manney et al. 1994).

e. Traveling planetary waves

Several distinct modes of traveling planetary waves have been observed in the SH stratosphere. In terms of wave amplitude and importance for the general circulation, the eastward-traveling wave 2 (discussed at the end of section 6.3e) is the largest. Another distinctive wave mode observed in the polar upper stratosphere is the so-called "4-day wave," discovered by Venne and Stanford (1979). Characteristics of this mode include

1) temperature-variance maxima confined to polar regions (centered near 60°–70°S) in the upper stratosphere and mesosphere (Venne and Stan-

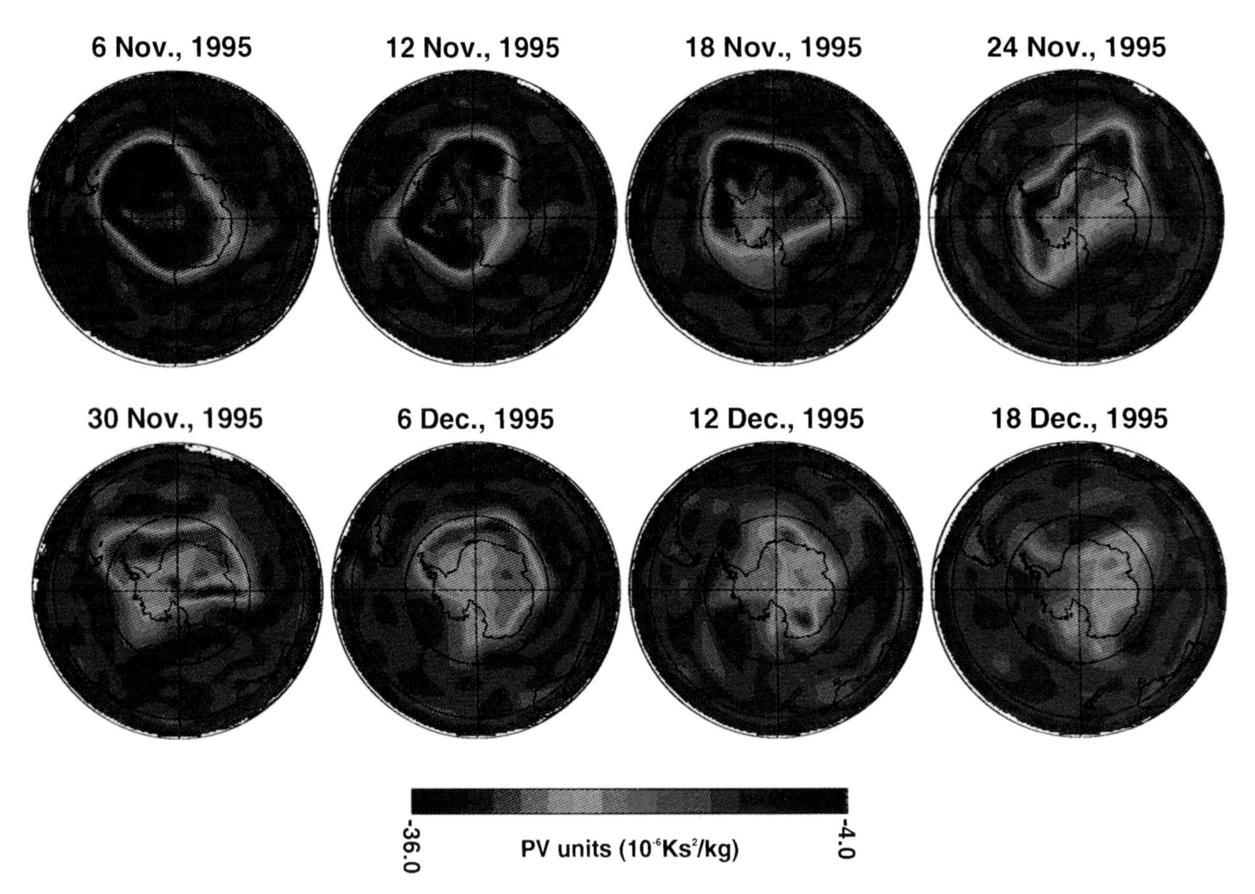

FIG. 6.21. Evolution of potential vorticity at 465 K during breakdown of the SH polar vortex in November–December 1995.

ford 1979, 1982; Prata 1984; Lait and Stanford 1988);

2) eastward-propagating power for zonal waves 1–4; reconstructed synoptic maps reveal one or more warm pools circling the pole with period near 4 days (Prata 1984; Lait and Stanford 1988; Allen et al. 1997). [An example is shown in Fig. 6.23, adapted from Lait and Stanford 1988.]; and
3) analysis of wave and background mean-flow structure strongly suggests instability of the polar night jet and/or mesospheric subtropical jet as the source of excitation for the 4-day wave (Hartmann 1983; Randel and Lait 1991; Manney 1991; Manney and Randel 1993; Lawrence and Randel 1996; Allen et al. 1997).

Another traveling wave with sharp spectral characteristics is the 2-day wave. This is a westward-propagating, zonal wave 3–4 oscillation observed in the upper stratosphere and mesosphere of the subtropics during summer. It has been documented in satellite observations by Rodgers and Prata (1981), Burks and Leovy (1986), Wu et al. (1993), Randel (1994), Limpasuvan and Leovy (1995), and Wu et al. (1996). One of the most interesting features of the 2-day wave is that it is observed in low to middle latitudes of both hemispheres, but only for a period of several weeks following summer solstice. Its origin has been discussed theoretically as being associated with a global normal mode (Salby 1981a,b) or due to dynamic instability of the summer easterly jet (Plumb 1983; Pfister 1985). Randel (1994) showed a combination of normal-mode structure and instability signature that suggests the 2-day wave is a near-resonant mode forced by dynamic instability.

Observational studies of satellite data have also shown evidence of westward-propagating, normal-mode Rossby waves in the upper stratosphere of the SH (Rodgers 1976; Hirota and Hirooka 1984; Venne 1989). These waves correspond to the free or resonant oscillations of the atmosphere and are characterized by 1) planetary-scale horizontal structure, 2) weak vertical phase tilts, and 3) regular westward propagation at (nearly) discrete frequencies, with typical periods of 5–15 days. These modes typically exhibit relatively small amplitudes and are interesting mainly because of their global coherence and similarity to calculated resonant modes (e.g., Salby 1981a).

6.5. Interannual variability

Most analyses of interannual variability in the stratosphere have focused on the NH, because of the relatively

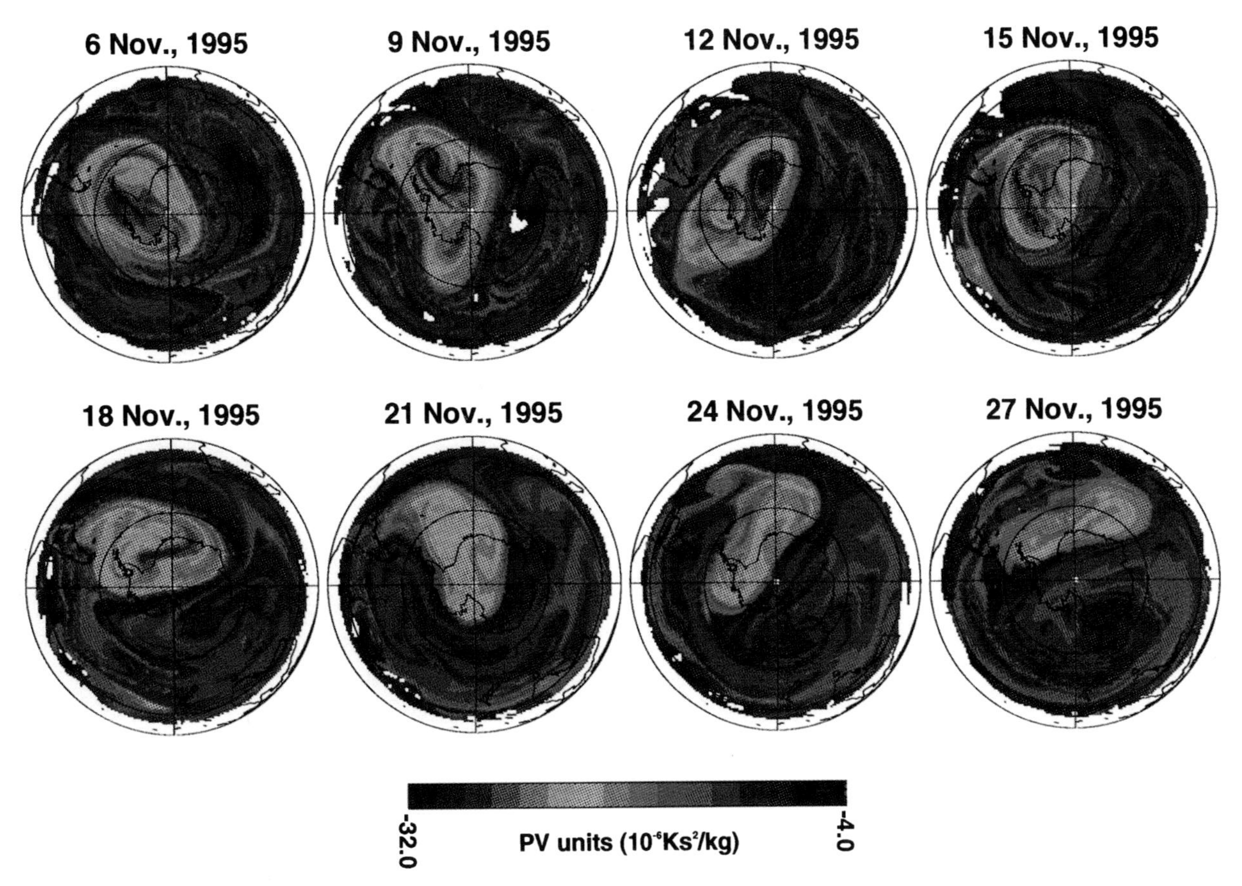

FIG. 6.22. Potential vorticity variations at 840 K during breakdown of the SH polar vortex in November 1995.

long record of observations compared to the SH. Labitzke (1982) has analyzed NH stratospheric analyses over 100–10 mb dating from 1957, while Baldwin and Holton (1988) and Dunkerton and Baldwin (1991) have shown results from NCEP data (up to 10 mb) using the weekly analyses available since 1964. The NCEP global data analyzed here span 17 years, providing the best record to date to analyze interannual stratospheric variability in the SH.

Several studies of NH stratospheric variability have focused on observing the response to specified external forcings. These include extratropical effects of the tropical QBO (Holton and Tan 1980, 1982; Labitzke 1982; Dunkerton and Baldwin 1991), the 11-year solar cycle (Labitzke and van Loon 1987, 1995), and ENSO (van Loon and Labitzke 1987; Baldwin and O'Sullivan 1995). Newman and Randel (1988) analyzed interannual variability in the SH stratosphere in relation to

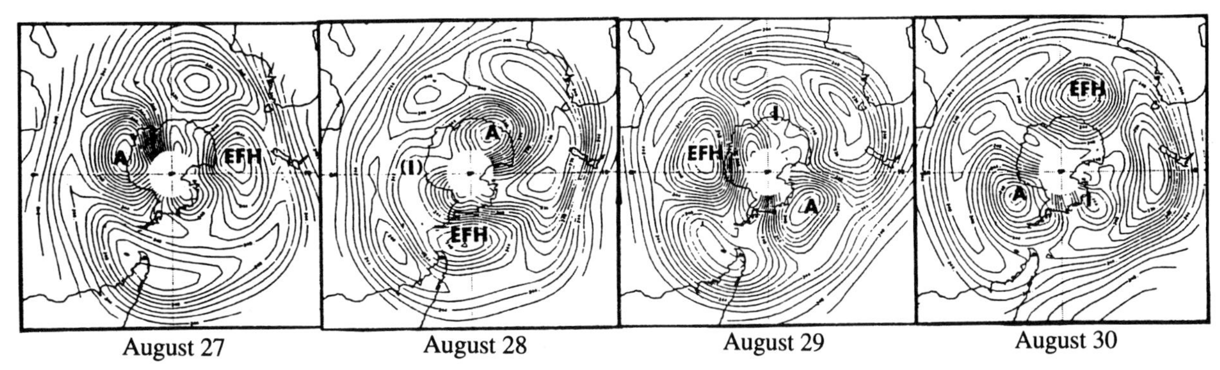

FIG. 6.23. Synoptic maps of polar temperatures in the upper stratosphere (near 45 km) during August 1981; contour interval is 2 K. Labeled features "A" and "EFH" are warm pools rotating about the pole with a period of 4 days. Feature A retains its identity for over seven complete revolutions about the pole (not shown here). [Adapted from Lait and Stanford (1988).]

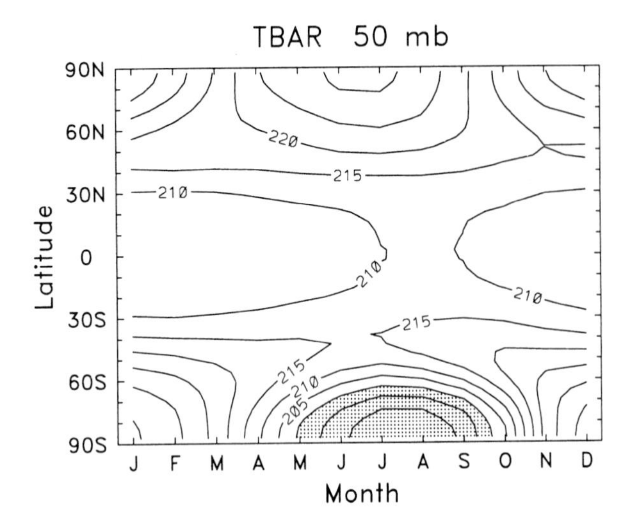

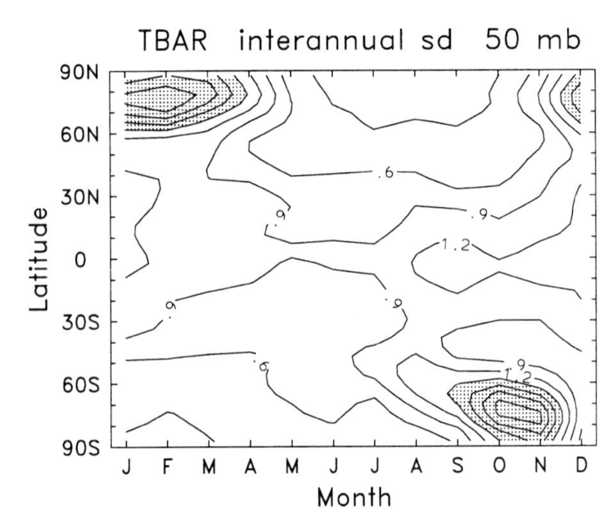

FIG. 6.24. Latitude–time sections at 50 mb of the zonal-mean temperature (top) and its interannual standard deviation [Eq. (8)] over 1979–95 (bottom-contour interval is 0.3 K).

ozone changes, and Hurrell and van Loon (1994) discuss long-term changes in the SH tropospheric circulation and their extension into the lower stratosphere. Our analyses here will discuss the observed QBO signal in the SH stratosphere, but we will not deal with the solar or ENSO components due to the relative shortness of the data record and confusion with effects of the major volcanic eruptions of El Chichon (1982) and Mount Pinatubo (1991).

Figure 6.24 shows latitude–time sections of the climatological mean and interannual standard deviation ($\sigma_{\rm int}$) of 50-mb monthly zonally averaged temperature from the NCEP data over 1979–95. The latter term is calculated as:

$$\sigma_{\rm int} = \left[\frac{1}{N}\sum_{i=1}^{N} \cos\,\phi(\bar{T}_i - \langle\bar{T}\rangle)^2\right]^{1/2}. \qquad (6.8)$$

Here, $\bar{T}_i$ is the monthly average temperature for year i, $\langle\bar{T}\rangle$ is the ensemble monthly average, N is the number of years available, and the cosine (ϕ = latitude) term accounts for geometrical effects (North et al. 1982). As a note, the $\sigma_{\rm int}$ estimates in the Tropics in Fig. 6.24 are probably underestimated (approximately by a factor of two), due to the inability of the NMC data to fully resolve the tropical QBO signal.

Interannual variability in the SH lower stratosphere exhibits a maximum during late winter and spring (August–November) poleward of 50°S—i.e., during the spring warming. Interannual variations are comparatively small during the rest of the year, in particular during midwinter. This variability in the SH is distinctive from the corresponding NH patterns, which exhibit strong polar variability throughout midwinter (December–March). The space–time maxima in interannual variability are similar to those for daily variability (see Fig. 6.14); much of the interannual variance is related to the presence or absence of large stratospheric warming events during a particular month. Interannual variability of zonal wind in the middle stratosphere shows similar seasonality to that in Fig. 6.24 but with the extratropical maxima shifted 10°–15° equatorward (associated with thermal wind balance).

Figure 6.25 shows time series of monthly mean temperatures during November at 100 mb for several Antarctic stations. November data are chosen because of the maximum interannual variability at this time seen in Fig. 6.24. These plots show data from the NCEP reanalysis (Kalnay et al. 1996), together with radiosonde data spanning approximately 1960–97, for a longer-term perspective on interannual variability. The monthly mean radiosonde data here have been constructed from the available daily station observations, using the method of Trenberth and Olson (1989), which minimizes sampling biases. Figure 6.25 shows a significant component of year-to-year see-saw behavior that is (statistically) related to the QBO (as discussed below). The long-term record also suggests that some significant cooling has occurred in the polar lower stratosphere since approximately 1980.

The altitude–time structure of the decadal-scale cooling over Antarctica is quantified in Fig. 6.26, which shows the antarctic temperature changes between the decades (1986–95 minus 1970–79), derived from radiosonde data (as in Fig. 6.25) (Randel and Wu 1998). These data show strong cooling (of order 6–10 K) in the lower stratosphere (~12–21 km), which maximizes in spring (October–December). Smaller-magnitude cooling (~1 K) persists throughout summer, while no significant changes are observed in winter. These decadal temperature changes exhibit similar space–time characteristics to antarctic ozone depletion since 1980. Furthermore, the observed cooling in Fig. 6.26 is in reasonable agreement with the ozone-hole temperature change calculations of Shine

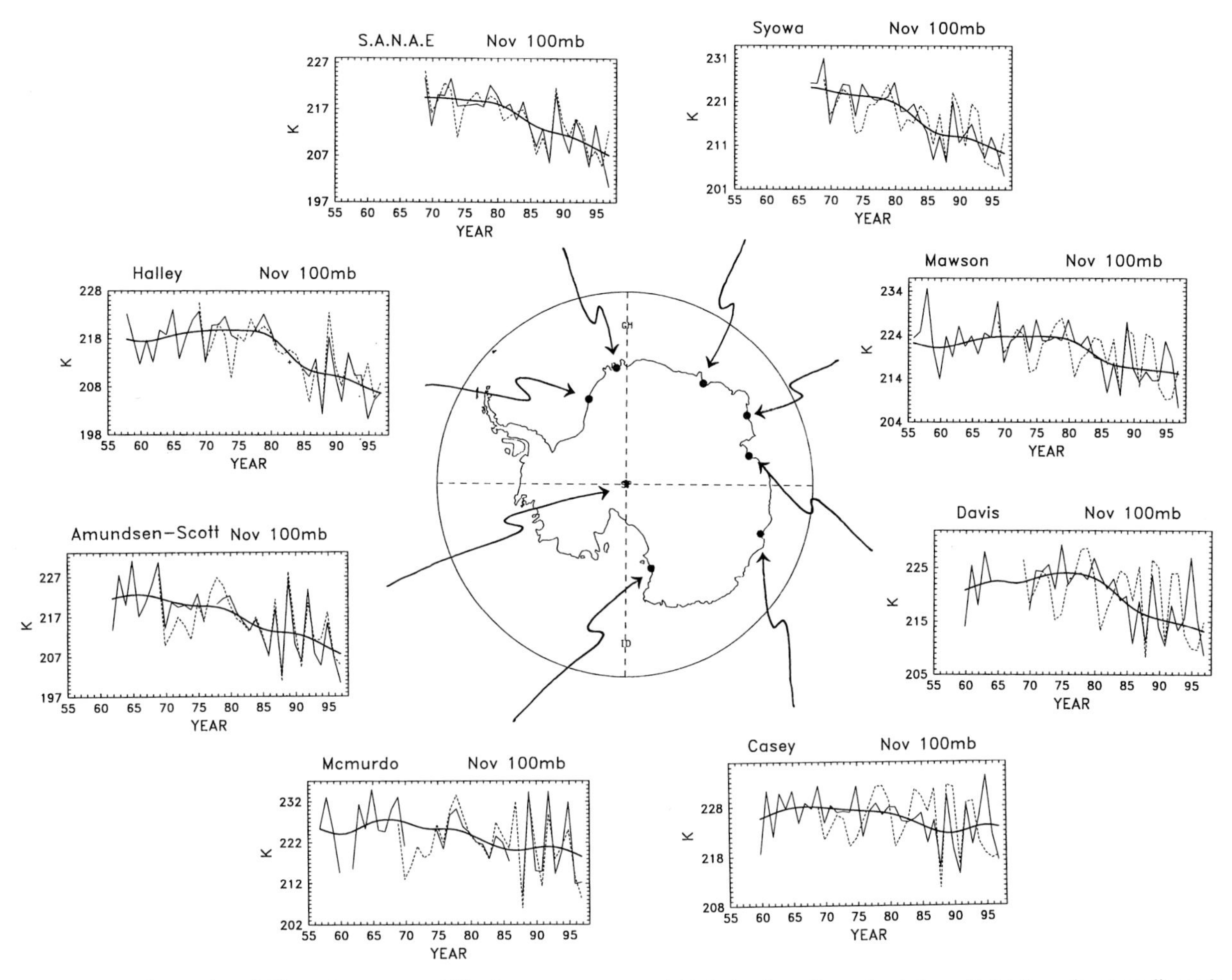

FIG. 6.25. Time series of November average 100-mb temperatures at eight Antarctic radiosonde stations. Solid lines denote radiosonde data, and dashed lines show data from NCEP reanalysis (Kalnay et al. 1996) at the radiosonde locations. The smooth curves in each panel show the decadal-scale variations at each location calculated from the radiosonde data. (From Randel and Wu 1998.)

(1986), Kiehl et al. (1988), Mahlman et al. (1994), and Ramaswamy et al. (1996). This strongly suggests that the observed polar cooling in spring is primarily a radiative response to the antarctic ozone hole. Waugh and Randel (1998) show a corresponding delay in the springtime vortex breakdown in the SH lower stratosphere (~2-week delay between the early 1980s and middle 1990s).

The seasonal evolution and interannual standard deviation of zonal-mean wind at 1 mb is shown in Fig. 6.27, illustrating interannual variability in the upper stratosphere (as with Fig. 6.24, the tropical variability may be underestimated in these NCEP-based estimates). Interannual variability in the SH upper stratosphere maximizes in midwinter near 30°S, on the equatorward flank of the climatological jet core. This SH midwinter variability maximum has been discussed by Shiotani et al. (1993) using a subset of the data here, and our discussion follows and extends their results. Figure 6.28 shows the seasonal march of

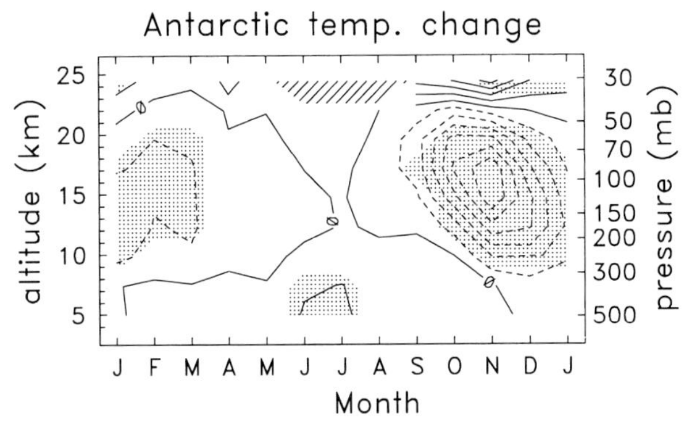

FIG. 6.26. Altitude–month profile of temperature differences over Antarctica between the decades (1986–95 minus 1970–79), derived from radiosonde data (see Fig. 6.25). Contour interval is 1 K, and shading denotes a 2-sigma statistically significant difference with respect to natural variability. Data are unavailable during midwinter at the uppermost levels. (From Randel and Wu 1998.)

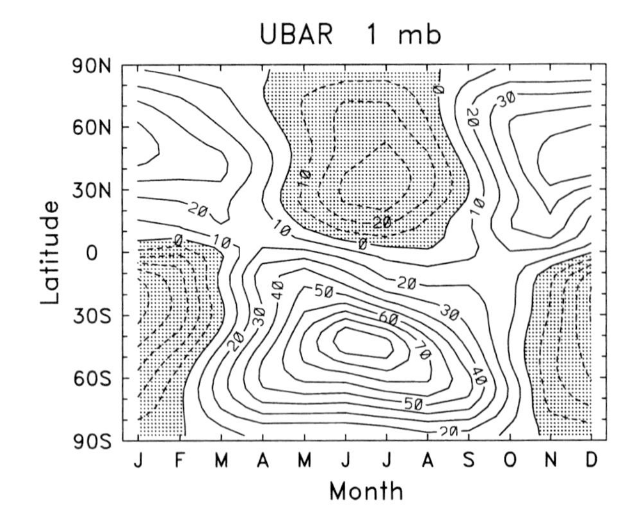

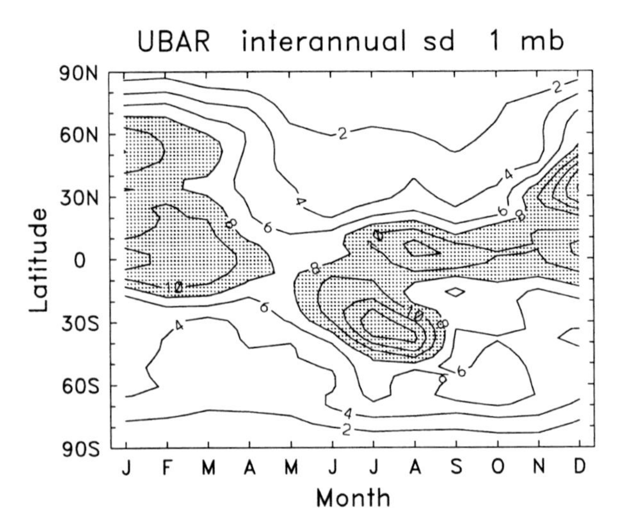

FIG. 6.27. Latitude–time sections at 1 mb of the zonal-mean zonal wind (top) and its interannual standard deviation [Eq. (8)] over 1979–95 (bottom-contour interval 2 m s^{-1}).

monthly mean zonal winds at 1 mb, 30°N and 30°S, for each year of the record here (1979–95); note the time axes are shifted for direct NH–SH seasonal comparisons. This figure illustrates the stronger zonal winds in the SH vs. NH subtropics and highlights the substantial interannual variability in the SH in midwinter (note the hint of similar variability in the December NH data). Time series of the SH anomalies versus year (not shown) reveal a see-saw pattern of positive and negative winds (with few near the long-term mean). These anomalies are not statistically correlated with the tropical QBO.

The spatial structure of the zonal-wind anomalies during July 1989 and 1992 (two opposite extremes) are shown in Fig. 6.29 (these years are noted with the light and heavy dashed lines in Fig. 6.28). The anomalies show a north–south dipole pattern, with centers near 30°–40°S and 60°–70°S, and extend throughout the middle and upper stratosphere. These anomalies correspond to a latitudinal shifting of the midwinter polar night jet; Shiotani et al. (1993) term these states the high-latitude jet and low-latitude jet. Shiotani et al. furthermore show evidence of systematic changes in planetary-wave structure during the respective regimes, such that larger wave amplitudes are observed in midwinter for the high-latitude-jet periods. The cause of this interannual variability is not well understood at present.

Extratropical influences of the QBO in the SH

The QBO dominates interannual variability in the tropical stratosphere. While it is not a dominant effect outside of the Tropics, statistical studies (primarily using the longer record of NH data) suggest that some component of middle–high-latitude variability is related to the QBO, in particular the strength of the NH winter polar vortex (e.g., Holton and Tan 1980, 1982; Dunkerton and Baldwin 1991). Evidence of an extra-

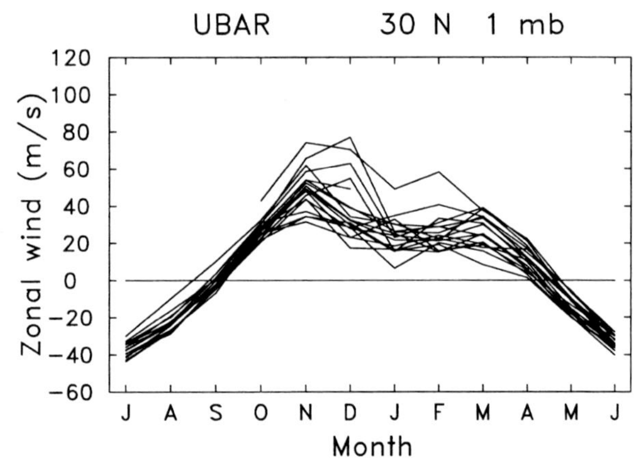

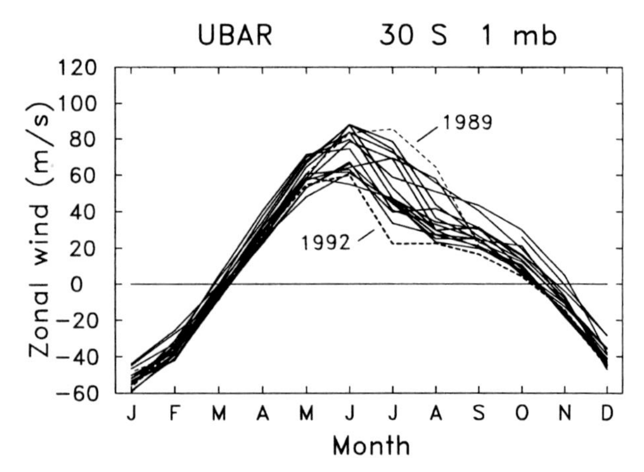

FIG. 6.28. Time traces of monthly mean zonal-mean wind at 1 mb, 30°N (top) and 30°S (bottom). Each line denotes an individual year during 1979–95.

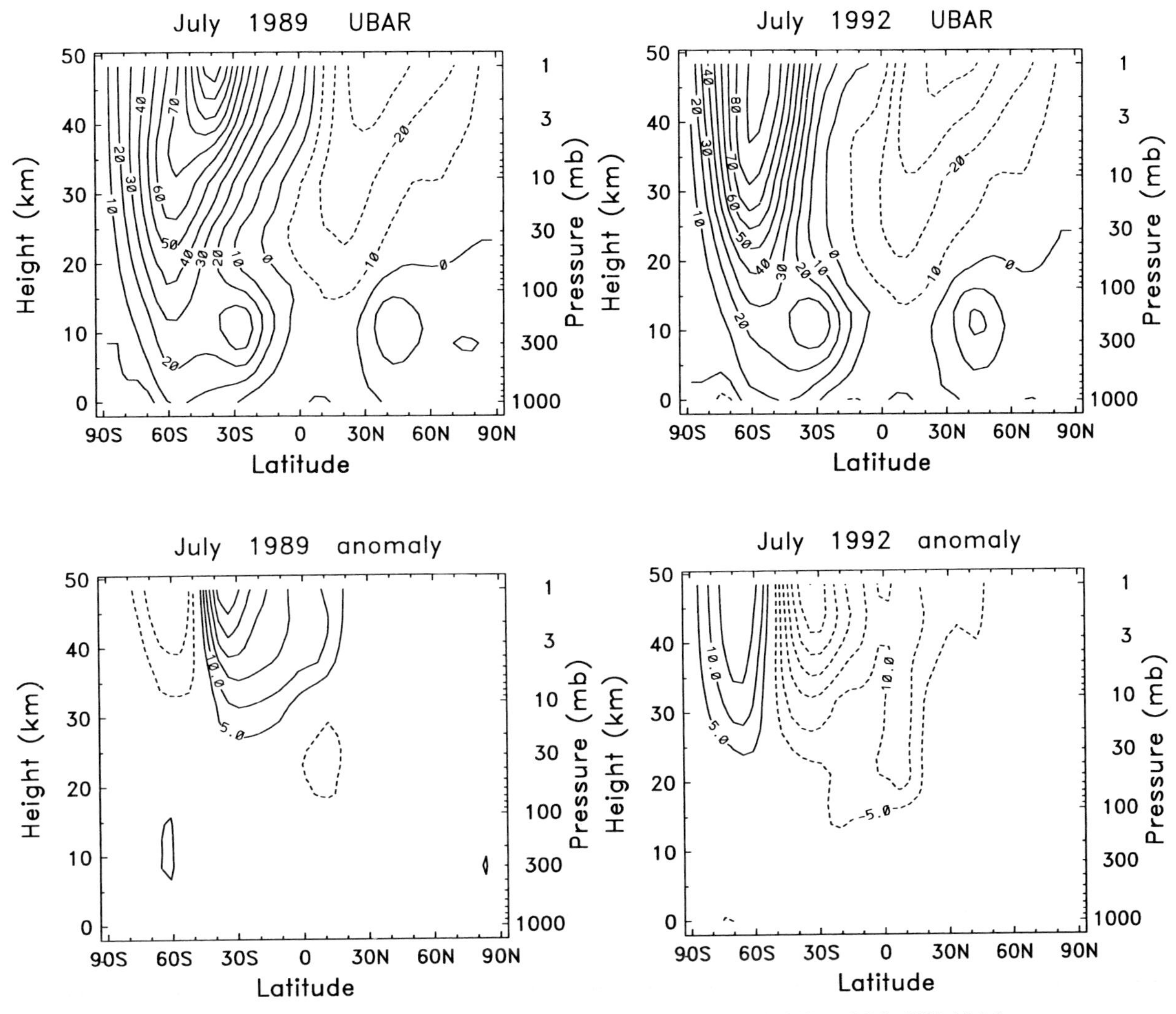

FIG. 6.29. Meridional cross sections of zonal-mean wind in July 1989 (left) and July 1992 (right). Top panels show full fields, and bottom panels are the respective anomalies.

tropical QBO signal is particularly clear in column ozone data and are observed in both the NH and SH (Bowman 1989; Lait et al. 1989; Tung and Yang 1994; Yang and Tung 1994). Randel and Cobb (1994) isolated extratropical QBO patterns in both column ozone and lower-stratosphere temperature using regression analyses tied to the observed QBO tropical winds. An update of their results for lower-stratosphere temperature (shown in Fig. 6.30) demonstrates that temperature variations in the midwinter subtropics and spring polar latitudes are statistically coherent with the tropical QBO winds. The modeling studies of Gray and Dunkerton (1990), O'Sullivan and Young (1993), O'Sullivan and Dunkerton (1994), and Buchart and Austin (1996) provide plausible mechanisms for transmitting the QBO signal to extratropics and are in qualitative agreement with these observations. Although the patterns in Fig. 6.30 are statistically significant, the QBO accounts for (at maximum) only 25%–35% of the interannual variance for seasonal anomalies in the time sample studied here; thus, the QBO is not a dominant mode of interannual variability in extratropics.

6.6. Ozone

Ozone is of prime importance in the stratosphere. Ozone absorbs solar ultraviolet radiation, providing the heat source for the temperature increase with altitude that defines the stratosphere [in the absence of ozone there would be no stratosphere, although the location of the tropopause itself is not very sensitive to the presence of ozone (Thuburn and Craig 1997)]. Ozone also provides the principal screening of this biologically harmful ultraviolet radiation. Thus, the analysis of the observed structure and variability of ozone is a central focus of stratospheric research.

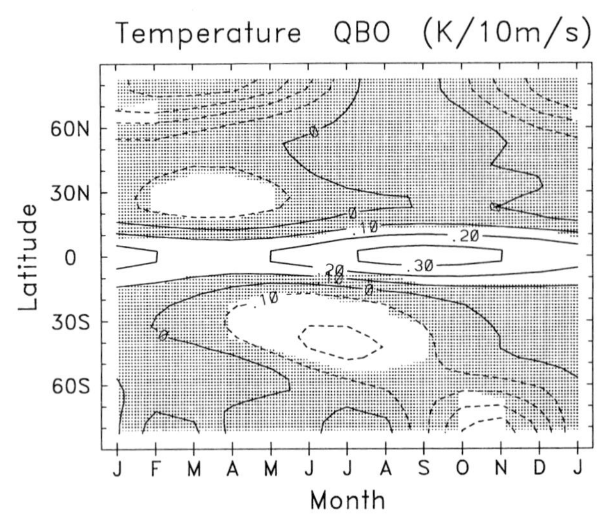

FIG. 6.30. Latitude–time section of the QBO-associated variation of lower stratospheric temperature, derived from regression analyses of data over 1979–98 (an update of the results presented in Randel and Cobb 1994). Units are K per 10 m s^{-1} of 30-mb equatorial zonal winds. Shading denotes that the associated correlations are not statistically significant.

The presence of ozone in the atmosphere was first deduced from observations of a sharp cutoff in the near ultraviolet end of the solar spectrum (Hartley 1880). Chapman (1930) provided a simple theoretical explanation for the formation and vertical structure of the ozone layer. The formation of ozone occurs by the photolysis of molecular oxygen by ultraviolet radiation (200–220 nm), followed by a combination of atomic and molecular oxygen:

$$O_2 + h\nu \rightarrow O + O \qquad (6.9a)$$

$$\frac{O + O_2 + M \rightarrow O_3 + M}{\text{net: } 3O_2 \rightarrow 2O_3}. \qquad (6.9b)$$

Here, M is a third body (typically a nitrogen or oxygen molecule) needed to conserve energy and momentum. Because the amount of ultraviolet radiation decreases downward from the top of the atmosphere while the amount of molecular oxygen decreases exponentially (with density) away from the earth's surface, the production of ozone must be a maximum in a layer at some intermediate height centered at low latitudes (where sunlight maximizes). Chapman originally believed that ozone destruction takes place by reaction of ozone with atomic oxygen. It is now known that this mechanism is secondary to ozone destruction by so-called catalytic cycles. Catalytic loss takes place by removing both ozone and atomic oxygen via:

$$Z + O_3 \rightarrow ZO + O_2 \qquad (6.10a)$$

$$\frac{O + ZO \rightarrow Z + O_2,}{\text{net: } O + O_3 \rightarrow 2O_2}, \qquad (6.10b)$$

where Z is principally reactive chlorine, bromine, nitrogen, or hydrogen species (additional catalytic cycles are important for winter polar-ozone losses, as discussed in the following sections). Ozone production primarily occurs in the Tropics, but observations show that the ozone concentration is greater in the high-latitude lower stratosphere than in the Tropics, thus showing that transport is a crucial factor in ozone behavior. Because this transport depends in turn on the heating distribution associated with ozone, detailed understanding involves the coupling of chemistry, radiation, and dynamics in the stratosphere.

a. Climatology of ozone

The strong wavelength dependence of ultraviolet light absorption by ozone allows deduction of ozone amount by surface and satellite measurements (the latter using ultraviolet radiation backscattered from the surface and clouds). One instrument that provided a long time series of global column-ozone measurements was the TOMS on the *Nimbus 7* satellite (spanning 1978–93; McPeters et al. 1993). Column ozone is the vertically integrated total amount of ozone, often measured in terms of Dobson units (DU), after the pioneering work of G. Dobson. One DU is approximately equal to 2.6×10^{22} molecules of ozone per square meter; typical observed values are of order 300 DU.

Figure 6.31 shows the climatological global structure of column ozone derived from TOMS data. Here,

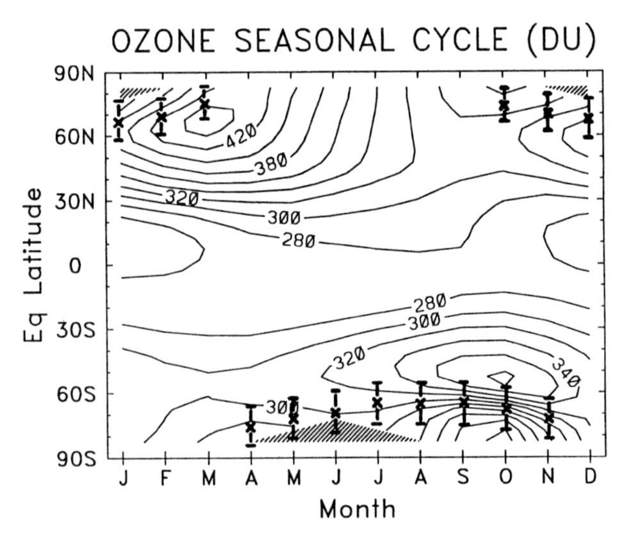

FIG. 6.31. Climatological column-ozone seasonal cycle, constructed using potential-vorticity coordinates (expressed in equivalent latitude). Contours are in Dobson units. The bracketed x's denote the statistical vortex edge and boundary region in the lower stratosphere (520 K). (From Randel and Wu 1995.)

TOMS data have been longitudinally averaged with respect to the polar-vortex structure in the lower stratosphere using potential vorticity coordinates (Buchart and Remsberg 1986; Schoeberl et al. 1992; Randel and Wu 1995); this allows direct comparisons of the NH and SH column-ozone structure in relation to the vortex (which is also indicated in Fig. 6.31). Column ozone is a minimum over the Equator, increasing toward higher (particularly winter) latitudes. There is a strong seasonal cycle in ozone in extratropics, with maximum amounts over high latitudes in winter and spring of the respective hemispheres. Winter–spring column-ozone amounts are substantially higher in the NH than in the SH, but similarly sized in summer. The observed middle–high-latitude maximum in column ozone, together with the seasonal maximum in winter–spring and the hemispheric differences, are related to global-scale transport from the tropical ozone source region by both mean meridional circulations and eddy processes. Note that the seasonal cycle and interhemispheric differences in residual mean circulation in Fig. 6.4 are consistent with these aspects of observed ozone structure (i.e., flow from Tropics to high latitudes in winter and stronger transport during winter in the NH than in the SH). Inspection of Fig. 6.31 shows that column ozone is a maximum over latitudes near the edge or slightly outside of the vortex in each hemisphere, peaking in spring (March in the NH and October in the SH). This aspect is much more pronounced in the SH, where a "collar," or maximum of ozone, is seen near 60°S. This maximum away from the pole is due to circulation effects, as the strongest downward velocity in the lower stratosphere occurs near the edge of the polar vortex, particularly in the SH (Schoeberl et al. 1992; Manney et al. 1994; clearly seen in Fig. 6.7c). Column ozone shows a relative minimum inside the vortex in the SH throughout winter–spring, especially pronounced in September–October. The deep spring minimum is mainly due to chemical depletion in the lower stratosphere associated with the ozone hole (discussed below), although a relative minimum in this region is a climatological feature related to the mean circulation (as discussed previously and observed in the NH winter in Fig. 6.31; see also Bojkov and Fioletov 1995, their Fig. 15).

The vertical profile of ozone and its seasonal evolution is shown in Fig. 6.32, based on HALOE data, combined with polar measurements from the microwave limb sounder (MLS) instrument on the UARS (Froidevaux et al. 1994; Manney et al. 1995). The quantity shown is ozone mixing ratio, in parts per million by volume. Note that column ozone (as shown above) is the vertical integral of mixing ratio times atmospheric density, which decreases exponentially upward, and hence the contribution to column ozone is heavily weighted from mixing ratios in the lower stratosphere in Fig. 6.32 [which is why column ozone is largest at high latitudes, even though the mixing ratio is largest in the middle stratosphere in the Tropics (the photochemical source region)]. There is a clear seasonal movement of the tropical mixing ratio maximum, following the position of the sun—over the SH subtropics in January and over NH subtropics in July. Relatively low values of ozone are found at high southern latitudes (inside the vortex) throughout SH winter. This is due to relatively weak poleward transport from tropical source latitudes across the vortex boundary; note the much-stronger high-latitude gradients of ozone in the SH versus the NH in winter (similar to the methane measurements seen in Fig. 6.8). Very low values are also observed in the polar lower stratosphere in October in Fig. 6.32, resulting from chemical ozone depletion (discussed below).

b. Ozone trends

Predictions of global ozone loss due to human activities were put forward by Johnston (1971) and Molina and Roland (1974), due to supersonic aircraft emissions and the release of chlorofluorocarbons, respectively. As nitrogen and chlorine compounds increase in the stratosphere, catalytic loss would also increase, and the new photochemical balance would shift ozone levels downward. The search for evidence of downward trends was inconclusive until the discovery of the antarctic ozone hole by Farman et al. (1985). Since that time, numerous observational analyses, based on both ground and satellite measurements, have established statistically significant decreases in total ozone over much of the globe (see WMO 1992; Stolarski et al. 1991; Stolarski et al. 1992; and the most recent updates in WMO 1995; Bojkov and Fioletov 1995; Harris et al. 1997; and McPeters et al. 1996). Figure 6.33 shows the latitude–season pattern of total ozone decreases during 1979–94 derived from TOMS data [similar to the results of Stolarski et al. (1991), but extended using data through 1994]. Similar results are derived using solar backscatter ultraviolet (BUV) data (Hollandsworth et al. 1995). Statistically significant decreases are observed in the SH at middle and high latitudes during spring (associated with the ozone hole) and extending into summer; relatively small losses are seen during autumn and winter. Large areas of ozone loss are observed in the NH midlatitudes during winter and spring; significant trends are not found in the Tropics. The majority of the column-ozone losses occur in the lower stratosphere. The SH ozone losses over midlatitudes during summer may be attributable to transport and persistence of ozone-poor air following breakup of the polar vortex (Atkinson et al. 1989; Sze et al. 1989; Prather et al. 1990; Cariolle et al. 1990; Mahlman et al. 1994; Brasseur et al. 1997).

Figure 6.34 shows a long record (1956–94) of monthly mean column-ozone anomalies at Mel-

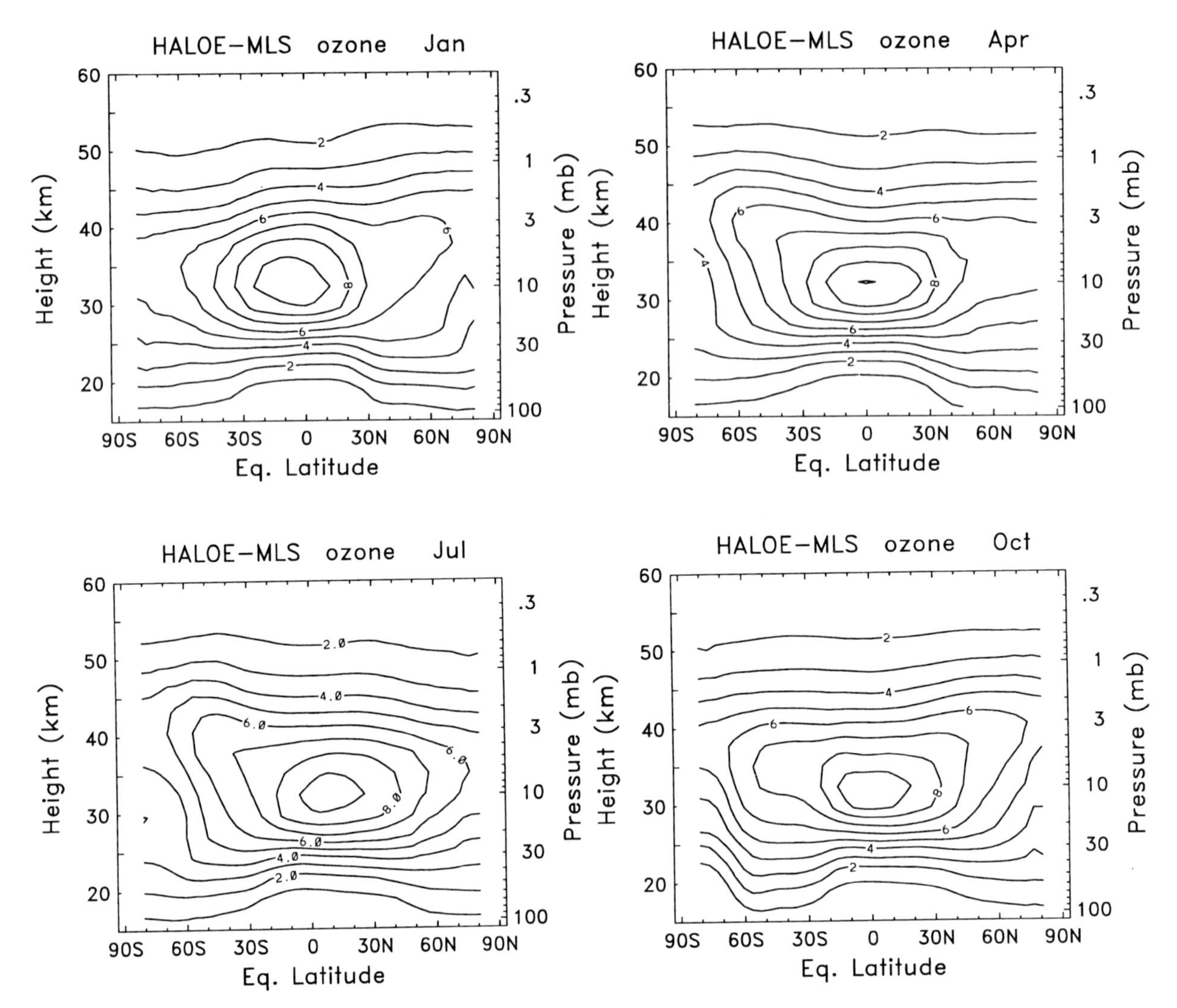

FIG. 6.32. Meridional cross sections of ozone mixing ratio in January, April, July, and October, derived from HALOE and MLS satellite data spanning 1991–96. Contour interval is 1 ppmv.

bourne, Australia (38°S), derived from ground-based Dobson spectrophotometer measurements (Atkinson and Easson 1989). The data in Fig. 6.34 have been deseasonalized by removing the mean annual cycle to highlight interannual variations and in turn divided by the long-term mean (317 DU) to obtain percentage variations. Additionally, interannual anomalies from TOMS data (over 1979–94) are superimposed in Fig. 6.34, and these show excellent agreement with the ground-based data for this overlap period. The data in Fig. 6.34 show month-to-month variations in column ozone at Melbourne of order ±4%, associated with natural meteorological variability inherent to the atmosphere. Additionally, there is an overall downward trend apparent in Fig. 6.34 (highlighted by the heavy solid line, which is a smoothed version of the monthly mean data), such that mean ozone values are 6%–8% lower in the 1990s than in the 1950s (approximately −2% per decade decrease). The trends are somewhat larger for the period 1979–94 (near −5% per decade). The enhanced ozone loss since approximately 1980 is a feature observed over extratropics of both hemispheres; hemispheric and global averages prior to this time show near constant values (Bojkov and Fioletov 1995), unlike the decreases seen during 1955–75 at Melbourne in Fig. 6.34. The SH contributed approximately 65% of the global 5% decline during 1980–95.

c. The ozone hole

1) OBSERVATIONS

The basic source, sink, and transport mechanisms affecting ozone were thought to have been well understood in the early 1980s, as evidenced by reasonable

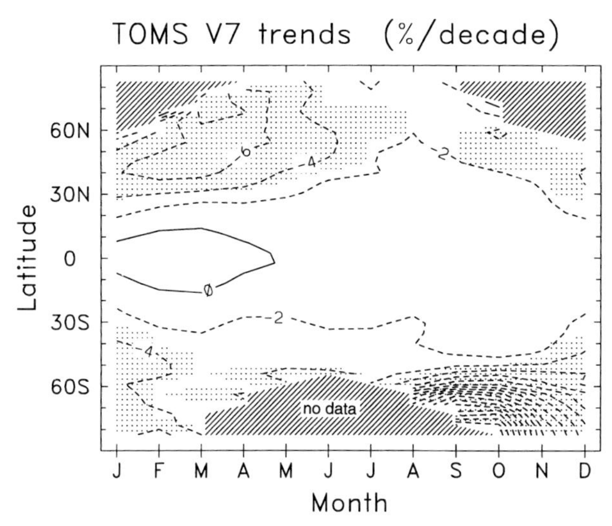

FIG. 6.33. Total ozone trends derived from TOMS Version 7 satellite data spanning 1979–94 (using combined *Nimbus 7* and *Meteor 3* data). Contour interval is 2% per decade. Shaded regions denote where the trends are statistically significant at the 2-sigma level.

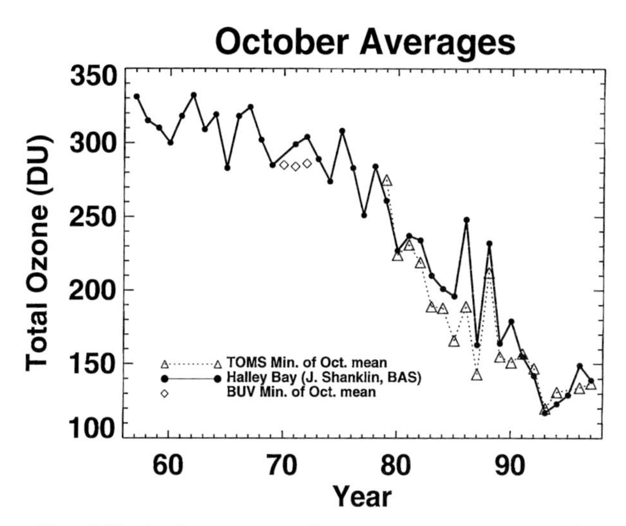

FIG. 6.35. October average column-ozone amounts over Halley Bay, Antarctica, from ground-based measurements. Also shown are minimum values of October column ozone from BUV data over 1970–72 and TOMS during 1979–97.

simulation of observed data (e.g., Garcia and Solomon 1983). This understanding was shaken by the unanticipated observations by Farman et al. (1985) of an approximate 50% decrease in column ozone over Antarctica during the decade 1975–85, but only during spring season (September–October). Figure 6.35 shows an updated version of the data presented in Farman et al. (see Jones and Shanklin 1995), showing October average column ozone measured by ground-based Dobson spectrophotometer over Halley Bay, Antarctica, over the years 1956–95. Values near 300 DU were observed prior to the mid-1970s, steadily declining to values below 150 DU in the 1990s.

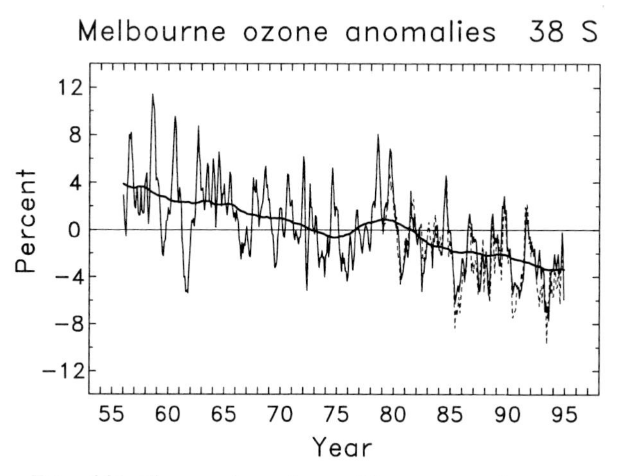

FIG. 6.34. Time series of monthly mean column-ozone anomalies from ground-based observations in Melbourne, Australia (38°S), spanning 1956–94 (light solid line), together with anomalies from TOMS data (dashed lines over 1979–94). Both data are deseasonalized and expressed as percent variations of the long-term mean. The heavy solid line is a smoothed version of the ground-based measurements, intended to highlight low-frequency variations.

Analysis of TOMS satellite data by Stolarski et al. (1986) showed that the depletion of ozone during SH spring occurred over a large area centered over the South Pole, comparable to the size of Antarctica. Figure 6.36 shows SH October monthly mean data from TOMS for the years 1979 and 1993–97, together with *Nimbus 5* BUV measurements from 1970 to 1972 (data from these instruments are included in Fig. 6.35, showing reasonable agreement with the ground-based measurements). Figure 6.36 illustrates the large ozone losses over polar regions between the 1970s and 1990s. The most dramatic losses occur inside the SH polar vortex (cf., the PV structure in Fig. 6.10), although midlatitude values also decrease (note the change in the collar region near 60°S in Fig. 6.36; see also Fig. 6.33). Note that a relative minimum in ozone over the pole was observed in the early 1970s (prior to the recent large depletions); this is the climatological minimum discussed above related to circulation and transport effects on ozone. The "ozone hole" refers not to this weak minimum, but to the relatively large depletions observed in comparing the earlier and later periods. Column-ozone values below 100 DU have been observed in the ozone hole since 1993 (Hofmann et al. 1994; Herman et al. 1995; Hofmann et al. 1997), and these are the lowest values ever observed anywhere on earth. Figure 6.37 shows the area of the ozone hole measured by TOMS, defined as the area inside the 220-DU contour during SH spring, illustrating rapid growth during the 1980s and approximately constant size during the 1990s (note that over 1979–

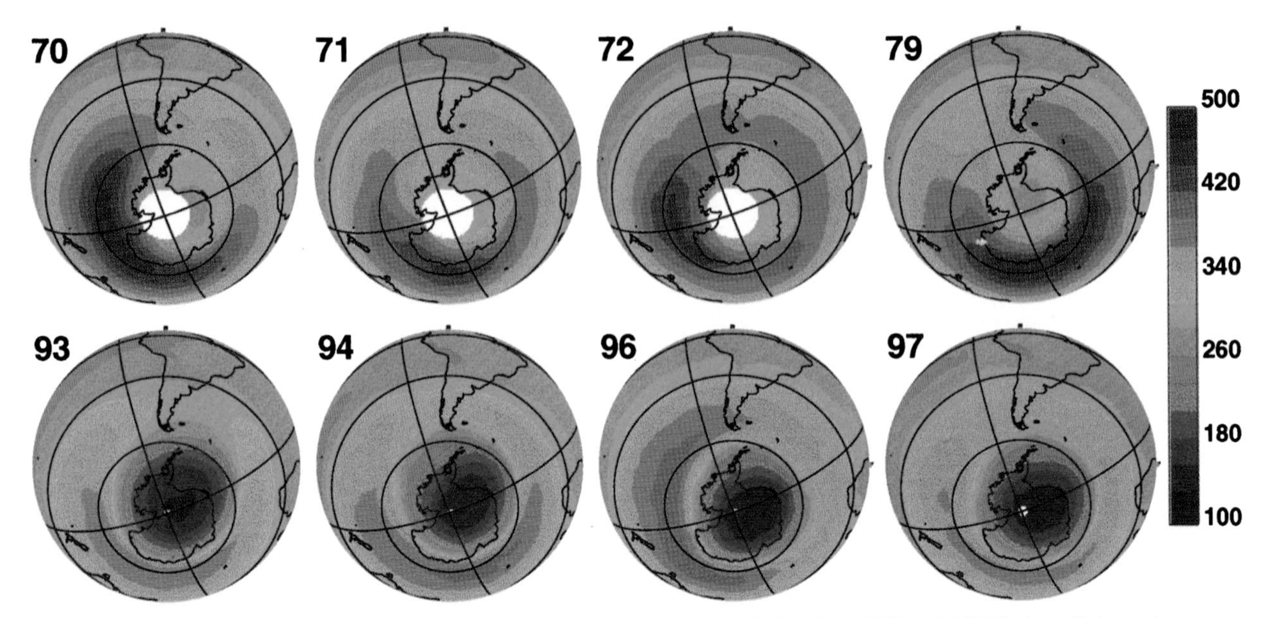

FIG. 6.36. Evolution of October average column ozone over the SH during 1970–97, from BUV and TOMS data. Color scale denotes values in DU. Note the strong depletion over Antarctica in the 1990s (the ozone hole) and reduced values over midlatitudes compared to the 1970s.

81, there was effectively zero area, with ozone levels below 220 DU). Jiang et al. (1996) present further details of ozone-hole interannual variability based on TOMS data.

The vertical profile of ozone depletion associated with the ozone hole is shown in Fig. 6.38, based on balloon-borne ozonesonde observation at the South Pole during 1992 and 1993 (Hofmann et al. 1994); the quantity shown is ozone partial pressure, proportional to mixing ratio times atmospheric density. The data in late August 1993 shows a column ozone amount near 270 DU, with a profile maximum over 15–20 km (approximately 100–50 mb). By the middle of October 1993, the ozone has substantially vanished over 13–20 km, and the column amount is 91 DU. Similar results are found for other years, demonstrating that the ozone hole is attributable to some process depleting ozone over this 13–20-km layer. Analyses of recent South Pole ozonesonde data compared with historical records (see WMO 1995 and Hofmann et al. 1997) clearly show that anomalous ozone loss begins in spring (September) and extends into early summer (December–January). Profiles show that this depletion occurs in the lower stratosphere, with strongest ozone losses occurring during the time period September–December.

2) THEORY OF CHEMICAL OZONE DEPLETION IN THE OZONE HOLE

The primary characteristics of the ozone hole (reviewed above) show strong ozone losses within the SH polar vortex in spring, primarily in the lower stratosphere. Departures of the column amount from climatological values near 300 DU are first evident in the mid-1970s (see Fig. 6.35); dramatic changes were first noted in the mid-1980s, and the ozone hole is now a regular feature in the SH stratosphere, with a near-complete loss of ozone in the lower stratosphere each year (Fig. 6.38). Although the causes of the ozone hole were strongly debated in the 1980s, a series of ground-based and aircraft measurement campaigns of the ozone-hole chemistry and dynamics have led to a scientific consensus that chemical ozone depletion associated with chlorine and bromine are responsible for the ozone hole (WMO 1992). Stratospheric chlorine primarily results from photochemical breakdown of chloroflurocarbons released in the troposphere from human activities, and the ozone hole is a remarkable signature of human climate influence on a planetary scale. Subsequent observations from aircraft and satellites, together with laboratory and modeling studies, have given better understanding to details of chemical ozone losses inside the ozone hole, as discussed in WMO (1995) and summarized briefly here.

The gas-phase catalytic cycles that destroy ozone over most latitudes are not effective over winter polar regions (particularly in the lower stratosphere, where the ozone hole occurs), because there is not enough sunlight to provide atomic oxygen [needed to complete the catalytic cycle in Eq. (6.10b)]. Rather, photochemical loss of ozone here is attributed to catalytic cycles involving enhanced levels of chlorine monoxide (ClO). The principal reactions involve the formation of the ClO dimer Cl_2O_2, followed by photolysis and thermal decomposition (Molina and Molina 1987):

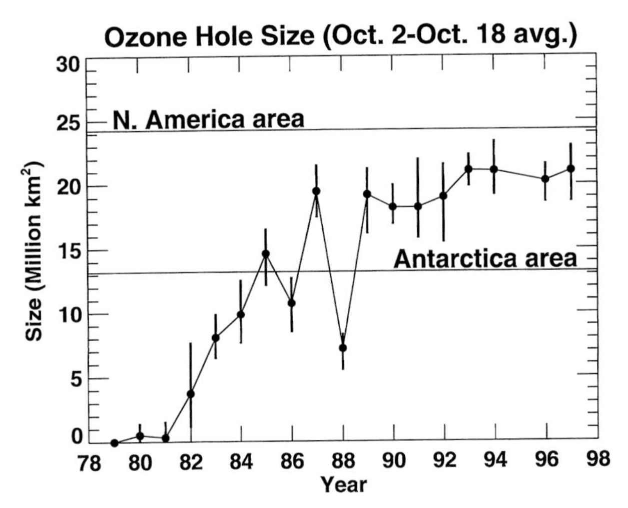

FIG. 6.37. Size of the antarctic ozone hole during 1979–97, as calculated by the area of column ozone with values under 220 DU (from TOMS data).

$$ClO + ClO + M \rightarrow Cl_2O_2 + M \quad (6.11a)$$

$$Cl_2O_2 + h\nu \rightarrow ClO_2 + Cl \quad (6.11b)$$

$$ClO_2 + M \rightarrow Cl + O_2 + M \quad (6.11c)$$

$$\frac{2(Cl + O_3 \rightarrow ClO + O_2)}{\text{net: } 2O_3 \rightarrow 3O_2}. \quad (6.11d)$$

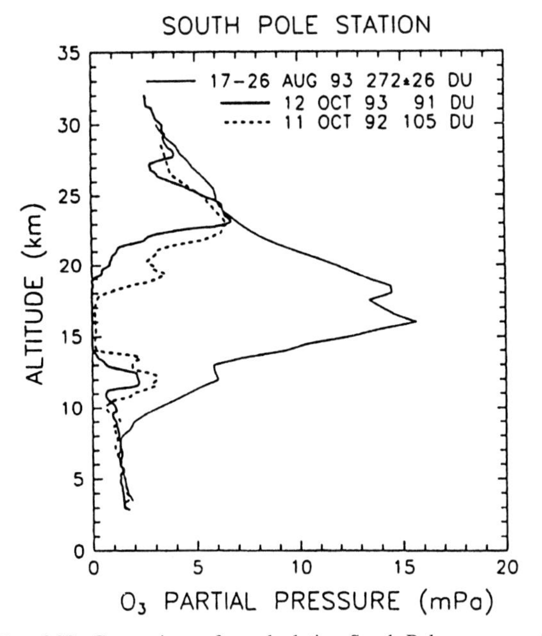

FIG. 6.38. Comparison of pre-depletion South Pole ozone profile in 1993, with the profile observed when total ozone reached a minimum in 1992 and 1993. (Adapted from Hofmann et al. 1994.)

Other reactions involve the interaction of ClO and bromine species (McElroy et al. 1986; see also Anderson et al. 1989; Solomon 1990; and WMO 1995). The Cl_2O_2 dimer reactions [Eq. (6.11)] account for approximately 80% of polar ozone losses (Solomon 1990). The rate-limiting step of these reactions is Eq. (6.11a), and the rate of chemical ozone loss is hence proportional to the amount of ClO squared (which helps explain the rapid increase in antarctic ozone depletion during the 1980s in response to the approximate linear increase in anthropogenic chlorine).

In order for these reactions to proceed, two important components must occur: 1) a change of chlorine from inactive, so-called "reservoir" species (namely $ClONO_2$ and HCl) into reactive chlorine (Cl_2, ClO, and Cl_2O_2); and 2) exposure to sunlight. These processes, which determine the location and timing of the ozone hole, are illustrated schematically in Fig. 6.39 [from WMO (1995), adapted from Webster et al. (1993)]. The key factor that occurs inside the SH polar vortex is the transformation from reservoir to reactive chlorine species. This occurs by chemical reactions occurring on the surfaces of particles (heterogeneous chemistry), rather than the more normal reaction of well-mixed gases (gas-phase or homogeneous chemistry) (Solomon et al. 1986a; Solomon 1988). One important heterogeneous reaction is

$$ClONO_2(g) + HCl(s) \rightarrow Cl_2(g) + HNO_3(s),$$

which converts the relatively benign reservoir species $ClONO_2$ and HCl into Cl_2, which can then be photolyzed by visible wavelength light (g, gas; s, solid).

The particles on which this and similar heterogeneous reactions occur are sulfate aerosols or polar stratospheric clouds (PSCs), which in turn form only at the very cold temperatures of the vortex interior (typically at temperatures below 195 K). It had long been recognized that such clouds form in the polar winter stratosphere (Stanford and Davis 1974), and that they occur with greater frequency and thickness in the Antarctic than in the Arctic (McCormick et al. 1982). The threshold formation and growth of PSC particles have been observed by balloon-borne measurements over Antarctica (Hofmann et al. 1989; Hofmann and Deshler 1989) and by aircraft measurements (Dye et al. 1990). A climatology of PSC observations derived from satellite measurements is shown in Fig. 6.40 (from Poole and Pitts 1994), indicating the frequent occurrence of PSCs inside the SH vortex in winter, primarily in the intensely cold lower stratosphere. Note the correspondence between the PSC probability in Fig. 6.40 and the polar temperature evolution in Fig. 6.5, and the much higher frequency of PSCs in the SH than in the NH due to the colder SH temperatures. The CLAES instrument on UARS was the first to obtain synoptic measurements (maps) of enhanced aerosol (PSC) levels over Antarctica, and

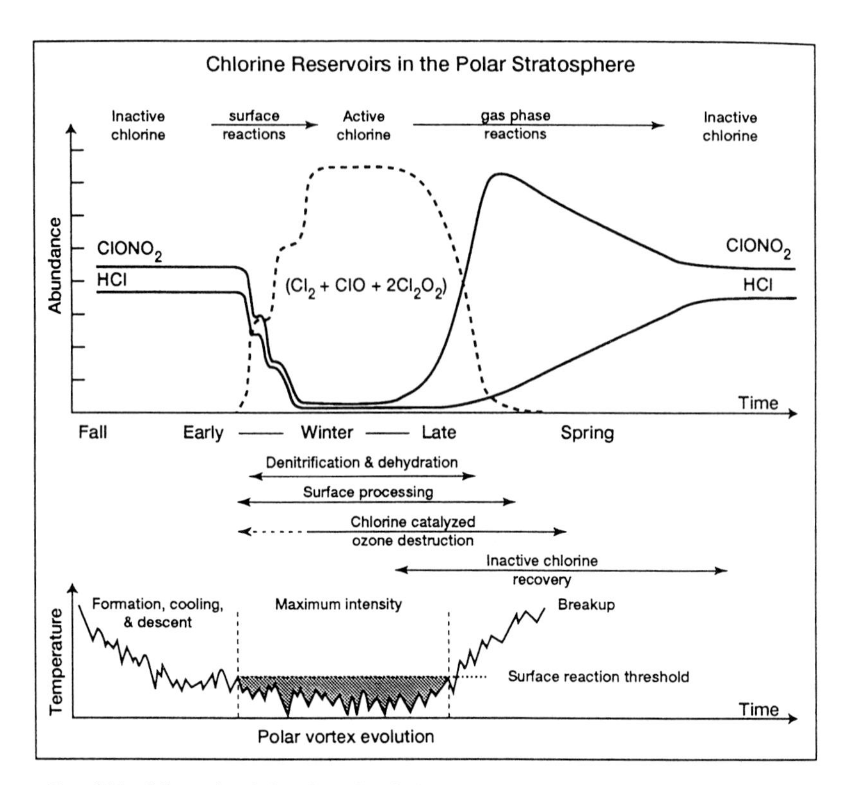

FIG. 6.39. Schematic of the photochemical and dynamical features of the polar regions related to ozone depletion. The upper panel represents the conversion of chlorine from inactive to active forms in winter in the lower stratosphere and the reformation of inactive forms in spring. The partitioning between the active chlorine species Cl_2, ClO, and Cl_2O_2 depends on exposure to sunlight after PSC processing. The corresponding stages of the polar vortex are indicated in the lower panel, where the temperature scale represents changes in the minimum polar temperatures in the lower stratosphere. (From WMO 1995; adapted from Webster et al. 1993.)

Fig. 6.41 shows examples of these data during SH winter 1992 (from Mergenthaler et al. 1997). Also shown in Fig. 6.41 are contours representing regions of temperature below 195 and 187 K (PSC threshold temperatures) from UKMO analyses. These CLAES data clearly show enhanced PSC amounts in polar regions throughout winter, generally coincident with regions of temperature below 195 K.

An additional important process in the SH involves the freezing out and sedimentation of nitric acid (HNO_3) in low-temperature PSCs. This results in an irreversible loss of reactive nitrogen inside the vortex (so-called denitrification), which in turn inhibits the reformation of the reservoir $ClONO_2$ from active chlorine (and hence maintains high levels of active chlorine). [A similar freezing out of water vapor inside the vortex leads to marked dehydration (Harwood et al. 1993; Tuck et al. 1993; Pierce et al. 1994), although this is not directly related to ozone chemistry.] By middle winter, most of the chlorine inside the vortex in the SH lower stratosphere is in the form of Cl_2 or Cl_2O_2, requiring only weak sunlight to initiate catalytic ozone loss.

When sunlight returns to the Antarctic in SH spring, ozone is rapidly destroyed in the lower stratosphere by catalytic cycles involving ClO and BrO (at the rate of approximately 1%–2% per day). This is why the ozone hole occurs in springtime. Some of the most compelling evidence for this sequence of events comes from direct observations of the ClO molecule (which occurs in very small amounts in the lower stratosphere away from aerosol- or PSC-processed air) and its anticorrelation with ozone. Figure 6.42 shows aircraft-based ClO and O_3 measurements covering a latitudinal sweep across the edge of the polar vortex (from Anderson et al. 1989). This shows a strong increase of ClO and decrease of O_3 inside the vortex (poleward of ~69°S) and strong anticorrelation for small-scale features near the vortex edge (over 67°–69°S). Figure 6.43 shows simultaneous observations of ClO and O_3 during September 1991 and 1992 from the MLS on UARS (Waters et al. 1993). These show strongly enhanced levels of ClO inside the SH polar vortex, with spatial patterns that are clearly mirrored in the low ozone values. Observations of other key chlorine and nitrogen species confirm the basic understanding

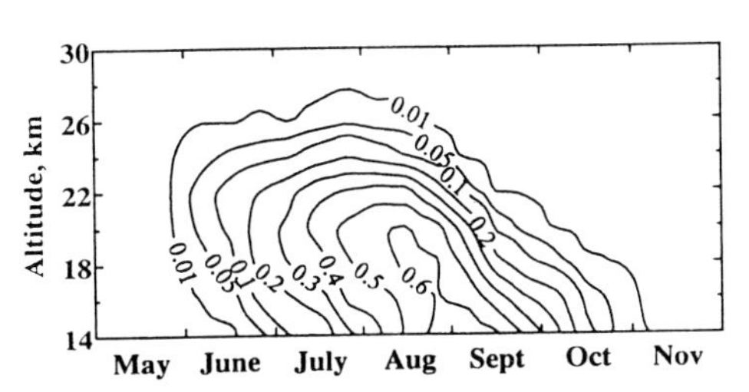

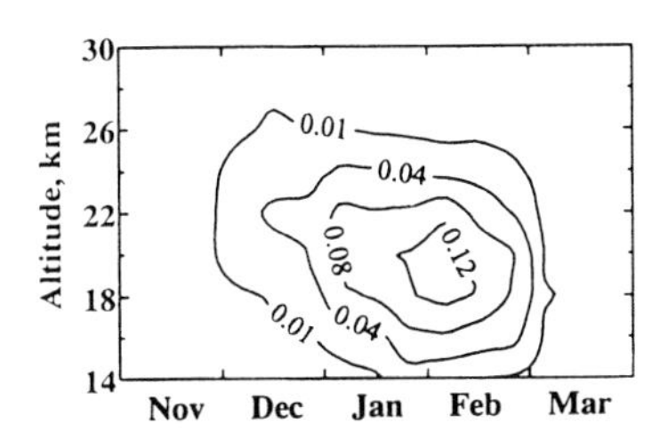

FIG. 6.40. PSC sighting probabilities derived from satellite data as function of altitude in the lower stratosphere during winter in the Antarctic (left) and Arctic (right). (From Poole and Pitts 1994.)

of the SH polar ozone budget (Solomon 1990; Santee et al. 1995; Douglass et al. 1995; Santee et al. 1996; Chipperfield et al. 1996). The most recent comprehensive modeling studies of antarctic ozone depletion, which incorporate comprehensive chemistry and aerosol microphysical parameterizations in realistic dynamical models (Brasseur et al. 1997; Portmann et al. 1996; Schoeberl et al. 1996), produce accurate simulations of observed ozone loss rates and reasonable estimates of observed interannual variability.

6.7. Looking to the future

Our study and understanding of the dynamics, radiation, and chemistry of the Southern Hemisphere stratosphere has increased considerably over the last decade, primarily motivated by the discovery of the antarctic ozone hole. Much of this rapid progress, however, resulted from the large body of observations compiled prior to the ozone hole discovery. These pre–ozone hole observations through the 1950s, 1960s, and 1970s in regions that appear to be pristine and that seemingly had little impact on human affairs justified the need for long-term monitoring and validated the foresight of the scientists who implemented these long-term observational programs.

There are a variety of open questions concerning the Southern Hemisphere stratosphere that require further research. The foremost question concerns the recovery of springtime antarctic ozone levels as stratospheric chlorine levels decrease in response to the regulations on chlorine-containing compounds, via the Montreal protocol and the amendments to that protocol. Continuous global monitoring of these substances now indicate that the total ozone-destroying potential peaked in the troposphere in 1994 and will peak in the stratosphere between 1997 and 1999 (Montzka et al. 1996). A slow healing process would then begin, with stratospheric chlorine levels reaching ~1980 values (associ-

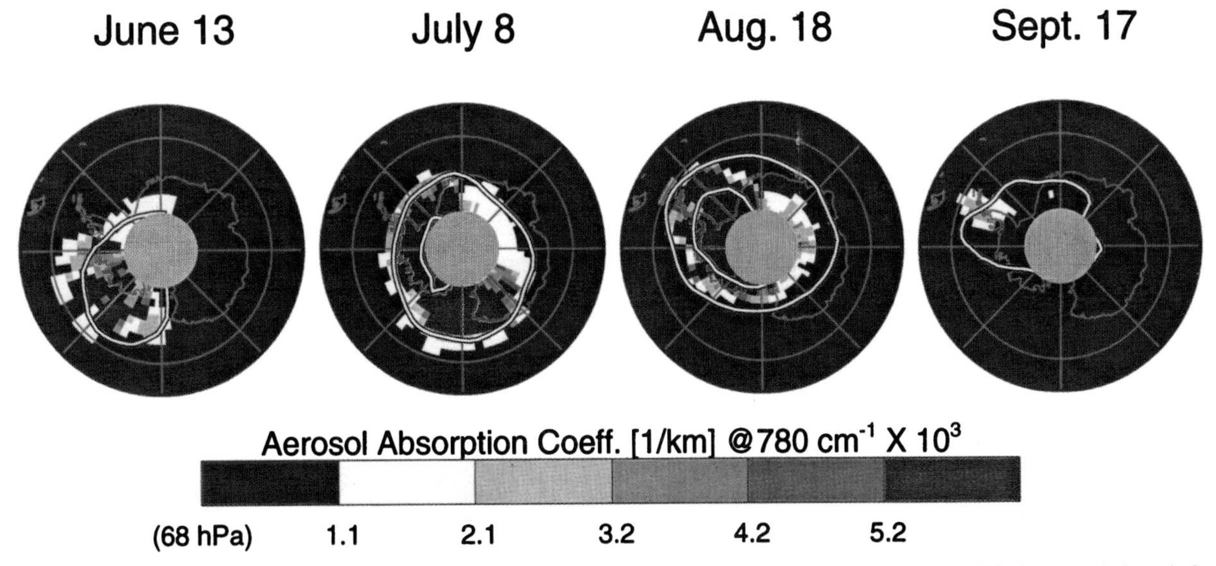

FIG. 6.41. Aerosol absorption coefficient at 68 mb (~19 km), measured by the CLAES instrument on UARS, for several days during SH winter 1992. The enhanced signals over polar regions are due to PSCs. White contours denote regions of UKMO temperature analyses below 195 and 187 K, respectively. No CLAES data are available over the pole (grey area). (Adapted from Mergenthaler et al. 1997.)

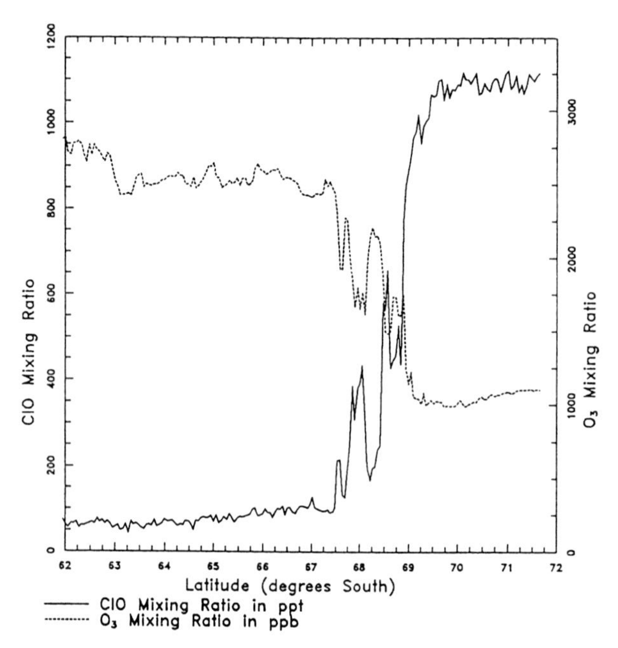

FIG. 6.42. Simultaneously observed ClO and O_3 obtained on 16 September 1987 by the ER-2 aircraft over the Antarctic. (From Anderson et al. 1989.)

ated with emergence of the ozone hole) in about the year 2050 (WMO 1995). Hofmann et al. (1997) propose that early healing may be detectable from ozonesonde data as early as 2004. Additional questions concern climate-change impacts on the southern stratosphere in response to increased greenhouse gas loading and ozone depletion. Ramaswamy et al. (1996) have recently shown that much of the long-term cooling of the lower stratosphere (see Figs. 6.25–6.26) is consistent with a radiative response to observed ozone depletion; moreover, the effect on lower-stratosphere temperatures of ozone losses during the last two decades outweighs the radiative effects of changes in other greenhouse gases over the past two centuries! The details of planetary-wave variability in the southern stratosphere require continued analysis and study for understanding the characteristics of the SH polar vortex and the polar-vortex breakup. Additionally, many aspects of trace constituent transport and variability remain to be explored in the SH stratosphere; one intriguing question concerns why the SH midlatitude lower stratosphere is substantially drier than that in the NH (Kelly et al. 1990; Tuck et al. 1993; Rosenlof et al. 1997). These questions point toward a continuation and improvement of observations in the SH, particularly at high latitudes.

6.8. Summary

The stratospheric polar vortex in the SH is colder, with more intense zonal winds and potential vorticity gradients, than that in the NH. The SH vortex is also more stable on a daily basis (with an absence of midwinter breakdowns) and persists longer into springtime than its NH counterpart. The fundamental reason for this difference is the smaller amplitude of large-scale wave forcing from the troposphere in the SH (due to absence of topography and continental heat sources found in the NH).

The intense polar vortex in the SH autumn, winter, and spring exerts a strong influence on transport and mixing of stratospheric trace constituents. There is minimal horizontal transport across the vortex edge, so that the vortex interior is effectively isolated from midlatitudes; this is clearly demonstrated in observations of long-lived constituents (see Figs. 6.8–6.9). The vortex structure is also a dominant factor in the seasonal evolution of ozone, both from a transport and chemical perspective (see also Schoeberl and Hartmann 1991).

The ozone hole is now a climatological feature of the SH stratosphere. There is a near-complete loss of ozone in the layer ~13–20 km throughout the polar vortex in spring. Observations of key chemical species (namely ClO) show conclusively that the antarctic ozone loss is due to chemical depletion, linked to the buildup of anthropogenic chlorine emissions. Chemical ozone loss in the antarctic ozone hole is tied to heterogeneous reactions occurring on the surface of PSCs. These PSCs occur only in the intense cold of the antarctic lower stratosphere; the return of sunlight in spring (completing the photochemical catalytic cycles) explains why the ozone hole occurs over Antarctica in spring. Analyses of long-term records show a substantial cooling of the SH lower stratosphere in spring (of order 6–10 K), and this is likely a radiative response to the depleted ozone levels. The antarctic stratosphere thus provides the most conclusive evidence of the effect of humans on the climate system.

Acknowledgments. A significant portion of this work was completed while WJR was on sabbatical leave at the Cooperative Research Centre for Southern Hemisphere Meteorology at Monash University, Melbourne; thanks to Prof. David Karoly and the CRC staff for hospitality and a professional work environment. Annita Allman (CRC) and Marilena Stone (NCAR) expertly prepared the manuscript. The National Center for Atmospheric Research is sponsored by the National Science Foundation; partial support was also provided to WJR by NASA grants W-16215 and W-18181. The authors thank Rolando Garcia, Kevin Hamilton, Jim Holton, Bill Mankin, Alan O'Neill, Mark Schoeberl, John Stanford, and Darryn Waugh for discussions throughout the course of the work and for comments on the manuscript, and Prof. Donald Johnson for a careful review of the text.

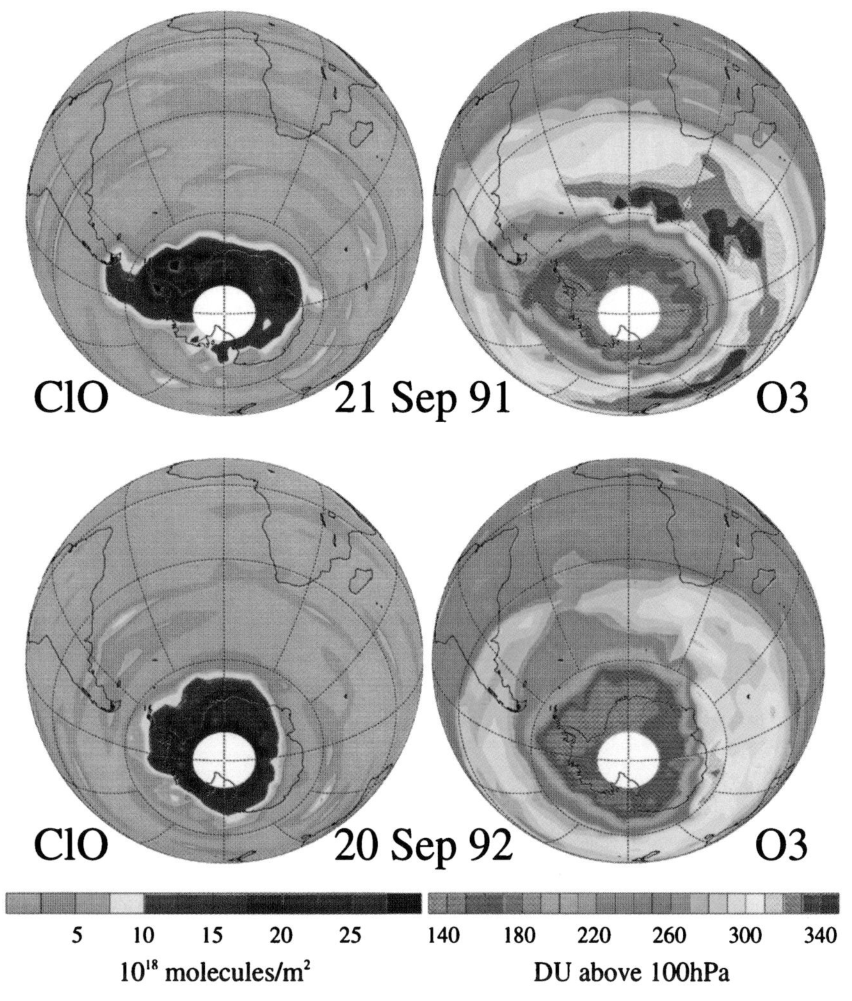

FIG. 6.43. Satellite measurements of ClO (left panels) and column ozone (right panels) during September 1991 and 1992, from the MLS instrument on UARS. (From Waters et al. 1993.)

APPENDIX
Notation

(u, v, w) — three-dimensional velocity components
$(\overline{v}^*, \overline{w}^*)$ — residual mean meridional circulation
T — temperature
θ — potential temperature [Eq. (6.6)]
(ϕ, λ) — latitude, longitude
z — log-pressure vertical coordinate [$z = -H \ln(p/p_s)$]
ρ_o — background atmospheric density ($\rho_o = \rho_{\text{surface}}\, e^{-z/H}$)
p — pressure
PV — potential vorticity [Eq. (6.7)]
N^2 — static stability parameter (Brunt–Väisälä frequency) $N^2 = (R/H)((\partial T/\partial z) + (2/7)(T/H))$
G — zonal momentum forcing [Eq. (6.1)]
$\mathbf{F}$ — Eliassen–Palm flux vector [Eq. (6.5)]
$\rho_o \nabla \cdot \boldsymbol{F}$ — EP flux divergence (large-scale zonal momentum forcing)
X — zonal momentum forcing from unresolved scales
Q — radiative heating/cooling rate
Ω — earth rotation rate ($7.3\ 10^{-5}$ s^{-1})
H — scale height (7 km)
a — earth radius ($6.37\ 10^6$ m)
R — gas constant (287 K^{-1} m^2 s^{-2})
$\bar{A}$ — zonal mean of A
A' — deviation of A from zonal mean

REFERENCES

Allen, D. R., J. L. Stanford, L. S. Elson, E. F. Fishbein, L. Froidevaux, and J. W. Waters, 1997: The 4-day wave as observed from the Upper Atmosphere Research Satellite microwave limb sounder. *J. Atmos. Sci.*, **54,** 420–434.

Anderson, J. G., W. H. Brune, S. A. Lloyd, D. W. Toohey, S. P. Sander, W. L. Starr, M. Loewenstein, and J. R. Podolske, 1989: Kinetics of O_3 destruction by ClO and BrO within the antarctic vortex: An analysis based on in situ ER-2 data. *J. Geophys. Res.*, **94,** 11 480–11 520.

Andrews, D. G., and M. E. McIntyre, 1978: Generalized Eliassen–Palm and Charney–Drazin theorems for waves on axisymmet-

ric mean flows in compressible atmospheres. *J. Atmos. Sci.*, **35,** 175–185.
———, J. R. Holton, and C. B. Leovy, 1987: *Middle Atmosphere Dynamics.* Academic Press, 489 pp.
Atkinson, R. J., and J. R. Easson, 1989: Reevaluation of the Australian total ozone data record. *Proc. Quadr. Ozone. Symp.*, Gottingen, Germany, A. Deepak, 168–171.
———, W. A. Matthews, P. A. Newman, and R. A. Plumb, 1989: Evidence of the mid-latitude impact of antarctic ozone depletion. *Nature*, **340,** 290–294.
Baldwin, M. P., and J. R. Holton, 1988: Climatology of the stratospheric polar vortex and planetary wave breaking. *J. Atmos. Sci.*, **45,** 1123–1142.
———, and D. O'Sullivan, 1995: Stratospheric effects of ENSO-related tropospheric circulation anomalies. *J. Climate.*, **8,** 649–667.
Bojkov, R. D., and V. E. Fioletov, 1995: Estimating the global ozone characteristics during the last 30 years. *J. Geophys. Res.*, **100,** 16 537–16 551.
Boville, B. A., 1995: Middle atmosphere version of CCM2 (MACCM2): Annual cycle and interannual variability. *J. Geophys. Res.*, **100,** 9017–9040.
Bowman, K. P., 1989: Global patterns of the quasi-biennial oscillation in total ozone. *J. Atmos. Sci.*, **46,** 3328–3343.
———, 1993a: Barotropic simulation of large-scale mixing in the antarctic polar vortex. *J. Atmos. Sci.*, **50,** 2901–2914.
———, 1993b: Large-scale isentropic mixing properties of the antarctic polar vortex from analyzed winds. *J. Geophys. Res.*, **98,** 23 013–23 027.
———, and N. J. Mangus, 1993: Observations of deformation and mixing of the total ozone field in the antarctic polar vortex. *J. Atmos. Sci.*, **50,** 2915–2921.
Boyd, J., 1976: The noninteraction of waves with the zonally averaged flow on a spherical earth and the interrelationship of eddy fluxes of energy, heat, and momentum. *J. Atmos. Sci.*, **33,** 2285–2291.
Brasseur, G. P., X. Tie, and P. J. Rasch, 1997: A three-dimensional simulation of the current and pre-industrial antarctic ozone and its impact on mid-latitudes and upper troposphere. *J. Geophys. Res.*, **102,** 8909–8930.
Buchart, N., and E. Remsberg, 1986: The area of the stratospheric polar vortex as a diagnostic for tracer transport on isentropic surfaces. *J. Atmos. Sci.*, **43,** 1319–1339.
———, and J. Austin, 1996: On the relationship between the quasi-biennial oscillation, total chlorine and the severity of the antarctic ozone hole. *Quart. J. Roy. Meteor. Soc.*, **122,** 183–217.
Burks, D., and C. B. Leovy, 1986: Planetary waves near the mesospheric easterly jet. *Geophys. Res. Lett.*, **13,** 193–196.
Cariolle, D., A. Lasserre-Bigorry, and J. F. Boyer, 1990: A general circulation model simulation of the springtime antarctic ozone decrease and its impact on mid-latitudes. *J. Geophys. Res.*, **95,** 1883–1898.
Chapman, S., 1930: On ozone and atomic oxygen in the upper atmosphere. *Philos. Mag.*, **10,** 369–376.
Charney, J. G., and P. G. Drazin, 1961: Propagation of planetary-scale disturbances from the lower into the upper atmosphere. *J. Geophys. Res.*, **66,** 83–109.
Chen, P., 1994: The permeability of the antarctic vortex edge. *J. Geophys. Res.*, **99,** 20 563–20 571.
———, J. R. Holton, A. O'Neill, and R. Swinbank, 1994: Quasi-horizontal transport and mixing in the antarctic stratosphere. *J. Geophys. Res.*, **99,** 16 851–16 855.
Chipperfield, M. P., M. L. Santee, L. Froidevaux, G. L. Manney, W. G. Read, J. W. Waters, A. E. Roche, and J. M. Russell, 1996: Analysis of UARS data in the southern polar vortex in September 1992 using a chemical transport model. *J. Geophys. Res.*, **101,** 18 861–18 881.
Dickinson, R. E., 1969: Theory of planetary wave–zonal flow interaction. *J. Atmos. Sci.*, **26,** 73–81.
Douglass, A. R., M. R. Schoeberl, R. S. Stolarski, J. W. Waters, J. M. Russell III, A. E. Roches, and S. T. Massie, 1995: Interhemispheric differences in springtime production of HCl and $ClONO_2$ in the polar vortices. *J. Geophys. Res.*, **100,** 13 967–13 978.
Dunkerton, T. J., 1978: On the mean meridional mass motions of the stratosphere and mesosphere. *J. Atmos. Sci.*, **35,** 2325–2333.
———, 1988: Body force circulation and the antarctic ozone minimum. *J. Atmos. Sci.*, **45,** 427–438.
———, and M. P. Baldwin, 1991: Modes of interannual variability in the stratosphere. *Geophys. Res. Lett.*, **19,** 49–52.
Dye, J., B. Gandrud, D. Baumgardner, K. Chan, G. Ferry, M. Loewenstein, K. Kelly, and J. Wilson, 1990: Observed particle evolution in the polar stratospheric cloud of January 24, 1989. *Geophys. Res. Lett.*, **17,** 413–416.
Edmon, H. J., B. J. Hoskins, and M. E. McIntyre, 1980: Eliassen–Palm cross sections for the troposphere. *J. Atmos. Sci.*, **37,** 2600–2616.
Eluszkiewicz, J., R. A. Plumb, and N. Nakamura, 1995: Dynamics of wintertime stratospheric transport in the Geophysical Fluid Dynamics Laboratory SKYHI general circulation model. *J. Geophys. Res.*, **100,** 20 883–20 900.
Farman, J. C., B. G. Gardiner, and J. D. Shanklin, 1985: Large losses of total ozone in Antarctica reveal seasonal ClO_x/NO_x interaction. *Nature*, **315,** 207–210.
Finger, F. G., M. E. Gelman, J. D. Wild, M. L. Chanin, A. Hauchecorne, and A. J. Miller, 1993: Evaluation of NMC upper-stratospheric temperature analyses using rocketsonde and lidar data. *Bull. Amer. Meteor. Soc.*, **74,** 789–799.
Fisher, M., A. O'Neill, and R. Sutton, 1993: Rapid descent of mesospheric air in the stratospheric polar vortex. *Geophys. Res. Lett.*, **20,** 1267–1270.
Froidevaux, L., J. W. Waters, W. G. Read, L. S. Elson, D. A. Flower, and R. F. Jarnot, 1994: Global ozone observations from the UARS MLS: An overview of zonal-mean results. *J. Atmos. Sci.*, **51,** 2846–2866.
Garcia, R. R., 1987: On the mean meridional circulation of the middle atmosphere. *J. Atmos. Sci.*, **44,** 3599–3609.
———, and S. Solomon, 1983: A numerical model of the zonally averaged dynamical and chemical structure of the middle atmosphere. *J. Geophys. Res.*, **88,** 1379–1400.
———, and B. A. Boville, 1994: "Downward control" of the mean meridional circulation and temperature distribution of the polar winter stratosphere. *J. Atmos. Sci.*, **51,** 2238–2245.
Geller, M. A., and M. F. Wu, 1987: Troposphere–stratosphere general circulation statistics. *Transport Processes in the Middle Atmosphere.* Reidel Publishers, 3–17.
Gelman, M. E., and R. M. Nagatani, 1977: Objective analyses of height and temperature at the 5-, 2- and 0.4-mb levels using meteorological rocketsonde and satellite radiation data. *Space Research*, **17,** 117–122.
———, A. J. Miller, K. W. Johnson, and R. M. Nagatani, 1986: Detection of long term trends in global stratospheric temperature from NMC analyses derived from NOAA satellite data. *Adv. Space Res.*, **6**(10), 17–26.
Gille, J. C., and L. V. Lyjak, 1987: Radiative heating and cooling rates in the middle atmosphere. *J. Atmos. Sci.*, **43,** 2215–2229.
Gray, L. T., and T. J. Dunkerton, 1990: The role of the seasonal cycle in the quasi-biennial oscillation in ozone. *J. Atmos. Sci.*, **47,** 2429–2451.
Harris, N. R. P., and Coauthors, 1997: Trends in stratospheric and tropospheric ozone. *J. Geophys. Res.*, **102,** 1571–1590.
Hartley, W. N., 1880: On the probable absorption of solar radiation by atmospheric ozone. *Chem. News*, **42,** 268–274.
Hartmann, D. L., 1976: The structure of the stratosphere in the Southern Hemisphere during late winter 1973 as observed by satellite. *J. Atmos. Sci.*, **33,** 1141–1154.
———, 1983: Barotropic instability of the polar night jet stream. *J. Atmos. Sci.*, **40,** 817–835.

———, C. R. Mechoso, and K. Yamazaki, 1984: Observations of wave-mean flow interaction in the Southern Hemisphere. *J. Atmos. Sci.*, **41,** 351–362.

———, K. R. Chan, B. L. Gary, M. R. Schoeberl, P. A. Newman, R. L. Martin, M. Loewenstein, J. R. Podolske, and S. E. Strahan, 1989: Potential vorticity and mixing in the south polar vortex during spring. *J. Geophys. Res.*, **94,** 11 625–11 640.

Harwood, R. S., 1975: The temperature structure of the Southern Hemisphere stratosphere August–October 1971. *Quart. J. Roy. Meteor. Soc.*, **101,** 75–91.

———, and Coauthors, 1993: Springtime stratospheric water vapor in the Southern Hemisphere as measured by MLS. *Geophys. Res. Lett.*, **20,** 1235–1238.

Haynes, P. H., C. J. Marks, M. E. McIntyre, T. G. Shepherd, and K. P. Shine, 1991: On the "downward control" of extratropical diabatic circulations by eddy-induced mean zonal forces. *J. Atmos. Sci.*, **48,** 651–678.

Herman, J. R., and Coauthors, 1995: *Meteor 3* total ozone mapping spectrometer observations of the 1993 ozone hole. *J. Geophys. Res.*, **100,** 2973–2984.

Hirota, I., and T. Hirooka, 1984: Normal-mode Rossby waves observed in the upper stratosphere. Part I: First symmetric modes of zonal wavenumbers 1 and 2. *J. Atmos. Sci.*, **41,** 1253–1267.

Hitchman, M. H., J. C. Gille, C. D. Rodgers, and G. P. Grasseur, 1989: The separated polar winter stratopause: A gravity wave–driven climatological feature. *J. Atmos. Sci.*, **46,** 410–422.

Hofmann, D. J., and T. Deshler, 1989: Comparison of stratospheric clouds in the Antarctic and Arctic. *Geophys. Res. Lett.*, **16,** 1429–1432.

———, J. M. Rosen, J. W. Harder, and J. V. Hereford, 1989: Balloon-borne measurements of aerosol, condensation nuclei, and cloud particles in the stratosphere at McMurdo Station, Antarctica, during the spring of 1987. *J. Geophys. Res.*, **94,** 11 253–11 269.

———, S. J. Oltmans, J. A. Lathrop, J. M. Harris, and H. Voemel, 1994: Record low ozone at the South Pole in the spring of 1993. *Geophys. Res. Lett.*, **21,** 421–424.

———, ———, J. M. Harris, B. J. Johnson, and J. A. Lathrop, 1997: Ten years of ozonesonde measurements at the South Pole: Implications for recovery of springtime antarctic ozone loss. *J. Geophys. Res.*, **102,** 8931–8943.

Hollandsworth, S. M., R. D. McPeters, L. E. Plynn, W. Planet, A. J. Miller, and S. Chandra, 1995: Ozone trends deduced from combined *Nimbus 7* SBUV and *NOAA 11* SBUV/2 data. *Geophys. Res. Lett.*, **22,** 905–908.

Holton, J. R., 1983: The influence of gravity wave breaking on the general circulation of the middle atmosphere. *J. Atmos. Sci.*, **40,** 2497–2507.

———, and H.-C. Tan, 1980: The influence of the equatorial quasi-biennial oscillation on the global circulation at 50 mb. *J. Atmos. Sci.*, **37,** 2200–2208.

———, and ———, 1982: The quasi-biennial oscillation in the Northern Hemisphere lower stratosphere. *J. Meteor. Soc. Japan*, **60,** 140–148.

———, and W.-K. Choi, 1988: Transport circulation deduced from SAMS trace species data. *J. Atmos. Sci.*, **45,** 1929–1939.

———, P. H. Haynes, M. E. McIntyre, A. R. Douglas, R. B. Rood, and L. Pfister, 1995: Stratosphere–troposphere exchange. *Rev. Geophys.*, **33,** 403–439.

Hoskins, B. J., M. E. McIntyre, and A. W. Robertson, 1985: On the use and significance of isentropic potential vorticity maps. *Quart. J. Roy. Meteor. Soc.*, **111,** 877–946.

Hurrell, J. W., and H. van Loon, 1994: A modulation of the atmospheric annual cycle in the Southern Hemisphere. *Tellus*, **46A,** 325–338.

Jiang, Y., Y. L. Yung, and R. W. Zurek, 1996: Decadal evolution of the antarctic ozone hole. *J. Geophys. Res.*, **101,** 8985–8999.

Johnston, H. S., 1971: Reduction of stratospheric ozone by nitrogen oxide catalysts from SST exhaust. *Science*, **173,** 517–522.

Jones, A. E., and J. D. Shanklin, 1995: Continued decline of total ozone over Halley, Antarctica since 1985. *Nature*, **376,** 409–411.

Jones, R. L., and J. A. Pyle, 1984: Observations of CH_4 and N_2O by the *NIMBUS-7* SAMS: A comparison with in-situ data and two-dimensional numerical model calculations. *J. Geophys. Res.*, **89,** 5263–5279.

Juckes, M. N., and M. E. McIntyre, 1987: A high resolution, one-layer model of breaking planetary waves in the stratosphere. *Nature*, **328,** 590–596.

Kalnay, E., and Coauthors, 1996: The NCEP/NCAR 40-year reanalysis project. *Bull. Amer. Meteor. Soc.*, **77,** 437–471.

Karoly, D. J., and B. J. Hoskins, 1982: Three-dimensional propagation of planetary waves. *J. Meteor. Soc. Japan*, **60,** 109–123.

Kelly, K. K., A. F. Tuck, L. E. Heidt, M. Loewenstein, J. F. Podolske, and S. E. Strahan, 1990: Comparison of ER-2 measurements of stratospheric water vapor between the 1987 antarctic and 1989 arctic airborne missions. *Geophys. Res. Lett.*, **17,** 465–468.

Kiehl, J. T., B. A. Boville, and B. P. Briegleb, 1988: Response of a general circulation model to a prescribed antarctic ozone hole. *Nature*, **332,** 501–504.

Kumer, J. B., J. L. Mergenthaler, and A. E. Roche, 1994: CLAES CH_4, N_2O and CCl_2F_2 (F12) global data. *Geophys. Res. Lett.*, **20,** 1239–1242.

Labitzke, K., 1982: On the interannual variability of the middle stratosphere during the northern winters. *J. Meteor. Soc. Japan*, **60,** 124–139.

———, and H. van Loon, 1972: The stratosphere in the Southern Hemisphere. *Meteorology of the Southern Hemisphere, Meteor. Monogr.*, No. 35, Amer. Meteor. Soc., 113–138.

———, and ———, 1987: Association between the 11-year solar cycle, the QBO and the atmosphere. Part I: The troposphere and the stratosphere in the Northern Hemisphere in winter. *J. Amer. Terr. Phys.*, **50,** 197–206.

———, and ———, 1995: Connection between the troposphere and stratosphere on a decadal time scale. *Tellus*, **47A,** 275–286.

Lahoz, W. A., and Coauthors, 1996: Vortex dynamics and the evolution of water vapor in the stratosphere of the Southern Hemisphere. *Quart. J. Roy. Meteor. Soc.*, **122,** 423–450.

Lait, L. R., and J. L. Stanford, 1988: Fast, long-lived features in the polar stratosphere. *J. Atmos. Sci.*, **45,** 3800–3809.

———, M. R. Schoeberl, and P. A. Newman, 1989: Quasi-biennial modulation of the antarctic ozone depletion. *J. Geophys. Res.*, **94,** 11 559–11 571.

Lawrence, B. N., and W. J. Randel, 1996: Variability in the mesosphere observed by the *NIMBUS 6* PMR. *J. Geophys. Res.*, **101,** 23 475–23 489.

Leovy, C. B., and P. J. Webster, 1976: Stratospheric long waves: Comparison of thermal structure in the Northern and Southern Hemispheres. *J. Atmos. Sci.*, **33,** 1624–1638.

———, C.-R. Sun, M. H. Hitchmann, E. E. Remsberg, J. M. Russell III, L. L. Gordley, J. M. Gille, and L. V. Lyjak, 1985: Transport of ozone in the middle stratosphere: Evidence for planetary-wave breaking. *J. Atmos. Sci.*, **42,** 230–244.

Limpasuvan, V., and C. B. Leovy, 1995: Observations of the two-day wave near the southern summer stratopause. *Geophys. Res. Lett.*, **22,** 2385–2388.

Lindzen, R. S., 1981: Turbulence and stress owing to gravity wave and tidal breakdown. *J. Geophys. Res.*, **86,** 9707–9714.

Mahlman, J. D., and L. J. Umscheid, 1987: Comprehensive modeling of the middle atmosphere: The influence of horizontal resolution. *Transport Processes in the Middle Atmosphere*, G. Visconti, Ed., D. Reidel Publishing Co., 251–256.

———, J. P. Pinto, and L. J. Umscheid, 1994: Transport, radiative, and dynamical effects of the antarctic ozone hole: A GFDL "SKYHI" model experiment. *J. Atmos. Sci.*, **51,** 489–508.

Manney, G. L., 1991: The stratospheric 4-day wave in NMC data. *J. Atmos. Sci.*, **48,** 1798–1811.

———, and W. J. Randel, 1993: Instability at the winter stratopause: A mechanism for the 4-day wave. *J. Atmos. Sci.*, **50,** 3928–3938.

———, C. R. Mechoso, L. S. Elson, and J. D. Farrara, 1991a: Planetary-scale waves in the Southern Hemisphere winter and early spring stratosphere: Stability analysis. *J. Atmos. Sci.*, **48,** 2509–2523.

———, J. D. Farrara, and G. R. Mechoso, 1991b: The behavior of wave 2 in the Southern Hemisphere stratosphere during late winter and early spring. *J. Atmos. Sci.*, **48,** 976–998.

———, R. W. Zurek, A. O'Neill, and R. Swinbank, 1994: On the motion of air through the stratospheric polar vortex. *J. Atmos. Sci.*, **51,** 2973–2994.

———, L. Froidevaux, J. W. Waters, and R. W. Zurek, 1995: Evolution of microwave limb sounder ozone and the polar vortex during winter. *J. Geophys. Res.*, **100,** 2953–2972.

———, R. Swinbank, S. T. Massie, M. E. Gelman, A. J. Miller, R. Nagatani, A. O'Neill, and R. W. Zurek, 1996: Comparison of UKMO and NMC stratospheric analyses during northern and southern winter. *J. Geophys. Res.*, **101,** 10 311–10 334.

Matsuno, T., 1970: Vertical propagation of stationary planetary waves in the winter Northern Hemisphere. *J. Atmos. Sci.*, **27,** 871–883.

McCormick, M. P., H. M. Steele, P. Hamill, W. P. Chu, and T. J. Swissler, 1982: Polar stratospheric cloud sightings by SAM II. *J. Atmos. Sci.*, **39,** 1387–1397.

McElroy, M. B., R. J. Salawitch, S. C. Wofsy, and J. A. Logan, 1986: Reductions of antarctic ozone due to synergistic interactions of chlorine and biomine. *Nature*, **321,** 759–762.

McIntyre, M. E., 1992: Atmospheric dynamics: Some fundamentals, with observational implications. *The Use of EOS for Studies of Atmospheric Physics*, J. C. Gille and G. Visconti, Eds., North–Holland, 313–386.

———, 1995: The stratospheric polar vortex and sub-vortex: Fluid dynamics and midlatitude ozone loss. *Phil. Trans. R. Soc. Lond.*, **352,** 227–240.

———, and T. N. Palmer, 1983: Breaking planetary waves in the stratosphere. *Nature*, **305,** 593–600.

———, and ———, 1984: The "surf zone" in the stratosphere. *J. Atmos. Terr. Phys.*, **46,** 825–849.

McPeters, R., and Coauthors, 1993: *NIMBUS-7* total ozone mapping spectrometer (TOMS) data products users guide. NASA Ref. Pub. 1323, 89 pp.

———, S. M. Hollandsworth, L. E. Flynn, J. R. Herman, and C. J. Seftor, 1996: Long-term ozone trends derived from the 16-year combined *Nimbus 7/Meteor 3* TOMS version 7 record. *Geophys. Res. Lett.*, **23,** 3699–3702.

Mechoso, C. R., and D. L. Hartmann, 1982: An observational study of traveling planetary waves in the Southern Hemisphere. *J. Atmos. Sci.*, **39,** 1921–1935.

———, ———, and J. D. Farrara, 1985: Climatology and interannual variability of wave, mean-flow interaction in the Southern Hemisphere. *J. Atmos. Sci.*, **42,** 2189–2206.

———, A. O'Neill, V. D. Pope, and J. D. Farrara, 1988: A study of the stratospheric final warming of 1982 in the Southern Hemisphere. *Quart. J. Roy. Meteor. Soc.*, **114,** 1365–1384.

Mergenthaler, J. L., J. B. Kumer, A. E. Roche, and S. T. Massie, 1997: Distribution of antarctic polar stratospheric clouds as seen by the CLAES experiment. *J. Geophys. Res.*, **102,** 19 161–19 170.

Molina, L. T., and M. J. Molina, 1987: Production of Cl_2O_2 from the self-reaction of the ClO radical. *J. Phys. Chem.*, **91,** 433–436.

Molina, M. J., and F. S. Rowland, 1974: Stratospheric sink for chlorofluoromethanes: Chlorine atom–catalyzed destruction of ozone. *Nature,* **249,** 810–812.

Montzka, S. A., J. H. Butler, R. C. Meyers, T. M. Thompson, T. H. Swanson, A. D. Clark, L. T. Locke, and J. W. Elkins, 1996: Decline in the tropospheric abundance of halogen from halocarbons: Implications for stratospheric ozone depletion. *Science*, **272,** 1318–1322.

Nash, E. R., P. A. Newman, J. E. Rosenfield, and M. R. Schoeberl, 1996: An objective determination of the polar vortex using Ertel's potential vorticity. *J. Geophys. Res.*, **101,** 9471–9478.

Newman, P. A., 1986: The final warming and polar vortex disappearance during the Southern Hemisphere spring. *Geophys. Res. Lett.,* **13,** 1228–1231.

———, and W. J. Randel, 1988: Coherent ozone-dynamical changes during the Southern Hemisphere spring, 1979–1986. *J. Geophys. Res.*, **93,** 12 585–12 606.

———, and M. R. Schoeberl, 1995: A reinterpretation of the data from the NASA stratosphere–troposphere exchange project. *Geophys. Res. Lett.*, **22,** 2501–2504.

North, G. R., F. J. Moeng, T. L. Bell, and R. F. Calahan, 1982: The latitude dependence of the variance of zonally averaged quantities. *Mon. Wea. Rev.*, **110,** 319–326.

Norton, W. A., 1994: Breaking Rossby waves in a model stratosphere diagnosed by a vortex-following coordinate system and a technique for advancing material contours. *J. Atmos. Sci.*, **51,** 654–673.

Olaguer, E. P., H. Yang, and K. K. Tung, 1992: A reexamination of the radiative balance of the stratosphere. *J. Atmos. Sci.*, **49,** 1242–1263.

O'Neill, A., and V. Pope, 1990: The seasonal evolution of the extra-tropical stratosphere in the Southern and Northern Hemisphere: Systematic changes in potential vorticity and the non-conservative effects of validation. *Dynamics, Transport and Photochemistry in the Middle Atmosphere of the Southern Hemisphere*, A. O'Neill, Ed., Kluwer, 33–54.

O'Sullivan, D., and R. E. Young, 1992: Modeling the quasi-biennial oscillation's effect on the winter stratospheric circulation. *J. Atmos. Sci.*, **49,** 2437–2448.

———, and T. J. Dunkerton, 1994: Seasonal development of the extratropical QBO in a numerical model of the middle atmosphere. *J. Atmos. Sci.*, **51,** 3706–3721.

Pfister, L., 1985: Baroclinic instability of easterly jets with applications to the summer mesosphere. *J. Atmos. Sci.*, **42,** 313–330.

Pierce, R. B., W. L. Grose, J. M. Russell III, and A. F. Tuck, 1994: Spring dehydration in the antarctic vortex observed by HALOE. *J. Atmos. Sci.*, **51,** 2931–2941.

Plumb, R. A., 1983: Baroclinic instability of the summer mesosphere: A mechanism for the quasi-two-day wave? *J. Atmos. Sci.*, **40,** 262–270.

———, 1989: On the seasonal cycle of stratospheric planetary waves. *Pure Appl. Geophys.*, **130,** 233–242.

Poole, L. R., and M. C. Pitts, 1994: Polar stratospheric cloud climatology based on Stratospheric Aerosol Measurement II observations from 1978 to 1989. *J. Geophys. Res.*, **99,** 13 083–13 089.

Portmann, R. W., S. Solomon, R. R. Garcia, L. W. Thomason, L. R. Poole, and M. P. McCormick, 1996: The role of aerosol variations in anthropogenic ozone depletion in the polar regions. *J. Geophys. Res.*, **101,** 22 991–23 006.

Prata, A. J., 1984: The 4-day wave. *J. Atmos. Sci.*, **41,** 150–155.

Prather, M., M. M. Garcia, R. Snuzzo, and D. Rind, 1990: Global impact of the antarctic ozone hole: Dynamical dilution with a three-dimensional chemical transport model. *J. Geophys. Res.*, **95,** 3449–3471.

Prusa, J. M., P. K. Smolarkiewicz, and R. R. Garcia, 1996: Propagation and breaking at high altitudes of gravity waves excited by tropospheric forcing. *J. Atmos. Sci.*, **53,** 2186–2216.

Ramaswamy, V., M. D. Schwarzkopf, and W. J. Randel, 1996: Fingerprint of ozone depletion in the spatial and temporal pattern of recent lower-stratospheric cooling. *Nature*, **382,** 616–618.

Randel, W. J., 1987: A study of planetary waves in the southern winter troposphere and stratosphere. Part I: Wave structure and vertical propagation. *J. Atmos. Sci.*, **44,** 917–935.

———, 1988: The seasonal evolution of planetary waves in the Southern Hemisphere stratosphere and troposphere. *Quart. J. Roy. Meteor. Soc.*, **114**, 1385–1409.

———, 1992: Global atmospheric circulation statistics, 1000–1 mb. NCAR Tech. Note, NCAR/TN-366 + STR, 256 pp.

———, 1994: Observations of the 2-day wave in NMC stratospheric analyses. *J. Atmos. Sci.*, **51**, 306–313.

———, and I. M. Held, 1991: Phase speed spectra of transient eddy fluxes and critical layer absorption. *J. Atmos. Sci.*, **48**, 688–697.

———, and L. R. Lait, 1991: Dynamics of the 4-day wave in the Southern Hemisphere polar stratosphere. *J. Atmos. Sci.*, **48**, 2496–2508.

———, and J. B. Cobb, 1994: Coherent variations of monthly mean total ozone and lower stratospheric temperature. *J. Geophys. Res.*, **99**, 5433–5447.

———, and F. Wu, 1995: TOMS total ozone trends in potential vorticity coordinates. *Geophys. Res. Lett.*, **22**, 683–686.

———, and ———, 1998: Cooling of the arctic and antarctic polar stratospheres due to ozone depletion. *J. Climate*, in press.

———, J. C. Gille, A. E. Roche, J. B. Kumer, J. L. Mergenthaler, J. W. Waters, E. F. Fishbein, and W. A. Lahoz, 1993: Stratospheric transport from the Tropics to midlatitudes by planetary-wave mixing. *Nature*, **365**, 533–535.

———, F. Wu, J. M. Russell III, and A. Roche, 1998: Seasonal cycles and interannual variability in stratospheric CH_4 and H_2O observed in UARS HALOE data. *J. Atmos. Sci.*, **55**, 163–185.

Rodgers, C. D., 1976: Evidence for the five-day wave in the stratosphere. *J. Atmos. Sci.*, **33**, 710–711.

———, and A. J. Prata, 1981: Evidence for a traveling two-day wave in the middle atmosphere. *J. Geophys. Res.*, **86**, 9661–9664.

Rong-jui, H., and K. Gambo, 1982: The response of a hemispheric multi-level model atmosphere to forcing by topography and stationary heat sources. *J. Meteor. Soc. Japan*, **60**, 93–108.

Rosenfield, J. E., M. R. Schoeberl, and M. A. Geller, 1987: A computation of the stratospheric residual circulation using an accurate radiative transfer model. *J. Atmos. Sci.*, **44**, 859–876.

———, P. A. Newman, and M. R. Schoeberl, 1994: Computations of diabatic descent in the stratospheric polar vortex. *J. Geophys. Res.*, **99**, 16 677–16 689.

Rosenlof, K. H., 1995: The seasonal cycle of the residual mean meridional circulation in the stratosphere. *J. Geophys. Res.*, **100**, 5173–5191.

———, A. F. Tuck, K. K. Kelly, J. M. Russell III, and M. P. McCormick, 1997: Hemispheric asymmetries in water vapor and inferences about transport in the lower stratosphere. *J. Geophys. Res.*, **102**, 13 213–13 234.

Russell, J. M. III, and Coauthors, 1993a: The Halogen Occulation Experiment. *J. Geophys. Res.*, **98**, 10 777–10 797.

———, A. F. Tuck, L. L. Gordley, J. H. Park, S. R. Drayson, J. E. Harries, R. J. Cicerone, and P. J. Crutzen, 1993b: HALOE antarctic observations in the spring of 1991. *Geophys. Res. Lett.*, **20**, 719–722.

Salby, M. L., 1981a: Rossby normal modes in nonuniform background conditions. Part II: Equinox and solstice conditions. *J. Atmos. Sci.*, **38**, 1827–1840.

———, 1981b: The 2-day wave in the middle atmosphere: Observations and theory. *J. Geophys. Res.*, **86**, 9654–9660.

Santee, M., W. Read, J. Waters, L. Froidevaux, G. Manney, D. Flower, R. Jarnot, R. Harwood, and G. Peckham, 1995: Interhemispheric differences in polar stratospheric HNO_3, H_2O, ClO, and O_3. *Science*, **267**, 849–852.

———, and Coauthors, 1996: Chlorine deactivation in the lower stratosphere polar regions during later winter: Results from UARS. *J. Geophys. Res.*, **101**, 18 835–18 859.

Schoeberl, M. R., and D. Hartmann, 1991: The dynamics of the stratospheric polar vortex and its relation to springtime ozone depletion. *Science*, **251**, 46–48.

———, and P. A. Newman, 1995: A multiple-level trajectory analysis of vortex filaments. *J. Geophys. Res.*, **100**, 25 801–25 815.

———, L. R. Lait, P. A. Newman, and J. E. Rosenfield, 1992: The structure of the polar vortex. *J. Geophys. Res.*, **97**, 7859–7882.

———, M. Luo, and J. E. Rosenfield, 1995: An analysis of the antarctic Halogen Occulation Experiment trace gas observations. *J. Geophys. Res.*, **100**, 5159–5172.

———, A. R. Douglass, S. R. Kawa, A. E. Dessler, P. A. Newman, R. S. Stolarski, A. E. Roche, J. W. Waters, and J. M. Russell III, 1996: Development of the antarctic ozone hole. *J. Geophys. Res.*, **101**, 20 909–20 924.

Shine, K. P., 1986: On the modelled thermal response of the antarctic stratosphere to a depletion of ozone. *Geophys. Res. Lett.*, **13**, 1331–1334.

———, 1987: The middle atmosphere in the absence of dynamic heat fluxes. *Quart. J. Roy. Meteor. Soc.*, **113**, 603–633.

Shiotani, M., and I. Hirota, 1985: Planetary wave-mean flow interaction in the stratosphere: A comparison between Northern and Southern Hemispheres. *Quart. J. Roy. Meteor. Soc.*, **111**, 309–334.

———, K. Kuroi, and I. Hirota, 1990: Eastward traveling waves in the Southern Hemisphere stratosphere during the spring of 1983. *Quart. J. Roy. Meteor. Soc.*, **116**, 913–927.

———, N. Shimoda, and I. Hirota, 1993: Interannual variability of the stratospheric circulation in the Southern Hemisphere. *Quart. J. Roy. Meteor. Soc.*, **119**, 531–546.

Solomon, S., 1988: The mystery of the antarctic ozone "hole." *Rev. Geophys.*, **26**, 131–148.

———, 1990: Progress towards a quantitative understanding of antarctic ozone depletion. *Nature*, **347**, 347–354.

———, R. R. Garcia, F. S. Rowland, and D. J. Wuebbles, 1986a: On the depletion of antarctic ozone. *Nature*, **321**, 755–758.

———, J. T. Kiehl, R. R. Garcia, and W. Grose, 1986b: Tracer transport by the diabatic circulation deduced from satellite observations. *J. Atmos. Sci.*, **43**, 1603–1617.

Stanford, J. L., and J. S. Davis, 1974: A century of stratospheric cloud reports: 1870–1972. *Bull. Amer. Meteor. Soc.*, **55**, 213–219.

———, J. R. Ziemke, and S. Y. Gao, 1993: Stratospheric circulation features deduced from SAMS constituent data. *J. Atmos. Sci.*, **50**, 226–246.

Stolarski, R. S., A. J. Krueger, M. R. Schoeberl, R. D. McPeters, P. A. Newman, and J. C. Alpert, 1986: *Nimbus 7* satellite measurements of the springtime antarctic ozone decrease. *Nature*, **322**, 808–811.

———, P. Bloomfield, R. McPeters, and J. Herman, 1991: Total ozone trends deduced from *Nimbus 7* TOMS data. *Geophys. Res. Lett.*, **18**, 1015–1018.

———, R. Bojkov, L. Bishop, C. Zerefos, J. Staehelin, and J. Zawodny, 1992: Measured trends in stratospheric ozone. *Science*, **256**, 342–349.

Sutton, R., 1994: Lagrangian flow in the middle atmosphere. *Quart. J. Roy. Meteor. Soc.*, **120**, 1299–1321.

Swinbank, R., and A. O'Neill, 1994: A stratosphere–troposphere data-assimilation system. *Mon. Wea. Rev.*, **122**, 686–702.

Sze, N. D., 1989: Antarctic ozone hole: Possible implications for ozone trends in the Southern Hemisphere. *J. Geophys. Res.*, **94**, 11 521–11 528.

Thuburn, J., and G. C. Craig, 1997: GCM tests of theories for the height of the tropopause. *J. Atmos. Sci.*, **54**, 869–882.

Trenberth, K. E., and J. G. Olson, 1989: Temperature trends at the South Pole and McMurdo Sound. *J. Climate*, **2**, 1196–1206.

Tuck, A. F., J. M. Russell III, and J. E. Harries, 1993: Stratospheric dryness: Antiphased dessication over Micronesia and Antarctica. *Geophys. Res. Lett.*, **20**, 1227–1230.

Tung, K. K., and H. Yang, 1994: Global QBO in circulation and ozone. Part I: Reexamination of observational evidence. *J. Atmos. Sci.*, **51**, 2699–2707.

van Loon, H., and K. Labitzke, 1987: The Southern Oscillation. Part V: The anomalies in the lower stratosphere of the Northern Hemisphere in winter and a comparison with the quasi-biennial oscillation. *Mon. Wea. Rev.*, **115,** 357–369.

Venne, D. E., 1989: Normal-mode Rossby waves observed in the wavenumber 1–5 geopotential fields of the stratosphere and troposphere. *J. Atmos. Sci.*, **46,** 1042–1056.

———, and J. L. Stanford, 1979: Observation of a 4-day temperature wave in the polar winter stratosphere. *J. Atmos. Sci.*, **36,** 2016–2019.

———, and ———, 1982: An observational study of high-latitude stratospheric planetary waves in winter. *J. Atmos. Sci.*, **39,** 1026–1034.

Waters, J. W., L. Froidevaux, W. G. Read, G. L. Manney, L. S. Elson, D. A. Flower, R. F. Jarnot, and R. S. Harwood, 1993: Stratospheric ClO and ozone from the microwave limb sounder on the Upper Atmosphere Research Satellite. *Nature*, **362,** 597–602.

Waugh, D. W., 1993a: Contour surgery simulation of a forced polar vortex. *J. Atmos. Sci.*, **50,** 714–730.

———, 1993b: Subtropical stratospheric mixing linked to disturbances in the polar vortices. *Nature*, **365,** 535–537.

———, and W. J. Randel, 1998: Climatology of arctic and antarctic polar vortices using elliptical diagnostics. *J. Atmos. Sci.*, in press.

Webster, G. R., and Coauthors, 1993: Chlorine chemistry on polar stratospheric cloud particles in the arctic winter. *Science*, **261,** 1130–1134.

WMO, 1992: Scientific assessment of ozone depletion: 1991. Global Ozone Research and Monitoring Project, World Meteorological Organization Rep. No. 25.

———, 1995: Scientific assessment of ozone depletion: 1994. Global Ozone Research and Monitoring Project, World Meteorological Organization Rep. No. 37.

Wu, D. L., P. B. Hays, W. R. Skinner, A. R. Marshall, M. D. Burrage, R. S. Lieberman, and D. A. Ortland, 1993: Observations of the quasi 2-day wave from the high resolution Doppler imager on UARS. *Geophys. Res. Lett.*, **20,** 1853–2856.

———, E. F. Fishbein, W. G. Read, and J. W. Waters, 1996: Excitation and evolution of the quasi-2-day wave observed in UARS/MLS temperature measurements. *J. Atmos. Sci.*, **51,** 728–738.

Yang, H., and K. K. Tung, 1994: Statistical significants and pattern of extratropical QBO in column ozone. *Geophys. Res. Lett.*, **21,** 2235–2238.

Yulaeva, E., J. R. Holton, and J. M. Wallace, 1994: On the cause of the annual cycle in the tropical lower stratospheric temperature. *J. Atmos. Sci.*, **51,** 169–174.

Chapter 7

The Role of the Oceans in Southern Hemisphere Climate

J. STUART GODFREY AND STEPHEN R. RINTOUL

CSIRO, Division of Marine Research, Hobart, Tasmania, Australia

7.1. Introduction

Atmospheric circulation is strongly influenced by details of the sea surface temperature. In many parts of the World Ocean, SST is fairly well approximated by a one-dimensional local balance, in which (at least on long-term mean) the SST adjusts locally until the losses due to latent and sensible heat and longwave radiation balance the incident shortwave radiation. There are large parts of the World Ocean, however, for which ocean currents affect SST quite strongly. Ocean currents connect regions of heat gain to regions of heat loss; heat gained from the atmosphere may be stored for many years and carried thousands of kilometers before being returned to the atmosphere. These currents are driven by the atmosphere, through surface winds or buoyancy fluxes. Thus, the atmosphere and the ocean interact strongly on one another, and the coupled system cannot be understood by considering either component in isolation. The aim of this chapter is to describe some of the physics underlying the observed patterns of sea surface temperature and heat flux, with an emphasis on the Southern Hemisphere.

While the emphasis on the Southern Hemisphere is appropriate in a book such as this one, the circulation of the ocean is global and cannot be discussed sensibly by focusing on the Southern Hemisphere alone. Because we cannot review all aspects of ocean circulation that may affect Southern Hemisphere meteorology, we have chosen to structure the chapter by focusing on the physical mechanisms responsible for maintaining the observed pattern of surface heat flux.

The surface heat-flux pattern is largely determined by the wind-driven circulation in the upper ocean. Strong poleward flow in subtropical western boundary currents carries warm water poleward, warming the overlying atmosphere. Along the Equator and some land boundaries, wind-driven divergence of surface waters causes upwelling of cold water and, hence, cooling of the atmosphere. Buoyancy-driven flows also play a part in maintaining the observed pattern of heat flux. In particular, the global overturning circulation associated with the formation of dense water in the North Atlantic is an efficient means of carrying heat in the meridional plane. To explain the observed patterns of SST and heat flux, we must consider the full three-dimensional circulation of the ocean.

The chapter begins with a description of the major features in the observed SST and surface heat-flux distributions. We then show how the lack of zonal symmetry in the observed heat-flux pattern reflects the dynamics of ocean currents, much of which can be understood using the simple physics of wind-driven circulation theory. In terms of circulation and heat transport, perhaps the most fundamental difference between the atmosphere and the ocean is the presence of continents to act as meridional boundaries in the ocean. Meridional walls can support zonal pressure gradients and, hence, geostrophic meridional flow. As a consequence, the mean overturning circulation plays a more important role in the meridional flux of heat than do transient eddies, in contrast to the atmosphere. We briefly describe the overturning circulation in the ocean and the relative contribution of ocean and atmosphere to the net poleward heat flux. The Southern Ocean is the one exception to the above: in the latitude of Drake Passage, there are no continental barriers. For this reason, the Southern Ocean provides the most direct oceanic analogue to atmospheric flows. We discuss the special dynamics of the Antarctic Circumpolar Current (ACC) and the pivotal role played by the ACC in the global overturning circulation as the primary means of exchange between the ocean basins. The Indonesian throughflow provides the only transfer of warm water between ocean basins and thus has a significant impact on climate. In the final section, we briefly discuss the link between ocean circulation and variations in climate.

We can only touch briefly on a number of these issues in this chapter. Readers interested in a more complete treatment of the dynamics of ocean circulation and air–sea interaction are referred to the textbook by Gill (1982). Tomczak and Godfrey (1994) provide a detailed description of the ocean currents mentioned here. Haidvogel and Bryan (1992) provide a useful review of ocean climate models and a brief description

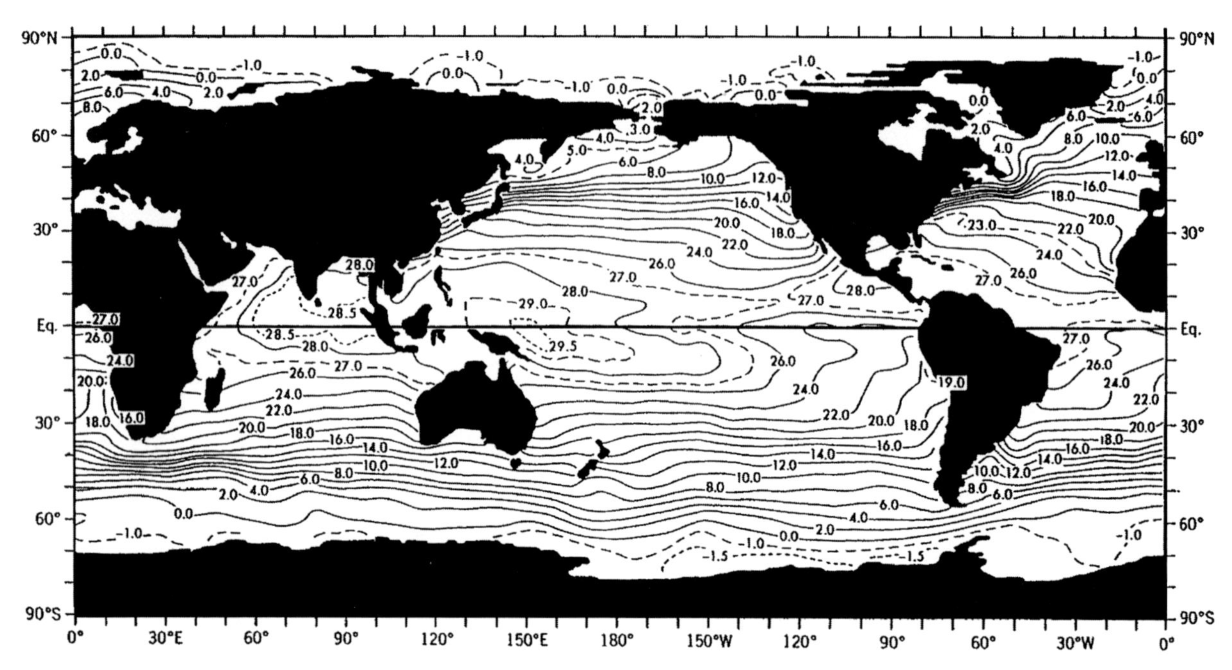

FIG. 7.1. Geographical distribution of annual mean SST. Adapted from Levitus (1982).

of coupled ocean–atmosphere models is given by Meehl in chapter 10 of this monograph.

7.2. Sea surface temperature and air–sea heat exchange

Figure 7.1 shows the observed annual mean pattern of SST. The poleward decrease in SST is, of course, expected, due to the poleward decrease of annual mean solar radiation. There are strong east–west and interbasin asymmetries, however: for example, the SST difference across the Pacific at 5°S is as much as 6°C. In the Tropics, the western sides of the Pacific and Atlantic are warmer than the eastern sides, while at northern high latitudes the zonal gradients are reversed, at least in the northern oceans. In the Indian Ocean, the warmest SSTs occur on the eastern side, rather than the western, and isotherms in the southeastern Indian Ocean tilt slightly poleward as they approach the coast, rather than equatorward as seen in Fig. 7.1 in the eastern Atlantic and Pacific. In the Southern Ocean, SSTs west of southern Chile are some 5°C warmer than at the same latitude in the mid-Atlantic. Meridional SST gradients are particularly strong south of South Africa.

To account for these zonal anomalies in annual mean SST, it is necessary to take ocean currents into account. The simplest model of the ocean and its role in climate is the "slab model," where the ocean is reduced to a thin, motionless layer of fixed depth that can act as a local store of heat. Because of the massive heat capacity of water—the upper few meters of the ocean can store as much heat as the entire atmosphere—such a model captures the ocean's ability to absorb and emit heat locally, and thus it can simulate the seasonal cycle of SST in much of the ocean. The resulting patterns of long-term mean SST, however, lack the zonal asymmetry seen in Fig. 7.1, because in reality the ocean not only stores but transports heat efficiently.

If the ocean really acted as a motionless slab, the annual mean pattern of SST would be set by the condition that the annual mean net heat flux into each point in the ocean must be zero. In fact, the ocean can absorb or release heat at any location in the long-term mean, so long as the global integral is zero in steady state. Wherever long-term mean heating (cooling) of the ocean occurs, there must be a mean inflow of cold (warm) water and removal of warmed (cooled) water; thus, the long-term mean net heat flux into the ocean is a better indicator of the role of ocean currents in modulating climate than the SST.

Figure 7.2a shows an estimate of the annual mean net heat flux into the ocean, from Oberhuber (1988). The main east–west asymmetries seen in Fig. 7.1 are accentuated in Fig. 7.2a. At a given latitude in Fig. 7.1, low (high) SSTs tend to correspond with heat flux into (out of) the ocean in Fig. 7.2a. There is strong heat gain in the eastern tropical Pacific and Atlantic and the western tropical Indian Ocean (regions of relatively low SST in Fig. 7.1), while strong heat loss is seen on the western sides of the North Pacific and North Atlantic, coinciding with zonal maxima in SST in Fig. 7.1. Further heat loss is seen in the subpolar North Atlantic. One place where high

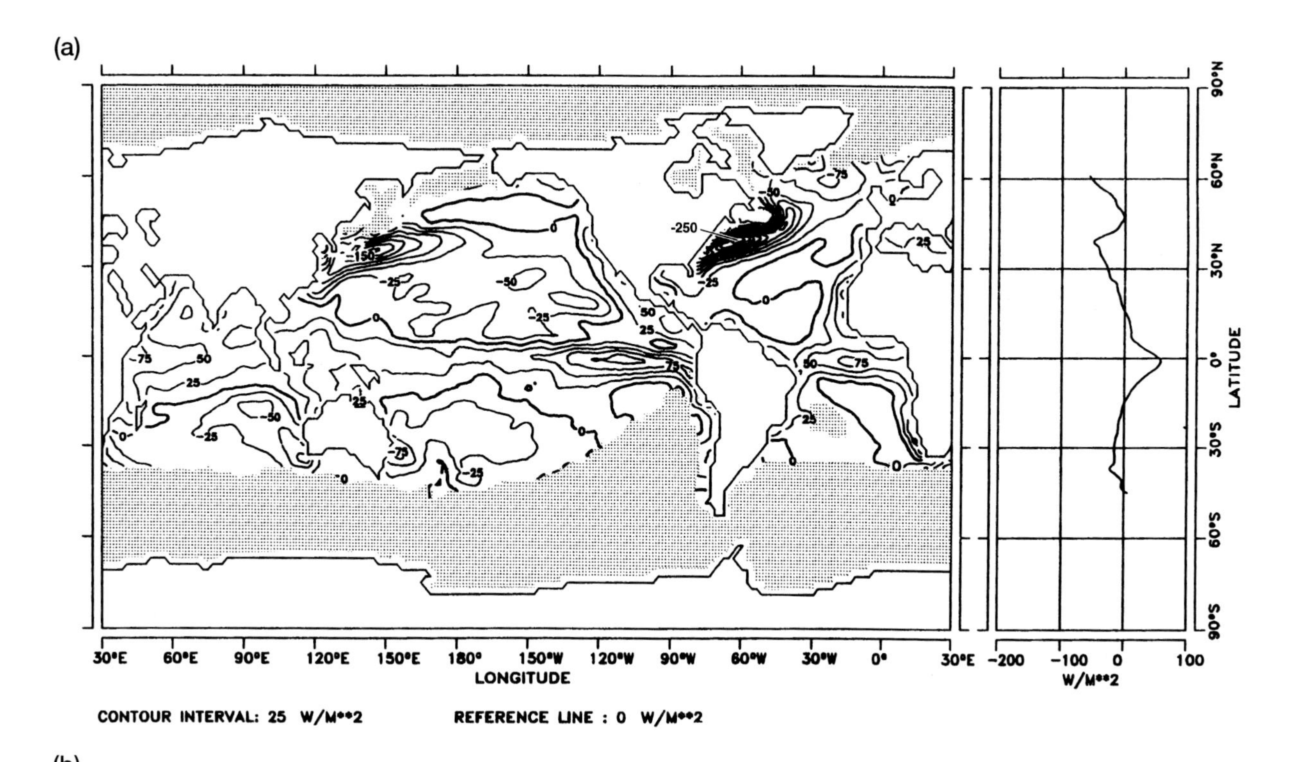

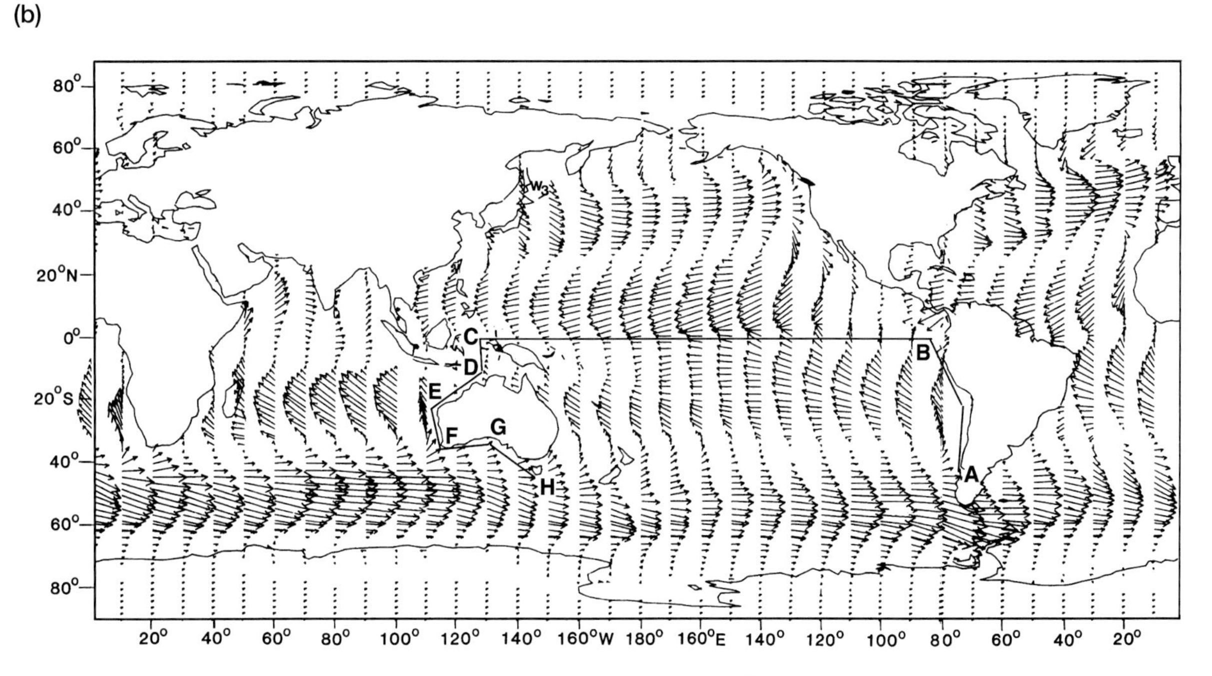

FIG. 7.2. (a) Annual mean net heat flux into the ocean. Contour interval 20 W m^{-2}. Positive (negative) values indicate heat gain (loss) by the ocean. Data are unavailable in stippled areas. From Oberhuber (1988). (b) Annual mean wind stress at the ocean surface. Largest vector is 0.2 N m^{-2}. From Hellerman and Rosenstein (1983).

SSTs coincide with heat flux into the water is the Indonesian region; this exception is thought to be due to the special role of tidal mixing (see below). The map in Fig. 7.2a has large gaps south of 30°S and in the southeast Pacific, due to the lack of merchant ship observations.

Even where shipping lanes are plentiful, the quantitative accuracy of estimates such as Fig. 7.2a is poor. Several such climatologies exist; they differ from one another by up to 50 W m^{-2} in places. This is due largely to sparse data, problems with the empirical algorithms used to infer heat flux from routine mer-

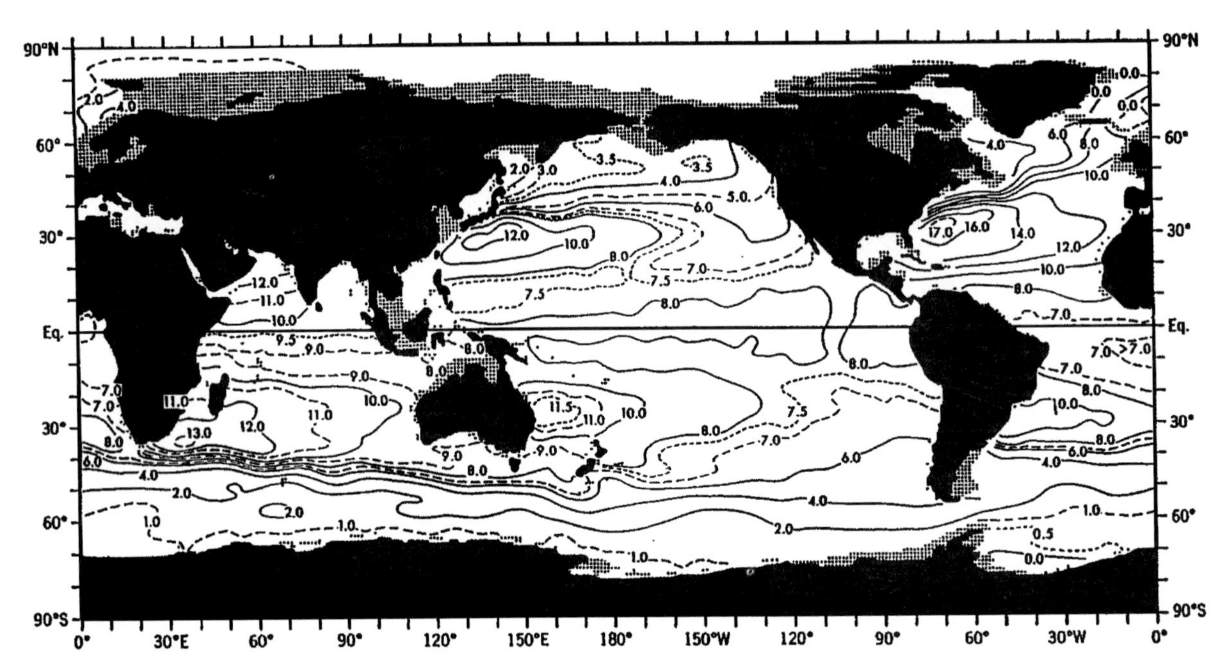

FIG. 7.3. Ocean temperature at 500 m, typical of the midthermocline in the midlatitude oceans. The strong zonal temperature gradient in each basin, near 30°N and 30°S, is generated by the Sverdrup (1947) mechanism. Adapted from Levitus (1982).

chant ship measurements, and the degree of spatial smoothing applied.

A starting point in understanding the ocean's role in climate variability is a good understanding of the ocean currents that maintain the annual mean heat flux seen in Fig. 7.2a. This chapter aims to describe the circulation features controlling the observed heat-flux pattern, emphasizing the Southern Hemisphere.

7.3. Ekman drifts and heat fluxes

While some of the flows associated with surface heat exchange are driven by the heat fluxes themselves, through buoyancy effects, most are generated by wind stresses. Figure 7.2b shows the annual mean wind stress over the World Ocean. Much of the pattern of Fig. 7.2a can be qualitatively understood by inspection of Fig. 7.2b, using simple principles of physical oceanography. The approach in the following is similar in spirit to that of Tomczak and Godfrey (1994).

The first of these principles is that a steady wind stress $\boldsymbol{\tau}$ directly induces a net surface "Ekman drift" of magnitude $\boldsymbol{\tau}/(\rho|f|)$ at right angles to the wind, where ρ is water density, and $f = 4\pi \sin(\theta)/T_d$ is the Coriolis parameter; the flow is to the left (right) of the wind in the Southern (Northern) Hemisphere. T_d is the length of the day, and θ the latitude.

Comparison of Fig. 7.2b with Fig. 7.1 shows that Ekman fluxes associated with the trade winds tend to carry water poleward, often down SST gradients. For example, Fig. 7.2b shows that, in the eastern Indian Ocean near 15°S, 110°E, the annual mean wind stress is about 0.1 N m^{-2} and the Ekman transport is directed southwestward, down a surface temperature gradient T_l (from Fig. 7.1) of about 0.4×10^{-4}°C m^{-1}. Assuming the Ekman transport is contained in the mixed layer, we would expect a heat loss of $\rho C_p(\tau/\rho|f(20°)|)T_l$ from the ocean, where C_p~4000 J kg^{-1} is the heat capacity of water. This is about 30 W m^{-2} and is probably the major cause of the net heat loss from this location seen in Fig. 7.2a. Similarly, Ekman fluxes are equatorward in the westerlies, causing water to move to higher temperatures; the ocean can gain heat here—though because the Coriolis parameter is larger at these latitudes, the effect is generally numerically smaller than in the trades.

Regions where the Ekman transport diverges, however, are often much stronger sources of oceanic heat gain, because they force cold subsurface water into the surface mixed layer. Inspection of Fig. 7.2b shows that strong equatorward winds occur along the west coasts of Africa and America, resulting in Ekman divergence. Similarly, easterly winds along the Equator induce Ekman transports that diverge on either side of the Equator; this upwells large volumes of water, because the Coriolis parameter is very small. Roughly speaking, the Ekman formula "works" outside of 2° from the Equator, so the upwelling associated with, say, an easterly wind stress of 0.03 N m^{-2} over a zonal extent L of 3000 km—typical of conditions in the mid-Pacific (see Fig. 7.2b)—is roughly $2\tau L/\rho f(2°)$ or 35×10^6 m^3 s^{-1}. If this water gains only 1°C between entering and leaving the 400-km × 3000-km enclosed area a, a net heat flux

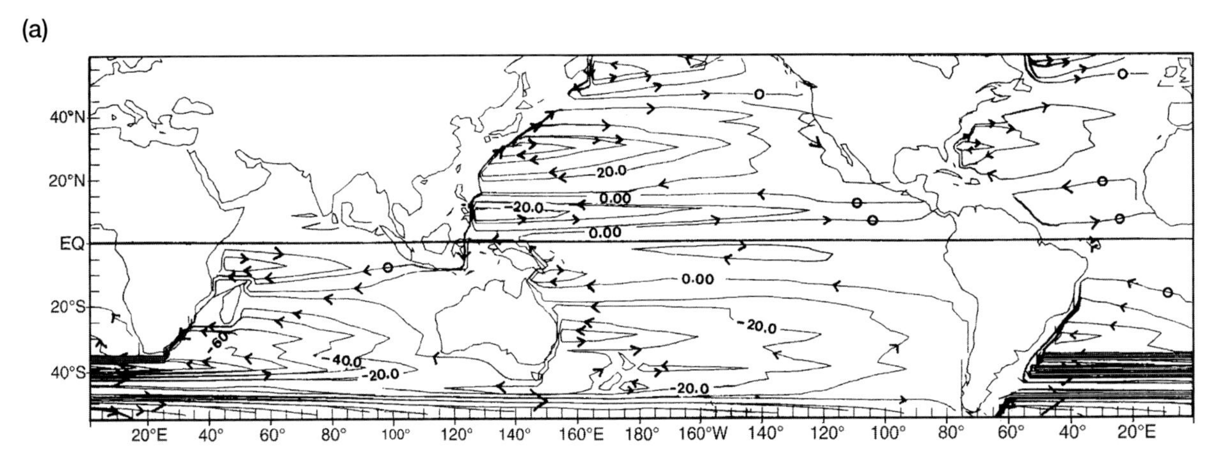

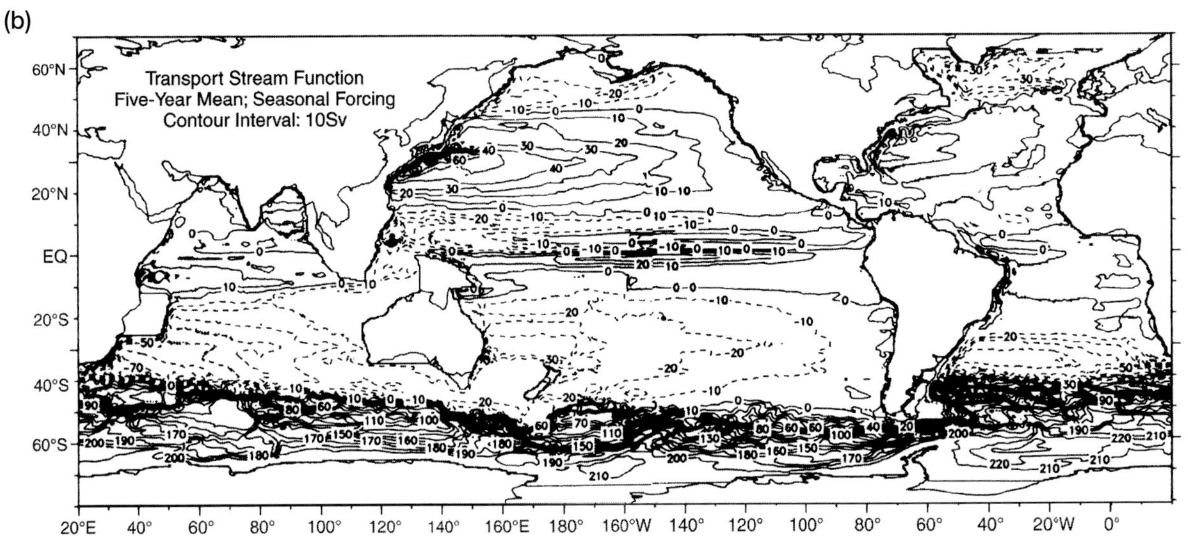

FIG. 7.4. (a) Annual mean mass-transport streamfunction for the World Ocean, calculated from the Sverdrup relation with Hellerman and Rosenstein annual mean wind stresses. Adapted from Godfrey (1989). (b) Annual mean mass-transport streamfunction for the World Ocean, from the fine-resolution model of Semtner and Chervin (1992) driven with Hellerman and Rosenstein seasonal mean wind stresses.

of $\rho C_p 2\tau L/\rho A[f(2°)]$ (1°C) is required through the surface. This is more than 100 W m^{-2}.

The third way in which winds induce heat fluxes through the ocean surface is that the convergence/divergence of Ekman drifts pushes the thermocline as a whole up and down. The motion of the thermocline generates oceanic Rossby waves; in the resulting steady state, the thermocline is deeper in the west, resulting in a zonal gradient of temperature (and density) (Fig. 7.3). This zonal gradient in turn generates meridional flows through the thermal wind relation:

$$f\frac{\partial v}{\partial z} = \frac{g}{\rho}\frac{\partial \rho}{\partial x}, \tag{7.1}$$

where $v(x,z)$ is the meridional velocity at depth z and longitude x, while g is the acceleration of gravity. The velocity v is generally weak below the thermocline. Assuming it to be zero below the thermocline and integrating upward, Sverdrup (1947) obtained the very simple "Sverdrup relation" for the total depth-integrated meridional flow in the interior **V** (i.e., the sum of the Ekman flow and the geostrophic flow in balance with the zonal gradient of Fig. 7.3):

$$\beta \mathbf{V} = \nabla \times \boldsymbol{\tau}/\rho = \frac{1}{\rho}\left(\frac{\partial \tau^{(y)}}{\partial x} - \frac{\partial \tau^{(x)}}{\partial y}\right). \tag{7.2}$$

Here $\beta = (1/R)\partial f/\partial \theta$ is the meridional gradient of the Coriolis parameter, R is the radius of the earth, and $(\boldsymbol{\tau}^{(x)}, \boldsymbol{\tau}^{(y)})$ are the components of wind stress in the (eastward, northward) directions (x,y). Inspection of Fig. 7.2b shows that between latitudes 25°–35° north and south of the Equator, $(\partial \tau^{(y)}/\partial x - \partial \tau^{(x)}/\partial y) \sim$

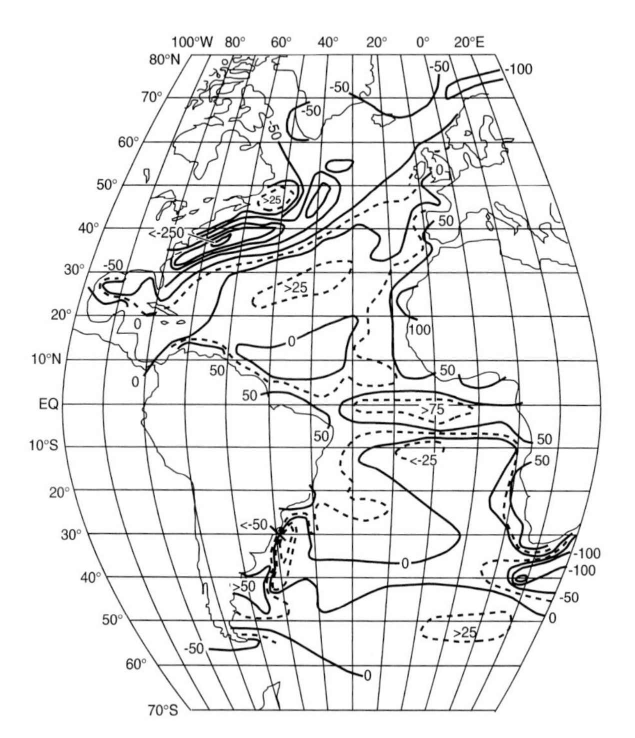

FIG. 7.5. Annual mean net heat flux into the Atlantic Ocean, from Bunker (1988). Contour interval 25 W m^{-2}.

$\partial \tau^{(x)}/\partial y$, and is negative (positive) at all longitudes in the Northern (Southern) Hemisphere. Thus, (7.2) predicts equatorward flows throughout these two latitude bands in the ocean interior. By mass continuity, the equatorward flow in the interior of the ocean must be balanced by poleward flow someplace where the simple Sverdrup dynamics do not hold—that is, near the boundaries of the basin, where friction and nonlinear effects can be important. Stommel (1948) showed that in order to remove the vorticity put in by the wind, the strong boundary currents must be on the western rather than the eastern side of the ocean basins and directed toward the pole in subtropical latitudes and toward the Equator in subpolar latitudes. The westward intensification can also be understood in terms of Rossby-wave propagation: Rossby waves carry the energy supplied by the wind over the basin interior westward to the western boundary, where it can be dissipated in strong flows along the western boundary.

7.4. The Sverdrup relation, overturning cells, and the strength of western boundary currents

Considering its great simplicity, the Sverdrup relationship holds quite well, even in the most sophisticated models of the World Ocean. For example, Fig. 7.4a shows the streamfunction for the World Ocean (north of Drake Passage) calculated from the Sverdrup relation and mass continuity alone, while Fig. 7.4b shows the streamfunction from the Semtner and Chervin (1992) fine-resolution global ocean model. There are large differences between Figs. 7.4a and 7.4b in the South Atlantic, where the Agulhas Current does not "retroflect" in the linear Sverdrup calculation and instead continues to the coast of South America as a zonal jet. Further, large differences are seen in the equatorial Pacific. Nonlinear terms are important in the momentum balance in each location (Ou and de Ruijter 1986; Godfrey et al. 1993). The "subtropical gyres" found between 15° and 45° latitude in each ocean basin, however—the equatorward Sverdrup flow in the interior and the compensating intense poleward

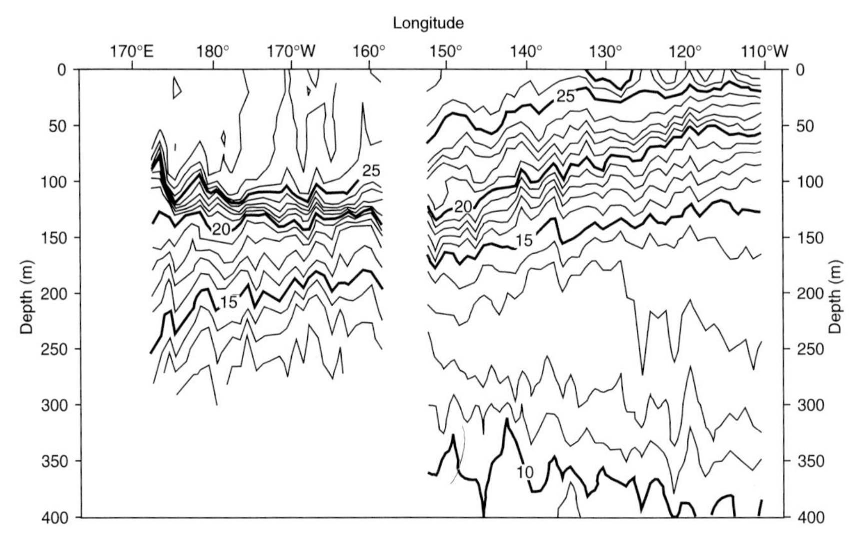

FIG. 7.6. A "snapshot" temperature section (°C) along the equatorial Pacific, from 170°E to 110°W; AXBT and XBT data, February and April–May 1979. From Halpern (1980).

flow along the western boundary—are well accounted for by the simple theory of the wind-driven circulation. As discussed in section 7.5, the wind-driven subtropical gyres largely control the pattern of surface heat flux seen in midlatitudes in Fig. 7.2a.

The Sverdrup relation is useful for understanding differences between ocean basins. For example, inspection of Fig. 7.2b shows that the wind-stress curl is clearly much weaker across the South Pacific Ocean than across the southern Indian Ocean, so it is not surprising that according to the Sverdrup relation the maximum net northward flow across the width of the South Pacific (near 30°S, excluding the western boundary current) is only 45×10^6 m^3 s^{-1}, whereas the flow across the much narrower southern Indian Ocean (near 32°S) reaches 51×10^6 m^3 s^{-1}. The northward flow across the South Atlantic reaches 36×10^6 m^3 s^{-1}, near 30°S.

If the Indonesian passages were closed, these flows would have to return southward in each basin, so the magnitudes of the three western boundary currents (the East Australian Current, Agulhas Current, and Brazil Current, respectively) would be given by the three northward flows just mentioned. In fact, the same depth-integrated momentum equations that lead to the Sverdrup relation predict a flow through the Indonesian passages of 16×10^6 m^3 s^{-1} (Godfrey 1989); this reduces the expected maximum strength of the East Australian Current from 45×10^6 to 29×10^6 m^3 s^{-1} and increases the expected strength of the Agulhas Current from 51×10^6 to 67×10^6 m^3 s^{-1}. Thus, if the throughflow were indeed as large as 16×10^6 m^3 s^{-1}, the Agulhas must carry much more heat poleward than the East Australian Current. The range of observational estimates of the throughflow mean magnitude remains quite wide, however: while most recent estimates are rather less than 16×10^6 m^3 s^{-1} (e.g., Toole and Warren 1993; Meyers et al. 1995; Wijffels 1993; Cresswell et al. 1993), Fieux et al. (1994) estimated a net transport of 22 Sv in the upper 500 m in 1989.

These wind-driven estimates of western boundary–current magnitudes must be further modified by consideration of the "overturning circulation." In addition to the Sverdrup flows, about 16×10^6 m^3 s^{-1} is observed to move northward in the Atlantic in the top 2000 m to compensate for the southward flow of North Atlantic deep water (NADW) (Rintoul 1991; Macdonald 1993; Schmitz and McCartney 1993; Dickson and Brown 1994). The extra northward current above 2000 m reduces the magnitude of the Brazil Current in the top 2000 m from about 30×10^6 down to 14×10^6 m^3 s^{-1} (Stommel 1958) and increases the strength of the Gulf Stream from about 37×10^6 to 53×10^6 m^3 s^{-1}. This brings the Gulf Stream (above 2000 m) to be comparable in magnitude to the Kuroshio, whose strength is estimated from the Sverdrup relation to be about 60×10^6 m^3 s^{-1}. (The transport of boundary currents such as the Gulf Stream and Kuroshio are further enhanced by tight recirculating gyres driven by nonlinear processes near the western boundary.)

Because deep water does not form in the North Pacific (Warren 1983), the Pacific has no analogue of the NADW buoyancy-driven current system, and the East Australian Current and the Kuroshio are quite well estimated by the wind-driven estimates just given. In the Indian Ocean, however, the conversion of deep to intermediate water (Toole and Warren 1993; see section 7.6) acts to increase the estimated magnitude of the Agulhas above 2000 m to some 94×10^6 m^3 s^{-1}. This estimate agrees fairly well with Toole and Warren's direct measurement of 85×10^6 m^3 s^{-1}. Thus, the Agulhas Current is by a large margin the strongest western boundary current in the Southern Hemisphere.

7.5. Currents and heat fluxes

Given the above brief description of the dynamics of wind-driven currents, we return to Fig. 7.2a and further explore the relationship between the surface heat-flux pattern and the main circulation features of the upper ocean, emphasizing the Southern Hemisphere.

a. Western boundary currents and heat fluxes

The most prominent features of Fig. 7.2a are the regions of strong oceanic heat loss in the western subtropical North Atlantic and North Pacific Oceans. Here, the warm water carried north in the western boundary currents of the subtropical gyres is strongly cooled by the atmosphere, particularly by outbreaks of cold, dry continental air in winter. The equatorward geostrophic flow of relatively cool water toward warmer climates in the gyre interior results in heat gain, but the magnitudes of the associated heat fluxes are quite small because the surface currents are slow.

Comparable phenomena might be expected in the outflows of the Southern Hemisphere western boundary currents. As discussed above, however, the western boundary currents of the South Pacific and South Atlantic subtropical gyres are weaker than their Northern Hemisphere counterparts and so carry less warm water south to supply heat to the atmosphere. For example, the maximum heat loss of 75 W m^{-2} over the East Australian Current in Fig. 7.2a is less than half the heat loss experienced over the Kuroshio.

The gaps in Fig. 7.2a make it difficult to compare the heat flux over the Brazil and Agulhas Currents to their Northern Hemisphere counterparts. Figure 7.5 shows annual heat gain over the Atlantic from Bunker (1988), who extended his heat-flux estimates to 60°S. Bunker's detailed map shows a number of features related to the circulation of the South Atlantic. The

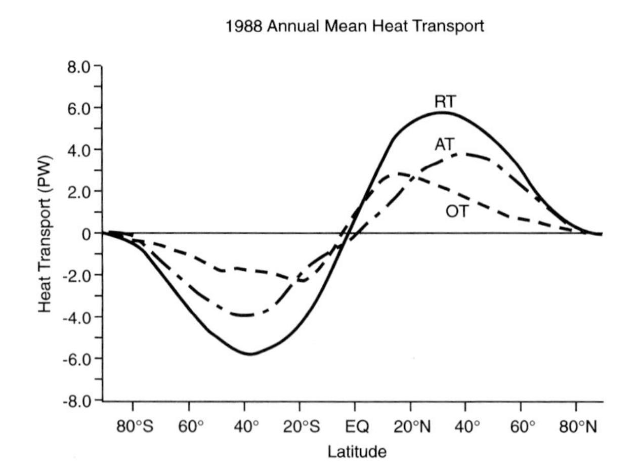

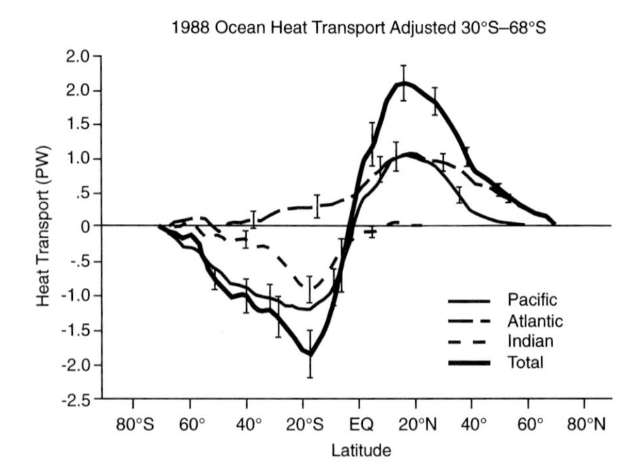

FIG. 7.7. (upper panel) The heavy full line shows the total annual mean northward heat transport for 1988 as a function of latitude, required to maintain the global pattern of radiative heat exchange at the top of the atmosphere, as observed by satellites. The light full line shows the estimated atmospheric heat transport, while the ocean heat transport (dashed line) is computed as a residual. Units are Petawatts (10^{15} W). From Trenberth and Solomon (1994). (lower panel) The poleward ocean transport in each ocean basin for 1988, summed over all oceans (total), as computed from the net heat flux through the surface, integrated from 65°N and adjusted south of 30°S to satisfy the constraint that the net heat flux into the ocean is zero. The effects of the Indonesian throughflow have been neglected in this calculation; they will enhance southward heat transport in the Indian Ocean with an equal and opposite effect in the Pacific. From Trenberth and Solomon (1994).

ocean loses heat from the warm waters of the Brazil Current between 30° and 45°S. South of 40°S and inshore of the Brazil Current, the ocean gains heat where cold water is carried equatorward by the Malvinas Current.

The largest heat loss in the South Atlantic—indeed, the largest in the Southern Hemisphere oceans—occurs south of Africa. Here, the Agulhas Current carries warm water poleward before "retroflecting" to flow eastward across the southern Indian Ocean. The Agulhas Current and its downstream extension, the South Indian Ocean Current (Stramma 1992), are the primary sources of heat for the Southern Ocean. Although no heat-flux estimates derived from observations extend into the high latitudes of the southern Indian Ocean, both the heat fluxes inferred from the ECMWF atmospheric model (Barnier et al. 1995) and the ocean model of Hirst and Godfrey (1993) show a region of oceanic heat loss over the Agulhas extension. (Although the models agree on the pattern of the heat flux, the heat loss in this region of the ocean model greatly exceeds that inferred from the atmospheric model.)

While the mass transport of the Agulhas Current is comparable in magnitude to that of the Gulf Stream and Kuroshio, the heat loss is less than half as intense. In part, this is because the warm waters of the Gulf Stream and Kuroshio experience extreme heat loss during winter when cold, dry continental air flows over these currents; there is no analogous flow of cold air in winter over the Southern Hemisphere boundary currents.

b. Eastern boundary flows and heat fluxes

Another striking feature of Fig. 7.2a is the presence of rather narrow (1000 km–wide) bands of heat input into the ocean off the west coast of the Americas, Europe, and Africa. The primary factor controlling these bands is wind-driven upwelling. It is apparent from Fig. 7.2b that the longshore component of the wind stress is equatorward along the eastern boundaries of both the Pacific and Atlantic, so the annual mean Ekman transport is offshore. Upwelling of subsurface water results; while the upwelling itself is confined to a few tens of kilometers next to the coast, the upwelled water takes several months to warm up, during which time it is advected about 1000 km offshore by the Ekman drift. An effect of comparable importance in generating the coastal bands is that, largely because of coastal upwelling, the temperature at the coast is lower than that 1000 km offshore—not just at the surface but throughout the top 200–300 m. The thermal wind relation (7.1) thus requires equatorward surface flows throughout this band, causing water to be advected toward warmer climates and thus to absorb more heat.

Upwelling-favorable winds also occur along the west coast of Australia. The ocean loses heat in this region (Fig. 7.2a), however, in contrast to other eastern boundary–current regimes. A plausible explanation has emerged from recent modeling studies (Godfrey and Weaver 1991; Hughes et al. 1992; Hirst and Godfrey 1993, 1994). The mechanism depends on the existence of the Indonesian passages. Because water flows through the passages from the Pacific to the Indian Ocean, a current known as the Indonesian

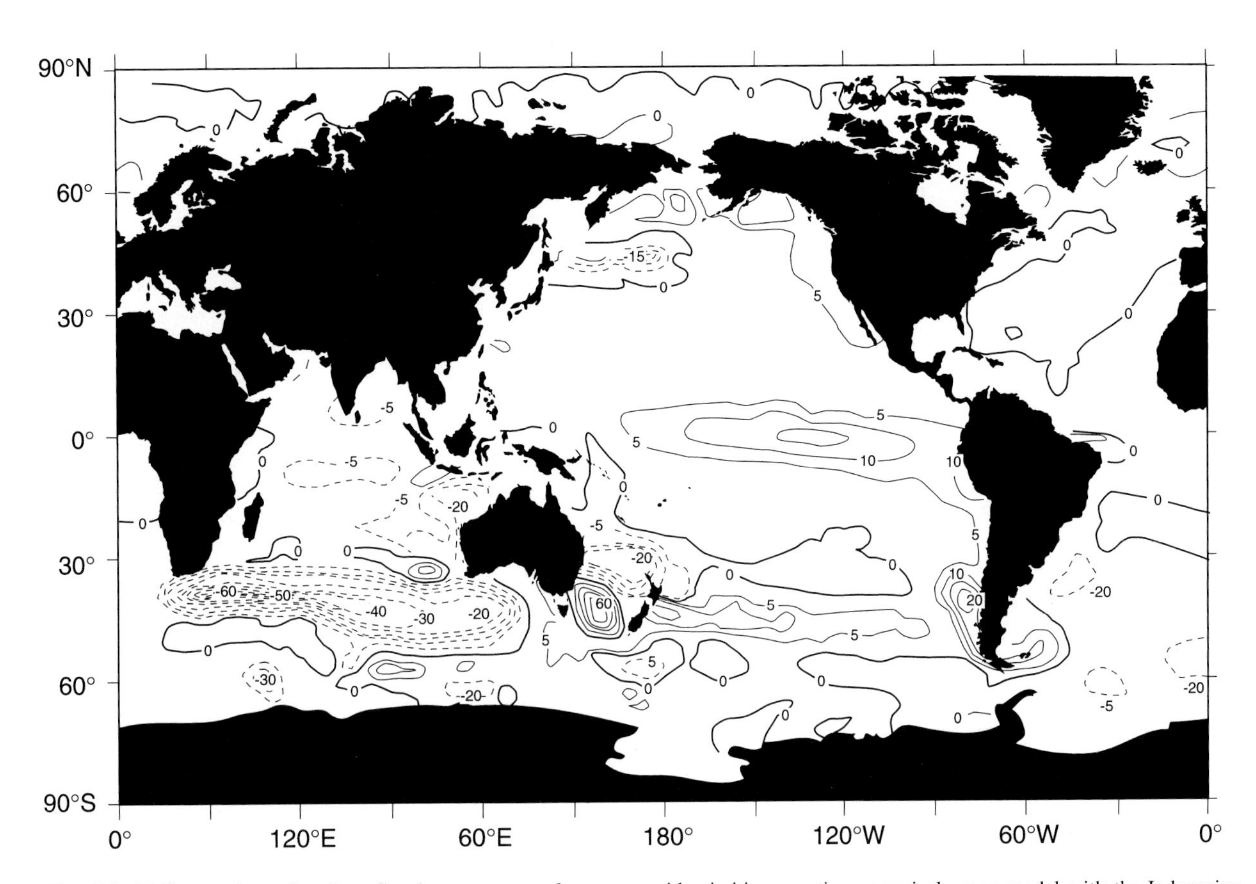

FIG. 7.8. Difference in surface heat flux between runs of a coarse-grid primitive equation numerical ocean model with the Indonesian passages open and closed (Hirst and Godfrey 1993). Both models were driven with Hellerman and Rosenstein (1983) annual mean winds and relaxed to a surface temperature close to the observed value. Contour interval is 5 W m^{-2}.

throughflow, the ocean thermal structure off northwestern Australia is similar to that of the western equatorial Pacific. As already noted, Ekman transports associated with the strong trade winds (Fig. 7.2b) carry the deep layers of warm water found in this region south into colder climates, where deep mixed layers form. The temperature in these mixed layers decreases steadily toward the south. The meridional temperature gradient in turn generates strong baroclinic currents through the thermal wind relation, with surface currents directed toward Australia in the top 200 m of the water column. The resulting onshore flow is strong enough to overwhelm the offshore Ekman drift. Furthermore, at the coast, the pressure gradients can no longer be balanced by geostrophic onshore flow; instead, they accelerate flow southward. An intense southward flow develops at the Australian shelf edge—the Leeuwin Current. It flows poleward, directly into the wind, resulting in more heat loss as the warm water encounters cooler climates. The reason this flow pattern occurs in the eastern Indian Ocean and not in the eastern Atlantic or Pacific is simply that the surface water in the eastern tropical Indian Ocean is much warmer than in the eastern parts of the other oceans, so heat loss and convective overturn occur when it is carried poleward.

c. Equatorial upwelling and heat fluxes

Bands of net ocean heating are seen along the Equator in Fig. 7.2a, in the eastern Atlantic and Pacific Oceans. As mentioned in section 7.3, these are due to upwelling, associated with the easterly winds along the Equator in both oceans (Fig. 7.2b). As in the coastal strip, horizontal advection by the Ekman transports carries the newly surfaced, cool water poleward, so the width of these bands is of order 1000 km—much wider than the region of actual upwelling.

Particularly in the Pacific, however, the region of strongest heat gain (about 105°W, Fig. 7.2a) is well to the east of the region of strongest easterly equatorial winds (about 160°W, Fig. 7.2b). The main reason is that along the equatorial Pacific, where the Coriolis parameter is zero, easterly winds blow surface warm water westward, piling up warm water in the western Pacific. The resulting zonal temperature gradients are very strong; Fig. 7.6 shows a "snapshot" in 1979. In models, temperatures are set by a very delicate balance in which vertical advection and diffusion and horizontal advection all play an active role; the temperature structure depends in an essential way on the values of vertical eddy diffusion and viscosity, which become large in the regions of low Richardson number between the westward surface current and the eastward equatorial undercurrent just beneath it (e.g., Pacanowski and Philander 1981). Numerical models have considerable difficulty in accurately simulating temperature patterns like Fig. 7.6 (e.g., Stockdale et al. 1993). It can be seen from Fig. 7.6, however, that vertical stratification is quite weak in the top 50 m, west of 140°W, and is much stronger east of that longitude. The mixed layer seldom reaches to 50 m in the western equatorial Pacific (e.g., Lukas and Lindstrom 1991), so the upwelled water will be cooler than the observed SST by only 1°C or less. This is not so in the eastern Pacific. Even with observed mixed-layer depths of order 20 m in this region, Fig. 7.6 suggests that the upwelled water will be 2°C or more colder than the SST. Thus, it is reasonable that heat fluxes into the equatorial region will be greater in the eastern than the central Pacific, despite the fact that the upwelling magnitude is greater in the central Pacific than the eastern.

The issue of how heat gained in the tropical oceans is redistributed to higher latitudes is a complicated one. The dynamics of the tropical oceans are rich and complex. The equatorial waveguide supports Kelvin and Rossby waves, which propagate much more quickly than Rossby waves at higher latitudes. As a result, the tropical oceans can respond more quickly to changes in wind forcing. Changes in the winds affect meridional heat transport directly by modifying the poleward transport of warm surface water in the Ekman layer. The system of zonal currents and countercurrents at low latitudes also responds to changes in wind stress on seasonal and longer timescales. Western boundary currents such as the North Brazil Current also play an important role in the meridional transfer of heat. The heat balance of the tropical oceans is maintained by a combination of equatorial waves, Ekman transport, vertical mixing, strong zonal flows in the interior, and strong boundary currents. Readers interested in a more complete discussion of how heat gained by the tropical oceans is carried to higher latitudes in each ocean basin are referred to Knox and Anderson (1985) and Philander (1990).

d. Tropical Indian Ocean

Returning to Fig. 7.2a, we note that the ocean gains considerable heat in the tropical Indian Ocean—but whereas this occurs on the eastern side of the ocean in the Atlantic and Pacific, it occurs primarily on the western side of the Indian Ocean.

Currents (and the monsoon winds that drive them) reverse direction each year in most parts of the tropical Indian Ocean, so the topic of how the cold water is supplied to the northern Indian Ocean and the warmed water is removed is a complex one. Two recent modeling studies, however, have provided a plausible picture of what the most important processes might be (McCreary et al. 1993; Wacogne and Pacanowski 1996). Once again, the annual mean winds of Fig. 7.2b provide at least a qualitative guide to the model behavior, despite the reversal of the winds over the year.

Briefly, the picture suggested by the models is as follows. The annual mean wind stresses of Fig. 7.2b are easterly in the southern Indian Ocean and westerly in the northern; consequently, the associated Ekman transports are southward both north and south of the Equator. Comparison of heat-flux climatologies like Fig. 7.2a with more detailed wind-stress maps like Fig. 7.2b suggests that the regions of strongest heat input into the tropical Indian Ocean, near the Somali, Arabian, and southern Indian coasts, are also regions of strongest wind-driven coastal upwelling. This upwelled water is then carried southward nearly to the Equator by Ekman transports, warming as it goes. A shallow (50 m–deep) cell develops across the Equator in one of the models (Wacogne and Pacanowski 1996), to provide mass continuity between the two regions of southward surface Ekman drift. The water carried south at the surface must be replaced from below. In both models, this occurs through a northward flow of water, originating partly from the southern Indian Ocean and partly from the Indonesian throughflow. This northward flow occurs in the Somali western boundary current, within the thermocline. The models cannot reproduce the deep circulation that Toole and Warren (1993) found (see section 7.6), so it is also possible that some of the deep inflow Toole and Warren found mixes right to the surface; this is another question that must await the completion of the World Ocean Circulation Experiment (WOCE) Indian Ocean campaign before it can be resolved.

The use of annual mean winds for any quantitative estimate of the heat fluxes in this region would be dangerous, but since the strong Asian summer monsoon winds are very like a stronger version of the annual mean winds of Fig. 7.2b, the true picture of the heat flux into the tropical Indian Ocean may be much as we have described, but with the main activity concentrated in northern summer.

e. Tidal mixing in the Indonesian region

As noted in section 7.3, the Indonesian region is unusual in that heat is strongly absorbed in a region that has very high SSTs. This is probably due to the action of internal tides. Very vigorous tidal flows occur in various parts of Indonesia, and the thermocline has been observed to rise and fall by tens of meters on tidal timescales (e.g., Ffield and Gordon

1992). Gordon (1986) noted that the water-mass structure changes quite markedly along the path of the Indonesian throughflow, not only at the surface but throughout the thermocline. He ascribed this to the action of vertical mixing by internal tides and found that an eddy diffusivity κ of about 10^{-4} m^2 s^{-1} was needed to account for the water-mass changes. Since the vertical temperature gradient T_z just below the mixed layer is typically 0.1°C m^{-1}, this implies a downward turbulent heat flux of $\rho C_p \kappa T_z$ in the upper thermocline of about 40 W m^{-2}—comparable to the observed excess of heat flux in the Indonesian region compared to neighboring regions (Fig. 7.2a). The very warm water of the western equatorial Pacific (Fig. 7.6) is believed to enter the tidal Indonesian region from south of the Philippines and pass through to the Indian Ocean; this passage of a warm-water flow through a localized region of intense tidal mixing allows the heat to be absorbed into an already warm part of the ocean.

f. Warm-pool heat budget

The above five subsections indicate how the ocean absorbs or releases heat in different parts of the globe, to give a qualitative account of the major features of Fig. 7.2a. It was also noted, however, that the heat fluxes of Fig. 7.2a are very uncertain. A concerted attempt has been made to reduce these uncertainties near the position of the global maximum of annual mean SST at 0°S, 155°E (Fig. 7.1), the site of the TOGA COARE experiment (Webster and Lukas 1992). This region was selected for intensive study, partly because atmospheric modeling studies have indicated that the atmosphere may be much more sensitive to SST anomalies here than elsewhere (e.g., Palmer and Mansfield 1984). The problems of uncertainties in heat-flux estimates are also particularly acute in this region. Godfrey and Lindstrom (1989) noted a range of between 20 and 80 W m^{-2} for estimates of annual mean heat flux into this region. Using ocean-based estimates of heat fluxes such as those discussed above, they suggested that the ocean was unable to absorb more than about 10 W m^{-2} in this region on long-term mean. Model studies by Gent (1991) tended to support this conclusion. Bradley et al. (1991, 1993) and Godfrey et al. (1991) combined direct heat-flux measurements with ocean heat-budget closures to confirm that the annual mean heat flux into this region is probably at or below the lower end of the 20–80 W m^{-2} range. They also identified problems with some of the heat-flux algorithms used. Part of the reason that the ocean absorbs so little heat is simply that the region lies near the SST maximum of the world—by definition, SST gradients are small and advection is also small. A second reason is that the region receives heavy rainfall. The base of the mixed layer is therefore often defined by a salinity gradient rather than a temperature gradient. The region with low temperature gradient and low diffusivity just below the mixed layer acts as a "barrier layer," preventing turbulent mixing of heat into the thermocline (e.g., Delcroix 1987; Lukas and Lindstrom 1991).

Weller and Anderson (1996) report accurate surface-flux measurements at a mooring at 1°45′S, 156°E—the center of the COARE intensive flux array. They show that a 1D mixed-layer model driven by these fluxes accounts surprisingly well for SST changes observed over the TOGA COARE experiment of November 1992 to February 1993. While horizontal advection is probably not negligible, problems with the mixed-layer model and the assumptions on water transparency are just as likely to be responsible for discrepancies between their modeled SST and observations.

Such "barrier layers" can have important consequences for ocean heat fluxes. For example, winds along the western Bay of Bengal are, on average, favorable to upwelling; one would therefore expect strong heat gain by the ocean here. During the upwelling season, however, there is also intense rainfall and runoff, so the upwelled water is saltier rather than cooler than the surface water, and little heat absorption takes place (Shetye et al. 1991).

g. Mixed-layer depths and subduction

We close this section by returning to the far Southern Hemisphere, where lack of shipping routes preclude direct estimates of surface heat fluxes from marine meteorological observations. The properties of ocean mixed layers provide one tool for improving our knowledge of heat fluxes in this region. Because of the massive heat capacity of water, poleward-flowing ocean currents such as the East Australian Current, Brazil Current, and Agulhas Current (or the Gulf Stream) lose heat over several years. A water column moving poleward in such a boundary current gains heat each summer, but then loses even more heat the following winter because the column finds itself farther poleward. As a result the winter mixed layer deepens year by year through convective overturn. This means that winter mixed-layer depth increases along the path of the current. In a qualitative way, the depth of the mixed layer indicates the amount of heat lost by the ocean along its circulation path. In contrast, a water column moving equatorward in the interior of the subtropical gyre will tend to gain heat over the annual cycle as it moves toward warmer latitudes. A new, shallow mixed layer of warmer water is formed over the deeper, cooler mixed layer formed during previous winters. The deeper layer is then isolated from the atmosphere and is said to be "subducted" (Woods and Barkmann 1986). The subduction of surface waters caused by variation of the seasonal cycle along the trajectory of a water column is the primary means by which the water of the ocean thermocline is renewed.

Thus, mixed-layer depths and temperatures, in combination with ocean currents, provide an integral measure of the mean heat flux into/out of the water column, and the very sparse ocean observations are nevertheless useful for constraining estimates of surface fluxes. Hydrographic sections completed as part of WOCE, in combination with ocean models containing recent improved algorithms for mixed-layer physics (e.g., Large et al. 1994), provide the best hope of improving and filling the gaps in maps such as Fig. 7.2a.

7.6. Meridional ocean heat transport

The general impression provided by Fig. 7.2a is that the oceans gain heat at low latitude and lose heat at higher latitudes, particularly over the western boundary currents. Indeed, to satisfy the annual mean radiation balance of the earth, the atmosphere and ocean together must transport heat from the Tropics, where the earth tends to gain heat, to the poles, where the earth tends to lose heat. The transport of heat in the atmosphere is discussed in chapter 2.

The meridional heat transport by the ocean can be estimated in several ways. Hydrographic sections across ocean basins can be used to measure the heat flux directly. Because it is generally not possible to measure the absolute velocity, additional assumptions are required to determine the net heat transport. In addition, measurements of the wind stress are needed to estimate the contribution of the Ekman flux. The calculation of Hall and Bryden (1982) is an example of the direct method of estimating ocean heat transport. Inverse methods can be used to refine the direct estimates by requiring the derived flows to satisfy additional constraints, such as mass conservation (e.g., Roemmich 1980).

A second, indirect technique is to estimate the oceanic heat flux as the difference between the observed or modeled meridional heat transport in the atmosphere and satellite measurements of the net energy flux leaving the top of the atmosphere (e.g., Carissimo et al. 1985; Hartmann et al. 1986; Trenberth and Solomon 1994; Keith 1995). Alternatively, estimates of surface heat flux such as shown in Fig. 7.2a can be integrated to determine how much heat the ocean must carry to support the observed heat transfer to the atmosphere. A difficulty with both the above techniques is that small biases in either the satellite measurements or the air–sea flux estimates can, when integrated over large areas, introduce large errors in the inferred ocean heat transports (Talley 1984).

Until recently, estimates of ocean heat flux derived from each of these methods have not agreed with each other (Bryden 1993). Ocean heat transports calculated as the difference between satellite and atmospheric measurements tend to be much larger than heat transports estimated from oceanographic data (Talley 1984). Recent studies by Trenberth and Solomon (1994) and Keith (1995), however, suggest that estimates based on the residual method now agree with direct ocean measurements to within the uncertainties in each. The convergence of results is primarily due to more accurate estimates of the energy leaving the top of the atmosphere from the Earth Radiation Budget Experiment satellites and better estimates of the atmospheric heat transport from ECMWF analyses than can be determined from observations, given the lack of atmospheric observations over the oceans. Direct heat-flux estimates from the global network of oceanographic sections collected during WOCE should further improve our understanding of the relative contributions of ocean and atmosphere to the meridional transport of heat.

As an illustration of our present understanding of global meridional heat flux, Fig. 7.7a shows the zonal integral of the meridional heat transport by the atmosphere and ocean, inferred from the ECMWF atmospheric model (Trenberth and Solomon 1994). (Figure 7.7a shows an example for a particular year, although the pattern is probably similar to the mean.) The ocean dominates the poleward transport of heat at low latitudes, while the reverse is true poleward of about 20° latitude. The pattern of meridional heat flux, however, is not the same in each ocean basin, as shown in Fig. 7.7b. The mechanism responsible for the meridional transport of heat varies from basin to basin as well.

In the North Atlantic, the dominant mechanism is the "overturning" or thermohaline circulation (Hall and Bryden 1982). Relatively warm water flows north in the Atlantic, cooling as it goes. Due to the high salinity of North Atlantic waters, cooling at high latitude makes the surface water dense enough to sink to depths greater than 2000 m. Southward flow of the newly formed NADW balances the northward flow of warmer water in the upper ocean.

Because of the large temperature contrast between the warm branch flowing north and the cold branch returning to the south, the overturning circulation associated with the formation of NADW is an efficient means of transporting heat northward. The large northward heat transport carried by this cell is responsible for the anomalous sign of the meridional heat flux in the South Atlantic (Fig. 7.7b): the northward heat flux due to the overturning circulation dominates the southward heat flux carried by the poleward flow of warm water in the Brazil Current.

In contrast, the meridional heat flux in the North Pacific is dominated by the more "horizontal" circulation of the wind-driven subtropical gyre (Bryden et al. 1991; Wilkin et al. 1995). Because the Pacific is relatively fresh compared to the North Atlantic, the surface waters in the North Pacific do not get dense enough to sink to great depth despite strong cooling by the atmosphere (Warren 1983). As a result, no overturning circulation exists in the Pacific. The poleward

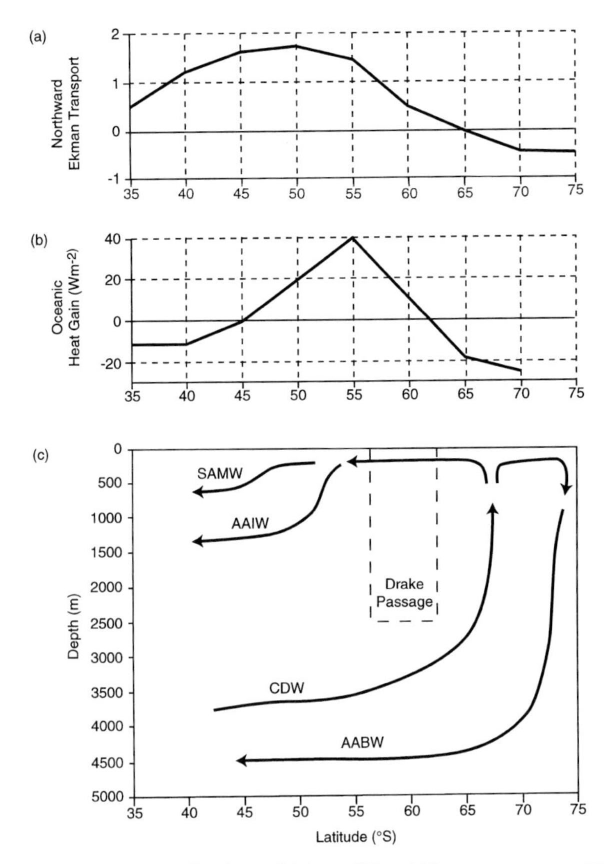

FIG. 7.9. Schematic view of (a) meridional Ekman transport, (b) oceanic heat gain (after Taylor et al. 1978), and (c) meridional transport of water masses in the Southern Ocean (after Sverdrup 1933). The divergent Ekman transport causes upwelling of circumpolar deep water (CDW). Some of the upwelled water is carried south, cooled, and converted to antarctic bottom water (AABW). The remainder is driven north beneath the westerlies, is warmed by the atmosphere, and ultimately feeds the formation of antarctic intermediate water (AAIW) and subantarctic mode water (SAMW). Note that no mean meridional flow can take place across the Drake Passage gap (shown as the dashed rectangle between 56° and 62°S), except in the Ekman layer.

heat flux results from poleward flow of warm water in the western boundary current (the Kuroshio) and a return flow of cooler (upper-ocean) water in the interior of the basin. Note the contrast in surface heat flux over the Atlantic and Pacific Oceans north of 50°N in Fig. 7.2a. Sinking of surface waters in the high-latitude North Atlantic draws a compensating flow of warm surface waters poleward, leading to large heat loss over the subpolar gyre; in the Pacific, the absence of sinking means less poleward flow of warm water and less heat loss to the atmosphere.

Due to the link provided by flow through the Indonesian passages, the meridional heat transport of the southern Indian and South Pacific basins must be considered together. The net heat flux is poleward, due to a combination of processes. The Indonesian throughflow is fed by cool water, which enters the South Pacific from the south and is gradually warmed within the basin before flowing through the Indonesian passages into the Indian Ocean. The warm throughflow water eventually flows south in the Agulhas and Leeuwin Currents. Equatorward flow of cool water in the Pacific compensated by poleward flow of warm water in the Indian carries a net poleward heat flux. A second contributor to the poleward heat flux in the Indian Ocean is the conversion of deep water to less dense intermediate waters (Toole and Warren 1993). In Toole and Warren's scheme, 27 Sv of relatively cold deep water flows north from the Southern Ocean into the Indian Ocean, where it is gradually converted through vigorous vertical mixing to warmer intermediate waters, which return to the south. This overturning in the Indian Ocean suggests that vertical-mixing processes at depth are more vigorous in the Indian Ocean than in the other two oceans—but if this is so, the reasons are not yet understood.

In each ocean, there is the possibility that ocean eddies play a role in poleward heat transport. There is little data for testing this idea critically. Bryan (1991a), however, in a review of poleward heat transport in a hierarchy of models, suggested that eddy heat transport tended to be important only in limited regions, such as the Southern Ocean. Elsewhere, changes in the mean flow tended to adjust to counter the effects of eddy heat transport.

7.7. Interbasin exchange

One particularly important factor that distinguishes the Southern Hemisphere oceans from their Northern Hemisphere counterparts is the possibility of exchange of mass, heat, and salt between the ocean basins. The Northern Hemisphere oceans are surrounded by land and nearly isolated from each other, except for a small flow from the Pacific to the Atlantic Ocean by way of the Arctic. By contrast, the ocean basins are linked at their southern end by the circumpolar Southern Ocean. The massive eastward flow of the ACC is the primary means by which mass, heat, and salt are carried from one basin to another. The flow from the Pacific to the Indian Ocean through the Indonesian passages, although much weaker than the ACC, also has a significant impact on the global redistribution of heat, as it provides the only interbasin exchange of warm water. The ACC and the Indonesian throughflow each play an important part in transforming the phenomenon of ocean heat transport from a regional to a global one.

a. Indonesian throughflow

This topic has been reviewed recently by Godfrey (1996), so discussion here will be brief. Much of the present conceptual basis for assessing the role of the Indonesian throughflow on ocean climate comes from a series of "thought experiments" with numerical ocean models, referred to in section 7.4b. In Hirst and Godfrey (1993), a global ocean GCM with 1° × 1° resolution and 14 vertical levels was driven with the winds of Fig. 7.2b. Heat fluxes were diagnosed by relaxing model surface temperature strongly toward an "observed" field that was close to observed SST. The ocean model was run to steady state in two configurations, one with the Indonesian passages open and another with them closed. Figure 7.8 shows the difference in the diagnosed heat fluxes between these two runs.

It is at first sight surprising that differences in surface heat flux occur in regions so far from the Indonesian passages, but a simple physical explanation can be found for most regions. First, the main difference in currents between the two runs consists of a westward flow across the Indian Ocean south of Indonesia and an eastward flow from south of Africa to Tasmania. These are connected by a southward western boundary current along Africa (enhanced Agulhas Current) and a northward flow along eastern Australia (reduced East Australian Current). The band of increased heat loss in the southern Indian Ocean occurs because the difference flow brings more warmed water down the Agulhas Current, which then loses heat in its three-year passage across the southern Indian Ocean. The intense band of heat gain off southeastern Australia corresponds to a reduced outflow of the East Australian Current. The region of increased heat loss off northwestern Australia, near 20°S, 100°E, comes from an increased temperature gradient here, resulting in an increase in the advection of heat by Ekman flow discussed in section 7.3. Further heat loss occurs along the Western Australian coast due to an increased Leeuwin Current. Opening the Indonesian gaps results in increased thermocline depth throughout the Indian Ocean and decreases in most of the rest of the World Ocean, to provide geostrophic balance for the currents just discussed. As a result, upwelled water is colder in the eastern tropical Pacific, leading to more heat gain there, and is warmer in the upwelling regions of the northern Indian Ocean, leading to reduced heat gains there. The region of heat gain off Chile is associated with a reduction in a Leeuwin Current–like flow pattern there. The authors were not able to account convincingly for the region of heat loss north of Japan, in Fig. 7.8.

Hirst and Godfrey (1994) considered the time-dependent problem of the response of these heat-flux patches to a sudden opening of the Indonesian throughflow. The time taken to establish these patches of heat-flux difference varied from a few months (in the case of the region off northwestern Australia) to years (for the patches in the southern Indian Ocean and southern East Australian Current) or decades (for the patch off southern Chile).

Gordon (1986) suggested that by linking the Indian and Pacific Oceans, the Indonesian throughflow might influence ocean circulation on even larger scales than indicated above. This issue is considered in more detail in section 7.8.

b. Southern Ocean

As might be anticipated from the gaps in Fig. 7.2a, the Southern Ocean is the most poorly observed region of the ocean. It is clear from the observations that do exist, however, that the Southern Ocean plays a particularly active role in maintaining the earth's climate. Many of the unique features of the Southern Ocean circulation result from the presence of Drake Passage, the only latitude band where there are no north–south continental barriers.

The atlases of Gordon et al. (1982) and Olbers et al. (1992) provide a good introduction to the hydrography of the Southern Ocean. The major circulation feature of the Southern Ocean is the ACC, which circles Antarctica from west to east. The ACC is the largest current system in the World Ocean, with a transport of about 130 Sv (Nowlin and Klinck 1986). The ACC is a deep-reaching current that appears to be strongly "steered" by the bottom topography. Marshall (1995) has recently shown that an inviscid, adiabatic flow that conserves density and potential vorticity in an ocean with realistic stratification will tend to be steered by, but not flow strictly along, the topography: the flow follows characteristics that lie between the lines of constant f/H found in a homogeneous ocean (where f is the Coriolis parameter and H the ocean depth) and the lines of constant latitude found in the limit of very strong stratification.

The lack of landmasses in the latitude band of Drake Passage means that there can be no net zonal pressure gradient above the shallowest bathymetry and, hence, no net geostrophic meridional flow. The impact of the Drake Passage gap on the global ocean circulation is profound, as demonstrated by the numerical experiments of Cox (1989). Cox compared three experiments: a global ocean with no gap at Drake Passage, driven by surface buoyancy fluxes but no wind stress; an ocean with an open Drake Passage (and similar forcing); and an experiment with wind forcing as well as buoyancy forcing. With no opening at Drake Passage, the global overturning is dominated by vigorous sinking (50 Sv) near Antarctica.

When a Drake Passage–like gap is introduced into the model, the poleward flow of warm water in the Southern Hemisphere is blocked by the geostrophic constraint in the gap, the overturning is much weaker (15 Sv), and water masses formed in the Northern Hemisphere account for a larger volume of the ocean, as observed. With the addition of realistic winds, the ACC strengthens, further isolating the high southern latitudes. Drake Passage in this way controls the distribution of water masses throughout much of the intermediate and deep ocean.

The present best estimate of the net oceanic heat loss south of the ACC is 3×10^{14} W, although the uncertainty in this number is large (Gordon and Taylor 1975). Because meridional geostrophic flow is not possible in the latitude band of Drake Passage, the poleward ocean heat flux required to balance the observed heat loss must be supplied by some other mechanism. De Szoeke and Levine (1981) demonstrated (by elimination of other mechanisms) that the poleward heat flux is likely accomplished by transient eddies. In this sense—no zonal boundaries, hence no net meridional geostrophic flow, and a dominant contribution by eddies to the meridional flux of heat—the Southern Ocean is the most direct oceanic analogue of atmospheric flows.

The nature of the dynamical balance of the ACC has been a focus of attention for several decades. Both the driving force of the ACC and the way in which the forcing is dissipated are not yet completely clear. Undoubtedly, the strong westerly winds play a part in driving the current, and indeed Sverdrup dynamics seems to explain much of the structure of the ACC north of Drake Passage (Stommel 1962; Baker 1982). In particular, the curl of the winds south of 50°S in Fig. 7.2a is negative, driving southward Sverdrup flow. Stommel (1962) suggested the Falklands Current flowing north along the east coast of South America provided the western boundary current required to balance the southward Sverdrup transport over the rest of the circumpolar ocean. Several authors have used output from the Fine Resolution Antarctic Model (FRAM) (FRAM Group 1991) to show that Stommel's suggestion seems to hold for this nearly eddy-resolving model (Saunders and Thompson 1993; Wells and de Cuevas 1995). Heat loss to the atmosphere near the Antarctic continent, however, likely contributes to the meridional pressure gradient that supports the strong eastward geostrophic flow of the ACC [e.g., as in the buoyancy-forced solutions of Cox (1989)].

The way in which the momentum put in by the wind is balanced has also been a topic of debate. Recent work has confirmed the early suggestion of Munk and Palmen (1951) that the torque exerted by the wind is balanced by form stress acting on the bottom topography. Johnson and Bryden (1989) emphasized the role of transient eddies in both the heat and momentum budget of the Southern Ocean: baroclinic instability results in downward eddy momentum flux as well as poleward eddy heat flux. Marshall et al. (1993) reached a similar conclusion through consideration of the potential vorticity dynamics of the ACC, noting in addition that transient eddies must act to transfer potential vorticity northward across the time–mean axis of the ACC. The overall balance between wind stress acting at the sea surface and form stress or bottom-pressure torque acting on the sea floor has been confirmed in high-resolution quasigeostrophic models (Treguier and McWilliams 1990; Wolff et al. 1991), channel models with simplified dynamics (Klinck 1992; Straub 1993), and in FRAM (Killworth and Nanneh 1994; Wells and de Cuevas 1995). Analyses of current meter measurements (Bryden and Heath 1985), altimeter data (Morrow et al. 1992), and the FRAM model (Killworth and Nanneh 1994) all agree that the contribution of the Reynolds stresses to the zonal momentum balance is small.

7.8. Air–sea–ice exchange and water-mass formation

The Southern Ocean is a region of intense air–sea exchange of momentum, energy, and dissolved gases. The interaction between the atmosphere and ocean modifies the water masses at the sea surface and drives vertical as well as horizontal motions.

Of particular importance for the ocean and its coupling to the atmosphere is the formation and melting of sea ice around the margin of Antarctica (see also chapter 4 in this monograph). In winter, the antarctic sea-ice pack occupies a larger area than the continent itself; most, but not all, of the sea ice melts in summer. When seawater freezes, most of the salt is left behind in the water underlying the ice. The rejection of brine during sea-ice formation plays an important part in the formation of dense water near Antarctica. The winds are essential to this process: a net release of salt over the annual cycle requires more ice formation than ice melt, so regions of net brine release correspond to regions where the wind tends to remove newly formed sea ice before it melts. Similarly, net melting of sea ice provides a source of freshwater that plays an important part in stabilizing the water column at high latitude.

Sea ice plays a number of other important roles in the climate system that can only be touched on briefly here. A continuous ice cover acts as an insulating lid, preventing the exchange of heat, moisture, and gases between the ocean and the atmosphere. Conversely, gaps in the ice cover such as leads or polynyas are areas of greatly enhanced exchange; for example, heat fluxes through leads can exceed 600 W m^{-2} (Andreas et al. 1979). Finally, ice has a much higher albedo than seawater. Changes in sea-ice extent can change the albedo of the earth as a whole, providing a potential positive climate feedback. Model studies suggest that interactions between sea ice and the thermohaline circulation can drive climate variability on timescales of decades and longer (Yang and Neelin 1993; Zhang et al. 1995). Readers interested in a more complete review of the role of sea ice in the climate system are referred to the review by Gow and Tucker (1990).

From the ocean's point of view, perhaps the most important impact of the antarctic sea ice is the release of brine. As mentioned above, the salt released during the formation of sea ice increases the density of shelf waters sufficiently to drive surface waters to sink to the sea floor. Interaction of seawater with the land ice flowing off the continent in the large ice shelves can also play a significant part in increasing the density of water near Antarctica. The dense water spills off the shelf and sinks, mixing as it goes, to form antarctic bottom water (AABW). AABW spreads northward to cool and ventilate a large fraction of the deep sea. The total formation rate of AABW is not well known. Most estimates suggest less than 10 Sv of dense water leaves the Antarctic continental shelf. The dense plumes entrain water as they sink, increasing the transport of AABW to perhaps 20 Sv or more. In other words, the Antarctic produces a volume of dense water that is about equal to or a little higher than the volume of dense water produced in the North Atlantic.

Near the axis of the ACC, two other important water masses are formed that ventilate the thermoclines of the Southern Hemisphere subtropical gyres. Antarctic intermediate water (AAIW) can be traced as a salinity minimum layer, with a temperature of 4°–5°C, from its source in the Southern Ocean into each of the Southern Ocean basins and indeed as far as the North Atlantic. The formation of AAIW is not completely understood, but probably involves a combination of convection (McCartney 1982) and subsurface mixing (Molinelli 1981; Piola and Georgi 1982). Subantarctic mode water (SAMW), produced by deep convection on the equatorward flank of the ACC, forms thick uniform layers, with temperatures in the range of 6°–13°C, that can also be traced from the Southern Ocean into the Southern Hemisphere gyres (McCartney 1977, 1982). Water masses formed in the Southern Ocean account for more than 50% of the volume of the World Ocean.

A simple schematic view of the meridional spreading of water masses is shown in Fig. 7.9c. Sverdrup (1933) first produced such a circulation scheme, based on tracing the spreading paths of property extrema such as the salinity maximum layer. Figure 7.9c shows deep water spreading south and upward in the Southern Ocean (this flow is deeper than the shallowest topography in the latitude band of Drake Passage, so zonal pressure gradients and meridional geostrophic flow are possible). The upwelling of deep water is driven by the divergent Ekman flux south of the westerly wind maximum shown in Fig. 7.9a (see also Fig. 7.2b). Part of the upwelled deep water moves south, is cooled, and is converted to denser AABW. The remainder is driven north in the Ekman layer and ultimately feeds the formation of AAIW and SAMW. In the conversion of upwelled deep water to warmer, less-dense intermediate water, the ocean gains heat from the atmosphere. This results in the rather surprising shape to the net annual air–sea heat-flux pattern seen in Fig. 7.9b. The northward drift of upwelled cold water in the region of divergent Ekman flux results in a band of oceanic heat gain, counter to the overall tendency toward oceanic heat loss at high latitude. The equatorward Ekman drift of cool surface water likely accounts for the region of heat gain near 50°S in Fig. 7.5.

Figure 7.9c is clearly a simplified view of the spreading in the meridional plane, and it is not clear to what extent the spreading paths deduced from water-mass tracers represent actual flow paths. Estimates of wind stress over the Southern Ocean, however, suggest that about $30 \times 10^6\ m^3\ s^{-1}$ of water upwells south of the ACC, which must be supplied by southward flow of deep water. To conserve mass, the southward flow of deep water must be balanced by an equal northward flow of AABW, AAIW, and SAMW. While Sverdrup's schematic view may not be accurate in detail, some circulation of this type must occur to balance the upwelling and downwelling driven by the strong westerly winds. Estimates of meridional flow based on heat and salt budgets south of the ACC give similar numbers (Gordon and Taylor 1975).

The circulation of the Southern Ocean in the meridional plane has been further illuminated by recent analyses of FRAM. Ocean models generally show a zonal-mean overturning streamfunction that differs significantly from the schematic picture shown in Fig. 7.9 (e.g., Manabe et al. 1990; Semtner and Chervin 1992). In particular, the models show an overturning cell consisting of northward flow in the Ekman layer balanced by deep southward flow, with strong upwelling and downwelling limbs connecting the two. This cell became known as the "Deacon cell" (Bryan 1991b), although Deacon may have been dismayed to have his name associated with the feature, for the following reason: the strong upwelling and downwelling limbs of the Deacon cell appear to cut across isopycnals, a pattern that is inconsistent with the observations of Deacon and subsequent investigators.

Doos and Webb (1994) used FRAM results to explain the apparent inconsistency. They showed that the strong upwelling and downwelling limbs of the Deacon cell result from taking zonal averages along depth surfaces. When the zonal integration is performed along density surfaces, the Deacon cell disappears (Hirst et al. 1996). Water parcels in the model flow along, not across, density surfaces, as expected. The Deacon cell results from the density convergence associated with correlations of zonal variations of density and meridional flow at constant depth (Karoly et al. 1996). In the Southern Ocean, these correlations are dominated by "standing eddies" or gyres, rather than transient eddies (e.g., at a given depth, northward boundary currents like the Malvinas Current off Argentina carry warmer water than the southward return flow in the ocean interior). The streamfunction formed by integrating on density surfaces is broadly similar to the pattern inferred by Sverdrup and Deacon from observations of water-mass properties. Thus, the Deacon cell in the Southern Ocean is analogous to the Ferrel cell in the atmosphere (Karoly et al. 1996), as discussed for the atmosphere in chapter 2.

Recent theoretical work on improved parameterization of eddies in ocean models has provided another perspective on the meridional circulation in the Southern Ocean (Gent and McWilliams 1990; Danabasoglu et al. 1994). Gent et al. (1995) show that the velocity at which tracers are advected is the sum of the large-scale velocity and an eddy-induced transport velocity that represents the effect of mesoscale eddies. The idea that tracers are not advected by the large-scale velocity alone is an old one in the meteorological community (e.g., Andrews et al. 1987; Plumb and Mahlman 1987), but has only recently been introduced in oceanography. Gent et al. (1995) estimate the strength of the meridional overturning streamfunction in the Southern Ocean associated with the eddy-induced transport velocity to be 18 Sv (for a particular choice of thickness diffusivity of 10^3 m^2 s^{-1}).

Formation of water masses in the Southern Ocean influences climate in several ways. First, the Southern Ocean water masses are the main way in which the abyssal and intermediate layers of the Southern Hemisphere oceans are ventilated. The rate at which these water masses are formed determines in part the rate at which heat and gases such as carbon dioxide are sequestered by the ocean.

A second important aspect of water-mass formation in the Southern Ocean is its connection to the global overturning cell associated with the formation of NADW. In recent years, increasing attention has been paid to the issue of how the overturning circulation is "closed": how is NADW converted to the relatively warm water flowing north in the Atlantic to feed the sites of deep-water formation? It is becoming clear that the Southern Ocean plays a major part in the transformation of NADW to thermocline water.

Vertical processes occurring in the ocean are generally too weak to observe directly. In the absence of clear sites of enhanced upwelling in the deep sea, the upwelling required to balance the sinking in the North Atlantic has been assumed to take place more or less uniformly over the World Ocean. This is the basis of the Stommel and Arons (1960) model of the abyssal circulation. Gordon (1986) also assumed that the upwelling balancing NADW formation was uniformly distributed around the globe, so that most of the conversion of deep to thermocline water took place in the Pacific, the ocean basin with the largest area. In Gordon's scheme, the warm water flowed into the Indian Ocean via the Indonesian passages, where it was joined by water upwelled in the Indian Ocean, and ultimately returned to the Atlantic around the southern tip of Africa. This scenario became known as the "warm-water path."

The warm-water path as described above was the basis of a now-famous schematic diagram of the "ocean conveyor" drawn up by Broecker (1987, 1991). The long, looping conveyor belt connecting the three ocean basins is a powerful image, which in a simple and direct way emphasizes the fact that the formation of dense water in the North Atlantic is part of a global circulation. Like all good schematics, some details were left out in order to illustrate the essence of the global overturning circulation in the clearest possible way. In particular, the nature of the upwelling branch of the overturning circulation, where the dense water produced in the North Atlantic is recycled into the less-dense water that returns to the Atlantic to close the loop, is not well represented in the ocean conveyor "logo" (Broecker 1991).

For example, Rintoul (1991) questioned the dominance of the warm-water path, showing instead that the export of NADW from the Atlantic was mostly compensated by cooler intermediate water entering the South Atlantic from the Pacific Ocean through Drake Passage. More recent studies have supported the idea that the "cold-water path" is the main way in which the global overturning circulation is closed (e.g., Gordon et al. 1992; Toggweiler and Samuels 1993a,b; Macdonald 1993; England and Garcon 1994; Schmitz 1995). The relatively cold water entering the Atlantic through Drake Passage gains heat and is made less dense as it flows north through the South Atlantic to compensate the export of NADW. The changes in heat and salt content of the northward-flowing water occur through heating by the atmosphere (Rintoul 1991), mixing with subtropical water in the Brazil–Malvinas confluence (Schmitz 1995), and mixing with Indian Ocean water. Some heat and salt from the Indian Ocean enters the Atlantic as part of a large-scale gyre that connects the South Atlantic and southern Indian basins (Gordon et al. 1992), and some enters in rings pinched off from the Agulhas Current (Gordon and Haxby 1990).

Note that the debate about "warm path" vs. "cold path" does not hinge on whether any water from the Indian Ocean enters the Atlantic—some Indian–Atlantic exchange certainly takes place. The question is really one of where the deep water gets converted to upper–layer water and the temperature of the upper branch. In the "warm-water path," water upwells from the abyss into the upper thermocline in the Indian and Pacific Oceans and enters the Atlantic as very warm (15°C; Gordon 1986) Indian Ocean water that flows north to the North Atlantic. The cold-water path is closed by converting deep water to intermediate water, primarily in the Southern Ocean (as described below). The review by Schmitz (1995) provides a good update of present ideas on the global overturning circulation or "ocean conveyor."

Broecker's ocean conveyor logo should not be interpreted as showing the main pathways associated with heat transport in the ocean. Most of the heat carried south in the subtropical Indian Ocean by the Agulhas

Current is lost to the atmosphere at higher latitudes within the Indian Ocean (in the large region of oceanic heat loss over the Agulhas extension in Fig. 7.2a). The unusual equatorward heat flux in the subtropical South Atlantic does not require an injection of warm water from the Indian Ocean; the water flowing north in the thermocline across 30°S is sufficiently warm relative to the southward-flowing deep water that the overturning circulation carries a significant equatorward heat flux (Rintoul 1991).

Whether the return flow compensating NADW export occurs via the warm or cold route, there is still the necessity to convert roughly 15–20 Sv of deep water to less-dense intermediate water somewhere in the ocean. The classical view is that the conversion takes place through vertical mixing and advection in the ocean interior (Stommel and Arons 1960; Munk 1966). Recent direct measurements of mixing in the ocean interior typically find values that are an order of magnitude too small to support this much upwelling (Toole et al. 1994). One possibility is that enhanced mixing near the boundary of the ocean is sufficient to close the heat and mass budget of the deep ocean (e.g., Polzin et al. 1997), although the number of direct measurements of near-boundary mixing is still small.

Another possibility is that the conversion is driven by air–sea fluxes, where the density layers corresponding to NADW outcrop in the Southern Ocean (as illustrated in Fig. 7.9). Toggweiler and Samuels (1993b), for example, showed that the radiocarbon distribution in the ocean was most consistent with conversion of deep to intermediate waters taking place in the Southern Ocean, rather than by upwelling through the main ocean thermocline at lower latitudes. The global ocean model of Hirst (1998) also suggests that this is what occurs. Hirst "tagged" water in the region of NADW formation and traced the flow of NADW through the model. Most of the NADW returned to the upper ocean in the Southern Ocean, where it was warmed and freshened by the atmosphere and converted to less-dense water, while being driven north by Ekman transports.

A number of interesting questions remain concerning the global overturning circulation. Although model and observational studies hint at the importance of the Southern Ocean in closing the loop, details of the closure are still sketchy. It is also not clear how the overturning circulation is driven. Is the cell "pushed" by the sinking of NADW, or "pulled" by the processes acting to convert deep water to less-dense water? What controls the strength of the overturning circulation? On the one hand, a number of models suggest that the presence of a fresh surface layer in the North Atlantic can shut down the formation of NADW, slowing the overturning (e.g., Bryan 1986; Manabe et al. 1991). On the other hand, Toggweiler and Samuels (1993a) show that the strength of the NADW overturning circulation scales with the strength of the wind stress (and northward Ekman drift) at the latitude of Drake Passage in their model. Undoubtedly, both wind and buoyancy forcing play some part in regulating the strength of the overturning circulation, but the details are not yet clear.

7.9. Influence of the oceans on climate change and variability

Due to the large thermal inertia of seawater, the ocean tends to respond slowly to changes in forcing. The atmosphere, in contrast, responds quickly and on timescales greater than a few weeks is in quasi-equilibrium with other components of the earth's climate system. As a consequence, changes in climate on timescales of years to centuries are largely controlled by ocean currents.

We have seen that the pattern of air–sea heat exchange is the result of convergence or divergence of heat carried by ocean currents, particularly the wind-driven circulation. Therefore, changes in the distribution of wind stress are likely to cause changes in ocean currents and changes in sea surface temperature (e.g., Sprintall et al. 1994). The atmosphere will respond to the change in its bottom boundary condition, perhaps leading to changes in winds that may feed back on ocean currents.

Another way to illustrate the impact of ocean circulation on climate change is to compare coupled models in which the ocean is represented as a motionless slab and coupled models with a full dynamical ocean model (Manabe et al. 1991). In the case of a slab ocean, the warming due to a doubling of carbon dioxide is roughly symmetric about the Equator. When carbon dioxide concentrations are gradually increased in an atmospheric model coupled to a dynamically active ocean, the warming in the Southern Hemisphere is delayed relative to the Northern Hemisphere. The delayed response in Southern Hemisphere surface temperatures is due to the deep penetration of heat in the Southern Ocean caused by deep mixing and water-mass formation. [Note, however, that the model of Manabe et al. appears to overestimate the extent of deep mixing in the Southern Ocean, hence possibly overestimating the delay in the Southern Hemisphere (England 1995)]. Coupled models are discussed in more detail in chapter 10.

Although it is straightforward to sketch mechanisms by which changes in forcing of the coupled ocean–atmosphere system may lead to changes in climate, we have not yet reached the point where we can make confident predictions of future climate. Nevertheless, much progress has been made in recent years in simulating the interactions between the ocean, atmosphere, and land surface that determine the earth's climate. It is not possible to review the large and rapidly expanding literature on coupled ocean–atmosphere modeling here (see chapter 10), but a few examples provide useful illustrations of the role of ocean currents in the coupled climate system.

a. The El Niño–Southern Oscillation phenomenon

ENSO was the focus of a major international program (TOGA) during the decade 1985–94. A textbook is available on ENSO (Philander 1990), so here our discussion is brief. The most dramatic effect of ENSO events is associated with disturbances to the delicate temperature balance in the equatorial Pacific referred to in section 7.5c. As described by Philander, equatorial temperature patterns like Fig. 7.6 vary strongly between El Niño and La Niña conditions—that is, between warm or cold conditions in the eastern equatorial Pacific. Figure 7.6 was obtained in 1979, a La Niña year; in an El Niño year, the near-surface thermal stratification is weaker in the east Pacific and SST rises, due to the action of downwelling equatorial Kelvin waves generated in the western Pacific by westerly winds. A positive feedback occurs: warmer SSTs in the eastern Pacific tend to induce central Pacific rain, which in turn causes more westerly winds in the west Pacific.

While the above statements would generally be agreed upon by ENSO researchers as being a major part of the ENSO mechanism, several aspects of ENSO remain to be explored. First, the magnitude of the throughflow varies strongly with ENSO, being stronger in La Niña conditions than in El Niño (Meyers 1996). This can affect the ENSO mechanism in two main ways. First, in a linear model it reduces the "reflectivity" of the western Pacific to ocean Rossby waves, impinging on it by about 20% (e.g., Clarke 1991); thus, a westerly wind burst will have a stronger warming effect than it would have if the throughflow were blocked. Second, it may change SST in the Indonesian region itself, both by reducing the magnitude of the Indonesian throughflow and by causing the thermocline to shallow. The latter effect will lead to greater tidal mixing, while the former will reduce the rate at which Pacific water advects through Indonesia; both will tend to cool Indonesian SSTs. These changes may potentially have a large effect on the atmosphere, because of their location near the world's SST maximum (Palmer and Mansfield 1984).

An intriguing question is whether the magnitude or frequency of ENSO are likely to change in a changing climate. Given the global impact of ENSO on interannual variations in climate, it is possible that one of the major effects of an enhanced greenhouse warming could be through changes in the nature of ENSO. To date, results from coarse resolution suggest that ENSO amplitude and frequency do not change dramatically with increasing CO_2, though mean SST gradient across the Pacific may change. This is currently a topic of intense research interest; Knutson et al. (1997) summarize recent results.

b. Indian Ocean SST dipole

Australian rainfall is significantly correlated with a pattern of SST anomalies known as the Indonesian–Indian Ocean SST dipole (IIOD) (Nicholls 1989; Drosdowsky 1993). One of the poles (pole A) lies between Indonesia and northwestern Australia. The other pole (pole B) is usually centered near 13°S, 80°E in the southern Indian Ocean. When pole A is anomalously warm and pole B is anomalously cool (known as warm polarity), Australia receives enhanced rainfall in a broad swath extending from northwestern Australia to Victoria. Low rainfall results from cool polarity. The anomalies are often first evident in autumn and become fully developed during winter. When cool polarity occurs at the same time as El Niño, as occurred in 1982–83 and in 1994, the drought in Australia can be extremely severe. When warm polarity occurs during El Niño, as in 1986 and 1991–92, the drought in Australia can be mild or may not occur at all.

Understanding the oceanography of the IIOD is still at an early phase. Qu et al. (1994) used expendable bathythermograph (XBT) data collected since 1983 and the results of an ocean GCM to document the currents in pole A and their influence on the variation of SST during an annual cycle. The study showed that SST in the area south of Java is warmed by the Indonesian throughflow and cooled by upwelling on the southern coast of Java. These two mechanisms are individually large terms in the heat budget, but they are balanced in the mean annual cycle in a way that produces relatively small changes of SST. Since the throughflow is driven primarily by Pacific winds and the upwelling primarily by local winds, interannual variations in the phase of the two mechanisms could upset the balance and generate SST anomalies. Further work is needed to determine the oceanographic processes responsible for SST variations at pole B.

c. Stability of the thermohaline circulation

In section 7.5 we discussed the significance of the overturning circulation in the meridional transport of heat. Modeling the overturning circulation places severe demands on global models, due to the long timescales involved. It is not possible to run a prognostic primitive equation model with even moderate resolution long enough for the deep circulation to reach equilibrium, given present computer resources. Given the importance of the thermohaline circulation to the poleward transport of heat in the ocean, there is a need for simpler models that are capable of long integration periods with which to explore the dynamics of the thermohaline circulation.

One approach to simplifying the primitive equations, thus permitting simulation of long-timescale phenomena, is to radically simplify the geometry of the ocean basins. For example, Stommel (1961) represented the global ocean as two well-mixed reservoirs connected by pipes. By varying the restoring time for temperature and salinity in the surface forcing, Stommel found up to three steady states, one of which was unstable.

Subsequent investigators, using a range of models including box models, two- and three-dimensional GCMs, and coupled ocean–atmosphere GCMs, have further explored the sensitivity of the thermohaline circulation to the surface forcing used to drive the model (Rooth 1982; Welander 1986; Bryan 1986; Marotzke et al. 1988; Manabe and Stouffer 1988; Maier-Reimer and Mikolajewicz 1989; Birchfield 1989; Weaver and Sarachik 1991; Zhang et al. 1993; and Tziperman et al. 1994). In particular, the models suggest that small changes in the freshwater forcing can lead to oscillatory solutions, stable and unstable states, and/or rapid transitions between states.

The existence of multiple equilibria or oscillations of the thermohaline circulation depends on the fundamental difference between thermal and haline forcing of the ocean (Marotzke and Willebrand 1991). The exchange of heat between the atmosphere and the ocean is closely related to the air–sea temperature difference; as a consequence, surface temperature anomalies tend to be damped by heat exchange with the atmosphere. In contrast, the freshwater flux between the atmosphere and the ocean surface does not depend on the surface salinity. Anomalies are not damped, the flow itself can strongly alter the salinity field, and the resulting feedback can force oscillatory behavior and transitions between equilibria.

The potential for multiple states of the thermohaline circulation—steady states, both stable and unstable, and oscillatory states—suggests that the internal dynamics of the ocean thermohaline circulation may drive climate variability on timescales of decades to millenia. The response of models of the thermohaline circulation to surface forcing, however, can be sensitive to details of the model design [e.g., choice of vertical and/or horizontal diffusivity (Winton and Sarachik 1993); vertical resolution (Weaver and Sarachik 1990, 1991); parameterization of deep convection (Marotzke 1991; Wright and Stocker 1991; Zhang et al. 1993); basin geometry (Zhang et al. 1992); and restoring time for the salinity boundary condition (Tziperman et al. 1994)]. In addition, there is some evidence that the ocean–atmosphere system is more stable than the simple models suggest. Although evidence from sediment and ice cores indicates the climate of the earth has undergone rapid changes in the past, the climate record is also characterized by periods of remarkable stability (e.g., Siegenthaler et al. 1988). Coupled models of the atmosphere and ocean also appear to be more stable than might be expected, based on the results of simple models (Gates et al. 1992).

7.10. Summary

The circulation of the atmosphere and ocean are coupled through the exchange of heat, moisture, and momentum at the sea surface. The annual mean air–sea heat flux varies strongly from place to place due to convergence or divergence of heat transport by ocean currents. Much of the pattern of air–sea heat flux is controlled by the wind-driven currents in the upper ocean. Upwelling and equatorward flow of cool water tends to cool the overlying atmosphere; poleward flow of warm water, as in the subtropical western boundary currents, tends to warm the atmosphere. In addition, the meridional heat transport carried by the global-scale overturning circulation associated with North Atlantic deep-water formation (and subsequent conversion to intermediate water and bottom water) helps to maintain the pattern of air–sea heat flux.

The ocean circulation is strongly controlled by the shape of the ocean basins. The presence of north–south barriers allows zonal pressure gradients and geostrophic meridional flows in each basin. As a result, advection by the mean meridional flows plays the dominant role in the meridional transport of heat, in contrast to the atmosphere, where eddy fluxes dominate. The exception to this rule is in the latitude band of Drake Passage, where the lack of land barriers prevents net meridional flow, and eddy fluxes of heat and momentum play a crucial role in the heat and momentum budget. The presence of Drake Passage also has a profound impact on the global distribution of water masses. A second, apparently inconsequential gap, the Indonesian passages, also has a significant impact on the regional and global ocean circulation and, hence, the pattern of atmosphere–ocean heat exchange.

The large heat capacity of the ocean relative to that of the atmosphere means that climate variability on timescales of years to centuries is largely controlled by ocean processes. The rate at which surface water is carried to the deep sea through vertical mixing and subduction of water masses determines the rate at which the ocean can store heat or gases such as carbon dioxide. The rate of water-mass formation and sinking in turn depends on the exchange of heat, moisture, and momentum with the atmosphere. Reliable regional climate predictions will require coupled ocean–atmosphere–land surface models that adequately represent these ocean processes.

Acknowledgments. This work is a component of the CSIRO Climate Change Research Program, which is partially supported by the Australian Department of Environment, Sports, and Territories. Michael Bessell assisted in the preparation of the figures.

REFERENCES

Andreas, E. L., C. A. Poulson, R. M. Williams, R. W. Lindsay, and J. A. Businger, 1979: The turbulent heat flux from arctic leads. *Bound.-Layer Meteor.,* **17,** 57–91.

Andrews, D. G., J. R. Holton, and C. B. Leovy, 1987: *Middle Atmosphere Dynamics.* Academic Press, 489 pp.

Baker, D. J., Jr., 1982: A note on Sverdrup balance in the Southern Ocean. *J. Mar. Res.,* **40** (suppl.), 21–26.

Barnier, B., L. Siefridt, and P. Marchasiello, 1995: Thermal forcing for a global ocean circulation model from a three-year climatology of ECMWF analyses. *J. Mar. Sys.,* **6,** 363–380.

Birchfield, G. E., 1989: A coupled ocean–atmosphere climate model: Temperature versus salinity effects on the thermohaline circulation. *Climate Dyn.,* **4,** 57–71.

Bradley, E. F., P. A. Coppin, and J. S. Godfrey, 1991: Measurements of sensible and latent heat fluxes in the western equatorial Pacific Ocean. *J. Geophys. Res.,* **96** (suppl.), 3375–3389.

———, J. S. Godfrey, P. A. Coppin, and J. A. Butt, 1993: Observations of net heat flux into the surface mixed layer of the western equatorial Pacific Ocean. *J. Geophys. Res.,* **98,** 22 521–22 532.

Broecker, W., 1987: The biggest chill. *Nat. Hist.,* **97,** 74–82.

———, 1991: The great ocean conveyor. *Oceanogr.,* **4,** 79–89.

Bryan, F., 1986: High-latitude salinity effects and interhemispheric thermohaline circulation. *Nature,* **323,** 301–304.

Bryan, K., 1991a: Poleward heat transport in the ocean: A review of a hierarchy of models of increasing resolution. *Tellus,* **43AB,** 104–115.

———, 1991b: Ocean circulation models. *Strategies for Future Climate Research,* M. Latif, Ed., Max-Planck Institut fur Meteorologie, 265–286.

Bryden, H. L., 1993: Ocean heat transport across 24°N latitude. *Interactions Between Global Climate Subsystems: The Legacy of Hann, Geophys. Monogr.,* No. 75, Intl. Union of Geodesy and Geophysics, 65–75.

———, and R. A. Heath, 1985: Energetic eddies at the northern edge of the Antarctic Circumpolar Current in the southwest Pacific. *Prog. Oceanogr.,* **14,** 65–87.

———, D. H. Roemmich, and J. A. Church, 1991: Ocean heat transport across 24°N in the Pacific. *Deep-Sea Res.,* **38,** 297–324.

Bunker, A. F., 1988: Surface energy fluxes of the South Atlantic Ocean. *Mon. Wea. Rev.,* **116,** 809–823.

Carissimo, B. C., A. H. Oort, and T. H. Vonder Haar, 1985: Estimating the meridional energy transports in the atmosphere and ocean. *J. Phys. Oceanogr.,* **15,** 82–91.

Clarke, A. J., 1991: On the reflection and transmission of low-frequency energy at the irregular western Pacific Ocean boundary. *J. Geophys. Res.,* **96** (suppl.), 3289–3305.

Cox, M. D., 1989: An idealized model of the World Ocean. Part 1: The global-scale water masses. *J. Phys. Oceanogr.,* **19,** 1730–1752.

Cresswell, G. R., A. Frische, J. Peterson, and D. Quadfasel, 1993: Circulation in the Timor Sea. *J. Geophys. Res.,* **98,** 14 379–14 389.

Danabasoglu, G., J. C. McWilliams, and P. R. Gent, 1994: The role of mesoscale tracer transports in the global ocean circulation. *Science,* **264,** 1123–1126.

de Szoeke, R. A., and M. D. Levine, 1981: The advective flux of heat by mean geostrophic motions in the Southern Ocean. *Deep-Sea Res.,* **28,** 1057–1085.

Delcroix, T., 1987: Net heat gain of the tropical Pacific Ocean computed from subsurface ocean data and wind stress data. *Deep-Sea Res.,* **34,** 33–44.

Dickson, R. R., and J. Brown, 1994: The production of North Atlantic deep water: Sources, rates, and pathways. *J. Geophys. Res.,* **99,** 12 319–12 341.

Doos, K., and D. J. Webb, 1994: The Deacon cell and other meridional cells of the Southern Ocean. *J. Phys. Oceanogr.,* **24,** 429–442.

Drosdowsky, W., 1993: Potential predictability of winter rainfall over southern and eastern Australia using Indian Ocean SST anomalies. *Aust. Meteor. Mag.,* **42,** 1–6.

England, M. H., 1995: Using chlorofluorocarbons to assess ocean climate models. *Geophys. Res. Lett.,* **22,** 3051–3054.

———, and V. C. Garcon, 1994: South Atlantic circulation in a World Ocean model. *Ann. Geophys.,* **12,** 812–825.

Ffield, A., and A. Gordon, 1992: Vertical mixing in the Indonesian thermocline. *J. Phys. Oceanogr.,* **22,** 184–198.

Fieux, M., C. Andrie, P., Delecluse, A. G. Ilahude, A. Kartavtseff, F. Mantisi, R. Molcard, and J. C. Swallow, 1994: Measurements within the Pacific–Indian Oceans throughflow region. *Deep-Sea Res.,* **41,** 1091–1130.

FRAM Group, 1991: An eddy-resolving model of the Southern Ocean. *Eos, Trans. Amer. Geophys. Union,* **72,** 169, 174–175.

Gates, W. L., J. F. B. Mitchell, G. J. Boer, U. Cubasch, and V. P. Meleshko, 1992: Climate modelling, climate prediction and model validation. *Climate Change 1992: The Supplementary Report to the IPCC Scientific Assessment,* J. T. Houghton, B. A. Callander, and S. K. Varney, Eds., Cambridge University Press, 97–134.

Gent, P. R., 1991: The heat budget of the TOGA COARE domain in an ocean model. *J. Geophys. Res.,* **96** (suppl), 3323–3330.

———, and J. C. McWilliams, 1990: Isopycnal mixing in ocean circulation models. *J. Phys. Oceanogr.,* **20,** 150–155.

———, J. Willebrand, T. J. McDougall, and J. C. McWilliams, 1995: Paremeterizing eddy-induced tracer transports in ocean circulation models. *J. Phys. Oceanogr.,* **25,** 463–474.

Gill, A. E., 1982: *Atmosphere–Ocean Dynamics.* Academic Press, 662 pp.

Godfrey, J. S., 1989: A Sverdrup model of the depth-integrated flow for the World Ocean allowing for island circulations. *Geophys. Astrophys. Fluid Dyn.,* **45,** 89–112.

———, 1996: The effect of the Indonesian throughflow on ocean circulation and heat exchange with the atmosphere: A review. *J. Geophys. Res.,* **101,** 12 217–12 238.

———, and E. J. Lindstrom, 1989: On the heat budget of the equatorial Pacific surface mixed layer. *J. Geophys. Res.,* **94,** 8007–8017.

———, and A. J. Weaver, 1991: Is the Leeuwin Current driven by Pacific heating and winds? *Progr. Oceanogr.,* **27,** 225–272.

———, M. Nunez, E. F. Bradley, P. A. Coppin, and E. J. Lindstrom, 1991: On the net surface heat flux into the western equatorial Pacific. *J. Geophys. Res.,* **96,** 3391–3400.

———, A. C. Hirst, and J. Wilkin, 1993: Why does the Indonesian throughflow appear to originate from the North Pacific? *J. Phys. Oceanogr.,* **23,** 1088–1098.

Gordon, A. L., 1986: Interocean exchange of thermocline water. *J. Geophys. Res.,* **91,** 5037–5046.

———, and H. W. Taylor, 1975: Heat and salt balance within the cold waters of the World Ocean. *Numerical Models of Ocean Circulation,* National Academy of Sciences, 54–56.

———, and W. F. Haxby, 1990: Agulhas eddies invade the South Atlantic: Evidence from GEOSAT atltimeter and shipboard conductivity–temperature–depth survey. *J. Geophys. Res.,* **95,** 3117–3125.

———, E. J. Molinelli, and T. N. Baker, 1982: *Southern Ocean Atlas,* Columbia University Press, 35 pp and 248 plates.

———, R. F. Weiss, W. M. Smethie Jr., and M. J. Warner, 1992: Thermocline and intermediate water communication between the South Atlantic and Indian Oceans. *J. Geophys. Res.,* **97,** 7223–7240.

Gow, A. J., and W. B. Tucker III, 1990: Sea ice in the polar regions. *Polar Oceanography, Part A: Physical Science,* W. O. Smith, Ed., Academic Press, 47–122.

Haidvogel, D. B., and F. O. Bryan, 1992: Ocean general circulation modeling. *Climate System Modeling,* K. E. Trenberth, Ed., Cambridge University Press, 371–413.

Hall, M. M., and H. L. Bryden, 1982: Direct estimates and mechanisms of ocean heat transport. *Deep-Sea Res.,* **29,** 339–360.
Halpern, D. A., 1980: Pacific equatorial temperature section from 172°E to 110°W during winter and spring 1979. *Deep-Sea Res.,* **27A,** 931–940.
Hartmann, D. L., V. Ramanathan, A. Berroir, and G. E. Hunt, 1986: Earth radiation budget data and climate research. *Rev. Geophys.,* **24,** 439–468.
Hellerman, S., and M. Rosenstein, 1983: Normal monthly wind stress over the World Ocean with error estimates. *J. Phys. Oceanogr.,* **13,** 1093–1104.
Hirst, A. C., 1998: On the fate of North Atlantic deep water as simulated by a global ocean GCM. *J. Geophys. Res.,* submitted.
———, and J. S. Godfrey, 1993: The role of the Indonesian throughflow in a global ocean GCM. *J. Phys. Oceanogr.,* **23,** 1057–1086.
———, and ———, 1994: The response of a global ocean GCM to a sudden change in the Indonesian throughflow. *J. Phys. Oceanogr.,* **24,** 1895–1910.
———, D. R. Jackett, and T. J. McDougall, 1996: The meridional overturning cells of a World Ocean model in neutral-surface coordinates. *J. Phys. Oceanogr.,* **26,** 775–791.
Hughes, T. M. C., A. J. Weaver, and J. S. Godfrey, 1992: Thermohaline forcing of the Indian Ocean by the Pacific Ocean. *Deep-Sea Res.,* **39,** 965–995.
Johnson, G. C., and H. L. Bryden, 1989: On the size of the Antarctic Circumpolar Current. *Deep-Sea Res.,* **36,** 39–53.
Karoly, D. J., P. C. McIntosh, P. Berrisford, T. J. McDougall, and A. C. Hirst, 1997: Similarities of the Deacon cell in the Southern Ocean and Ferrel cells in the atmosphere. *Quart. J. Roy. Meteor. Soc.,* **123,** 519–526.
Keith, D. W., 1995: Meridional energy transport—Uncertainty in zonal means. *Tellus,* **47A,** 30–44.
Killworth, P. D., and M. M. Nanneh, 1994: Isopycnal momentum budget of the Antarctic Circumpolar Current in the fine-resolution Antarctic model. *J. Phys. Oceanogr.,* **24,** 1201–1223.
Klinck, J. M., 1992: Effects of wind, density and bathymetry on a one-layer Southern Ocean model. *J. Geophys. Res.,* **97,** 20 179–20 189.
Knox, R. A., and D. L. T. Anderson, 1985: Recent advances in the study of the low-latitude ocean circulation. *Prog. Oceanogr.,* **14,** 259–317.
Knutson, T. R., S. Manabe, and D. Gu, 1997: Simulated ENSO in a global coupled ocean–atmosphere model: Multidecadal amplitude modulation and CO_2 sensitivity. *J. Climate,* **10,** 138–161.
Large, W. G., J. C. McWilliams, and S. C. Doney, 1994: Oceanic vertical mixing: A review and a model with a nonlocal boundary layer parameterization. *Rev. Geophys.,* **32,** 363–404.
Levitus, S., 1982: Climatological atlas of the World Ocean. NOAA Prof. Paper No. 13, 173 pp.
Lukas, R., and E. Lindstrom, 1991: The mixed layer of the western equatorial Pacific Ocean. *J. Geophys. Res.,* **96** (suppl.), 3343–3357.
Macdonald, A. M., 1993: Property fluxes at 30°S and their implications for the Pacific–Indian throughflow and global heat budget. *J. Geophys. Res.,* **98,** 6851–6868.
Maier-Reimer, E., and U. Mikolajewicz, 1989: Experiments with an OGCM on the cause of the Younger Dryas. Max-Planck-Institut fur Meteorologie Rep. No. 39, Hamburg.
Manabe, S., and R. J. Stouffer, 1988: Two stable equilibria of a coupled ocean–atmosphere model. *J. Climate,* **1,** 841–866.
———, K. Bryan, and M. J. Spelman, 1990: Transient response of a global ocean–atmosphere model to a doubling of atmospheric carbon dioxide. *J. Phys. Oceanogr.,* **20,** 722–749.
———, M. J. Spelman, and R. J. Stouffer, 1991: Transient responses of a coupled ocean–atmosphere model to gradual changes of atmospheric CO_2. Part 1: Annual mean response. *J. Climate,* **4,** 785–818.
Marotzke, J., 1991: Influence of convective adjustment on the stability of the thermohaline circulation. *J. Phys. Oceanogr.,* **21,** 903–907.
———, and J. Willebrand, 1991: Multiple equilibria of the global thermohaline circulation. *J. Phys. Oceanogr.,* **21,** 1372–1385.
———, P. Welander, and J. Willebrand, 1988: Instability and multiple steady states in a meridional-plane model of the thermohaline circulation. *Tellus,* **40A,** 162–172.
Marshall, D., 1995: Topographic steering of the Antarctic Circumpolar Current. *J. Phys. Oceanogr.,* **25,** 1636–1650.
Marshall, J., D. Olbers, H. Ross, and D. Wolf-Gladrow, 1993: Potential vorticity constraints on the dynamics and hydrography of the Southern Ocean. *J. Phys. Oceanogr.,* **23,** 465–487.
McCartney, M., 1977: Subantarctic mode water. *Deep-Sea Res.,* **24** (suppl.), 103–119.
———, 1982: The subtropical recirculation of mode waters. *J. Mar. Res.,* **40** (suppl.), 427–464.
McCreary, J. P., P. K. Kundu, and R. L. Molinari, 1993: A numerical investigation of dynamics, thermodynamics and mixed-layer processes in the Indian Ocean. *Prog. Oceanogr.,* **31,** 181–244.
Meyers, G., 1996: Variations of Indonesian throughflow and the El Niño Southern Oscillation. *J. Geophys. Res.,* **101,** 12 255–12 264.
———, R. J. Bailey, and A. P. Worby, 1995: Volume transport of Indonesian throughflow. *Deep-Sea Res.,* **42,** 1163–1174.
Molinelli, E. J., 1981: The antarctic influence on antarctic intermediate water. *J. Mar. Res.,* **39,** 267–293.
Morrow, R., R. Coleman, J. Church, and D. Chelton, 1994: Surface eddy momentum flux and velocity variances in the Southern Ocean from GEOSAT altimetry. *J. Phys. Oceanogr.,* **24,** 2050–2071.
Munk, W. H., Abyssal recipes. 1966: *Deep-Sea Res.,* **13,** 207–230.
———, and E. Palmen, 1951: Note on the dynamics of the Antarctic Circumpolar Current. *Tellus,* **3,** 53–55.
Nicholls, N., 1989: Sea surface temperature and Australian rainfall. *J. Climate,* **2,** 965–973.
Nowlin, W. D., Jr., and J. M. Klinck, 1986: The physics of the Antarctic Circumpolar Current. *Rev. Geophys.,* **24,** 469–491.
Oberhuber, J. M., 1988: An atlas based on the COADS data set: The budget of heat, buoyancy and turbulent kinetic energy at the surface of the global ocean. Max-Planck Institut fur Meteorologie Rep. No. 15.
Olbers, D., V. Gouretski, G. Seiss, and J. Schroter, 1992: *Hydrographic Atlas of the Southern Ocean,* Alfred Wegener Institute, 17 pp. and 82 plates.
Ou, H. W., and W. P. M. de Ruijter, 1986: Separation of an inertial boundary current from a curved coastline. *J. Phys. Oceanogr.,* **16,** 280–289.
Pacanowski, R., and S. G. H. Philander, 1981: Parameterization of vertical mixing in numerical models of tropical oceans. *J. Phys. Oceanogr.,* **11,** 1443–1451.
Palmer, T. N., and D. A. Mansfield, 1984: Response of two atmospheric general circulation models to sea-surface temperature anomalies in the tropical east and west Pacific. *Nature,* **310,** 483–485.
Philander, S. G. H., 1990: *El Niño, La Niña and the Southern Oscillation.* Academic Press, 289 pp.
Piola, A., and D. T. Georgi, 1982: Circumpolar properties of antarctic intermediate water and subantarctic mode water. *Deep-Sea Res.,* **29,** 687–711.
Plumb, R. A., and J. D. Mahlman, 1987: The zonally averaged transport characteristics of the GFDL general circulation–transport model. *J. Atmos. Sci.,* **47,** 1825–1836.
Polzin, K. L., J. M. Toole, J. R. Ledwell, and R. W. Schmitt, 1997: Spatial variability of turbulent mixing in the abyssal ocean. *Science,* **276,** 93–96.

Qu, T., G. Meyers, S. J. Godfrey, and D. Hu, 1994: Ocean dynamics in the region between Australia and Indonesia and its influence on the variation of SST in a global GCM. *J. Geophys. Res.*, **99,** 18 433–18 446.

Rintoul, S. R., 1991: South Atlantic interbasin exchange. *J. Geophys. Res.*, **96,** 2675–2692.

Roemmich, D., 1980: Estimation of meridional heat flux in the North Atlantic by inverse methods. *J. Phys. Oceanogr.*, **10,** 1972–1983.

Rooth, C., 1982: Hydrology and ocean circulation. *Prog. Oceanogr.*, **11,** 131–149.

Saunders, P. M., and S. R. Thompson, 1993: Transport, heat, and freshwater fluxes within a diagnostic numerical model (FRAM). *J. Phys. Oceanogr.*, **22,** 452–464.

Schmitz, W. J., 1995: On the interbasin scale thermohaline circulation. *Rev. Geophys.*, **33,** 151–174.

———, and M. S. McCartney, 1993: On the North Atlantic circulation. *Rev. Geophys.*, **31,** 29–49.

Semtner, A. J., and R. M. Chervin, 1992: Ocean general circulation from an eddy-resolving model. *J. Geophys. Res.*, **97,** 5493–5550.

Shetye, S. R., S. S. C. Shenoi, A. D. Gouveia, G. S. Michael, D. Sundar, and G. Nampoothiri, 1991: Wind-driven coastal upwelling along the western boundary of the Bay of Bengal during the southwest monsoon. *Cont. Shelf Res.*, **11,** 1397–1408.

Siegenthaler, U., H. Friedli, H. Loetscher, E. Moor, A. Neftel, H. Oeschger, and B. Stauffer, 1988: Stable-isotope ratios and concentration of CO_2 in air from polar ice cores. *Ann. Glaciol.*, **10,** 1–6.

Sprintall, J., D. Roemmich, B. Stanton, and R. Bailey, 1994: Regional climate variability and ocean heat transport in the south-west Pacific Ocean. *J. Geophys. Res.*, **100,** 15 865–15 871.

Stockdale, T., D. Anderson, M. Davey, P. Delecluise, A. Kattenberg, Y. Kitamura, M. Latif, and T. Yamagata, 1993: TOGA Numerical Experimentation Group intercomparison of tropical ocean GCM's. World Climate Research Programme WCRP-79, WMO/TD 545, 90 pp.

Stommel, H., 1948: The westward intensification of wind-driven ocean currents. *Eos, Trans. Amer. Geophys. Union,* **29,** 202–206.

———, 1958: *The Gulf Stream: A Physical and Dynamical Description.* University of California Press and Cambridge University Press, 202 pp.

———, 1961: Thermohaline convection with two stable regimes of flow. *Tellus,* **13,** 224–230.

———, 1962: An analogy to the Antarctic Circumpolar Current. *J. Mar. Res.*, **20,** 92–96.

———, and A. B. Arons, 1960: On the abyssal circulation of the World Ocean. I. Stationary planetary flow patterns on a sphere. *Deep-Sea Res.*, **6,** 140–154.

Stramma, L., 1992: The South Indian Ocean Current. *J. Phys. Oceanogr.*, **22,** 421–430.

Straub, D. N., 1993: On the transport and angular momentum balance of channel models of the Antarctic Circumpolar Current. *J. Phys. Oceanogr.*, **22,** 776–782.

Sverdrup, H. U., 1933: On vertical circulation in the ocean due to action of the wind with application to conditions within the Antarctic Circumpolar Current. *Discovery Rep.*, **7,** 139–170.

———, 1947: Wind driven currents in a baroclinic ocean, with application to the equatorial currents in the eastern Pacific. *Proc. Natl. Acad. Sci.*, **33,** 318–326.

Talley, L. D., 1984: Meridional heat transport in the Pacific Ocean. *J. Phys. Oceanogr.*, **14,** 231–241.

Taylor, H. W., A. L. Gordon, and E. Molinelli, 1978: Climatic characteristics of the Antarctic Polar Front Zone. *J. Geophys. Res.*, **83,** 4572–4578.

Toggweiler, J. R., and B. Samuels, 1993a: Is the magnitude of the deep outflow from the Atlantic Ocean actually governed by Southern Hemisphere winds? *The Global Carbon Cycle,* NATO ASI Series Vol. 15, M. Heimann, Ed., Springer-Verlag, 303–331.

———, and ———, 1993b: New radiocarbon constraints on the upwelling of abyssal water to the ocean's surface. *The Global Carbon Cycle,* NATO ASI Series Vol. 15, M. Heimann, Ed., Springer-Verlag, 333–366.

Tomczak, M., and J. S. Godfrey, 1994: *Regional Oceanography: An Introduction.* Pergamon Press, 422 pp.

Toole, J. M., and B. A. Warren, 1993: A hydrographic section across the subtropical south Indian Ocean. *Deep-Sea Res.*, **40,** 1973–2019.

———, K. L. Polzin, and R. W. Schmitt, 1994: Estimates of diapycnal mixing in the abyssal ocean. *Science,* **264,** 1120–1123.

Treguier, A. M., and J. C. McWilliams, 1990: Topographic influences on wind-driven, stratified flow in a beta-plane channel: An idealized model for the Antarctic Circumpolar Current. *J. Phys. Oceanogr.*, **20,** 321–343.

Trenberth, K. E., and A. Solomon, 1994: The global heat balance: Heat transports in the atmosphere and ocean. *Climate Dyn.*, **10,** 107–134.

Tziperman, E., J. R. Toggweiler, Y. Feliks, and K. Bryan, 1994: Instability of the thermohaline circulation with respect to mixed boundary conditions: Is it really a problem for realistic models? *J. Phys. Oceanogr.*, **24,** 217–232.

Wacogne, S., and R. C. Pacanowski, 1996: Seasonal heat transport in the tropical Indian Ocean. *J. Phys. Oceanogr.*, **26,** 2666–2699.

Warren, B. A., 1983: Why is no deep water formed in the North Pacific? *J. Mar. Res.*, **41,** 327–347.

Weaver, A. J., and E. S. Sarachik, 1990: On the importance of vertical resolution in certain ocean circulation models. *J. Phys. Oceanogr.*, **20,** 600–609.

———, and ———, 1991: The role of mixed boundary conditions in numerical models of the ocean's climate. *J. Phys. Oceanogr.*, **21,** 1470–1493.

Webster, P. J., and R. Lukas, 1992: TOGA COARE: The Coupled Ocean–Atmosphere Response Experiment. *Bull. Amer. Meteor. Soc.*, **73,** 1377–1416.

Welander, P., 1986: Thermohaline effects in the ocean circulation and related simple models. *Large-Scale Transport Processes in Oceans and Atmosphere,* J. Willebrand and D. L. T. Anderson, Eds., Reidel, 163–200.

Weller, R. A., and S. P. Anderson, 1996: Surface meteorology and air–sea fluxes in the western equatorial Pacific warm pool during the TOGA Coupled Ocean–Atmosphere Response Experiment. *J. Climate,* **9,** 1959–1990.

Wells, N. C., and B. A. de Cuevas, 1995: Depth-integrated vorticity budget of the Southern Ocean from a general circulation model. *J. Phys. Oceanogr.*, **25,** 2569–2582.

Wijffels, S. E., 1993: Exchanges between hemispheres and gyres: A direct approach to the mean circulation of the equatorial Pacific. Ph.D. thesis, Massachusetts Institute of Technology, [WHOI-93-42], 267 pp.

Wilkin, J. L., J. V. Mansbridge, and J. S. Godfrey, 1995: Pacific Ocean heat transport at 24°N in a high-resolution global model. *J. Phys. Oceanogr.*, **25,** 2204–2214.

Winton, M., and E. S. Sarachik, 1993: Thermohaline oscillations induced by strong steady salinity forcing of ocean general circulation models. *J. Phys. Oceanogr.*, **23,** 1389–1410.

Wolff, J. O., E. Maier-Raimer, and D. J. Olbers, 1991: Wind-driven flow over topography in a zonal beta plane channel: A quasi-geostrophic model of the Antarctic Circumpolar Current. *J. Phys. Oceanogr.*, **21,** 236–264.

Woods, J. D., and W. Barkmann, 1986: The response of the upper ocean to solar heating. I: The mixed layer. *Quart. J. Roy. Meteor. Soc.*, **471,** 1–27.

Wright, D. G., and T. F. Stocker, 1991: A zonally averaged ocean model for the thermohaline circulation. Part I: Model development and flow dynamics. *J. Phys. Oceanogr.,* **21,** 1713–1724.
Yang, J., and J. D. Neelin, 1993: Sea ice interactions with the thermohaline circulation. *Geophys. Res. Lett.,* **20,** 217–220.
Zhang, S., C. A. Lin, and R. J. Greatbatch, 1992: A thermocline model for ocean–climate studies. *J. Mar. Res.,* **50,** 99–124.
———, R. J. Greatbatch, and C. A. Lin, 1993: A reexamination of the polar halocline catastrophe and implications for coupled ocean–atmosphere modeling. *J. Phys. Oceanogr.,* **23,** 287–299.
———, C. A. Lin, and R. J. Greatbatch, 1995: A decadal oscillation due to the coupling between an ocean circulation model and a thermodynamic sea-ice model. *J. Mar. Res.,* **53,** 79–106.

Chapter 8

Interannual and Intraseasonal Variability in the Southern Hemisphere

GEORGE N. KILADIS

Aeronomy Laboratory, ERL, NOAA, Boulder, Colorado

KINGTSE C. MO

Climate Prediction Center, NMC/NWS/NOAA, Washington, DC

8.1. Introduction

The aim of this chapter is to examine the temporal and spatial variability of the troposphere over the Southern Hemisphere and, in the case of El Niño/ Southern Oscillation, link it to fluctuations in the tropical oceans as well. In the time since the first volume on Southern Hemisphere meteorology appeared in 1972, it has been possible to examine in much more detail the variability of the SH atmosphere, due to the vast improvements in the amount and quality of data that have since become available. Despite this progress, vast regions of the southern oceans remain very poorly observed. Fortunately, advances in remote sensing by satellite have done much to fill in some of the existing gaps in the ship- and ground-based observing network.

We deal specifically here with interannual (IA) and intraseasonal (IS) variability, with respect to the mean state of the atmosphere as outlined in chapters 1 and 2. As discussed in those chapters, over most regions the SH atmosphere displays a lower amplitude mean seasonal cycle than does the Northern Hemisphere, where continental effects result in a much larger first annual harmonic in the meridional temperature gradient and circulation strength. This is also reflected in less of a seasonal dependence in the variability of the SH troposphere. Despite the dominance of ocean over the SH, however, continental effects, orography, and forcing from the Tropics still produce substantial zonal asymmetries in both the time-mean flow and statistics of IA and IS variability.

We will refer to the time mean at a given point in the seasonal cycle as the climatological or "basic" state, whereas time-mean local deviations from the zonal mean will be referred to as the "stationary" waves (sometimes called "standing" waves in the literature). At any given moment, the atmosphere will deviate from its climatological state. Here, anomalies from the mean seasonal cycle of the atmosphere will be referred to as "perturbations" or "transients." These perturbations occur over all temporal scales, from high-frequency turbulence on up to fluctuations involving decadal and longer-term climate variability. In this chapter, IA perturbations refer to year-to-year variability. These can be calculated, for example, by simply subtracting individual seasonal means from long-term seasonal averages or by digital filtering in the time domain, with appropriate weighting functions depending on the desired response. For seasonal averaging, we use the DJF, MAM, etc., definitions, which differ from the "oceanic" (e.g., JFM, etc.) convention used in chapters 1 and 2. As far as the statistics presented here are concerned, however, only very slight differences in the results are to be noted between these conventions.

Intraseasonal variability here will refer to subseasonal fluctuations in roughly the 6- to 70-day period range. Higher-frequency fluctuations involving the synoptic timescale, generally considered as disturbances due to individual weather systems and with periods less than about 6 days, are discussed in chapters 3, 4, and 6. Decadal and longer timescale variability is covered in chapters 9 and 10.

Unless otherwise noted, the calculations performed specifically for this chapter were carried out using NCEP/NCAR reanalysis data (Kalnay et al. 1996). The NCEP/NCAR reanalysis product provides a consistently assimilated dataset of the state of the global atmosphere at 28 levels in the vertical, as well as at the surface. The temporal resolution is four times daily, and here we use the gridded pressure level set with a spatial resolution of 2.5° latitude/longitude at 17 levels in the vertical (1000–10 hPa). At the time of this writing, data from 1 January 1958 through 31 December 1997 were available.

8.2. Interannual variability

a. General features

To a much greater extent than the NH, the principal modes of variability in the SH are dominated by

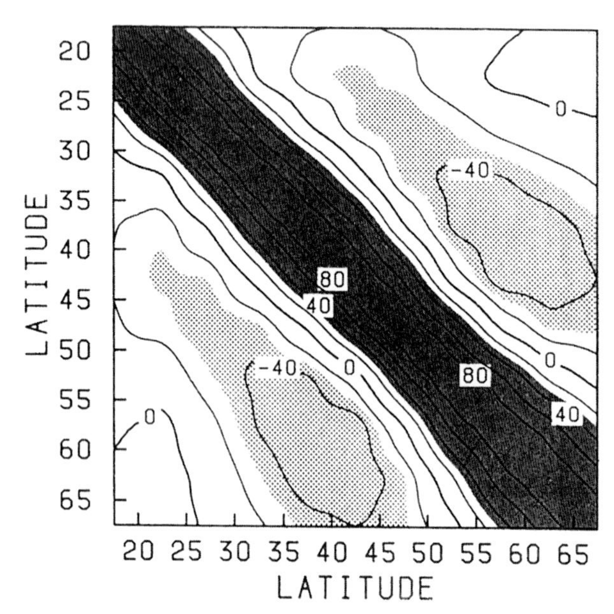

FIG. 8.1. Correlations (×100) of daily geostrophic zonal-wind anomalies at 500 hPa between pairs of latitudes over the Southern Hemisphere, using the Australian analyses from 1 August 1968 to 8 August 1983 (from Kidson 1988a).

zonally symmetric, equivalent barotropic structures with zonal-wind anomalies opposing each other between intervals of roughly 20° latitude (Trenberth 1979, 1981; Rogers and van Loon 1982; Kidson 1986, 1988a). For example, Fig. 8.1 shows the correlation of 500-hPa geostrophic zonal-mean westerly anomalies between pairs of latitudes over the SH using daily Australian analyses from 1 August 1968 through 7 August 1983 (Kidson 1988a). Kidson notes that the magnitude of the correlations increases as the data are averaged over longer intervals, so while the opposition is present over a wide frequency range, the strongest relationships are on the IA timescale.

Trenberth (1981) demonstrated that the largest contribution to the zonal mean westerly fluctuations on monthly to IA timescales was from the Australasian and western Pacific sectors from 120°E to 155°W, reflecting alternating extensions and retractions of the subtropical jet. It was shown that these zonal-mean zonal-wind anomalies display a pronounced tendency to migrate poleward over time (Swanson and Trenberth 1981), a phenomenon that has continued to be observed in more recent data (Dickey et al. 1992) and on IS timescales as well (e.g., Weickmann et al. 1997). Figure 8.2 plots, by latitude, the zonally averaged, 200-hPa zonal-wind anomalies from 1968 through 1997. The propagation of anomalies from low to high latitudes, often symmetric about the Equator, is quite evident in both hemispheres. There are several periods when the subtropical zonal-wind anomalies oppose those at both equatorial and high latitudes. We will return to these signals when we relate a portion of this variability to ENSO.

One way to objectively obtain preferred modes of variability is to perform an empirical orthogonal function (EOF) analysis, which can be interpreted as preferred spatial patterns of covariability across a given domain (see Mo and Ghil 1987). In general, the explained variance (and the spatial scale) comprising each EOF becomes smaller as the order of the EOF increases. The variability associated with each EOF pattern, however, may be spread across several frequency bands (Lau et al. 1994; Mo and Kousky 1993). Therefore, temporal filtering is often effectively employed to separate modes at different timescales prior to EOF analysis.

Kidson (1988b) performed an EOF analysis of fluctuations in 500-hPa height over the SH for all seasons with periods greater than 50 days, utilizing seven years of ECMWF data. It was found that this period range essentially represented IA variability, although we will see below that IS quasi-stationary states and blocking are also captured by this filtering. Even after seasonally stratifying the data, Kidson (1988b) and Rogers and van Loon (1982) also obtained rather similar EOFs for both JJA and DJF, so the leading modes of IA variability appear to change little throughout the year.

The first four unrotated EOFs of 500-hPa height anomalies over the SH with periods longer than 50 days were calculated using the NCEP/NCAR reanalysis data for the 1973–95 period. These EOFs were obtained from the filtered, detrended data on a 5° grid using a covariance matrix, after removing the first two harmonics of the mean annual cycle. The EOFs were calculated both for the entire record and for individual seasons. Results were similar enough that we only present the EOFs for all seasons together here. The patterns are nearly identical to those obtained by Rogers and van Loon (1982) and Kidson (1988a), and they are quite robust in that they are reproducible with only minor differences in completely independent samples.

The leading EOF (Fig. 8.3a), which explains nearly 25% of the total variance, has a strong zonally symmetric component, showing an out-of-phase relationship between heights in a latitude belt just south of 40°S and heights over the subtropics and over Antarctica poleward of 60°S. This "high-latitude mode" (Karoly 1990) is also dominant in the studies of Mo and White (1985), Mo (1986), Mo and Ghil (1987), Kousky and Bell (1992), and Kidson and Sinclair (1995). The zonal-mean geostrophic wind anomalies associated with this EOF would lead to the maximum correlation between the zonal-mean westerlies at about 40° and 60°S, as observed in Fig. 8.1. Apart from zonal symmetry, a zonal wave 3 is also evident, especially in the height field around Antarctica. Larger amplitudes over the western Pacific and Indian Oceans

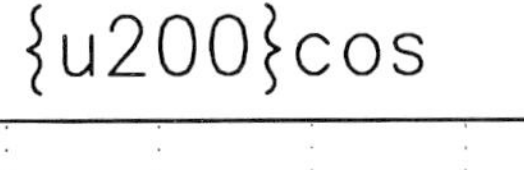

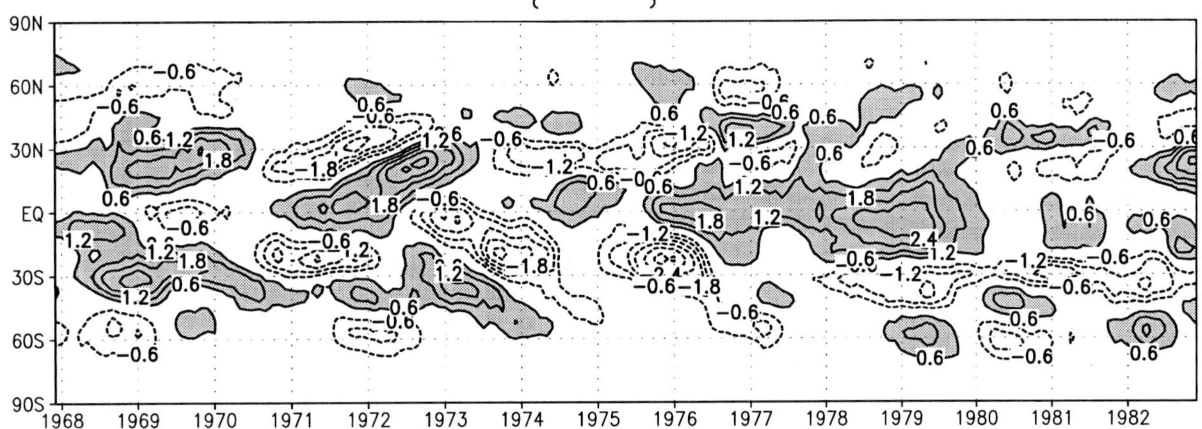

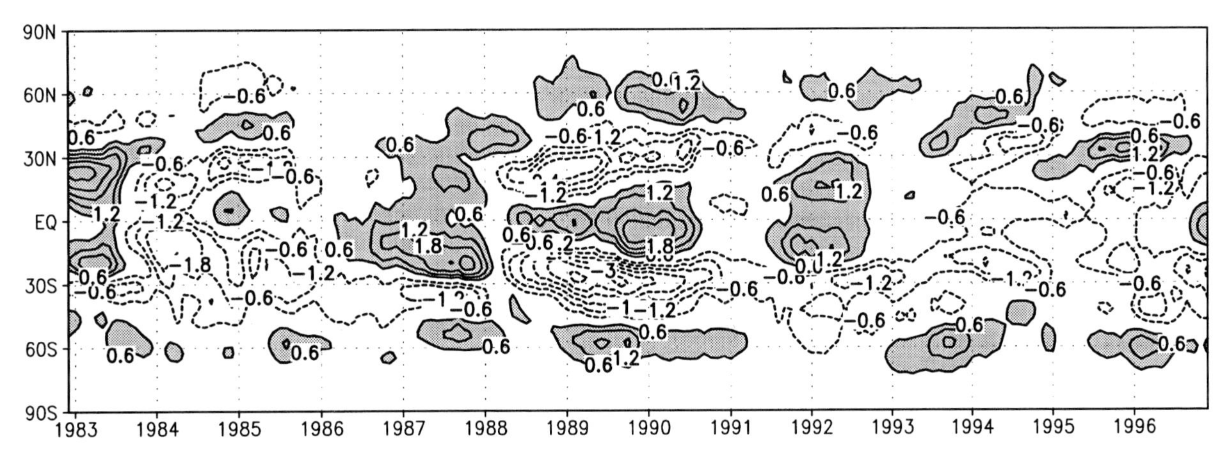

FIG. 8.2. 200-hPa zonal-mean zonal-wind anomalies by latitude from 1968 through 1997, weighted by the cosine of latitude. Data are 5-month running means. Contour interval is 0.6 m s^{-1}. Negative contours are dashed, and the zero contour has been omitted.

also lead to a projection onto wavenumber 1 in the midlatitudes.

In its positive phase (as in Fig. 8.3a), EOF 1 is characterized by an equatorward shift of the polar jet and weakening of the westerlies around the Antarctic coast, while in the opposite phase the climatological double-jet structure is more prominent (Kidson and Sinclair 1995). Sinclair et al. (1997) found that this pattern modulates cyclone activity between middle and high latitudes, with increased (decreased) westerlies near 55°–60°S accompanied by more (fewer) cyclones in the circumpolar region and fewer (more) in middle latitudes. Trenberth (1984a) demonstrated a reversal in the high-latitude mode between the winter of 1979, when the double jet was prominent, and the winter of 1980, when a single zonal-mean jet was shifted to the north of its climatological position. A somewhat similar reversal was also noted between the summers of 1976/77 and 1978/79.

The projection of EOF 1 onto the vertical distribution of zonal-mean zonal wind shows that the out-of-phase relationship between fluctuations centered on 35° and 58°S continues with little phase tilt or change in amplitude with height between the surface and 100 hPa (see Kidson 1988b). Szeredi and Karoly (1987a,b) also show a nearly equivalent barotropic structure in the leading mode of zonal-mean meridional wind above 850 hPa, with a latitudinal distribution indicative of the eddy poleward flux of momentum associated with the zonal wind maxima (see also chapter 2; Trenberth 1987; Karoly 1990).

EOF 1 also appears in sea level pressure (Rogers and van Loon 1982; Mo and White 1985; Sinclair et al. 1997) and 200-hPa streamfunction (Lau et al. 1994), and is the dominant mode in EOFs calculated separately for summer and winter (not shown). It will be shown below that the pattern displays a strong connection to tropical convection on both IA and IS

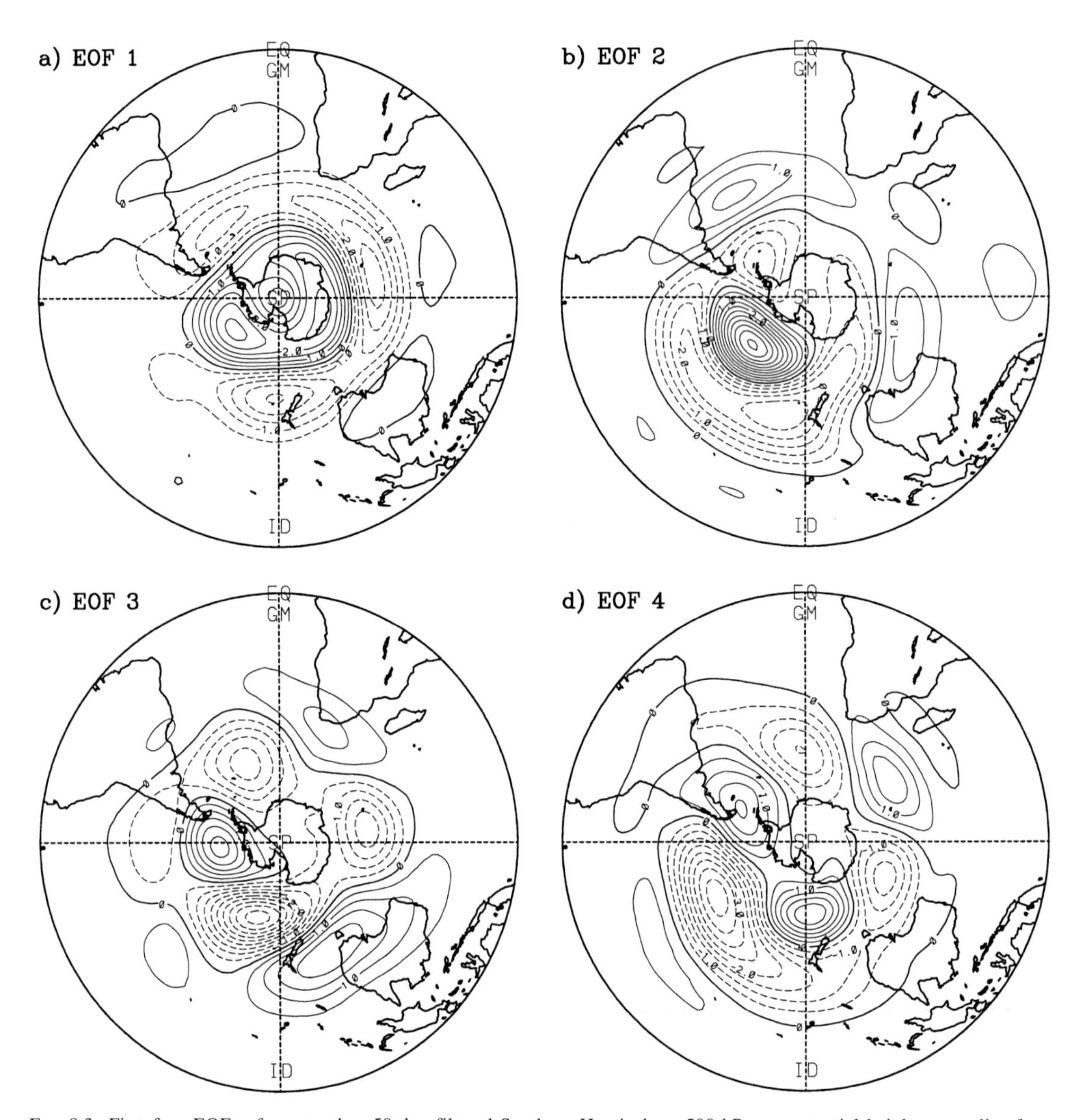

FIG. 8.3. First four EOFs of greater-than-50-day filtered Southern Hemisphere 500-hPa geopotential height anomalies from 1973 through 1995. Contour interval is 0.5 units. Negative contours are dashed.

timescales. In particular, zonally symmetric circulations in the SH appear to have a strong connection to ENSO events, although non-ENSO years can also be dominated by this mode (e.g., van Loon and Rogers 1981; Trenberth 1984a).

EOF 2 (Fig. 8.3b), which explains 12.5% of the variance, has a strong zonal wavenumber 1 component, with its largest perturbations in the Pacific sector. This pattern represents a distinct wave train, starting with a strengthening or weakening of the South Pacific trough to the east of New Zealand, with an opposing anomaly located to the north of West Antarctica and two downstream centers to the north of the Antarctic peninsula and along the east coast of South America. The center over the Bellingshausen Sea is a preferred region of persistent anomalies and blocking (see section 8.3b); thus, the pattern is also representative of low-frequency fluctuations on the IS as well as the IA timescales. EOF 2 is present throughout the year but does show some seasonal dependence, being most common in the winter and least common in summer. Once again, it appears that this pattern has a relationship to tropical convection, although it can also persist through seasons with only weak anomalies in the Tropics (Kidson 1988b).

EOF 3 (Fig. 8.3c) explains slightly less variance than EOF 2 (10.2%), and depicts a wavenumber 3 pattern with maximum amplitude between 50° and 60°S, along with a center farther north over the Tasman Sea. The centers again attain maximum strength over the Pacific sector and are in quadrature with those of EOF 2, although Kidson (1988b) did not find any evidence that this represents a traveling oscillation. EOF 3 is also the dominant teleconnection pattern during JJA found by Mo and White (1985) using monthly data, but is present with lower amplitude the rest of the year as well. Similar wavenumber 3 patterns appear in EOFs of sea level pressure (not shown) and are associated with blocking to the southeast of New Zealand (Sinclair et al. 1997). The wave train–like nature of these modes is strongly suggestive of the propagation of Rossby-wave energy with a source region in the subtropics. We will see below that wavenumber 3 patterns over the SH are dominant modes over a broad range of timescales, and that they have a strong relationship to ENSO.

EOF 4 (Fig. 8.3d) also shows a pronounced wavenumber 3 pattern, with some wave 1 component as well, but explains much less variance (6.5%) than the three leading EOFs. This pattern essentially represents the amplification/suppression of the climatological stationary waves over the SH, since throughout the year there are mean troughs and ridges near the positions of the corresponding negative and positive centers in Fig. 8.3d (van Loon and Jenne 1972; see Fig. 1.20). Unlike the first three patterns, EOF 4 does not show up as a leading mode during summer, but instead a wavenumber 4 pattern is more evident (not shown).

b. El Niño/Southern Oscillation

1) General features

A large portion of the IA variability of the atmosphere in both hemispheres is closely linked to that in the tropical oceans. By far, the most important mode of IA variability of the ocean–atmosphere system is due to ENSO. The surface pressure signal of ENSO was studied extensively by Walker in the early twentieth century (e.g., Walker and Bliss 1932). Walker noted that Indian monsoon rainfall was related to a surface pressure "seesaw" between stations in the western and eastern South Pacific, a phenomenon he termed the "Southern Oscillation." The El Niño portion of ENSO, which originally referred to a warming of SST along the Peruvian coast, was linked to the SO and warming of the central equatorial Pacific SST in the late 1950s and 1960s (e.g., Berlage 1957; Bjerknes 1966). Research on ENSO, and thus its literature, is extensive, and we can only give a brief overview of the phenomenon here before discussing its influence on circulation, temperature, and precipitation over the SH. For in-depth treatments of the dynamics, impacts, and historical aspects of ENSO, we refer the reader to the volumes by Philander (1990), Diaz and Markgraf (1992), Glantz et al. (1991), and Allan et al. (1996), as well as the references cited in the following sections.

The "center of action" of ENSO is in the tropical Pacific, where interactions between SST, convection, and wind stress drive the coupling between the ocean and atmosphere. In the mean, southeasterly trade winds associated with the South Pacific high dominate the equatorial flow in the central and eastern Pacific (see chapters 1 and 3). The effect of this wind stress, along with the change in sign of the Coriolis parameter at the Equator, causes the surface waters in the ocean to flow to the left of the wind south of the Equator and to the right of the wind to the north of the Equator. This leads to divergence in the upper oceanic layer, resulting in the upward transport of colder subsurface water, a process referred to as upwelling.

Upwelling is also prevalent along the eastern boundaries of the oceans, due to equatorward wind stress, which causes divergence within the surface ocean drift away from the coastline. This factor, combined with the northward transport of cold water in the Peru current, gives rise to a large area of relatively cold water extending westward in a wedge-shaped region bounded on the north by the equatorial region and on the east by the coast of South America. The stabilizing effect of the relatively cool SST on the boundary layer results in low precipitation and extensive low stratus cloud cover over this region, extending westward along the Equator in the "equatorial dry zone." These factors also favor the location of off-equatorial convergence zones over the open ocean of the Pacific.

The trade winds and upwelling weaken toward the western Pacific, and consequently, SST increases as the so-called "warm pool" of the western Pacific and eastern Indian Ocean is approached. A key to the forcing of the atmosphere by the ocean is found in the relationship between SST and deep convection over the Tropics. In general, the heaviest rainfall over the tropical oceans occurs in regions of SST greater than 28°C, although it is the SST gradient that is primarily responsible for moisture convergence (Lindzen and Nigam 1987; Trenberth 1991). Thus, the most widespread region of heavy rainfall in the Tropics is centered on the warm pool, although the continental effects of the monsoon circulations shift the center of this maximum toward northern Australia during southern summer and toward southern Asia during southern winter (Meehl 1987; Webster 1994; see chapter 3). In the mean, relatively low surface pressure over the warm pool and higher pressure over the dry zone drives the trade-wind circulation, which in turn maintains the SST gradient across the Pacific (Philander 1990).

According to most theoretical models of ENSO, an event develops as a coupled instability caused by the feedback between SST, convection, and wind-stress forcing. This coupling can be used to explain both the warm and cold extreme states of ENSO. A "warm event" is associated with El Niño conditions, while the opposite phase, here termed a "cold event," is characterized by below-normal SST in the central and eastern equatorial Pacific and along the coast of South America. Both extremes are associated with substantial global climatic anomalies.

The initiation of an ENSO event may be caused by various factors, which may actually differ from event to event and be related to random forcing at high frequencies. The "delayed oscillator" concept suggests that the timing of ENSO variability is caused by the alteration of propagating, equatorially trapped oceanic waves reflecting off the boundaries of the Pacific (Philander et al. 1984; Battisti and Hirst 1989). In this scenario, a warm event is initiated as downwelling equatorial Kelvin waves, progressing eastward at about 2.5 m s^{-1}, causing an increase in SST over the central and eastern equatorial Pacific. This results in the development of anomalous convection to the east of the warm pool. As convection sets up near the date line, it produces westerly surface winds within the equatorial waveguide, which generate more downwelling Kelvin waves and anomalous eastward oceanic flow along the Equator. This further elevates the SST to the east, so that the anomaly grows (Kessler and McPhaden 1995). In effect, this instability results in the transfer of a portion of the heat content of the warm pool to the eastern Pacific.

As the downwelling Kelvin waves progress eastward, they also generate westward-propagating off-equatorial Rossby waves. These take 6–12 months to reflect off the western boundary of the Pacific and produce upwelling Kelvin waves, which then propagate eastward along the Equator, this time cooling the SST. It is believed that the wave dynamics set the timescale for the transition between warm and cold events, although the waxing and waning of the actual heat content of the equatorial oceanic mixed layer also appear to be an important factor in determining the amplitude and timing of ENSO events (e.g., Philander 1990).

Along with the higher SST and enhanced convection in the central and eastern Pacific during warm events, there is a lowering of SLP over the eastern Pacific and an increase over Australasia. The pressure seesaw pattern between Australia and the eastern South Pacific allows the use of station pressure differences to document SO fluctuations back to the late nineteenth century. There are several different variations of the Southern Oscillation Index, but the most commonly used rely on the normalized monthly mean surface pressure difference between Tahiti (17.5°S, 149.6°W) in the South Pacific and Darwin (12.4°S, 130.9°E) in Australia (Chen 1982; Trenberth 1984b; Ropelewski and Jones 1987; Allan et al. 1991). The time for the recurrence of ENSO events averages about 36–48 months (van Loon and Madden 1981; Rasmusson and Carpenter 1982; Chen 1982; Trenberth 1976, 1984b), but there are periods when the system shows little variability or stays in one state for several years (Trenberth and Shea 1987; Wang and Wang 1996).

One intriguing aspect of ENSO concerns the occasionally strong reversal in the ocean–atmosphere state in adjacent years (van Loon and Shea 1985, 1987; Meehl 1987; Kiladis and van Loon 1988; Kiladis and Diaz 1989; Rasmusson et al. 1990; Ropelewski et al. 1992; Shen and Lau 1995). This "biennial tendency" of ENSO appears to be a manifestation of the fact that an extreme in the tropical climate system sets up a necessary (but not always sufficient) state for a transition to the opposite extreme in the following year. It is also related to the strong coupling of ENSO to the seasonal cycle (Trenberth and Shea 1987) and appears to be associated with a portion of the biennial peak in variance of SH circulation (Trenberth 1975).

2) Warm- and cold-event teleconnections over the Southern Hemisphere

(i) 500-hPa heights

To illustrate the anomalous circulation pattern over the SH during ENSO extremes, we composited 500-hPa height anomalies for warm and cold events back to 1958 from the list compiled by Kiladis and van Loon (1988). The objective criteria used to select these events allowed for the inclusion of the more-recent cold and warm events. Composite anomalies were calculated for JJA and SON of the years during which the events started (year 0), through DJF and MAM of the following year (year + 1). The warm-event year 0's used in this compositing are 1957, 1963, 1965, 1969, 1972, 1976, 1982, 1986, 1991, and 1997. The cold events are 1964, 1970, 1973, 1975, 1988, and 1995. Only the available seasons for the 1957 and 1997 warm events were included in the composites. The patterns obtained for both extremes are quite robust in that they are reproducible to a large degree by splitting the data record into two independent samples. The warm-event patterns also compare well with those obtained by van Loon and Madden (1981), van Loon and Shea (1987), and Karoly (1989) using independent analyses and different samples.

Sea level pressure patterns (not shown) are similar to the 500-hPa height anomalies over the extratropics, although a slight westward tilt with height is seen in many regions (Karoly 1989). This vertical structure is generally consistent with the equivalent barotropic nature typical of extratropical IA circulation anomalies. At low latitudes, however, there is most often a sign reversal between the lower- and upper-tropospheric pressure perturbations, as is typical of the vertical structure associated with tropical diabatic heating anomalies.

In Fig. 8.4, we reproduce the seasonal composite 500-hPa height anomalies for warm events. In JJA (Fig. 8.4a), a region of anomalously low pressure extends along 40°S over the Pacific from New Zealand eastward to 90°W. Poleward of this trough there is a

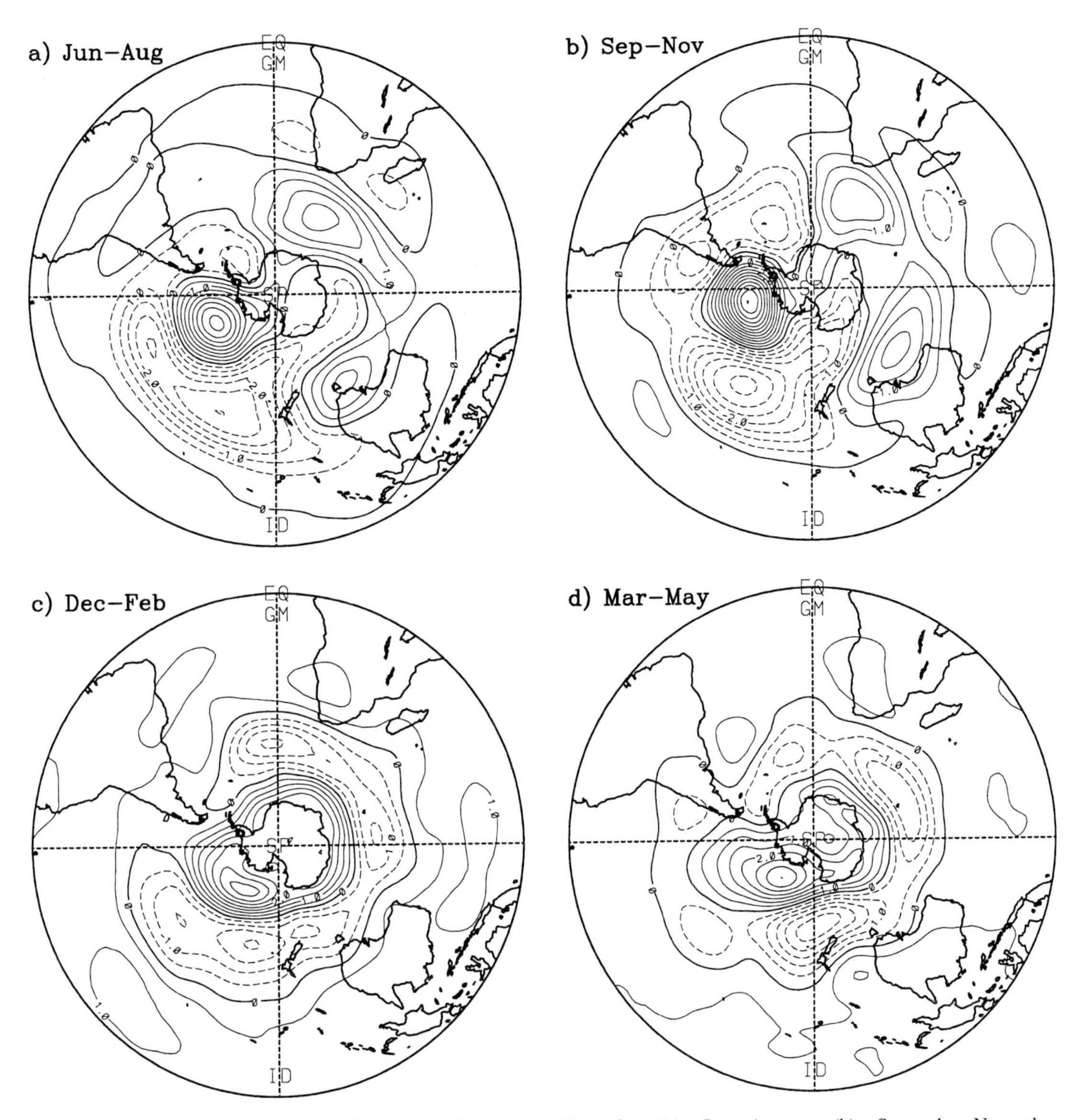

FIG. 8.4. Composite warm-event 500-hPa height anomalies for (a) June–August, (b) September–November, (c) December–February, and (d) March–May. Contour interval is 0.5 dam. Negative contours are dashed.

strong positive anomaly over the Bellingshausen Sea, along with anomalous ridging to the south of Australia and Africa. This pattern has elements of both IA EOFs 2 and 3 described in the previous section (Fig. 8.3), and obviously has a substantial projection onto wavenumber 3 in middle latitudes. The developing strong trough over the Pacific was discussed in some length by van Loon (1984). This feature and the downstream ridge strongly resemble the signature of wintertime persistent anomalies associated with blocking over the Bellingshausen Sea (discussed in section 8.3b), consistent with the predisposition toward blocking over this area during warm events, and vice versa during cold events (Rutllant and Fuenzalida 1991). Karoly (1989) showed that this pattern was somewhat variable within his warm-event sample and attributed this to the large spread in SST anomalies during this developing stage of individual events.

The SON pattern (Fig. 8.4b) is quite similar to Fig. 8.4a, but significantly more amplified. The strengthening of the anomalous Pacific trough results in a suppression of the usual seasonal cycle over this sector, since in the mean pressure normally rises over the South Pacific into the southern spring (van Loon 1984).

The pattern becomes zonally symmetric during summer (Fig. 8.4c), with negative anomalies between 35° and 50°S surrounding above-normal heights over Antarctica and the adjacent ocean. The Pacific trough is still strong, as are troughs over the South Atlantic and South Indian Oceans. This pattern strongly resembles

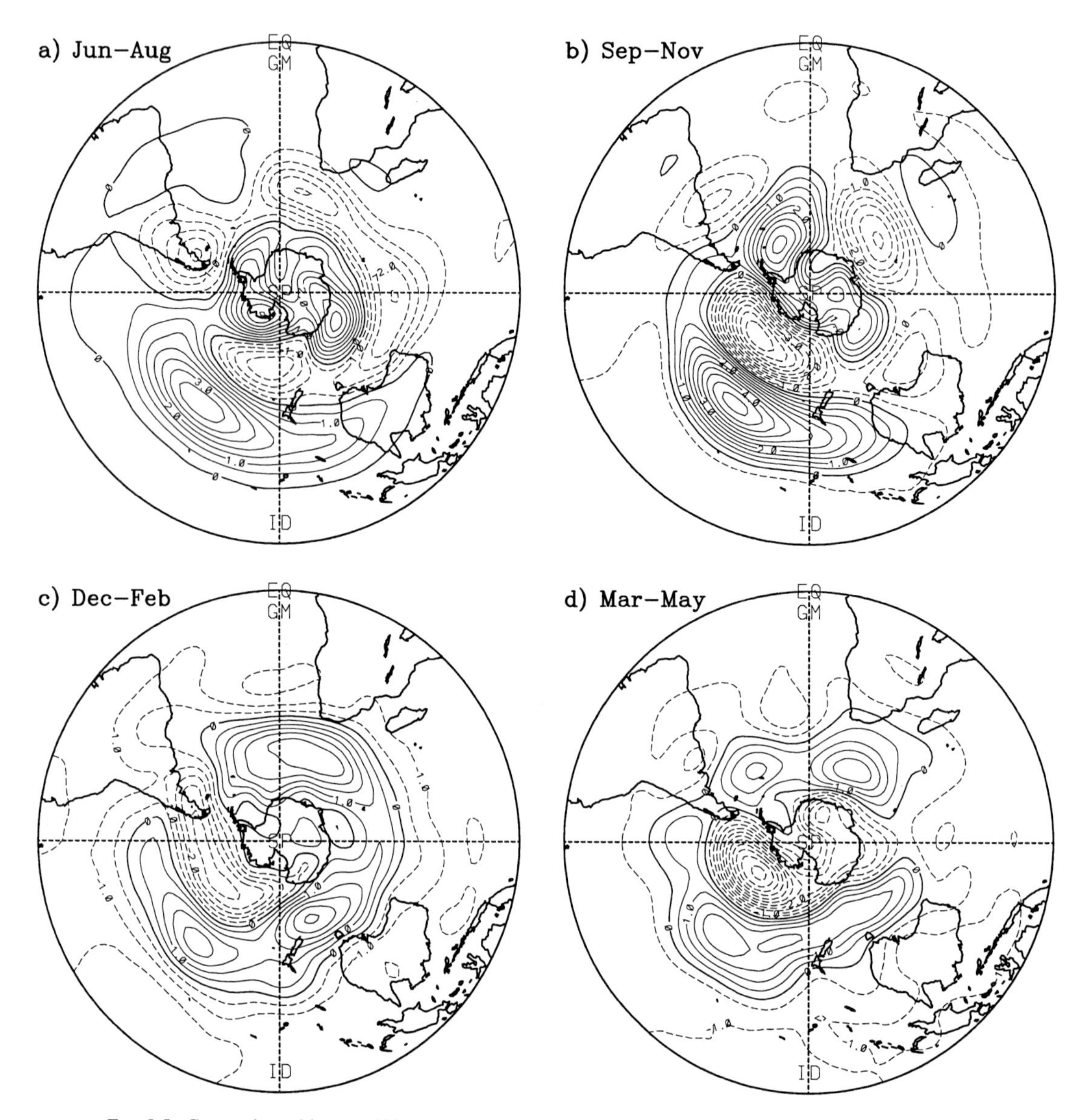

FIG. 8.5. Composite cold-event 500-hPa height anomalies for (a) June–August, (b) September–November, (c) December–February, and (d) March–May. Contour interval is 0.5 dam. Negative contours are dashed.

the high-latitude mode of IA EOF 1 (Fig. 8.3a) and was shown by Karoly (1989) to be quite stable from event to event. By the following MAM, the high-latitude mode remains strong (Fig. 8.4d), with pressure remaining high over Antarctica, although the troughs in Fig. 8.4c have retrogressed somewhat. Despite the differences in Fig. 8.4, we note that the Pacific trough and the Bellingshausen Sea ridge are present throughout the seasonal cycle.

During the cold-event phase of the SO, the pressure anomalies are in most regions opposite to those during warm events (Fig. 8.5). The weak Pacific trough is a very robust feature in all seasons, as is anomalous troughing over the Bellingshausen Sea from spring through fall. The reversal between warm and cold events is least evident during JJA (Fig. 8.5a), but there is a strong opposition to warm events in the high latitudes during DJF and MAM in cold events (Figs. 8.5c,d). Wavenumber 1 and 3 patterns are still prominent in all seasons. The lack of a perfect reversal in the pressure patterns between warm and cold events, though likely due in part to sampling, may also be due to a real nonlinearity in the atmospheric response to the presence versus the absence of large-scale convection over the central and eastern tropical Pacific.

(ii) 200-hPa zonal wind

In his pioneering work on the SO, Bjerknes (1966) noted a persistently strong Aleutian low during warm events, resulting in stronger-than-normal westerlies

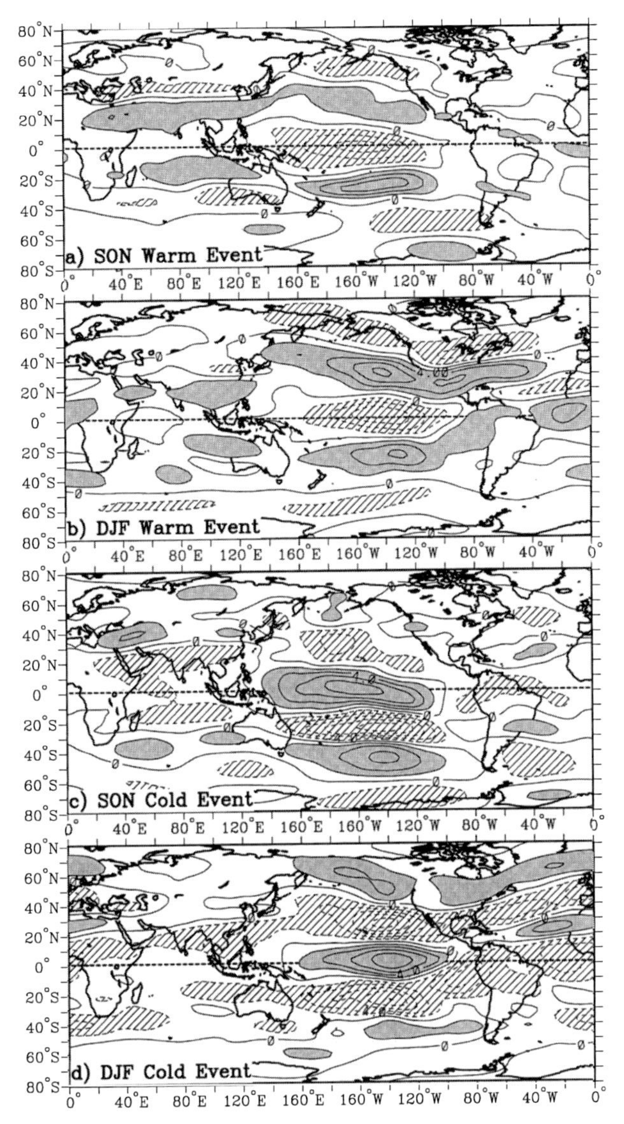

FIG. 8.6. Composite 200-hPa zonal-wind anomalies for (a) September–November warm events, (b) December–February warm events, (c) September–November cold events, and (d) December–February cold events. Contour interval is 2 m s^{-1}. Negative contours are dashed. Positive anomalies greater than 2 m s^{-1} are shaded; negative anomalies less than −2 m s^{-1} are hatched.

over the North Pacific, especially during winter. He reasoned that an enhanced Hadley circulation, driven by strong convection over the central Pacific, can maintain a strong subtropical jet stream by the large-scale meridional flux of westerly momentum. Here, we examine global zonal-wind anomalies at 200 hPa to determine whether similar arguments can be used to account for wind patterns at jet level over the SH during warm and cold events.

Figure 8.6 shows 200-hPa zonal-wind anomalies during SON and DJF for warm and cold events, calculated from the same sample of years as used in Figs. 8.4 and 8.5. Results are similar, but with lower amplitude, for JJA and MAM. Over the SH, the bulk of the 200-hPa zonal-wind signals are in line with the implied 500-hPa geostrophic wind anomalies, due to the pressure fields in Figs. 8.4 and 8.5. To a remarkable extent, the cold-event anomalies are in most regions opposite in sign to the warm-event signals. Anomalies are most pronounced over the Pacific sector, consistent with the region of largest forcing by diabatic heating displacements as envisioned by Bjerknes. The patterns are strikingly symmetric about the Equator, especially over the Pacific but to some extent over the Atlantic and Indian sectors as well. These symmetries about the Equator have been shown to be preferred modes of IA variability of the global circulation in the upper troposphere (see Fig. 1.34), and it appears that tropical convection is the primary forcing for these signals (Arkin 1982; Mo and Kousky 1993; Hsu 1994).

Over the Pacific, the warm-event composite has easterly anomalies along the Equator and along 40°–50°N and 40°–50°S (Figs. 8.6a,b). Enhanced westerlies are observed in the subtropics of both hemispheres and along the coast of Antarctica. These patterns essentially represent an equatorward movement of the climatological storm track and strengthening of the Pacific subtropical jet during warm events, as well as a more pronounced double jet in cold events as manifested by a strong polar-front jet in the Pacific subantarctic latitudes in Figs. 8.6c and 8.6d. Corresponding shifts in the frequency of cyclones related to ENSO were found by Sinclair et al. (1997), in the sense that increases in cyclone frequency were found along the enhanced westerlies in Fig. 8.6, especially during winter.

In the Atlantic, the anomalies show a tendency to be opposite to those over the Pacific for a given latitude. Although the equatorial anomalies over the Indian sector are weak, the symmetric pattern resembles the pattern over the Pacific with an equatorward shift, with a tendency for strong (weak) westerlies centered on 15°N and 15°S during warm (cold) events, and opposite anomalies poleward of these. An equatorward movement of the climatological westerly jet at 45°S over the Atlantic/Indian Oceans during DJF is also evident during warm events (compare Fig. 8.5c with Fig. 1.30) and is representative of the fluctuations in the high-latitude mode discussed above.

Many of the zonal-wind perturbations described above occurred during the 1986–89 ENSO cycle (Vincent et al. 1995). Chen et al. (1996) pointed out the change in the zonal-mean zonal-flow regime from a strong single jet in the 1986–87 warm event to a double-jet structure during the 1988–89 warm event, with strong flow around Antarctica. These changes were related to anomalies in the vorticity budget by Mo and Rasmusson (1993) and Rasmusson and Mo (1993). Sardeshmukh and Hoskins (1985, 1988) also

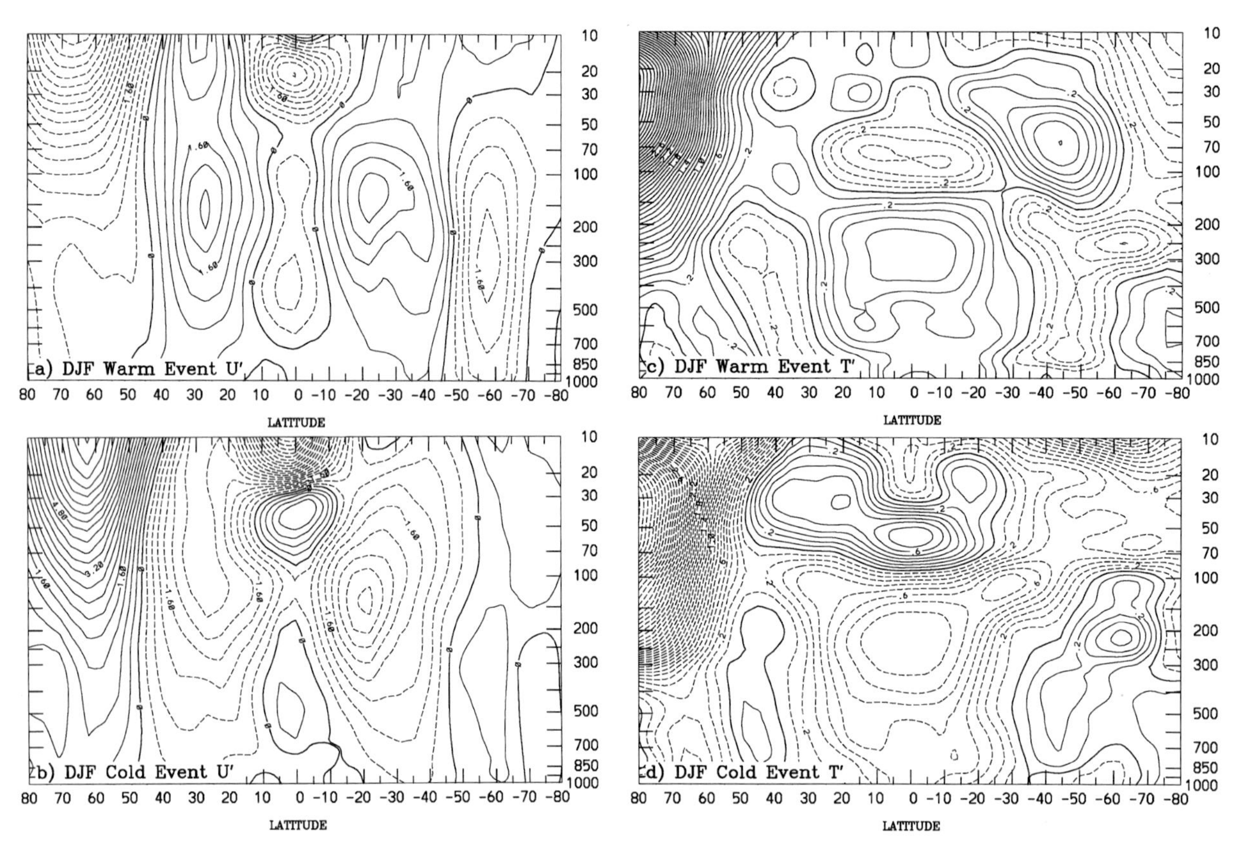

FIG. 8.7. Latitude–pressure diagrams of composite zonal-mean anomalies of (a) December through February zonal wind during warm events; (b) December through February zonal wind during cold events; (c) December through February temperature during warm events; and (d) December through February temperature during cold events. Contour interval is 0.4 m s^{-1} in (a) and (b), and 0.1°C in (c) and (d). Negative contours are dashed. Pressure is shown in hPa on the vertical scale.

showed that the large anomalies in tropospheric zonal wind during the 1982–83 and 1986–87 warm events could be accounted for by altered forcing of the rotational flow by the divergent circulation, as originally implied by Bjerknes. This outflow produces anomalous subtropical anticyclonic circulations poleward of the convection in both hemispheres. The opposing downstream response over the Atlantic can be understood as a planetary-scale Rossby response to the subtropical Pacific circulations (Sardeshmukh and Hoskins 1988).

The more substantial DJF anomalies in the SH, when compared to those in the NH during its summer, are likely related to the fact that the forcing of the Rossby-wave sources in the subtropics by the divergent wind is stronger in the former. This is due to the fact that the westerlies and associated absolute-vorticity gradient are stronger and closer to the upper-level divergent wind forced by the tropical convection during summer in the SH than during summer in the NH (see Fig. 1.30). It is still necessary, however, to take into account the effect of feedback by the transients, especially within the storm track (e.g., Trenberth 1987; Karoly 1990), in order to completely account for the strengthened subtropical jets during warm events (see Held et al. 1989; Hoerling and Ting 1994).

(iii) Zonally symmetric signals

Karoly (1989) analyzed the zonal-mean anomalies in wind, temperature, and pressure over the SH associated with warm events from the surface up to 100 hPa and found primarily equivalent barotropic structures from polar regions well into the subtropics. Figure 8.7 shows a similar analysis up to 10 hPa from 80°N to 80°S for DJF zonal-mean zonal wind during warm events (Fig. 8.7a), cold events (Fig. 8.7b), and for temperature during warm (Fig. 8.7c) and cold events (Fig. 8.7d). Comparison of Figs. 8.7a and 8.7b with Figs. 8.6b and 8.6d reveals that the zonal-mean signatures are dominated by the large anomalies over the Pacific sector. This is in line with the results of Trenberth (1981) and Kidson (1988b), who examined zonal-mean zonal-wind variability over the SH using smaller samples. Poleward of about 15°N and 15°S, the wind anomalies are of the same sign from the surface well into the stratosphere, with little tilt with height apparent. During warm events (Fig. 8.7a), strengthened westerly flow is found centered on 28°N and 23°S, extending to within 10° of the Equator and out to 45° latitude in both hemispheres. These westerlies peak at just below 100 hPa and represent an equatorward displacement and overall strengthening of the subtropical jets. Poleward of these jets, deep easterly anomalies dominate from the surface into the stratosphere, while westerlies are again found below 150 hPa over Antarctica. The midlatitude SH easterly signal peaks at a lower level than the westerly anomalies in the subtropics, but in the NH the easterly anomalies continue to increase in strength with height to at least 10 hPa at around 65°N.

The weaker-than-normal stratospheric polar vortex in the NH winter during warm events was documented in detail by van Loon and Labitzke (1987; see also chapter 6) and is seen in every warm event back to 1958 except for 1963, 1982, and 1991, which were accompanied by major volcanic eruptions affecting the stratosphere (see Labitzke and van Loon 1989). In contrast, the influence of warm events on the SH stratosphere appears to be weak, even during southern winter (not shown). The apparent strong equatorial stratospheric signal in Figs. 8.7a and 8.7b is likely due to sampling of the large anomalies associated with the stratospheric QBO (see chapter 6), as no consistent signal of these zonal winds was found above 100 hPa when individual events were examined.

Apart from the equatorial stratospheric signals just noted, zonal-mean zonal-wind anomalies during cold events are to a large extent the inverse of the warm-event signals. The strengthened stratospheric polar vortex in the NH and stronger westerlies over the Southern Ocean at around 60°S are robust, appearing to some degree in each individual cold event (not shown).

Many examples of these zonal-mean zonal-flow signatures during warm and cold events can be found in Fig. 8.2. The simultaneous strengthening and weakening of the subtropical flow in both hemispheres, in opposition to the flow along the Equator, is particularly evident. These anomalies give rise to variations in the angular momentum of the atmosphere during ENSO, which is observed to increase during warm events and decrease during cold events (e.g., Rosen et al. 1984).

(iv) Precipitation patterns during ENSO extremes

The anomalous global precipitation patterns during ENSO extremes were studied by Ropelewski and Halpert (1987, 1989), Kiladis and van Loon (1988), Kiladis and Diaz (1986, 1989), Lau and Sheu (1988), and Barnston and Smith (1996), among many others. Here we present a brief summary of those results and refer the reader to chapter 3 and other studies for more specific details of regional signals. Most of these signals show a reversal in sign between warm and cold events, so the response to ENSO is roughly linear in that sense (Kiladis and Diaz 1989; Ropelewski and Halpert 1989).

Figure 8.8a shows the composite DJF precipitation anomalies computed for a large sample of warm events back to 1877 (see Kiladis and van Loon 1988) from a gridded dataset assembled from land stations by Eischeid et al. (1991). Localized signals may not be seen at the resolution of this analysis, and although the magnitudes may be misrepresented due to sampling, the large-scale pattern is characteristic of the major warm-event anomalies over the SH during this season. The cold-event composite (not shown) shows virtually the opposite pattern with similar amplitudes at most locations, confirming the roughly linear relationship to equatorial Pacific SST anomalies. Here, we discuss the warm-event composite signals, with the understanding that cold events will tend to have opposite anomalies in general.

The large-scale atmospheric signal in the tropical Pacific during ENSO is manifested as a displacement of the convergence zones. Both the ITCZ and the SPCZ move equatorward toward positive SST anomalies during a warm event, while the SPCZ also moves eastward to a large extent (Trenberth 1976). Available island station data (e.g., Kiladis and van Loon 1988) and satellite data for more recent events confirm that precipitation greatly increases along the entire equatorial dry zone in the central and eastern Pacific, connecting the two regions of positive anomalies over the west Pacific and coastal South America in Fig. 8.8a. Regions normally under the influence of the ITCZ and SPCZ, such as Micronesia and, in the South Pacific, Fiji, Samoa, and New Caledonia, then experience below-normal rainfalls (Donguy and Henin 1980; Morliere and Rebert 1986). At the same time, there is a weakening of the monsoon and fewer-than-normal tropical storms over Australasia and the eastern Indian Ocean (e.g., Evans and Allan 1992; Basher and Zheng 1995). There is also evidence that precipitation is enhanced over the western Indian Ocean, when a strong signal over northern Madagascar and the Mascarene Islands is observed in Fig. 8.8a.

The tendency for drought conditions over northern and eastern Australia during warm events has been studied extensively (e.g., Nicholls 1988; Allan 1991; Drosdowsky and Williams 1991), while the increase in flood frequency during cold events is also an important impact in Australia (e.g., Nicholls 1991). Detailed examination of the rainfall events in eastern Australia by Nicholls and Kariko (1993) indicates that ENSO affects rainfall mainly by influencing both the number and intensity of events, which is also seen in terms of the cyclone frequency over the continent (Sinclair et al. 1997). Figure 8.8a indicates generally wetter conditions over New Zealand during DJF; however, the interplay between the seasonality of the flow and orography of New Zealand leads to complex details in ENSO precipitation signals there (Gordon 1986; Mul-

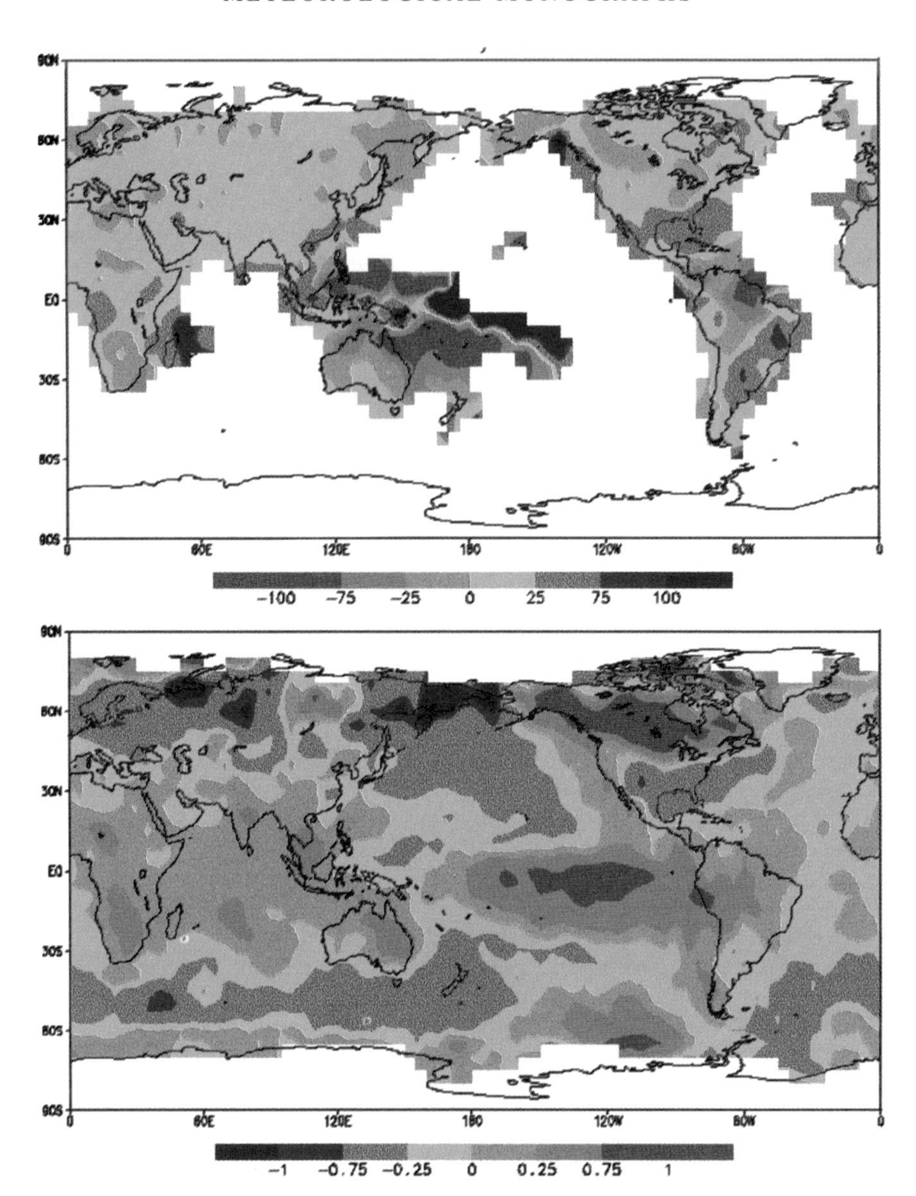

FIG. 8.8. Composite warm-event anomalies calculated from a 25-event sample back to 1877 (see Kiladis and Diaz 1989). Data are on a 5° global grid (Eischeid et al. 1991). (top) precipitation, in mm month^{-1}; (bottom) temperature, in °C.

lan 1995). For example, enhanced southwesterly flow during summer and spring (Figs. 8.4c,d) leads to heavier orographic precipitation along the west coast of the South Island but drier-than-normal conditions in the lee of the Southern Alps.

In South America, as SST increases along the coast of Ecuador and northern Peru, extensive flooding is often experienced, with above-normal precipitation over the mountains of northern and central Peru but drier-than-normal conditions over the southern Peruvian Andes (Enfield 1992). Dry conditions also extend eastward into the Amazon basin. A relationship between drought in northeastern Brazil and ENSO has been found by Hastenrath and Heller (1977) and Ropelewski and Halpert (1987). These rainfall deficits appear to be due to increased atmospheric subsidence over northern South America associated with the northward displacement of the Atlantic ITCZ and meridional circulation during warm event years (Kousky et al. 1984; see chapter 3).

The effect of an anomalous trough during JJA and SON in Figs. 8.4a and 8.4b over southern South America is associated with enhanced frontal activity and wetter-than-normal conditions over southern Chile and Argentina (Aceituno 1988; 1989), leading to a signal of increased snowcover over the southern Andes (Cerveny et al. 1987). These wet conditions persist, but with lower amplitude, into the following summer (Fig. 8.8a).

Southeastern Africa, south of 15°S, also shows drier-than-normal conditions during warm events. This tendency for drought occurs during the main DJF rainy season and lasts into the following MAM season, and appears to extend out into southern Madagascar (Jury et al. 1994, 1995; Lindesay 1988; see chapter 3). The signal lies in the region of expected anomalous subsidence on the equatorward side of westerly anomalies just south of the continent (see Figs. 8.4c,d and 8.6b). At the same time, the equatorial eastern African region, which encompasses parts of Kenya, Uganda, Rwanda, Burundi, and Tanzania, tends to experience greater-than-normal rainfall. This dipole pattern of rainfall over Africa has also been documented by Janowiak (1988), Nicholson (1986), and Nicholson and Entekhabi (1986). The equatorial eastern African signal first appears during SON and appears to be the western extension of a large region of heavy rainfall covering the western Indian Ocean at this stage of a warm event, as reflected in the strong signal in the Seychelles during both phases of ENSO (Kiladis and Diaz 1989). It is important to emphasize, however, that patterns of African precipitation are complex in detail and only a relatively small portion of the overall variability of the Africa rainfall can be ascribed to ENSO (e.g., Tyson 1986; Janowiak 1988).

(v) Temperature patterns during ENSO extremes

The zonal-mean temperature signals during DJF warm and cold events were shown in Figs. 8.7c and 8.7d. The warming of the low-latitude troposphere out to about 30° latitude away from the Equator is a well-documented feature of warm events (e.g., Newell and Weare 1976; Horel and Wallace 1981; Yulaeva and Wallace 1994); corresponding cooling is noted for cold events in Fig. 8.7d. Beyond this, in warm (cold) events there are bands of cooling (warming) in the midlatitudes of both hemispheres that exhibit poleward tilts with height. As with the zonal wind, the largest signals of all are located in the NH polar stratosphere, with correspondingly small and inconsistent signals over Antarctica in all seasons. Overall, the tropospheric temperature anomalies are in phase with the zonally symmetric DJF 500-hPa height patterns in Figs. 8.4c and 8.5c and are of the right sense to produce a thermal wind balance of the zonal winds in Figs. 8.7a and 8.7b.

The distribution of surface-temperature signals associated with warm events during DJF is mapped in Fig. 8.8b. As in Fig. 8.8a, the temperature anomalies were obtained from a gridded set of station data, with SST from ship observations used as a proxy for surface air temperature over the oceans, giving near-global coverage. The far-field surface temperature lags the Pacific SST signal by about three to six months (Kiladis and Diaz 1989; Halpert and Ropelewski 1992), as it does in the rest of the troposphere (Angell 1981; Horel and Wallace 1981; Yulaeva and Wallace 1994). Over the drought regions of Africa, Australasia, and South America, it is quite likely that an increase in subsidence and shortwave radiation is responsible for the above-normal air temperature signals. This can exacerbate the impacts of reduced precipitation by increasing evaporation, worsening the effect of drought on crop production and fire potential.

Since surface air temperature closely follows the local SST over most regions of the oceans, it is necessary to explain the forcing of SST anomalies in order to explain anomalies in surface air temperature over the bulk of the SH. The observed warming in the central and eastern equatorial Pacific can be understood to be a direct result of the decrease in upwelling over those regions. The extension of this warming along the west coast of the Americas is partially a result of the poleward propagation of downwelling coastal Kelvin waves initially striking the coast along the Equator (e.g., Philander 1990).

Over the rest of the tropical oceans, it has been hypothesized that shortwave radiation may be important in forcing the above-normal SST (e.g., Kiladis and Diaz 1989), which would give rise to the observed temporal phase lag with the eastern Pacific. This would seem to be the case over the Indian Ocean and Australasia, where drought conditions persist late into a warm event despite the high SST there (e.g., Nicholls 1984). Van Loon and Shea (1987) point out, however, that over the SPCZ region meridional advection is also in the right sense to produce a warming of the SST during DJF of a warm event.

The zonal-mean signal of below-normal temperatures between 30°S and the Antarctic coast has contributions from every sector except the southeastern Pacific (see also Barnston and Smith 1996). At least part of this signal may be related to westerly wind stress forcing of the ocean surface, which would produce lower-than-normal SST due to latent and sensible heat fluxes, as well as mechanical mixing within the upper layer of the ocean (e.g., Lau and Nath 1996). It is worth noting that these temperature signals, as well as those precipitation signals discussed above, appear to show a fairly linear relationship with the Pacific equatorial SST, with cold events having patterns of the opposite sign with similar magnitude in most regions (Kiladis and Diaz 1989; Ropelewski and Halpert 1989).

c. Other sources of interannual variability

Forcing leading to IA and longer-timescale climatic variability can be categorized as either "internal" and "external" (e.g., Diaz and Kiladis 1996). ENSO is an example of an "internal" source of climate variability; that is, one that is due to instabilities within the ocean–atmosphere system. Solar variability and volcanism are examples of external climate forcings, which are completely removed from feedback from the climate system.

Apart from ENSO, there is evidence that SST fluctuations outside the tropical Pacific are related to IA variability over the SH. For example, observations show that rainfall over northeastern Brazil, the Nordeste region, is related to SSTs in the Atlantic Ocean (e.g., Hastenrath 1978; see chapter 3). Numerical experiments performed by Moura and Shukla (1981), Mechoso and Lyons (1988), and Mechoso et al. (1990) support the hypothesis that the atmosphere is being forced by the SST anomalies. SST anomalies in the South Atlantic and Indian Oceans have also been implicated in precipitation variability over southern Africa (e.g., Mason 1995). Feedback between the atmosphere and sea ice around Antarctica is also a possible source of internal forcing in the SH (e.g., Simmonds and Jacka 1995; see chapter 4).

The stratospheric QBO (see chapter 6) is another example of internal variability. Trenberth (1980) found that little, if any, of the extratropical variance in the SH is related to the QBO, although Yasunari (1989) and Gray et al. (1992) suggest some loose connection between the QBO and tropical signals. There is also some evidence for QBO-associated signals around the southern African region (Jury et al. 1994; see chapter 3).

The 11-yr solar cycle has been shown to be related to stratospheric and tropospheric anomalies in the NH (e.g., van Loon and Labitzke 1987, 1988), but little is known of its influence on the SH. Large volcanic eruptions produce a cooling of the surface temperature (e.g., Dutton and Christy 1992) and increase in stratospheric temperature (Labitzke and van Loon 1989). The net result of such external forcing, combined with SST and cryospheric feedbacks, could conceivably produce substantial SH variability on IA timescales, although much more work is needed in this area to assess the extent of these effects.

8.3. Intraseasonal variability

a. General patterns

Overall, there is less pronounced geographical dependence and seasonal variability in SH extratropical transient kinetic-energy activity when compared to the NH. This is due to more substantial orographic forcing and land–sea temperature contrasts in the NH, producing larger-amplitude stationary waves. Nevertheless, the distribution of continental heat sources and sinks, tropical heating, and orography are not negligible in the SH (Kalnay and Mo 1986; Quintanar and Mechoso 1995a,b), and these result in considerable zonal asymmetries in statistics of both the time-mean and transient circulations (e.g., Trenberth 1982; see chapter 1).

Most of the extratropical IS variability in circulation over the SH occurs in the westerly belt between 40° and 60°S (Jones and Simmonds 1993), with the latitude of maximum variability shifting from the equatorward side of this latitude range at 200 hPa poleward into the circumpolar trough at the surface (see Figs. 1.14, 1.24, 1.31, and 1.33). At lower latitudes, low-frequency IS variability is dominated by the eastward-propagating Madden–Julian oscillation on the 30–60-day timescale, which has primarily a zonal wavenumber 0 through 2 spatial scale. At higher frequencies, a host of equatorially trapped disturbances are important at low latitudes (Wheeler and Kiladis 1999). These include equatorially trapped Rossby waves (Kiladis and Wheeler 1995) and mixed Rossby–gravity waves (e.g., Liebmann and Hendon 1990), which in the past have been identified as "easterly waves" (see chapter 3). The MJO in particular displays important relationships to extratropical IS variability over the SH (see following sections).

Space–time Fourier analysis has been used extensively to examine IS spatial and temporal variability in the troposphere of the SH extratropics. In contrast to the NH, which has substantial energy in westward-propagating "ultra-long" wavenumbers 1 and 2 (e.g., Fraedrich and Böttger 1978; Madden 1979), SH planetary-wave energy is more concentrated in higher-wavenumber, quasi-stationary and eastward-propagating features (Fraedrich and Kietzig 1983).

In SH summer, the bulk of the variability is due to propagating medium-scale (wavenumber 4–7) features (Randel and Stanford 1983, 1985a,b), with wave 5 dominant for extended periods (Salby 1982; Hamilton 1983; Chen et al. 1987). In winter, power shifts to wavenumbers 1 through 4, with waves 3 and 4 most dominant (Mechoso and Hartmann 1982; Randel 1987; Randel et al. 1987). During all seasons the transient wave energy peaks near the jet level at around 300 hPa, but, in winter, vertical propagation of wavenumber 1 and 2 energy into the stratospheric westerlies is evident (Randel 1987; see chapter 6). Higher-wavenumber energy decays very rapidly above the jet.

Wavenumber 3 has a maximum in stationary variance during winter (Hansen et al. 1989), while wavenumber 4 patterns in this season are mainly eastward-traveling modes associated with baroclinic waves at higher latitudes (Randel 1987). These phase-propagation speeds are qualitatively consistent with the dispersion relationship for barotropic Rossby waves when Doppler shifted by the background wind, with shorter waves generally propagating eastward faster than longer waves, as expected (Fraedrich and Kietzig 1983).

Zonal-mean variability on the IS timescale in the SH troposphere is in many respects similar to patterns of IA variability. For example, the correlation as a function of latitude of zonal-mean zonal wind in Fig. 8.1, whereby zonally symmetric signals of zonal wind alternate at roughly 20°-latitude intervals, holds well into the IS range (Kidson 1988a). Several studies have identified distinct vacillations in these zonal-wind signals during SH winter (e.g., Webster and Keller 1975), whereby the flow periodically goes from more zonally symmetric to highly amplified. These zonal-index transitions are very rapid compared to the persistence of the regimes themselves (Yoden et al. 1987; Hartmann 1995), with the transitions between and maintenance of each regime attributable to wave–mean flow interactions (Hartmann et al. 1984; Karoly 1990; Yu and Hartmann 1993).

When more zonally symmetric, there is a tendency for a broad westerly double-jet structure, with a primary maximum at about 55°S and a secondary jet at around 25°S. The more amplified flow has a single zonal-wind maximum centered at 40°S and strong wavenumber 1 and weak wavenumber 3 components, as opposed to the more dominant wavenumber 3 in the zonally symmetric flow case (Hansen and Sutera 1988). Hansen and Sutera (1991) found evidence that the switching between these two regimes leads to a bimodal distribution in the probability density for wavenumber 3 amplitude, which dominates the variance during most SH winter seasons. This implies two stable flow regimes and that the average state of the circulation is not the most probable state. Such behavior gives the impression of multiple stable regimes of an almost intransitive system (e.g., Mo and Ghil 1987). The bimodality found by Hansen and Sutera (1988, 1991), however, appears to be very sensitive to the parameters and samples chosen to represent it and may not be statistically significant (Nitsche et al. 1994). Despite a rather broad preferred timescale of around 18 to 30 days for these vacillations, the actual spectrum of such zonal-mean zonal winds has statistically significant peaks only on the seasonal to IA timescale (Kidson 1988a), much of which is related to the ENSO variability described in the previous section.

As in section 8.2, a convenient starting point in the discussion of spatial patterns of IS variability in the SH is an EOF analysis of 500-hPa heights. EOFs were calculated as in the previous section, except using data filtered to retain only fluctuations on the 10–50-day timescale. These modes can thus be viewed as preferred IS perturbations superimposed on the lower-frequency IA patterns in Fig. 8.3. Our results are very similar to those obtained from a shorter sample by Kidson (1991) using ECMWF analyses. Broadly similar patterns were also obtained by Ghil and Mo (1991), Farrara et al. (1989), Mechoso et al. (1991), Lau et al. (1994), and Kidson and Sinclair (1995), although some differences can be accounted for by the fact that those studies included lower-frequency fluctuations in their analyses. Kidson shows that the 10–50-day band accounts for more than 40% of the total daily variance, exclusive of the seasonal cycle, in 500-hPa height over much of the mid- and high latitudes of the SH. Since the leading modes show little seasonal dependence, results will only be discussed for the EOFs obtained using the entire data record.

The first two IS EOFs (Figs. 8.9a,b) describe 11.5% and 9.8% of the variance, respectively, and represent two zonal wavenumber 4 patterns nearly in quadrature. The centers of these wave trains extend from south of Australia into the Pacific, where they reach maximum amplitude and are farthest poleward, and then equatorward again over the Atlantic. As with the IA EOFs 2 and 3 (Figs. 8.3b,c), the IS EOFs 1 and 2 are also suggestive of Rossby wave trains. The IS EOFs represent an eastward progressive mode, however, with preferred periods in both the 16-day and 22–26-day ranges (Kidson 1991; Ghil and Mo 1991; Lau et al. 1994), whereas the IA modes are essentially quasi-stationary wavenumber 3 patterns.

Intraseasonal EOFs 1 and 2 are most dominant during winter, and their 2–4-week timescale is consistent with the dominance of propagating wave 4 variance during this season. These patterns still have substantial amplitude in summer as well, but they are the second and third EOFs in that season. As we would expect for fluctuations with periods greater than 10 days, the vertical structure shown is nearly equivalent barotropic, with only a slight westward tilt with height (Kidson 1991). By comparison, Mechoso and Hartmann (1982) show that eastward-propagating wavenumber 4 patterns, which dominated the SH midlatitudes during the austral winter of 1979, had a vertical structure more consistent with baroclinic waves.

The third and fourth EOFs (Figs. 8.9c,d) are also in spatial quadrature. They represent a mix of waves 3 and 4 in midlatitudes, and their zonally symmetric components display an exchange of mass between Antarctica and lower latitudes. Thus, these patterns are partly a manifestation of an IS high-latitude mode like that dominant on IA timescales. In summer, IS EOF 3 is much more zonally symmetric (not shown) and is very similar to the IA EOF 1 of Fig. 8.3a. As with the first two EOFs, and the IA fluctuations discussed in section 8.2a, EOFs 3 and 4 have a nearly equivalent barotropic structure.

There are distinct differences, as well as similarities, between the IA EOFs described in section 8.2a and the IS results. Both timescales display a dominance of wave 3 and 4 structures, with fluctuations in similar locations on a broad range of timescales. For example, IA EOF 3 (Fig. 8.3c) and IS EOF 4 (Fig. 8.9d) are quite similar, with centers in the southern Indian Ocean and to the southeast of New Zealand, just poleward of the climatological-mean, zonally asymmetric 500-hPa heights (Fig. 1.20). This suggests a periodic amplification/suppression of the stationary waves in these sectors. Mo and van Loon (1984) and Mo and Higgins (1997) comment on the fact that the amplitude of wave 3 has substantial secular variability, being much less prominent in the

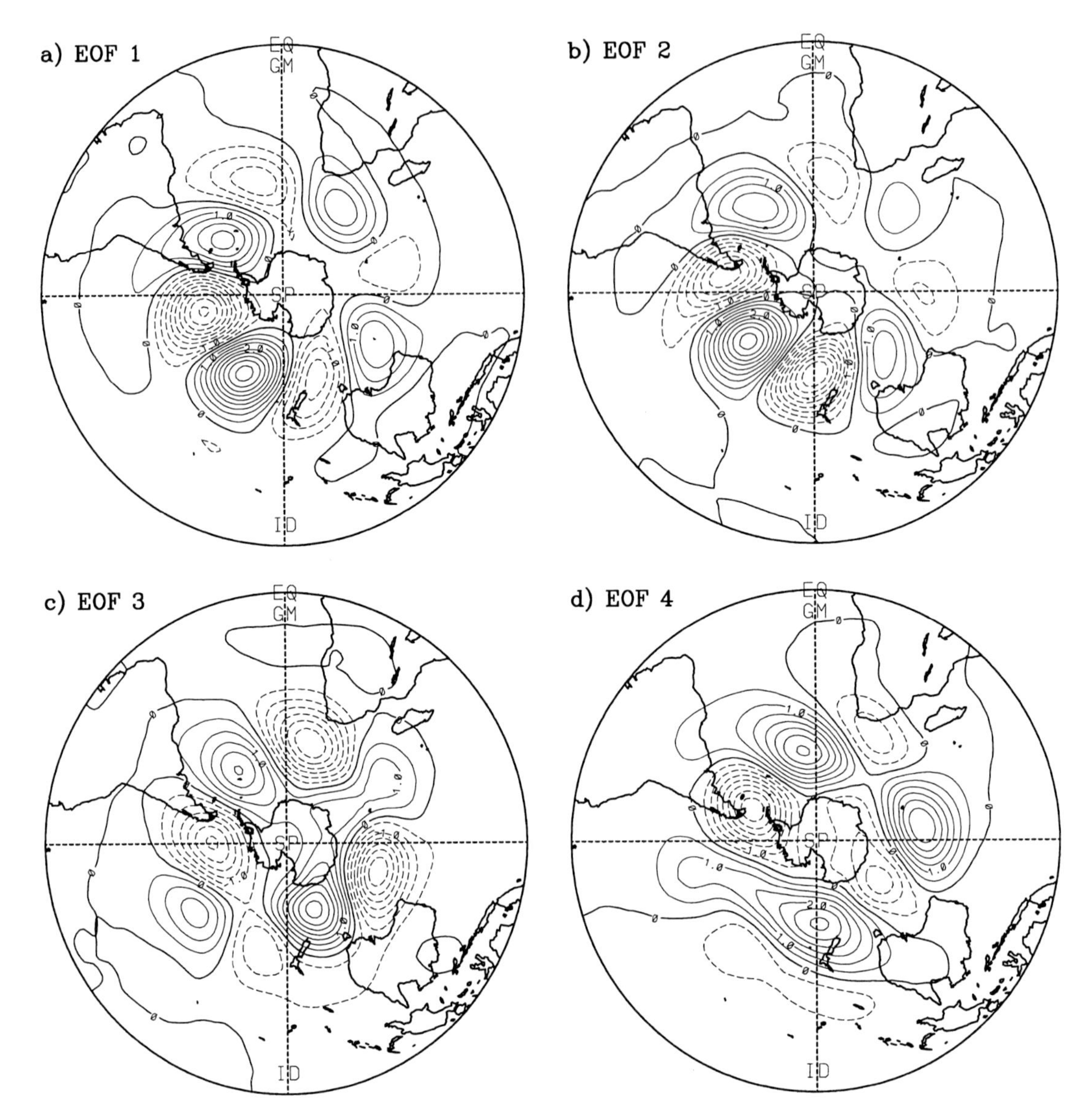

FIG. 8.9. First four EOFs of 10–50-day filtered Southern Hemisphere 500-hPa geopotential height anomalies from 1973 through 1995. Contour interval is 0.5 units. Negative contours are dashed.

1970s and stronger over the recent decade. These variations also appear to carry over into the IS variability and may be related to changes in the SH semiannual cycle (van Loon et al. 1993; Hurrell and van Loon 1994), but more work is needed to fully assess these potential scale interactions.

In addition to zonal wavenumbers 3 and 4, wavenumber 5 patterns also appear in summer EOFs (Kidson 1991). Salby (1982) and Hamilton (1983) documented pronounced wavenumber 5 variability during the FGGE year 1979. This wave shows a regular eastward phase progression with periods on the order of 10 days, but it also has a stationary component on the timescale of 8–15 days. Randel and Stanford (1983) reviewed the intermediate-scale waves in the SH, substantiated their equivalent barotropic structure, and demonstrated the extent of their interactions with the zonal-mean flow.

Kalnay et al. (1986) reported on the presence of large-amplitude stationary Rossby waves of zonal wavenumber 7–scale between 20° and 40°S in the South Pacific and east of South America during January 1979. GCM experiments suggested that the stationary waves were maintained by tropical heating and that subtropical convection was important in sustaining their large amplitude and phase. Intraseasonal tropical convection as a potential forcing of the SH circulation is discussed further in section 8.3d.

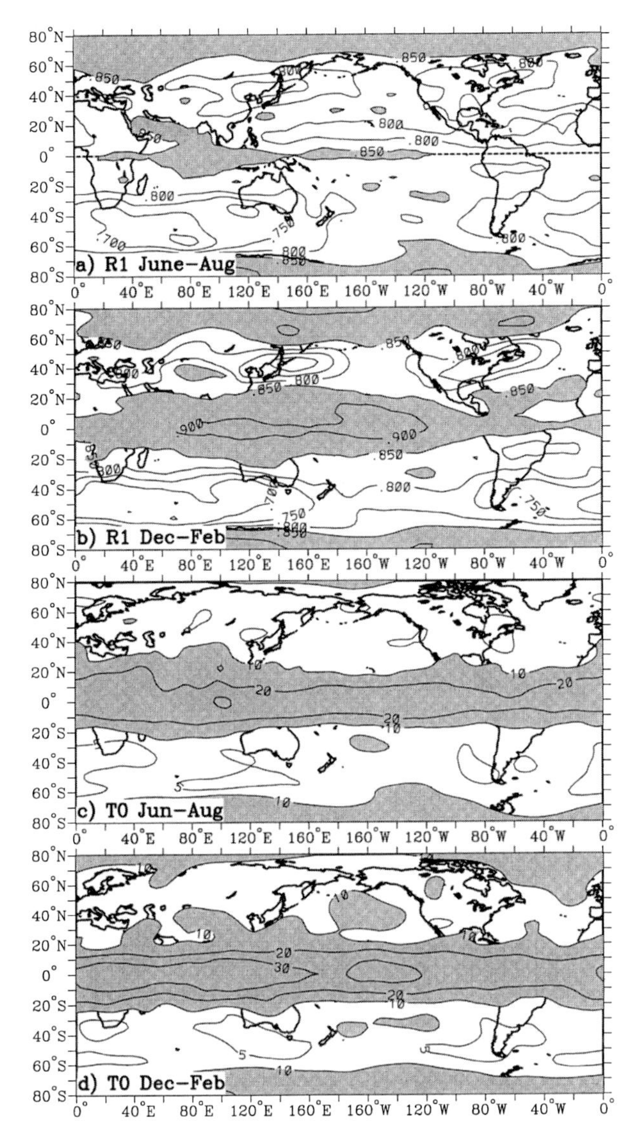

FIG. 8.10. (a,b) Lag 1 autocorrelations (R1) for daily 500-hPa height anomalies from 1973 through 1995 for (a) June through August, and (b) December through February. Contour interval is 0.05, with values above 0.85 shaded. (c,d) The estimate (T_0) of the effective time between independent samples based on a first-order autoregressive model (see text) for (c) June through August, and (d) December through February. Contours are shown for 5, 10, and 20 days. Values above 10 days are shaded.

b. Persistent anomalies and blocking in the Southern Hemisphere

When compared to the NH, anomalous flows in the SH are, in general, less persistent and of lower amplitude (Trenberth 1985; Mo 1986). Trenberth (1985) calculated various measures of persistence for the SH using Australian analyses from 1972 to 1980. Two of these are shown for global fields using the NCEP reanalysis data from 1973 through 1996. Figures 8.10a and 8.10b show the lag 1 autocorrelations (R1) for daily 500-hPa height anomalies for JJA and DJF. Also shown is an estimate (T_0) of the effective time between independent samples (Figs. 8.10c,d), based on a first-order autoregressive model. While the patterns in Fig. 8.10 are similar to Trenberth's over the SH, our values for both T_0 and the autocorrelation at lag 1 day are higher, which is likely due to the fact that we have used the more-homogeneous NCEP/NCAR reanalysis product.

The highest values of day-to-day persistence as measured by the R1 statistic are over the Tropics and polar regions of both hemispheres, with the lowest values in the storm tracks. Values of T_0 exceeding 20 days over the entire equatorial zone are partly due to IA forcing from ENSO, and the locally high values of T_0 over the extratropical central North and South Pacific during DJF are also likely to be a consequence of this forcing. Low storm-track values can be attributed to baroclinic development and rapid advection. As shown by the areas covered by R1 greater than 0.75, there is less persistence over the SH overall. Thus, the average value for T_0 in the midlatitudes is about 5–6 days for the SH and about 8–9 days in the NH (Gutzler and Mo 1983; Shukla and Gutzler 1983). This is due to the lower-amplitude waves and more-zonal circulation over the SH in general.

A somewhat different measure of persistence is shown in Fig. 8.11, which uses the methodology employed for the NH by Dole and Gordon (1983) and depends on the amplitude of the anomaly itself. These maps show the number of cases when anomalies at a given location, greater in magnitude than a prespecified threshold, persist for a given time. Figures 8.11a and 8.11b map the number of anomalies greater than 150 m at 500 hPa lasting for 1–3 days, for May–September and November–March. In both seasons, maxima are located along 50°S, extending from the South Atlantic eastward to the south of Australia and New Zealand. These patterns resemble 2–8-day bandpass-filtered variance fields of 500-hPa heights (Trenberth 1982; Trenberth and Mo 1985), and so can be interpreted as large-amplitude storm-track fluctuations due to baroclinic waves, which have maximum activity in these regions (e.g., Sinclair 1994; Sinclair et al. 1997; see chapter 10). Winter values are slightly higher, when the baroclinic waves are more amplified, but this difference is much smaller than that observed for similar seasonal statistics in the NH.

The spatial pattern of persistent anomalies greater than 100 m that lasted for 7 days or more show very different patterns (Figs. 8.11c,d) and to a large extent represent large-amplitude cut-off lows and blocking highs. While the distribution tends to be similar for both positive and negative departures (not shown), Trenberth and Mo (1985) found pronounced skewness in the overall frequency of 500-hPa geopotential height, which was a function of latitude. As in the NH, these statistics indicate that blocking anticyclones are

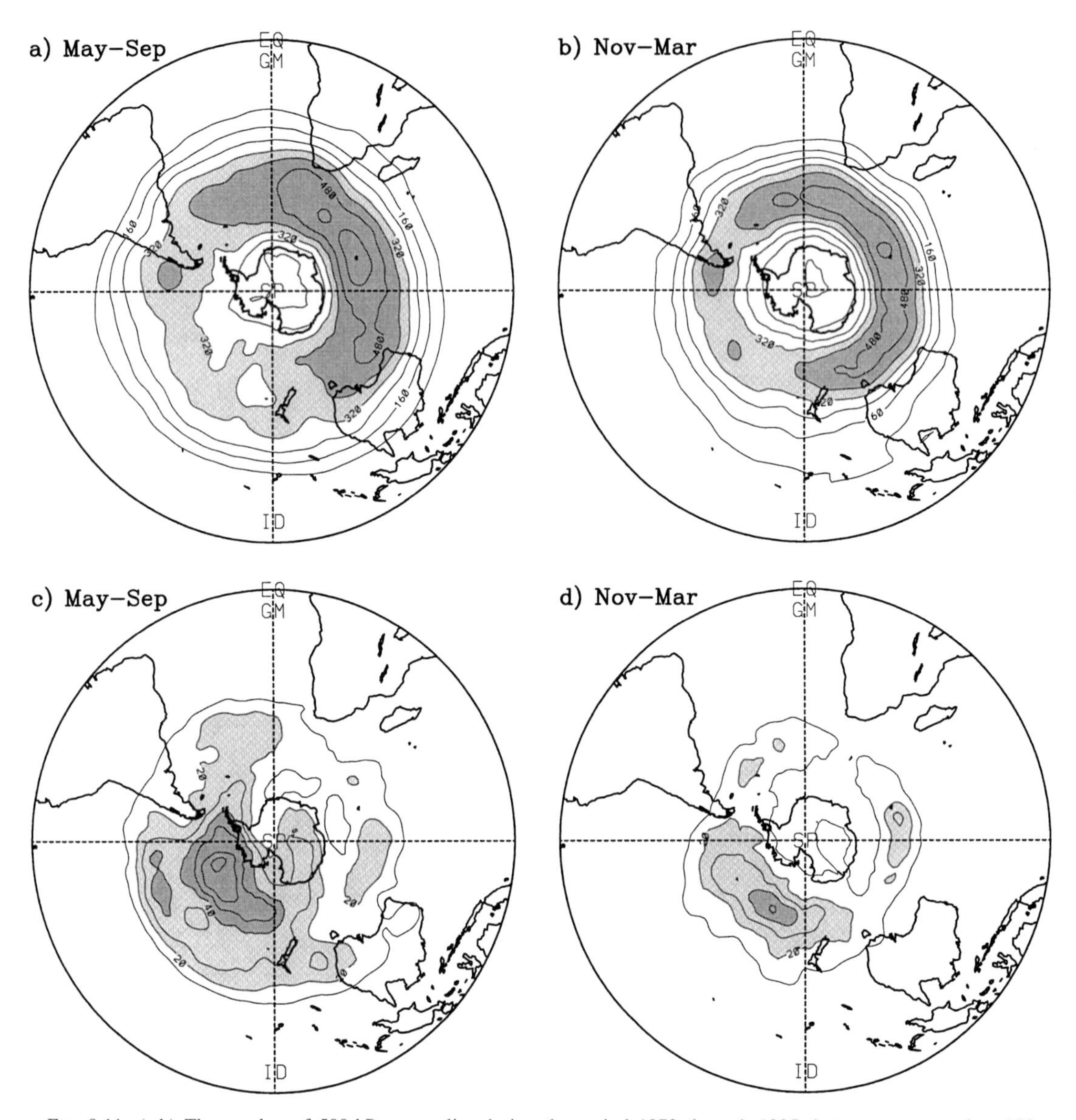

FIG. 8.11. (a,b) The number of 500-hPa anomalies during the period 1973 through 1995 that were greater than 150 m and lasted for 1–3 days for (a) May–September and (b) November–March. Contour interval is 80 events. Values between 320 and 400 events are shaded lightly; values of above 400 events have dark shading. (c,d) The number of 500-hPa anomalies during the period 1973 through 1995 that were greater than 100 m and lasted for greater than 10 days for (c) May–September and (d) November–March. Contour interval is 10 events. Values between 20 and 40 events are shaded lightly; values of above 40 events have dark shading.

most common on the poleward side of the polar-front jet, with cut-off lows most common on the equatorward side.

Figure 8.11c implies that there should be a primary maximum in wintertime persistent anomalies over the South Pacific centered at around 60°S, 120°W, which is also confirmed in the studies of Sinclair (1996), Sinclair et al. (1997), and Mo and Higgins (1998). In summer, the primary region shifts westward to southeast of New Zealand (Fig. 8.11d). Other regions of frequent persistence, present in both seasons, are located over the southern Indian Ocean, to the southwest of Australia, and southeast of South America.

The patterns in Figs. 8.11c and 8.11d are consistent with the locations of most-frequent blocking over the SH identified in recent studies by Sinclair (1996) and Mo and Higgins (1998), but differ somewhat from earlier studies (e.g., van Loon 1956; Trenberth and Mo 1985; Lejenäs 1984, 1988). These differences may be due to the varying methods used to identify blocks, the representativeness of the datasets used, or real secular variability in the locations and frequencies of blocks.

All studies find that persistent anomalies are most common in winter, with one maximum of activity located southeast of New Zealand near 60°S, 160°W. This is a region where the climatological split in the jets favors the formation of blocks. It is also known that this region is a favored location for the formation of cut-off lows (see Taljaard 1972), although there are few recent studies of these events in the SH. As with persistent anomalies, the duration of blocking is shorter and its amplitude is less on average in comparison to the NH (van Loon 1956; Coughlan 1983). This may be attributed to the presence of the generally stronger westerlies at middle and high latitudes of the SH, which makes blocks more vulnerable to the effects of transient disturbances.

There is considerable interest in whether persistent anomalies, and blocks in particular, are connected to hemisphere-wide planetary-scale fluctuations. In the NH, blocking seems to be mainly a local phenomenon, with little tendency for a block in one region to be connected to blocking in another area (Lejenäs and Økland 1983). This also appears to be the case in the SH, and although statistics show that locally zonal wave 3 is most consistently involved in SH blocks, they often appear to be part of wave trains that do not necessarily have a zonal orientation (Trenberth and Mo 1985). At times, multiple blocks involving wave 3 do occur, although this is not common.

As an example of persistent anomalies tied to hemispheric wave trains, Mo and Higgins (1998) studied the flow associated with large-amplitude wintertime anomalies in the two favored regions of the South Pacific identified above. The pattern for anomalies at 60°S, 120°W is nearly identical to the predominantly wavenumber 3 IA EOF 2 in Fig. 8.3b, which in turn strongly resembles the composite warm-event circulations during JJA and SON in Figs. 8.4a and 8.4b. This is also consistent with the eastward displacement of the subtropical jet and strengthening of the westerlies along the Antarctic coast in Figs. 8.6a and 8.6b, which in essence results in an eastward shift and enhancement of the climatological split flow near New Zealand (Chen et al. 1996). Above-normal heights over the Bellingshausen Sea are a signature of the tendency for blocking over this area during warm events (Rutllant and Fuenzalida 1991).

Mo and Higgins found that the pattern associated with persistent anomalies at 60°S, 160°W is virtually the same pattern as in Fig. 8.3b, except shifted 40° longitude to the west. These patterns are dominated by wave 3, with an orientation consistent with propagation of energy out of the Tropics near Australasia, more zonally through the South Pacific, then equatorward again in the South American/Atlantic sector. We show in section 8.3d that a portion of this variability can be related to IS variability in tropical convection.

Since there is strong geographical dependence in the locations of most-frequent blocking in both hemispheres, zonal asymmetries in the basic state must play an important role in determining where blocks should preferentially form. Thermal and orographic forcing is responsible for the maintenance of the stationary waves (see chapters 1 and 2); thus, the interplay between this forcing and transient wave activity is likely important for the initiation of blocks. Baines (1983) concluded that blocking in the Australian region is primarily a locally forced phenomenon, rather than a hemispheric one, and reviewed the possible initiation and maintenance mechanisms. These were baroclinic instability influenced by local topography or thermal forcing (e.g., Frederiksen 1982) and a "modon" mechanism (McWilliams 1980) aided by thermal forcing associated with zonally asymmetric SSTs.

It appears that a variety of possible forcings can be responsible for persistent anomalies. Hoskins et al. (1983) note that the split in the zonal flow near New Zealand is supported by the eddy fluxes of vorticity, with the transients tending to force westerlies between 40° and 60°S and easterlies to the north. During blocking, this feedback appears to be amplified. Trenberth (1986a,b) examined in detail a blocking episode near New Zealand during the winter of 1979 and found that the transients maintained the block by systematically accelerating the westerlies in the storm track to the south of the block, while decelerating the westerlies at the split of the jet. These results support the key role of synoptic eddies in maintaining blocks (Shutts 1983).

Mo et al. (1987) performed GCM experiments to study the mechanisms for the June 1982 block and concluded that the most important forcing maintaining that particular event was heating related to land–sea contrast. Berbery and Núñez (1989) analyzed the blocking episode during June 1985 over the South American sector. Using a hemispheric shallow-water model, they were able to simulate the mechanism of wave amplification by local resonance between Rossby waves generated over the Pacific and the downstream Andes mountains. This mechanism was proposed by Kalnay and Merkine (1981) to explain the frequent occurrence of blocking downstream of the Andes over the South Atlantic Ocean.

c. Mechanisms responsible for SH teleconnections

The IA and IS variability described in the previous sections is superimposed on the climatological seasonal cycle described in chapter 1. Over the last two decades, significant progress has been made toward understanding the behavior of transient fluctuations with respect to the basic-state circulation by considering the propagation of linear barotropic Rossby waves (e.g., Hoskins and Karoly 1981; Hoskins et al. 1983). This involves examination of the refractive-index properties of the slowly varying large-scale flow through use of the so-called Wentzel–Kramers–Brillouin (WKB) approximation. Some important implications predicted by the theory include the following.

1) In a westerly, constant angular-velocity flow on a sphere (i.e., superrotation), Rossby-wave energy disperses eastward along great circle routes.
2) Rossby waves of a typical IS spatial scale should be trapped along westerly jets, with the jet acting as a waveguide, or preferred path, for propagation.
3) The region where the zonal flow is equal to the zonal phase speed of a Rossby wave represents a critical line, through which wave propagation is precluded.

Manifestations of Rossby-wave propagation can be seen in the plots of IA and IS EOFs in Figs. 8.3 and 8.9, where, in many cases, wave trains originating near Australia follow at first a poleward path across the Pacific, then an equatorward arc into the Atlantic sector. Deviations from a great-circle route can occur due to refraction by the basic state, development due to barotropic and baroclinic instability, and modification of the waves by orography, diabatic effects, and other sources of dissipation or generation. The theory can be extended to three-dimensional flow to account for vertical propagation (Karoly and Hoskins 1982; Karoly 1983) and generalized further through the use of the Eliasson–Palm flux (Edmon et al. 1980; Trenberth 1986b; see chapter 2).

It follows from point 3 above that Rossby-wave energy should be unable to propagate from regions of westerly flow into regions of easterlies (Webster and Holton 1982). During much of the year, such critical lines are observed along the margin of the equatorial upper-level easterlies. Cross-equatorial Rossby-wave propagation should be possible, however, in sectors such as the eastern equatorial Pacific and Atlantic, where westerlies extend from the extratropics of both hemispheres across the Equator during austral summer (see Fig. 1.30a). Such interhemispheric propagation is indeed observed during this season (Kiladis and Weickmann 1992b; Hsu and Lin 1992; Tomas and Webster 1994; Kiladis 1998) and represents a cross-hemispheric source for IS variability.

Several studies have compared preferred teleconnections in the SH with that expected from WKB theory and simulated in linear barotropic models. Using one-point correlation maps Berbery et al. (1992) and Ambrizzi et al. (1995) obtained teleconnection patterns for austral winter that are nearly identical to the IS EOFs in Fig. 8.9. They identified the wintertime polar and subtropical jets as waveguides and the South Atlantic and eastern South Pacific as preferred regions for equatorward propagation (Ambrizzi and Hoskins 1997). The polar-jet waveguide is quite evident in Figs. 8.3 and 8.9, although variance within the subtropical jet is best viewed in upper-tropospheric fields (e.g., Fig. 1.33). In summer, there is a single waveguide along the polar jet, with the strongest teleconnectivity along 45°S from 90°W to around 90°E, where the jet is strongest (Hsu and Lin 1992; Hoskins and Ambrizzi 1993; Ambrizzi and Hoskins 1997). Frequent troughs and associated frontal zones propagating into the Tropics from the SH westerlies are seen over Australia, the western South Pacific, South America, and southern Africa during this season, giving rise to the episodic cloud bands, such as the SPCZ, connecting the Tropics and extratropics in satellite imagery (e.g., Streten 1978; Kiladis et al. 1989; Kiladis and Weickmann 1992a, 1997; Kodama 1993; see chapter 3).

d. Intraseasonal tropical–extratropical interaction over the SH

The discussion in section 8.2 showed that a large portion of the IA variability of SH circulation is related to shifts in tropical convection associated with ENSO. A significant portion of the IS variability is also known to be related to tropical convection. Evidence supports the notion that the relationship is two-way; that is, in some cases shifts in tropical convection force associated changes in circulation, while at other times the convection appears to be related to forcing by disturbances originating in the extratropics.

1) The Madden–Julian oscillation

We will not attempt to summarize here the vast literature on the MJO, but give only a broad description of the disturbance and focus on its signals over the SH. Detailed descriptions of the observational aspects of the MJO can be found in Knutson and Weickmann (1987), Lau and Chan (1988), Rui and Wang (1990), Hendon and Salby (1994), and Madden and Julian (1994).

The MJO is an eastward-propagating disturbance that has its maximum amplitude within the Tropics of the eastern Indian to the western Pacific Oceans. The "oscillation" was discovered by Madden and Julian (1971, 1972) through the spectral analysis of wind and pressure from a network of tropical radiosonde stations. It has a broad period range maximizing between about 41 and 53 days, and most of the spatial amplitude of the wave is contained within zonal wavenumbers 0 through 2 (Hendon and Salby 1994). In particular, the wave 0 or zonal-mean component of the MJO has significant amplitude and is strongly connected to fluctuations in both the total and latitudinal distribution of IS angular momentum of the atmosphere (Weickmann et al. 1997). Large perturbations in convection and circulation within the Tropics are associated with the disturbance, but significant systematic extratropical circulation signals in both hemispheres are also observed during its evolution. There is still much work to be done concerning the origin and maintenance of the MJO. Madden and Julian (1994) give a good summary of the competing theories that attempt to explain MJO dynamics.

In the following sections, we describe the broad features of the "typical" MJO in the Tropics, then relate these to far-field signals in the SH. It must be kept in mind that the evolution of each MJO will differ significantly from this canonical description, with the large-scale convective signal the most robust feature and the large-scale circulation less consistent from event to event.

(i) Convection related to the MJO

Tropical convective activity is often monitored through the use of satellite-derived outgoing longwave radiation, which can detect high, cold cloud tops associated with precipitating convective clouds. Using the timing convention of Madden and Julian (1972) and Knutson and Weickmann (1987), the MJO cycle begins with the appearance of anomalously low OLR over the central Indian Ocean, which increases in intensity as it moves eastward at 5–10 m s^{-1}. After several days, the cloudiness begins to increase over the western Pacific as it decreases over the Indian sector. Thus, the eastward movement of the MJO has some characteristics of a standing oscillation between two poles over the eastern Indian and the western Pacific Oceans. The path of the MJO is primarily through the monsoon regions and so is significantly affected by the seasonal cycle. During southern winter, the convective envelope moves at first poleward over India from the equatorial Indian Ocean, then across southeast Asia to the Philippine region, while in southern summer it moves more directly eastward across Indonesia and then into the SPCZ. During summer, a modulation of both the onset and Australian monsoon activity by the MJO is often apparent (McBride 1987; Hendon and Liebmann 1990; see chapter 3). Usually, the convective signal becomes weak over the equatorial dry zone of the Pacific, although in later stages a signal of enhanced cloudiness is often observed almost simultaneously over the South American and African continents, with enhanced activity in the South American convergence zone (see Kodama 1993) apparent in southern summer (e.g., Knutson and Weickmann 1987). A new MJO cycle begins as convection dies out over South America and Africa and begins to increase again over the Indian Ocean.

Several aspects of the convective envelope deserve mention. It is important to realize that, while the MJO disturbance itself moves eastward, it modulates a host of higher-frequency disturbances that individually can propagate westward or eastward (e.g., Nakazawa 1988; Lau et al. 1991; Hendon and Liebmann 1994; see chapter 3). During some years, a well-defined MJO may be followed by long periods of relative inactivity, when the oscillation is practically absent. Some of the IA variability in the MJO is related to ENSO (Lau and Chan 1988). During warm events, when SST is higher than normal in the Pacific to the east of the date line, the envelope is observed to move into the central Pacific closer to the Equator, reflecting the equatorward shift in the positions of the Pacific convergence zones (see section 8.2 above). During the intense 1982–83 warm event, the OLR signal associated with the MJO was displaced well eastward of its normal domain and propagated along the Equator to the South American coast (e.g., Lau and Chan 1988).

(ii) Circulation associated with the MJO

In Fig. 8.12 we show the OLR and circulation anomalies during May–September and November–March associated with the phase of the MJO when convection is over the western tropical Pacific. These patterns were obtained using the regression technique outlined in Kiladis and Weickmann (1992a), and complete details of this approach can be found in that study. Briefly, 30–70-day filtered OLR in a selected 20°-box base region is regressed against deseasonalized u and v components of the wind at each global grid point. This procedure yields a separate regression equation between OLR in the base region and the wind component for each global grid point. The linear dependence of the circulation can then be mapped by applying the regression equation for each grid point using an arbitrary deviation in OLR at the base region as the independent variable. The correlation coefficient enables the statistical significance of the local linear relationship between OLR in the base region and the dependent variable at any given grid point to be assessed. Wind vectors on the figures depict where the linear relationship is locally significant at the 95% level. The OLR predictor is also regressed against OLR perturbations at all other grid points. Lagged regression relationships can be used to examine the evolution of the convective and circulation signals over time.

The 30–70-day filtering is quite effective at isolating the MJO signal, with little contamination from other disturbances (Kiladis and Weickmann 1992a). The circulations are mapped for a negative deviation in OLR and are thus meant to represent those associated with deep convection at the base region. In individual cases, an extensive region of OLR less than 220 W m^{-2} is observed within the MJO convective envelope. This corresponds to an OLR anomaly of roughly -30 W m^{-2} with respect to the long-term mean within the domain of the disturbance, so we scale the regressions to this value to produce realistic amplitudes of the circulations associated with a typical MJO.

Figures 8.12a and 8.12b show that MJO convection during southern summer is more extensive and farther south, and during both seasons is embedded in upper-

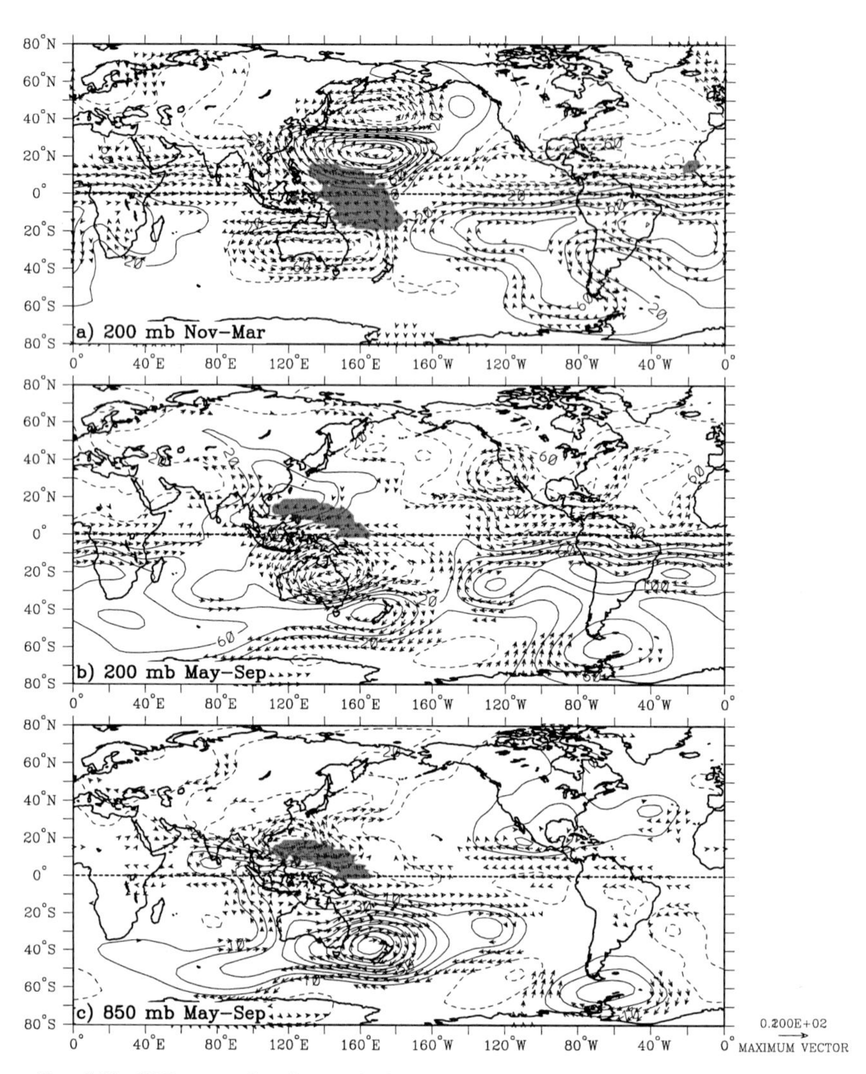

FIG. 8.12. OLR, streamfunction, and wind anomalies associated with a 30–70-day filtered, area-averaged OLR anomaly of −30.0 W m^{-2}, calculated over the period 1979–1995: (a) 200 hPa, November–March; (b) 200 hPa, May–September; (c) 850 hPa, May–September. Contour interval is 20 × 10^5 m^2 s^{-1} in (a) and (b), and 10 × 10^5 m^2 s^{-1} in (c). Negative contours are dashed. The OLR base region is 15°S–5°N, 140°–160°E in (a), and 5°S–15°N, 140°–160°E in (b) and (c). Shading denotes OLR anomalies less than −10 W m^{-2}. Only those wind vectors locally significant at the 95% level are shown. The largest vectors are about 10 m s^{-1}.

level easterly anomalies. An extensive reach of equatorial westerly anomalies dominates most of the globe outside the convective region. Flanking the convection are upper-level anticyclonic perturbations in the subtropics of both hemispheres. Comparison between Figs. 8.12b and 8.12c shows that the vertical structure of the circulation is baroclinic at low latitudes, with low-level westerlies underlying upper easterlies near the Equator. The vertical structure transitions to more-nearly equivalent barotropic in the extratropics, with like-signed perturbations through the depth of the troposphere.

In both Figs. 8.12a and 8.12b, convection in the equatorial western Pacific tends to be accompanied by an anomalous anticyclonic circulation over Australia and enhanced westerlies to the south. In southern winter, a wave train is seen emanating from the Australian anticyclone, crossing the South Pacific, and then heading northward over the South American/Atlantic sector. In Fig. 8.13, the wintertime SH 500-hPa height anomalies corresponding to the same phase of the MJO in Figs. 8.12b and 8.12c are mapped over the SH. While there is little evidence of the Australian anticyclone at this level, the extratropical features can be seen to comprise a wave train similar to those of IS EOFs 1 and 2 in Figs. 8.9a and 8.9b, with somewhat different phasing. Figure 8.13 also corresponds well to the pattern associated with persistent anomalies in the

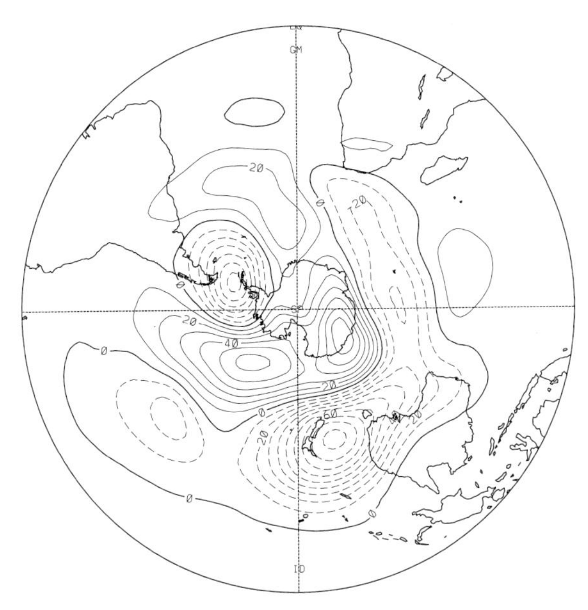

FIG. 8.13. 500-hPa geopotential height during May–September, corresponding to Fig. 8.12b. Contour interval is 10 m. Negative contours are dashed.

preferred region of 60°S, 120°W (Mo and Higgins 1998). Thus, a substantial portion of the IS variability in the SH winter appears to be related to the MJO, although Mo and Higgins concluded that, based on the relatively weak correlations with OLR, the persistent anomalies must in many occasions be generated internally and are not exclusively dependent on tropical forcing.

The strong relationship between tropical convection and the Australian subtropical jet on IS timescales has been noted in several studies (Nogues-Paegle and Zhen 1987; Nogues-Paegle and Mo 1988; Hurrell and Vincent 1990, 1991; Schrage and Vincent 1996). The nature of this linkage on MJO timescales has been likened to that occurring during ENSO (Sardeshmukh and Hoskins 1985, 1988; Rasmusson and Mo 1993). Berbery and Nogues-Paegle (1993) and Tyrrell et al. (1996) demonstrate that the anticyclonic anomaly over Australia is a direct result of upper-level divergent outflow from anomalous convection blowing across the strong mean gradient in absolute vorticity along the SH westerlies. The resulting anticyclonic vorticity tendency creates a subtropical Rossby-wave source over Australia, which is then followed by the dispersion of energy eastward as predicted by barotropic Rossby-wave theory (e.g., Grimm and Silva Dias 1995). This is supported by the lagged regressions for the period surrounding that in Figs. 8.12b and 8.13 (not shown), which show downstream centers strengthening over time at the expense of those upstream.

2) SUBMONTHLY CIRCULATION VARIABILITY RELATED TO TROPICAL CONVECTION

Kiladis and Weickmann (1992a,b; 1997) studied the relationships between tropical convection and large-scale circulation at 200 and 850 hPa on submonthly (6–30-day) timescales. For the most part, at these higher-frequency IS timescales the teleconnection patterns are more localized in the regions of convection when compared to the MJO and are dominated by spatial scales centered on zonal wavenumbers 4 through 6. In general, submonthly tropical convection in regions of tropical easterlies is associated with outflow into winter-hemisphere subtropical anticyclones and more symmetric signals of twin anticyclones during the transition seasons.

An example of some particularly strong connections over the Indian and Australasian sectors during southern winter is shown in Fig. 8.14 (based on results in Kiladis and Weickmann 1997). These patterns were obtained in the same manner as those in Figs. 8.12 and 8.13, except that, in this case, June–August less-than-30-day high-pass–filtered OLR data were used as a predictor in the regressions. In Fig. 8.14a, convection in the Arabian Sea is accompanied by upper-level outflow across the Equator into the SH and associated with an anomalous anticyclone and intensification of the subtropical westerly jet over the southern Indian Ocean. Downstream, a wave train extending across Australia and into the South Pacific is seen, although these features are not statistically significant. Shifting the base region eastward to southeast Asia (Fig. 8.14b) shifts the outflow pattern and downstream anticyclone by a similar distance. For convection to the east of the Philippines (Fig. 8.14c), the outflow is associated with an anticyclonic perturbation to the east of Australia and poleward shift and intensification of the core of the subtropical jet at this location.

These results indicate that Asian monsoon convection can significantly affect SH circulation on submonthly timescales. Submonthly convection produces outflow and subtropical anticyclonic perturbations in the SH, which appear tied to downstream Rossby wave trains. A vorticity budget of the time evolution of the pattern in Fig. 8.14c (not shown) reveals that much of the northerly flow emanating from the region of the heating is divergent, and that the advection of absolute vorticity of the basic state by the perturbation divergent flow contributes strongly to the maintenance of the subtropical anticyclone. Thus, the relationship discussed above between tropical convection, Rossby-wave sources, and jet fluctuations over the Australian–Pacific sector on IA and MJO timescales is also present at submonthly timescales and appears to involve similar forcing mechanisms. The sensitivity of the SH to forcing by convection is likely due to the close proximity of the subtropical westerlies to sources

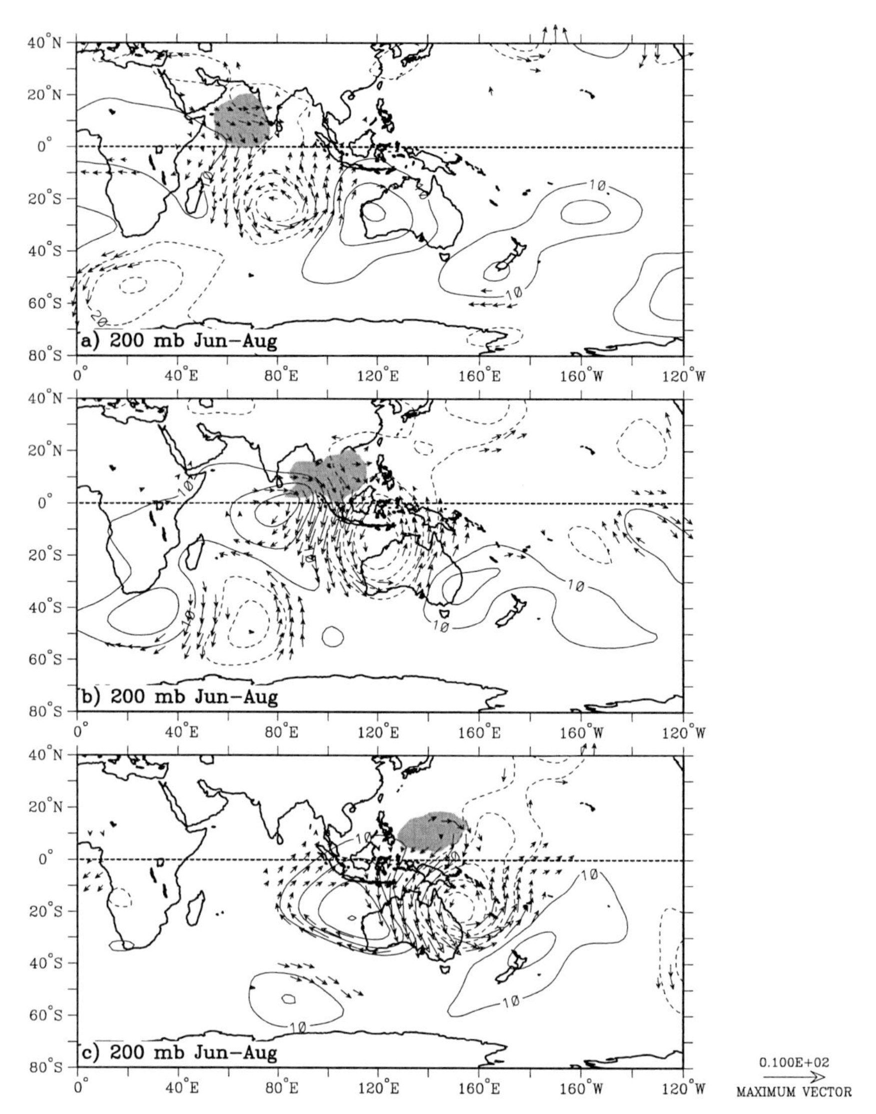

FIG. 8.14. 200-hPa streamfunction, wind, and OLR perturbations associated with a −30 W m^{-2} perturbation in less-than-30-day filtered OLR in the base regions: (a) 5°–15°N, 60°–70°E; (b) 5°–15°N, 100°–110°E; and (c) 5°–15°N, 140°–150°E. Contour interval is 10 × 10^{5} m^{2} s^{-1}. Negative contours are dashed. Shading denotes OLR anomalies less than −10 W m^{-2}. Only those wind vectors locally significant at the 95% level are shown. The largest vectors are about 5 m s^{-1}.

of tropical forcing in most sectors of that hemisphere (Kiladis and Weickmann 1997).

e. Other sources of intraseasonal circulation variability

As we have emphasized, the interplay between tropical convection and atmospheric circulation is one important source of internal atmospheric variability. In addition, the atmosphere is also characterized by a large amount of variability that is present solely due to the interaction between various scales of the circulation. This variability would be present even in a dry atmosphere without any variable coupling to the boundary. The importance of such internal variability can be assessed by performing extended GCM runs with constant boundary forcing (see chapter 11). Using this approach, it has been estimated that such internal variability accounts for up to 80% of the total variance in some regions (e.g., Lau 1992).

The source for much of the internal IS variability over the extratropics is generally considered to be barotropic and baroclinic instability (see Frederiksen and Webster 1988; Young and Houben 1989). Considerable progress has been made in the understanding of this variability through the method of normal-mode analysis. For the atmosphere, this involves linearizing a given set of the equations of motion about a basic state and isolating the most unstable growing patterns through an expansion in terms of normal modes (Frederiksen 1992). Such an analysis of a relatively simple two-level primitive equation model isolates the locations of SH storm tracks and downstream regions of maximum low-frequency variability with quite good fidelity (Frederiksen and Frederiksen 1993). In particular, the signatures of many of the EOF patterns in Figs. 8.3 and 8.9 have been simulated in these simple models, indicating that instability of the basic state is responsible for much of the observed variability in the SH extratropics.

Of course, nonlinear interactions between the various spatial and temporal scales of the circulation are also responsible for the growth and decay of perturbations (Young and Houben 1989). The feedback of synoptic-scale transients onto the large-scale flow appears to be particularly important for the generation of low-frequency zonal-flow variability along the storm tracks of the SH (e.g., Trenberth 1987; Karoly 1990; Akahori and Yoden 1997). Yu and Hartmann (1993) found low-frequency zonal-flow vacillations, which persisted out to seasonal timescales in a primitive equation model of the SH. Once these flows formed, they were maintained primarily by barotropic wave–mean flow interactions.

At somewhat higher frequencies, variability in the strength of the zonal flow in the South Pacific jet with a period between one and two weeks was observed by Ko and Vincent (1995) during four SH summers. While some of these jet strengthenings can be related to tropical convection, a large number appear to be generated through local baroclinic instability (Vincent et al. 1997). Their timescale is similar to a dominant timescale of NH Asian jet fluctuations and happens to be close to the period of the organization of synoptic-scale eddies into baroclinic "wave packets" observed by Lee and Held (1993) during SH summer. The wave packets are comprised of a group of baroclinic waves whose amplitude is strongly modulated in the zonal direction by an envelope of wave activity. The envelopes propagate quite clearly around the globe at 45°S for many cycles, with a mean period of around 12 days. It was pointed out that such envelopes may have implications for medium-range weather forecasting. As such, the more zonally symmetric and less-baroclinic SH atmosphere provides a "cleaner" laboratory than the NH for the study of atmospheric variability encompassing the synoptic to interannual timescales.

Acknowledgments. We thank K.-M. Lau, Harry van Loon, Matthew Wheeler, John Kidson, Mark Sinclair, James Renwick, and Don Johnson for valuable discussions and comments on earlier drafts of this chapter.

REFERENCES

Aceituno, P., 1988: On the functioning of the Southern Oscillation in the South American sector. Part I: Surface climate. *Mon. Wea. Rev.,* **116,** 505–524.

———, 1989: On the functioning of the Southern Oscillation in the South American sector. Part II: Upper air circulation. *J. Climate,* **2,** 341–355.

Akahori, K., and S. Yoden, 1997: Zonal-flow vacillation and bimodality of baroclinic eddy life cycles in a simple global circulation model. *J. Atmos. Sci.,* **54,** 2349–2361.

Allan, R. J., 1991: Australasia. *Teleconnections Linking Worldwide Climate Anomalies.* Cambridge University Press, M. H. Glantz, R. W. Katz, and N. Nicholls, Eds., 73–120.

———, N. Nicholls, P. D. Jones, and I. J. Butterworth, 1991: A further extension of the Tahiti–Darwin SOI, early ENSO events, and Darwin pressure. *J. Climate,* **4,** 743–749.

———, J. Lindesay, and D. Parker, 1996: *El Niño Southern Oscillation and Climatic Variability.* CSIRO Publishing, 405 pp.

Ambrizzi, T., and B. J. Hoskins, 1997: Stationary Rossby-wave propagation in a baroclinic atmosphere. *Quart. J. Roy. Meteor. Soc.,* **123,** 919–928.

———, ———, and H.-H. Hsu, 1995: Rossby-wave propagation and teleconnection patterns in the austral winter. *J. Atmos. Sci.,* **52,** 3661–3672.

Angell, J. K., 1981: Comparison of variations in atmospheric quantities with sea surface temperature variations in the equatorial eastern Pacific. *Mon. Wea. Rev.,* **109,** 230–243.

Arkin, P. A., 1982: The relationship between interannual variability in the 200-mb tropical wind field and the Southern Oscillation. *Mon. Wea. Rev.,* **110,** 1393–1401.

Baines, P. G., 1983: A survey of blocking mechanisms with application to the Australian region. *Aust. Meteor. Mag.,* **31,** 27–36.

Barnston, A. G., and T. M. Smith, 1996: Specification and prediction of global surface temperature and precipitation from global SST using CCA. *J. Climate,* **9,** 2660–2697.

Basher, R. E., and X. Zheng, 1995: Tropical cyclones in the southwest Pacific: Spatial patterns and relationships to Southern Oscillation and sea surface temperature. *J. Climate,* **8,** 1249–1260.

Battisti, D. S., and A. C. Hirst, 1989: Interannual variability in a tropical atmosphere/ocean model: Influence of the basic state, ocean geometry, and nonlinearity. *J. Atmos. Sci.,* **46,** 1687–1712.

Berbery, E. H., and M. N. Núñez, 1989: An observational and numerical study of blocking episodes near South America. *J. Climate,* **2,** 1352–1361.

———, and J. Nogues-Paegle, 1993: Intraseasonal interactions between the Tropics and extratropics in the Southern Hemisphere. *J. Atmos. Sci.,* **50,** 1950–1965.

———, ———, and J. D. Horel, 1992: Wavelike Southern Hemisphere extratropical teleconnections. *J. Atmos. Sci.,* **49,** 155–177.

Berlage, H. P., 1957: Fluctuations of the general atmospheric circulation of more than one year, their nature and prognostic value. *K. Ned. Meteorol. Inst., Meded. Verh.* **69,** 152 pp.

Bjerknes, J., 1966: A possible response of the atmospheric Hadley circulation to equatorial anomalies of ocean temperature. *Tellus,* **18,** 820–829.

Cerveny, R. S., R. R. Skeeter, and K. F. Dewey, 1987: A preliminary investigation of the relationship between South American snow cover and the Southern Oscillation. *Mon. Wea. Rev.,* **115,** 620–623.

Chen, B., S. R. Smith, and D. H. Bromwich, 1996: Evolution of the tropospheric split jet over the South Pacific Ocean during the 1986–89 ENSO cycle. *Mon. Wea. Rev.,* **124,** 1711–1731.

Chen, T. C., M. C. Yen, and D. P. Nune, 1987: Dynamic aspects of the Southern Hemisphere medium scale waves during the summer season. *J. Meteor. Soc. Japan,* **65,** 401–421.

Chen, W. Y., 1982: Assessment of Southern Oscillation sea level pressure indices. *Mon. Wea. Rev.,* **110,** 800–807.
Coughlan, M. J., 1983: A comparison climatology of blocking action in the two hemispheres. *Aust. Meteor. Mag.,* **31,** 3–31.
Diaz, H. F., and V. Markgraf, 1992: *El Niño.* Cambridge University Press, 476 pp.
———, and G. N. Kiladis, 1995: Climatic variability on decadal to century time scales. *World Survey of Climatology,* Vol. 16, A. Henderson-Sellers, Ed., 608 pp.
Dickey, J. O., S. L. Marcus, and R. Hide, 1992: Global propagation of interannual fluctuations in atmospheric angular momentum. *Nature,* **357,** 484–488.
Dole, R. M., and N. D. Gordon, 1983: Persistent anomalies of the extratropical Northern Hemisphere wintertime circulation: Geographical distribution and regional persistence characteristics. *Mon. Wea. Rev.,* **111,** 1567–1586.
Donguy, J. R., and C. Henin, 1980: Climatic teleconnections in the western South Pacific with El Niño phenomenon. *J. Phys. Oceanogr.,* **10,** 1952–1958.
Drosdowsky, W., and M. Williams, 1991: The Southern Oscillation in the Australian region. Part I: Anomalies at the extremes of the oscillation. *J. Climate,* **4,** 619–638.
Dutton, E. G., and J. R. Christy, 1992: Solar radiative forcing at selected locations and evidence for global lower tropospheric cooling following the eruptions of El Chichón and Pinatubo. *Geophys. Res. Lett.,* **19,** 2313–2316.
Edmon, H. J., B. J. Hoskins, and M. E. McIntyre, 1980: Eliassen–Palm cross sections for the troposphere. *J. Atmos. Sci.,* **37,** 2600–2616.
Eischeid, J. K., H. F. Diaz, R. S. Bradley, and P. D. Jones, 1991: A comprehensive precipitation data set for global land areas. U.S. Department of Energy Rep. No. DOE/ER-69017T-H1, 81 pp.
Enfield, D. B., 1992: Historical and prehistorical overview of El Niño/Southern Oscillation. *El Niño,* H. F. Diaz and V. Markgraf, Eds., Cambridge University Press, 476 pp.
Evans, J. L., and R. J. Allan, 1992: El Niño/Southern Oscillation modification to the structure of the monsoon and tropical cyclone activity in the Australasian region. *Int. J. Climatol.,* **12,** 611–623.
Farrara, J. D., M. Ghil, C. R. Mechoso, and K. C. Mo, 1989: Empirical orthogonal functions and multiple flow regimes in the Southern Hemisphere winter. *J. Atmos. Sci.,* **46,** 3219–3223.
Fraedrich, K., and H. Böttger, 1978: A wavenumber-frequency analysis of the 500-mb geopotential at 50°N. *J. Atmos. Sci.,* **35,** 745–750.
———, and E. Kietzig, 1983: Statistical analysis and wavenumber-frequency spectra of the 500-mb geopotential along 50°S. *J. Atmos. Sci.,* **40,** 1037–1045.
Frederiksen, J. S., 1982: A unified three-dimensional instability theory of the onset of blocking and cyclogenesis. *J. Atmos. Sci.,* **39,** 969–987.
———, 1992: Towards a unified instability theory of large-scale atmospheric disturbances. *Trends Atmos. Sci.,* **1,** 239–261.
———, and P. J. Webster, 1988: Alternative theories of atmospheric teleconnections and low-frequency fluctuations. *Rev. Geophys.,* **26,** 459–494.
———, and C. S. Frederiksen, 1993: Southern Hemisphere storm tracks, blocking, and low-frequency anomalies in a primitive equation model. *J. Atmos. Sci.,* **50,** 3148–3163.
Ghil, M., and K. Mo, 1991: Intraseasonal oscillations in the global atmosphere. Part II: Southern Hemisphere. *J. Atmos. Sci.,* **48,** 780–790.
Glantz, M. H., R. W. Katz, and N. Nicholls, Eds., 1991: *Teleconnections Linking Worldwide Climate Anomalies.* Cambridge University Press, 535 pp.
Gordon, N. D., 1986: The Southern Oscillation and New Zealand weather. *Mon. Wea. Rev.,* **114,** 371–387.
Gray, W. M., J. D. Sheaffer, and J. Knaff, 1992: Influence of the stratospheric QBO on ENSO variability. *J. Meteor. Soc. Japan,* **70,** 975–994.
Grimm, A. M., and P. L. Silva Dias, 1995: Analysis of tropical–extratropical interactions with influence functions of a barotropic model. *J. Atmos. Sci.,* **52,** 3538–3555.
Gutzler, D. S., and K. C. Mo, 1983: Autocorrelation of Northern Hemisphere geopotential heights. *Mon. Wea. Rev.,* **111,** 155–164.
Halpert, M. S., and C. F. Ropelewski, 1992: Surface temperature patterns associated with the Southern Oscillation. *J. Climate,* **5,** 577–593.
Hamilton, K., 1983: Aspects of wave behavior in the mid- and upper-troposphere of the Southern Hemisphere. *Atmos.–Ocean,* **21,** 40–54.
Hansen, A. R., and A. Sutera, 1988: Planetary-wave amplitude bimodality in the Southern Hemisphere. *J. Atmos. Sci.,* **45,** 952–964.
———, and ———, 1991: Planetary-scale flow regimes in midlatitudes of the Southern Hemisphere. *J. Atmos. Sci.,* **48,** 952–964.
———, ———, and D. E. Venne, 1989: An examination of midlatitude power spectra: Evidence for standing variance and the signature of El Niño. *Tellus,* **41A,** 371–384.
Hartmann, D. L., 1995: A PV view of zonal-flow vacillation. *J. Atmos. Sci.,* **52,** 2561–2576.
———, C. R. Mechoso, and K. Yamazaki, 1984: Observations of wave–mean flow interaction in the Southern Hemisphere. *J. Atmos. Sci.,* **41,** 351–362.
Hastenrath, S., 1978: On modes of tropical circulation and climate anomalies. *J. Atmos. Sci.,* **35,** 2222–2231.
———, and L. Heller, 1977: Dynamics of climatic hazards in northeast Brazil. *Quart. J. Roy. Meteor. Soc.,* **103,** 77–92.
Held, I. M., S. W. Lyons, and S. Nigam, 1989: Transients and the extratropical response to El Niño. *J. Atmos. Sci.,* **46,** 163–174.
Hendon, H. H., and B. Liebmann, 1990: A composite study of onset of the Australian summer monsoon. *J. Atmos. Sci.,* **47,** 2227–2240.
———, and ———, 1994: Organization of convection within the Madden–Julian oscillation. *J. Geophys. Res.,* **99,** 8073–8083.
———, and M. L. Salby, 1994: The life cycle of the Madden–Julian oscillation. *J. Atmos. Sci.,* **51,** 2225–2237.
Hoerling, M. P., and M. Ting, 1994: Organization of extratropical transients during El Niño. *J. Climate,* **7,** 745–766.
Horel, J. D., and J. M. Wallace, 1981: Planetary-scale atmospheric phenomena associated with the Southern Oscillation. *Mon. Wea. Rev.,* **109,** 813–829.
Hoskins, B. J., and D. Karoly, 1981: The steady, linear response of a spherical atmosphere to thermal and orographic forcing. *J. Atmos. Sci.,* **38,** 1179–1196.
———, and T. Ambrizzi, 1993: Rossby-wave propagation on a realistic longitudinally varying flow. *J. Atmos. Sci.,* **50,** 1661–1671.
———, I. N. James, and G. H. White, 1983: The shape, propagation, and mean-flow interaction of large-scale weather systems. *J. Atmos. Sci.,* **40,** 1595–1612.
Hsu, H.-H., 1994: Relationship between tropical heating and global circulation: Interannual variability. *J. Geophys. Res.,* **99,** 10 473–10 489.
———, and S.-H. Lin, 1992: Global teleconnections in the 250-mb streamfunction field during the Northern Hemisphere winter. *Mon. Wea. Rev.,* **120,** 1169–1190.
Hurrell, J. W., and D. G. Vincent, 1990: Relationship between tropical heating and subtropical westerly maxima in the Southern Hemisphere during SOP-1, FGGE. *J. Climate,* **3,** 751–768.
———, and ———, 1991: On the maintenance of short-term subtropical wind maxima in the Southern Hemisphere during SOP-1, FGGE. *J. Climate,* **4,** 1009–1022.

———, and H. van Loon, 1994: A modulation of the atmospheric annual cycle in the Southern Hemisphere. *Tellus,* **46A,** 325–338.

Janowiak, J. E., 1988: An investigation of interannual rainfall variability in Africa. *J. Climate,* **1,** 240–255.

Jones, D. A., and I. Simmonds, 1993: Time and space spectral analyses of Southern Hemisphere sea level pressure variability. *Mon. Wea. Rev.,* **121,** 661–672.

Jury, M. R., C. McQueen, and K. Levey, 1994: SOI and QBO signals in the African region. *Theor. Appl. Climatol.,* **50,** 103–115.

———, B. A. Parker, N. Raholijao, and A. Nassor, 1995: Variability of summer rainfall over Madagascar: Climatic determinants at interannual scales. *Int. J. Climatol.,* **15,** 1323–1332.

Kalnay, E., and L.-O. Merkine, 1981: A simple mechanism for blocking. *J. Atmos. Sci.,* **38,** 2077–2091.

———, and K. C. Mo, 1986: Mechanistic experiments to determine the origin of short-scale Southern Hemisphere stationary Rossby waves. *Adv. Geophys.,* **29,** 415–442.

———, ———, and J. Paegle, 1986: Large-amplitude, short-scale stationary waves in the Southern Hemisphere: Observations and mechanistic experiments to determine their origin. *J. Atmos. Sci.,* **43,** 252–275.

———, and Coauthors, 1996: The NCEP/NCAR 40-year reanalysis project. *Bull. Amer. Meteor. Soc.,* **77,** 437–471.

Karoly, D. J., 1983: Rossby wave propagation in a barotropic atmosphere. *Dyn. Atmos. Oceans,* **7,** 111–125.

———, 1989: Southern Hemisphere circulation features associated with El Niño–Southern Oscillation events. *J. Climate,* **2,** 1239–1252.

———, 1990: The role of transient eddies in low frequency zonal variations in the Southern Hemisphere circulation. *Tellus,* **42A,** 41–50.

———, and B. J. Hoskins, 1982: Three-dimensional propagation of planetary waves. *J. Meteor. Soc. Japan,* **60,** 109–123.

Kessler, W. S., and M. J. McPhaden, 1995: Ocean equatorial waves and the 1991–93 El Niño. *J. Climate,* **8,** 1757–1774.

Kidson, J. W., 1986: Index cycles in the Southern Hemisphere during the Global Weather Experiment. *Mon. Wea. Rev.,* **114,** 1654–1663.

———, 1988a: Indices of the Southern Hemisphere zonal wind. *J. Climate,* **1,** 183–194.

———, 1988b: Interannual variations in the Southern Hemisphere circulation. *J. Climate,* **1,** 1177–1198.

———, 1991: Intraseasonal variations in the Southern Hemisphere circulation. *J. Climate,* **4,** 939–953.

———, and M. R. Sinclair, 1995: The influence of persistent anomalies on Southern Hemisphere storm tracks. *J. Climate,* **8,** 1938–1950.

Kiladis, G. N., 1998: Observations of Rossby waves linked to convection over the eastern tropical Pacific. *J. Atmos. Sci.,* **55,** 321–339.

———, and H. F. Diaz, 1986: A documentation of the ENSO event of 1877–78 and a comparison with 1982–83. *Mon. Wea. Rev.,* **144,** 1035–1047.

———, and H. van Loon, 1988: The Southern Oscillation. Part VII: Meteorological anomalies over the Indian and Pacific sectors associated with the extremes of the oscillation. *Mon. Wea. Rev.,* **116,** 120–136.

———, and H. F. Diaz, 1989: Global climatic anomalies associated with extremes in the Southern Oscillation. *J. Climate,* **2,** 1069–1090.

———, and K. M. Weickmann, 1992a: Circulation anomalies associated with tropical convection during northern winter. *Mon. Wea. Rev.,* **120,** 1900–1923.

———, and ———, 1992b: Extratropical forcing of tropical Pacific convection during northern winter. *Mon. Wea. Rev.,* **120,** 1924–1938.

———, and M. Wheeler, 1995: Horizontal and vertical structure of observed tropospheric equatorial Rossby waves. *J. Geophys. Res.,* **100,** 22 981–22 997.

———, and K. M. Weickmann, 1997: Horizontal structure and seasonality of large-scale circulations associated with submonthly tropical convention. *Mon. Wea. Rev.,* **125,** 1997–2013.

———, H. von Storch, and H. van Loon, 1989: Origin of the South Pacific convergence zone. *J. Climate,* **2,** 1161–1171.

Knutson, T. R., and K. M. Weickmann, 1987: 30–60-day atmospheric oscillations: Composite life cycles of convection and circulation anomalies. *Mon. Wea. Rev.,* **115,** 1407–1436.

Ko, K.-C., and D. G. Vincent, 1995: A composite study of the quasi-periodic subtropical wind maxima over the South Pacific during November 1984–April 1985. *J. Climate,* **8,** 579–588.

Kodama, Y.-M., 1993: Large-scale common features of subtropical precipitation zones (the Baiu frontal zone, the SPCZ, and the SACZ). Part II: Conditions of the circulations for generating the STCZs. *J. Meteor. Soc. Japan,* **71,** 581–610.

Kousky, V. E., and G. D. Bell, 1992: *Atlas of Southern Hemisphere 500-mb Teleconnection Patterns Derived from National Meteorological Center Analyses.* NOAA Atlas No. 9, Dept. of Commerce, 90 pp.

———, M. T. Kayano, and I. F. A. Cavalcanti, 1984: A review of the Southern Oscillation: Oceanic–atmospheric circulation changes and related rainfall anomalies. *Tellus,* **36A,** 490–502.

Labitzke, K., and H. van Loon, 1989: The Southern Oscillation. Part IX: The influence of volcanic eruptions on the Southern Oscillation in the stratosphere. *J. Climate,* **2,** 1223–1226.

Lau, K. M., and P. H. Chan, 1988: Intraseasonal and interannual variations of tropical convection: A possible link between the 40–50-day oscillation and ENSO? *J. Atmos. Sci.,* **45,** 506–521.

———, and P. J. Sheu, 1988: Annual cycle, quasi-biennial oscillation and Southern Oscillation in global precipitation. *J. Geophys. Res.,* **93,** 10 975–10 988.

———, T. Nakazawa, and C. H. Sui, 1991: Observations of cloud cluster hierarchies over the tropical western Pacific. *J. Geophys. Res.,* **96,** 3197–3208.

———, P.-J. Sheu, and I.-S. Kang, 1994: Multiscale low-frequency circulation modes in the global atmosphere. *J. Atmos. Sci.,* **51,** 1169–1193.

Lau, N.-C., 1992: Climate variability simulated in GCMs. *Climate System Modelling,* K. Trenberth, Ed., Cambridge University Press, 788 pp.

———, and M. J. Nath, 1996: The role of the "atmospheric bridge" in linking tropical Pacific ENSO events to extratropical SST anomalies. *J. Climate,* **9,** 2036–2057.

Lee, S., and I. M. Held, 1993: Baroclinic wave packets in models and observations. *J. Atmos. Sci.,* **50,** 1413–1428.

Lejenäs, H., 1984: Characteristics of Southern Hemisphere blocking as determined from a time series of observational data. *Quart. J. Roy. Meteor. Soc.,* **110,** 967–979.

———, 1988: Southern Hemisphere planetary-scale waves and blocking. *J. Meteor. Soc. Japan,* **66,** 777–781.

———, and H. Økland, 1983: Characteristics of Northern Hemisphere blocking as determined from a long time series of observational data. *Tellus,* **35A,** 350–362.

Liebmann, B., and H. H. Hendon, 1990: Synoptic-scale disturbances near the Equator. *J. Atmos. Sci.,* **47,** 1463–1479.

Lindesay, J. A., 1988: Southern African rainfall, the Southern Oscillation, and a Southern Hemisphere semi-annual cycle. *Int. J. Climatol.,* **8,** 17–30.

Lindzen, R. S., and S. Nigam, 1987: On the role of sea surface temperature gradients in forcing low-level winds and convergence in the Tropics. *J. Atmos. Sci.,* **44,** 2418–2436.

Madden, R. A., 1979: Observations of large-scale traveling Rossby waves. *Rev. Geophys. Space Phys.,* **17,** 1935–1949.

———, and P. Julian, 1971: Detection of a 40–50-day oscillation in the zonal wind. *J. Atmos. Sci.,* **28,** 702–708.

———, and ———, 1972: Description of global-scale circulation cells in the Tropics with a 40–50-day period. *J. Atmos. Sci.,* **29,** 1109–1123.

———, and ———, 1994: Observations of the 40–50-day tropical oscillation—A review. *Mon. Wea. Rev.,* **122,** 814–837.

Mason, S. J., 1995: Sea-surface temperature–South African rainfall associations, 1910–1989. *Int. J. Climatol.,* **15,** 119–135.

McBride, J. L., 1987: The Australian summer monsoon. *Monsoon Meteorology,* C.-P. Chang and T. N. Krishnamurti, Eds., Oxford University Press, 544 pp.

McWilliams, J. C., 1980: An application of equivalent modons to atmospheric blocking. *Dyn. Atmos. Oceans,* **5,** 43–66.

Mechoso, C. R., and D. L. Hartmann, 1982: An observational study of traveling planetary waves in the Southern Hemisphere. *J. Atmos. Sci.,* **39,** 1921–1935.

———, and S. W. Lyons, 1988: On the atmospheric response to sea surface temperature anomalies associated with the Atlantic warm event during 1984. *J. Climate,* **1,** 422–428.

———, ———, and J. A. Spahr, 1990: The impact of sea surface temperature anomalies on the rainfall over northeast Brazil. *J. Climate,* **3,** 812–826.

———, J. D. Farrara, and M. Ghil, 1991: Intraseasonal variability of the winter circulation in the Southern Hemisphere atmosphere. *J. Atmos. Sci.,* **48,** 1387–1404.

Meehl, G. A., 1987: The annual cycle and interannual variability in the tropical Pacific and Indian Ocean regions. *Mon. Wea. Rev.,* **115,** 27–50.

Mo, K. C., 1986: Quasi-stationary states in the Southern Hemisphere. *Mon. Wea. Rev.,* **114,** 808–823.

———, and H. van Loon, 1984: Some aspects of the interannual variation of mean monthly sea level pressure on the Southern Hemisphere. *J. Geophys. Res.,* **89**(D6), 9541–9546.

———, and G. H. White, 1985: Teleconnections in the Southern Hemisphere. *Mon. Wea. Rev.,* **113,** 22–37.

———, and M. Ghil, 1987: Statistics and dynamics of persistent anomalies. *J. Atmos. Sci.,* **44,** 877–901.

———, and V. E. Kousky, 1993: Further analysis of the relationship between circulation anomaly patterns and tropical convection. *J. Geophys. Res.,* **98**(D3), 5103–5113.

———, and E. M. Rasmusson, 1993: The 200-mb vorticity budget during 1986–89 as revealed by NMC analyses. *J. Climate,* **6,** 577–594.

———, and R. W. Higgins, 1997: Planetary waves in the Southern Hemisphere and linkage to the Tropics. NCAR Tech. Note TN-433 + PROC, 287 pp.

———, and ———, 1998: The Pacific–South American modes and tropical convection during the Southern Hemisphere winter. *Mon. Wea. Rev.,* **126,** 1581–1596.

———, J. Pfaendtner, and E. Kalnay, 1987: A GCM study of the maintenance of the June 1982 blocking in the Southern Hemisphere. *J. Atmos. Sci.,* **44,** 1123–1142.

Morliere, A., and J. P. Rebert, 1986: Rainfall shortage and El Niño–Southern Oscillation in New Caledonia, southwestern Pacific. *Mon. Wea. Rev.,* **114,** 1131–1137.

Moura, A. D., and J. Shukla, 1981: On the dynamics of droughts in northern Brazil: Observation, theory and numerical experiments with a general circulation model. *J. Atmos. Sci.,* **38,** 2653–2675.

Mullan, A. B., 1995: On the linearity and stability of Southern Oscillation–climate relationships for New Zealand. *Int. J. Climatol.,* **15,** 1365–1386.

Nakazawa, T., 1988: Tropical super clusters within intraseasonal variations over the western Pacific. *J. Meteor. Soc. Japan,* **66,** 823–839.

Newell, R. E., and B. C. Weare, 1976: Factors governing tropospheric mean temperature. *Nature,* **194,** 1413–1414.

Nicholls, N., 1984: The Southern Oscillation and Indonesian sea surface temperature. *Mon. Wea. Rev.,* **112,** 424–432.

———, 1988: El Niño–Southern Oscillation and rainfall variability. *J. Climate,* **1,** 418–421.

———, 1991: Teleconnections and health. *Teleconnections Linking Worldwide Climate Anomalies,* M. H. Glantz, R. W. Katz, and N. Nicholls, Eds., Cambridge University Press, 535 pp.

———, and A. Kariko, 1993: East Australian rainfall events: Interannual variations, trends, and relationships with the Southern Oscillation. *J. Climate,* **6,** 1141–1152.

Nicholson, S. E., 1986: The nature of rainfall variability in Africa south of the Equator. *J. Climatol.,* **6,** 515–530.

———, and D. Entekhabi, 1986: The quasi-periodic behavior of rainfall variability in Africa and its relationship to the Southern Oscillation. *Arch. Meteor. Geophys. Bioklimatol.,* **34,** 311–348.

Nitsche, G., J. M. Wallace, and C. Kooperberg, 1994: Is there evidence for multiple equilibria in planetary-wave amplitude statistics? *J. Atmos. Sci.,* **51,** 314–322.

Nogues-Paegle, J., and Z. Zhen, 1987: The Australian subtropical jet during the second observing period of the Global Weather Experiment. *J. Atmos. Sci.,* **44,** 2277–2289.

———, and K. C. Mo, 1988: Transient response of the Southern Hemisphere subtropical jet to tropical forcing. *J. Atmos. Sci.,* **45,** 1493–1508.

Philander, S. G. H., 1990: *El Niño, La Niña, and the Southern Oscillation.* Academic Press, 293 pp.

———, T. Yamagata, and R. C. Pacanowski, 1984; Unstable air–sea interactions in the Tropics. *J. Atmos. Sci.,* **41,** 604–613.

Quintanar, A. I., and C. R. Mechoso, 1995a: Quasi-stationary waves in the Southern Hemisphere. Part I: Observational data. *J. Climate,* **8,** 2659–2672.

———, and ———, 1995b: Quasi-stationary waves in the Southern Hemisphere. Part II: Generation mechanisms. *J. Climate,* **8,** 2673–2690.

Randel, W. J., 1987: A study of planetary waves in the southern winter troposphere and stratosphere. Part I: Wave structure and vertical propagation. *J. Atmos. Sci.,* **44,** 917–935.

———, and J. L. Stanford, 1983: Structure of medium-scale atmospheric waves in the Southern Hemisphere summer. *J. Atmos. Sci.,* **40,** 2312–2318.

———, and ———, 1985a: An observational study of medium-scale wave dynamics in the Southern Hemisphere summer. Part I: Wave structure and energetics. *J. Atmos. Sci.,* **42,** 1172–1188.

———, and ———, 1985b: The observed life cycle of a baroclinic instability. *J. Atmos. Sci.,* **42,** 1172–1188.

———, D. E. Stevens, and J. L. Stanford, 1987: A study of planetary waves in the southern winter troposphere and stratosphere. Part II: Life cycles. *J. Atmos. Sci.,* **44,** 936–949.

Rasmusson, E. M., and T. H. Carpenter, 1982: Variations in tropical sea surface temperature and surface wind fields associated with the Southern Oscillation/El Niño. *Mon. Wea. Rev.,* **110,** 354–384.

———, and K. Mo, 1993: Linkages between 200-mb tropical and extratropical circulation anomalies during the 1986–89 ENSO cycle. *J. Climate,* **6,** 577–594.

———, X. Wang, and C. E. Ropelewski, 1990: The biennial component of ENSO variability. *J. Marine Sys.,* **1,** 71–96.

Rogers, J. C., and H. van Loon, 1982: Spatial variability of sea level pressure and 500-mb height anomalies over the Southern Hemisphere. *Mon. Wea. Rev.,* **110,** 1135–1392.

Ropelewski, C. F., and M. S. Halpert, 1987: Global and regional-scale precipitation patterns associated with the El Niño/Southern Oscillation. *Mon. Wea. Rev.,* **115,** 1606–1626.

———, and P. D. Jones, 1987: An extension of the Tahiti–Darwin Southern Oscillation index. *Mon. Wea. Rev.,* **115,** 2161–2165.

———, and M. S. Halpert, 1989: Precipitation patterns associated with the high-index phase of the Southern Oscillation. *J. Climate,* **2,** 268–284.

———, ———, and X. Wang, 1992: Observed tropospheric biennial variability and its relationship to the Southern Oscillation. *J. Climate,* **5,** 594–614.

Rosen, R. D., D. A. Salstein, T. M. Eubanks, J. O. Dickey, and J. A. Steppe, 1984: An El Niño signal in atmospheric angular momentum and earth rotation. *Science,* **225,** 411–414.

Rui, H., and B. Wang, 1990: Development characteristics and dynamic structure of tropical intraseasonal convection anomalies. *J. Atmos. Sci.,* **47,** 357–379.

Rutllant, J., and H. Fuenzalida, 1991: Synoptic aspects of the central Chile rainfall variability associated with the Southern Oscillation. *Int. J. Climatol.,* **11,** 63–76.

Salby, M. L., 1982. A ubiquitous wavenumber 5 anomaly in the Southern Hemisphere during FGGE. *Mon. Wea. Rev.,* **110,** 1712–1720.

Sardeshmukh, P. D., and B. J. Hoskins, 1985: Vorticity balances in the Tropics during the 1982–83 El Niño–Southern Oscillation event. *Quart. J. Roy. Meteor. Soc.,* **111,** 261–278.

———, and ———, 1988: The generation of global rotational flow by steady idealized tropical divergence. *J. Atmos. Sci.,* **45,** 1228–1251.

Schrage, J. M., and D. G. Vincent, 1996: Tropical convection on 7–21-day timescales over the western Pacific. *J. Climate,* **9,** 587–607.

Shen, S., and K.-M. Lau, 1995: Biennial oscillation associated with the east Asian summer monsoon and tropical sea surface temperatures. *J. Meteor. Soc. Japan,* **73,** 105–124.

Shukla, J., and D. Gutzler, 1983: Interannual variability and predictability of 500-mb geopotential heights over the Northern Hemisphere. *Mon. Wea. Rev.,* **111,** 1273–1279.

Shutts, G. J., 1983: The propagation of eddies in different jet streams: Eddy vorticity forcing of blocking flow fields. *Quart. J. Roy. Meteor. Soc.,* **109,** 737–761.

Simmonds, I., and T. H. Jacka, 1995: Relationships between the interannual variability of Antarctic sea ice and the Southern Oscillation. *J. Climate,* **8,** 637–647.

Sinclair, M. R., 1994: An objective cyclone climatology for the Southern Hemisphere. *Mon. Wea. Rev.,* **122,** 2239–2256.

———, 1996: A climatology of anticyclones and blocking for the Southern Hemisphere. *Mon. Wea. Rev.,* **124,** 245–263.

———, J. A. Renwick, and J. W. Kidson, 1997: Low-frequency variability of Southern Hemisphere sea level pressure and weather system activity. *Mon. Wea. Rev.,* **125,** 2531–2543.

Streten, N. A., 1978: A quasi-periodicity in the motion of the South Pacific cloud band. *Mon. Wea. Rev.,* **106,** 1211–1214.

Swanson, G. S., and K. E. Trenberth, 1981: Interannual variability in the Southern Hemisphere troposphere. *Mon. Wea. Rev.,* **109,** 1890–1897.

Szeredi, I., and D. Karoly, 1987a: The vertical structure of monthly fluctuations of the Southern Hemisphere troposphere. *Aust. Meteor. Mag.,* **35,** 19–30.

———, and ———, 1987b: The horizontal structure of monthly fluctuations of the Southern Hemisphere troposphere from station data. *Aust. Meteor. Mag.,* **35,** 119–129.

Taljaard, J. J., 1972: Synoptic meteorology of the Southern Hemisphere. *Meteorology of the Southern Hemisphere, Meteor. Monogr.,* No. 35, Amer. Meteor. Soc., 263 pp.

Tomas, R. A., and P. J. Webster, 1994: Horizontal and vertical structure of cross-equatorial wave propagation. *J. Atmos. Sci.,* **51,** 1417–1430.

Trenberth, K. E., 1975: A quasi-biennial standing wave in the Southern Hemisphere and interrelations with sea surface temperature. *Quart. J. Roy. Meteor. Soc.,* **101,** 55–74.

———, 1976: Spatial and temporal variations of the Southern Oscillation. *Quart. J. Roy. Meteor. Soc.,* **102,** 639–653.

———, 1979: Interannual variability of the 500-mb zonal-mean flow in the Southern Hemisphere. *Mon. Wea. Rev.,* **107,** 1515–1524.

———, 1980: Atmospheric quasi-biennial oscillations. *Mon. Wea. Rev.,* **108,** 1370–1377.

———, 1981: Interannual variability of the Southern Hemisphere circulation: Regional characteristics. *Mon. Wea. Rev.,* **109,** 127–136.

———, 1982: Seasonality in Southern Hemisphere eddy statistics at 500 mb. *J. Atmos. Sci.,* **39,** 2507–2520.

———, 1984a: Interannual variability of the Southern Hemisphere circulation: Representativeness of the year of the Global Weather Experiment. *Mon. Wea. Rev.,* **112,** 108–123.

———, 1984b: Signal versus noise in the Southern Oscillation. *Mon. Wea. Rev.,* **112,** 326–332.

———, 1985: Persistence of daily geopotential height over the Southern Hemisphere. *Mon. Wea. Rev.,* **113,** 38–53.

———, 1986a: The signature of a blocking episode on the general circulation in the Southern Hemisphere. *J. Atmos. Sci.,* **43,** 2061–2069.

———, 1986b: An assessment of the impact of transient eddies on the zonal flow during a blocking episode using localized Eliassen–Palm flux diagnostics. *J. Atmos. Sci.,* **43,** 2070–2087.

———, 1987: The role of eddies in maintaining the westerlies in the Southern Hemisphere winter. *J. Atmos. Sci.,* **44,** 1498–1508.

———, 1991: General characteristics of El Niño–Southern Oscillation. *Teleconnections Linking Worldwide Climate Anomalies,* M. H. Glantz, R. W. Katz, and N. Nicholls, Eds., Cambridge University Press, 13–42.

———, and K. C. Mo, 1985: Blocking in the Southern Hemisphere. *Mon. Wea. Rev.,* **133,** 38–53.

———, and D. J. Shea, 1987: On the evolution of the Southern Oscillation. *Mon. Wea. Rev.,* **115,** 3078–3096.

Tyrrell, G. C., D. J. Karoly, and J. L. McBride, 1996: Links between tropical convection and variations of the extratropical circulation during TOGA COARE. *J. Atmos. Sci.,* **53,** 2735–2748.

Tyson, P. D., 1986: *Climatic Change and Variability in Southern Africa.* Oxford University Press, 220 pp.

van Loon, H., 1956: Blocking action in the Southern Hemisphere. *Notos,* **5,** 171–177.

———, 1984: The Southern Oscillation. Part III: Associations with the trades and with the trough in the westerlies of the South Pacific Ocean. *Mon. Wea. Rev.,* **112,** 947–954.

———, and R. L. Jenne, 1972: The zonal harmonic standing waves in the Southern Hemisphere. *J. Geophys. Res.,* **77,** 992–1003.

———, and R. A. Madden, 1981: The Southern Oscillation. Part I: Global associations with pressure and temperature in northern winter. *Mon. Wea. Rev.,* **109,** 1150–1162.

———, and J. C. Rogers, 1981: Remarks on the circulation over the Southern Hemisphere in FGGE and on its relation to the phases of the Southern Oscillation. *Mon. Wea. Rev.,* **109,** 2255–2259.

———, and D. J. Shea, 1985: The Southern Oscillation. Part IV: The precursors south of 15°S to the extremes of the oscillation. *Mon. Wea. Rev.,* **113,** 2063–2074.

———, and K. Labitzke, 1987: The Southern Oscillation. Part V: The anomalies in the lower stratosphere of the Northern Hemisphere in winter and a comparison with the quasi-biennial oscillation. *Mon. Wea. Rev.,* **115,** 357–369.

———, and D. J. Shea, 1987: The Southern Oscillation. Part VI: Anomalies of sea level pressure on the Southern Hemisphere and of Pacific sea surface temperature during the development of a warm event. *Mon. Wea. Rev.,* **115,** 370–379.

———, and K. Labitzke, 1988: Association between the 11-year solar cycle, the QBO, and the atmosphere. Part II: Surface and 700 mb in the Northern Hemisphere in winter. *J. Climate,* **1,** 905–920.

———, J. W. Kidson, and A. B. Mullan, 1993: Decadal variation of the annual cycle in the Australian dataset. *J. Climate,* **6,** 1227–1231.

Vincent, D. G., J. M. Schrage, and L. D. Sliwinski, 1995: Recent climatology of kinematic variables in the TOGA COARE region. *J. Climate,* **8,** 1296–1308.

———, K.-C. Ko, and J. M. Schrage, 1997: Subtropical jet streaks over the South Pacific. *Mon. Wea. Rev.,* **125,** 438–447.

Walker, G. T., and E. W. Bliss, 1932: World Weather V. *Mem. Roy. Meteor. Soc.,* **4,** 53–84.

Wang, B., and Y. Wang, 1996: Temporal structure of the Southern Oscillation as revealed by waveform and wavelet analysis. *J. Climate,* **9,** 1586–1598.

Webster, P. J., 1994: The role of hydrological processes in ocean–atmosphere interactions. *Rev. Geophys.,* **32,** 427–476.

———, and J. L. Keller, 1975: Atmospheric variations: Vacillations and index cycles. *J. Atmos. Sci.,* **32,** 1283–1300.

———, and J. R. Holton, 1982: Cross-equatorial response to middle-latitude forcing in a zonally varying basic state. *J. Atmos. Sci.,* **39,** 722–733.

Weickmann, K. M., G. N. Kiladis, and P. D. Sardeshmukh, 1997: The dynamics of intraseasonal atmospheric angular momentum oscillations. *J. Atmos. Sci.,* **54,** 1445–1461.

Wheeler, M., and G. N. Kiladis, 1999: Convectively coupled equatorial waves: Analysis of clouds and temperature in the wavenumber-frequency domain. *J. Atmos. Sci.,* **56,** 374–399.

Yasunari, T., 1989: A possible link of the QBOs between the stratosphere, troposphere, and sea surface temperature in the Tropics. *J. Meteor. Soc. Japan,* **67,** 483–493.

Yoden, S., M. Shiotani, and I. Hirota, 1987: Multiple planetary flow regimes in the Southern Hemisphere. *J. Meteor. Soc. Japan,* **64,** 571–585.

Young, R. E., and H. Houben, 1989: Dynamics of planetary-scale waves during Southern Hemisphere winter. *J. Atmos. Sci.,* **46,** 1365–1383.

Yu, J.-I., and D. L. Hartmann, 1993: Zonal-flow vacillation and eddy forcing in a simple GCM of the atmosphere. *J. Atmos. Sci.,* **50,** 3244–3259.

Yulaeva, E., and J. M. Wallace, 1994: The signature of ENSO in global temperature and precipitation fields derived from the microwave sounding unit. *J. Climate,* **7,** 1719–1736.

Chapter 9

Climatic Change and Long-Term Climatic Variability

PHILIP D. JONES
Climatic Research Unit, University of East Anglia, Norwich, United Kingdom

ROBERT J. ALLAN
Division of Atmospheric Research, CSIRO, Aspendale, Victoria, Australia

9.1. Introduction

This chapter reviews both climatic change and climatic variability in the Southern Hemisphere. While its scope concentrates on the changes and processes that have occurred during the twentieth century, particularly since 1950, it also attempts to bring together much paleoclimatic data and consider whether "pseudo-cyclic" phenomena, such as the El Niño–Southern Oscillation, have influences on longer timescales.

Few studies concentrate on this hemisphere, partly because of the lack of both long instrumental and, to some extent, paleoclimatic data. Here, we bring together much of the available data, attempting to link the evidence with current ideas about teleconnections. The most important generator of these teleconnections is ENSO, and few regions in the Southern Hemisphere are unaffected by its influence. Antarctica may be one less-affected region, but the shortness of the record and the great interannual variability of the climate may be hiding the clues.

ENSO embodies large-scale ocean–atmosphere interactions and is responsible for widespread environmental impacts, with many of the most prominent effects being found in the Southern Hemisphere. Much is still to be resolved concerning ENSO characteristics (viz. manifestations, onset, cessation, duration, frequency, and magnitude) and the physical mechanisms underlying them. Even less is known about other, lower-frequency fluctuations in climate, particularly those operating on decadal to century timescales. Such fluctuations may have considerable ramifications not only for climate patterns that shape the evolution of marine and terrestrial environments, but also for the modulation of the climatic regime in which ENSO dynamics operate. In fact, until recently, variations in ENSO characteristics and longer-period fluctuations in the climate system have tended to be studied in isolation. This has changed, with indications of possible interactions between such scales of variability both illustrating the complexity of the climate system and revealing deficiencies in dynamical understanding of ocean–atmosphere processes.

Many factors of the Southern Hemisphere are quite different from the Northern Hemisphere. The large oceanic part means that few regions develop very continental climates, and the three main mid- to low-latitude continental regions have strong relationships with one another, both on the ENSO and longer timescales. In the Antarctic, ENSO relationships are often weak, and the length of the record means that it is difficult to assess longer-timescale similarities. Antarctica is clearly "in phase" with the rest of the world on the Ice Age timescale, but it will take further improvements to the paleoclimatic database before any similarities in rates of change on the 50- to 100-yr timescale can be quantitatively addressed.

This chapter reviews what is known about climatic change in this hemisphere, particularly over the last century and, to a lesser extent, the last few millennia, linking the changes to the variability in recurrence intervals of ENSO events.

9.2. Climatic change in the Southern Hemisphere

a. Background

Instrumental records in the Southern Hemisphere rarely extend back into the nineteenth century, except in a few regions. Their length is considerably shorter, therefore, than comparable NH records, which extend back to the late eighteenth century over most of Europe and many midlatitude and some tropical coastlines. Thus, to consider climatic change in many SH regions on the decadal to century timescale, it is vital, even more so than in the NH, to make use of paleoclimatic reconstructions.

Many different types of paleoclimatic or proxy climatic reconstructions have been attempted throughout the world. The principles behind most of the techniques are discussed in Bradley (1985) and Berglund (1986). The applicability of the techniques obviously is dependent on the availability of material or evidence: ice cores are restricted to high-latitude and high-altitude ice caps; trees to the mid- to high latitudes, where annual rings are regularly laid down; corals to shallow tropical seas; pollen and other indicators in lacustrine cores and lake levels to mid- to high-latitude regions; etc. Although many indicators give past-temperature indications, subtropical and tropical proxy records over terrestrial areas will more likely provide evidence of past wetter or drier periods.

As their name implies, all proxy records are some form of indicator of the past climate. It must always be remembered, however, that they are surrogates. They will only explain a fraction of the true climatic variance and may only respond to a specific season of the year (e.g., trees to the growing season) (Jones and Bradley 1992). At best, therefore, the proxy indicators are annual in their resolution (not to be confused with a response to the conditions during the whole year, however defined). More importantly, however, although a reconstruction may explain 40% of the variance compared to an instrumental record for an independent verification period, this variance has only been calculated using interannual data for, at best, the last hundred years. The inherent assumption made is that this variance applies to all timescales over the period of the reconstruction. The assumption regarding the nature of this variance can only possibly be assessed by using long adjacent reconstructions made with different proxies. A related issue in paleoclimatology concerns the quality of the proxy information over the length of the record. If this varies with frequency, the quality of the reconstruction must vary, and this may be mostly manifest in the variations at the lowest frequencies (see, e.g., Briffa et al. 1996 and Jones and Briffa 1996).

Despite the above caveats, paleoclimatology provides our only means of assessing past climatic changes and variability on century and longer timescales. In the discussion of the climate of the SH, we also assume that each reconstruction is only applicable to the region for which it has been developed. For most of the SH, this will only be the region within a 2000-km radius (Briffa and Jones 1993 and Jones et al. 1997). In other words, a single reconstruction is only of local significance, just as a single instrumental record would be indicative of only a local region. As the discussion moves from certain to less-certain evidence, it begins with the instrumental evidence extending to gradually less well–resolved proxy evidence.

b. The Instrumental Period (1850 to the present)

1) History and homogeneity

Land-based instrumental recording of the climate began between 1830 and 1860 at a few coastal sites in South America, southern Africa, Australia, New Zealand, Indonesia, and a few smaller islands. Gradually, sites were developed inland until by 1900 most of the SH continents, with the exception of Antarctica, had some form of measurements. The development of the fledgling recording networks, often colonially run, has been documented in some instances (e.g., Salinger et al. 1995). The colonial nature has led to some of the early data being held only in Europe (see, e.g., Ropelewski and Jones 1987 and Können et al. 1998). Extensive weather and climate recording in the Antarctic had to wait until the International Geophysical Year in 1957/58, although sporadic records have been taken by expeditions to the continent since the turn of the century (Jones 1990 and Venter 1957). All types of data are subject to biases that affect long-term homogeneity (see later).

Marine recording also began in the early to mid-nineteenth century, on the same ships that brought and supplied the European settlers. Available records are confined to the major shipping routes from Europe to South America and from Europe around Africa to Australasia. Even now there are few records in the higher-latitude Southern Ocean and in the island-free southeastern Pacific Ocean. In these regions, the only source of information for weather forecasting and climatic studies is satellite and buoy data (Reynolds and Marsico 1993 and Guymer and Le Marshall 1980).

Bringing together all the records from the mid-nineteenth century is not a simple task. Some of the land records have been compiled in datasets such as World Weather Records (WWR) and the Global Historic Climate Network (GHCN) (see references in Jones et al. 1986, Jones 1994, and Vose et al. 1992), while much of the marine data has been entered into the Comprehensive Ocean Atmosphere Data Set (Woodruff et al. 1987) or the UKMO's marine data bank (Bottomley et al. 1990 and Parker et al. 1995). Both the land and marine datasets are of variable data quality, and much effort has been expended both in improving the number of available records and in homogeneity assessment. Land-based data are affected by site and screen design changes, methods used to calculate mean temperatures, observation time changes, and urbanization (Jones et al. 1986; Jones et al. 1990; Jones et al. 1991; Rhoades and Salinger 1993; Peterson and Easterling 1994; Easterling and Peterson 1995; Salinger et al. 1995; and Torok 1996). Marine variables, such as SST, are affected both by changes in recording practices, particularly from the use of uninsulated (canvas) buckets prior to 1941 to the use of engine-intake measurements and better-insulated buckets (Jones et al. 1991 and Parker et al. 1994), and by changes to shipping routes. Atmospheric pressure, air temperature, cloudiness, and wind measurements on-board ships are also affected by systematic and random biases (see section 9.36). Corrections to the marine datasets have been attempted and docu-

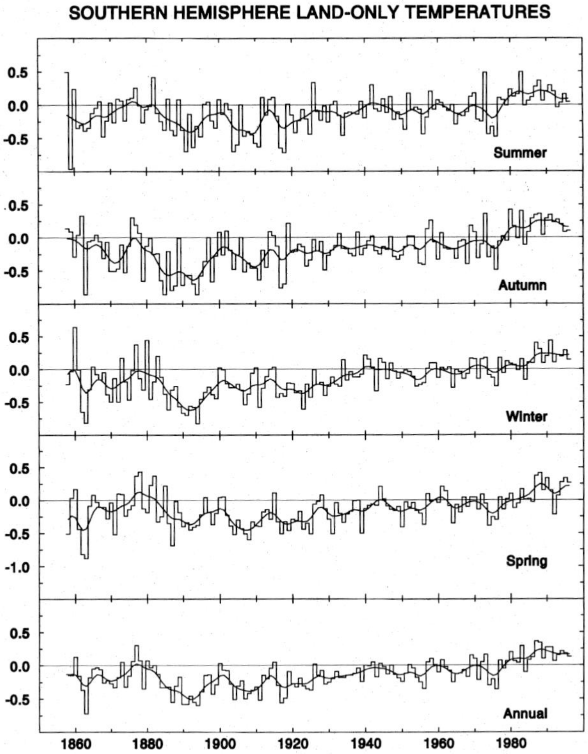

SOUTHERN HEMISPHERE LAND-ONLY TEMPERATURES

FIG. 9.1. Seasonal and annual average temperatures over SH land areas, 1858–1996. Data come from Jones (1994) and are expressed as anomalies from 1961 to 1990. Seasons are standard three-month climatological seasons (summer being December to February, etc.). The smooth lines in this and subsequent figures up to Fig. 9.7 are 10-yr Gaussian filters.

mented (see, e.g., Bottomley et al. 1990; Ward 1992; Parker et al. 1994, and Folland and Parker 1995).

2) HEMISPHERIC ANALYSES

We show three sets of seasonal temperature analyses: one for land-only regions (Fig. 9.1) of the hemisphere, one for marine-only (Fig. 9.2) areas, and one combined (Fig. 9.3). The data for Fig. 9.1 come from the latest land-based dataset of Jones (1994), where a 5° × 5° grid-box temperature anomaly dataset has been developed. The box averages are produced by simple averages of the monthly station temperature anomalies, from 1961 to 1990, of each station in the box. A station cannot influence more than one box. The hemispheric series is the weighted average of available boxes. At best, the coverage is between 20% and 25% of the surface area of the hemisphere. Antarctic data is available only since 1957, and the period since then constitutes the best hemispheric coverage. The sparse coverage in earlier years might be expected to bias the years prior to the 1950s. Various attempts to assess this influence have been undertaken [by Jones et al. (1986), Hansen and Lebedeff (1987), and Jones (1994); see also Trenberth et al. (1992), Madden et al. (1993), and Jones et al. (1997)]. The basic conclusion is that while the earlier years may be less reliable individually, the additional errors are not serially correlated and so will have less effect on estimates of decadal mean temperatures. Additional evidence that the lack of Antarctic data prior to 1957 does not seriously affect hemispheric estimates on the decadal and longer timescales comes from interhemispheric agreements on these timescales. As a consequence, the observed long-term temperature change is a robust measure of the rise in air temperature on the hemispheric scale over the last 140 years.

Along with comparable series for the NH, the series are the cornerstone of observational evidence of a gradual warming of the climate. In this regard, they have been comprehensively reviewed—for example, by the Intergovernmental Panel on Climate Change (see Folland et al. 1990, 1992; and Nicholls et al. 1996a). Here, we highlight the main conclusions. Figure 9.1 shows a warming of about 0.4°C since the mid-nineteenth century, common to all seasons. The greater variability in the nineteenth century is due to the sparse coverage at the time. The warming is almost monotonic since the coldest decades of the 1880s and 1890s and amounts to nearly 0.6°C over the 1901 to 1995 period. The warming over the whole series is less because of the relative warmth of the first two

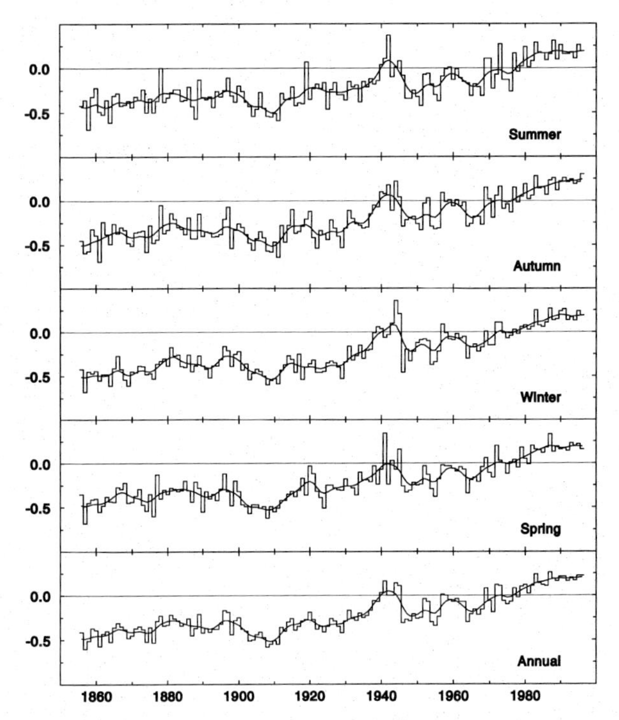

FIG. 9.2. Seasonal and annual average SSTs over the SH oceans, 1856–1996. Data come from Parker et al. (1994) and are expressed as anomalies from 1961 to 1990.

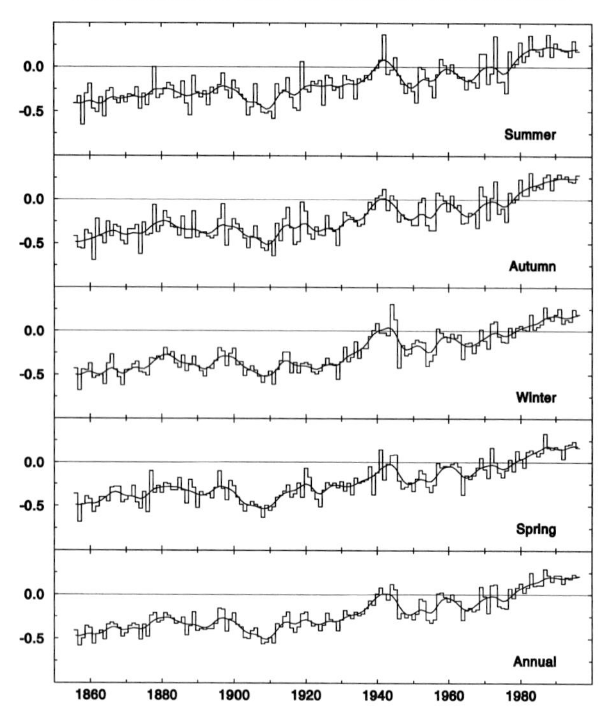

FIG. 9.3. Seasonal and annual average temperatures from land and marine regions of the SH, 1858–1996. Data come from a combination of Jones (1994) and Parker et al. (1995) and are expressed as anomalies from 1961 to 1990.

decades, the 1860s and 1870s. In contrast to the NH (see Jones 1994), year-to-year variability seems slightly greater during the summer and autumn compared to the winter and spring seasons. This is a common feature of many station records in the 20° to 60°S zone.

Figure 9.2 is based on SST anomalies from the Hadley Center of the UKMO (Parker et al. 1995). The series shows similar features to the land data but considerably less year-to-year variability, making the trends this century appear clearer. Temperatures began rising in 1910 and, apart from a marked cooling in the mid-1940s, have continued to rise to the early 1990s. As with the land series, year-to-year variability is greater in summer and autumn compared to winter and spring in contrast to the NH. Figure 9.3 is based on the combined land and marine records (see also Jones and Briffa 1992 and Parker et al. 1994). In all seasons, the rise in temperatures appears monotonic since the turn of the century and conforms to linear trends much more convincingly than in the NH.

3) CONTINENTAL ANALYSES

Hemispheric analyses hide much of the regional detail evident in the individual grid-box data series. For other important variables of interest, such as precipitation, they tell us very little about the changes that are occurring and the interrelationships between variables. This situation is partially redressed for temperature by mapping the temperature anomalies by decade (see, for example, Jones and Briffa 1992 and Parker et al. 1994).

Air temperature records over the SH in the global maps of Parker et al. (1994) show regions of deficient coverage, particularly the large gaps over the interior of continents such as Africa and South America up until the 1930s–1940s. If temperature responses in the most recent and relatively "data-rich" decades are, to any degree, representative of the historical period, then for annual averages at least, patterns over land tend to show up as extensions of the more dominant signals over the ocean basins.

In this section, we present temperature and precipitation series for a number of regions of the hemisphere (for definition, see Fig. 9.4). For precipitation (Fig. 9.5), they are the major continental grain-growing regions, while for temperature they are the same regions as well as averages for the four main continents (Figs. 9.6 and 9.7). All series are shown as calendar-year averages. To understand linkages with circulation features, it is necessary to consider seasonal or even monthly totals (see section 9.3b1). All temperature series have been developed from the Jones (1994) dataset, while those for precipitation data come from the updated Hulme (1994) dataset. The data sources are based on WWR, GHCN, additional information available to the Climatic Research Unit, and, during the 1990s, climate reports routinely issued by national meteorological agencies. Precipitation series were produced by areally weighting each $5° \times 5°$ grid-box series expressed as standardized departures. The resulting series were transformed back to millimeter departures using the average region mean and standard deviation (see Jones and Hulme 1996 for more discussion). Regional series for some areas in the southwest Pacific have been shown by Salinger et al. (1995).

In Fig. 9.5, precipitation variations for the seven regions show both long- and short-term trends and variations. In Australasia there is little long-term trend, although totals over New Zealand (see also Salinger et al. 1992) and part of southwestern Australia (see also Pittock 1975 and Allan and Haylock 1993) have declined slightly, while southeastern and northern Australia show slight increases. Year-to-year variability is clearly greatest in northern Australia. An examination of decadal to multidecadal fluctuations in Figs. 9.5 and 9.6 shows varying coherency amongst the regions. As expected, relationships are generally most marked among spatially proximate locations under the influence of similar climate regimens. For precipitation, Australian regions in the southeast and north of the continent show some in-phase consistency, while the southeast and southwest areas are

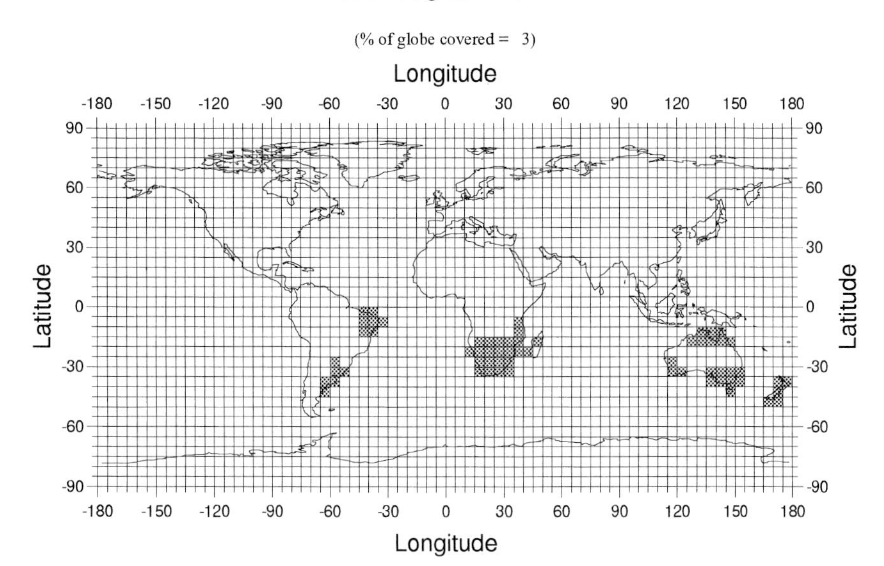

FIG. 9.4. Areas of the SH selected for regional precipitation and temperature analyses.

more strongly out of phase. In the two other continental regions, different features emerge: northeast Brazil shows marked decadal-scale variability, particularly at the beginning and end of the record (see also Ward 1994), while southern Africa shows its greatest variability in the recent 20 years, with a clear trend to drier conditions (see also Allan et al. 1995). In the Rio de la Plata basin encompassing eastern Argentina, Uruguay, and southern Brazil, an upward trend in precipitation is evident throughout the twentieth century. The discharge of the river has increased by over 20% between 1944–70 and 1971–92 (Garcia and Vargas 1998). Variations between continents are best represented by generally out-of-phase relationships between southwestern Australia and New Zealand (primarily after 1920) (as inferred from circulation patterns in Allan and Haylock 1993) and in-phase relationships involving both northeastern and southeastern Australia and southern Africa.

Temperature variations in Figs. 9.6 and 9.7 tend to show more decadal- to century-timescale trends in most regions. The late 1980s/early 1990s are among the warmest years of the records, although 1992 and 1993 have clearly been exceptionally cool in New Zealand and the Antarctic. The century-timescale increase in temperatures appears greatest in southern Africa, but is hardly noticeable in Australasia. As with the hemispheric averages, the coldest decades were the 1880s and 1890s. For air temperature, intercontinental relationships are, to varying degrees, in phase between northeastern Australia and northeastern Brazil, southern Africa and eastern Argentina/Uruguay/southeast Brazil (from 1900 to 1980), and southeastern and northern Australia and New Zealand (since the 1950s), and out of phase between New Zealand and eastern Argentina/Uruguay/southeast Brazil (since 1900). Other time series involving air temperature show marked shifts in relationships between being in and out of phase over parts of the observational record (e.g., northeastern Australia and southern Africa). Similar shifts are found in precipitation links between some of the above regions. There is need for care, however, when evaluating such links, because all of the time series discussed here are for average annual conditions and thus may mask important relationships found only on seasonal timescales.

In some parts, temperature analyses have been undertaken on a regional/national basis using more comprehensive datasets than are available internationally. In Australia, Torok (1996) has homogenized over 200 stations, many requiring a number of adjustments. Between 1908 and 1992, this series is highly correlated ($r = 0.94$) with the series in Fig. 9.7. This result indicates that the Jones (1994) dataset and the associated earlier homogeneity exercise can give reliable results on regional to continental scales, even for the relatively small number of stations regularly exchanged internationally. For more local studies, however, it will be necessary to resort to national datasets, such as that developed by Torok. Before 1908, Australian records are affected by non-standard screens (Nicholls et al. 1996b) and require urgent correction.

Further confirmation of the pervasiveness of the large-scale temperature increases comes from the com-

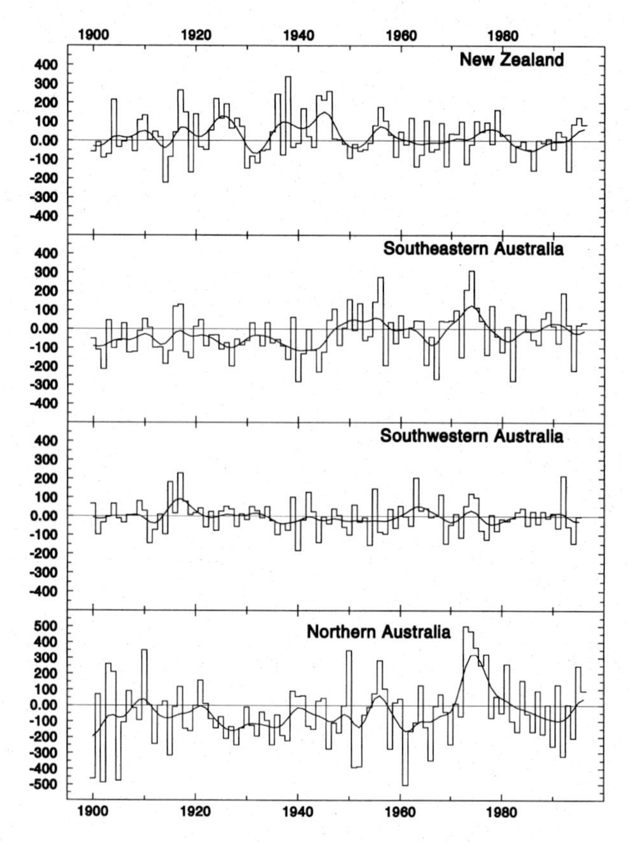

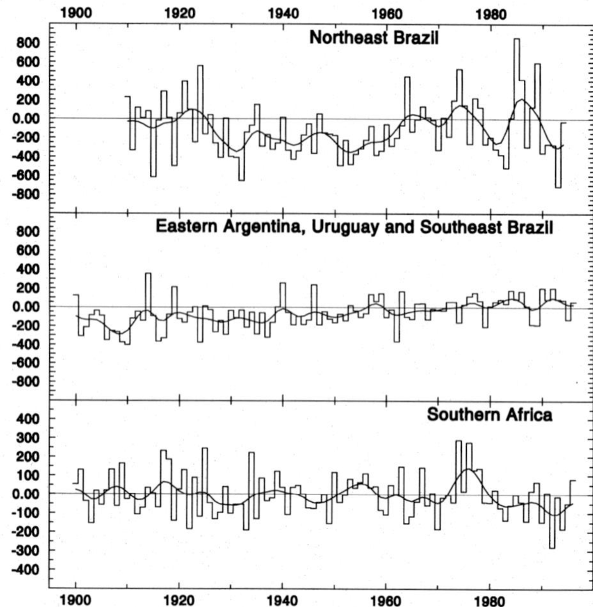

FIG. 9.5. Annual time series of precipitation averages for seven selected regions of the SH. Regions are defined in Fig. 9.4. Data come from the updated 5° × 5° grid-box precipitation series produced by Hulme (1994). Averages are expressed as mm anomalies from 1951 to 1980.

parison of the totally independent (in terms of sources and homogeneity assessments) land and marine datasets. Agreement is not only good on the hemispheric scale, as evidenced by Fig. 9.1 and 9.2, but also for much smaller regions. Folland and Salinger (1995) compare the land temperatures for New Zealand, developed from seven stations (all available to the Jones 1994 dataset) to the SST average from the surrounding ocean. Over 1871 to 1993, the two series correlate to 0.82, with the land series warming by 1.1°C since 1900 and the marine series by 0.7°C. One of the main reasons for the good agreement is the use of consistent Stevenson screens since 1871 (Salinger et al. 1995). In South America, recent regional studies for Chile and Argentina (Rosenbluth et al. 1997) indicate warmings in agreement with the grid-box data.

Recent evidence, principally in the NH, that there has been a reduction in the diurnal temperature range (DTR) over some land areas of the world (Karl et al. 1993) during the last 40 years is beginning to be more fully assessed in the SH (Easterling et al. 1997). Over New Zealand and the southwest Pacific (Salinger et al. 1993; Salinger 1995; and Salinger and Jones 1996), decreases in DTR have been observed, but they are small and regionally specific, while Antarctica (Jones 1995) shows no reductions. Eastern Australia and South Africa (Plummer et al. 1995; Lough 1995; and Karl et al. 1993) and Chile (Rosenbluth et al. 1997) show larger and significant reductions in DTR since 1950.

c. Proxy records

1) INTRODUCTION

As with the instrumental data, the availability of paleoclimatic information is considerably sparser than for the NH. The land area is smaller, leaving few areas in the mid- to high-latitude regions for information from terrestrial indicators such as tree rings. In low-latitude regions, the development of coral records is beginning to provide important new information about climate change. Ice caps are restricted in the Tropics to a couple of locations in South America—one in Kenya and a potential site in Irian Jaya, Indonesia. Ice layers in the Antarctic provide high-resolution information from coastal cores, especially from the peninsula region and Law Dome in the Australian sector and long, less-resolved records from the interior cores at Vostok, Byrd, and the South Pole.

For convenience, we will consider proxy records as being of two types: high resolution (annually resolved, spanning up to the last 2 to 3 millennia) and low resolution (50–100-yr resolution, spanning the last 1 to 2 million yr). The latter are important for

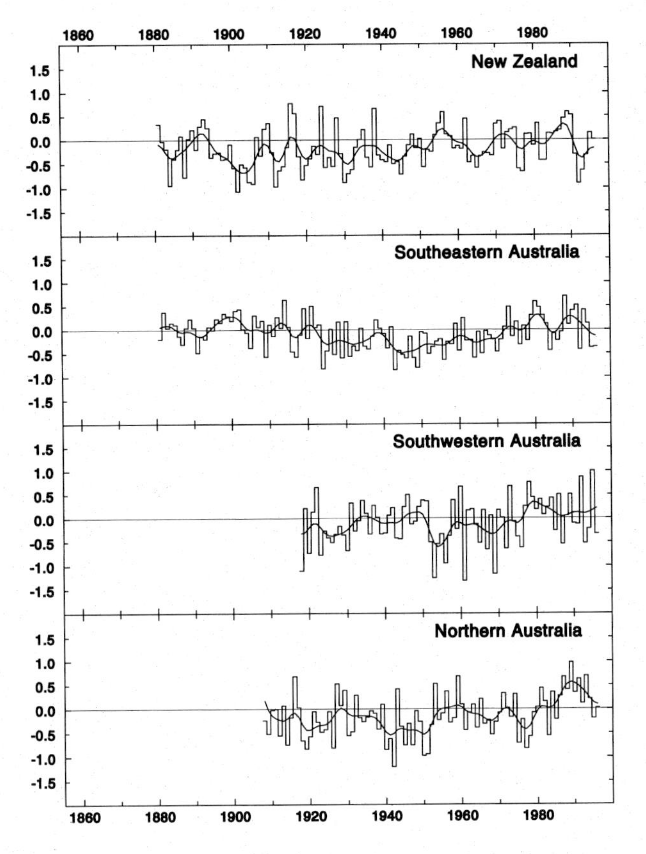

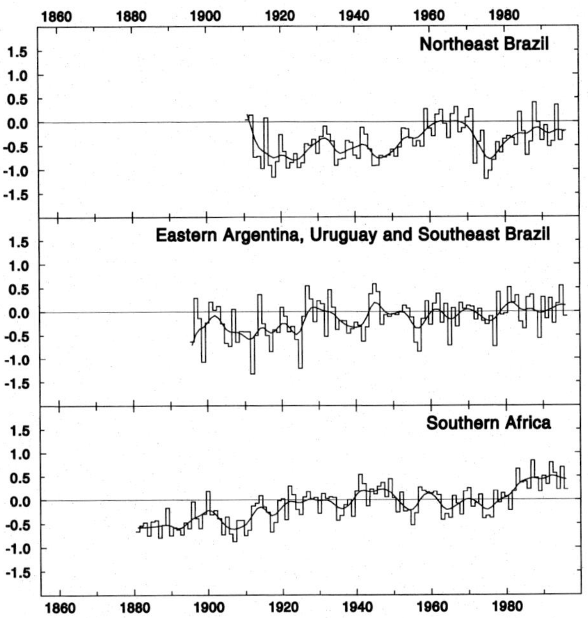

FIG. 9.6. Annual time series of temperature for the same seven selected regions of the SH as Fig. 9.5. Data comes from the 5° × 5° grid-box temperature anomaly series produced by Jones (1994). Averages are expressed as °C anomalies from 1961 to 1990.

what they tell us about the millennial-timescale course of climate change. Information from the central, deeper cores can be compared with deep-sea cores in the ocean. This aspect is considered only briefly in this chapter. High-resolution information from the former is vital for extending our knowledge of change and variability on decadal to century timescales and our understanding of how representative of the climate system the last 140 years of instrumental data have been. One question of relevance is the variability of ENSO events over the last few thousand years. Have they changed in their magnitude, duration, or frequency of occurrence significantly over this time frame, for example?

Despite improvements in paleoclimatology in the SH, there are still too few reconstructions to attempt to synthesize a record equivalent to that for the NH, as has been done by Bradley and Jones (1993). Few of the records, the notable exception being the Peruvian ice core of Quelccaya, show features such as the Little Ice Age and medieval warm period apparent in most paleoclimatic reconstructions from the North Atlantic region [see Bradley and Jones (1993) and Hughes and Diaz (1994) for more discussion]. This highlights the need for continued development and eventual combination of records from the four main landmasses.

2) HIGH RESOLUTION

(i) Tree-ring records

Until very recently, there were few dendroclimatic studies in the SH. Initial work seemed to indicate that the material was difficult to cross-date and that the tree ring–climate relationships were weak (LaMarche 1975; see also Scott 1972). The situation has been dramatically reversed during the last decade, through patient and detailed sampling strategies, and some of the important advances in the field have come in the SH (Cook 1995). In Figs. 9.8a and 9.8b, we show a number of the reconstructions on two timescales—the last 3000 and the last 1000 years.

In South America, two millennium-long reconstructions of summer (DJF) temperature have been made using ring-width data from Alerce (Fitzroya cupressoides) trees. One is located on the western side of the Andes at Lenca, south-central Chile (Lara and Villalba 1993), while the other is on the eastern side at Alerce, Andean Argentina (Villalba 1990). Both reconstructions show little evidence of warming during the present century (in accord with the instrumental evidence for the summer season) and little century-scale variability over the last 1000 years. This latter fact may be an artifact of the tree-ring standardization (removal of the age-related trends, including probably

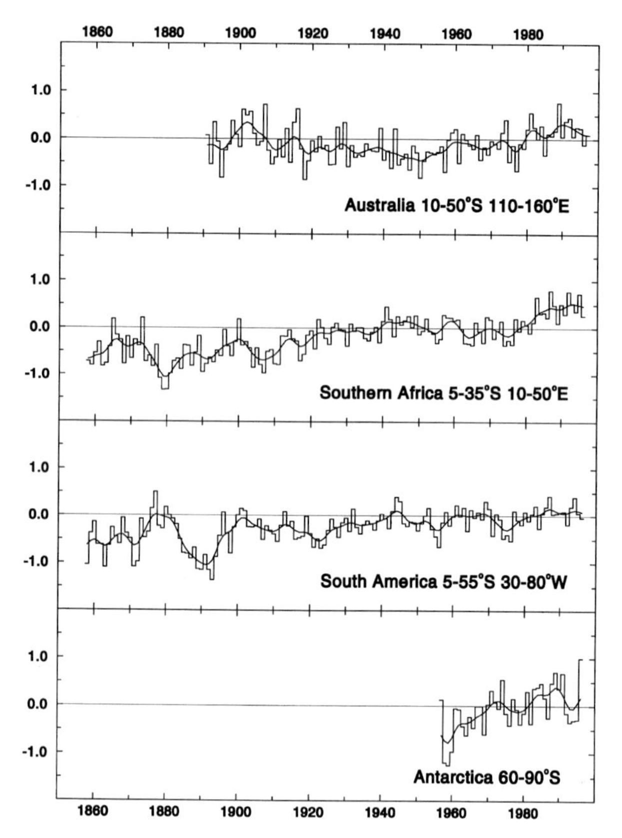

FIG. 9.7. Annual time series of average temperatures from the four main continents of the SH. Data comes from the 5° × 5° grid-box temperature anomaly series produced by Jones (1994). Averages are expressed as °C anomalies from 1961 to 1990.

all variability due to climate on century and longer timescales; see Briffa et al. 1996 for more discussion of this issue), and further work, both on this species and others in the region, such as Pilgerodendron, is clearly warranted. As shown in Figs. 9.8a and 9.8b, the two long reconstructions are not always in very good agreement, even though they attempt to reconstruct the same instrumental temperature series (based on three sites between 40° and 43°S on the eastern side of the Andes). Elsewhere in South America (south of 20°S), river flow, precipitation, and other temperature reconstructions have been produced, but they are only a maximum of 500 yr in length (see Boninsegna 1992 and Villalba 1994 for reviews, and Villalba et al. 1997).

The Villalba et al. (1997) work is the first to attempt to link tree-ring reconstructions in southern South America with those in New Zealand, using the out-of-phase relationship between pressure data in the two regions. This pressure link has been termed the Trans Polar Index (see Pittock 1980a and Karoly et al. 1996). The link gives greater incentive to the tree-ring work being undertaken from Nothofagus in Tierra del Fuego using material buried in peat bogs during the last 8000 years (Roig et al. 1996).

In the Australasian region, temperature reconstructions have been produced using ring-width data from huon pine (Lagarostrobos franklinii) near Lake Johnston, western Tasmania (Cook et al. 1991; Cook et al. 1992; Cook et al. 1996). The Tasmanian records extend back nearly 4000 years and have the potential to be extended much further. This record is important for the fact that the twentieth century is the warmest century except for a short period around 300 B.C. In New Zealand, nationwide temperature and zonal and meridional pressure indices have been reconstructed back to the early 1700s (Norton et al. 1989 and Salinger et al. 1994). Extension to about 1500 should be possible. Temperature reconstructions for Stewart Island, New Zealand, are beginning to be produced at this time (D'Arrigo et al. 1995). Potential for long reconstructions similar to those in Tasmania exists in the North Island of New Zealand from kauri trees (Agathis australis), but has yet to be realized (see Norton and Palmer 1992 for a review for the Australasian region).

Although other land regions where trees grow are tropical, reconstruction of precipitation may be possible in a few regions where there are clearly defined single rainy seasons. Such an example is Indonesia, where teak trees (Tectona grandis L.F.) appear to offer the greatest potential (Jacoby and D'Arrigo 1990 and Palmer and Murphy 1993). In this region, the latest studies have built on earlier work undertaken by Berlage (1931) and de Boer (1951).

Southern Hemisphere dendroclimatic studies have until recently been produced solely from tree-ring widths. Variations in maximum latewood density in NH conifers have produced some of the best dendroclimatic reconstructions (see Schweingruber and Briffa 1996 for a review). Preliminary work, however, in the SH appears to indicate little benefit is gained from densitometric analyses (Schweingruber and Briffa 1996).

(ii) Coral records

Coral growth (in scleratinian corals) involves the accretion of skeletal carbonate. As the process involves temperature-dependent fractionation, the resulting carbonate is enriched in the heavier (^{18}O) isotope at lower water temperatures (Fairbridge and Dodge 1979 and Cole 1996). In regions with a strong seasonal cycle, variations in δ^{18}O will be recorded through the growth axis of the coral (Dunbar and Wellington 1981). In Fig. 9.9, we show a number of the reconstructions from the Pacific Ocean over the last 400 years.

A 366-yr δ^{18}O record from the Galapagos Islands has been produced (Dunbar et al. 1994; Dunbar et al. 1996). El Niño events in this region are identifiable as large negative δ^{18}O anomalies as water temperature rises during the events. Temperatures vary interannu-

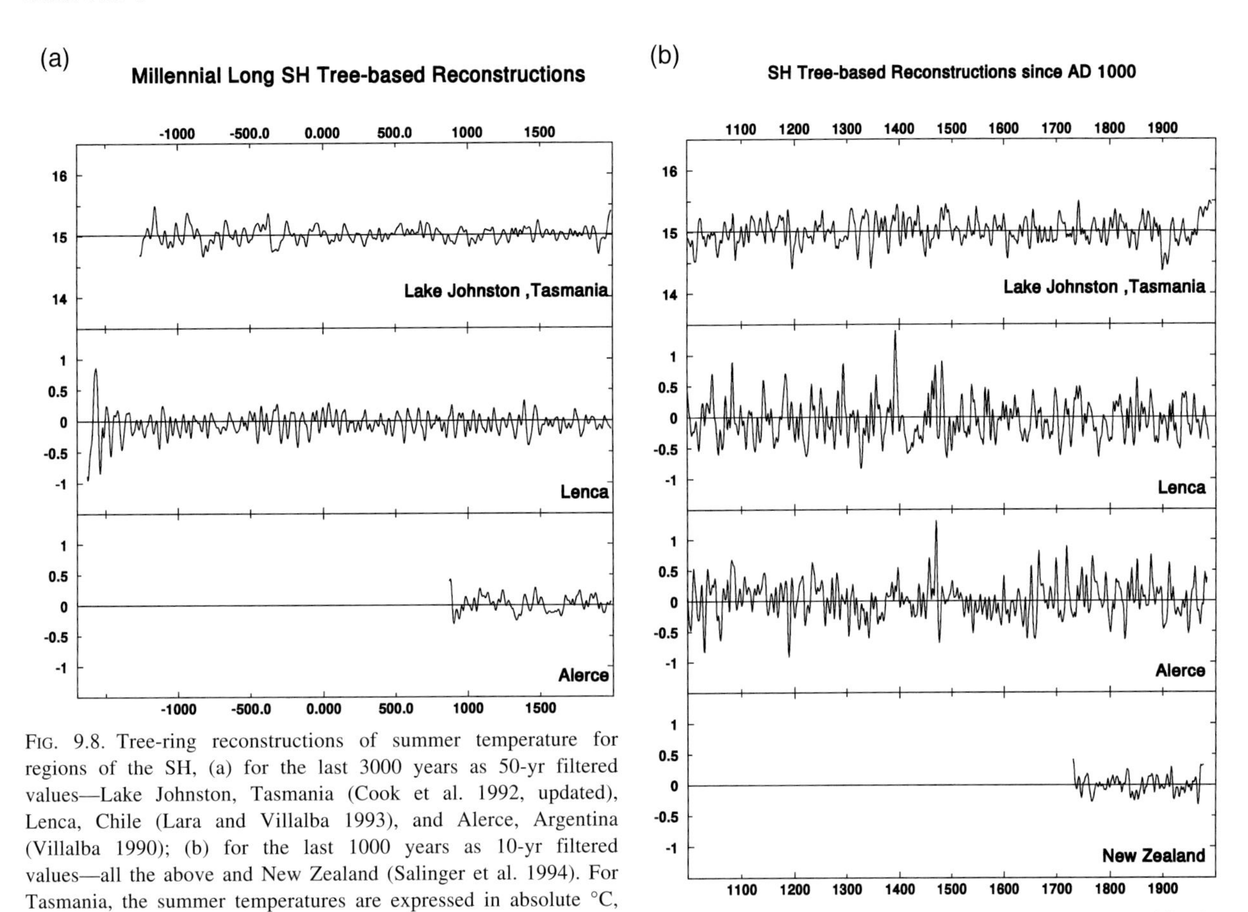

FIG. 9.8. Tree-ring reconstructions of summer temperature for regions of the SH, (a) for the last 3000 years as 50-yr filtered values—Lake Johnston, Tasmania (Cook et al. 1992, updated), Lenca, Chile (Lara and Villalba 1993), and Alerce, Argentina (Villalba 1990); (b) for the last 1000 years as 10-yr filtered values—all the above and New Zealand (Salinger et al. 1994). For Tasmania, the summer temperatures are expressed in absolute °C, while for the other sites they are in anomaly units. Details of the calibration and verification periods are given in the original publications and in Cook (1995).

ally by up to 2.0°C between extreme years, and it is possible to reconstruct subannual time series. The reconstruction shows warmer SSTs during the seventeenth, nineteenth, and twentieth centuries and cooler ones during the eighteenth.

On the western side of the Pacific, corals from the Great Barrier Reef, eastern Australia, have been examined using a number of indicators [$\delta^{18}O$, fluorescence, calcification rates, etc.—see Dunbar and Cole (1993) and Cole (1996) for a discussion of the type of information available in corals]. Reconstructions of SST have been made from a number of coral cores in the region (Druffel and Griffin 1993 and Lough et al. 1996). The records indicate an increase in temperature since the 1740s, with an associated increase in the seasonal range. River runoff for the Burdekin River in Queensland has been produced back to the 1640s from coral fluorescence. This latter reconstruction is a good proxy for Queensland-wide precipitation (Lough 1995) and the extended record shows that the heavier rains of the 1970s (see Fig. 9.5) were not that unusual in a longer-term context. Further development in coral reconstructions is expected in the Indian Ocean, as well as the Pacific, during the next few years.

(iii) Ice-core records

Several ice cores have been drilled in both high-elevation, low-latitude sites in Peru and at higher-latitude sites in the Antarctic Peninsula and the Antarctic continent itself. Evidence from each ice cap comes from a variety of indices in the layers—oxygen isotopes, accumulation rates, percent of dust particles, acidity, etc. (see Bradley 1985 for more details). Figure 9.10 shows a number of the ice-core isotopic temperature series over the last 1000 years. We will discuss each of the three groups in turn.

Ice caps in the Peruvian Andes provide the longest records from proxy sources in all of the lower-latitude regions of the SH. Two long records are available in the area: Quelccaya in southern Peru (Thompson 1992, 1996) and Huascarán in northern Peru (Thompson et al. 1995). The latter record (as yet only preliminarily analyzed) is likely to provide the better and certainly longer record. Quelccaya is 5670 m and Huascarán is ~6000 m in height. The Quelccaya isotopic and snow-accumulation records from two cores indicate a cool period from 1520 to 1880, marked also by an increase in the seasonality of

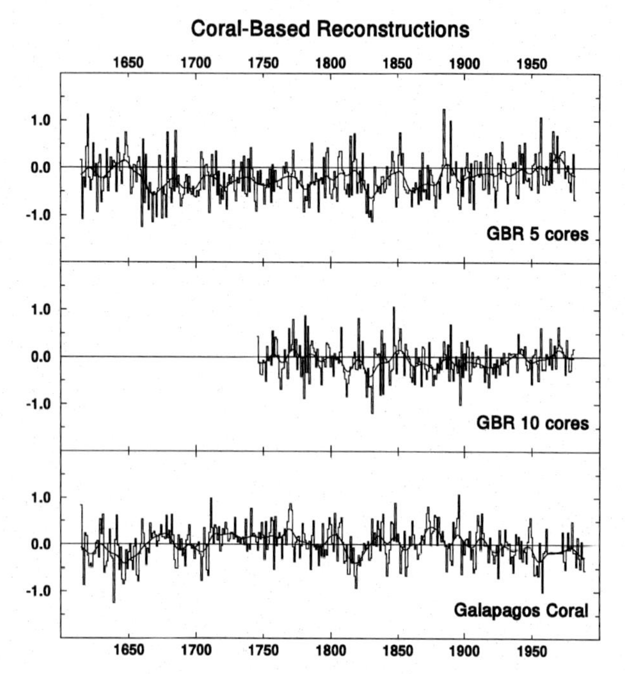

FIG. 9.9. Coral reconstructions for annual temperatures for the last 400 years: two reconstructions from the Great Barrier Reef, Australia (Lough et al. 1996) and one from the Galapagos Islands (Dunbar et al. 1994, 1996). The temperatures are expressed as anomalies from periods given in the original publications and in Cook (1995). All series have been filtered with a 20-yr Gaussian filter.

oxygen isotope variations. Alternations between dry and wet periods are marked in the accumulation rates. Wetter periods occurred between 750 and 1000 and between 1500 to 1700.

Evidence from Peru (Thompson et al. 1993) and eastern Africa (Hastenrath 1992) and west Irian Jaya (Thompson 1995, personal communication) points to a recent pronounced recession of high-elevation tropical ice caps. In some regions there has been so much melting that the annual isotope cycle is not being produced. Thompson et al. (1993) notes that low-latitude, high-altitude ice caps show both evidence of Little Ice Age cooling and twentieth-century warming more strongly than polar ice caps, suggesting they are more sensitive to large-scale climatic changes. Many GCMs suggest that low-latitude, high-elevation sites may experience the largest temperature increases as a result of future greenhouse gas–induced climate changes.

Ice-core evidence from the Antarctic Peninsula is discussed in Peel (1992), Mosley-Thompson (1992), and Thompson et al. (1994). Ice cores have been drilled at a number of sites (James Ross Island, 64°S; Dolleman Island, 70°S; Dyer Plateau, 71°S; and Siple, 76°S). The relationships between isotopic analyses and station temperatures never explain more than 25% of the instrumental variance, although the stations are all coastal sites on both sides of the peninsula. The relationships between isotopic content and temperature are clearly related to changes in source material, particularly when a polynya is evident in the Weddell Sea (Jones et al. 1993). The better indicator of temperature as opposed to source changes may come from the more continental sites such as Siple. The longest record is also found at Siple, as accumulation here is least. The record does not indicate any evidence of a protracted cold period in the sixteenth to nineteenth centuries. All the Antarctic Peninsula records clearly show warmer conditions in the early nineteenth century, with cooler conditions in the late nineteenth century, although the timing and magnitude of the temperature drop varies from 1845 at Dyer Plateau to near 1890 at James Ross Island. All the cores in this region show a strong warming since 1950, which is also shown in the instrumental records (King 1994 and Jones 1995).

Most ice cores from the Antarctic continent do not have sufficient accumulation to identify clearly annual layers before the twentieth century. Dating is therefore a severe problem, and possible circularity is apparent when dust/acidity layers are used. These need to be categorically related to specific volcanic events before they can be used. Even if the event can be identified, it is still not known how long it took the volcanic material to reach the site. Cores exist for a number of sites: South Pole, Dome C, Filchner–Ronne Ice Shelf, Plateau Law Dome, and Mizuho (see Mosley-Thompson 1992, for details). The best dated of these are Law Dome (Morgan 1985 and Morgan and van Ommen 1997), South Pole (Mosley-Thompson 1992), and Plateau (Mosley-Thompson 1996). The ice-core resolution is so good at Law Dome that a seasonal record since 1300 has been derived (Morgan and van Ommen 1997). None of these records show indications of long-term changes over this period, except for a tendency to lower isotopic values (cooler temperatures) at Law Dome and Plateau in the autumns and winters of the late eighteenth and early nineteenth centuries.

3) SOUTHERN HEMISPHERE SYNTHESES DURING THE LAST FEW THOUSAND YEARS

Although they have often revealed signs of longer-term climatic fluctuations, proxy data have generally been examined in a local to regional context. Few studies have attempted, or been able to call on enough data to produce, more hemispherical overviews of climatic variability. Works that have attempted to develop a broader SH picture of climatic change and variability have focused on a particular phenomenon (e.g., ENSO) as resolved in one proxy type (e.g., tree rings; Lough and Fritts 1990); the pattern of climatic response in one or two proxy type(s) (Thompson et al. 1992; Villalba 1993; Cook 1995; Villalba et al. 1997);

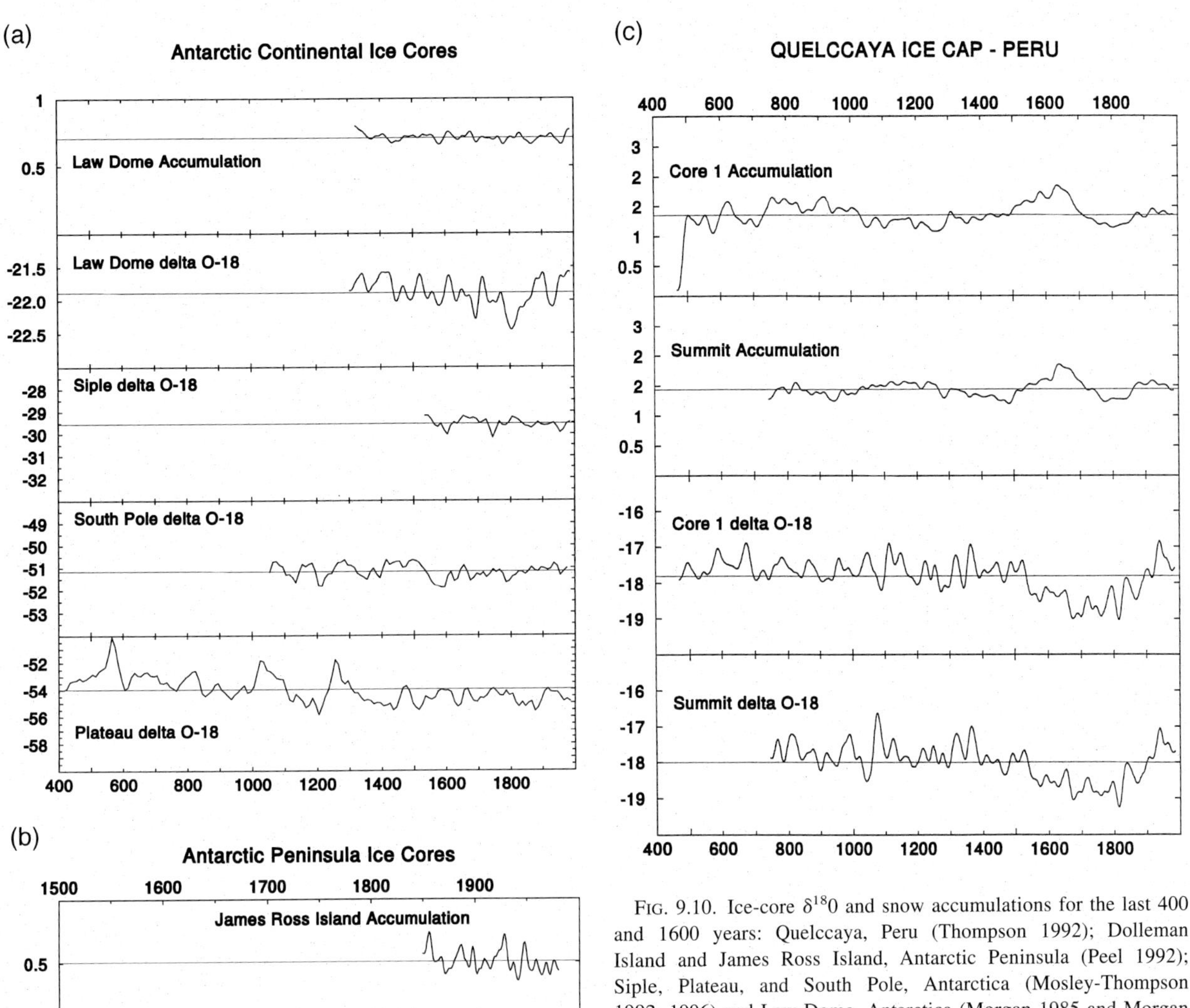

FIG. 9.10. Ice-core $\delta^{18}0$ and snow accumulations for the last 400 and 1600 years: Quelccaya, Peru (Thompson 1992); Dolleman Island and James Ross Island, Antarctic Peninsula (Peel 1992); Siple, Plateau, and South Pole, Antarctica (Mosley-Thompson 1992, 1996) and Law Dome, Antarctica (Morgan 1985 and Morgan and van Ommen 1997). Snow accumulations are in ice equivalent (meters) per year, while the isotope records (delta 0–18 or delta D) imply warmer temperatures for lighter (less negative) isotopic values and vice versa. Antarctic continental and Quelccaya cores are filtered with a 50-yr Gaussian filter in Figs. 9.10a and 9.10c, while the Antarctic Peninsula cores are filtered with a 10-yr Gaussian filter in Fig. 9.10b.

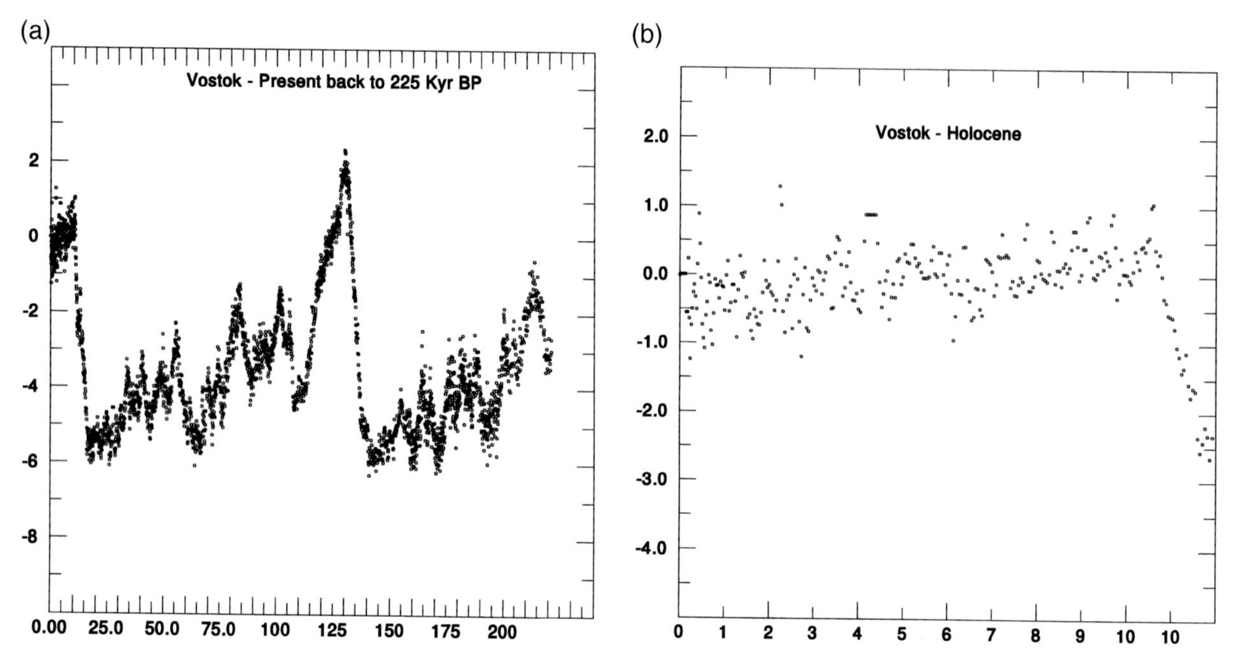

FIG. 9.11. Ice-core δD for the last 220 000 years from Vostok, Antarctica. Data are from Jouzel et al. (1993). Data are plotted for (a) the whole period and (b) the last 11 000 years. The isotopic data are expressed as °C anomalies (based on the spatial dependence between isotopic values and temperature observed over Antarctica and Greenland) from the average instrumental temperature observed at Vostok.

one phenomena as it affects a particular region of the hemisphere over time in proxy data (Enfield 1989; Enfield and Cid 1991; Martin et al. 1993; Ortlieb and Machare 1993; Villalba 1994); or all available proxy evidence on timescales of millennia for the Holocene period (Markgraf et al. 1992). More recently, Diaz and Pulwarty (1994), Whetton and Rutherford (1994), Whetton et al. (1996), Stahle et al. (1998), and Allan and D'Arrigo (1998) have attempted to resolve characteristics of ENSO phases and longer climatic fluctuations in several proxy data sources of high resolution across the Indo-Pacific region. The quality and quantity of high-resolution proxy indicators, however, has to date restricted efforts to provide a detailed study of climatic change and variability on decadal to millennial timescales. As pointed out earlier, attempts at syntheses stress the need for more work to be performed, particularly in the SH (Bradley and Jones 1993; Hughes and Diaz 1994; and Crowley and Kim 1993). The most consistent features are generally cool conditions during the nineteenth century and the warmest since about 1950.

4) MILLENNIAL-TIMESCALE PROXY RECORDS

Proxy records on this timescale come from the Vostok ice core and from deep-sea cores in the main oceanic basins. Recent intercomparisons of the two data sources show remarkable similarities (Martinson et al. 1987; Lorius et al. 1992; and Pichon et al. 1992). Here we just show the Vostok isotopically (from δD) inferred temperatures (from Jouzel et al. 1993) for the last 230 000 and the last 11 000 years in Fig. 9.11. The last 11 000 years, which encompass the entire Holocene period, has been the longest and most stable warm period of the last 230 000 years. Indeed, the present temperature level at Vostok has occurred for only a small fraction of the last 230 000 years. The most striking feature of Fig. 9.11a is the very rapid warming to the interglacial state at about 125 000 and 15 000 years ago. Deeper drilling at Vostok will be able to indicate whether this rapid transition occurred after earlier ice ages.

9.3. Circulation and ENSO variability in the Southern Hemisphere

a. Background

In general, the bulk of research on climatic variability has dealt with either locally representative series from single or, at best, a small number of sites, or more regional averages (see section 9.2). Such efforts are important, but in order to improve understanding of the climate system, validate model simulations, and ultimately resolve physical mechanisms, there is an increasing need to obtain a hemispherical to global perspective of the problem. A few studies are now in a position to begin to assess spatially extensive fields capable of resolving the nature and variability of atmospheric and oceanic circulation features and patterns over the historical instrumental period. Some of the initial findings of this research are reviewed in this section. Overall, however, most studies have been able to do little more than document the observational patterns and speculate as to the physical processes involved. This has often been expressed only in terms of interannual variability and long-term trends being related to ENSO and possible enhanced greenhouse warming, respectively.

b. Circulation records

In addition to recording longer-term trends and changes in various temperature (both land and marine) and precipitation data, the time series in section 9.2b exhibit interannual to multidecadal fluctuations. From the historical instrumental record, fluctuations in climatic features that will assist in understanding the above processes can also be examined using time series of mean sea level pressure, wind components, and cloudiness.

Recent examinations of regional- to hemispheric-mean cloudiness variations are found in the papers of London et al. (1991), Henderson-Sellers (1992), and Parungo et al. (1994). Detailed studies over countries in the SH are restricted to those of P. A. Jones (1991) and Jones and Henderson-Sellers (1992) for Australia. Efforts to accumulate and assess data essential to longer-term examinations of the spatial patterns of global-cloudiness fluctuations are best represented by the atlases of Warren et al. (1986, 1988) and the update of Hahn et al. (1994). Problems with the reliability and representativeness of such data are discussed in the above compilations. As a result, analyses to date have been restricted generally to data since 1930, and then only consider total cloudiness. The spatial distribution is most sparse over the oceanic regions, outside the major historical shipping lanes. In particular, regions of the southeastern Pacific and higher latitudes of the Southern Ocean are often data void.

Historical surface to near-surface wind reconstructions have been assembled specifically for various regions and ocean basins by Bigg (1992, 1993), Bigg et al. (1987), Cooper et al. (1989), Shriver (1993), Ward (1991, 1992), and Whysall et al. (1987). Other studies have incorporated wind fields in multiparameter analyses of atmospheric and oceanic patterns (Allan et al. 1995; Gu and Philander 1995; Parthasarathy et al. 1991; Wang and Ropelewski 1995). As with all other long-term variables, concerns have been raised about the quality and representativeness of such data and the trends and fluctuations deduced from them. In particular, major changes in observer and national practices, instrument type, positioning, and data distribution have meant that wind reconstructions are particularly difficult to work with (Cardone et al. 1990; Postmentier et al. 1989; Wright 1988). As with cloud data, only sparse information is available over the southeastern Pacific and higher latitudes of the Southern Ocean.

Until recently, SH MSLP coverage, capable of considering interdecadal variations, has been restricted to gridded fields of portions of the hemisphere (Barnett and Jones 1992; P. D. Jones 1991; Jones and Wigley 1988). Lately, MSLP has begun to receive more attention in terms of quality control and data reconstruction, to the extent that, along with land and marine (SST) temperatures and precipitation, spatial fields of near-global to global coverage spanning the last 100–120 years are becoming available (Allan 1993). Such coverage, together with local to regional time series, is examined for the SH. Despite these efforts, the available database is both shorter and, in most regions, of a poorer quality than in the NH. This will soon change to a considerable degree with the completion of a full global MSLP data compilation covering the period 1871–1994 (Allan et al. 1996).

1) REGIONAL ANALYSES

Section 9.2 showed that the pattern of decadal to multidecadal fluctuations over the SH varies in time and space, and such nature is also evident from an evaluation of specific regional studies. This can be seen in a number of papers, ranging from the early efforts covering major climatic regimes and features in both hemispheres by Kraus (1954, 1955a,b, 1956), through to the works of Fletcher et al. (1982) on low-frequency Southern Ocean–Antarctic variability; Nicholson and Entekhabi (1986), Hulme (1992), and Nicholson (1993) on long-term African rainfall variability; Nitta and Yoshimura (1993) on continental to global air temperatures; Smith et al. (1994) on regional to global SST patterns; local to regional evaluations of Queensland rainfall patterns by Lough (1991, 1992, 1993); and Yu and Neil (1991, 1993) on southwestern and southeastern Australian rainfall trends. Similar findings have come from specific studies of ENSO variability over the instrumental period (e.g., Allan 1993; Diaz and Pulwarty 1994; Elliott and Angell 1987, 1988; Enfield 1988, 1989; Enfield and Cid 1991; Fu 1986; Fu et al. 1986; Lough and Fritts 1990; Ortlieb and Machare 1993; Trenberth and Shea 1987; Whetton et al. 1990; Whetton and Rutherford 1994; Whetton et al. 1996). A more extensive documentation of such research, together with some of the most detailed SH regional syntheses, can be found in Allan et al. (1995), Karoly et al. (1996), and Salinger et al. (1996).

Interannual to multidecadal fluctuations of precipitation and land air temperatures in Figs. 9.5 and 9.6 are from regions that generally display a high degree of sensitivity to ENSO phases (Ropelewski and Halpert 1987, 1989). There is now considerable evidence that ENSO influences are not only manifest by interactions involving quasibiennial (QBO) and lower-frequency signals within the broad 2- to 7-yr time frame (Barnett 1991; Rasmussen et al. 1990; Ropelewski et al. 1992), but are also modulated by decadal–multidecadal fluctuations in the climate system (Jinghua et al. 1988; Jingxi et al. 1990; Allan 1993; Gu and Philander 1995; Shriver 1993; Wang and Ropelewski 1995; Allan et al. 1996). Recent research has begun to explore this contention further.

Much of the incentive to focus on lower-frequency variations in the climate system and ENSO has been because of the presence of a persistent El Niño sequence from 1990/91 until 1995 (Trenberth and Hoar 1996; Harrison and Larkin 1997; Rajagopalan et al. 1997; Allan and D'Arrigo 1998), together with indications of a possible abrupt climatic change over the Pacific basin since the mid-1970s (e.g., Nitta 1993; Graham et al. 1994; Graham 1994; Kleeman et al. 1996; Latif and Barnett 1994a,b; Miller et al. 1994a,b; and Trenberth and Hurrell 1994). Studies investigating this abrupt change involve a mixture of observational data and model simulations that have gained additional support from long century to millennial runs of GCM experiments working toward resolving the physical mechanisms underlying climatic variability (Barnett et al. 1992; Hunt and Davies 1997; Miller and Del Genio 1994; von Storch 1994). Most of this research, however, has had a heavy emphasis on conditions over the North Pacific and their implications for North America. Although a Pacific focus for natural climatic variability appears to be dominant in most of the above GCM simulations, there is evidence for wider tropical involvement and a potential for influences to extend into the SH through teleconnection patterns and modulations of the principal wave-train modes. Nevertheless, care must be taken with the interpretation of these preliminary results, particularly as they are very sensitive to the distribution and parameterizations of clouds in the GCMs (Miller and Del Genio 1994).

Historical and proxy data findings provide further insights. A study of major ENSO-sensitive variables in a number of Indo-Pacific regions by Diaz and Pulwarty (1994), indicates that while there is very significant coherency in responses on 2- to 6-yr time frames, such structure is considerably diminished when examining the signal encompassed by the broad 11- to 150-yr band. Detailed analysis of the full range of dominant frequencies in global surface temperature fields by Mann and Park (1993, 1994a,b), indicates that multidecadal variability patterns (centering around 15–18 years) appear to resemble the near-global temperature response associated with the major interannual ENSO signal (see Halpert and Ropelewski 1992). Further evaluation of this type of evidence is made in the course of this section, through an analysis of climatic variables in ENSO-sensitive regions across the SH.

Aspects of the linkage between interannual ENSO characteristics and lower-frequency decadal–multidecadal fluctuations can be seen from an analysis of the dominant QBO (18–35 months) and lower-frequency (LF) (approximately 3–7 yr) signals inherent in ENSO. The Southern Oscillation Index is filtered in the 18–88 month band and in the QBO (18–35 months) and LF (32–88 months) bands in Fig. 9.12. When taken together with similar analyses of central equatorial Pacific–eastern equatorial Pacific (CEP–EEP) SST and CEP precipitation series of Wright (1989), this diagram shows strong coherency in responses between these indices and that ENSO phases (El Niño and La Niña), characterized by the superposition of QBO and LF signals, occur with varying frequency, magnitude, and duration, giving rise to the concept of a "family" of El Niño and La Niña events. Also shown in Fig. 9.12 are longer and coherent decadal–multidecadal variations in signal frequency and magnitude.

Evidence that QBO and LF interactions may result in a range of ENSO phases has included indications that extensive sequences of El Niño and La Niña episodes result from the dominance of a strong LF signal in the climate system. This is particularly relevant in light of the recent predominance of a recurrent El Niño sequence. Figure 9.13 shows the detailed response of the SOI and CEP–EEP SST during all of the recurrent or persistent ENSO sequences that have been recorded in the historical instrumental record. When such persistent ENSO sequences are put into context with other ENSO responses in Fig. 9.12, it can be seen that such events are not as common as the more "typical" shorter-duration events. Nevertheless, they have important consequences for climatic patterns, and any advance in understanding the physical processes underlying the QBO and LF signals is vital to improving ENSO dynamics.

As noted in studies described earlier, even LF signals in the climate system over the Pacific basin may provide a broader "envelope" that acts to modulate the ENSO phenomenon. The most pronounced low-frequency feature in Fig. 9.12 is the tendency for QBO and, in particular, LF signals to be strong from the late 1800s up to the 1910s and again since the 1950s, while showing a marked weakening during the 1920s and 1930s. This low-frequency aspect is a coherent feature of the time series of all three ENSO indices examined. In fact, fluctuations of the above type display strong multidecadal characteristics in line with the type of response noted in Mann and Park (1993, 1994a,b). Interestingly, coherent patterns at these frequencies are missing from ENSO-sensitive regions in the study by Diaz and Pulwarty (1994), perhaps because they are swamped by the large width of the band (11 to 150 yr) used in that paper to define all low-frequency signals. Much effort is underway to resolve the nature of decadal–multidecadal variability, particularly with a focus on the potential source of

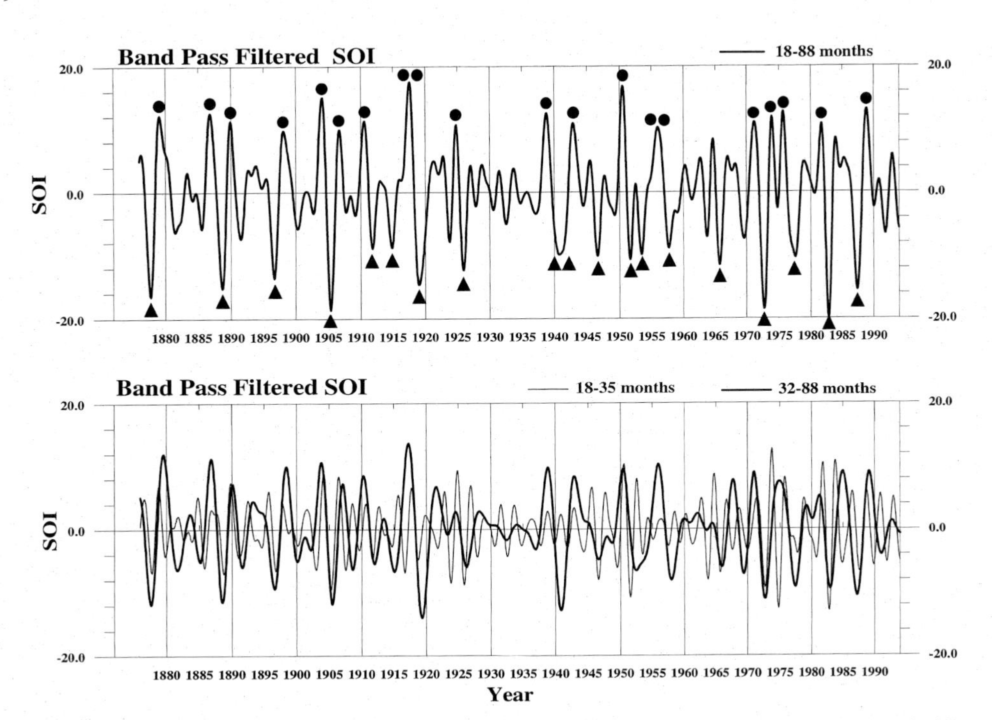

FIG. 9.12. Bandpass-filtered SOI from 1876 to 1992. Top panel shows the SOI filtered in the 18–88-month band, and the bottom panel shows the SOI filtered in the 18–35 (light line weight) and the 32–88-(heavy line weight) month bands. The solid dots and triangles in the top panel indicate major La Niña and El Niño phases of the ENSO phenomenon, respectively.

such fluctuations in the relatively slower response of the oceans to various forcings.

Climatic fluctuations of the decadal to multidecadal type have also been found in regional studies of atmospheric and/or oceanic circulation patterns. Interannual relationships involving ENSO and rainfall over parts of the west coast of South America have been related through changes in the mean latitude of the subtropical anticyclone near the Chilean coast by Rubin (1955), Pittock (1980a,b), Aceituno (1988), Rutllant and Fuenzalida (1991), Ortlieb and Machare (1993), and Villalba (1994). Quinn and Neal (1983), in their discussion of El Niño characteristics in the South American region, detailed evidence suggestive of long-term variations in the anticyclone over the southeastern Pacific Ocean. They attributed changes in the pattern of rainfall in north to central Chile to this climatic feature. In reconstructing SH midlatitude MSLP over parts of this century, (P. D.) Jones (1991) indicates a significant decrease in the central pressure of the South Pacific anticyclone from 1951 to 1985. Interestingly, the bulk of this downward trend in MSLP is found to varying degrees in all seasons in the period since the mid-1970s. This is in line with the suggested abrupt change in ENSO circulation features over the Pacific Ocean basin discussed earlier.

Over southern Africa, low-frequency periodicities in rainfall of around 10–12, 16–20, and greater than 20 years have been noted by Lindesay (1984), Tyson (1986), Mason (1990), and Mason and Tyson (1992). Efforts to link such fluctuations to physical features of the climate system have focused primarily on SSTs, particularly over the South Atlantic Ocean (Mason 1990; Mason and Tyson 1992). A strengthening of southerly flow in both wind and pressure fields over the northern Mozambique Channel during the austral winter since the mid-1940s is documented by Bigg (1992).

Most research on climatic patterns over the tropical South Atlantic Ocean has generally focused on interannual variability associated with ENSO-like, or possibly ENSO-related, patterns (e.g., Polonsky 1994; Delecluse et al. 1994) or lower-frequency responses in SST that are part of a wider signal that leads to major rainfall fluctuations in parts of Africa (particularly the Sahel) and South America (especially northeastern Brazil) (e.g., Folland et al. 1986; Lough 1986; Palmer 1986; Hastenrath 1990; Lamb and Peppler 1992; Ward 1994). Distinct low-frequency modes in South Atlantic SSTs on interdecadal timescales have been reported recently in EOF analyses by Kawamura (1994). Fluctuations in the South Atlantic anticyclone are found in seasonal examinations of SH midlatitude MSLP by (P. D.) Jones (1991). Of the circulations described in Jones, only the South Atlantic anticyclone shows a distinct increase in MSLP, particularly since the mid-1970s (a statistically significant 0.6-hPa rise in MSLP since 1951).

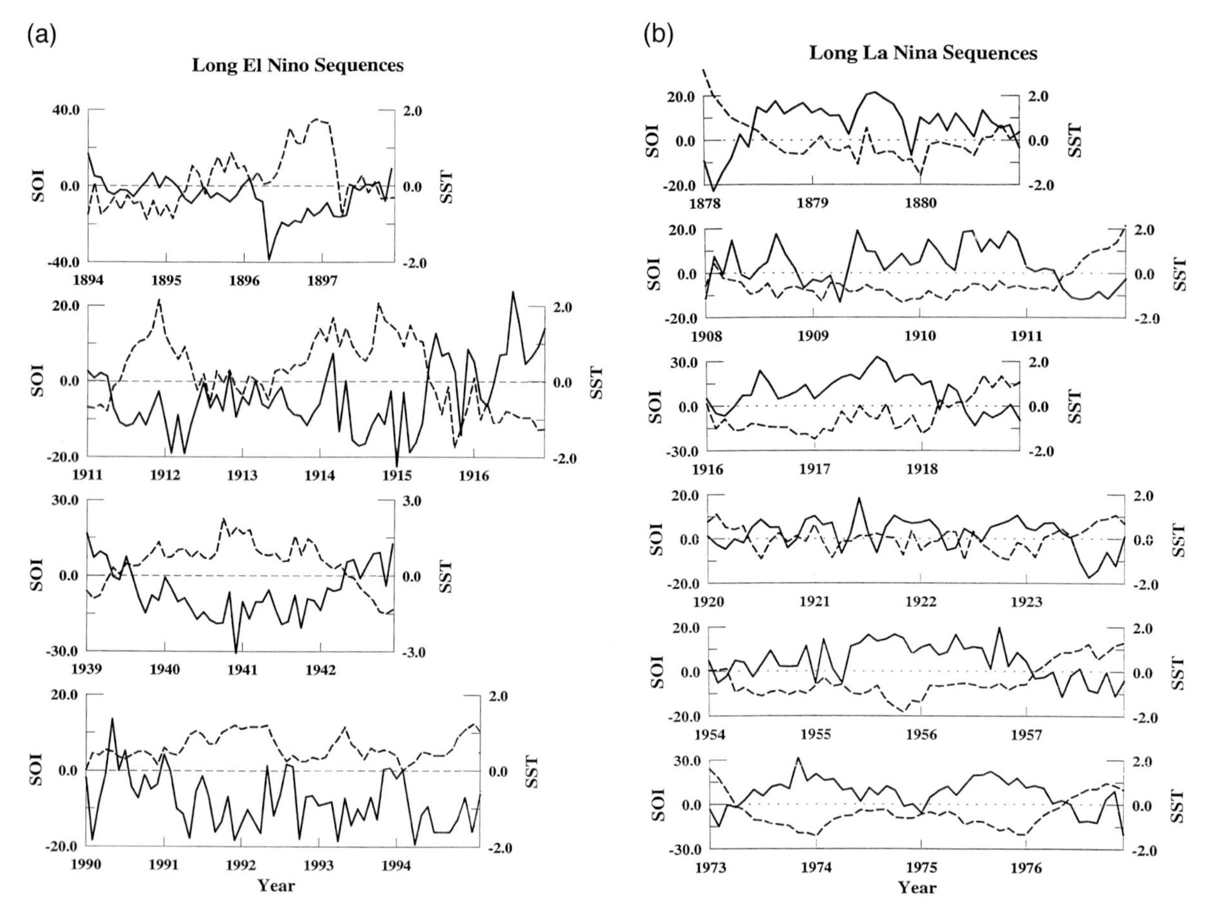

FIG. 9.13. (a) Persistent El Niño events defined from raw SOI and CEP–EEP rainfall values since 1876. (b) Persistent La Niña events defined from raw SOI and CEP–EEP rainfall values since 1876. SOI traces given by solid lines, CEP–EEP rainfall traces given by dashed lines.

Over the Indian Ocean, Jones also examined the time series of seasonal MSLP but found no significant trends in the central pressure of the South Indian Ocean anticyclone. There was, however, a significant positive correlation between the time series of the central pressure of the South Atlantic and southern Indian Ocean anticyclones. Further insight has been gained from broader studies focusing on a number of climatic features in the major ocean basins. One recent example of such research is the paper by Allan et al. (1995) on multidecadal fluctuations over the Indian Ocean basin. Although focusing only on the austral summer (December to February), Allan et al. (1995) detail low-frequency variations in the strength of the semipermanent anticyclone over the southern Indian Ocean. Evidence of a weaker anticyclone during the first half of this century, with the most distinct change centered on the 1920s and 1930s, is supported by coherent responses in SST, wind, MSLP, and cloudiness fields. Low-frequency warmings and coolings of SSTs are found to be most pronounced and robust along the outflow region of the Agulhas Current in the southern Indian Ocean. It is interesting to note that the region used in (P. D.) Jones (1991) to define grid points contributing to the central pressure of the southern Indian Ocean anticyclone is the least representative of its mean position in the austral summer. This may explain why a significant fluctuation in the central pressure of the above feature is found in Allan et al. (1995), but not in Jones. Specific GCM experiments forcing an oceanic GCM by the type of circulation changes over the Indian Ocean observed in Allan et al. (1995), and by additional modulations of Pacific Ocean wind fields, have been reported by Reason et al. (1996). Initial results of this work suggest that local Indian Ocean forcing alone is not enough to produce the observed long-term SST response.

Over Australasia/Oceania, the recent studies of Fitzharris et al. (1992), Allan and Haylock (1993), and Salinger et al. (1996) are among the most comprehensive attempts to document and understand climatic fluctuations. Emphasizing the New Zealand situation, Fitzharris et al. show that austral summer and winter MSLP patterns have varied considerably, with changes in the subtropical anticyclonic belt in the Tasman Sea/New Zealand region being most pronounced. Such modulations of circulation in both of the seasonal extremes are marked by MSLP anomaly patterns indicative of fluctuations in the longwave pattern over this part of the SH. Using the (P. D.) Jones (1991) dataset during the austral winter, Allan and Haylock find a similar result when analyzing low-pass–filtered (> 25 yr) MSLP patterns. In particular, both studies reveal a generally out-of-phrase relationship between MSLP anomalies immediately to the south-southwest of Australia and those just to the southeast of New Zealand. This configuration slowly changes sign, from a negative/positive to a positive/negative alignment over the period from 1911 to 1989. When seen against the average MSLP pattern for austral winter, this changing alignment of anomalies has maxima and minima that occur in locations associated with the major longwave trough and ridge features in the Oceania region. The net result of these changes in circulation has been a downward trend in austral winter rainfall and westerly flow across southwestern Australia over the last 40-odd years, with a marked increase in westerly zonal flow over New Zealand being most prominent since the 1980s. These changes have led to a general retreat of New Zealand glaciers to about 1980 and a recent re-advance (Fitzharris et al. 1992).

More decadal- to multidecadal-type patterns over Australia during austral winter are also resolved in the study of Allan and Haylock (1993), in their analysis of 7–20-yr fluctuations in the climate. These are manifest by an out-of-phase relationship between MSLP over the continent, associated with fluctuations in the central pressure and longitudinal structure of the winter anticyclone, and MSLP at high latitudes over the Southern Ocean. This response is linked to wet and dry extremes in southwestern Australia and, in MSLP anomaly fields, is related to changes in the longwave trough to the south-southwest of Australia.

For the austral summer, Fitzharris et al. (1992) resolve a pronounced fluctuation in MSLP over New Zealand, from a negative anomaly during 1911–49, to a positive anomaly from 1950 to 1979, and then back to a negative anomaly during 1980–89. This alternation resulted in major changes in the dominant wind regime and, hence, glacier termini over New Zealand. Lesser variations in flow during the above period are evident over the southeastern margins of the Australian continent.

A broader focus examining the southwest Pacific Ocean and embracing both atmospheric and marine data is given by Salinger et al. (1996). They show that time series of seasonal MSLP off both the west and east coasts of Australia at 40°S confirm much of the above findings. In particular, there is further evidence that major low-frequency circulation fluctuations involve periods of amplification and suppression of the longwave trough pattern in the region. There is no suggestion, however, that this involves displacements of the latitude of the anticyclone over the Australian continent during the austral winter, as suggested by Pittock et al. (1980a,b).

There has been considerable argument regarding data quality in contentions about the validity of claims that there are distinctive and significant low-frequency patterns, fluctuations, and a trend in the trade-wind field over the tropical Pacific Ocean (Bigg et al. 1987; Cooper et al. 1989; Shriver 1993; Ward 1991, 1992; Whysall et al. 1987; Cardone et al. 1990; Postmentier et al. 1989; Wright 1988; Inoue and Bigg 1995). In general, those claiming real variations in the wind field indicate a tendency for a greater proportion of westerly anomalies during the 1940s and 1960s and a more even distribution in anomalies during the 1920s, 1950s, and 1970s. The 1930s showed a propensity for more easterly anomalies. Claims that have indicated links between these apparent low-frequency features and more interannual ENSO fluctuations have been questioned by Rasmusson (1987). Nevertheless, wind-field changes have had some support from indications of increased central MSLP in the South Pacific anticyclone [(P. D.) Jones 1991, detailed earlier] and ridging of this feature into the central equatorial Pacific (Cooper et al. 1989; Inoue and Bigg 1995). A broad trend indicative of a strengthening of the Pacific trade-wind regime by about 1 m s^{-1} up until the mid-1970s is also noted in Cooper et al. (1989).

Fluctuations in circulation and climatic features at middle to higher latitudes of the SH have been detailed by Hurrell and van Loon (1994), Enomoto (1991), Carleton (1989), and van Loon et al. (1993). Only the period since the 1950s can be studied in this region. Even then there are problems with data quality. Enomoto (1991), in particular, uses highly dubious data sources. Such studies have shown changes in zonal asymmetry, the semiannual wave in MSLP and wind fields, the timing of the breakdown of the tropospheric polar vortex, MSLP in the circumpolar trough, and relationships between sea-ice extent and circulation indices since the mid- to late 1970s.

A more holistic perspective of the patterns of variability and their coherence among prominent ENSO-sensitive regions can be gleaned from an analysis of the complete spatial structure of these features over the entire SH.

2) Analyses of the whole Southern Hemisphere

Research efforts in this area are best represented by studies examining global surface temperature (land and/or marine) and MSLP patterns. Examples of the former (e.g., Parker et al. 1994; Kawamura 1994; Nitta and Yamada 1989) all show that decadal to multidecadal variability is generally manifest at middle to higher latitudes in both hemispheres. Over the oceans, monthly mean SSTs show the most pronounced broad-scale signals across the Indian, midlatitude North Pacific, and North and South Atlantic Oceans. Inter-

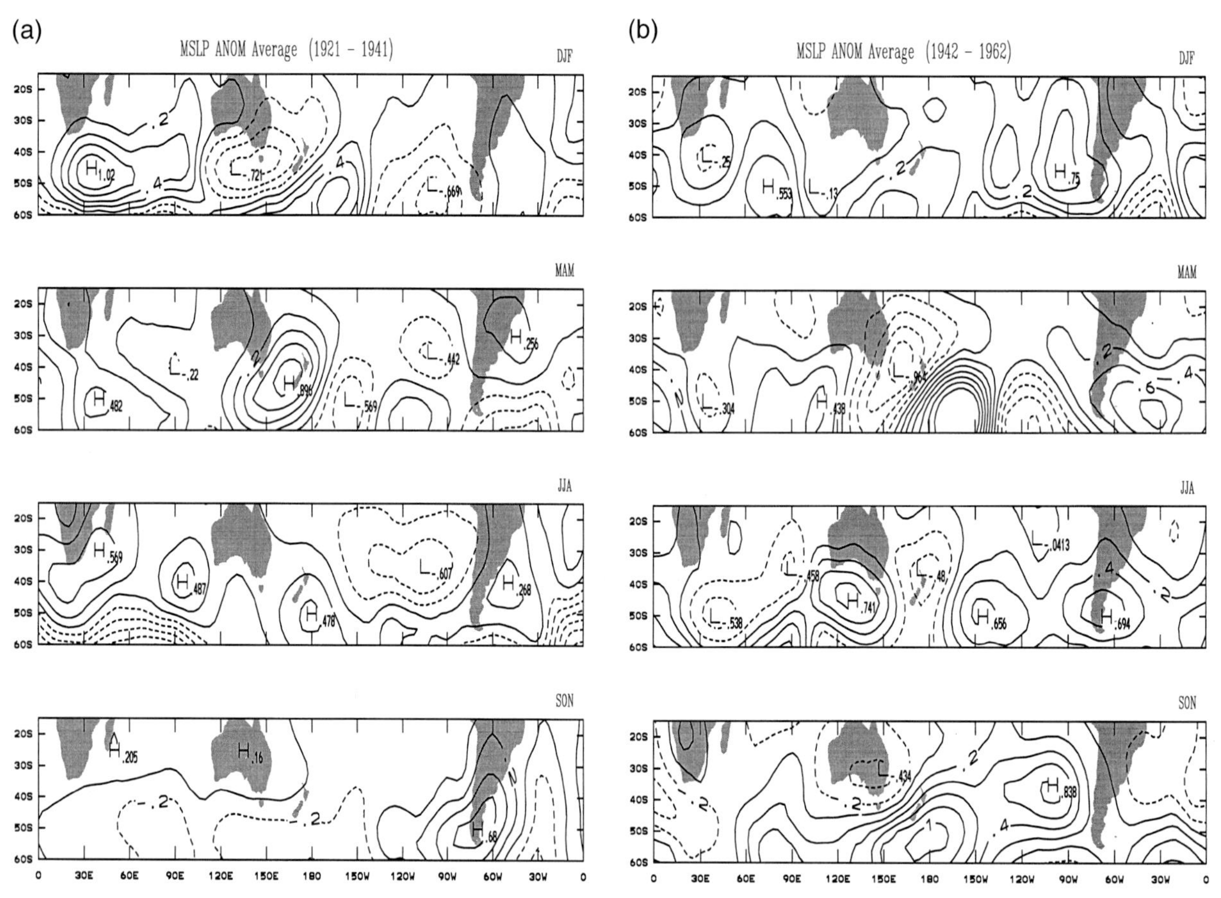

FIG. 9.14. (a) MSLP anomalies during each season (DJF, MAM, JJA, SON) for the 1921–41 epoch. (b) As in (a), except for the 1942–62 epoch. (c) As in (a), except for the 1963–83 epoch. (d) MSLP during each season for the full period of data from 1912 to 1985. All pressure values given in hPa. Contours in (a) to (c) every 0.2 hPa, and in (d) every 4 hPa.

estingly, Kawamura suggests that there are two separate SST modes involved with this scale of low-frequency variability, one between the Indian and Pacific Oceans and the other involving the North and South Atlantic Ocean alone. The above paper and that of Nitta and Yamada, however, examine only data since the 1950s and focus on annual mean conditions. The study of Parker et al. provides more information, by addressing historical global surface temperature data since the 1880s and by providing seasonal resolution from the 1930s. For the SH overall, and within the constraints of data reliability and sparsity, the major SST fluctuations on decadal to multidecadal timescales are found across the southern midlatitudes of the ocean basins. This is especially evident over the southern Indian Ocean, a region that appears to be particularly sensitive to climatic fluctuations (Allan et al. 1995 and Reason et al. 1996).

Supporting evidence from studies of other climatic variables is basically restricted to efforts to resolve patterns in Southern Hemisphere MSLP observations (Allan 1993; Karoly et al. 1996). This approach tends to confirm the suggestion, noted in the analysis of filtered signals of ENSO variables, that lower-frequency MSLP fluctuations show reduced variance and a different structure in the epoch between the two World Wars than both prior to and since that period. In their study of rotated EOF modes in Southern Hemisphere MSLP, Karoly et al. (1996) also address the question of the nature of the so called "Trans-Polar Index" (TPI), identified and discussed by Pittock (1980a,b, 1984). The TPI, which measures the zonal wavenumber 1 asymmetry of the SH circumpolar vortex via the difference in MSLP between Hobart (Australia) and Stanley (Falkland Islands), has been shown to have little relationship to the various measures of ENSO. In fact, despite being resolved by the second rotated EOF in Karoly et al. (1996), relatively little use has been made of the TPI in climate studies. Nevertheless, the above study suggests that the TPI is perhaps indicative of a low-frequency climatic mode that is independent of the decadal to multidecadal mode associated with ENSO. Stronger relationships on interannual timescales (3.3–3.6 yr) between the TPI and ENSO in the austral summer are noted in Villalba et al. (1997). More-detailed resolution of such climatic

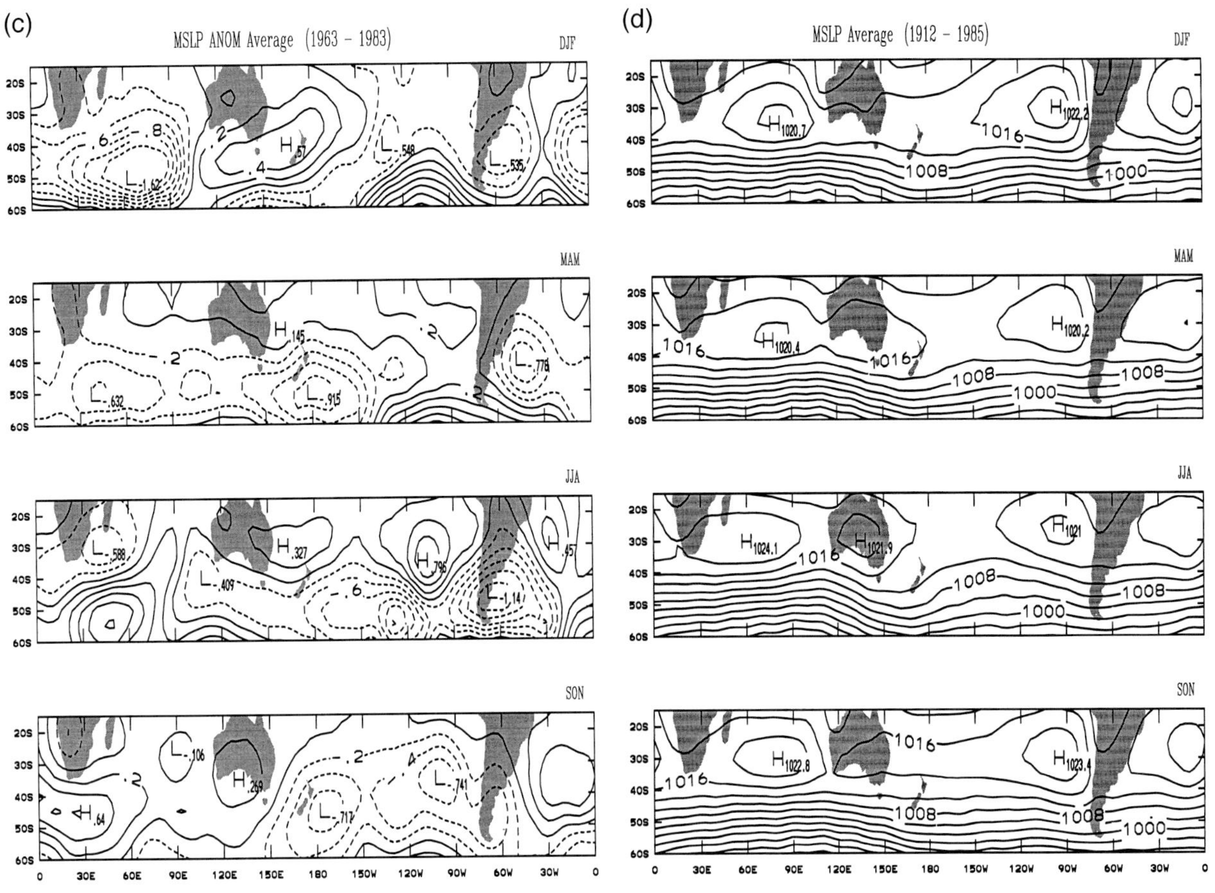

FIG. 9.14. Continued.

fluctuations requires gridded fields of SST, MSLP, and ENSO teleconnections (Allan 1993). In this analysis, the essence of SH variability is captured by the patterns of MSLP and SST on a seasonal basis in Figs. 9.14 and 9.15. Given the nature of the multidecadal signals observed in Fig. 9.12, and to facilitate easy intercomparison, the length of reduced ENSO—the activity period in the 1920s to 1930s (21 yr)—is used to determine equal intervals in the record. The MSLP anomalies examined are from (P. D.) Jones (1991) for the subtropical to higher latitudes of the SH, while SST anomalies are derived from the global sea-ice and sea surface temperature (GISST) dataset (Parker et al. 1994). MSLP data cover 21-yr epochs since 1921, while GISST is a longer compilation with 21-yr epochs shown since 1900 (overlapping from 1921 onward with the MSLP data).

An examination of the MSLP anomalies in Fig. 9.14 shows that, in general, the most prominent features are found at middle to higher latitudes over the SH. Anomalies in the southern African to Indian Ocean sector tend to be centered between 40° and 60°S, suggestive of changes in the circumpolar vortex and longwave structure. In the austral summer (DJF), autumn (MAM), and winter (JJA) seasons, there is an indication of a distinct alternation in the MSLP anomaly pattern over the three epochs shown. Few MSLP anomalies are found to be associated directly with the semipermanent anticyclone in the southern Indian Ocean in any season. As mentioned previously, however, this anticyclonic feature was found to display considerable multidecadal variability during the austral summer in the multiparameter studies by Allan et al. (1995). Such differences in these analyses may result from no shipborne MSLP observations being used in the (P. D.) Jones (1991) data. A general reduction in data reliability over SH oceanic basins as a consequence of the reliance on land- and island-based data is noted by Jones in the above paper. The atlas of Allan et al. (1996) should provide global MSLP coverage that can help to resolve such problems.

Over Australasia, MSLP anomalies are mostly found immediately to the south-southwest of the Australian continent or in the Tasman Sea to New Zealand region. There is no distinct evidence for latitudinal displacements of the high-pressure belt between the epochs, but rather suggestions of longitudinal variability and fluctuations in the longwave troughs to the southwest and southeast of Australia. As in the Indian Ocean, there are indications of distinct alternations in

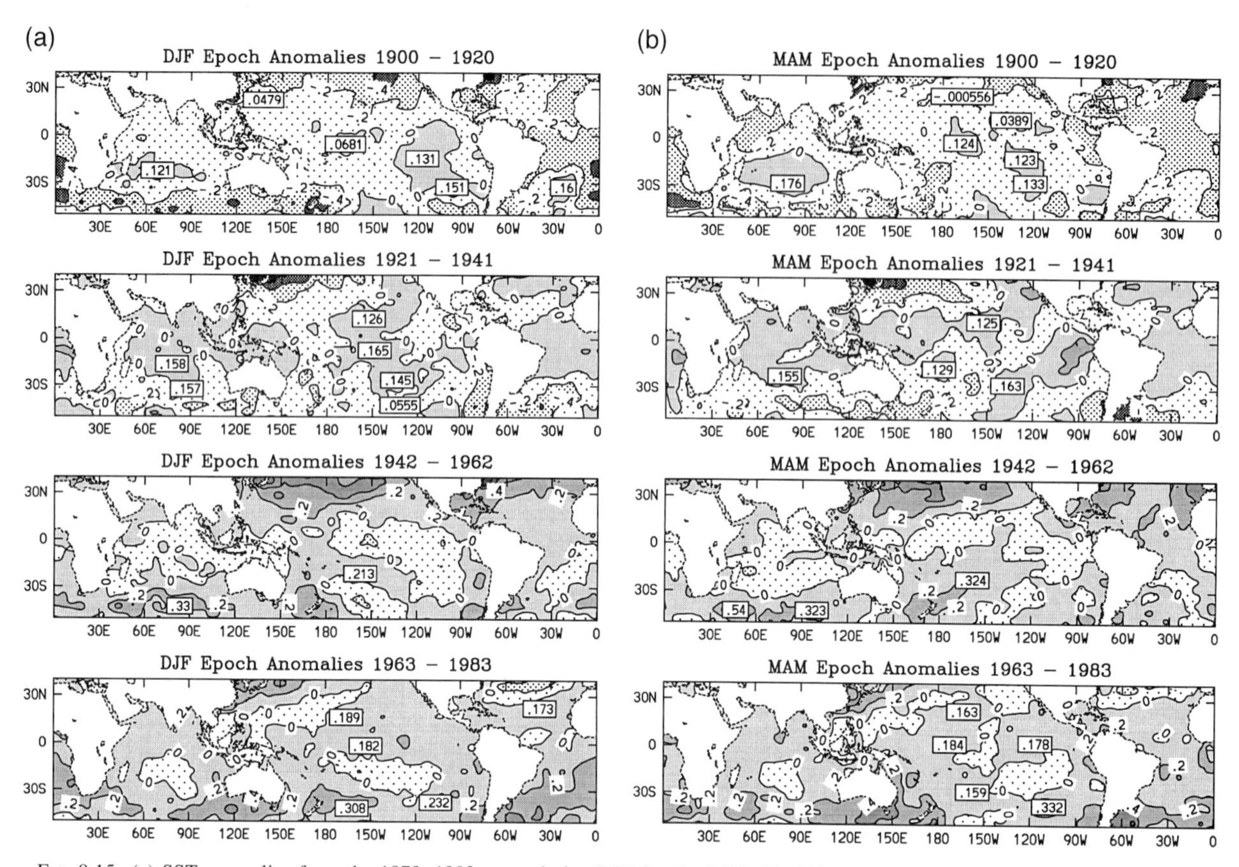

FIG. 9.15. (a) SST anomalies from the 1879–1983 mean during DJF for the 1900–20, 1921–41, 1942–62, and 1963–83 epochs. (b) As in (a), except during MAM. (c) As in (a), except for JJA. (d) As in (a), except for SON. Positive SST anomalies are shown by shading; negative SST anomalies are shown by dot stippling. Contours are every 0.2°C.

MSLP anomalies. Relationships between the southwest Australian and Tasman Sea–New Zealand troughs have been noted earlier and can be seen clearly in Fig. 9.14. As discussed in Salinger et al. (1996), these troughs are positively and significantly correlated in austral spring (SON) and summer seasons, and poorly related during the remainder of the year. Despite encroaching into the oceanic regimes around Australia and New Zealand, these longwave trough features are consistent with analyzed seasonal patterns deduced from synoptic charts produced by the South African Weather Bureau and the Australian Bureau of Meteorology [(P. D.) Jones 1991].

Over the Pacific Ocean to South American sector, Fig. 9.14 provides the most definitive evidence for changes in the latitudinal position of a major climatic feature. The southeastern Pacific anticyclone appears to be the most mobile closed circulation in the midlatitude regions of the SH. During the epochs shown, it is found to wax and wane in magnitude and be displaced latitudinally and longitudinally. As mentioned earlier, however, the sparsity of oceanic data in the Jones MSLP set must be considered when assessing the nature of the southeastern Pacific anticyclone. As has been shown by a number of compilations of atmospheric or oceanic data (viz. Bottomley et al. 1990; Parker et al. 1994), the southeastern Pacific Ocean is one of the most data-sparse regions of the globe. Nevertheless, examinations of coastal Chilean data and island stations, in studies detailed earlier, tend to support much of the above contentions about the variability of anticyclonic circulation in this region.

In general, the epoch analysis of MSLP anomalies and mean conditions in Fig. 9.14 indicates that the long-term Southern Hemisphere MSLP patterns, indicative of atmospheric circulation, have undergone periods of higher and lower intensities, with pronounced variations in the longwave patterns and circumpolar vortex structure. The recent 1963–83 epoch provides the most solid evidence for a period of more meridional SH circulation. On the other hand, the 1921–41 epoch has a more zonal circulation pattern around the hemisphere. Further verification of such findings will be available through the new, near-global MSLP datasets using both terrestrial and shipborne observations (Allan et al. 1996).

Epoch SST anomalies for each season since 1900 are shown in Fig. 9.15. As noted earlier in this subsection, SST anomalies on these multidecadal timescales show a tendency to be most prominent at mid-

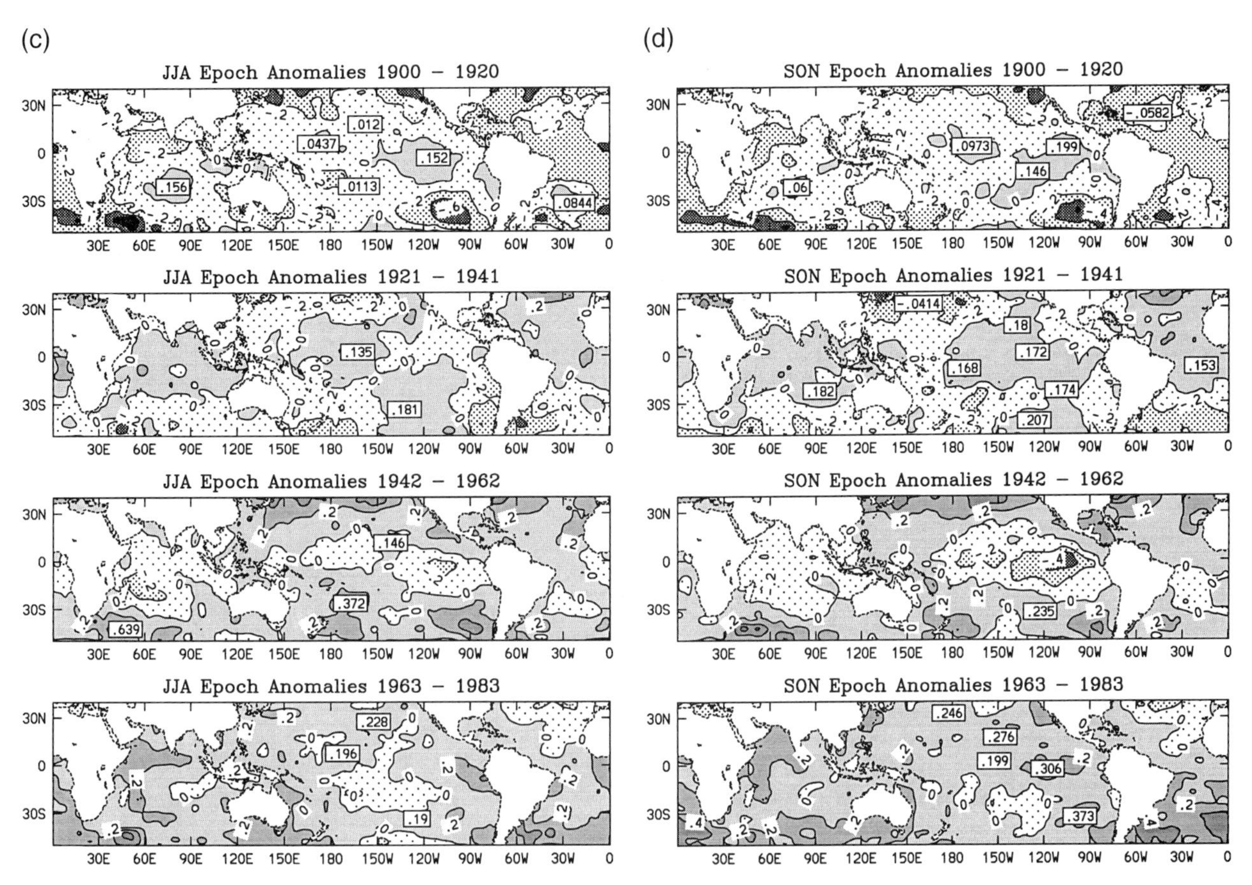

FIG. 9.15. Continued.

to higher latitudes in either hemisphere. Over the SH, the strongest anomalies are found across the southern margins of the Indian and Atlantic Oceans and, to a lesser degree, the Pacific Ocean in all seasons. In general, anomalies in the above locations are found to be negative in the 1900–20 and 1921–41 epochs and positive in the 1942–62 and 1963–83 epochs. As is also evident, such general patterns of cooling and warming are part of wider responses in each of the ocean basins. This has a structure that is typically one of the opposite, but weaker, anomalies in tropical to subtropical latitudes, to those at mid- to higher latitudes. Anomaly patterns of the type described above would appear to be suggestive of an organized, low-frequency forcing on the oceans.

An examination of both the MSLP and SST anomaly patterns in Figs. 9.14 and 9.15 since 1921 doesn't provide an immediately visible link suggestive of ocean–atmosphere interaction mechanisms. The broad changes in the zonality of the circumpolar flow, however, noted earlier, do coincide with changes in the SST at mid- to higher latitudes. As noted in Allan et al. (1995), analyses of the changes in a number of oceanic and atmospheric variables show only a coherency in broad responses and require careful modeling simulations to determine whether physical forcings operate to produce the observed patterns (Allan et al. 1996). Thus, careful observational studies are needed to resolve the types of changes apparent in oceanic and atmospheric variables, with the reality of the linkage being examined through modeling studies. What is apparent when longer records are analyzed is that the climate system is not only modulated by interannual forcings but also by forcing operating on decadal–multidecadal time frames. The unraveling of such natural fluctuations from those expected from an anthropogenic source is an even more challenging task.

9.4. Conclusions

Evidence from the instrumental period and earlier paleoclimatic sources indicates that the climate of the Southern Hemisphere has varied over the last few thousand years. Since the mid-nineteenth century, the average temperature has risen by about 0.6°C, the rise being much more like a linear trend than in the Northern Hemisphere. All seasons show a similar rise in temperature. Although most regions show the rise in temperature, the characteristics of most regional temperature and precipitation series relate to their location with respect to the ENSO phenomenon. Both in-phase and out-of-phase relationships are evident on the interannual and decadal timescales. Few regions in the hemisphere seem immune from the ENSO influence. Although claims have been made, the influence is weak in South America south of 40°S, Antarctica, and southwestern parts of Africa.

Each new ENSO event provides new insights into the phenomena. The most recent extended negative phase is probably the longest yet recorded. Not only is our understanding of the linkages and teleconnections on the interannual timescale being improved, but recent work is highlighting how decadal-scale fluctuations in ENSO are also manifest in different regions of the hemisphere. The twentieth century began with a period of strong ENSO activity between 1900 and 1920, being followed by one of much reduced activity between 1921 and 1941. In this, the main ENSO events were generally smaller in magnitude and there was a greater length of time between events. The frequency and magnitude of events then increased between 1942 and 1962 to the levels between 1963 and 1983. Recent ENSO variability has probably been the greatest this century, comparable to the levels of the late nineteenth century.

Analyses of the available historical instrumental records of atmospheric and oceanic circulation parameters over the Southern Hemisphere indicate that the most significant changes have occurred with regard to zonality of atmospheric flow and fluctuations in mid- to higher-latitude SST. During the periods 1900–20 and 1921–41, there was a tendency for the strongest signal to be of cooler SSTs at mid- to higher latitudes of the southern margins of the Indian, Pacific, and Atlantic Oceans. Atmospheric circulation data available during the 1921–41 period show more zonal circulation around the Southern Hemisphere. For the later 1942–62 and 1963–83 epochs, the reverse situation applies, with warmer SSTs at mid- to higher latitudes of the southern margins of the Indian, Pacific, and Atlantic Oceans and more meridional hemispheric circulation. The atmospheric circulation changes across the midlatitudes of the Southern Hemisphere tend to be manifest by fluctuations in the magnitude of the longwave troughs to the west and east of the Australian continent and southern Africa, and through fluctuations in the position and magnitude of the southeastern Pacific anticyclone. Analyses of the austral winter anticyclone over Australia show only variations in magnitude between the various epochs. A specific situation with regard to austral winter rainfall over southwestern Australia appears to be related to fluctuations involving the longwave trough to the southwest of Australia and the continental anticyclone and those relating to the longwave pattern in the wider Oceania region.

ENSO relationships with the observed multidecadal fluctuations of the climate system take the form of waxings and wanings in the magnitude of the QBO and LF components of the phenomenon. The 1921–41 period shows much weaker ENSO characteristics compared to the stronger, robust ENSO signal in the 1963–83 epoch. More data are needed to resolve any possible tendency for persistent ENSO sequences of either phase to be more predominant in times of reduced ENSO signal.

The major challenge for future research is to provide some understanding of the physical processes underlying interannual to multidecadal fluctuations in the climate system. This is beginning to occur, as improved observational datasets become available and are used in conjunction with the results of various GCM simulations of climatic features and phenomena.

Paleoclimatic data is our main hope of extending the climatic records back in time to see how unusual the twentieth century has been in a longer context. The dearth of land regions means that paleoclimatology in the hemisphere is only just beginning to challenge some of the concepts about past warm and cold periods apparent in Northern Hemisphere– and, particularly, North Atlantic–based records. The Southern Hemisphere is endowed with three types of records: coral records from the tropical oceans; dendroclimatic records from Tasmania, New Zealand, and southern South America; and ice-core records from Antarctica and a few tropical ice caps, particularly in South America. Historical or documentary records are generally not of relevance here except in some parts of South America. The coral records will be particularly important in studying ENSO variability over the last 500 years. The challenge to paleoclimatologists in this hemisphere is to analyze all the records independently of studies in the Northern Hemisphere, particularly for the time frame of the last few thousand years.

Acknowledgments. The authors wish to thank David Karoly and Neville Nicholls for comments on earlier drafts of this chapter. We also thank many individuals for the provision of the many climatic and paleoclimatic series, specifically Alberto Aristarain, Ed Cook, Rosanne D'Arrigo, Rob Dunbar, Mike Hulme, Jean Jouzel, Antonio Lara, Janice Lough, Vin Morgan, Ellen Mosley-Thompson, David Peel, Lonnie Thompson, and Ricardo Villalba. P. D. Jones is supported by the U.S. Department of Energy, Atmospheric and Climate Research Division, under Grant No. DE-FG02-86ER60397. R. J. Allan is supported jointly through the CSIRO "Climate Variability and Impacts Programme" and "Climate Change and Impacts Programme," and funded in part by the Australian Federal Government's National Greenhouse Research Programme of the Australian Greenhouse Office, and the State Government of Queensland.

REFERENCES

Aceituno, P., 1988: On the functioning of the Southern Oscillation in the South American sector. Part I: Surface climate. *Mon. Wea. Rev.,* **116,** 505–524.

Allan, R. J., 1993: Historical fluctuations in ENSO and teleconnection structure since 1879: Near-global patterns. *Quat. Aust.,* **11,** 17–27.

———, and M. R. Haylock, 1993: Circulation features associated with the winter rainfall decrease in southwestern Australia. *J. Climate,* **6,** 1356–1367.

———, and R. D. D'Arrigo, 1998: "Persistent" ENSO sequences: How unusual was the recent El Nino? *The Holocene,* in press.

———, J. A. Lindesay, and C. J. C. Reason, 1995: Multidecadal variability in the climate system over the Indian Ocean region during the austral summer. *J. Climate,* **8,** 1853–1873.

———, ———, and D. E. Parker, 1996: *The El Niño Southern Oscillation and Climatic Variability.* CSIRO Publishing, 405 pp.

Barnett, T. P., 1991: The interaction of multiple timescales in the tropical climate system. *J. Climate,* **4,** 269–285.

———, and P. D. Jones, 1992: Intercomparison of two different Southern Hemisphere sea level pressure datasets. *J. Climate,* **5,** 92–99.

———, A. D. Del Genio, and R. A. Ruedy, 1992: Unforced decadal fluctuations in a coupled model of the atmosphere and ocean mixed layer. *J. Geophys. Res.,* **97,** 7341–7354.

Berglund, B. E., Ed., 1986: *Handbook of Holocene Paleoecology and Paleohydrology.* Wiley, 869 pp.

Berlage, H. P., 1931: Over het verband tusschen de dikte derjaaringen van djatiboomen (*Tectona grandis L.F.*) en den regenval op Java. [On the relationship between thickness of tree rings of Djati (teak) trees and rainfall in Java.] *Tectona,* **24,** 939–953 (in Dutch).

Bigg, G. R., 1992: Validation of trends in the surface wind field over the Mozambique Channel. *Int. J. Climatol.,* **12,** 829–838.

———, 1993: Comparison of coastal wind and pressure trends over the tropical Atlantic: 1946–1987. *Int. J. Climatol.,* **13,** 411–421.

———, K. D. B. Whysall, and A. S. Cooper, 1987: Long-term trends and a major climate anomaly in the tropical Pacific wind-field. *Trop. Ocean–Atmos. News.,* **37,** 1–2.

Boninsegna, J. A., 1992: South American dendroclimatological records. *Climate Since A.D. 1500,* R. S. Bradley and P. D. Jones, Eds., Routledge, 446–462.

Bottomley, M., C. K. Folland, J. Hsiung, R. E. Newell, and D. E. Parker, 1990: *Global Ocean Surface Temperature Atlas (GOSTA).* Meteorological Office and Massachusetts Institute of Technology, 313 plates, 20 pp.

Bradley, R. S., 1985: *Quaternary Paleoclimatology: Methods of Paleoclimatic Reconstruction.* Allen and Unwin, 472 pp.

———, and P. D. Jones, 1993: "Little Ice Age" summer temperature variations: Their nature and relevance to recent global warming trends. *The Holocene,* **3,** 367–376.

Briffa, K. R., and P. D. Jones, 1993: Global surface air temperature variations during the twentieth century: Part 2, Implications for large-scale high-frequency paleoclimatic studies. *The Holocene,* **3,** 82–93.

———, ———, F. H. Schweingruber, W. Karlén, and S. G. Shiyatov, 1996: Tree-ring variables as proxy-climatic indicators: Problems with low-frequency signals. *Climatic Variability and Forcing Mechanisms of the Last 2000 Years,* P. D. Jones, R. S. Bradley, and J. Jouzel, Eds., Springer-Verlag, 9–41.

Cardone, V. J., J. G. Greenwood, and M. Cane, 1990: On trends in historical marine wind data. *J. Climate,* **3,** 113–127.

Carleton, A. M., 1989: Antarctic sea-ice relationships with indices of the atmospheric circulation of the Southern Hemisphere. *Climate Dyn.,* **3,** 207–220.

Cole, J., 1996: Coral records of climatic change: Understanding past variability in the tropical ocean–atmosphere. *Climatic Variability and Forcing Mechanisms of the Last 2000 Years,* P. D. Jones, R. S. Bradley, and J. Jouzel, Eds., Springer-Verlag, 331–353.

Cook, E. R., 1995: Temperature histories from tree rings and corals. *Climate Dyn.,* **11,** 211–222.

———, T. Bird, M. Peterson, M. Barbetti, B. Buckley, R. D. D'Arrigo, R. Francey, and P. Jans, 1991: Climatic change in Tasmania inferred from a 1089-year tree-ring chronology of subalpine huon pine. *Science,* **253,** 1266–1268.

———, ———, ———, ———, ———, ———, ———, 1992: Climatic change over the last millenium in Tasmania reconstructed from tree-rings. *The Holocene,* **2,** 285–217.

———, B. M. Buckley, and R. D. D'Arrigo, 1996: Inter-decadal climate oscillations in the Tasmanian sector of the Southern Hemisphere: Evidence from tree rings over the past three millennia. *Climatic Variability and Forcing Mechanisms of the Last 2000 Years,* P. D. Jones, R. S. Bradley, and J. Jouzel, Eds., Springer-Verlag, 141–160.

Cooper, N. S., K. D. B. Whysall, and G. R. Bigg, 1989: Recent decadal climate variations in the tropical Pacific. *Int. J. Climatol.,* **9,** 221–242.

Crowley, T. J., and K.-Y. Kim, 1993: Towards development of a strategy for determining the origin of decadal-centennial scale climate variability. *Quat. Sci. Rev.,* **12,** 375–385.

D'Arrigo, R. D., B. M. Buckley, E. R. Cook, and W. S. Wagner, 1995: Temperature sensitive tree-ring width chronologies of pink pine (Halocarpus biformis) from Stewart Island, New Zealand. *Paleogeogr. Paleoclimatol. Paleoecol.,* **119,** 293–300.

de Boer, H. J., 1951: Tree-ring measurements and weather fluctuations in Java from A.D. 1514. *Proc. Konink. Neder. Akad. Weterschappen,* **B54,** 194 (in Dutch).

Delecluse, P., J. Servain, C. Levy, K. Arpe, and L. Bengtsson, 1994: On the connection between the 1984 Atlantic warm event and the 1982–1983 ENSO. *Tellus,* **46A,** 448–464.

Diaz, H. F., and R. S. Pulwarty, 1994: An analysis of the time scales of variability in centuries-long ENSO-sensitive records in the last 1000 years. *Clim. Change,* **26,** 317–342.

Druffel, E. R. M., and S. Griffin, 1993: Large variations of surface ocean radiocarbon: Evidence of circulation changes in the southwest Pacific. *J. Geophys. Res.,* **98,** 20 249–20 259.

Dunbar, R. B., and G. M. Wellington, 1981: Stable isotopes in a branching coral monitor seasonal temperature variation. *Nature,* **293,** 453–455.

———, and J. E. Cole, 1993: *Coral Records of Ocean–Atmosphere Variability.* NOAA Climate and Global Change Program Special Rep. No. 10, 37 pp.

———, G. M. Wellington, M. W. Colgan, and P. W. Glynn, 1994: Eastern Pacific sea surface temperature since 1600 A.D. The $\delta^{18}0$ record of climate variability in the Galapagos corals. *Paleooceanogr.,* **9,** 291–315.

———, B. K. Linsley, and G. M. Wellington, 1996: Eastern Pacific corals monitor El Niño/Southern Oscillation, precipitation and sea surface temperature variability over the past three centuries. *Climatic Variability and Forcing Mechanisms of the Last 2000 Years,* P. D. Jones, R. S. Bradley, and J. Jouzel, Eds., Springer-Verlag, 373–405.

Easterling, D. R., and T. C. Peterson, 1995: A new method for detecting undocumented discontinuities in climatological time series. *Int. J. Climatol.,* **15,** 367–377.

———, and Coauthors, 1997: Maximum and minimum temperature trends for the globe. *Science,* **277,** 364–367.

Elliott, W. P., and J. K. Angell, 1987: The relation between Indian monsoon rainfall, the Southern Oscillation, and hemispheric air and sea temperature: 1884–1984. *J. Climate Appl. Meteor.,* **26,** 943–948.

———, and ———, 1988: Evidence for changes in Southern Oscillation relationships during the last 100 years. *J. Climate,* **1,** 729–737.

Enfield, D. B., 1988: Is El Niño becoming more common? *Oceanogr. Mag.,* **1**(2), 23–27, 59.

———, 1989: El Niño, past and present. *Rev. Geophys.,* **27**(1), 159–187.

———, and L. S. Cid, 1991: Low-frequency changes in El Niño–Southern Oscillation. *J. Climate,* **4,** 1137–1146.

Enomoto, H., 1991: Fluctuations of snow accumulation in the Antarctic and sea level pressure in the Southern Hemisphere in the last 100 years. *Clim. Change,* **18,** 67–87.

Fairbridge, R. G., and R. E. Dodge, 1979: Annual periodicity of the 0-18/0-16 and C-13/C-12 ratios in the coral *Mantastrea annularis. Geochim. Cosmochim. Acta,* **43,** 1009–1020.

Fitzharris, B. B., J. E. Hay, and P. D. Jones, 1992: Behaviour of New Zealand glaciers and atmospheric circulation changes over the past 130 years. *The Holocene,* **2,** 97–106.

Fletcher, J. O., U. Radok, and R. Slutz, 1982: Climatic signals of the Antarctic Ocean. *J. Geophys. Res.,* **87,** 4269–4276.

Folland, C. K., and D. E. Parker, 1995: Correction of instrumental biases in historical sea surface temperature data. *Quart. J. Roy. Meteor. Soc.,* **121,** 319–367.

———, and M. J. Salinger, 1995: Surface temperature trends and variations in New Zealand and the surrounding ocean, 1871–1993. *Int. J. Climatol.,* **15,** 1195–1218.

———, T. N. Palmer, and D. E. Parker, 1986: Sahel rainfall and worldwide sea temperatures, 1901–85. *Nature,* **320,** 602–607.

———, T. R. Karl, and K. Ya. Vinnikov, 1990: Observed climate variations and change. *Climatic Change: The IPCC Scientific Assessment,* J. T. Houghton, G. J. Jenkins, and J. J. Ephraums, Eds., Cambridge University Press, 195–238.

———, ———, N. Nicholls, B. S. Nyenzi, D. E. Parker, and K. Ya. Vinnkov, 1992: Observed climate variability and change. *The Supplementary Report to the IPCC Scientific Assessment,* J. T. Houghton, B. A. Callander, and S. K. Varney, Eds., Cambridge University Press, 135–170.

Fu, C., 1986: A review of studies of El Niño–Southern Oscillation phenomenon associated with interannual climate variability. *Chinese J. Atmos. Sci.,* **11,** 237–253.

———, H. F. Diaz, and J. O. Fletcher, 1986: Characteristics of the response of sea surface temperature in the central Pacific associated with warm episodes of the Southern Oscillation. *Mon. Wea. Rev.,* **114,** 1716–1738.

Garcia, N. O., and W. M. Vargas, 1998: The temporal climatic variability in the Rio de la Plata basin displayed by the river discharges. *Clim. Change,* **38,** 359–379.

Graham, N. E., 1994: Decadal-scale climate variability in the tropical and North Pacific during the 1970s and 1980s: Observations and model results. *Climate Dyn.,* **10,** 135–162.

———, T. P. Barnett, R. Wilde, M. Ponater, and S. Schubert, 1994: On the roles of tropical and midlatitude SSTs in forcing interannual to interdecadal variability in the winter Northern Hemisphere circulation. *J. Climate,* **7,** 1416–1441.

Gu, D., and S. G. H. Philander, 1995: Secular changes of annual and interannual variability in the Tropics during the past century. *J. Climate,* **8,** 864–876.

Guymer, L. B., and J. F. Le Marshall, 1980: The impact of FGGE buoy data on Southern Hemisphere analysis. *Aust. Meteor. Mag.,* **28,** 19–42.

Hahn, C. J., S. G. Warren, and J. London, 1994: Climatological data for clouds over the globe from surface observations: The total cloud edition. CDIAC NDP-026A, ORNL/CDIAC-72. [Available from Carbon Dioxide Information Analysis Center, Oak Ridge, TN 37831-6335.]

Halpert, M. S., and C. F. Ropelewski, 1992: Surface temperature patterns associated with the Southern Oscillation. *J. Climate,* **5,** 577–593.

Hansen, J., and S. Lebedeff, 1987: Global trends of measured surface air temperature. *J. Geophys. Res.,* **92,** 13 345–13 372.

Harrison, D. E., and N. K. Larkin, 1997: Darwin sea level pressure, 1876–1996: Evidence for climate change? *Geophys. Res. Lett.,* **24,** 1779–1782.

Hastenrath, S., 1990: Decadal-scale changes of the circulation in the tropical Atlantic sector associated with Sahel drought. *Int. J. Climatol.,* **10,** 459–472.

———, 1992: Greenhouse indicators in Kenya. *Nature,* **355,** 503–504.

Henderson-Sellers, A., 1992: Continental cloudiness changes this century. *Geo. J.,* **27,** 255–262.

Hughes, M. K., and H. F. Diaz, 1994: Was there a medieval warm period, and if so, where and when. *Clim. Change,* **27,** 255–262.

Hulme, M., 1992: Rainfall changes in Africa: 1931–1960 to 1961–1990. *Int. J. Climatol.,* **12,** 685–699.

———, 1994: Validation of large-scale precipitation fields in general circulation models. *Global Precipitations and Climate Change,* M. Desbois and F. Désalmand, Eds., Springer-Verlag, 387–405.

Hunt, B. G., and H. L. Davies, 1997: Mechanism of multi-decadal climatic variability in a global climatic model. *Int. J. Climatol.,* **17,** 565–580.

Hurrell, J. W., and H. van Loon, 1994: A modulation of the atmospheric annual cycle in the Southern Hemisphere. *Tellus,* **46A,** 325–338.

Inoue, M., and G. R. Bigg, 1995: Trends in wind and sea-level pressure in the tropical Pacific Ocean for the period 1950–1979. *Int. J. Climatol.,* **15,** 35–52.

Jacoby, G. C., and R. D. D'Arrigo, 1990: Teak (Tectona grandis L.F.): A tropical tree of large-scale dendroclimatic potential. *Dendrochronol.,* **8,** 83–98.

Jinghua, Y., C. Longxun, and W. Gu, 1988: The propagation characteristics of interannual low-frequency oscillations in the tropical air–sea system. *Adv. Atmos. Sci.,* **5,** 405–420.

Jingxi, L., D. Yihui, and G. Shiying, 1990: Interannual low-frequency oscillations of meridional winds over the equatorial Indian–Pacific Oceans. *Acta Meteor. Sinica,* **4,** 586–597.

Jones, P. A., 1991: Historical records of cloud cover for Australia. *Aust. Meteor. Mag.,* **39,** 181–189.

———, and A. Henderson-Sellers, 1992: Historical records of cloudiness and sunshine in Australia. *J. Climate,* **5,** 260–267.

Jones, P. D., 1990: Antarctic temperatures over the present century—A study of the early expedition record. *J. Climate,* **3,** 1193–1203.

———, 1991: Southern Hemisphere sea-level pressure data: An analysis and reconstructions back to 1951 and 1911. *Int. J. Climatol.,* **11,** 585–607.

———, 1994: Hemispheric surface air temperature variations: A reanalysis and an update to 1993. *J. Climate,* **7,** 1794–1802.

———, 1995: Recent variations in mean temperature and the diurnal temperature range in the Antarctic. *Geophys. Res. Lett.,* **22,** 1345–1348.

———, and T. M. L. Wigley, 1988: Antarctic gridded sea level pressure data: An analysis and reconstruction back to 1957. *J. Climate,* **1,** 1199–1220.

———, and R. S. Bradley, 1992: Climatic variations over the last 500 years. *Climate Since A.D. 1500,* R. S. Bradley and P. D. Jones, Eds., Routledge, 649–665.

———, and K. R. Briffa, 1992: Global surface air temperature variations during the twentieth century: Part 1, Spatial, temporal and seasonal details. *The Holocene,* **2,** 165–179.

———, and ———, 1996: What can the instrumental record tell us about longer timescale paleoclimatic reconstructions. *Climatic Variations and Forcing Mechanisms of the Last 2000 Years,* P. D. Jones, R. S. Bradley, and J. Jouzel, Eds., Springer-Verlag, 625–644.

———, and M. Hulme, 1996: Methods for calculating regional precipitation and temperature time series. *Int. J. Climatol.,* **16,** 361–377.

———, S. C. B. Raper, and T. M. L. Wigley, 1986: Southern Hemisphere surface air temperature variations, 1851–1984. *J. Climate Appl. Meteor.,* **25,** 1213–1230.

———, P. Ya. Groisman, M. Coughlan, N. Plummer, W.-C. Wang, and T. R. Karl, 1990: Assessment of urbanization effects in time series of surface air temperatures over land. *Nature,* **347,** 169–172.

———, T. M. L. Wigley, and G. Farmer, 1991: Marine and land temperature data sets: A comparison and a look at recent trends. *Greenhouse-Gas-Induced Climatic Change: A Critical*

Appraisal of Simulations and Observations, M. E. Schlesinger, Ed., Elsevier, 153–172.

———, R. Marsh, T. M. L. Wigley, and D. A. Peel, 1993: Decadal timescale links between Antarctic Peninsula ice-core oxygen-18, deuterium and temperature. *The Holocene,* **3,** 14–26.

———, T. J. Osborn, and K. R. Briffa, 1997: Estimating sampling errors in large-scale temperature averages. *J. Climate,* **10,** 2548–2568.

Jouzel, J., and Coauthors, 1993: Extending the Vostok ice record of paleoclimate to the penultimate glacial period. *Nature,* **364,** 407–412.

Karl, T. R., and Coauthors, 1993: Asymmetric trends of daily maximum and minimum temperature: Empirical evidence and possible causes. *Bull. Amer. Meteor. Soc.,* **74,** 1007–1023.

Karoly, D. J., P. Hope, and P. D. Jones, 1996: Decadal variations of the Southern Hemisphere circulation. *Int. J. Climatol.,* **16,** 723–738.

Kawamura, R., 1994: A rotated EOF analysis of global sea surface temperature variability with interannual and interdecadal scales. *J. Phys. Oceanogr.,* **24,** 707–715.

King, J. C., 1994: Recent climate variability in the vicinity of the Antarctic Peninsula. *Int. J. Climatol.,* **14,** 357–369.

Kleeman, R., R. Colman, and S. B. Power, 1996: A recent change in the mean state of the Pacific Ocean: Observational evidence, atmospheric response and implications for coupled modeling. *J. Geophys. Res.,* **101,** 20 483–20 499.

Können, G. P., P. D. Jones, M. H. Kaltofen, and R. J. Allan, 1998: Pre-1866 extensions of the Southern Oscillation Index using early Indonesian and Tahitian meteorological readings. *J. Climate,* **11,** 2325–2339.

Kraus, E. B., 1954: Secular changes in the rainfall regime of S.E. Australia. *Quart. J. Roy. Meteor. Soc.,* **80,** 591–601.

———, 1955a: Secular changes of tropical rainfall regimes. *Quart. J. Roy. Meteor. Soc.,* **81,** 198–210.

———, 1955b: Secular changes of east-coast rainfall regimes. *Quart. J. Roy. Meteor. Soc.,* **81,** 430–439.

———, 1956: Secular changes in the standing circulation. *Quart. J. Roy. Meteor. Soc.,* **82,** 289–300.

LaMarche, V. C. Jr., 1975: Potential of tree-rings for reconstruction of past climatic variations in the Southern Hemisphere. *Proc. WMO/IAMAP Symp. on Long-Term Climatic Fluctuations,* WMO, 21–30.

Lamb, P. J., and R. A. Peppler, 1992: Further case studies of tropical Atlantic surface atmospheric and oceanic patterns associated with sub-Saharan drought. *J. Climate,* **5,** 476–488.

Lara, A., and R. Villalba, 1993: A 3620-year temperature record from *Fitzroya cupressoides* tree rings in southern South America. *Science,* **260,** 1104–1106.

Latif, M., and T. P. Barnett, 1994a: Causes of decadal climate variability over the North Pacific and North America. Max-Planck-Institut fur Meteorologie, Rep. No. 141, 17 pp.

———, and ———, 1994b: Causes of decadal climate variability over the North Pacific and North America. *Science,* **266,** 634–637.

Lindesay, J. A., 1984: Spatial and temporal rainfall variability over South Africa, 1963 to 1981. *S. Afr. Geogr. J.,* **66,** 168–175.

London, J., S. G. Warren, and C. J. Hahn, 1991: Thirty-year trend of observed greenhouse clouds over the tropical oceans. *Adv. Space Res.,* **11,** 45–49.

Lorius, C., J. Jouzel, and D. Raynauld, 1992: The ice-core record: Past archive of the climate and signpost to the future. *Phil. Trans. Roy. Soc. London B,* **388,** 227–234.

Lough, J. M., 1986: Tropical Atlantic sea surface temperature and rainfall variations in sub-Saharan Africa. *Mon. Wea. Rev.,* **114,** 561–570.

———, 1991: Rainfall variations in Queensland, Australia: 1891–1986. *Int. J. Climatol.,* **11,** 745–768.

———, 1992: Variations of sea-surface temperatures off north-eastern Australia and associations with rainfall in Queensland: 1956–1987. *Int. J. Climatol.,* **12,** 765–782.

———, 1993: Variations of some seasonal rainfall characteristics in Queensland, Australia: 1921–1987. *Int. J. Climatol.,* **13,** 391–409.

———, 1995: Temperature variations in a tropical–subtropical environment: Queensland, Australia, 1910–1987. *Int. J. Climatol.,* **15,** 77–95.

———, and H. C. Fritts, 1990: Historical aspects of El Nino/Southern Oscillation—Information from tree rings. *Global Ecological Consequences of the 1982–83 El Nino–Southern Oscillation,* P. W. Glynn, Ed., Elsevier Oceanography Series, 285–321.

———, D. J. Barnes, and R. B. Taylor, 1996: The potential of massive corals for the study of high-resolution climate variation in the past millennium. *Climatic Variations and Forcing Mechanisms of the Last 2000 Years,* P. D. Jones, R. S. Bradley, and J. Jouzel, Eds., Springer-Verlag, 355–369.

Madden, R. A., D. J. Shea, G. W. Branstator, J. J. Tribbia, and R. O. Weber, 1993: The effects of imperfect spatial and temporal sampling on estimates of the global mean temperature: Experiments with model data. *J. Climate,* **6,** 1057–1066.

Mann, M. E., and J. Park, 1993: Spatial correlations of interdecadal variation in global surface temperatures. *Geophys. Res. Lett.,* **20,** 1055–1058.

———, and ———, 1994a: Globally correlated variability in surface temperatures. Preprints, *6th Conf. on Climate Variations,* Nashville, TN, Amer. Meteor. Soc., 297–301.

———, and ———, 1994b: Global-scale modes of surface temperature variability on interannual to century timescales. *J. Geophys. Res.,* **99,** 25 819–25 833.

Markgraf, V., J. R. Dodson, A. P. Kershaw, M. McGlone, and N. Nicholls, 1992: Evolution of late Pleistocene and Holocene climates in the circum-South Pacific land areas. *Climate Dyn.,* **6,** 193–211.

Martin, L., M. Fournier, P. Mourguiart, A. Sifeddine, B. Turcq, M. L. Absy, and J-M. Flexor, 1993: Southern Oscillation signal in South American palaeoclimatic data of the last 7000 years. *Quat. Res.,* **39,** 338–346.

Martinson, D. G., N. G. Pisias, J. D. Hays, J. Imbrie, T. C. Moore, and N. J. Shackleton, 1987: Age dating and the orbital theory of the ice ages: Development of a high resolution 0 to 300,000 year chronostratigraphy. *Quat. Res.,* **27,** 477–482.

Mason, S. J., 1990: Temporal variability of sea surface temperatures around Southern Africa: A possible forcing mechanism for the 18-year rainfall oscillation? *S. Afr. J. Sci.,* **86,** 243–252.

———, and P. D. Tyson, 1992: The modulation of sea surface temperature and rainfall associations over southern Africa with solar activity and the quasi-biennial oscillation. *J. Geophys. Res.,* **97,** 5847–5856.

Miller, A. J., D. R. Cayan, T. P. Barnett, N. E. Graham, and J. M. Oberhuber, 1994a: Interdecadal variability of the Pacific Ocean: Model response to observed heat flux and wind stress anomalies. *Climate Dyn.,* **9,** 287–302.

———, ———, ———, ———, and ———, 1994b: The 1976–77 climate shift of the Pacific Ocean. *Oceanogr.,* **7,** 21–26.

Miller, R. L., and A. D. Del Genio, 1994: Tropical cloud feedbacks and natural variability of climate. *J. Climate,* **7,** 1388–1402.

Morgan, V., 1985: An oxygen isotope climatic record from Law Dome, Antarctica. *Clim. Change,* **7,** 415–426.

———, and T. D. van Ommen, 1997: Seasonality in late Holocene climate from ice core records. *The Holocene,* **7,** 351–354.

Mosley-Thompson, E., 1992: Paleoenvironmental conditions in Antarctica since A.D. 1500: Ice core evidence. *Climate Since A.D. 1500,* R. S. Bradley and P. D. Jones, Eds., Routledge, 572–591.

———, 1996: Rapid Holocene climate changes recorded in an East Antarctica ice core. *Climatic Variations and Forcing Mechanisms of the Last 2000 Years,* P. D. Jones, R. S. Bradley, and J. Jouzel, Eds., Springer-Verlag, 263–279.

Nicholls, N., G. V. Gruza, J. Jouzel, T. R. Karl, L. Ogallo, and D. E. Parker, 1996a: Observed climate variability and change. *Climate Change 1995: The Science of Climate Change,* J. T. Houghton, L. G. Meira Filho, B. A. Callander, N. Harris, A.

Kattenburg, and K. Maskell, Eds., Cambridge University Press, 133–192.

———, R. Tapp, K. Burrows, and D. Richards, 1996b: Historical thermometer exposures in Australia. *Int. J. Climatol.,* **16,** 705–710.

Nicholson, S. E., 1993: An overview of African rainfall fluctuations of the last decade. *J. Climate,* **6,** 1463–1466.

———, and D. Entekhabi, 1986: Quasi-periodic behaviour of rainfall variability in Africa. *Arch. Meteor. Geophys. Bioklimatol. Ser. A,* **34,** 311–348.

Nitta, T., 1993: Interannual and decadal scale variations of atmospheric temperature and circulations. *Climate Variability, Proc. Int. Workshop on Climate Variabilities,* Beijing, China, National Natural Science Foundation of China and Cosponsors, 15–22.

———, and S. Yamada, 1989: Recent warming of tropical sea surface temperature and its relationship to the Northern Hemisphere circulation. *J. Meteor. Soc. Japan,* **67,** 375–383.

———, and J. Yoshimura, 1993: Trends and interannual and interdecadal variations of global land surface air temperature. *J. Meteor. Soc. Japan,* **71,** 367–374.

Norton, D. A., and J. G. Palmer, 1992: Dendroclimatic evidence from Australasia. *Climate Since A.D. 1500,* R. S. Bradley and P. D. Jones, Eds., Routledge, 463–482.

———, K. R. Briffa, and M. J. Salinger, 1989: Reconstruction of New Zealand summer temperature to 1730 A.D. using dendroclimatic techniques. *Int. J. Climatol.,* **9,** 633–644.

Ortlieb, L., and J. Machare, 1993: Former El Niño events: Records from western South America. *Global Planet. Change,* **7,** 181–202.

Palmer, J. G., and J. G. Murphy, 1993: An extended tree-ring chronology (teak) from Java. *Proc. Konink. Neder. Akad. Wetten,* **B92,** 241–257.

Palmer, T. N., 1986: Influence of the Atlantic, Pacific and Indian Oceans on Sahel rainfall. *Nature,* **322,** 251–253.

Parker, D. E., P. D. Jones, C. K. Folland, and A. Bevan, 1994: Interdecadal changes of surface temperature since the late nineteenth century. *J. Geophys. Res.,* **99,** 14 373–14 399.

———, C. K. Folland, and M. Jackson, 1995: Marine surface temperature: Observed variations and data requirements. *Clim. Change,* **31,** 559–600.

Parthasarathy, K., K. R. Kumar, and A. A. Munot, 1991: Evidence of secular variations in Indian monsoon rainfall–circulation relationships. *J. Climate,* **4,** 927–938.

Parungo, F., J. F. Boatman, H. Sievering, S. W. Wilkison, and B. B. Hicks, 1994: Trends in global marine cloudiness and anthropogenic sulfur. *J. Climate,* **7,** 434–440.

Peel, D. A., 1992: Ice core evidence from the Antarctic Peninsula regions. *Climate Since A.D. 1500,* R. S. Bradley and P. D. Jones, Eds., Routledge, 549–571.

Peterson, T. C., and D. R. Easterling, 1994: Creation of homogeneous composite climatological reference series. *Int. J. Climatol.,* **14,** 671–680.

Pichon, J. J., L. D. Labeyrie, G. Bareille, M. Labracherie, J. Duprat, and J. Jouzel, 1992: Surface water temperature changes in the high latitudes of the Southern Hemisphere over the last glacial–interglacial cycle. *Paleooceanogr.,* **7,** 289–318.

Pittock, A. B., 1975: Climatic change and the patterns of variation in Australian rainfall. *Search,* **6,** 498–504.

———, 1980a: Patterns of climatic variation in Argentina and Chile. I. Precipitation, 1931–1960. *Mon. Wea. Rev.,* **108,** 1347–1361.

———, 1980b: Patterns of climatic variation in Argentina and Chile. II. Temperature, 1931–1960. *Mon. Wea. Rev.,* **108,** 1362–1369.

———, 1984: On the reality, stability, and usefulness of Southern Hemisphere teleconnections. *Aust. Meteor. Mag.,* **32,** 75–82.

Plummer, N., Z. Lin, and S. Torok, 1995: Recent changes in the diurnal temperature range over Australia. *Atmos. Res.,* **37,** 79–86.

Polonsky, A. B., 1994: Comparative study of the Pacific ENSO event of 1991–92 and the Atlantic ENSO-like event of 1991. *Aust. J. Mar. Freshwater Res.,* **45,** 705–725.

Postmentier, E. S., M. A. Cane, and S. E. Zebiak, 1989: Tropical Pacific climate trends since 1960. *J. Climate,* **2,** 731–736.

Quinn, W. H., and V. T. Neal, 1983: Long-term variations in the Southern Oscillation, El Niño and Chilean subtropical rainfall. *Fish. Bull.,* **81,** 363–374.

Rajagopalan, B., L. Upmanu, and M. A. Cane, 1997: Anomalous ENSO occurrences: An alternate view. *J. Climate,* **10,** 2351–2357.

Rasmusson, E. M., 1987: Tropical Pacific variations. *Nature,* **327,** 192.

———, X. Wang, and C. F. Ropelewski, 1990: The biennial component of ENSO variability. *J. Mar. Sys.,* **1,** 71–96.

Reason, C. J. C., R. J. Allan, and J. A. Lindesay, 1996: Dynamical response of the oceanic circulation and temperature to interdecadal variability in the surface winds over the Indian Ocean. *J. Climate,* **9,** 97–114.

Reynolds, R. W., and D. C. Marsico, 1993: An improved real-time global SST analysis. *J. Climate,* **6,** 114–119.

Rhoades, D. A., and M. J. Salinger, 1993: Adjustment of temperature and rainfall records for site changes. *Int. J. Climatol.,* **13,** 899–913.

Roig, F. A., C. Roig, J. Rabassa, and J. Boninsegna, 1996: Fuegian floating tree-ring chronology from sub-fossil *Nothofagus* wood. *The Holocene,* **6,** 469–476.

Ropelewski, C. F., and M. S. Halpert, 1987: Global and regional-scale precipitation patterns associated with the El Niño/Southern Oscillation. *Mon. Wea. Rev.,* **115,** 1606–1626.

———, and P. D. Jones, 1987: An extension of the Tahiti–Darwin Southern Oscillation Index. *Mon. Wea. Rev.,* **115,** 2161–2165.

———, and M. S. Halpert, 1989: Precipitation patterns associated with the high index phase of Southern Oscillation. *J. Climate,* **2,** 268–284.

———, ———, and X. Wang, 1992: Observed tropospheric biennial variability and its relationship to the Southern Oscillation. *J. Climate,* **5,** 594–614.

Rosenbluth, B., H. F. Fuenzalida, and P. Aceituno, 1997: Recent temperature variations along Chile. *Int. J. Climatol.,* **17,** 67–85.

Rubin, M. J., 1955: An analysis of pressure anomalies in the Southern Hemisphere. *Notos,* **4,** 11–16.

Rutllant, J., and H. Fuenzalida, 1991: Synoptic aspects of the central Chile rainfall variability associated with the Southern Oscillation. *Int. J. Climatol.,* **11,** 63–76.

Salinger, M. J., 1995: Southwest Pacific temperatures: Trends in maximum and minimum temperatures. *Atmos. Res.,* **37,** 87–100.

———, and P. D. Jones, 1996: Southern Hemisphere climate: The modern record. *Proc. Roy. Soc. Tasmania,* **130,** 101–107.

———, R. McGann, L. Coutts, B. Collen, and E. Fouhy, 1992: *Rainfall Trends in New Zealand and Outlying Islands, 1920–1990.* New Zealand Meteorological Service, 33 pp.

———, J. Hay, R. McGann, and B. Fitzharris, 1993: Southwest Pacific temperatures: Diurnal and seasonal trends. *Geophys. Res. Lett.,* **20,** 935–938.

———, J. G. Palmer, P. D. Jones, and K. R. Briffa, 1994: Reconstructions of New Zealand climate indices back to A.D. 1731 using dendroclimatic techniques: Some preliminary results. *Int. J. Climatol.,* **14,** 1135–1149.

———, R. E. Basher, B. B. Fitzharris, J. E. Hay, P. D. Jones, J. P. MacVeigh, and I. Schmidely-Leleu, 1995: Climate trends in the southwest Pacific. *Int. J. Climatol.,* **15,** 285–302.

———, and Coauthors, 1996: Observed variability and change in climate and sea-level in Oceania. *Greenhouse: Coping with the Climate Change,* W. J. Bourma, G. Pearman, and M. Manning, Eds., CSIRO Publishing, 100–126.

Schweingruber, F. H., and K. R. Briffa, 1996: Tree-ring density networks for climate reconstruction. *Climatic Variability and*

Forcing Mechanisms of the Last 2000 Years, P. D. Jones, R. S. Bradley, and J. Jouzel, Eds., Springer-Verlag, 43–65.

Scott, D., 1972: Correlations between tree-ring width and climate in two areas of New Zealand. *J. Roy. Soc. N.Z.,* **2,** 545–560.

Shriver, J. F., 1993: Interdecadal variability of the equatorial Pacific Ocean and atmosphere: 1930–1989. Tech. Rep., Mesoscale Air–Sea Interaction Group, Florida State University, 97 pp. [Available from the Mesoscale Air–Sea Interaction Group, Florida State University, Tallahassee, FL 32306, USA.]

Smith, T. M., R. W. Reynolds, and C. F. Ropelewski, 1994: Optimal averaging of seasonal sea surface temperatures and associated confidence intervals (1860–1989). *J. Climate,* **7,** 949–964.

Stahle, D. W., and Coauthors, 1997: Experimental multiproxy reconstructions of the Southern Oscillation. *Bull. Amer. Meteor. Soc.,* in press.

Thompson, L. G., 1992: Ice core evidence from Peru and China. *Climate Since A.D. 1500,* R. S. Bradley and P. D. Jones, Eds., Routledge, 517–548.

———, 1996: Climatic changes for the last 2000 years inferred from ice-core evidence in tropical ice cores. *Climatic Variations and Forcing Mechanisms of the Last 2000 Years,* P. D. Jones, R. S. Bradley, and J. Jouzel, Eds., Springer-Verlag, 281–295.

———, E. Mosley-Thompson, and P. A. Thompson, 1992: Reconstructing interannual climate variability from tropical and subtropical ice-core records. *El Niño: Historical and Paleoclimatic Aspects of the Southern Oscillation,* H. F. Diaz and V. Markgraf, Eds., Cambridge University Press, 295–322.

———, ———, M. Davis, P.-N. Lin, T. Yao, M. Dyurgerov, and J. Dai, 1993: "Recent warming" ice core evidence from tropical ice cores with emphasis on central Asia. *Global Planet. Change,* **7,** 145–156.

———, D. A. Peel, E. Mosley-Thompson, R. Mulvaney, J. Dai, P.-N. Lin, M. E. Davis, and C. F. Raymond, 1994: Climate since A.D. 1510 on Dyer Plateau, Antarctic Peninsula: Evidence for recent climatic change. *Ann. Glaciol.,* **20,** 420–426.

———, E. Mosley-Thompson, M. E. Davis, P.-N. Lin, K. A. Henderson, J. Cole-Dai, J. F. Bolzan, and K.-B. Lin, 1995: Late glacial stage and Holocene tropical ice core records from Huascarán, Peru. *Science,* **269,** 46–50.

Torok, S., 1996: The development of a high quality historical temperature data base for Australia. Ph.D. thesis, University of Melbourne, 298 pp.

Trenberth, K. E., and D. J. Shea, 1987: On the evolution of the Southern Oscillation. *Mon. Wea. Rev.,* **115,** 3078–3096.

———, and J. W. Hurrell, 1994: Decadal atmosphere–ocean variations in the Pacific. *Climate Dyn.,* **9,** 303–319.

———, and T. T. Hoar, 1996: The 1990–1995 El Niño–Southern Oscillation event: Longest on record. *Geophys. Res. Lett.,* **23,** 57–60.

———, J. R. Christy, and J. W. Hurrell, 1992: Monitoring global monthly mean surface temperatures. *J. Climate,* **5,** 1405–1423.

Tyson, P. D., 1986: *Climatic Change and Variability in Southern Africa.* Oxford University Press, 220 pp.

van Loon, H., J. W. Kidson, and A. B. Mullan, 1993: Decadal variation of the annual cycle in the Australian dataset. *J. Climate,* **6,** 1227–1231.

Venter, R. J., 1957: Sources of meteorological data for the Antarctic. *Meteorology of the Antarctic,* M. P. van Rooy, Ed., Weather Bureau (Pretoria, South Africa), 17–38.

Villalba, R., 1990: Climatic fluctuations in northern Patagonia during the last 1000 years as inferred from tree-ring growth. *Quat. Res.,* **34,** 346–360.

———, 1993: Tree-ring and glacial evidence for the medieval warm epoch and the Little Ice Age in southern South America. *Clim. Change,* **26,** 183–197.

———, 1994: Dendroclimatology: A Southern Hemisphere perspective. Paper prepared for the Southern Hemisphere Paleo- and Neoclimates Workshop, Proj No. 341/IGCP/IUGS/UNESCO, Mendoza, Argentina, 41 pp. [Available from Departmento de Dendrocronologia e Historia Ambientol, IANIGLA, CONICET, C.C.330, 5500 Mendoza, Argentina.]

———, E. R. Cook, R. D. D'Arrigo, G. C. Jacoby, P. D. Jones, M. J. Salinger, and J. Palmer, 1997: Sea-level pressure variability around Antarctica since A.D. 1750 inferred from subantarctic tree-ring records. *Climate Dyn.,* **13,** 375–390.

von Storch, J.-S., 1994: Interdecadal variability in a global coupled model. *Tellus,* **46A,** 419–432.

Vose, R. S., R. L. Schmoyer, P. M. Steurer, T. C. Peterson, R. Heim, T. R. Karl, and J. Eischeid, 1992: The Global Historical Climatology Network: Long-term monthly temperature precipitation, sea-level pressure and station pressure data. NDP-041, ORNL/CDIAC-53, 99 pp. [Available from Oak Ridge National Laboratory, Oak Ridge, Tennessee.]

Wang, X. L., and C. F. Ropelewski, 1995: An assessment of ENSO-scale secular variability. *J. Climate,* **8,** 1584–1599.

Ward, M. N., 1991: Worldwide ocean–atmosphere surface fields in Sahel wet and dry years using provisionally corrected surface wind data. Climate Research Tech. Note No. 20, 26 pp. [Available from the National Meteorological Library, Meteorological Office, Bracknell, Berkshire, UK.]

———, 1992: Provisionally corrected surface wind data, worldwide ocean–atmosphere surface winds, and Sahelian rainfall variability. *J. Climate,* **5,** 454–475.

———, 1994: Tropical North African rainfall and worldwide monthly to multidecadal climate variations. Ph.D. thesis, University of Reading, 313 pp.

Warren, S. G., C. J. Hahn, J. London, R. M. Chervin, and R. L. Jenne, 1986: Global distribution of total cloud cover and cloud type over land. NCAR Tech. Note TN-273+STR, 29 pp + 199 maps.

———, ———, ———, ———, and ———, 1988: Global distribution of total cloud cover and cloud type amounts over the ocean. NCAR Tech. Note TN-317+STR, 42 pp + 170 maps.

Whetton, P. H., and I. Rutherford, 1994: El Niño–Southern Oscillation teleconnections in the Eastern Hemisphere over the last 500 years. *Clim. Change,* **28,** 221–253.

———, D. Adamson, and M. A. J. Williams, 1990: Rainfall and river flow variability in Africa, Australia and east Asia linked to El Niño–Southern Oscillation events. *Lessons for Human Survival: Nature's Record,* P. Bishop, Ed., Quaternary Geological Society, 71–82.

———, R. J. Allan, and I. Rutherford, 1996: Historical ENSO teleconnections in the Eastern Hemisphere: Comparison with latest El Niño series of Quinn. *Clim. Change,* **32,** 103–109.

Whysall, K. D. B., N. S. Cooper, and G. R. Bigg, 1987: Long-term changes in the tropical Pacific surface wind field. *Nature,* **327,** 216–219.

Woodruff, S. D., R. J. Slutz, R. J. Jenne, and P. M. Steurer, 1987: A comprehensive ocean–atmosphere dataset. *Bull. Amer. Meteor. Soc.,* **68,** 1239–1250.

Wright, P. B., 1988: On the reality of climatic changes in wind over the Pacific. *Int. J. Climatol.,* **8,** 521–527.

———, 1989: Homogenized long-period Southern Oscillation indices. *Int. J. Climatol.,* **9,** 33–34.

Yu, B., and D. T. Neil, 1991: Global warming and regional rainfall: The difference between average and high intensity rainfalls. *Int. J. Climatol.,* **11,** 653–661.

———, and ———, 1993: Long-term variations in regional rainfall in the south-west of Western Australia and the difference between average and high intensity rainfalls. *Int. J. Climatol.,* **13,** 77–88.

Chapter 10

Climate Modeling*

GERALD A. MEEHL

*National Center for Atmospheric Research,** Boulder, Colorado*

10.1. Introduction

The purpose of this chapter is to describe current Southern Hemisphere climate model simulation capabilities in terms of seasonal mean quantities, the annual cycle, interannual and longer timescale variability, and possible future climate change. Climate modeling is defined here to include global climate simulations with general circulation models. Since the timescales will include monthly to seasonal to interannual to interdecadal and longer, the subject matter of this chapter is distinct from numerical weather prediction studies on shorter-than-monthly timescales. The climate models covered in this chapter involve spatial resolutions from about 2° × 2° to 5° × 5°. A number of the climate model results in this chapter will include some type of interactive ocean surface as well as specified SST experiments.

A somewhat different class of climate model than the GCMs discussed in this chapter involves embedding a regional high-resolution climate model in a global GCM (McGregor et al. 1993). This class of model is generally able to improve regional details (e.g., over a region the size of Australia) of the global GCM, even though errors in the global GCM can cause problems with the embedded model's regional simulation (Walsh and McGregor 1995). A detailed discussion of results from models of this class is beyond the scope of the present chapter, but it should be mentioned that such models are used not only to better simulate present-day climate over specific regions, but are also used in climate-change studies (McGregor and Walsh 1994). Still another class of climate model that is beyond the scope of the present chapter involves limited-domain coupled models. These are usually configured for the tropical Pacific region at high resolution and are often used as the basis for ENSO forecasting (e.g., Zebiak and Cane 1987).

The hierarchy of climate model configurations to be discussed in this chapter is given in Fig. 10.1. At the top is the representation of an atmospheric GCM with specified climatological or time-evolving SSTs. Second from top in Fig. 10.1 is a configuration with an atmospheric GCM coupled to a simple nondynamic slab ocean. In this type of model, sometimes called a "mixed-layer model" (e.g., IPCC 1990), the mixed-layer or slab ocean is of the order of 50 to 100 m deep and allows the model to be run with a seasonal cycle (the slab of water provides for a crude accounting of seasonal heat storage at each grid point, which, together with the surface energy balance, computes SSTs interactively with the atmospheric GCM). These models also usually have simple interactive thermodynamic sea–ice representations. The third type of model configuration is a so-called global coupled ocean–atmosphere GCM, sometimes simply termed "coupled model" (e.g., Washington and Meehl 1989). These models are potentially capable of representing dynamically coupled processes of the observed climate system. They are also the most computationally intensive because (as depicted in Fig. 10.1, bottom) the atmospheric GCM is coupled to a global ocean GCM. Thus, the atmospheric GCM passes net heat flux, net freshwater flux, and wind stress to the ocean and sea–ice components, and SST and sea–ice distribution is passed, in turn, back to the atmospheric model. SSTs are computed interactively between the atmosphere and ocean GCMs, with the ocean GCM accounting explicitly for heat storage, advection, and diffusion to contribute to the surface energy balance and eventual SST calculation. A brief history of the development of coupled models is given by Meehl (1990a), and a description of some of the details of coupled models in terms of coupling details, spin-up issues, climate drift, and other aspects is given by Meehl (1992). Issues involved with the mid-1990s state of the art of coupled models is provided by Meehl (1995).

---

* Portions of this manuscript were supported by the Office of Health and Environmental Research of the U.S. Department of Energy as part of its Carbon Dioxide Research Program.

** The National Center for Atmospheric Research is sponsored by the National Science Foundation.

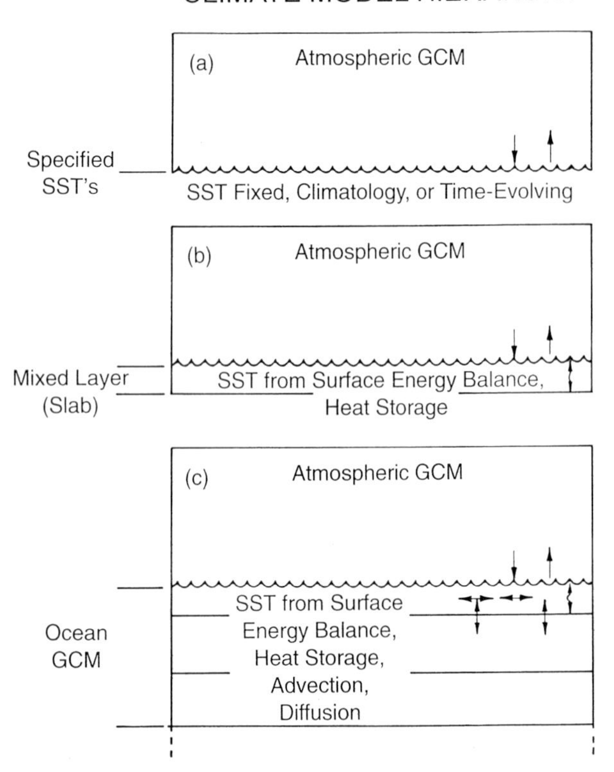

FIG. 10.1. Schematic of climate model hierarchy: (a) atmospheric GCM with specified climatological SSTs, (b) atmospheric GCM coupled to a simple mixed-layer or slab ocean, and (c) atmospheric GCM coupled to an ocean GCM.

In a coupled model, the component models are imperfect and, when coupled together interactively, the coupled simulation with an ocean formulation reflects errors in the component models (Meehl 1990a). Thus, the coupled simulation will contain systematic errors that are a product of model errors, as well as errors in coupled interactions not well represented by the coupling of the component models (e.g., Meehl 1989, 1997b). In some models, these errors remain in the simulations, and sensitivities and processes are studied in the context of those errors. In other models, these errors at the air–sea interface are corrected to produce a simulation that closely resembles the present-day SSTs (Meehl 1992). This technique is variously called "q-flux" for mixed-layer models and "flux correction" or "flux adjustment" for coupled models (these techniques are discussed in detail in Meehl 1992). Briefly, ocean and atmosphere components are run separately with observed boundary forcing from the other medium to calibrate appropriate correction terms. When the components are coupled, these corrections are added in at each grid point and time step. A flux-corrected model produces a simulation that more closely resembles present-day climate, even though all the errors in such models are present (but not visible), just as they are in the noncorrected versions.

For example, Fig. 10.2 shows SST differences between a non–flux-corrected coupled-model integration and a flux-corrected calibration integration with the same model (Murphy 1995). The high-latitude southern oceans tend to be too warm by about 1°–2°C,

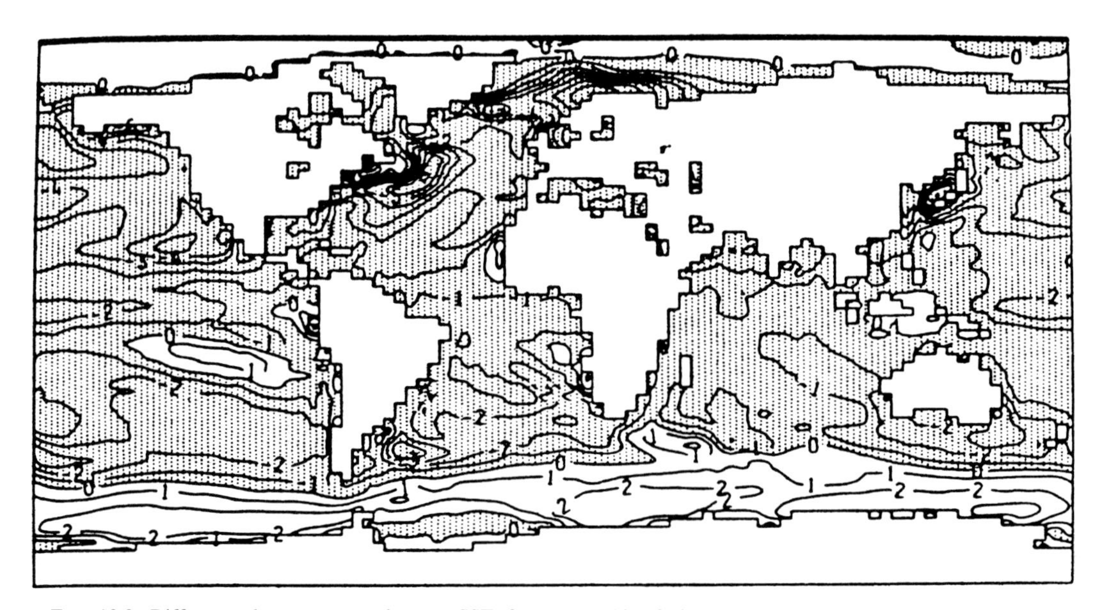

FIG. 10.2. Difference between annual-mean SST from year 10 of the coupled-model integration with no flux adjustment and the smoothed initial state, showing systematic SST errors in the coupled-model, 10-yr average. Contour interval is 1°C; negative values indicating areas where the model is colder than observations are stippled (Murphy 1995).

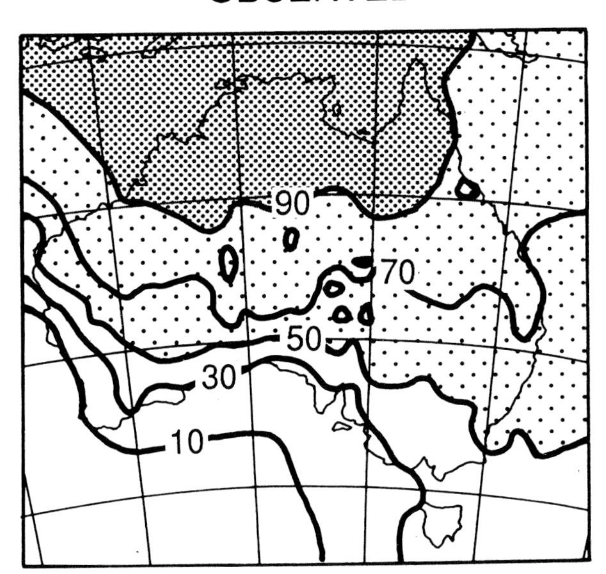

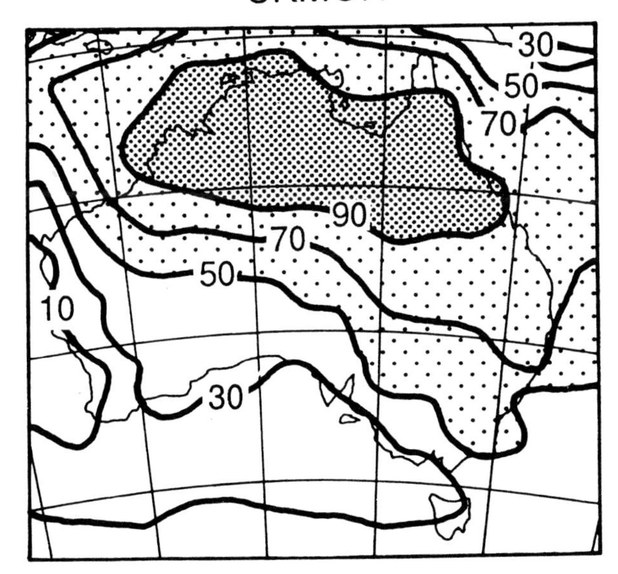

FIG. 10.3. Rainfall seasonality (DJF rainfall as a percentage of the DJF plus JJA average) over Australia for (a) observed, (b) UKMO high-resolution (2.5 × 3.75) climate model (Whetton et al. 1996c).

while the Tropics and southern subtropical oceans are too cold by about the same amount. As suggested by Fig. 10.2, the consequent reduction of baroclinicity implied by these SST errors has ramifications for the Southern Hemisphere midlatitude circulation produced

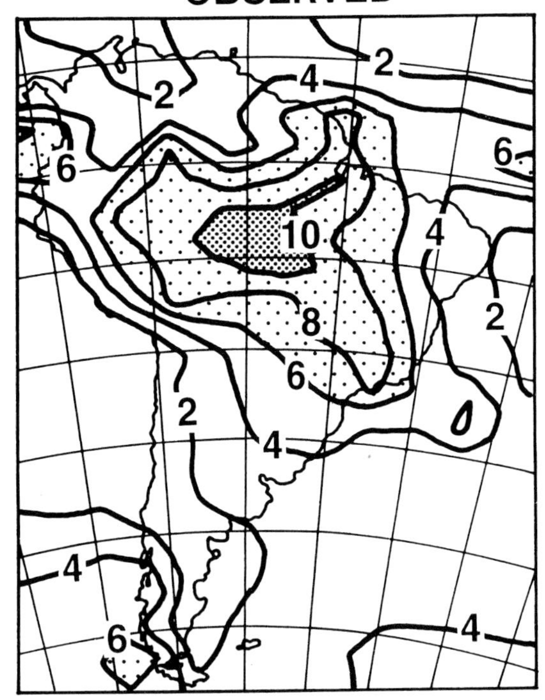

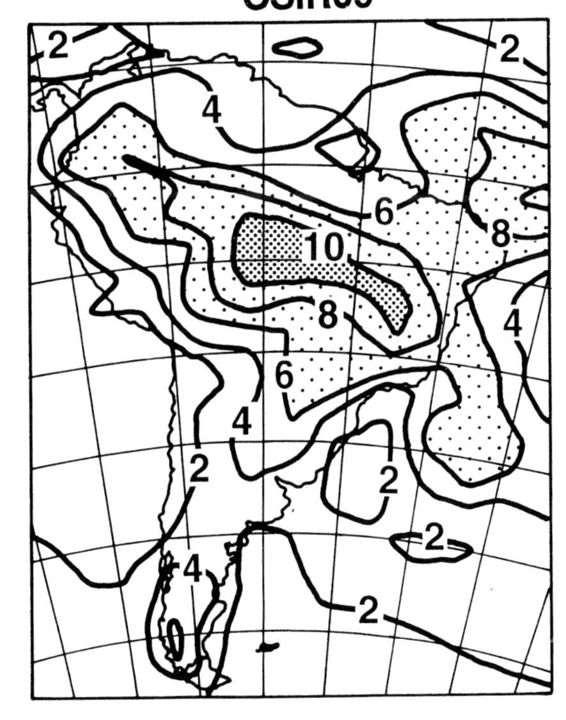

FIG. 10.4. Mean DJF precipitation in mm day^{-1} for (a) observed and (b) CSIRO (R21 resolution) climate model (Whetton et al. 1996c).

by these models. All groups do not use flux correction, because of suggestions that flux correction could adversely affect certain nonlinear interacting processes in the coupled climate system (e.g., Neelin and Dijkstra 1995). A clearer indication of model variability and sensitivity may be produced by non–flux-corrected models. Yet, since their base-state climate simulations are removed from observations, there could be questions as to how such nonlinear processes are being represented in non–flux-corrected models. The solution to this problem (discussed in Meehl 1995) is to improve model components to remove such systematic errors. Nevertheless, the present mix of flux-corrected and non–flux-corrected models requires that this issue be kept in mind when viewing climate model results.

Concerning the ocean model components in coupled models, further details on results from ocean observations and ocean GCM studies are given in chapter 7. In a number of cases, appropriate observed atmospheric quantities are provided here to compare with the climate model results, but the detailed discussions of the atmospheric observations are contained in previous chapters. In general, "low-resolution" models refer to a general class that includes roughly 500-km resolution (e.g., R15, T21), while "high-resolution" models include about 250-km resolution (e.g., R21, R30, T42). We begin with a description of typical features of mean climate simulated by climate models, including a variety of features discussed in previous chapters. Then, simulations of seasonal-cycle features are followed by results from modeling experiments examining aspects of interannual and longer timescale variability (to the multicentury timescale). Finally, results from climate model simulations of anthropogenic climate change in the Southern Hemisphere are presented.

10.2. Simulation of seasonal mean climate

This section will focus chiefly on rainfall regimes and dynamically linked circulation-related features of climate models. Further description of typical simulation errors in other variables such as temperature are given in IPCC (1996).

a. Tropical rainfall regimes

In the Southern Hemisphere Tropics, there are three continental rainfall regimes and three oceanic. Climate models can generally represent many features of these tropical rainfall regimes. Figure 10.3a shows the Australian summer monsoon-season rainfall as a percent of combined winter and summer rainfall, for observations that can be compared to a similar quantity from a high-resolution climate model in Fig. 10.3b (Whetton et al. 1996c). In general, the model is able to capture the tropical northern Australian summer-monsoon rainfall maximum that accounts for most of the annual

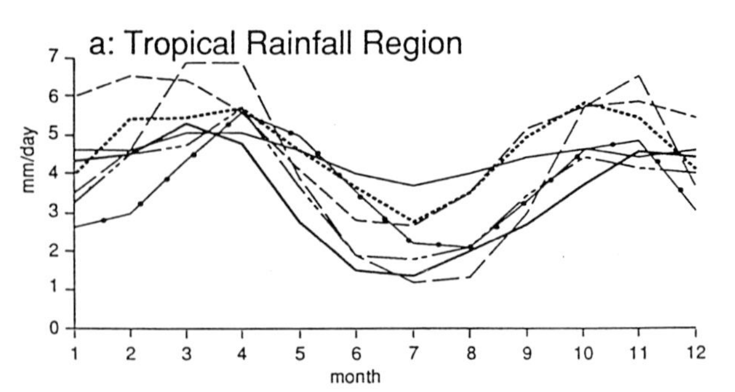

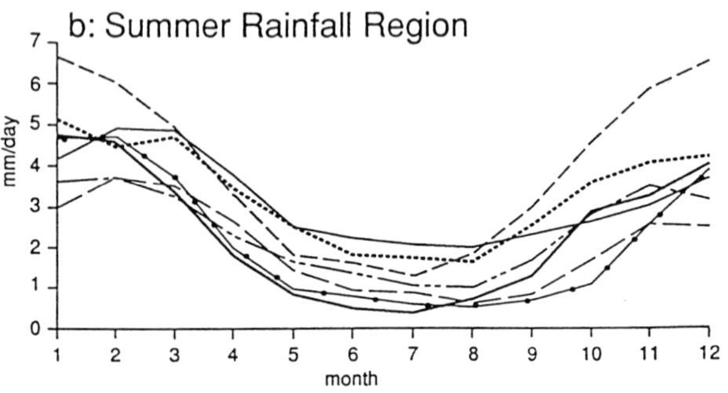

FIG. 10.5. Observed (solid line with dots) and simulated (for six coarse-resolution climate models) African monthly average-area average-precipitation rates for (a) tropical (land points over Africa between 5°N and 10°S) and (b) summer (land points over Africa between 10° and 30°S) rainfall regions in mm day^{-1} (Joubert 1994).

rainfall in that region. Average summer precipitation from observations for South America is shown in Fig. 10.4a, and a similar plot in Fig. 10.4b from another high-resolution climate model demonstrates that this model can simulate the rainfall maximum over the Amazon Basin as well as the South Atlantic convergence zone (SACZ) trailing to the southeast of the continent (Whetton et al. 1996c). Whetton et al. also pointed out less-successful model simulations of these quantities in both low- and high-resolution climate models. Almost all models can simulate some representation of the land precipitation regimes over Australia and South America, even though the complicated topography over South America presents problems for many models.

The third continental precipitation regime is over southern Africa. An indication of model-to-model variability is shown in Fig. 10.5 for the seasonality of tropical rainfall (top, area-averaged over about 5°N–10°S) and summer rainfall (bottom, 10°–30°S) for seven low-resolution climate models. All are able to simulate the gross features of seasonality of rainfall but with varying degrees of consistency with regard to amount (Joubert 1995).

The southern tropical oceanic rainfall regimes are the South Pacific convergence zone (SPCZ); SACZ, mentioned above; and the South Indian convergence zone (SICZ). These are noted in Fig. 10.6a for the observations and in Fig. 10.6b for a low-resolution

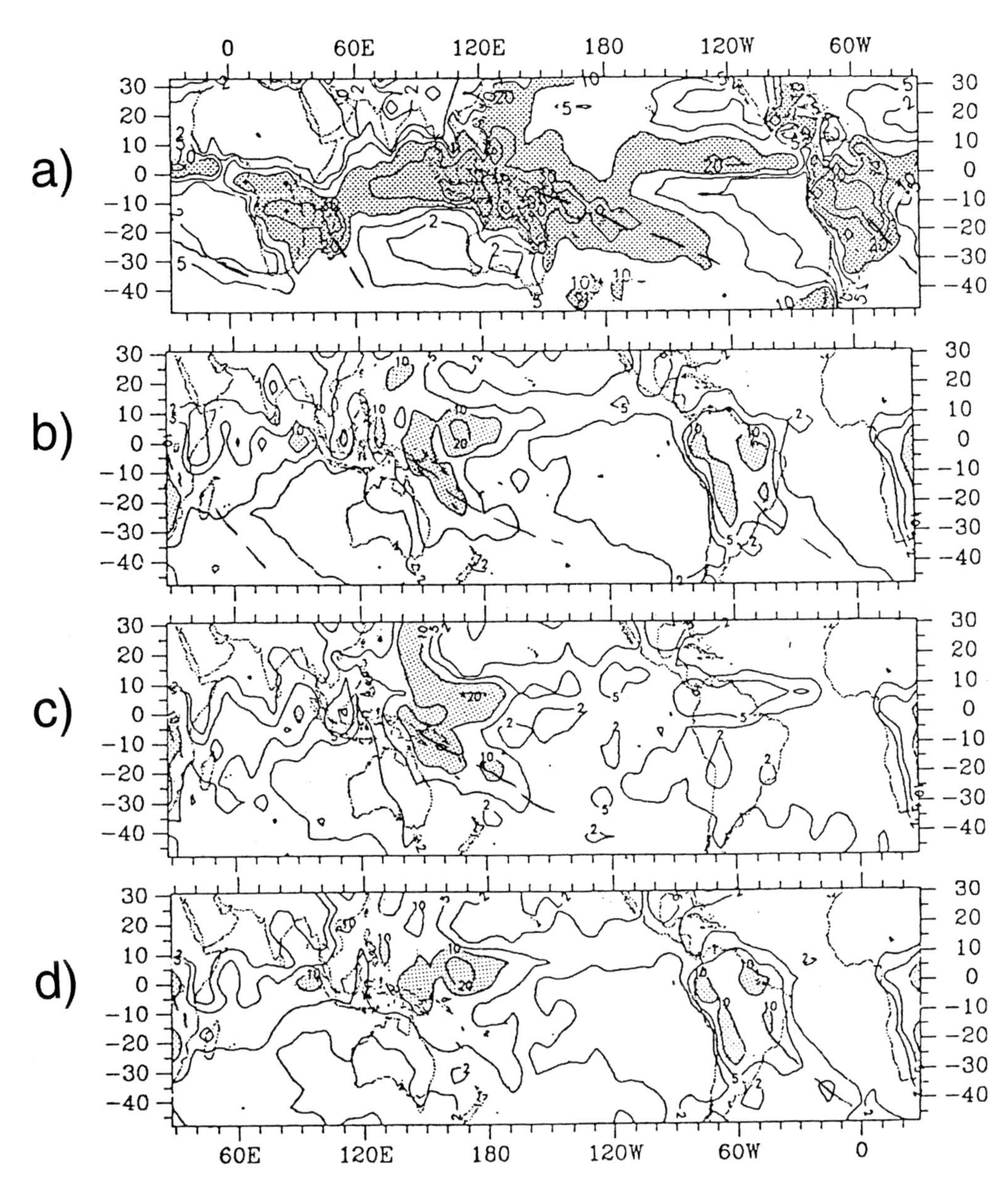

FIG. 10.6. Mean January precipitation (cm month^{-1}): (a) observed; (b) control experiment with the Max Planck Institute for Meteorology T21 climate model; (c) experiment with South America removed; and (d) experiment with Australia removed. Areas with more than 10 cm month^{-1} are stippled (Kiladis et al. 1989).

climate model. As noted for the African precipitation simulations in Fig. 10.5, most models can capture the essence of many tropical precipitation features but not necessarily the magnitudes. This is the case for the model in Fig. 10.6b. All three oceanic convergence zones are represented by precipitation maxima, but amounts are less than the observed in Fig. 10.6a.

Given that most climate models can qualitatively represent the major features of Southern Hemisphere tropical precipitation, the goal then is to use such models to provide insight into the workings of the climate system. An interesting example of such an experiment to determine the role of the Australian and South American continents in the simulation of the SPCZ is shown in Figs. 10.6c and 10.6d (Kiladis et al. 1989). In these two experiments, South America was first removed from the climate model and replaced by water; then South America was returned and Australia removed. For the experiment with South America removed (Fig. 10.6c), the SPCZ is virtually unaffected, but the SACZ has disappeared. When Australia is removed and South America is returned (Fig. 10.6d), the SPCZ is only weakly present and the SACZ has reappeared. This suggests

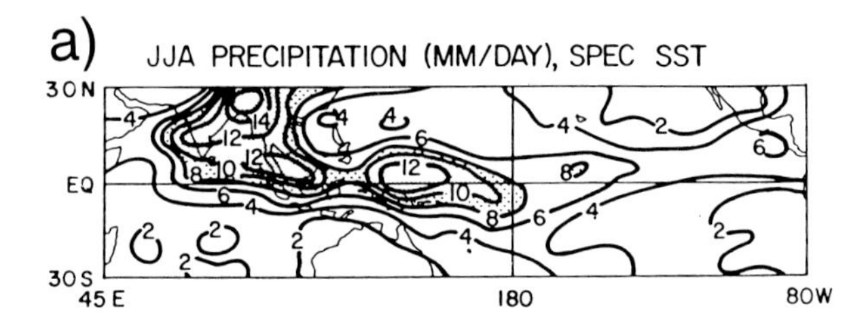

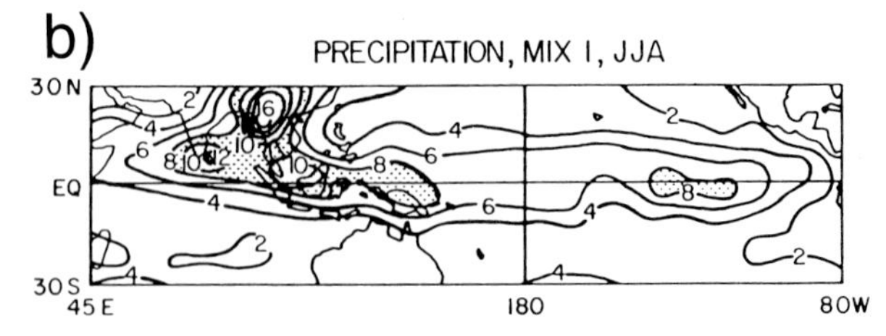

FIG. 10.7. (a) Observed JJA precipitation from a climate model with specified observed SSTs (SPEC SST), and (b) JJA precipitation with the same atmospheric model coupled to a nondynamic slab or mixed-layer ocean. Units are mm day^{-1} (Meehl 1989).

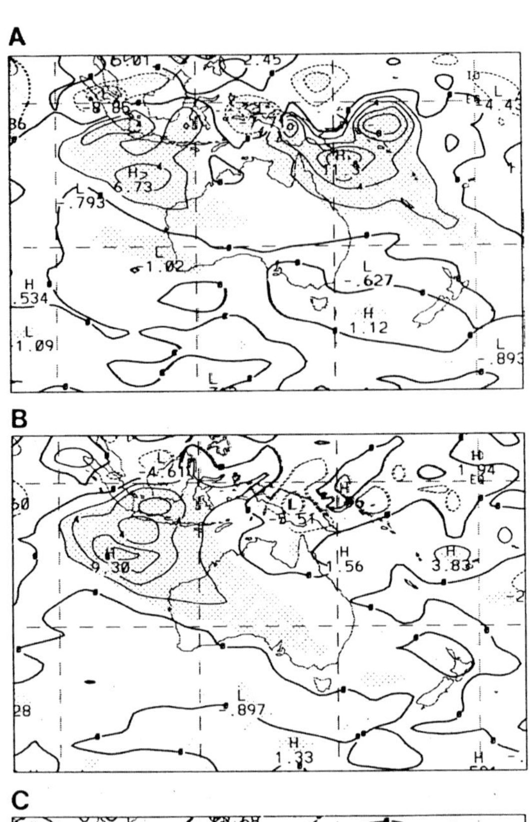

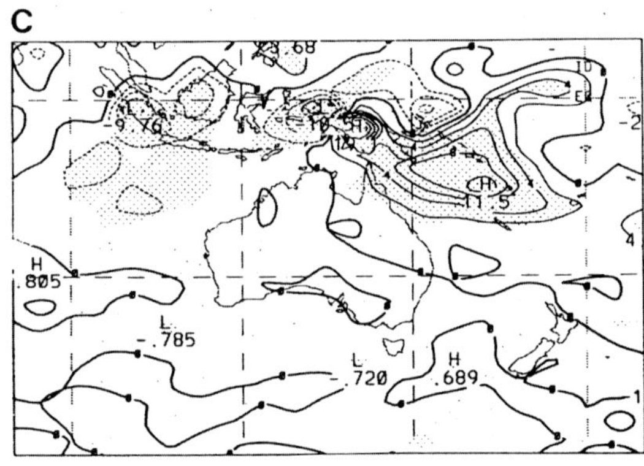

FIG. 10.8. Differences between the precipitation rate for (a) experiment with warm SST anomalies both to the northwest and northeast of Australia, (b) experiment with warm SST anomalies only to the northwest of Australia, and (c) experiment with warm SST anomalies only to the northeast of Australia, and the control simulation. Contour interval is 2 mm day^{-1}; negative contours are dashed. Regions of differences significant at the 95% confidence level are stippled.

that the continental influence in interacting with the midlatitude westerlies is essential in producing the SPCZ and SACZ.

The role of the Pacific SST distribution is also important for the SPCZ. A different coarse-grid climate model can represent the SPCZ with observed SSTs (Fig. 10.7a), but when the model is coupled to a slab ocean without flux correction, the tropical SSTs become zonally oriented with warmest SSTs in the equatorial Tropics across the Pacific. The resulting precipitation simulation (Fig. 10.7b) shows that the tropical precipitation in the Pacific follows the warmest SSTs, and the SPCZ has disappeared. The role of tropical SSTs in the position of the SPCZ has been noted in chapter 8 such that during ENSO events the SPCZ shifts northeastward, toward the warm SST anomalies in the central equatorial Pacific.

Another example of a climate model being used as a tool to attempt to understand tropical rainfall regimes and anomalies is shown in Fig. 10.8. This was an experiment with a climate model to study the effects of SST anomalies to the northwest and northeast of Australia (Simmonds 1990a). The rainfall anomalies generated by the model when there were warm SST anomalies both to the northwest and northeast of Australia (Fig. 10.8a), just to the northwest (Fig. 10.8b), and just to the northeast (Fig. 10.8c) show local increases over the warm SST anomalies. Yet only the anomalies to the northwest have a significant effect in increasing rainfall over continental Australia. A similar experiment was performed for the southern African region, where southwestern Indian Ocean SST anomalies appeared to exert a greater influence over summer rainfall variability for that region compared to South Atlantic SST anomalies (Lindesay and Smith 1993). Additional simulations of tropical rainfall regimes affected by ENSO in climate models are discussed in section 10.3.

b. Extratropical circulation

Climate model simulation of extratropical Southern Hemisphere circulation has steadily improved since the early 1980s (Simmonds 1990b). For example,

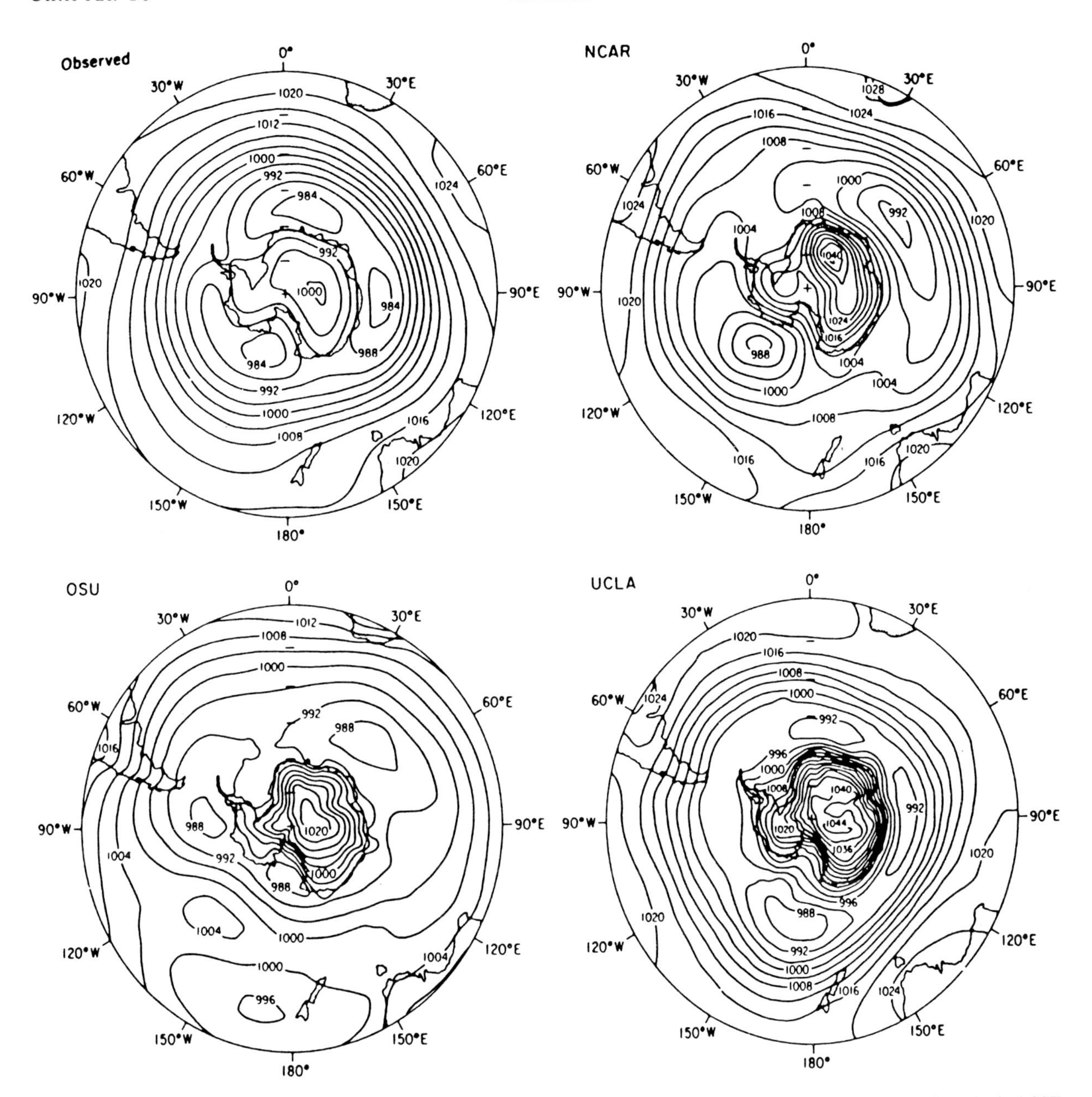

FIG. 10.9. Mean SLP for southern winter for (a) observed, and early 1980s–vintage coarse-grid climate models with climatological SSTs from (b) NCAR, (c) OSU, and (d) UCLA. The contour interval is 4 hPa (Schlesinger 1984).

climate models of the early 1980s had consistent difficulties in representing the correct depth and position of the circumpolar trough around Antarctica with a wave 3 structure (Fig. 10.9, after Schlesinger 1984; the dominant wave 3 structure reflects Southern Hemisphere land–ocean distribution with three ocean basins separated by continents). By the late 1980s, climate models were capable of capturing most elements of the circumpolar trough and the wave 3 structure of the extratropical circulation (Fig. 10.10, after Simmonds et al. 1988). Yet some basic features of extratropical climate, such as rainfall, suffer from questionable verification data from observed datasets. An ensemble of zonal-mean GCM rainfall data and a satellite-derived rainfall estimate suggest that there is somewhat greater zonal-mean rainfall during JJA (2–4 mm day^{-1}) compared to DJF (about 1–3 mm day^{-1}) between 30° and 70°S, while two other observed datasets indicate almost the opposite seasonality (Whetton 1997).

What has become clear, however, is that model resolution is very important for the simulation of southern extratropical climate. This is illustrated in Fig. 10.11 for a model intercomparison between a

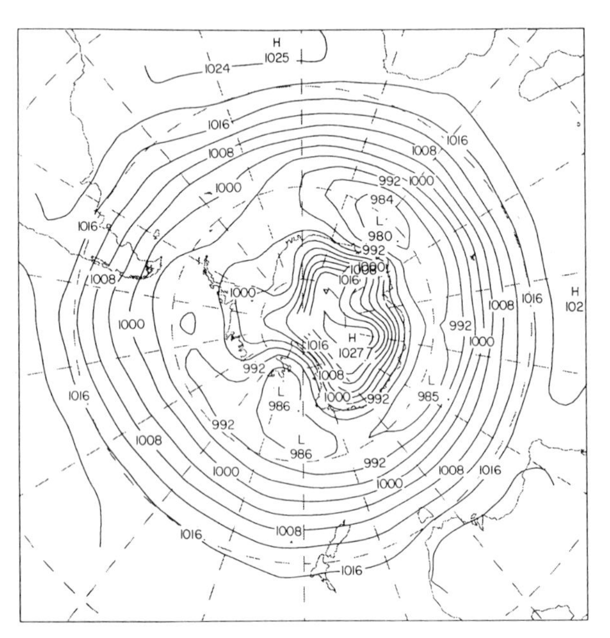

FIG. 10.10. Mean SLP for southern winter for a late 1980s–vintage coarse-grid climate model from Melbourne University. Contour interval is 4 hPa (Simmonds et al. 1988).

number of high- and low-resolution climate models run with observed SSTs (Boer et al. 1992). The zonal-mean sea level pressure plots reveal much less agreement between coarse-resolution models and observations in both winter and summer (Fig. 10.11, left side), while high-resolution models have much improved simulations of SLP in the Tropics and subtropics and a deepening and southward shift of the circumpolar trough in both seasons (Fig. 10.11, right side). South of about 70°S, all models show less agreement, indicating difficulties in correctly simulating the high-altitude climate over Antarctica.

A further analysis of a wide variety of GCMs from the Atmospheric Model Intercomparison Project (AMIP; see Gates 1995) shows that model resolution of greater than about 4° by 4° seems to be necessary for accurate simulation of the position and intensity of the circumpolar trough (Chen and Bromwich 1995). In some cases, however, high-resolution models actually overshoot reality and produce westerlies that are stronger than observed (Whetton et al. 1996c).

The effect of increasing resolution in a single model is shown in Fig. 10.12 (Boville 1991). The same climate model was run at four resolutions. As resolution is progressively increased from top to bottom of that figure, the three subtropical highs in the southern oceans become more organized and stronger, and the circumpolar trough deepens and shifts farther south, in closer agreement with the observations in Fig. 10.9a.

Part of the role of resolution involves better representation of the Antarctic and Andean topography. Previous modeling studies have shown the importance of the elevated Antarctic topography for the simulation of Southern Hemisphere climate (Mechoso 1981). Higher-resolution climate models better resolve the steep escarpment of Antarctica and allow the lows in the circumpolar trough to move farther south (see discussion of the Antarctic topography and the general circulation in chapter 2). A second role played by Antarctic topography is in producing the katabatic wind regime. Climate models that represent the Antarctic topography well are able to simulate the strong observed katabatic winds that originate over the antarctic ice dome (Fig. 10.13). Topographically induced katabatic winds have been shown to be important in simulating the circumpolar trough (Parrish et al. 1994). In Fig. 10.14, a climate model with Antarctic topography produces a much better simulation of the circumpolar trough (Fig. 10.14a) compared to a version of the same model without Antarctic topography (Fig. 10.14b), partly due to the absence of an organized katabatic wind regime in the version with a flat Antarctica.

Another important consideration in simulating the extratropical climate of the Southern Hemisphere is correct representation of the overall tropospheric temperature structure (Meehl and Albrecht 1988). In a climate model with simple convective adjustment, where there was inadequate heating of the tropical troposphere and a reduced Equator-to-pole midtropospheric temperature gradient, the circumpolar trough in January was weak and located too far equatorward (Fig. 10.15a). In the same model with a mass flux convective scheme that had the effect of producing a much-improved tropospheric temperature simulation with a steeper Equator-to-pole temperature gradient, the extratropical circulation was dramatically improved (Fig. 10.15b). The subtropical highs over oceans intensify as do lows over the continents, and the circumpolar trough deepens and moves poleward in better agreement with the observations in Fig. 10.9a.

The general improvements involved with a better representation of the tropospheric temperature structure and Antarctic topography are also reflected in another feature that many coarse-grid climate models have found difficult to simulate—the upper-troposphere double jet in the Pacific sector (e.g., Kinter et al. 1988; Mullan and McAvaney 1995). A climate model with a good representation of meridional temperature gradients and Antarctic topography is capable of producing a reasonable simulation of the double jet (Fig. 10.16, from Kitoh et al. 1990). This figure shows the relationship between steep subtropical and high-latitude temperature gradients and the position and intensity of the subtropical and high-latitude jets, respectively. Kitoh (1994) has used this model to study the role of the Tropics in producing the double-jet structure in the Pacific sector. It was demonstrated

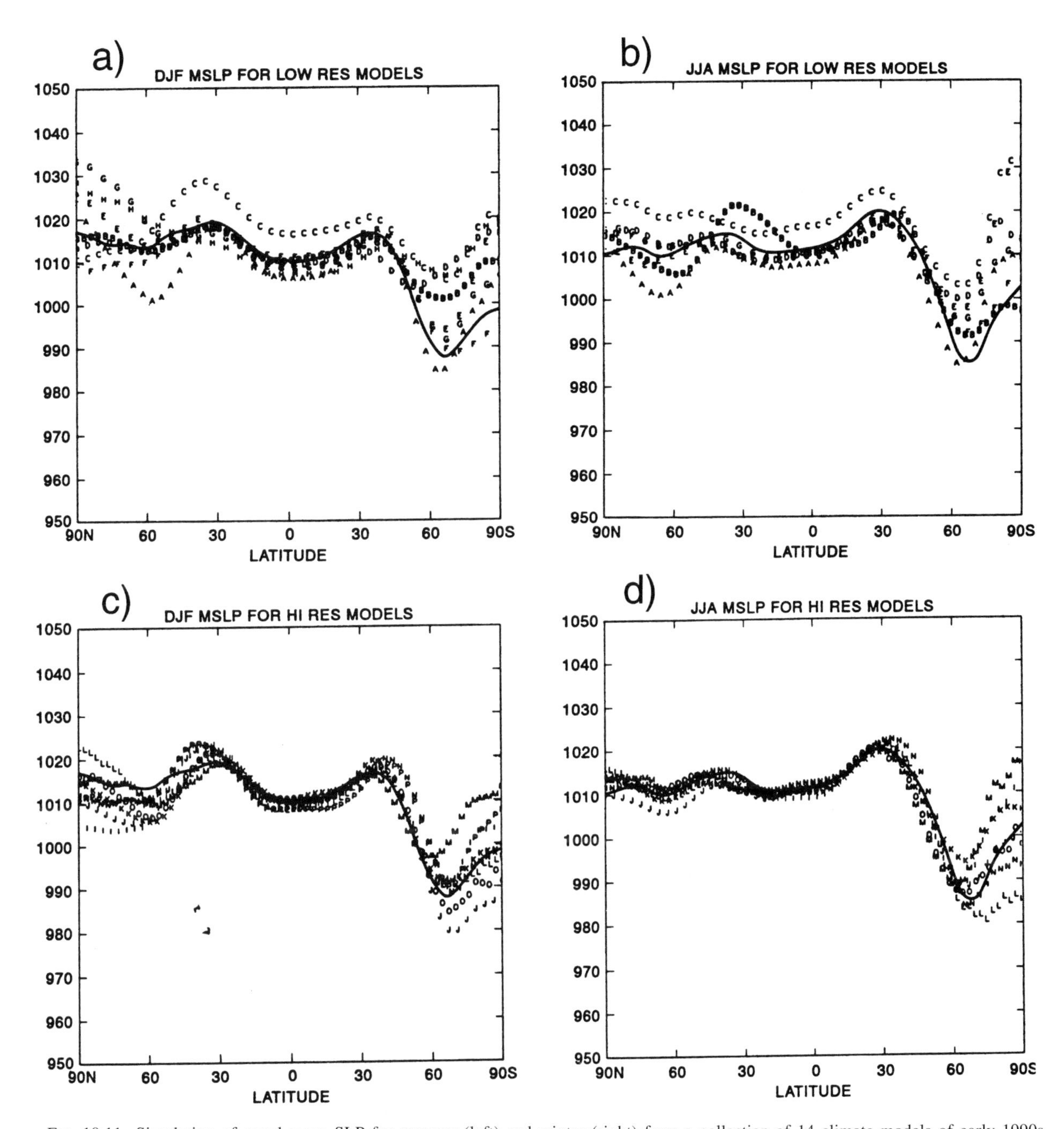

FIG. 10.11. Simulation of zonal-mean SLP for summer (left) and winter (right) from a collection of 14 climate models of early 1990s vintage run with observed SSTs, separated into low-resolution (top) and high-resolution (bottom) classes. Models are letters; observations are solid lines (Boer et al. 1992).

that tropical convective heating in the Pacific maintains the subtropical part of the double jet, and the high-latitude jet is associated with a wave train forced by convective heating northwest of Australia (Fig. 10.17a for the observations, and Fig. 10.17b for the model). These connections have been noted in simple climate models as well (e.g., Karoly et al. 1989). It is also notable that coarse-grid models that are not capable of simulating well the position and intensity of the circumpolar trough or the double jet also have difficulty in properly simulating the stationary-wave flux (Yang and Gutowski 1994), as shown in Fig. 10.17. Variability simulated in association with the double jet in climate models will be discussed in section 10.3.

Climate models are also capable of reproducing aspects of Southern Hemisphere extratropical atmospheric dynamics on various timescales consistent with the basic climate variables shown so far. For

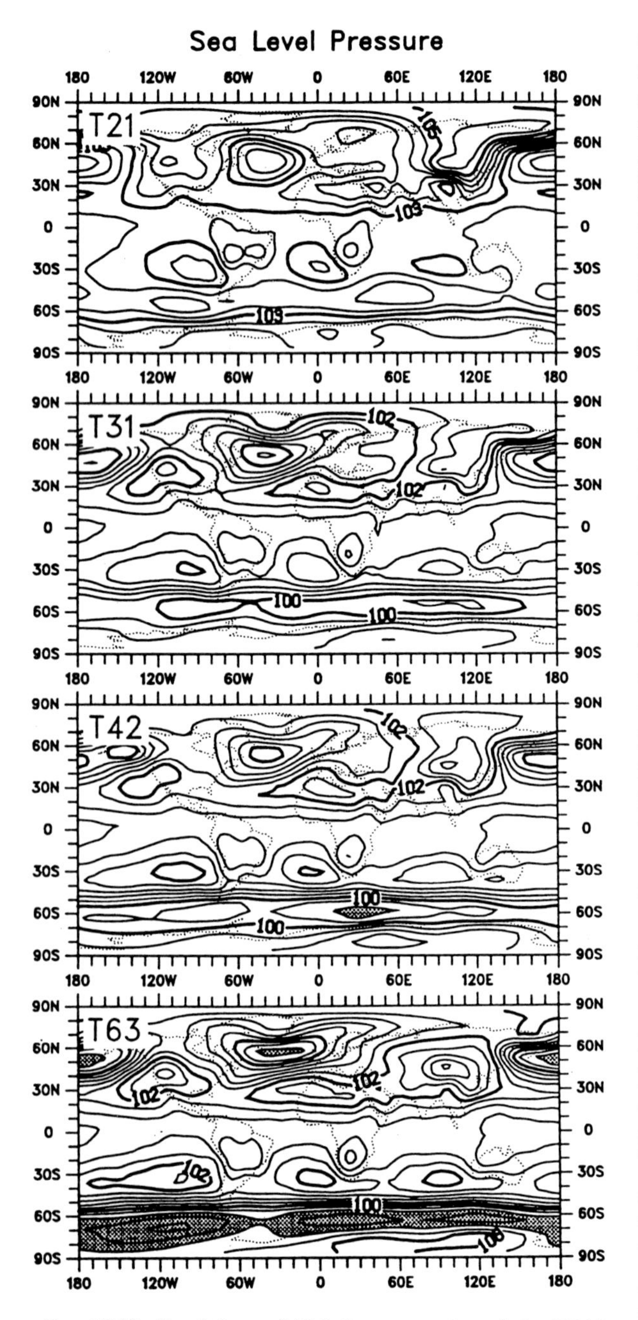

FIG. 10.12. Simulations of SLP from a version of the NCAR Community Climate Model at resolutions varying from low (top) to high (bottom). Multiply contour labels by 10 to get SLP in hPa. Contour interval is 5 hPa, and SLP areas less than 990 hPa are stippled (Boville 1991).

example, Fig. 10.18 shows cyclone density and cyclogenesis for observations (Jones and Simmonds 1993) and a climate model simulation (Simmonds and Wu 1993) for the winter season. Both cyclone density and cyclogenesis show maxima near 60°S in the region of the circumpolar trough, with three relative maxima in the southern Indian, Pacific, and Atlantic sectors. This representation of synoptic timescale variability can be compared with another one, shown in Fig. 10.19, that is the standard deviation of bandpass-filtered, 500-mb geopotential height (filtered to retain the 2.5- to 6-day band timescale; Murphy 1995) in winter and summer. Similar to Fig. 10.18, greatest synoptic timescale activity occurs in the region of the circumpolar trough near 60°S.

Measures of January zonal-mean transient eddy momentum and heat flux for observations and two model versions are shown in Fig. 10.20 (Meehl and Albrecht 1991). The "standard" model includes moist convective adjustment and the "hybrid" includes a mass-flux convective scheme that has the effect of improving the Southern Hemisphere simulation, as shown in Fig. 10.15. Both model versions agree with the observations of transient eddy heat flux in that maximum values occur in the lower troposphere, with an increase of magnitude in the model with the mass-flux convective scheme compared to the model version with convective adjustment. Similarly, both model versions show maximum values of transient eddy momentum flux in the upper troposphere, with eddy momentum-flux convergence near 55°S. Consistent with previous results from these experiments, the mass-flux convective scheme in the "hybrid" experiment produces a stronger (and closer to observations) transient eddy momentum-flux convergence near that latitude.

Persistent atmospheric-circulation anomalies with multiple-day timescales, termed "blocking," have also been studied in climate models (Bates and Meehl 1986). For example, numbers of 500-mb persistent height anomalies indicative of blocking events per 10 years in the observations (an anomaly of 100 m lasting at least 5 days in 128-day-long winter and summer seasons, from Swanson and Trenberth 1982; see also Trenberth and Mo 1985 and Trenberth 1986) are shown in Figs. 10.21a and 10.21c. Using this criteria, we see that there are greater numbers of midlatitude blocking events in winter (Fig. 10.21c) than summer (Fig. 10.21a). During winter, centers of blocking occur in the southern Pacific sector near New Zealand, with other centers in the South Atlantic and southern Indian Ocean sectors. Similar centers of blocking action appear in summer but with fewer numbers. Using a slightly more stringent criteria for a mixed-layer model (an anomaly of 150 m lasting at least seven days in three-month seasons), the model shows qualitative similarity with the observations. As could be expected from the more stringent criteria, there are fewer blocks tabulated from the model, but the centers of action are in similar locations (though the New Zealand sector is not particularly well simulated) and there are more blocks in the model in winter than summer.

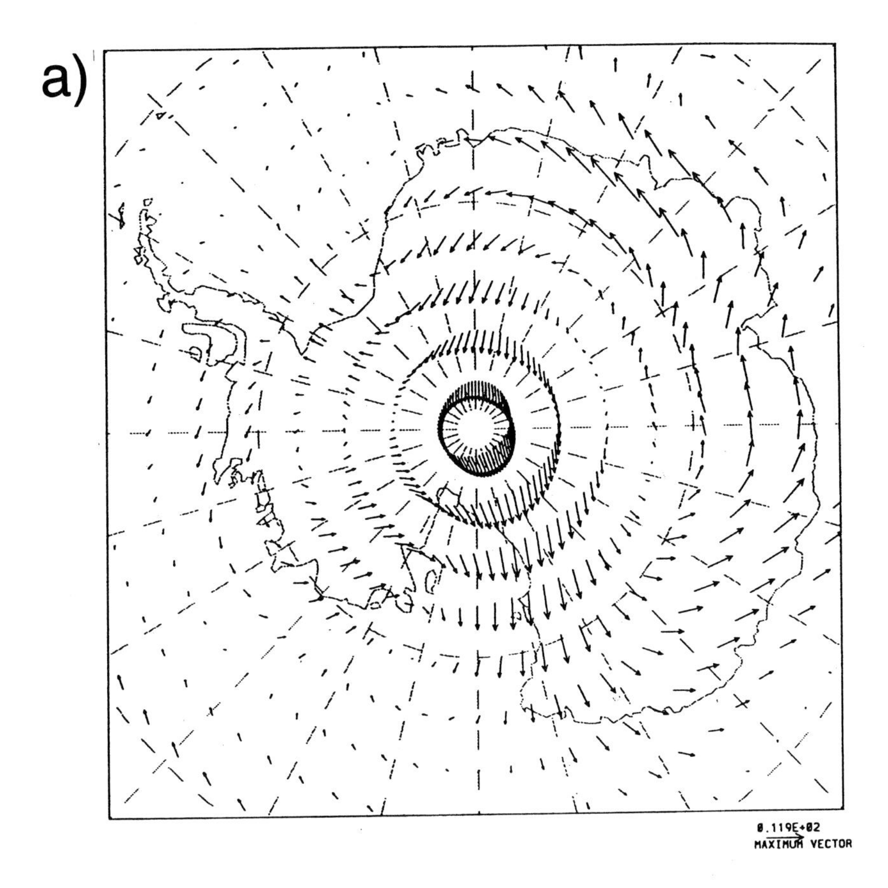

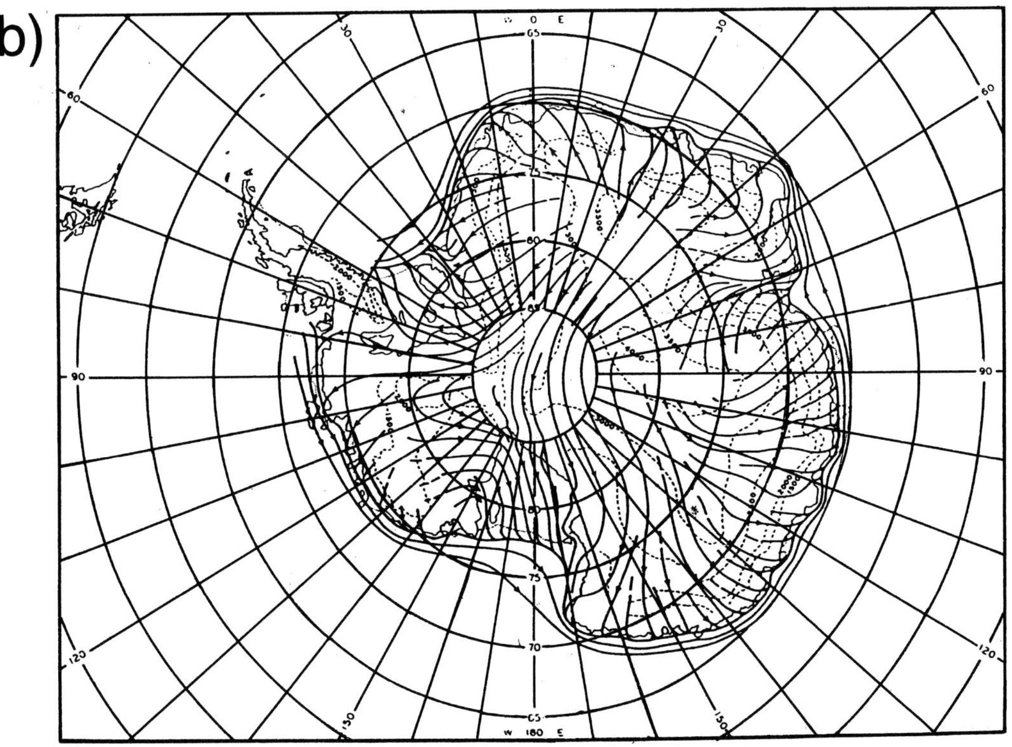

FIG. 10.13. Low-level vector wind field from (a) the July simulation of the Melbourne University climate model (Simmonds et al. 1988) and (b) observations (after Simmonds 1990b).

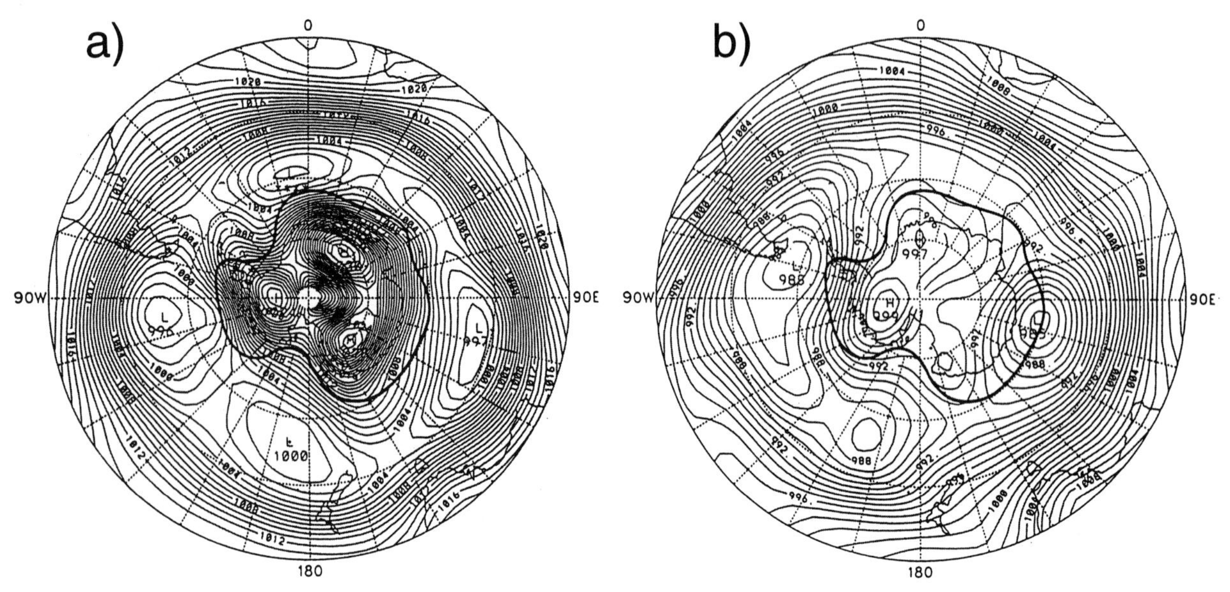

FIG. 10.14. July SLP for (a) model that includes Antarctic topography; and (b) same model without Antarctic topography. Contour interval is 1 hPa (Parrish et al. 1994).

c. Antarctic sea ice

As noted above, both mixed-layer and coupled climate models have historically included simple thermodynamic sea-ice formulations (Meehl 1992). Recently, coupled models are including dynamic sea-ice model components in addition to the thermodynamic formulations (e.g., Meehl and Washington 1995; Washington and Meehl 1996). Meanwhile, models with specified sea-ice extent and fraction have been used to study sensitivity of climate to changes in sea-ice parameters. For example, Simmonds and Wu (1993) progressively increased the specified open-water fraction of antarctic sea ice in a series of climate model experiments. Results from two of those simulations are shown in Fig. 10.22a for sea-ice open-water fraction of 0.80 and 1.00, respectively (compare to observed and model-simulated cyclone densities in Fig. 10.18). Stippled areas in Fig. 10.22 indicate that the greater

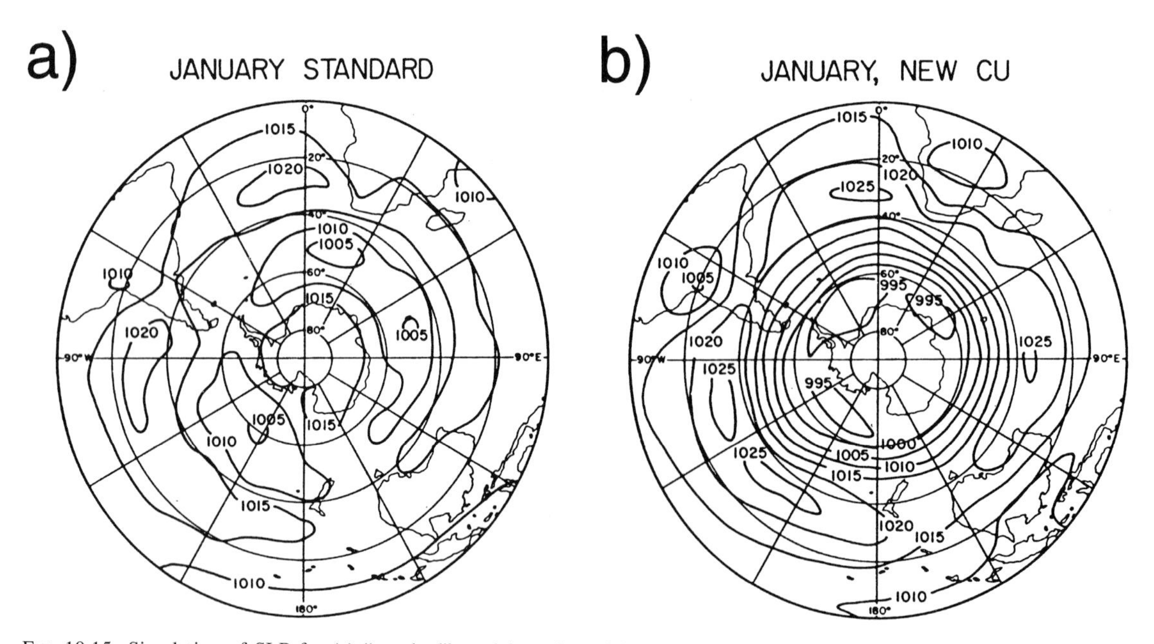

FIG. 10.15. Simulation of SLP for (a) "standard" model version with convective adjustment, and (b) "new cu" model version with mass-flux convection scheme. Contour interval is 5 hPa (Meehl and Albrecht 1988).

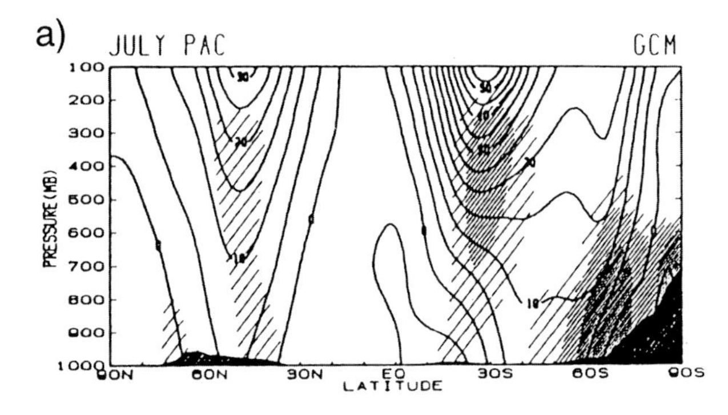

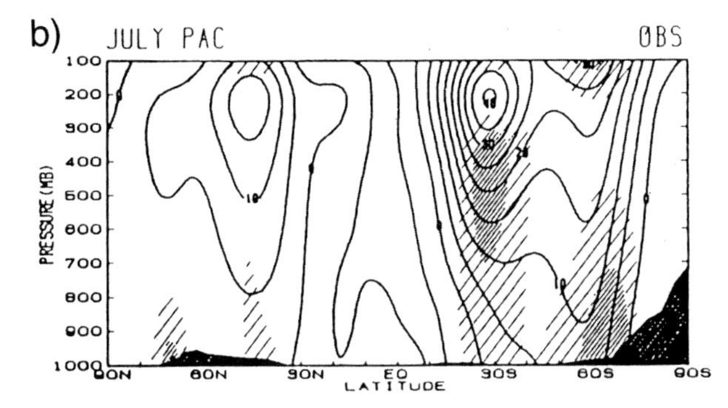

FIG. 10.16. Cross sections of (a) model-simulated and (b) observed zonal wind for July, averaged for the Pacific sector (120°E–60°W). Contour interval is 5 m s^{-1}. Light, medium, and heavy shading indicates regions with meridional temperature gradient greater than 2, 3, and 5 K per 4 degrees of latitude, respectively.

the open-water fraction, the greater the cyclone density at high latitudes and somewhat less in the midlatitudes. These changes are reflected by zonal-mean wind differences in Fig. 10.23 from Simmonds and Budd (1991). The southward shift of cyclonic activity noted in Fig. 10.22 is characterized by a weakening of westerlies in midlatitudes (negative values near 50°S in Fig. 10.23) and strengthening at higher latitudes (positive values in Fig. 10.23, roughly near 65°S). There are indications in Fig. 10.23 that changes in antarctic sea-ice parameters influence the global circulation with a weakening of high-latitude westerlies in the Northern Hemisphere as well.

The concept that alterations of sea ice can influence global climate sensitivity has been demonstrated repeatedly in carbon dioxide (CO_2) climate-change experiments with climate models (e.g., IPCC 1990, 1992, 1996), but also holds true for simulation of present-day climate. Results from a sensitivity study illustrating this point are shown in Figs. 10.24 and 10.25 (Meehl and Washington 1990). In the former, sea-ice extent in the original simulation of present-day climate with a mixed-layer model with thermodynamic sea ice is shown for DJF and JJA and in a revised experiment, where a simple change to the threshold for ice-albedo reduction has been made to account for increased melt ponds and leads in melting sea ice. There are retreats of sea ice in the revised experiment control case in both the Arctic and Antarctic (comparing the solid-line ice boundary between revised and original experiments in Fig. 10.24), with about a 1°-warmer globally averaged surface air temperature for present-day climate. Biggest changes are evident in a sensitivity experiment where CO_2 is doubled in the model. The sea ice melts back farther in the revised than in the original experiment in both hemispheres (comparing the dashed-line ice limits in Fig. 10.24), showing greater sensitivity to the change in external forcing. Thus, these alterations in the sea-ice parameterization also alter the global climate sensitivity. Globally averaged surface air temperature warming due to the change in external forcing is 3.5°C in the original experiment, but is enhanced in the revised sea-ice experiment to 4.0°C with a corresponding increase of globally averaged precipitation of 7.12% to 7.25%. Figure 10.25 shows zonal-mean temperature differences for the sensitivity experiment, with increased CO_2 for the revised and original sea-ice formulations. Magnitudes of the temperature differences are greatest in the upper tropical troposphere in both seasons (comparing Figs. 10.25c and 10.25d to 10.25a and 10.25b) and at high southern latitudes in summer (comparing Fig. 10.25a and 10.25c).

d. The seasonal cycle: The semiannual oscillation at mid- and high southern latitudes

The semiannual oscillation has been characterized as a twice-annual expansion and weakening of the circumpolar trough from March to June and September to December, and as a contraction and intensification from June to September and December to March. It is associated with similar fluctuations of tropospheric temperature gradients, heights, pressure, and winds at middle and high latitudes (e.g., van Loon 1967; Meehl 1991; see also chapters 1, 2, and 4). In spite of apparent recent changes in its characteristics [Hurrell and van Loon (1994) show a reduction of magnitude of the SAO starting about 1979, which Meehl et al. (1998) link to changes in the seasonal cycle of SSTs near 50°S], the SAO has been a consistent aspect of the seasonal cycle in the Southern Hemisphere and something that climate models should be able to simulate. As with other aspects of Southern Hemisphere circulation, climate models have had improving success simulating the SAO, while earlier coarse-grid models had mixed success (Xu et al. 1990). Simulations of the annual cycle of monthly mean SLP intensity (Fig. 10.26a) and position (Fig. 10.26b) of the circumpolar trough show the typical features of coarse-grid models noted in section 10.2b. These include a circumpolar trough that is too weak and too far equatorward. Of the models shown, the NCAR model

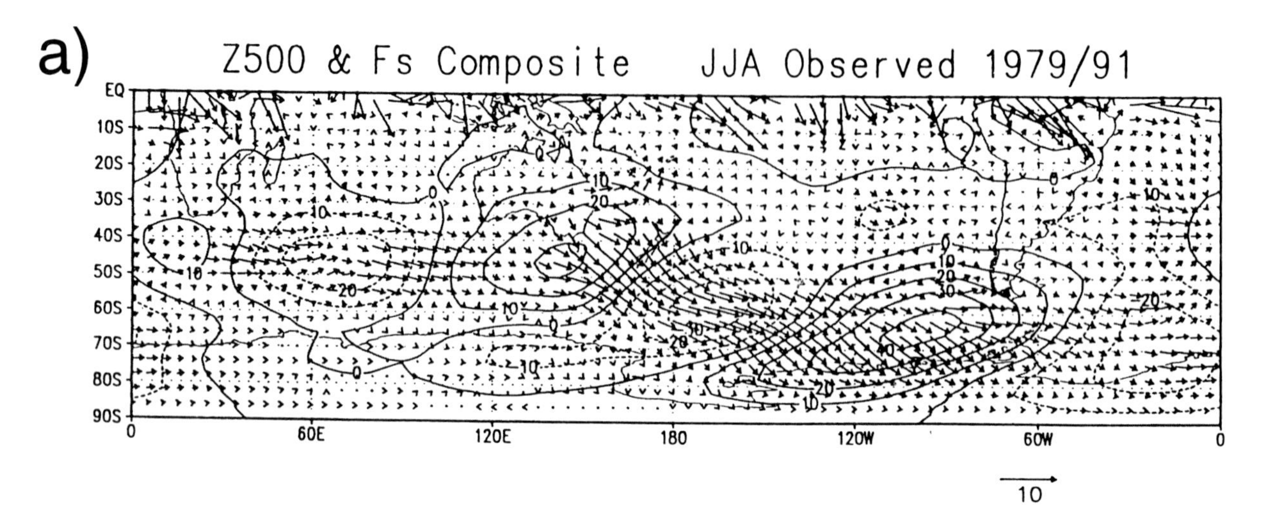

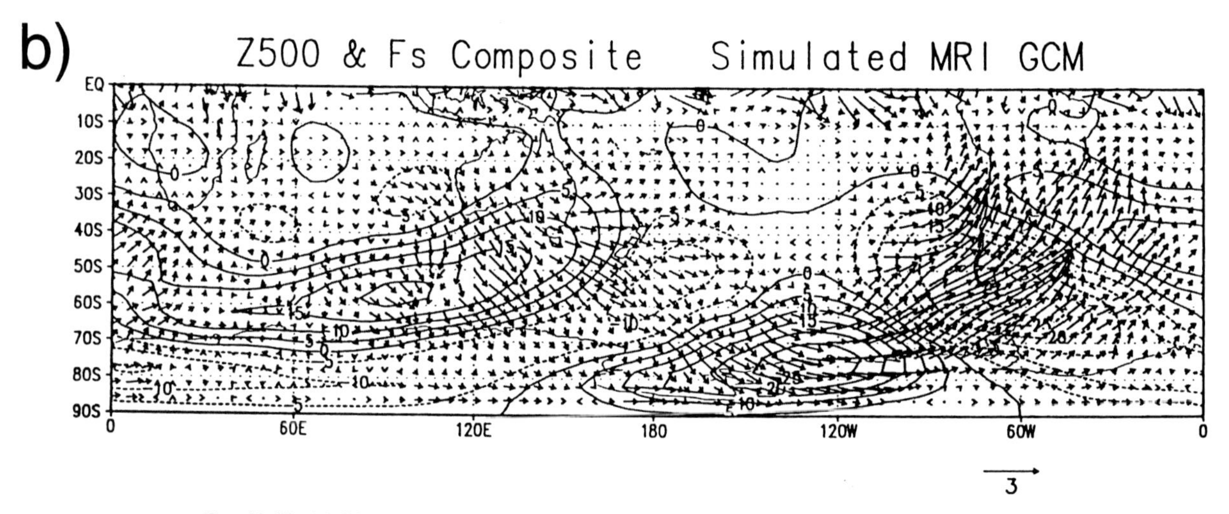

FIG. 10.17. (a) Observed and (b) simulated composite of the geopotential height anomaly and stationary-wave activity flux at 500 hPa for JJA (Kitoh 1994).

seems to show a small-magnitude SAO compared to the observations, with circumpolar trough intensification in February and September (Fig. 10.26a) and contraction around February–April and again near October–November. A more recent assessment of the simulation of the SAO from the AMIP GCMs shows that many models can simulate some features of the SAO (Mullan and McAvaney 1995), but SAO biases in higher-resolution models are generally less than in the low-resolution models (Chen and Bromwich 1995).

An SAO index devised by van Loon (1967) of the 500-mb temperature difference between 50° and 65°S is shown in Fig. 10.27a from the Australian analyses from May 1972 to May 1984. There is a dominant second harmonic with peaks in March and September, illustrating the twice-annual intensification of the midtropospheric temperature gradient characteristic of the SAO. The NCAR model shown in Fig. 10.26 was run with observed SSTs (denoted SPEC SST in Fig. 10.27b) and with a non–flux-corrected mixed-layer ocean (MIX 1 in Fig. 10.27c). As suggested in Fig. 10.26, this model simulates a weak SAO, with the amplitude of the second harmonic of the 500-mb temperature index (explaining 46% of the total variance of the monthly mean values) of larger amplitude than the first harmonic (explaining 38% of the variance, Fig. 10.27b). The mixed-layer model produces a different annual cycle of SSTs in the circumpolar ocean, and the result is a reduction of the SAO second harmonic compared to the first harmonic (Fig. 10.27c). This modeling study suggests that the seasonal cycle of SSTs near 50°S is crucial for simulating the amplitude of the SAO.

Results from another model that has had some success with simulating a weak SAO (Kitoh et al. 1990) are shown in Fig. 10.28a in comparison with observations from the NMC analyses for a period

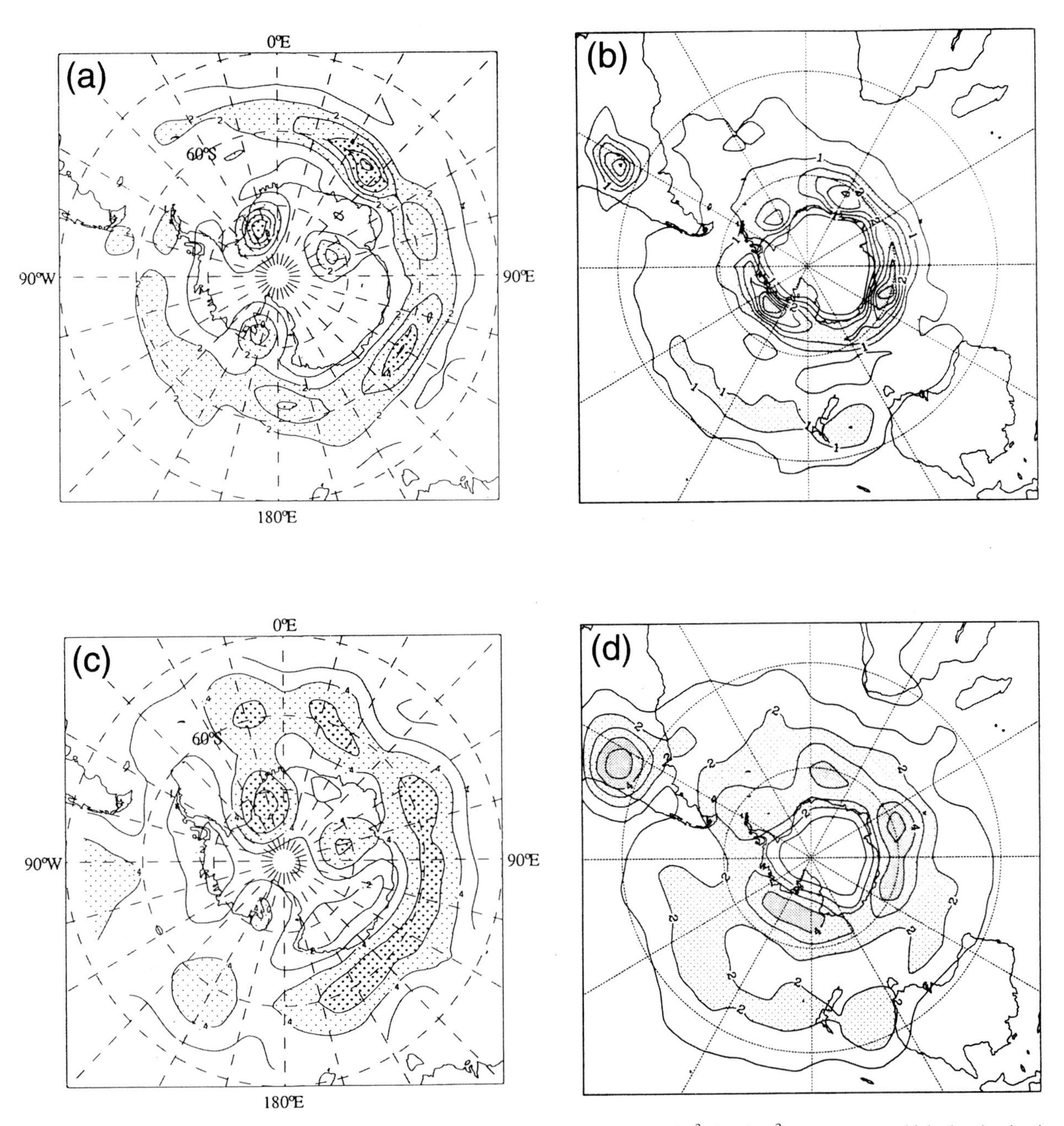

FIG. 10.18. Distribution of (a) simulated July cyclone density (units are 1.0×10^{-3} deg lat^{-2})—areas over which the density is greater than 2 and 4 are covered by light and heavy stippling, respectively; (b) observed winter cyclone density (units are 0.5×10^{-3} deg lat^{-2})—areas over which the density is greater than 1 and 3 are covered by light and heavy stippling, respectively; (c) simulated July cyclogenesis (units are 0.1×10^{-3} cyclones deg lat^{-2} day^{-1})—areas over which the rate of cyclogenesis is greater than 0.4 and 0.5 are covered by light and heavy stippling, respectively; and (d) observed winter cyclogenesis (units are 0.1×10^{-3} cyclones deg lat^{-2} day^{-1})—areas over which the rate of cyclogenesis is greater than 0.2 and 0.4 are covered by light and heavy stippling, respectively. Note different area projections between computed (left) and observed (right) (model values from Simmonds and Wu 1993; observations from Jones and Simmonds 1993).

(1979–1987) when the SAO amplitude was reduced in the observed system (Hurrell and van Loon 1994). Nevertheless, there is evidence of areas where the phase arrows indicate a dominant SAO with March and September maxima near 50°S in the observations in the southern Indian, Pacific, and Atlantic Ocean sectors (i.e., phase arrows pointing to the northeast in Fig. 10.28b). In the model, there are also similar areas to the observations where the SAO has comparable phase and dominates the annual harmonic of SLP (Fig. 10.28a).

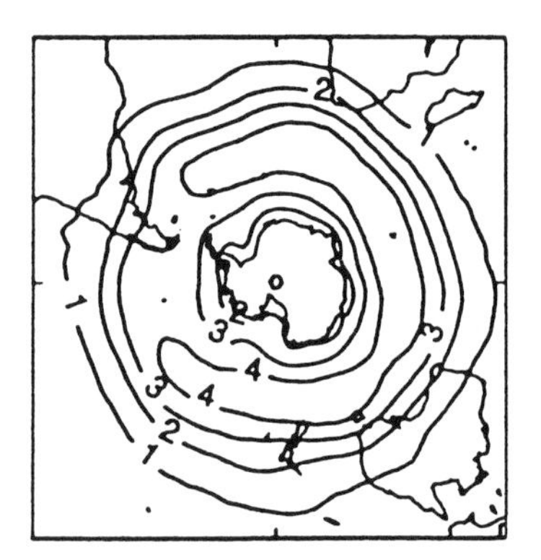

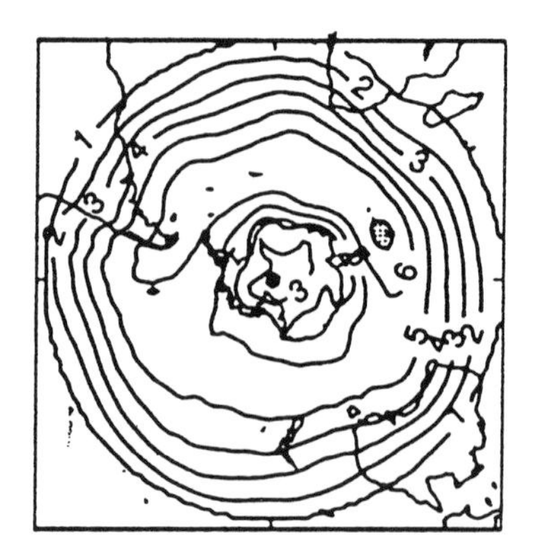

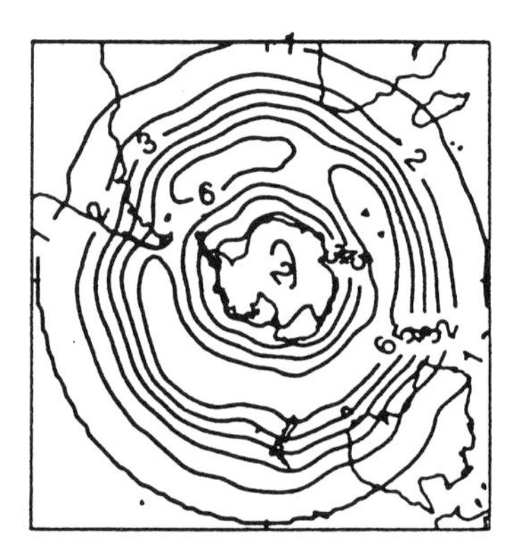

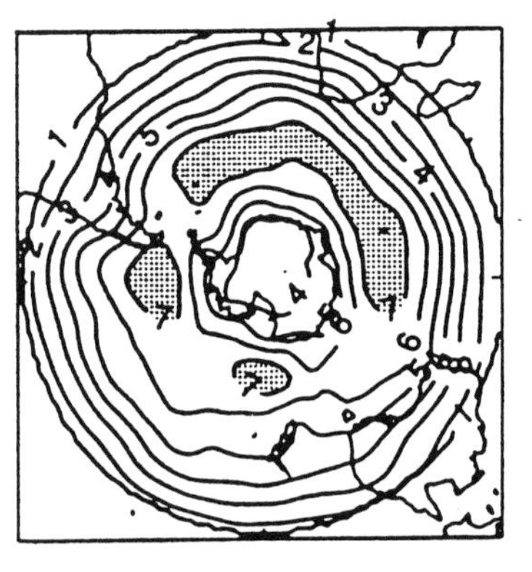

FIG. 10.19. Standard deviation of bandpass-filtered (2.5- to 6-day), 500-hPa geopotential height, 10-yr averages from a global coupled climate model with flux correction from the Hadley Centre (top), and mean observed distribution for 1983–92 (bottom), for DJF (left) and JJA (right) (Murphy 1995).

As noted above, almost all aspects of simulated Southern Hemisphere climate seem to improve with more-recent model versions as well as increased model resolution, and the SAO is further evidence of this (Bromwich et al. 1995). Figure 10.29 illustrates the previously noted feature of coarse-resolution models simulating the circumpolar trough that is too weak and too far north. Yet, the SAO is evident in terms of the seasonal cycle of the strength of the trough (Fig. 10.29a) for the ensemble of both the low- and high-resolution models. The greatest improvement in the SAO simulation with the high-resolution models appears in the position of the trough (Fig. 10.29b). As in the observations, the high-resolution models simulate a contraction of the trough in March–April and November near the correct latitudes.

10.3. Interannual and longer timescale variability

Modeling studies of variability at interannual and longer timescales have been undertaken with specified anomalous SSTs representative of extremes in interannual variability (e.g., ENSO conditions); with specified time series of observed SSTs over some period of time; and with coupled models that generate their own dynamically interactive ocean–atmosphere variability. These three types of experiments are reviewed next.

a. Single-season specified SST experiments

The first type of experiment, with anomalous ENSO-like SST anomalies specified in the tropical Pacific, demonstrated that GCMs of the early 1980s were capable of simulating midlatitude teleconnections commonly observed in association with ENSO events in the Tropics (e.g., Rowntree 1972; Blackmon et al. 1983; Shukla and Wallace 1983; Fennessy et al. 1985; Palmer and Mansfield 1986a,b). Subsequently, this type of experiment has been used to test model sensitivity. For example, Fig. 10.30 shows results from sensitivity experiments with a model run in perpetual January mode with a specified ENSO-like

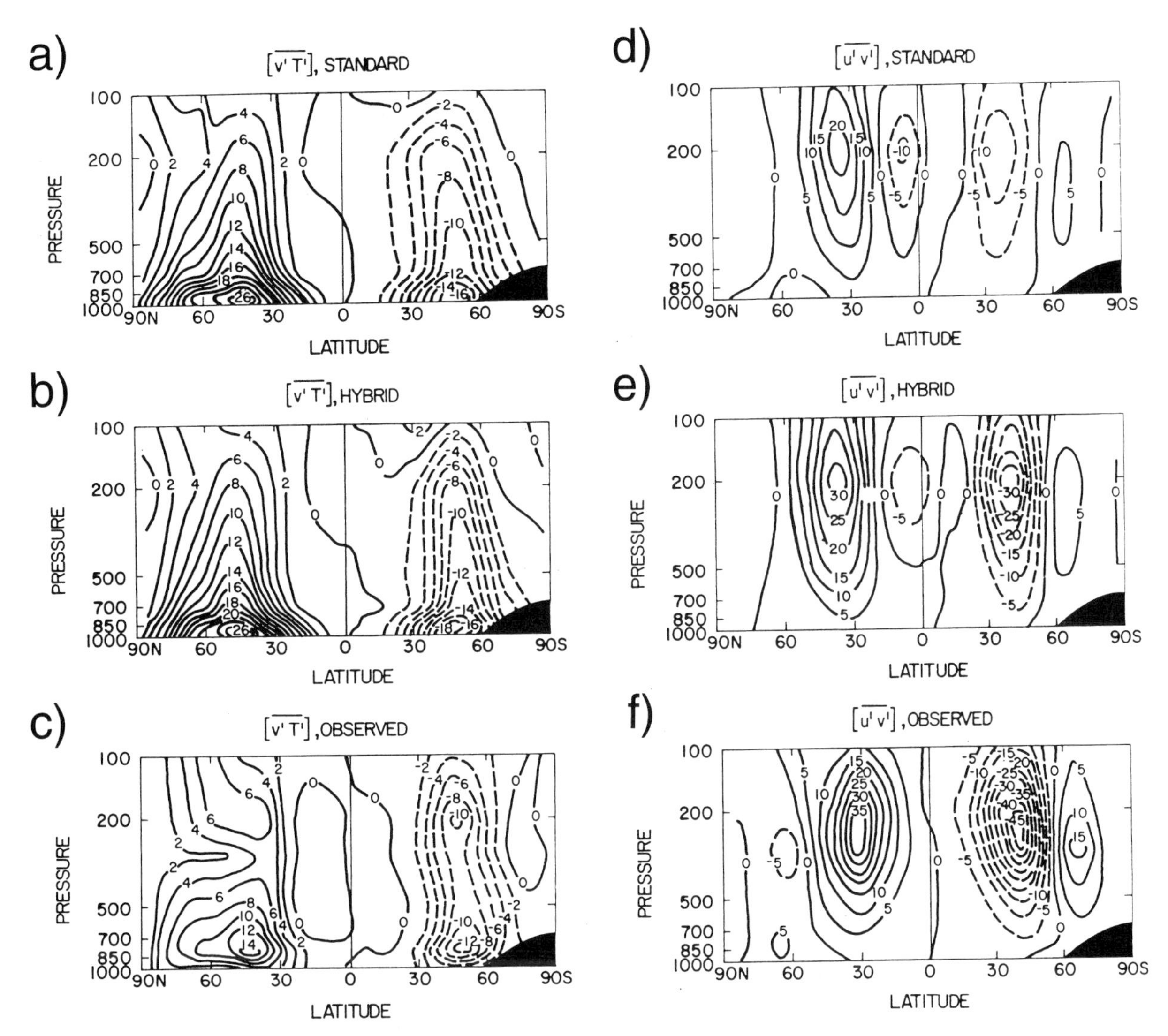

FIG. 10.20. January zonal-mean transient eddy heat flux (°C m s^{-1}) for (a) "standard" model experiment with convective adjustment; (b) "hybrid" model experiment with mass-flux convective scheme; and (c) observed from ECMWF analyses, 1979–87. (d) Same as (a) except for zonal-mean transient eddy momentum flux; (e) same as (b) except for transient eddy momentum flux; and (f) same as (c) except for transient eddy momentum flux (Meehl and Albrecht 1991).

SST anomaly in the eastern tropical Pacific (Meehl and Albrecht 1991). The changes in u-component wind are shown for an experiment with standard convective adjustment (Fig. 10.30a), the same model with a mass-flux convective scheme (Fig. 10.30b), and observations from a composite difference from the ECMWF analyses that includes two ENSO events (Fig. 10.30c). The model simulations in Figs. 10.30a and 10.30b both agree with the observations in showing a strengthening of the southern subtropical jet (positive differences in the upper troposphere near 30°S) and a weakening of the midlatitude westerlies (negative differences near 55°S). These experiments were run to test model sensitivity to the type of convective scheme used, with the mass-flux convective scheme (which has the effect of warming and moistening the upper tropical troposphere) producing more realistic response to an ENSO SST anomaly in the Southern Hemisphere with less success in the Northern Hemisphere.

A somewhat different type of GCM SST anomaly experiment showed linkages between SSTs in the southwest Indian and South Atlantic Oceans and interannual summer rainfall variability over southern Africa, with the Indian Ocean SSTs exerting the greater influence (Lindesay and Smith 1993).

b. Time-evolving specified SST experiments

With basic model capability established in terms of midlatitude response to a fixed tropical SST anomaly

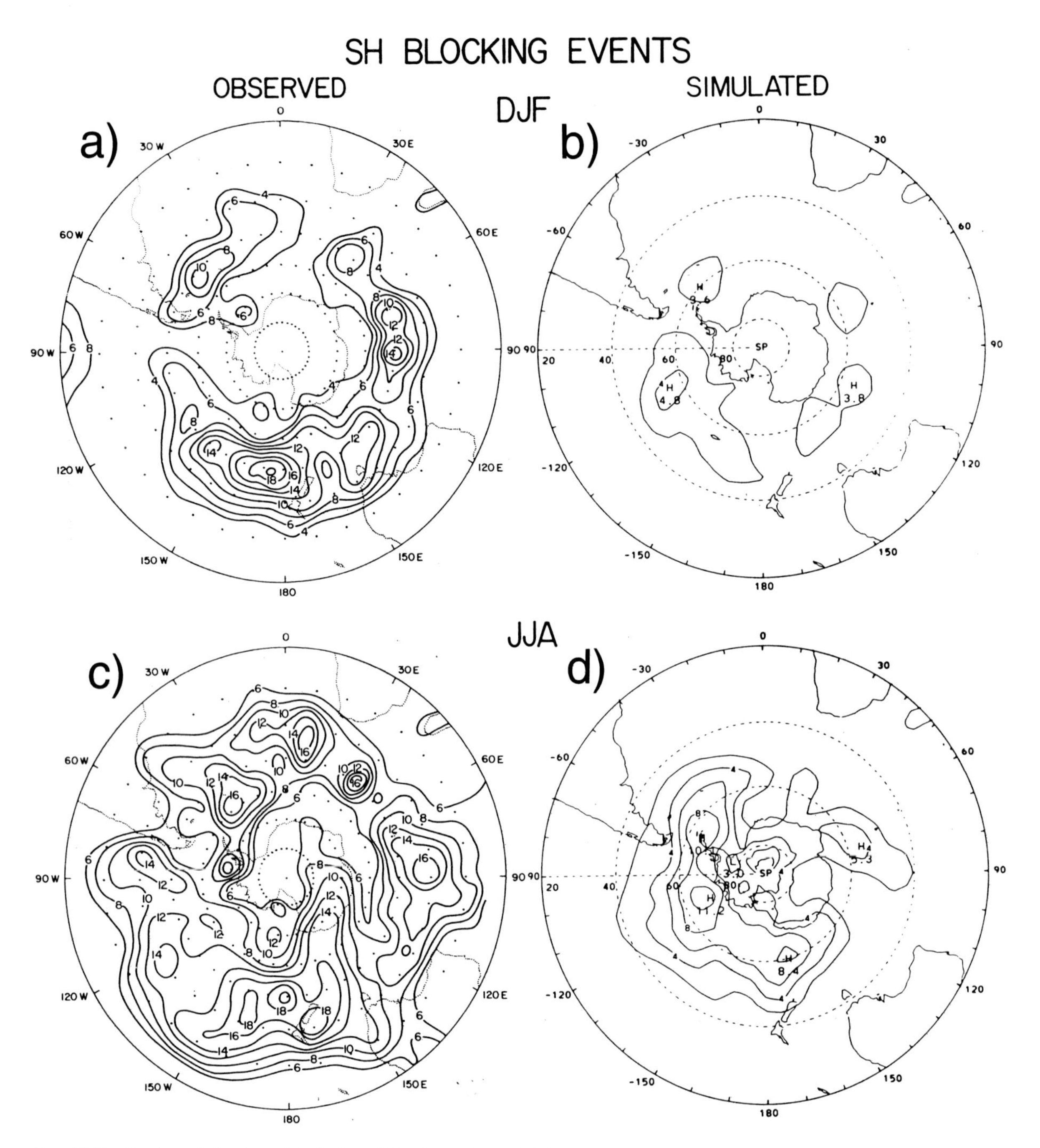

FIG. 10.21. Number of Southern Hemisphere 500-mb persistent height anomalies indicative of blocking events per 10 years in the observations (6 November to 11 March season) and per 15 years in the climate model DJF season: (a) observed, (b) simulated in the climate model with a mixed-layer ocean and no flux correction, (c) observed for 15 May to 19 September, and (d) simulated for JJA. Criteria for blocks are somewhat more stringent for the model compared to the observations—see text (Bates and Meehl 1986).

in experiments like those cited above, it was of interest to determine if a climate model could respond realistically to time-evolving SST forcing. For example, Lau (1985) used the month-to-month time series of observed SSTs over the tropical Pacific for the period 1962–76 (which included three ENSO events) to assess this capability in a climate model. The composite response for the three ENSO events in that time series for 500-mb height anomalies compared to observed 500-mb height anomalies (from Karoly 1989) is shown in Fig. 10.31. The model is in qualitative agreement with the observations, in that 500-mb heights generally

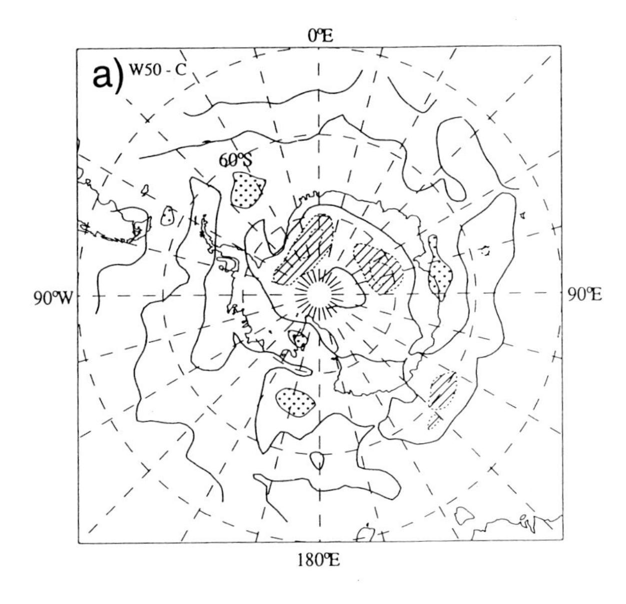

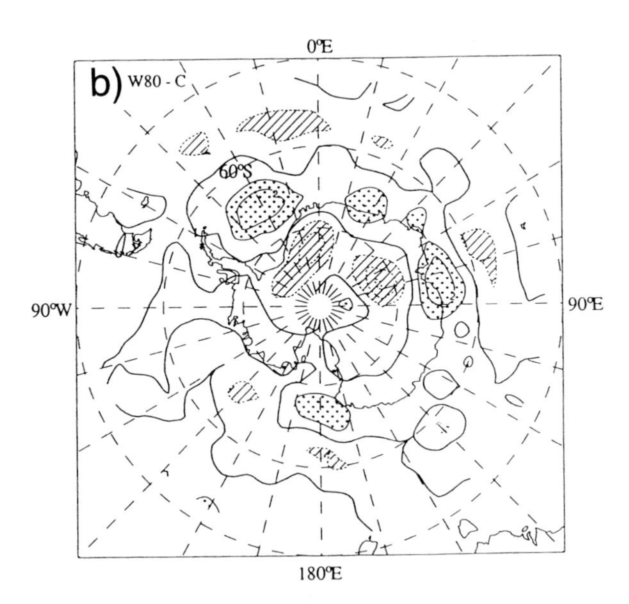

FIG. 10.22. Anomalies of cyclone density (defined as in Fig. 10.18a) for model experiments with (a) 50% open water in sea-ice area and (b) 80% open water in sea-ice area. Areas of cyclone density increase greater than 1 are stippled; areas of decrease less than −1 are hatched (Simmonds and Wu 1993).

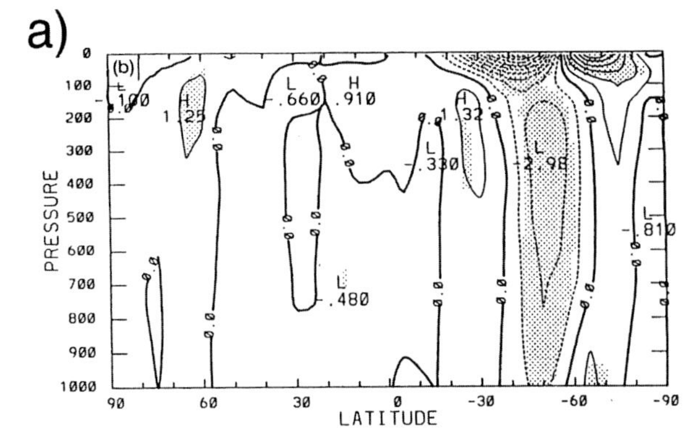

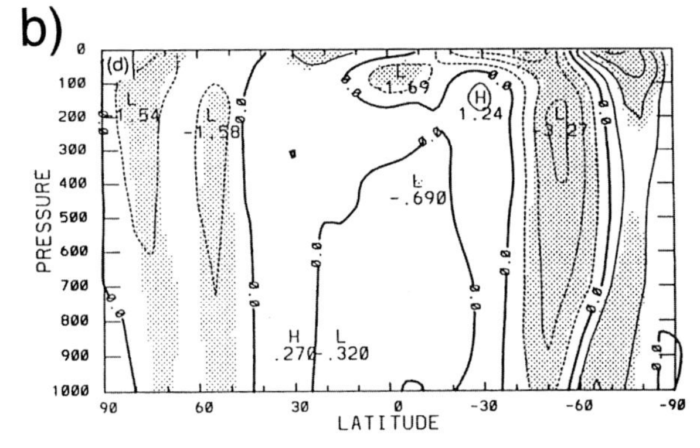

FIG. 10.23. Difference between the zonally averaged zonal wind for model experiments, with (a) 50% open water in sea-ice area and (b) 80% open water in sea-ice area. Zonal-wind decreases have dashed contours, and regions where differences are significant at the 95% confidence level are stippled (Simmonds and Budd 1991).

fall over the southern midlatitudes (with maximum negative values corresponding to the three southern oceans) and rise at higher latitudes.

More recently, a number of models have been integrated with the observed time series of global SSTs from 1979 to 1988 as part of AMIP. To assess the response of the time-evolving upper-level winds, Bromwich et al. (1995) have plotted the zonally averaged wind at 200 mb in the 180°–120°W sector for observations represented by the ECMWF analyses (Fig. 10.32a) for the period 1980–90 and the average of winds at that level for five high-resolution models for 1979–88 (Fig. 10.32b). As noted above, the high-resolution class of models generally better represents southern climate and, in particular, the double jet. In Fig. 10.32, the interannual out-of-phase relationship between the subtropical and high-latitude jet is simulated by the models, with the subtropical jet intensifying during ENSO events (1982–83, 1986–87) and the high-latitude jet weakening during those periods.

Still longer multidecadal integrations with specified observed time-evolving global SSTs have demonstrated atmospheric GCM simulation skill in reproducing decadal-timescale climate fluctuations over the globe and Southern Hemisphere, in general, and Australia in particular (Frederiksen et al. 1995).

c. Global coupled-model variability

In the late 1980s and early 1990s, it was established that the multidecadal global coupled GCMs being run at that time were capable of producing ENSO-like interannual variability purely due to the dynamical coupling between model components (e.g., Sperber et

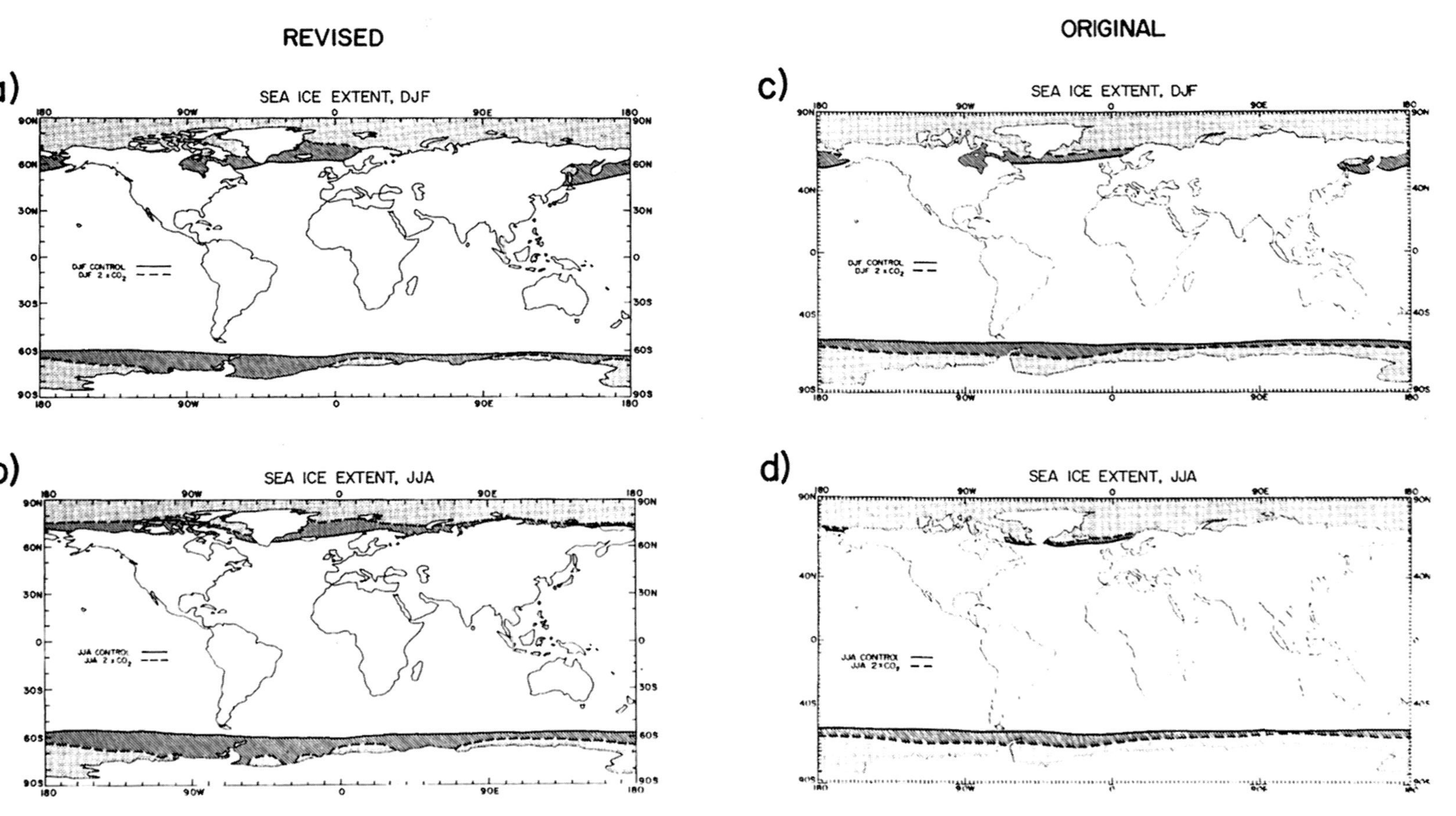

FIG. 10.24. Sea-ice extent (limit of ice 0.2 m thick) for the control case (solid lines) and the 2 × CO_2 case (dashed lines) for (a) revised sea-ice albedo experiment, DJF; (b) as in (a) but for JJA; (c) original sea-ice albedo experiment, DJF; (d) as in (c) except for JJA (Meehl and Washington 1990).

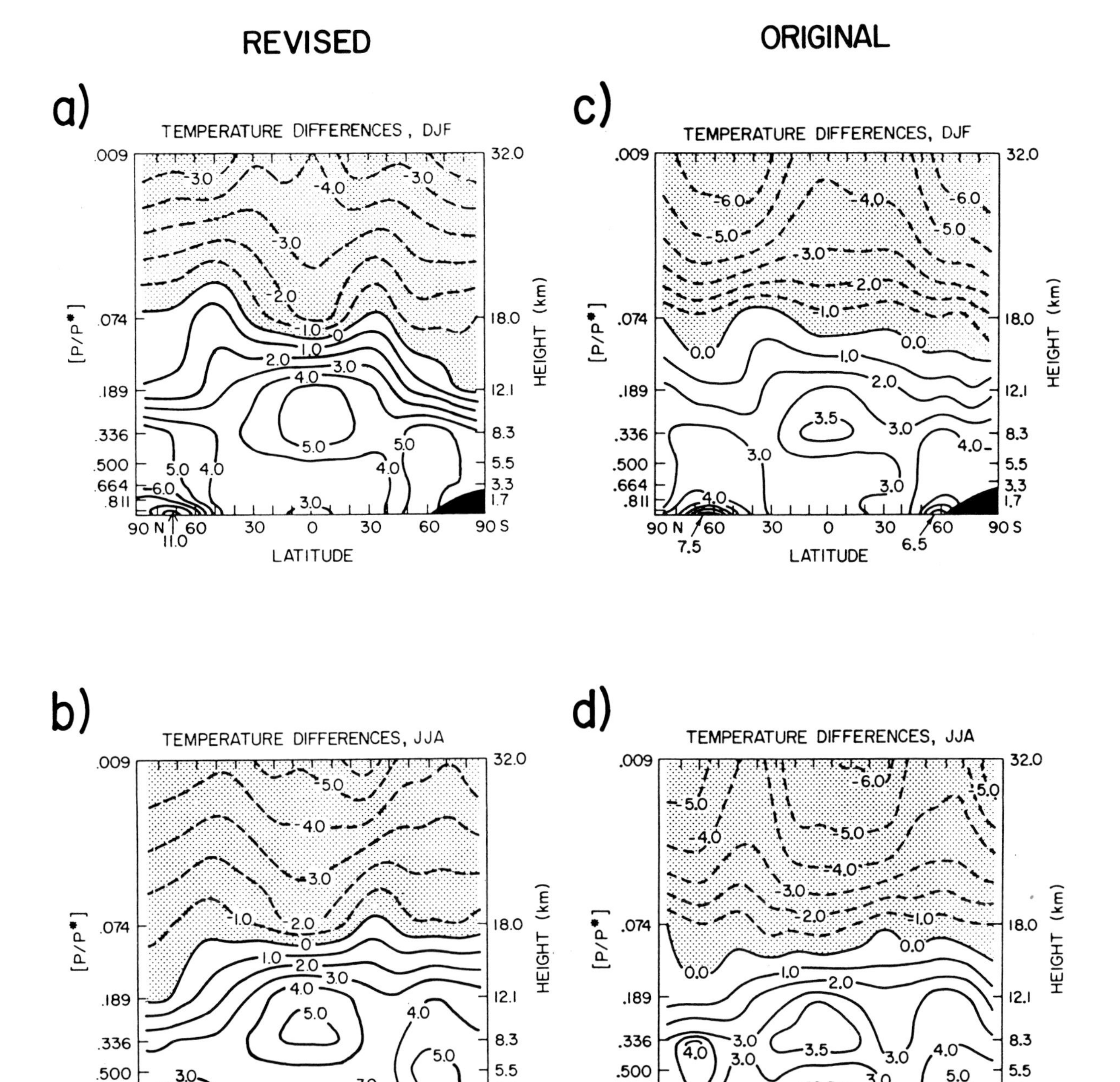

FIG. 10.25. Zonal-mean temperature differences (°C), 2 × CO_2 minus control: (a) DJF revised sea-ice albedo experiment; (b) same as in (a) but for JJA; (c) original sea-ice albedo experiment, DJF; and (d) as in (c) except for JJA (Meehl and Washington 1990).

al. 1987; Meehl 1990b; Lau et al. 1992; Nagai et al. 1992; Tett 1995). Before this time, limited-domain coupled models of the tropical Pacific basin had been able to produce ENSO-like oscillations (Neelin et al. 1992). With ENSO-like signals appearing in the global coupled models, truly global-scale response of the climate system to ENSO generated in these models could be studied. For example, the observed large-scale pattern of the Southern Oscillation (with opposition of SLP anomalies between the Pacific and Australasian sectors) could be produced in such coarse-grid global coupled models (Fig. 10.33 from Sperber et al. 1987). ENSO-like signals from a 20-yr time series produced by a global coarse-grid coupled model

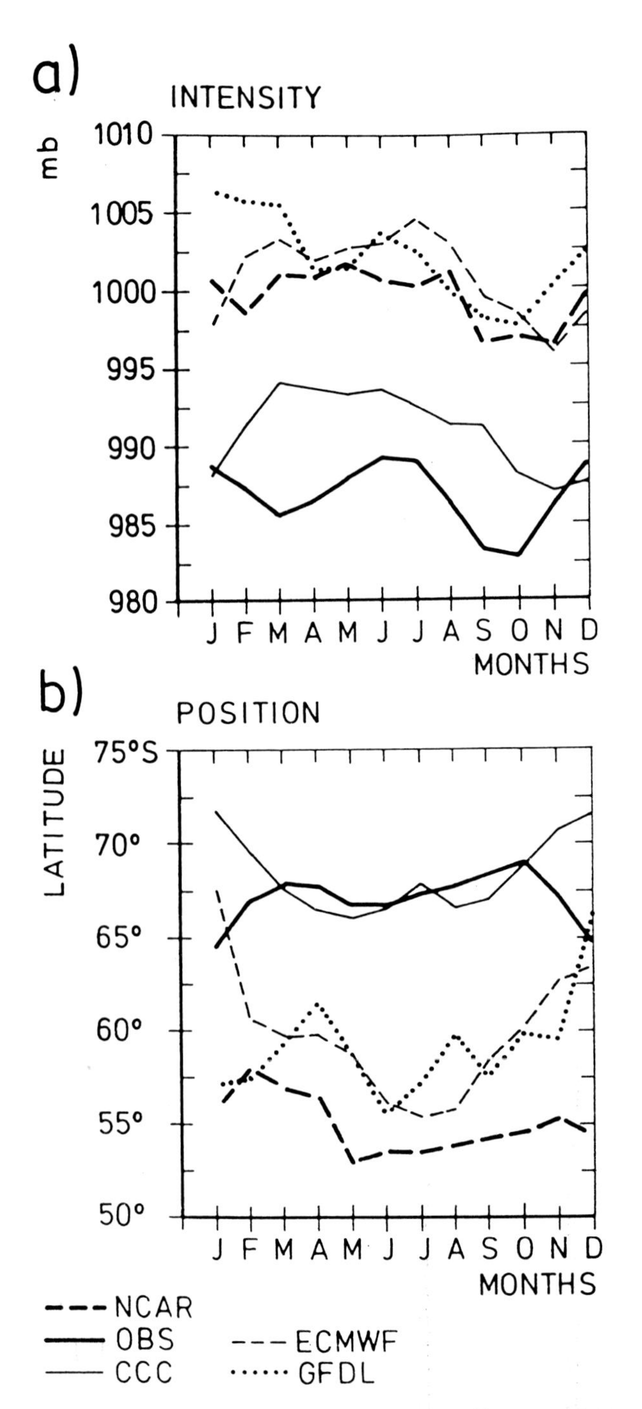

FIG. 10.26. Time series of monthly mean zonally averaged (a) intensity (mb) and (b) position (degrees south) of the circumpolar trough in the observations (solid heavy line) and four coarse-grid climate models run with specified climatological SSTs (Xu et al. 1990).

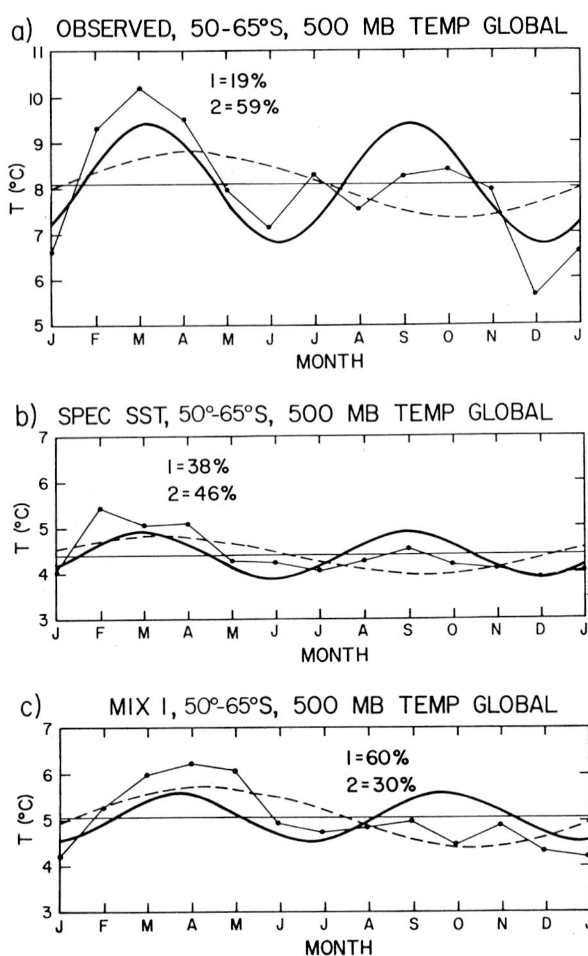

FIG. 10.27. Annual cycle of global zonal-mean 500-mb temperature differences (°C), 50°S minus 65°S, for (a) long-term mean observations (1972–84), (b) "SPEC SST" experiment with specified observed climatological SSTs, and (c) "MIX 1" experiment with the same atmospheric model coupled to a nondynamic slab or mixed-layer ocean formulation. Dots connected by thin line indicate monthly mean values. Dashed line is first harmonic; solid heavy line is second harmonic. Percentages indicate amount of total variance of monthly mean values accounted for by first and second harmonics (Meehl 1991).

from Lau et al. (1992) show warm SSTs in the central equatorial Pacific coincident with westerly surface wind anomalies, negative SOI, greater precipitation, and westerly anomaly ocean surface currents (Fig. 10.34). Results from all of these models show that the ENSO-like oscillations are of lower magnitude by typically half compared to observations, and only a subset of coupled processes thought to be responsible for ENSO in the observed climate system is represented (Neelin et al. 1992).

The global patterns of SST and SLP anomalies generated from a composite of warm events minus cold events (with the sign of the anomalies representative of a warm event) in a global coupled GCM are shown in Figs. 10.35a and 10.35b, respectively (Meehl 1990b). Observed SST anomalies from a typical warm event (December 1987) are shown in Fig. 10.35c, and composite SLP anomalies from observed warm events are shown in Fig. 10.35d (from van Loon 1986). The

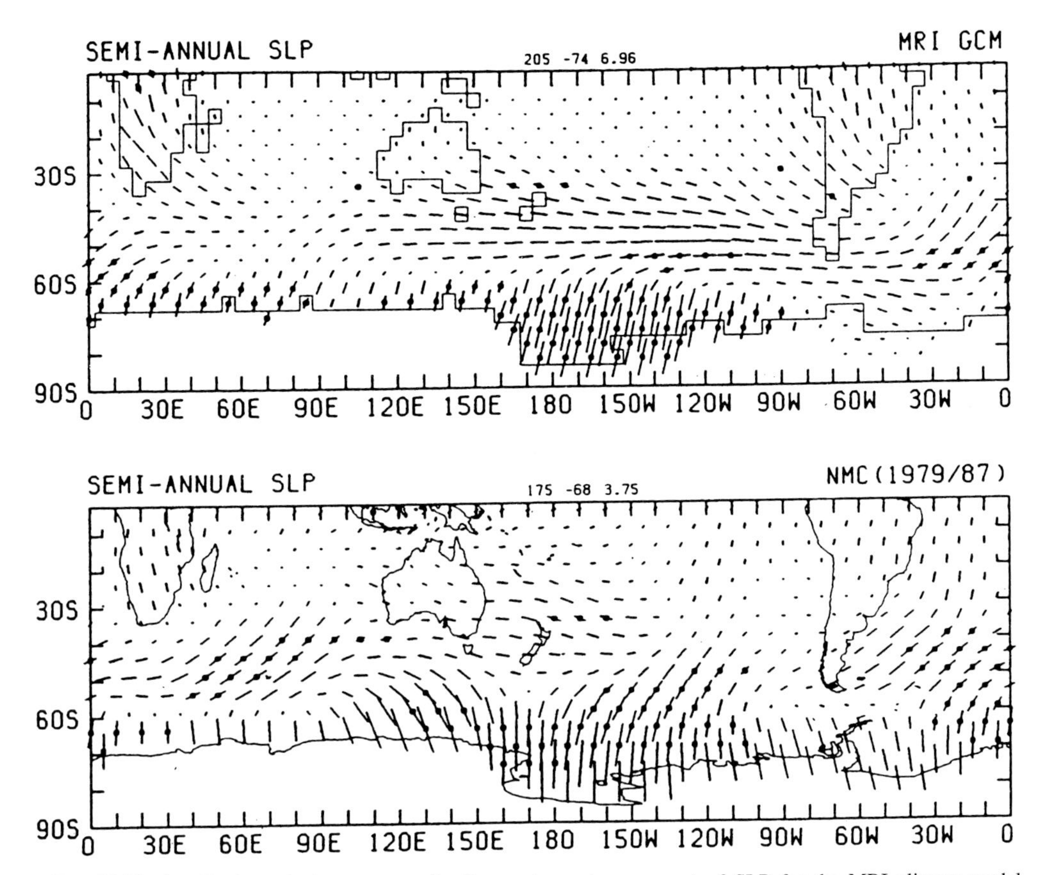

FIG. 10.28. Amplitude and phase arrows for the semiannual component of SLP for the MRI climate model (top) and NMC analyses from 1979–87 (bottom). Length of arrow indicates relative amplitude. Arrow pointing straight north has first maximum 1 January; one pointing straight east has first maximum 1 April, etc. Length of arrows is normalized with the largest value of 6.96 mb at 155°W, 74°S for the model, and 3.75 mb at 175°E, 68°S for the observations. Dots on arrows indicate that the semiannual amplitude at that grid point exceeds the amplitude of the annual harmonic (Kitoh et al. 1990).

model produces warm SST anomalies associated with negative SLP anomalies in the equatorial Pacific, as observed. Positive SLP anomalies are seen over Australasia, northeastern South America, and southern Africa, also as observed. There are negative SLP anomalies in the model and observations in the Indian Ocean near 40°S and in the eastern Pacific at that same latitude. There is less correspondence between model and observations at high southern latitudes, possibly due to the systematic errors at mid- and high southern latitudes in this non–flux-corrected coupled model (Whetton et al. 1996a).

In spite of the systematic errors, this coupled model can reproduce the large-scale tropical moisture anomalies associated with warm events. Figure 10.35e shows observed composite ENSO precipitation anomalies for southern summer and can be compared to area-averaged composite soil moisture anomalies (cm) for the tropical continental areas of southern Africa, Australasia, and northeastern South America from the coupled model in Fig. 10.35f. These areas are particularly at risk for moisture deficits during ENSO warm events (Fig. 10.35e). As seen by the correspondence between the model results in Fig. 10.35f and the observations in Fig. 10.35e, the coupled model is able to simulate moisture deficits in each of those areas for model-generated warm events.

Biennial signals involving coupled air–sea anomalies in the tropical Indian and Pacific regions, termed the tropospheric biennial oscillation (TBO), have been identified in observations and a global coupled model, with extremes in these signals manifested as warm and cold events in the Southern Oscillation (Meehl 1994, 1997a). These tropical air–sea anomalies that have identifiable links to the Southern Hemisphere midlatitude circulation (Meehl 1987; Kiladis and van Loon 1988) appear to be associated with midlatitude circulation anomalies in the Northern Hemisphere that influence the south Asian monsoon. Figure 10.36 is a schematic representing these interactions. SST anomalies of opposite sign in the Australian monsoon and tropical Pacific regions are associated with convective heating anomalies (i.e., a weak Australian monsoon is associated with cool SSTs to the north of Australia and

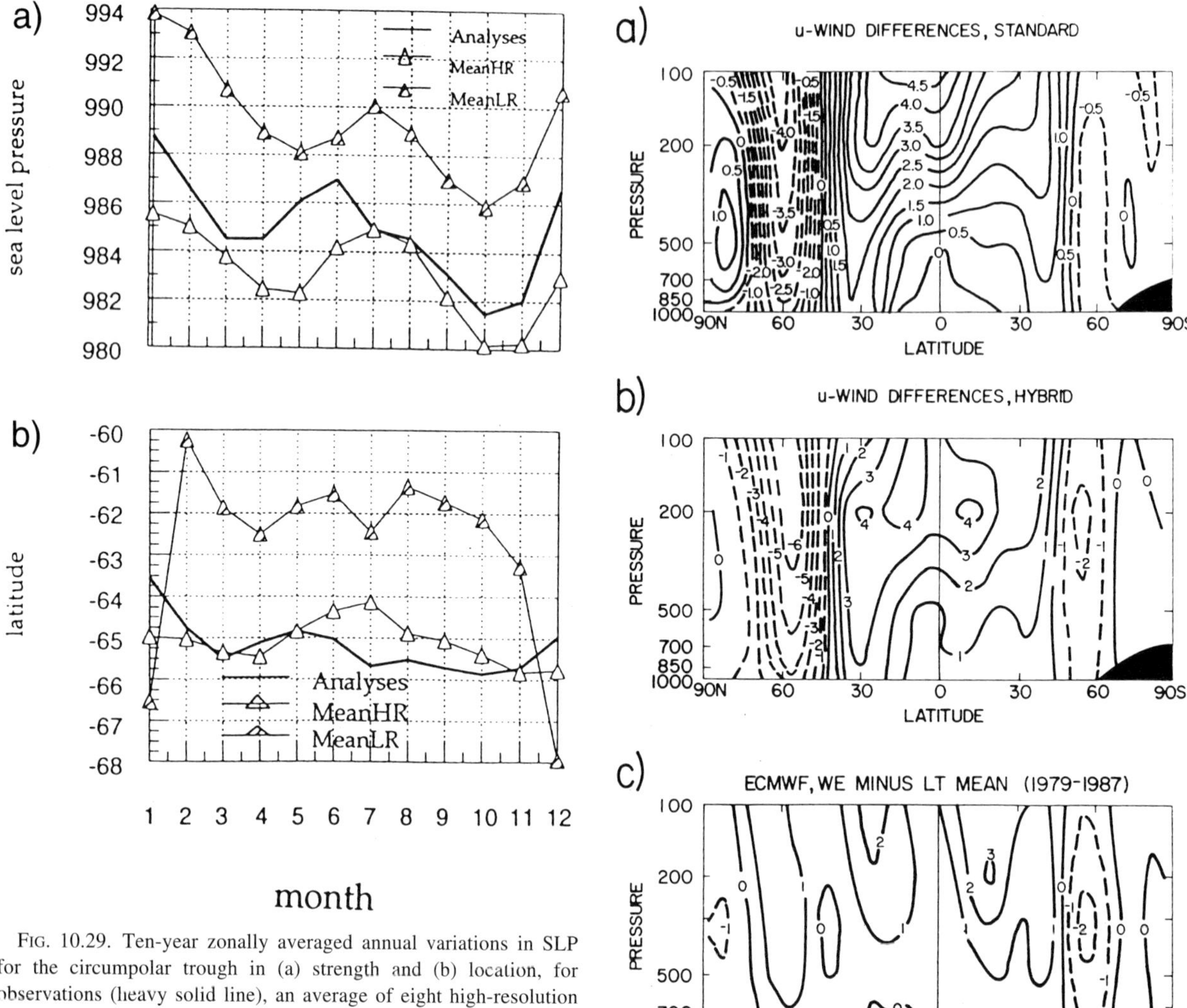

FIG. 10.29. Ten-year zonally averaged annual variations in SLP for the circumpolar trough in (a) strength and (b) location, for observations (heavy solid line), an average of eight high-resolution climate models (clear triangles), and an average of nine low-resolution models (hatched triangles) (Bromwich et al. 1995).

FIG. 10.30. Zonal-mean u-component wind differences (m s^{-1}) for warm-event composite minus control for (a) "standard" experiment with convective adjustment, (b) "hybrid" experiment with mass-flux convective scheme, and (c) observations, two warm events from the ECMWF analyses, 1979–87 (Meehl and Albrecht 1991).

warm SSTs to the east; Fig. 10.36a). The associated air–sea interaction alters the SST anomaly pattern near Australia as well as in the tropical Indian and Pacific Oceans, and northern midlatitude circulation anomalies contribute to a subsequent strong south Asian monsoon (Fig. 10.36b). In the following southern summer, the warm SSTs north of Australia are associated with a strong Australian monsoon with cool SSTs in the Indian and Pacific (Fig. 10.36c). This is followed by a weak south Asian monsoon (Fig. 10.36d), partly due to the convective heating anomalies in the Australian region the southern summer before.

The characteristics of interannual variability of lower-troposphere temperature in a global coupled GCM have been compared to observations as represented by microwave sounding unit (MSU) satellite data (Meehl et al. 1994). Figure 10.37 shows comparable measures of zonal average lower-tropospheric temperature for the MSU data (at left) and coupled model (at right) at 60°N, Equator, and 60°S. The general characteristic of the midlatitude variability is reproduced by the coupled model, with greater interannual variance (heavy solid line) at 60°N compared to 60°S in both model (right) and observations (left). At the Equator, the ENSO-like variability discussed above is present in the coupled model at center right, with somewhat smaller amplitude than the observations at center left.

A geographic representation of interannual variability of lower-tropospheric temperature in a global coupled model compared to observations, as represented by the MSU satellite data, is shown in Fig. 10.38 (Meehl et al. 1994). As suggested by Fig. 10.37, the coupled model correctly characterizes less variability

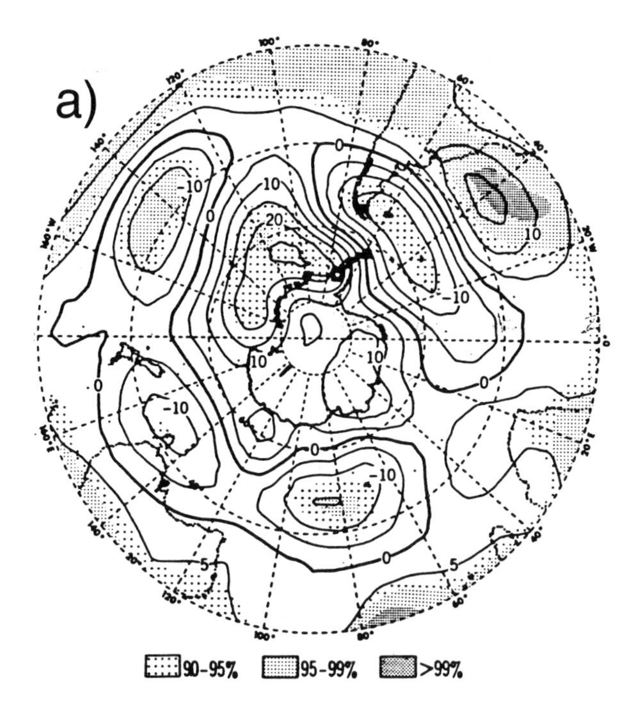

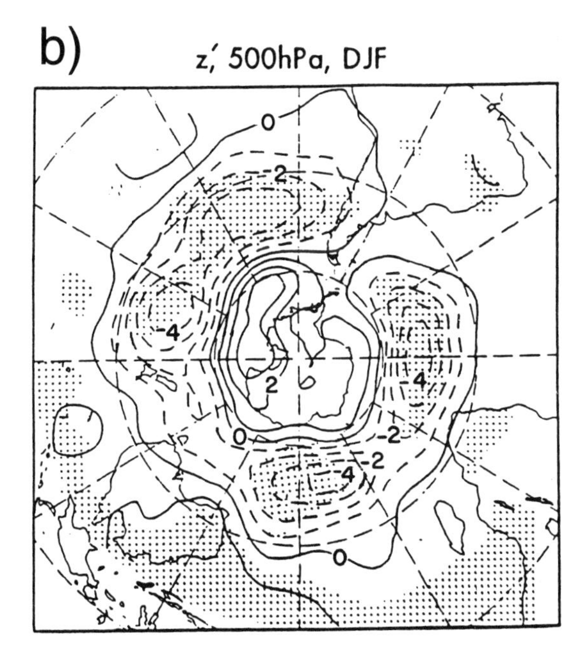

FIG. 10.31. (a) Composite charts of the model-generated departure from 30-yr climatology of 500-mb height, as constructed using simulated data for the period of December (year 0) to February (year +1) in six individual ENSO episodes from a climate model run with the observed time evolution of tropical Pacific SSTs from 1962 to 1976. Contour interval is 5 m. Light, medium, and dark shading indicates anomalies exceeding the 90%, 95%, and 99% significance levels, respectively (Lau 1985). (b) Observed composite winter anomalies for three ENSO events for 500-mb height. Anomalies significant at the 95% significance level are shaded (Karoly 1989).

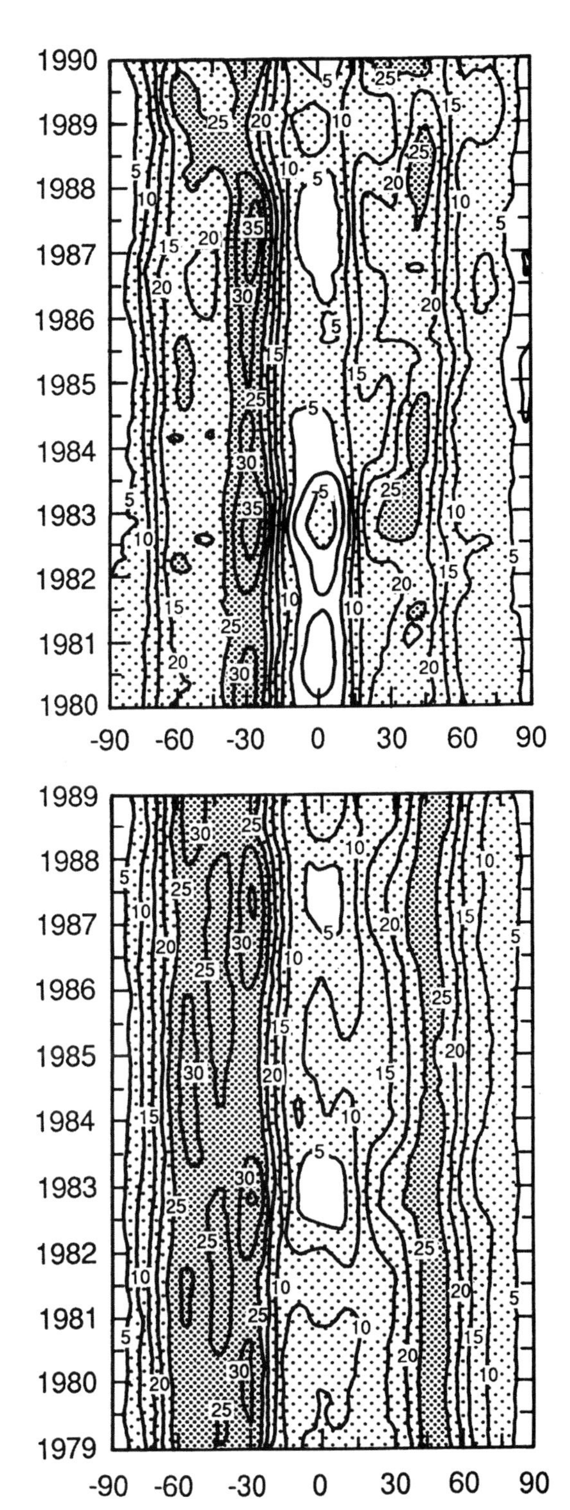

FIG. 10.32. Hovmöller diagram of zonally averaged zonal wind in Pacific sector (180° to 120°W) at 200 mb: (a) ECMWF analysis from January 1980 through December 1989; (b) simulations of five high-resolution climate models run with the observed time-evolving global distribution of SSTs from January 1979 through December 1988 (Bromwich et al. 1995).

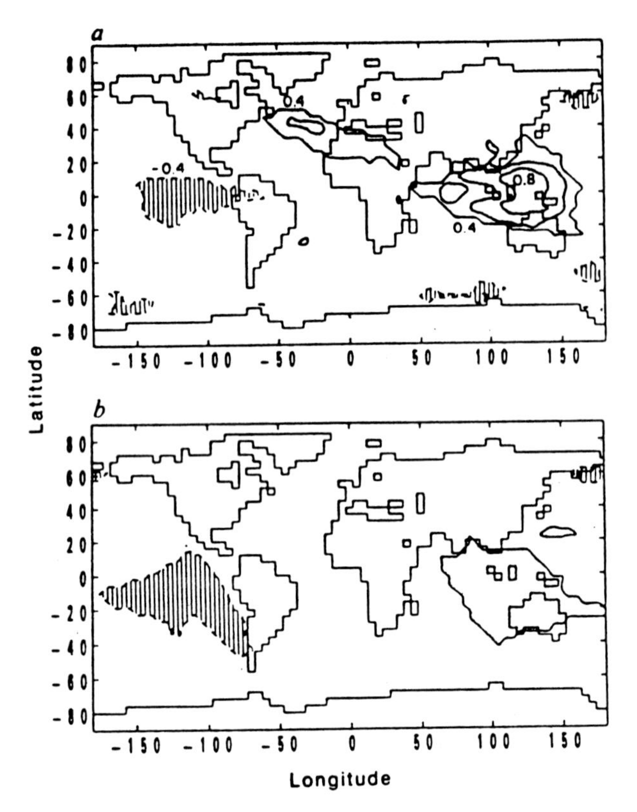

FIG. 10.33. (a) Correlation of SLP at a point near Darwin, Australia, with all other points in the global coupled model; (b) same as (a) except for observed correlations. First contour is 0.4 or −0.4 (the latter areas hatched), and contour interval is 0.2 (Sperber et al. 1987).

in the southern midlatitudes compared to northern midlatitudes, with greater variability over Antarctica compared to the southern midlatitude circumpolar oceans. The ratio of variances in Fig. 10.38c, model divided by observations, shows that the model is reasonably close to the observations in the Tropics and subtropics but underestimates the interannual variability poleward of about 60°S over Antarctica.

Global coupled models run for longer periods of time can provide estimates of variability at longer timescales. Observed variability of 5-yr seasonal averages is shown in Figs. 10.39a and 10.39b for summer and winter, and similar values are shown from a global coupled model in Figs. 10.39c and 10.39d (Meehl et al. 1993b). The observations, where available, suggest greater variability at higher southern latitudes. This is certainly the case for the coupled model. In particular, 5-yr timescale variability is enhanced in the model in the winter hemisphere, probably in association with coupled ocean–sea-ice variability at those timescales. These same features have been noted in the global coupled model of Manabe and Stouffer (1996).

Decadal timescale variability in global coupled models has shown distinct Southern Hemisphere signatures. For example, von Storch (1994) showed two leading modes of interdecadal variability (Fig. 10.40) from a coarse-grid global coupled model. One has a dominant tropical Pacific signal with connections to high southern latitudes (Fig. 10.40a). The other is a southern midlatitude mode. These results from the model suggest that the southern oceans could carry a significant part of global decadal variability signals. In that same study, von Storch showed a Pacific mode with a 17-yr period with signals that rotate around the South Pacific gyre on that timescale (see discussion of observed longer-period variability in chapter 9).

Bjerknes (1964) proposed a north–south oscillation on decadal timescales connecting SST anomalies of opposite sign between the North and South Atlantic. Bryan and Stouffer (1991) suggested that this oscillation could be a product of decadal variability of the thermohaline circulation. They cited results from a global coupled model experiment by Manabe and Stouffer (1988) that showed SST anomalies between two climate states, one with an active thermohaline circulation and one without (Fig. 10.41). This experiment shows a clear north–south signal, not only in the Atlantic but between hemispheres, such that the Northern Hemisphere is out of phase with the Southern Hemisphere in relation to the strength of the thermohaline circulation.

Multicentury integrations of coupled ocean–atmosphere climate models have indicated the nature of long-timescale variability that can be compared with paleoclimate proxy data (see discussion in chapter 9). Manabe and Stouffer (1996) calculated 25-yr timescale variability of surface air temperature from such a coupled-model integration in comparison to their mixed-layer model and the atmospheric model run with observed climatological monthly mean SSTs (Fig. 10.42). There is enhanced variability on the 25-yr timescale at high southern latitudes in the coupled model (Fig. 10.42a) compared to less variability in the mixed-layer model on this timescale (Fig. 10.42b) and much less in the specified SST model (Fig. 10.42c). This suggests that ocean dynamics in the southern oceans contribute to 25-yr timescale variability. This result is consistent with the study by von Storch presented in Fig. 10.40. To illustrate the timescales of this very low–frequency variability in the southern oceans, Manabe and Stouffer (1996) computed spectra for a grid point at 60.75°S, 142.5°W from their coupled and mixed-layer models (Fig. 10.42d). The coupled model shows enhanced variability on all timescales compared to the mixed-layer model, but the very long–timescale variability is enhanced even more at timescales greater than a century. This result points to the possible important role of the southern oceans in very long–timescale variability of the global climate system.

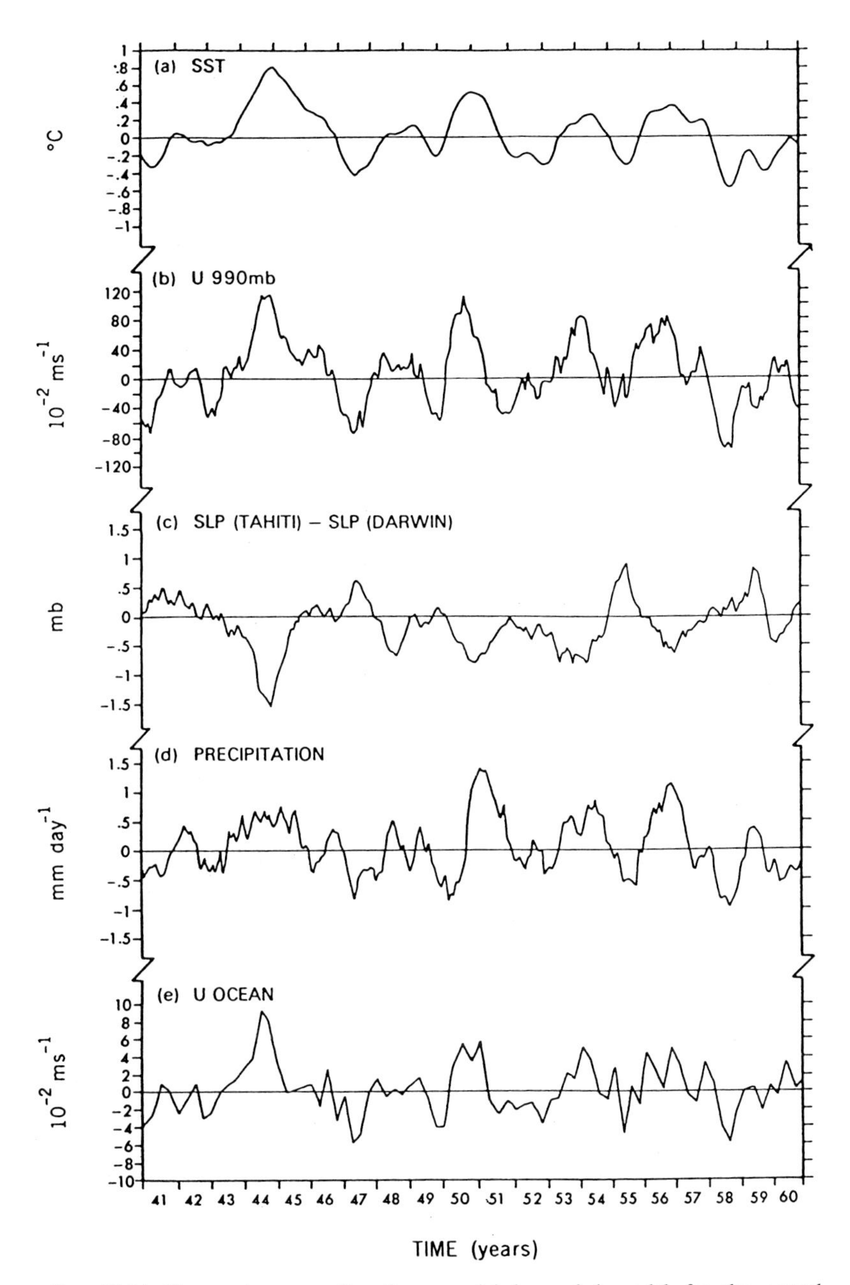

FIG. 10.34. Time-series anomalies from a global coupled model for the central equatorial Pacific areas [6.75°S to 6.75°N, 165°E to 165°W in (a), (b), and (d); 4.5°S to 4.5°N, 165°E to 165°W in (e)] for (a) SST; (b) zonal wind at 990 mb; (c) SLP difference between grid points near Tahiti and Darwin; (d) precipitation; and (e) zonal current speed of the top ocean layer. Values in (a)–(d) are seven-month running means; seasonal mean values are plotted in (e) (Lau et al. 1992).

10.4. Simulated climate change in the Southern Hemisphere

a. Results from mixed-layer models

Most of the climate model results to date that have been used for regional climate-change scenarios for the Southern Hemisphere have relied on mixed-layer model experiments. A typical experiment with this class of model involves a control integration with present amounts of CO_2 and a doubled CO_2 experiment run to equilibrium (i.e., a state of little drift of globally averaged temperature). Difference maps of $2 \times CO_2$ minus control usually show warming everywhere, with somewhat greater values over continents.

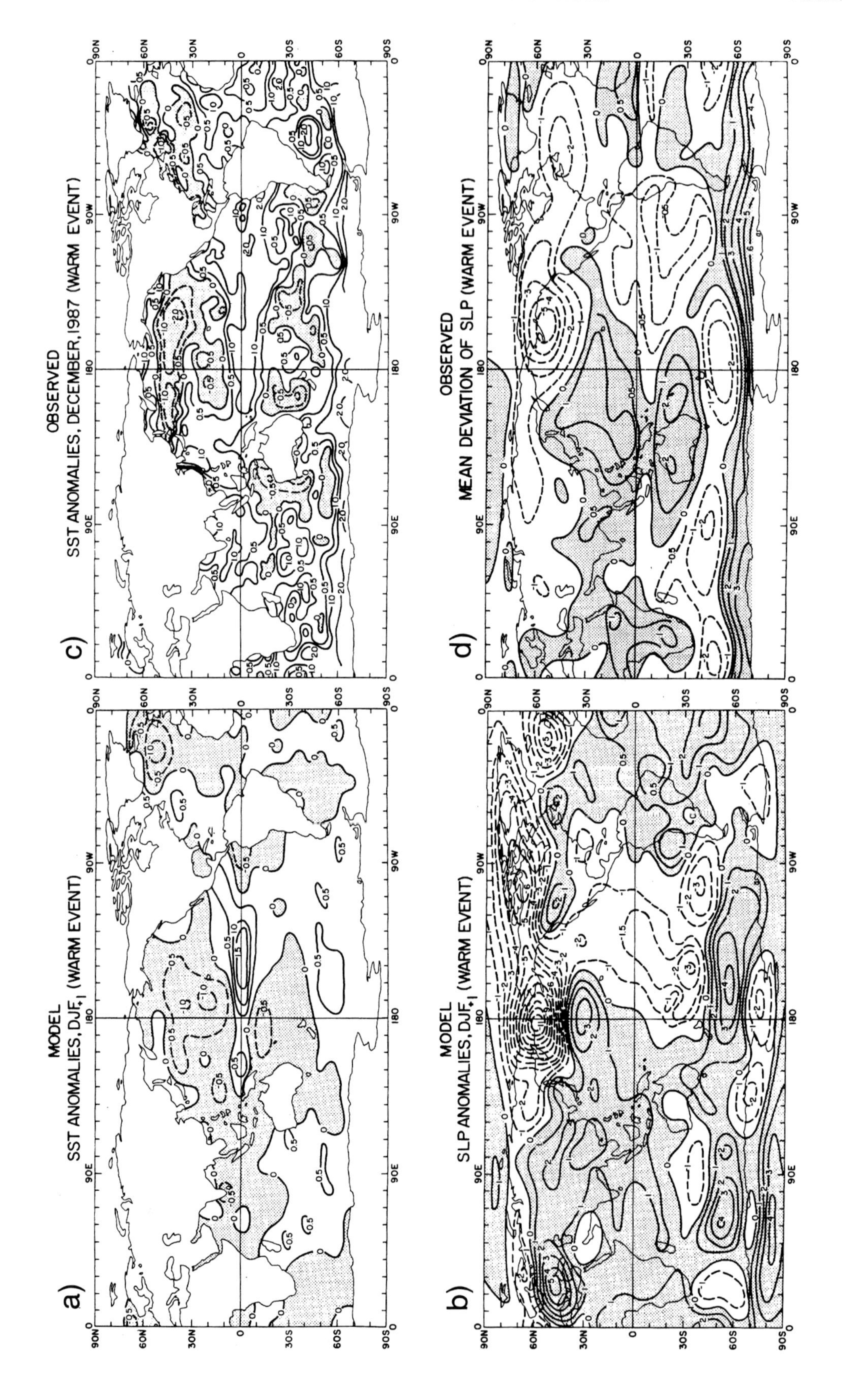
a)
MODEL
SST ANOMALIES, DJF+1 (WARM EVENT)
b)
MODEL
SLP ANOMALIES, DJF+1 (WARM EVENT)
c)
OBSERVED
SST ANOMALIES, DECEMBER, 1987 (WARM EVENT)
d)
OBSERVED
MEAN DEVIATION OF SLP (WARM EVENT)

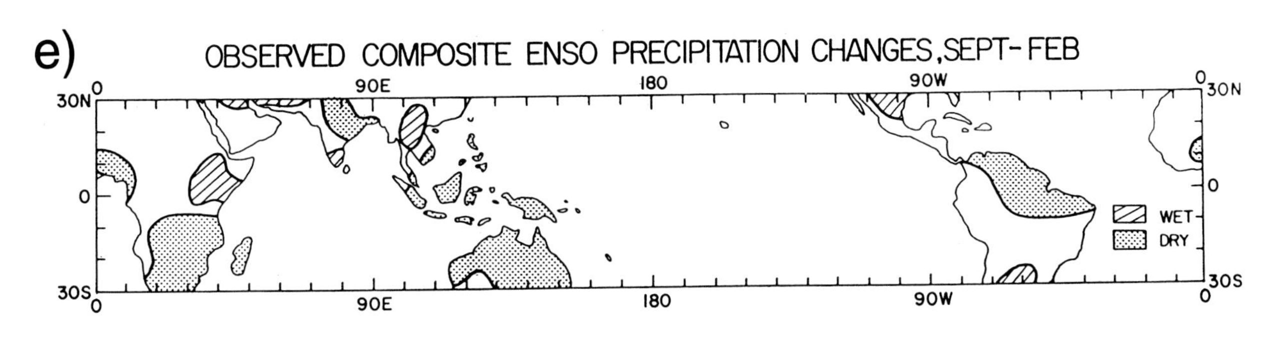

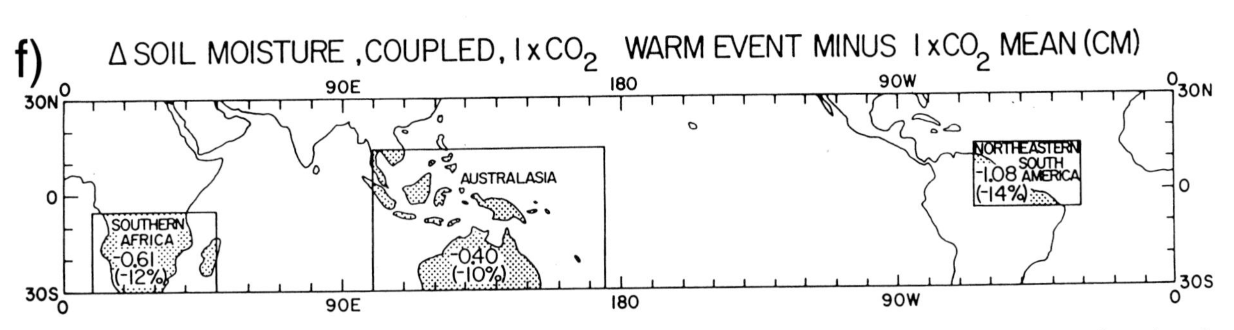

FIG. 10.35. Geographic plots from a global coupled model for (a) SST differences for warm- minus cold-event composites, sign of differences is representative of a warm event in the central equatorial Pacific; (b) same as (a) except for SLP; (c) observed global blend SST anomalies from the Climate Analysis Center for December 1987—this period is representative of a warm event in the central equatorial Pacific; (d) mean SLP anomalies for observed warm events [parts (a)–(d) from Meehl 1990b]; (e) observed composite precipitation anomalies, stippling indicates areas susceptible to moisture deficits during warm events; (f) area-averaged soil moisture differences (cm) for land points from the coupled model for the areas indicated for composite warm events, percentage changes of the differences are given in parentheses [parts (e) and (f) from Meehl et al. 1993a].

Large warming is seen in the winter hemisphere high latitudes due to ice-albedo feedback (IPCC 1990).

The standard climate-change scenario for the Southern Hemisphere produced by mixed-layer models (IPCC 1990) is almost-uniform warming of a couple of degrees over the oceans, several degrees of warming over the southern continents, and large warming (15°C and more) during winter near Antarctica due to the retreat of sea ice. In these equilibrium integrations, there is a general decrease of soil moisture over the southern continents, with some models showing increases of precipitation and soil moisture in the Australian monsoon region. These results for mean climate change have generally been reconfirmed in the later coupled models, the major exception being smaller-amplitude warming at high southern latitudes in the coupled models (even though those results could be underestimated due to the likelihood of excessive mixing in the ocean GCMs in the circumpolar ocean, discussed later). Thus, even though the lack of ocean dynamics precludes studying changes of certain types of interannual variability in mixed-layer models (e.g., ENSO), and the time evolution of such changes cannot be determined from equilibrium experiments, results from mixed-layer model experiments may still provide adequate first-order indications of climate-change scenarios for certain regions of the Southern Hemisphere (Whetton et al. 1996a). For example, mixed-layer model results suggest that the southern African region could become warmer and that tropical, subtropical, and midlatitude circulation systems could weaken and shift southward (Joubert 1994). There are also indications that there could be an increase in the probability of dry years over large areas of southern Africa (Joubert and Mason 1996), with increases in both the frequency and intensity of extreme daily rainfall events (Mason and Joubert 1997). Often, contradictory results from mixed-layer model experiments have made the establishment of climate-change scenarios for South America difficult, but there does seem to be a tendency for the models to show more frequent, intensive, and extensive precipitation deficits for that region with increased CO_2 in the future (Burgos et al. 1991).

Concerning changes in Australian precipitation, Ryan et al. (1992) showed that the equatorial Australian monsoon shear line did not shift in a mixed-layer model with increased CO_2, but the monsoon precipitation increased somewhat (Fig. 10.43). An examination of summer- and winter-region precipitation over Australia (Whetton et al. 1994) showed a nearly year-round increase in summer-region precipitation in

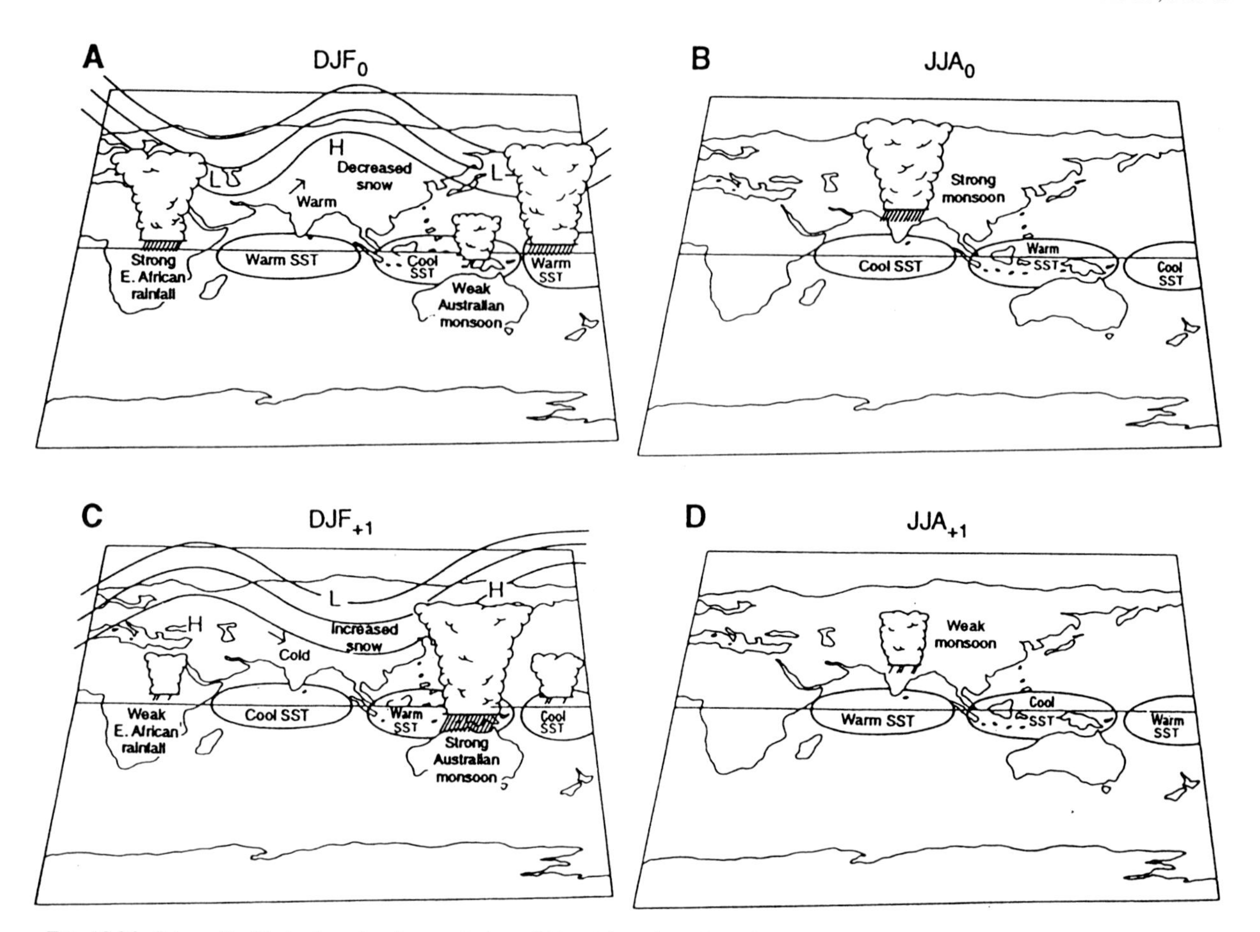

FIG. 10.36. Schematic illustration showing evolution of biennial cycle and tropical–midlatitude interaction noted in a global coupled model and observations for (a) summer of a weak Australian monsoon, (b) winter after the weak Australian monsoon showing a strong Asian monsoon, (c) summer of a strong Australian monsoon, and (d) the following winter and weak south Asian monsoon (Meehl 1994).

northern Australia with increased CO_2 in a mixed-layer model (Fig. 10.44a), and a nearly year-round decrease of winter-region precipitation in southern Australia (Fig. 10.44b). Whetton et al. (1994) also noted that these model-produced changes were similar to recent observed changes of precipitation in those two regions (Figs. 10.44c,d). In mixed-layer model studies of precipitation over Australia, Pittock et al. (1991), Gordon et al. (1992), and Whetton et al. (1993) documented a tendency for heavy rainfall events to increase, with a decrease of return period for such heavy rain events (Fig. 10.45). For New Zealand, climate-change scenarios derived from GCM experiments suggest mostly warmer and wetter conditions in the future due to increased CO_2 (Whetton et al. 1996b).

In the extratropics, changes in blocking activity (shown for the present-day climate simulated by a mixed-layer model in Fig. 10.21) in Fig. 10.46 show general decreases in both summer and winter with increased CO_2. Because of the relatively small surface warming at high southern latitudes in coupled models, however, compared to large warming at those latitudes in mixed-layer models such as this one (discussed later), this result will have to be revisited with the later generation of models with improved ocean-mixing schemes.

Mixed-layer models have also been used to study changes in interannual variability simulated by those models, keeping in mind that those models are not capable of producing ENSO variability. Mearns (1993), Gordon and Hunt (1994), and Liang et al. (1995) used mixed-layer models and confirmed some of the earlier mixed-layer model results that showed some areas of increased interannual temperature variability in the Tropics and subtropics, with decreases at higher latitudes.

b. Results from coupled ocean–atmosphere GCMs

Publication of results from global coupled–climate model experiments with gradually increasing amounts of CO_2 began in the late 1980s and early 1990s. These experiments confirmed some results from the mixed-layer models. For example, warming from the coupled models (summarized by IPCC 1992, 1996) occurs over

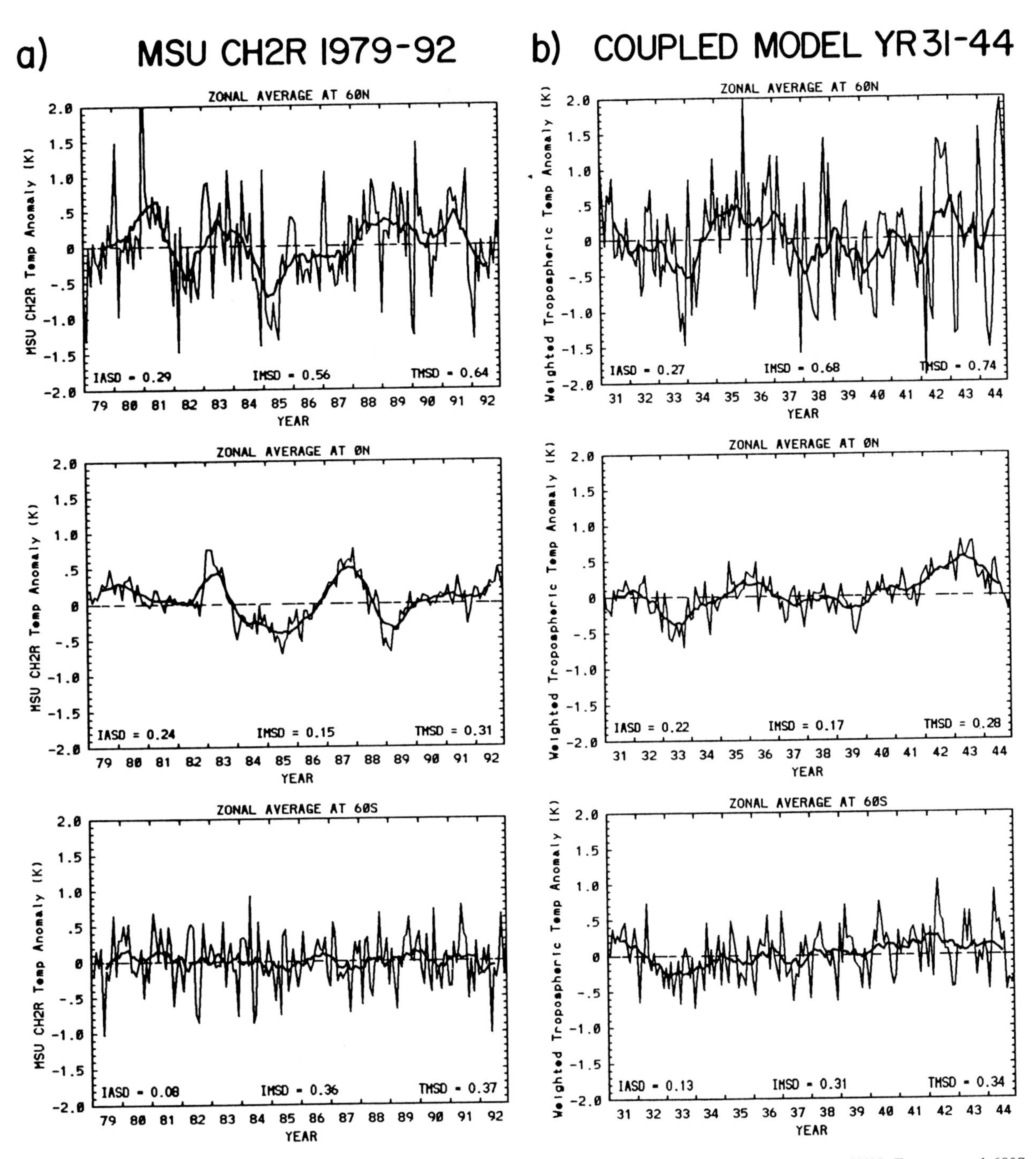

FIG. 10.37. Time series of the zonal average monthly (thin line) lower-tropospheric temperature anomalies at 60°N, Equator, and 60°S in °C from observations represented by MSU CH2R satellite data (left) and the coupled model (right). The thick line is the 12-month running mean to represent interannual variability. IASD = interannual standard deviation; IMSD = intermonthly standard deviation; and TMSD = total monthly standard deviation (Meehl et al. 1994).

the tropical and subtropical oceans, and somewhat greater warming occurs over the continents. Also similar to the mixed-layer model results, there appear to be small mean decreases of soil moisture over most of the southern continental areas. Some of the results from the coupled models, however, have changed previous perceptions of simulated CO_2 climate change in the Southern Hemisphere in terms of the consequences of including ocean dynamics as well as the more realistic gradual increases of CO_2 in these experiments. As we have seen earlier, the coupled models are able to simulate some aspects of coupled climate

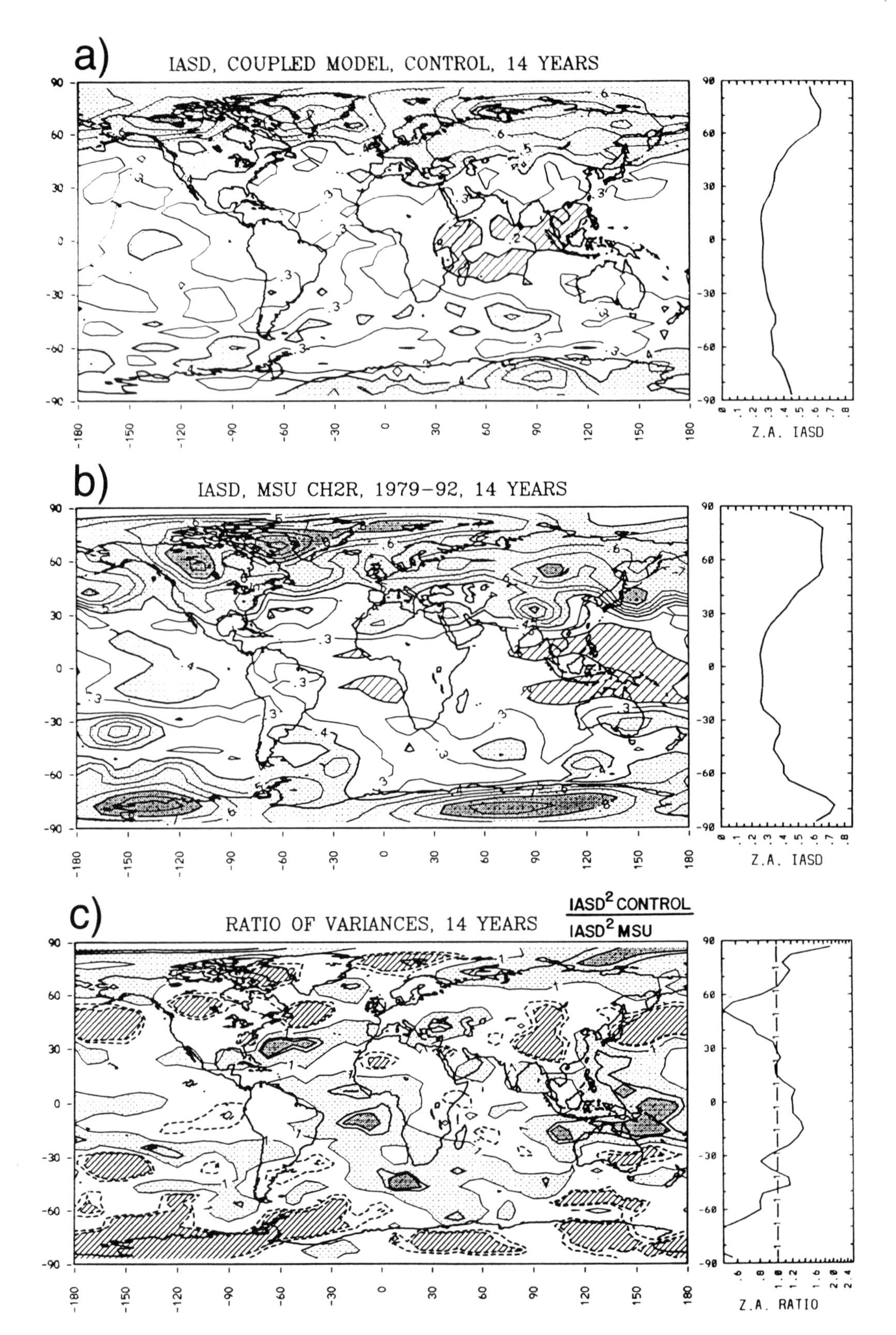
a)
IASD, COUPLED MODEL, CONTROL, 14 YEARS
Z.A. IASD
b)
IASD, MSU CH2R, 1979–92, 14 YEARS
Z.A. IASD
c)
RATIO OF VARIANCES, 14 YEARS
IASD2 CONTROL
IASD2 MSU
Z.A. RATIO

variability at ENSO and longer timescales, and changes in this variability due to increased CO_2 can be studied in these models.

Figure 10.47 shows the time evolution of zonal-mean decadal warming from two of these coarse-grid coupled models (Meehl et al. 1993b; Stouffer et al. 1989). The NCAR model uses a CO_2 transient increase of 1% per year linear without flux correction, while the Geophysical Fluid Dynamics Laboratory (GFDL) uses 1% per year compound with heat and freshwater-flux correction. Therefore, CO_2 is increasing more rapidly in the GFDL experiment. In any case, general features of the warming are similar between the two models. These characteristics are mirrored in the other coupled models that have been run in this same configuration with varying degrees of flux correction (IPCC 1992; Colman et al. 1995). Warming in the Southern Hemisphere lags that in the Northern Hemisphere at high latitudes. There is little warming at southern high latitudes in winter because the ice-albedo feedback that played such a large role in high-latitude Southern Hemisphere warming in the mixed-layer models is not a factor in those areas in the coupled models. This is due to the deep mixing in the circumpolar Southern Ocean associated with the Deacon cell, though the strength of this mixing is not entirely clear (see discussion later in this chapter).

Figure 10.47 also shows that, as the oceans in the southern midlatitudes warm, the delayed warming of the high-latitude southern oceans causes an increase of Equator-to-pole temperature gradient. This is associated with an increase in baroclinicity and zonal-wind stress near about 55°S (Fig. 10.48, taken from Washington and Meehl 1989; Manabe et al. 1991). This results in even more vigorous mixing of the circumpolar ocean and a stronger Deacon cell in these models, thus further delaying southern high-latitude warming. In Fig. 10.49, the zonal-mean meridional streamfunction from the GFDL model shows a weakening of the overturning in the northern midlatitudes from about 15 to 10 Sv, but in the southern midlatitudes, where there is a slight increase in wind stress as noted in Fig. 10.48, the Deacon cell strengthens somewhat (note greater area encompassed by the 15-Sv contour near 50°S in Fig. 10.49b in the increased CO_2 experiment compared to the control experiment in Fig. 10.49a) and antarctic bottom-water formation is reduced (note values of over -10 Sv next to Antarctica near 70°S in Fig. 10.49a, weakening to around -5 Sv in Fig. 10.49b). As noted earlier, the mixed-layer models all showed large warming and sea-ice retreat due to ice-albedo feedback in the Antarctic region. In the coupled model shown above from GFDL, antarctic sea-ice decreases in the Atlantic sector in summer with increased CO_2 and actually thickens somewhat in most other regions (Fig. 10.50).

It has been noted, however, that the simplistic representation of the Deacon cell in the meridional streamfunction as a large, uniform, strong deep-mixing circulation (as in Fig. 10.49, derived from integration on depth levels) is quite different (and considerably less apparent) when integrating along density surfaces (Karoly et al. 1997). There have also been suggestions, based on observed ocean tracers such as chlorofluorocarbons, that the deep mixing associated with the Deacon cell noted in model simulations (such as those in Fig. 10.49) is overestimated (England 1995). If this is the case, ocean models that include improved mixing schemes could show less deep mixing in the vicinity of the Deacon cell and greater warming of the circumpolar southern oceans, which would more resemble the earlier mixed-layer model climate-change results (Whetton et al. 1996a). For example, such differences in simulated climate change due to increased CO_2 between mixed-layer models and coupled models have been noted for the southern African area, with greater normalized warming in coupled models than in mixed-layer models over subcontinental southern Africa, while mixed-layer models show greater normalized warming in the high-latitude Southern Ocean (Joubert and Tyson 1996).

Several studies with global coupled ocean–atmosphere GCMs (Knutson and Manabe 1995; Tokioka et al. 1995; Tett 1995; Meehl and Washington 1996) have shown a differential mean warming of SSTs across the tropical Pacific due to increased CO_2, with the eastern Pacific warming faster than the western Pacific (Fig. 10.51a). This is also the pattern of SST anomalies associated with ENSO events. Consequently, the mean precipitation change due to an increase of CO_2 resembles precipitation anomalies associated with present-day ENSO events in those models (Meehl and Washington 1996). That is, relatively greater mean increases in precipitation occur in an increased CO_2 climate over the equatorial Pacific, with relatively fewer increases or some decreases in the ITCZ and SPCZ regions, as well as in the far western Pacific over Indonesia and northern Australia (Fig. 10.51b). The consequences of the similarity in

FIG. 10.38. (a) Interannual standard deviation (IASD) of the coupled-model lower-tropospheric temperature for a 14-yr period in the model integration; different 14-yr periods show similar patterns. (b) IASD for 14 years of observations of lower-troposphere temperature from MSU satellite data; for panels (a) and (b) contour interval is 0.1°C, light shading for 0.8°C, and cross-hatching for IASD less than 0.2°C; the zonal average of the IASD is shown on the right. (c) The ratio of the interannual variances (i.e., the square of the interannual standard deviations) of the coupled-model lower-troposphere temperature divided by the MSU temperature. Light stippling indicates a ratio greater than 1, dark shading indicates greater variability in the model significant at the 95% level, and cross-hatching indicates less variability in the model significant at the 95% level. The zonal average of the ratio is shown to the right (Meehl et al. 1994).

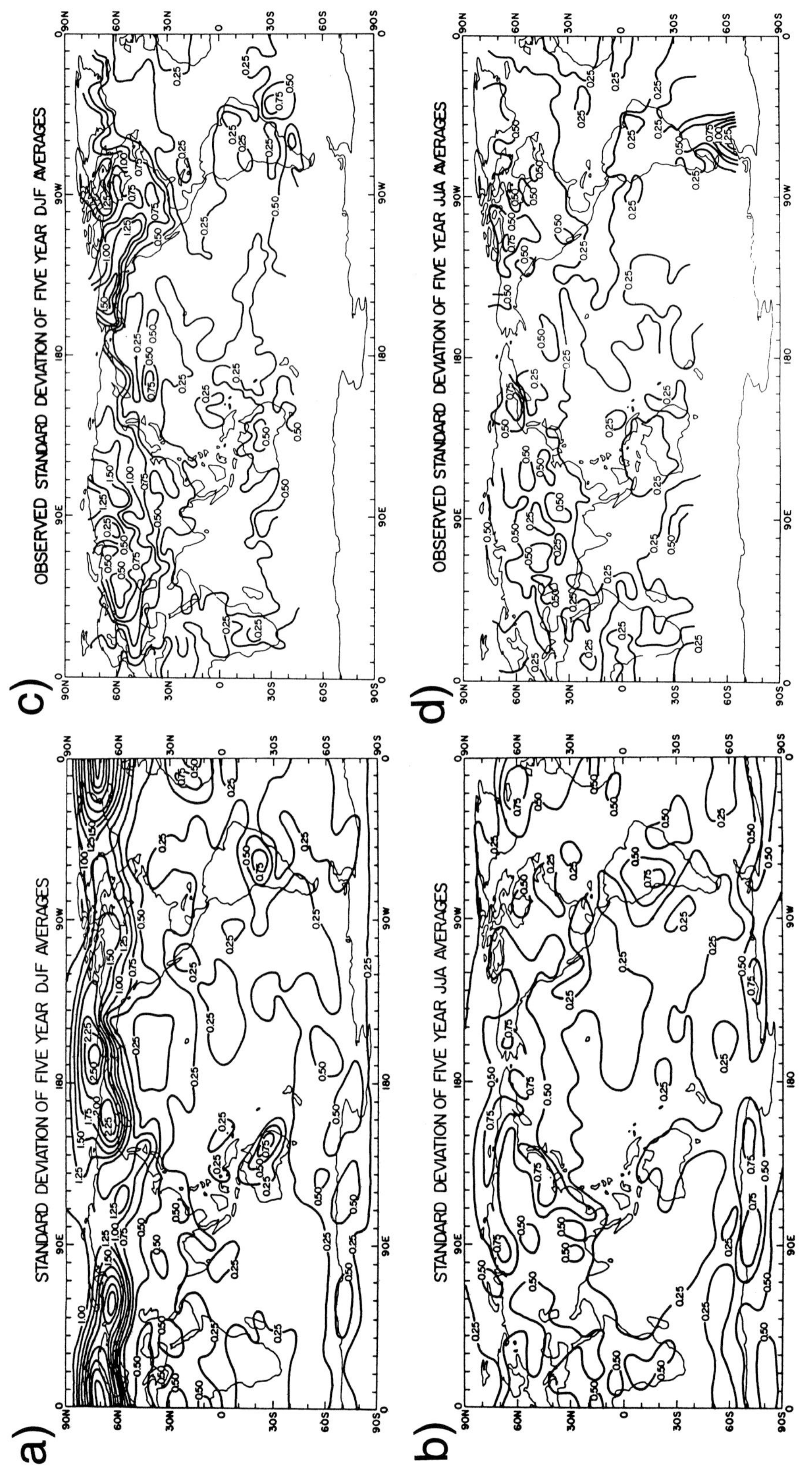

FIG. 10.39. Standard deviations (°C) of 5-yr seasonal mean surface air temperature for (a) coupled model DJF, (b) coupled model JJA, (c) observations (1945–84) for DJF, and (d) observations (1945–84) for JJA (Meehl et al. 1993b).

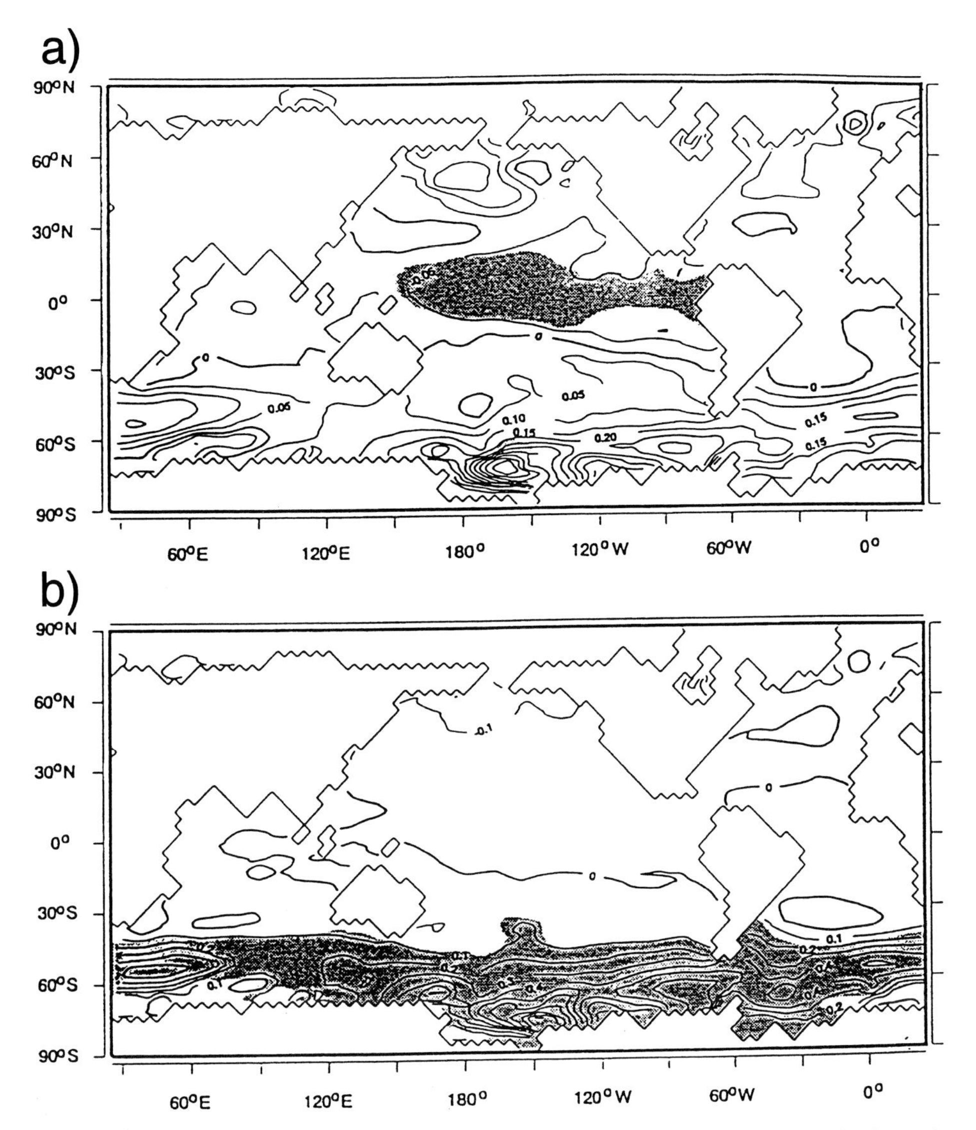

FIG. 10.40. SST anomalies related to the first two modes of variability, filtered to retain timescales between 7 and 50 yr from a 325-yr integration of a coarse-grid coupled model. Shaded areas indicate regions where the variance explained is larger than 10% (von Storch 1994).

precipitation anomaly patterns of ENSO events and CO_2 climate change are discussed later.

c. Changes in variability in coupled models

Since the instrumental record of the earth's climate is only about a century long, one way to assess time–space variability on longer timescales is through the use of paleoclimate data (e.g., Barnett et al. 1996; also see discussion in chapter 9). Coupled models also provide the means to produce estimates of "natural" century-timescale variability of the climate system (e.g., Knutson and Manabe 1994; Stouffer et al. 1994). Such simulations are used to quantify long-timescale variability in climate-change detection studies (e.g., Karoly et al. 1994).

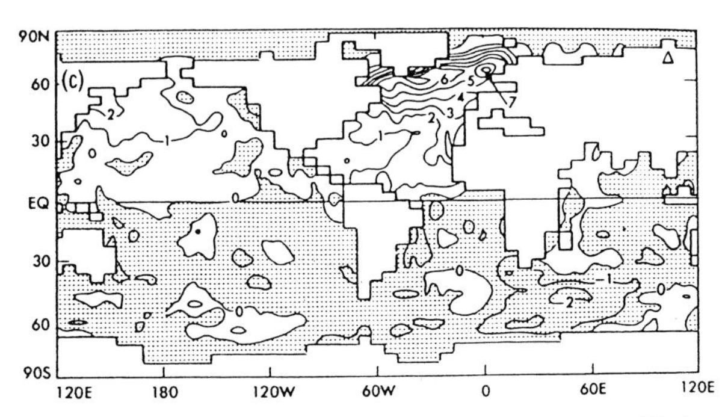

FIG. 10.41. SST difference between two model equilibrium climates found in a coupled model, one with an active thermohaline circulation and the other without (Manabe and Stouffer 1988).

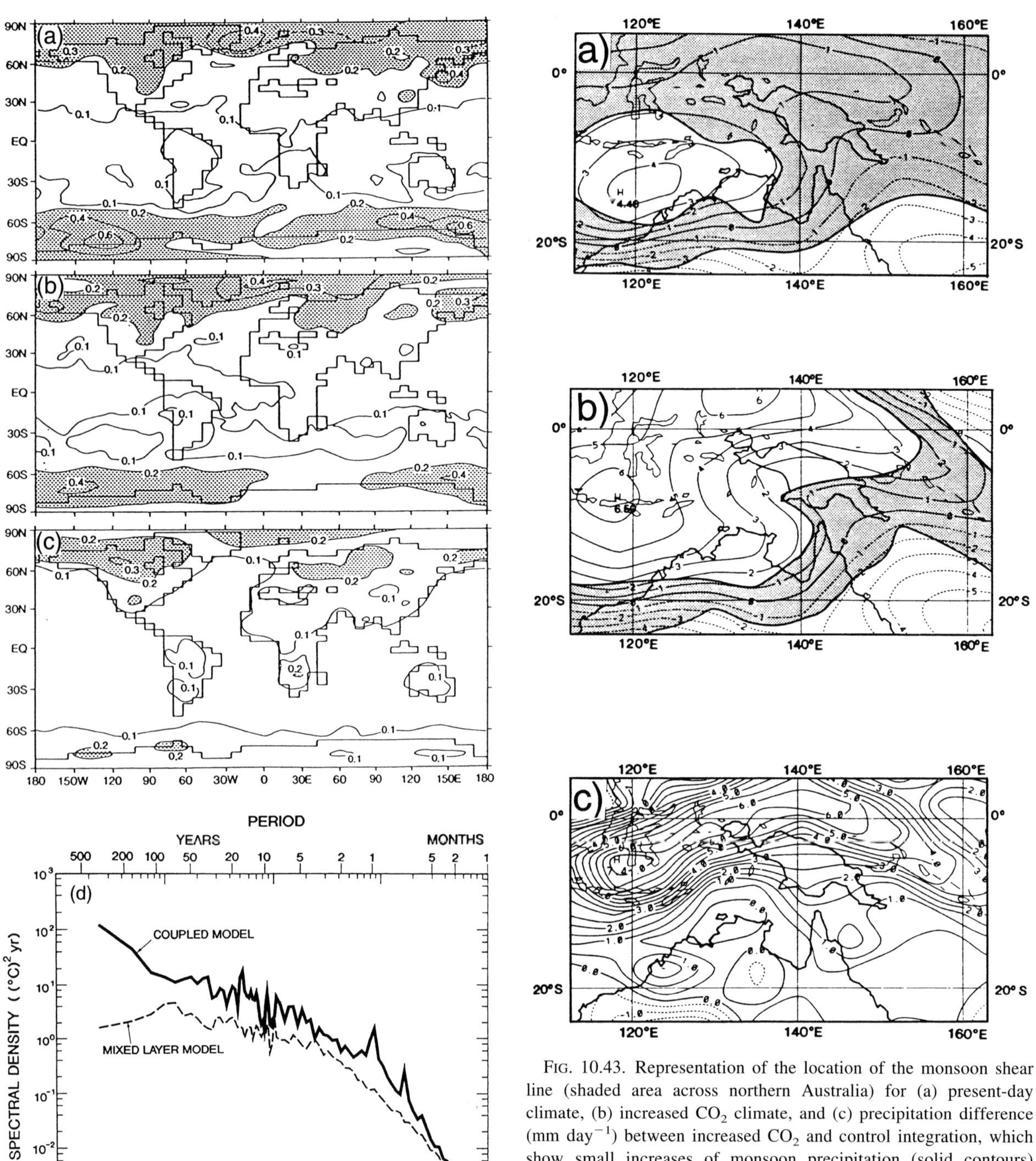

FIG. 10.42. Geographical distribution of the standard deviation of 25-yr mean surface air temperature anomaly (°C) for (a) coupled model 1000-yr integration; (b) mixed-layer model 1000-yr integration; (c) specified climatological SST 500-yr integration; (d) power spectra of monthly mean SST anomaly at 60.75°S, 142.5°W from the 1000-yr integration of the coupled model (solid line) and 1000-yr integration of the mixed-layer model (dashed line) (Manabe and Stouffer 1996).

FIG. 10.43. Representation of the location of the monsoon shear line (shaded area across northern Australia) for (a) present-day climate, (b) increased CO_2 climate, and (c) precipitation difference (mm day^{-1}) between increased CO_2 and control integration, which show small increases of monsoon precipitation (solid contours) over northern Australia land areas (Ryan et al. 1992).

Key questions to be addressed with the coupled models are what changes there may be, if any, to interannual variability associated with ENSO from an increase of CO_2, and how precipitation patterns could be affected. Several model simulations with coarse-grid global coupled models show that ENSO-like events in those models continue to occur with an increase of CO_2 in the model atmosphere with either similar-amplitude SST variability in the tropical east-

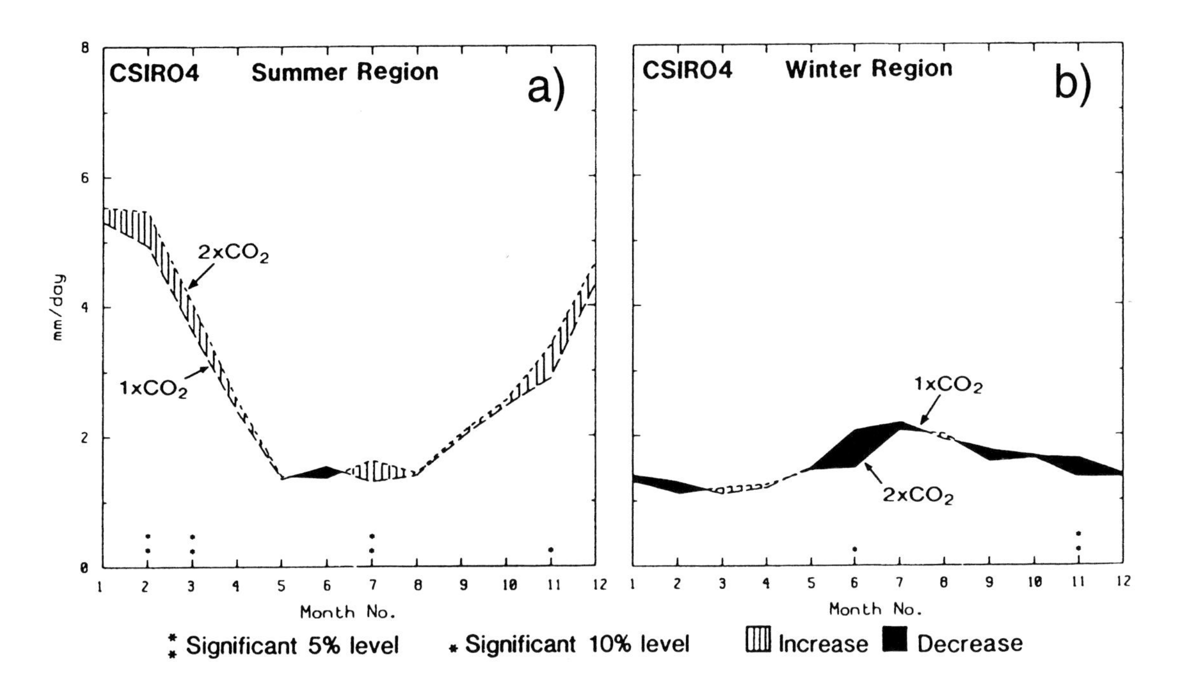

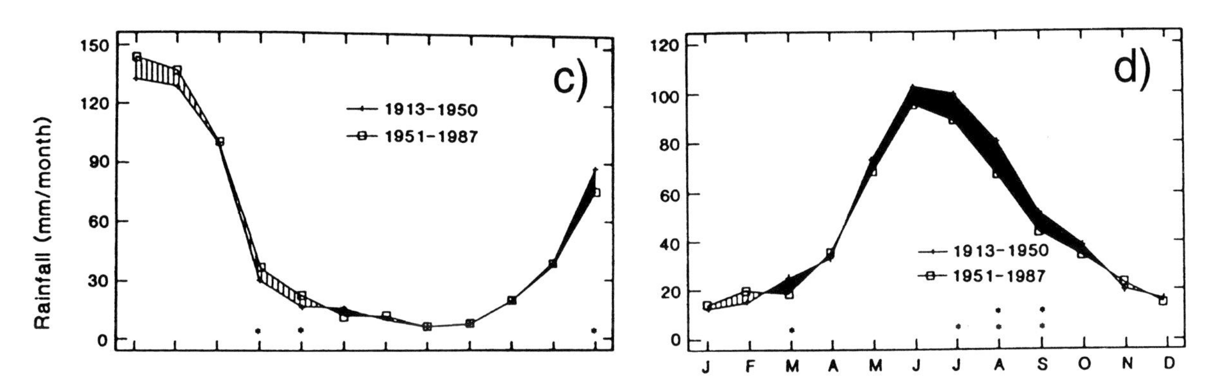

FIG. 10.44. Change in rainfall (increases are hatched, decreases are solid) through the 12 months of the year in a mixed-layer model for doubled CO_2 for (a) summer (November–April) and (b) winter (May–October) regions in northern and southern Australia, respectively. (c) and (d) As in (a) and (b) except for observed average rainfall for each of two periods (1913–50 and 1951–87) for summer and winter rainfall regions (after Whetton et al. 1996).

ern Pacific or somewhat of a decrease (Meehl et al. 1993a; Knutson and Manabe 1994; Tett 1995; Tokioka et al. 1995). Those models show, however, that there may be some alterations to effects associated with ENSO. Meehl et al. (1993a) showed that SST variability over the tropical Pacific is not appreciably changed for the relatively short time series considered, but precipitation variability associated with ENSO is enhanced, particularly over the tropical continents in southern summer. That is, with ENSO-like events in the increased CO_2 climate, anomalously wet areas become wetter and dry areas drier. This is illustrated in Fig. 10.52 for the coupled-model ENSO-like simulations discussed in Fig. 10.35 for tropical continental regions at risk for moisture deficits in present-day ENSO events. With increased CO_2 in that coupled model, there are enhanced precipitation deficits associated with ENSO events, with increased CO_2 in two of the three areas—Australasia and northeastern South America. Meehl et al. (1993a) noted that, following the earlier suggestions of Rind et al. (1989), an increase of mean surface temperature could result in enhanced precipitation variability due to processes associated with the nonlinear relationship between surface temperature and evaporation. A subsequent coupled model experiment (Fig. 10.51) indicates that since the mean climate-change signal from an increase of CO_2 resembles the precipitation anomaly pattern from a typical ENSO event, more severe precipitation deficits would occur in future ENSO events over

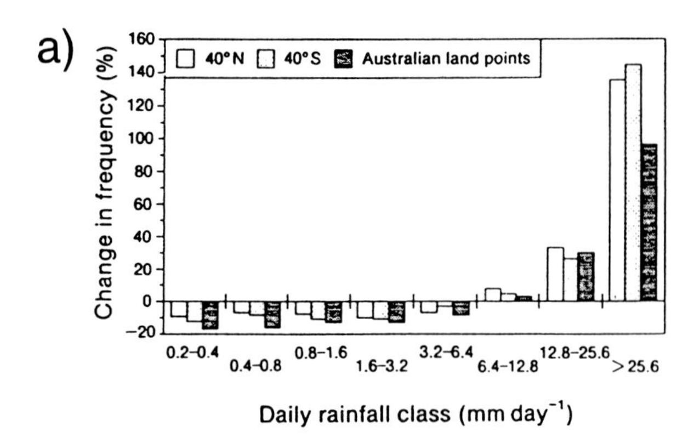

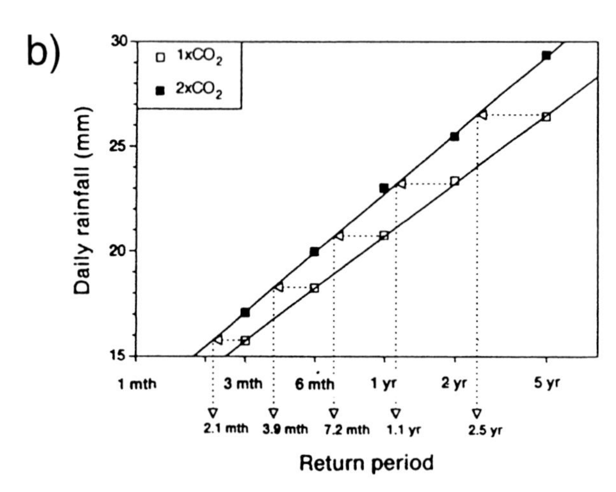

FIG. 10.45. (a) Changes in the frequency of occurrence of daily rainfall classes with doubled CO_2 in a mixed-layer model, and (b) daily rainfall amount vs. return period in Australia, as simulated by a mixed-layer model for both control and doubled CO_2 integrations (Pittock et al. 1991).

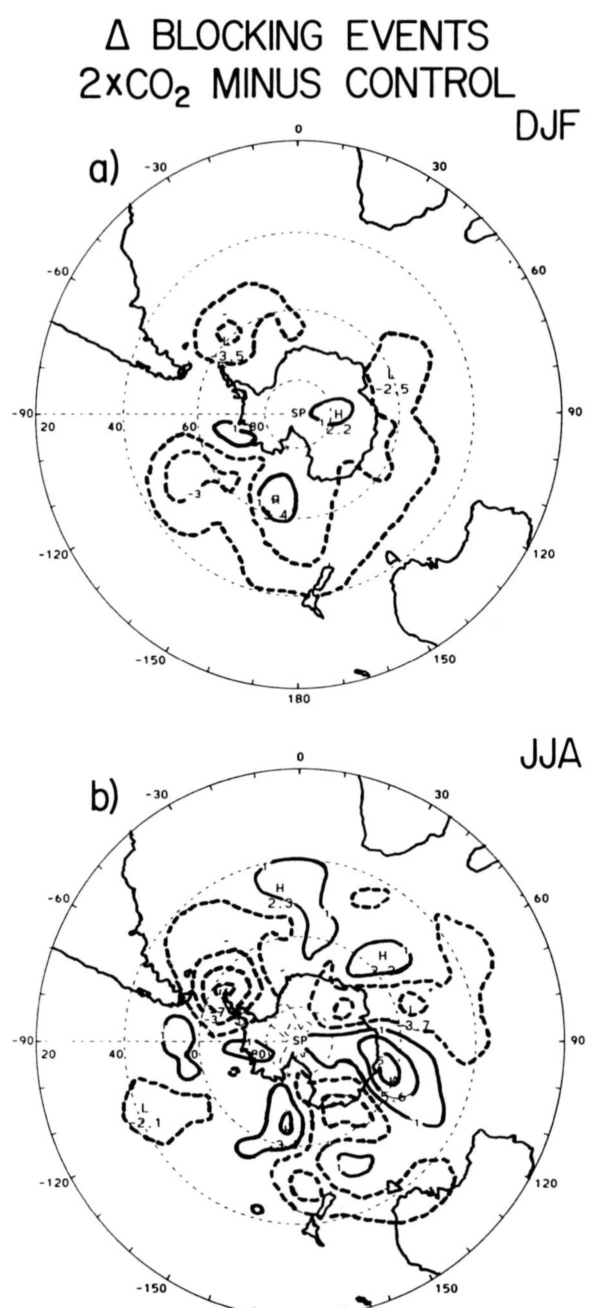

FIG. 10.46. Differences between the number of blocking events per 15 years in a mixed-layer model, 2 × CO_2 minus control, during (a) DJF and (b) JJA (Bates and Meehl 1986).

Australasia due to the juxtaposition of same-sign anomalies, with implications for management of freshwater resources in that region (Meehl 1996).

Meehl et al. (1993a) also noted that the changed mean basic state of the extratropical circulation due to increased CO_2 (altered tropospheric temperature and wind structure) was associated with an alteration of the midlatitude teleconnections involved with ENSO-like events in the tropical Pacific.

Knutson and Manabe (1994) noted a decrease in interannual SST variability over the tropical Pacific on the 2- to 15-yr timescale in a multicentury GFDL coupled-model integration with increased CO_2, and a small enhancement of precipitation variability on that same timescale (Fig. 10.53). Tett (1995) and Tokioka et al. (1995) showed that there were no significant alterations of SST variability over the tropical Pacific associated with ENSO-like events in their coupled models, but Tett suggested that there may be small enhancements of precipitation variability. A recent study with an atmospheric GCM with SSTs specified to simulate both an ENSO event and CO_2 warming suggests that signals associated with intensification of hydrological effects may be relatively small, even though that model was not producing those changes interactively (Smith et al. 1997).

The possibility for warming of SSTs with implications for changes in tropical cyclones has also been

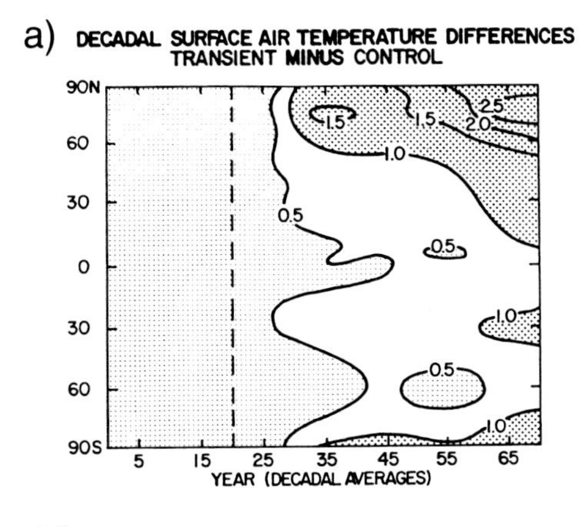

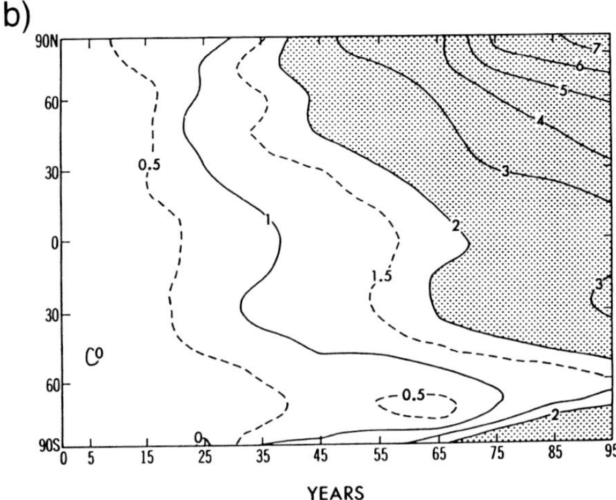

FIG. 10.47. Zonal-mean decadal mean differences, increased CO_2 minus control, for two coupled models with gradually increasing amounts of CO_2: (a) the NCAR model without flux correction (Meehl et al. 1993b) and (b) the GFDL model with flux correction (Stouffer et al. 1989).

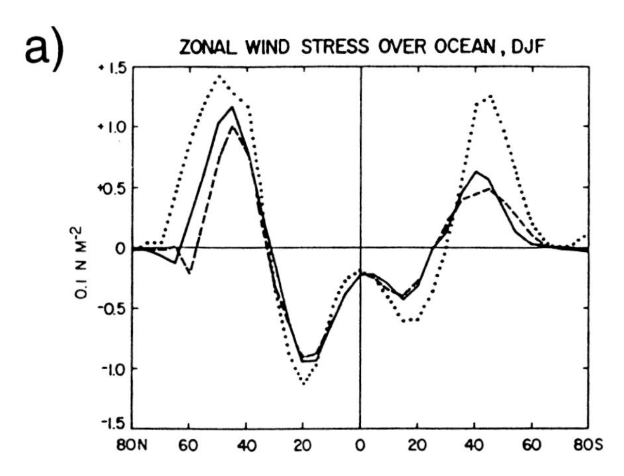

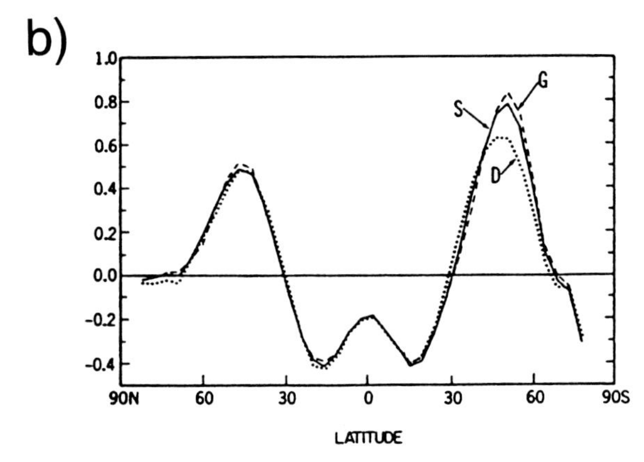

FIG. 10.48. Zonally averaged zonal-wind stress over ocean (DJF, 0.1 N m^{-2}) from (a) NCAR coupled model without flux correction for control integration (solid line), 2 × CO_2 integration (dashed line), and observed; and (b) GFDL coupled model with flux correction for control (labeled S, solid line), increased CO_2 experiment (labeled G, dashed line), and decreased CO_2 experiment (labeled D, dotted line) (Manabe et al. 1991). Note in both (a) and (b) that the dashed line rises above the solid line near 50°S, indicating increased zonal-wind stress at that latitude with increased CO_2 in the coupled models.

suggested (e.g., Nicholls 1991; Evans 1993). These changes, however, are more circumstantial, and it has been pointed out that we must await higher-resolution coupled GCMs for better simulation of phenomena on those space scales before more quantitative results can be produced (Lighthill et al. 1994).

Concerning the larger-scale aspects in changes of interannual variability, Meehl et al. (1994) mainly showed increases of tropical interannual variability associated with the inclusion of ENSO-like phenomena in the coupled model, as well as small increases of interannual variability in the southern midlatitudes (Fig. 10.54).

10.5. Summary

There has been a marked improvement in the capabilities of climate models to simulate Southern Hemisphere climate over the past decade. This has been made possible by improved model formulations in physics and parameterizations. Additionally, there has been increased ability to run models at higher resolution that better resolves Antarctic and Andean topography. This has proven critical for improvements of mid- and high-latitude southern climate.

Though studies described in this chapter give general model characteristics and features in relation to their simulations of the Southern Hemisphere climate, recent activities should shed further light on the accuracy of the models. For example, AMIP, designed to intercompare simulation characteristics of many atmospheric GCMs currently in use today (Gates 1992, 1995), presents the prospect of evaluating features of these models in a uniform way (e.g., Joubert 1997). Additionally, the reanalysis project underway at NCEP and NCAR, as well as a similar effort at the ECMWF, is providing uniform, observation-based datasets for

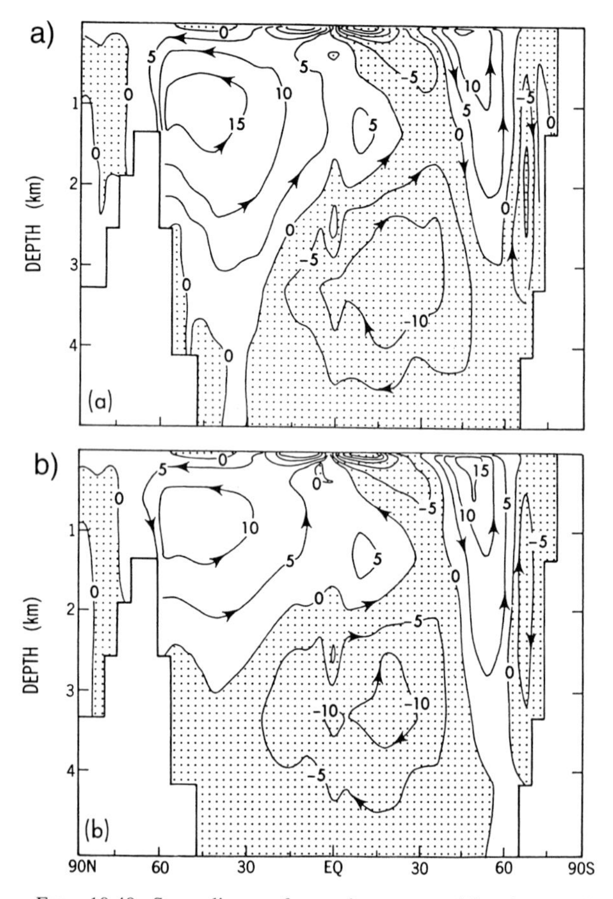

FIG. 10.49. Streamlines of zonal-mean meridional oceanic circulation (in Sv) from a coupled model for (a) control integration with present-day CO_2 and (b) increased CO_2 experiment. Note strengthened Deacon cell in the increased CO_2 experiment in panel (b) (Stouffer et al. 1989).

model evaluation and analysis (Kalnay et al. 1996; Gibson et al. 1996). The increasingly important role played by global coupled GCMs is being addressed in an international intercomparison exercise called the Coupled Model Intercomparison Project (Meehl et al. 1997). Some early indications of the coupled models' simulation capabilities have been provided in an intercomparison of a number of global coupled models as part of the Intergovernmental Panel on Climatic Change assessment (IPCC 1996). The net effect of these various model intercomparisons, together with the uniform observation-based datasets from the reanalysis efforts, is to provide better evaluations of the accuracy of climate-model simulations of the Southern Hemisphere climate.

Climate models run with observed SSTs or models with interactive oceans that use flux correction (to adjust SSTs so they are close to observed values) are able to satisfactorily simulate many features of Southern Hemisphere climate, including the continental and oceanic rainfall regimes, the circumpolar trough, the upper-tropospheric double jet in the Pacific sector, and the semiannual oscillation at middle and high southern latitudes. Climate-modeling studies have also shown that the simulated climate system appears to be very sensitive to variations of sea ice. This sensitivity has appeared in studies that have altered details of sea-ice parameterizations, as well as in experiments with a change in external forcing, such as increased CO_2.

The most recent advance in terms of climate modeling has involved multidecadal to multicentury integrations with global coupled ocean–atmosphere GCMs. These models, though coarse grid to date, have shown a surprising ability to reproduce many aspects of variability in the coupled climate system on a variety of time and space scales. For example, these models reproduce some aspects of ENSO and the associated teleconnections to the southern mid- and high latitudes. Additionally, interannual, 5-yr, and decadal timescale variability from such models compares favorably with observations. Due to the compounding of model errors when imperfect components are coupled, coupled models tend to have errors in their climate simulations. Some groups correct these errors (called flux correction or flux adjustment) and others do not. As model components improve, the need for using flux correction will lessen, since the coupled simulation will be closer to the observed climate system. Yet many of the processes that produce coupled variability appear in present-day coupled models, both with and without flux correction. This suggests that certain modes of variability in the climate system must be very fundamental and quite robust. Thus, such models have been used to study changes in that simulated variability due, for example, to increased CO_2.

The inclusion of the radiative forcing from sulfate aerosols has been shown to be important for simulating the past 100 years of climate evolution, as well as for future climate change (IPCC 1996). Most of these effects, however, are tied to large industrial areas of the Northern Hemisphere, with comparatively little contribution from industrial activity in the Southern Hemisphere over the past 100 years. Future increases in sulfate aerosol production over areas of southeastern Australia, South America, and South Africa, however, will have some bearing on the magnitude of future temperature increases in those regions and must be kept in mind for subsequent climate-model simulations (IPCC 1996).

Climate-modeling studies have highlighted the importance of the southern circumpolar oceans in terms of a response to CO_2-induced warming that depends critically on the representation of ocean mixing. Coupled-model experiments have also shown that the circumpolar oceans are very important for decadal- and longer-timescale variability, and ocean GCMs must have appropriate mixing schemes to accurately simulate these features. Paleoclimate data provide some estimates of this longer-timescale variability, and coupled models are playing a valuable role in

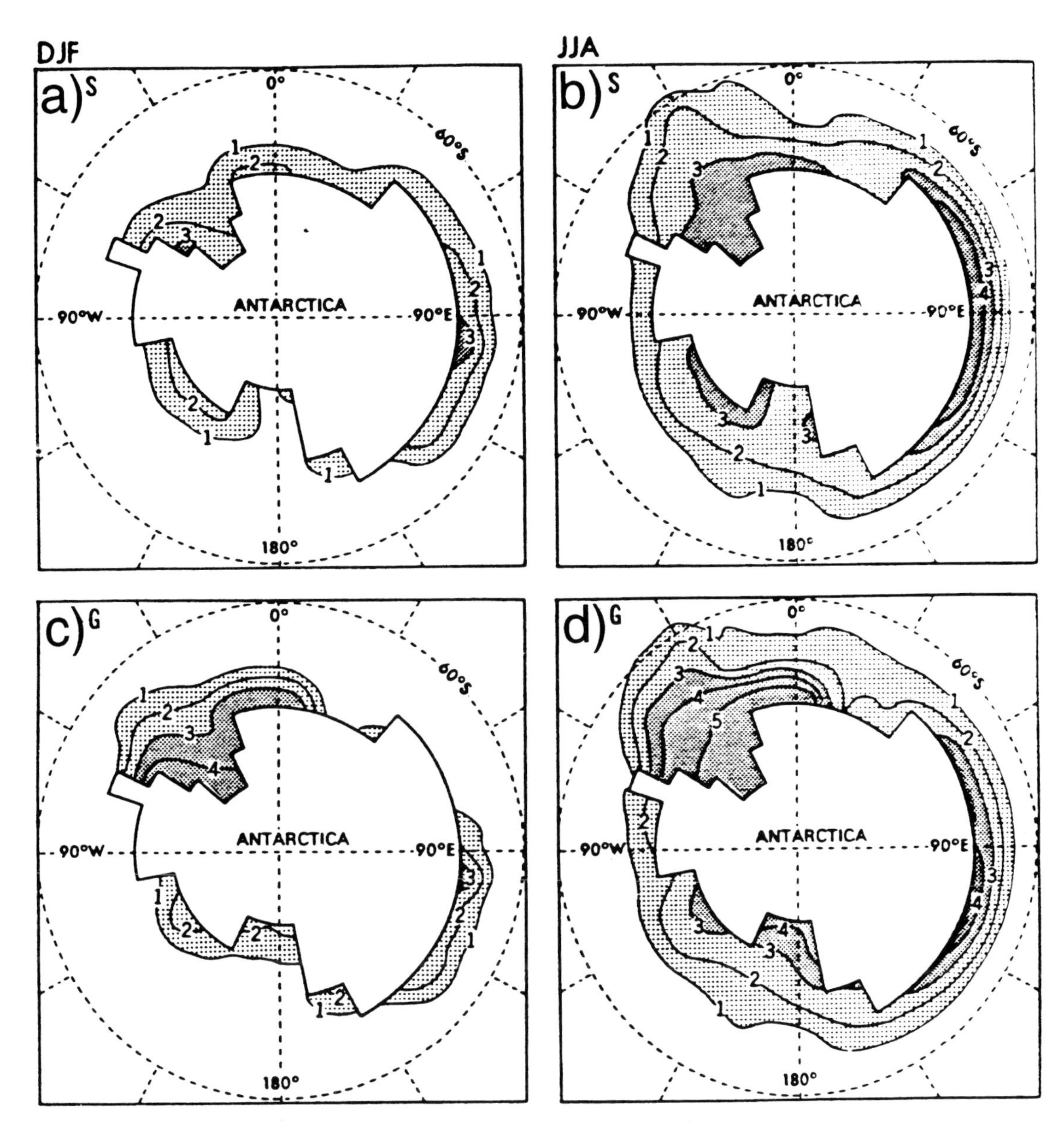

FIG. 10.50. Geographical distribution of seasonal mean sea-ice thickness from a coupled model with flux correction: (a) control integration, DJF; (b) control integration, JJA; (c) increased CO_2 experiment, DJF; (d) increased CO_2 experiment, JJA. Light and dark shadings cover the regions where the thickness of sea ice exceeds 1 and 3 m, respectively (Manabe et al. 1992).

attempting to quantify so-called "natural" variability in the climate system. Multi-thousand-year integrations with global coupled models are being used to estimate such long-timescale variability to assist in climate-change detection studies. Since the southern oceans seem to be pivotal in climate system variability, future studies with coupled models will focus even more on the Southern Hemisphere for understanding present-day climate, as well as possible future climate change.

REFERENCES

Barnett, T. P., B. D. Santer, P. D. Jones, R. S. Bradley, and K. R. Briffa, 1996: Estimates of low-frequency natural variability in near-surface air temperature. *Holocene,* **6,** 255–263.

Bates, G. T., and G. A. Meehl, 1986: The effect of CO_2 concentration on the frequency of blocking in a general circulation model coupled to a simple mixed-layer ocean model. *Mon. Wea. Rev.,* **114,** 687–701.

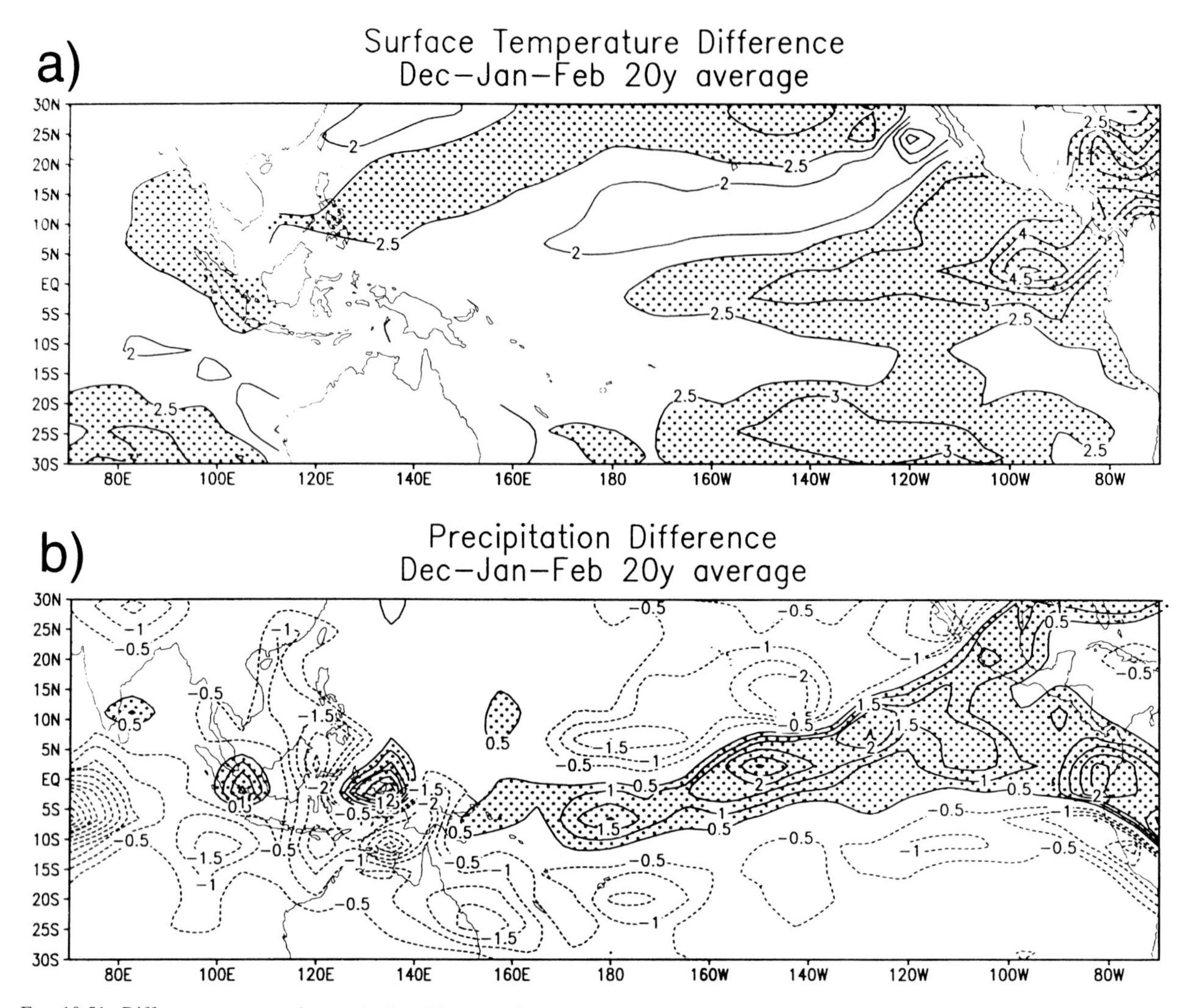

FIG. 10.51. Differences computed over the last 20 years of a transient CO_2 increase experiment minus the control integration from the second-generation NCAR coupled model, DJF: (a) SST and (b) precipitation (after Meehl and Washington 1996).

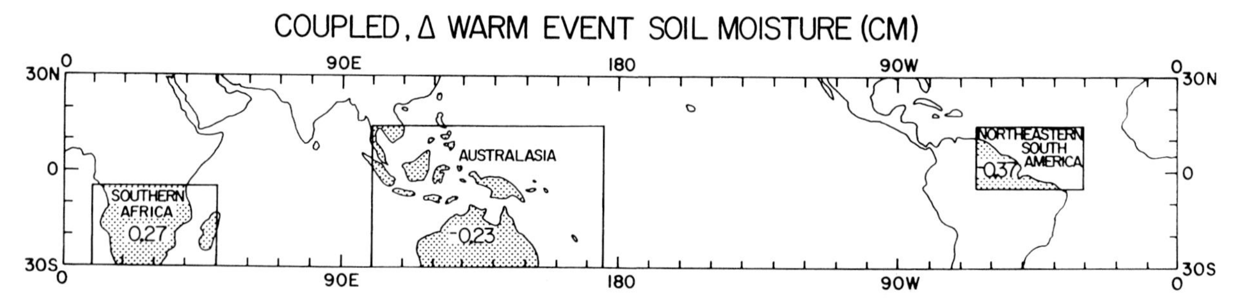

FIG. 10.52. Area-averaged soil moisture differences (cm) for land points from the coupled model, increased CO_2 warm events minus control-case warm events, for the areas indicated in Fig. 10.35e that are susceptible to moisture deficits during warm events. Negative differences for Australasia and northeastern South America indicate that moisture deficits in those land areas are more severe during increased CO_2 warm events (Meehl et al. 1993a).

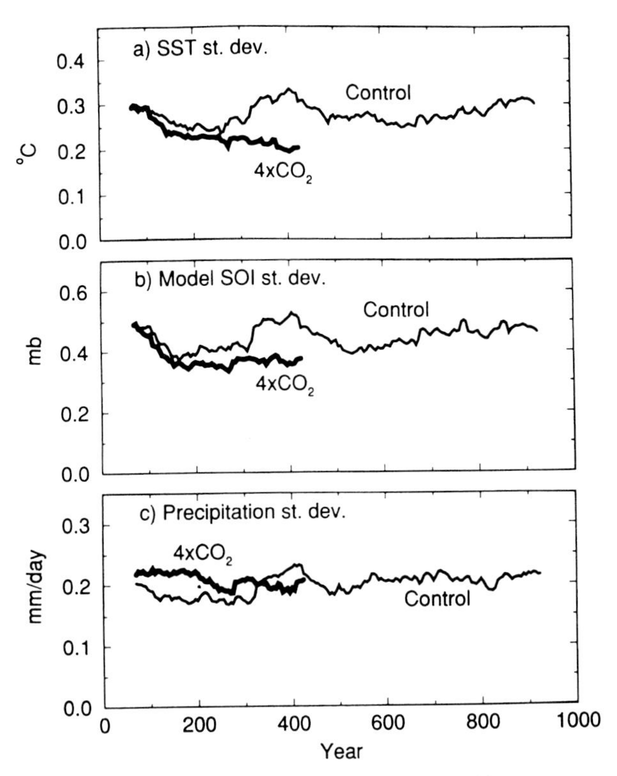

FIG. 10.53. Standard deviations of overlapping 100-yr segments of bandpass (2- to 15-yr) filtered anomalies from 4 × CO_2 and control experiments. (a) SST (7°N–7°S, 173°E–120°W) in °C; (b) SOI; and (c) precipitation (11°N–11°S, 173°E–113°W) in mm day^{-1} (Knutson and Manabe 1994).

Bjerknes, J., 1964: Atlantic air–sea interaction. *Advances in Geophysics*. Academic Press, 1–82.

Blackmon, M. L., J. E. Geisler, and E. J. Pitcher, 1983: A general circulation model study of January climate anomaly patterns associated with interannual variation of equatorial Pacific sea surface temperatures. *J. Atmos. Sci.*, **40,** 1410–1425.

Boer, G. J., and Coauthors, 1992: Some results from an intercomparison of the climates simulated by 14 atmospheric general circulation models. *J. Geophys. Res.*, **97,** 12 771–12 786.

Boville, B. B., 1991: Sensitivity of simulated climate to model resolution. *J. Climate,* **4,** 469–485.

Bromwich, D. H., B. Chen, and X. Pan, 1995: Intercomparison of simulated polar climates by global climate models. Preprints, *Fourth Conf. on Polar Meteorology and Oceanography,* Dallas, TX, Amer. Meteor. Soc., (J9)14–(J9)19.

Bryan, K., and R. Stouffer, 1991: A note on Bjerknes' hypothesis for North Atlantic variability. *J. Mar. Sys.,* **1,** 229–241.

Burgos, J. J., H. F. Ponce, and L. C. B. Molion, 1991: Climate change predictions for South America. *Clim. Change,* **18,** 223–239.

Chen, B., and D. H. Bromwich, 1995: High latitude pressure patterns simulated by global climate models. *Proc. First Int. AMIP Scientific Conf.,* WCRP-92, WMO/TD-No. 732, World Climate Research Programme, 439–444.

Colman, R., S. Power, B. McAvaney, and R. Dahni, 1995: A non–flux-corrected transient CO_2 experiment using the BMRC coupled A/OGCM. *Geophys. Res. Lett.,* **22,** 3047–3050.

England, M. H., 1995: Using chlorofluorocarbons to assess ocean climate models. *Geophys. Res. Lett.,* **22,** 3051–3054.

Evans, J., 1993: Sensitivity of tropical cyclone intensity to sea surface temperature. *J. Climate,* **6,** 1133–1140.

Fennessy, M. J., L. Marx, and J. Shukla, 1985: General circulation model sensitivity to 1982–83 equatorial Pacific sea surface temperature anomalies. *Mon. Wea. Rev.,* **113,** 858–864.

Frederiksen, C. S., P. Indusekaran, R. Balgovind, D. P. Rowell, and C. K. Folland, 1995: Simulations of Australian climate variability: The role of global SSTs. *Proc. First Int. AMIP Scientific Conf.,* WCRP-92, WMO/TD-No. 732, World Climate Research Programme, 413–418.

Gates, W. L., 1992: AMIP: The atmospheric model intercomparison project. *Bull. Amer. Meteor. Soc.,* **73,** 1962–1970.

———, Ed., 1995: *Proc. First Int. AMIP Scientific Conf.,* WCRP-92, WMO/TD-No. 732, World Climate Research Programme, 529 pp.

Gibson, R., P. Kallberg, and S. Uppala, 1996: The ECMWF Reanalysis (ERA) Project. *ECMWF Newsletter,* **73,** 7–17.

Gordon, H. B., and B. G. Hunt, 1994: Climatic variability within an equilibrium greenhouse simulation. *Climate Dyn.,* **9,** 195–212.

———, P. H. Whetton, A. B. Pittock, A. M. Fowler, and M. R. Haylock, 1992: Simulated changes in daily rainfall intensity due to the enhanced greenhouse effect: Implications for extreme rainfall events. *Climate Dyn.,* **8,** 83–102.

Hurrell, J. W., and H. van Loon, 1994: A modulation of the atmospheric annual cycle in the Southern Hemisphere. *Tellus,* **46A,** 325–338.

IPCC, 1990: *Climate Change: The IPCC Scientific Assessment.* J. T. Houghton, G. J. Jenkins, and J. J. Ephraums, Eds., Cambridge University Press, 366 pp.

———, 1992: *Climate Change 1992: The Supplementary Report to the IPCC Scientific Assessment.* J. T. Houghton, B. A. Callander, and S. K. Varney, Eds., Cambridge University Press, 200 pp.

———, 1996: *Climate Change 1995: The Science of Climate Change.* J. T. Houghton, L. G. Meira Filho, B. A. Callander, N. Harris, A. Kattenberg, and K. Maskell, Eds., Cambridge University Press, 572 pp.

Jones, D. A., and I. Simmonds, 1993: A climatology of Southern Hemisphere extratropical cyclones. *Climate Dyn.,* **9,** 131–145.

Joubert, A., 1994: Simulations of southern African climate by carly-generation general circulation models. *Water S. A.,* **20,** 315–322.

———, 1995: Simulations of southern African climate by early generation general circulation models. *S. African J. Sci.,* **91,** 85–91.

———, 1997: AMIP simulations of atmospheric circulation over southern Africa. *Int. J. Climatol.,* **17,** 1129–1154.

———, and S. J. Mason, 1996: Droughts over southern Africa in a doubled-CO_2 climate. *Int. J. Climatol.,* **16,** 1149–1156.

———, and P. D. Tyson, 1996: Equilibrium and fully-coupled GCM simulations of future southern African climates. *S. African J. Sci.,* **92,** 471–484.

Kalnay, E., and Coauthors, 1996: The NCEP/NCAR Reanalysis Project. *Bull. Amer. Meteor. Soc.,* **77,** 437–471.

Karoly, D. J., 1989: Southern Hemisphere circulation features associated with El Niño–Southern Oscillation events. *J. Climate,* **2,** 1239–1252.

———, R. A. Plumb, and M. Ting, 1989: Examples of the horizontal propagation of quasi-stationary waves. *J. Atmos. Sci.,* **46,** 2802–2811.

———, J. A. Cohen, G. A. Meehl, J. F. B. Mitchell, A. H. Oort, R. J. Stouffer, and R. T. Wetherald, 1994: An example of fingerprint detection of greenhouse climate change. *Climate Dyn.,* **10,** 97–105.

———, P. C. McIntosh, P. Berrisford, T. J. McDougall, and A. C. Hirst, 1997: Similarities of the Deacon cell in the Southern Ocean and the Ferrel cells in the atmosphere. *Quart. J. Roy. Meteor. Soc.,* **123,** 519–526.

Kiladis, G. N., and H. van Loon, 1988: The Southern Oscillation. Part VII: Meteorological anomalies over the Indian and Pacific sectors associated with the extremes of the oscillation. *Mon. Wea. Rev.,* **116,** 120–136.

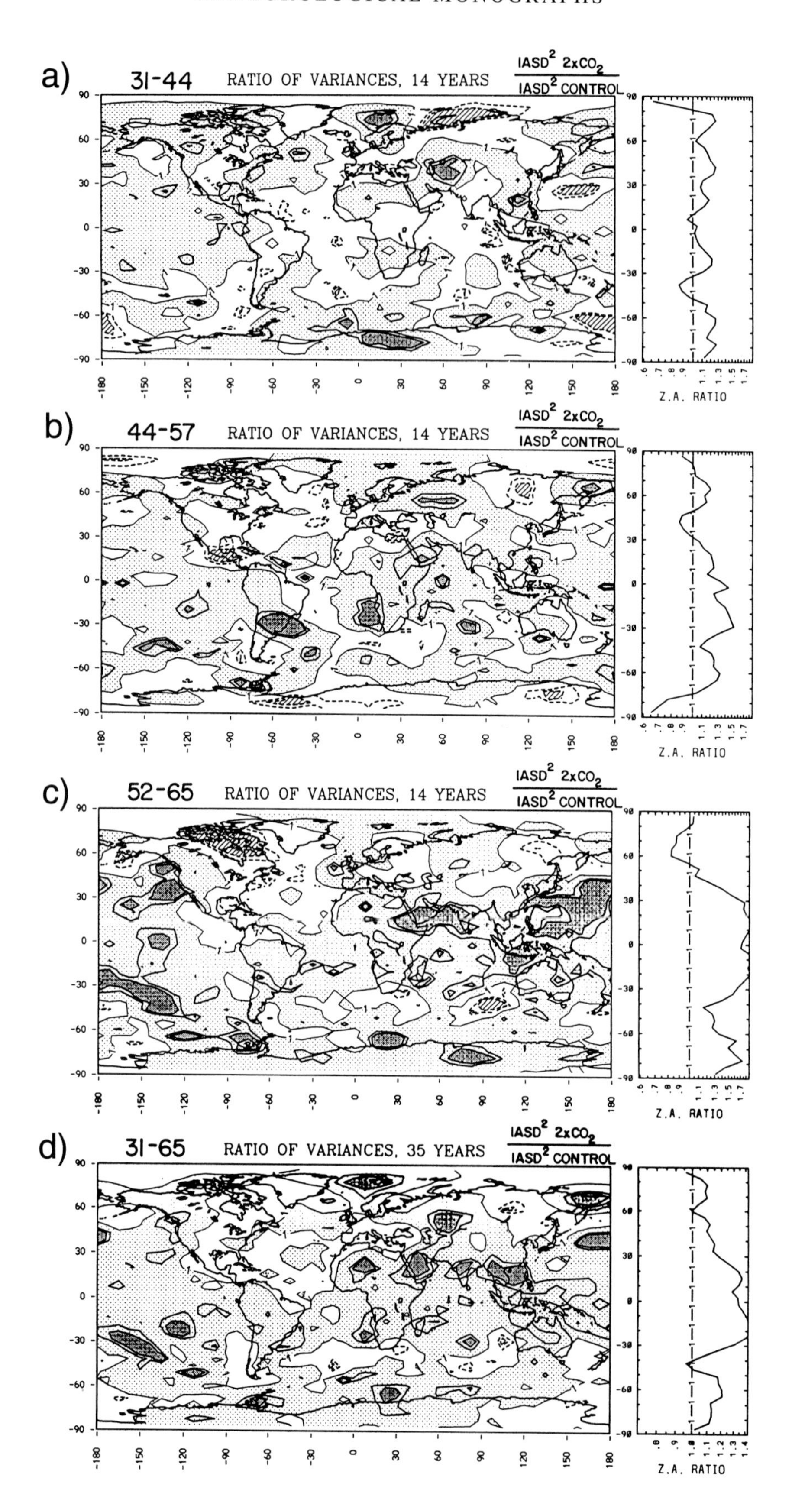
a)
31-44 RATIO OF VARIANCES, 14 YEARS
IASD² 2xCO₂ / IASD² CONTROL
Z.A. RATIO
b)
44-57 RATIO OF VARIANCES, 14 YEARS
IASD² 2xCO₂ / IASD² CONTROL
Z.A. RATIO
c)
52-65 RATIO OF VARIANCES, 14 YEARS
IASD² 2xCO₂ / IASD² CONTROL
Z.A. RATIO
d)
31-65 RATIO OF VARIANCES, 35 YEARS
IASD² 2xCO₂ / IASD² CONTROL
Z.A. RATIO

———, H. von Storch, and H. van Loon, 1989: Origin of the South Pacific convergence zone. *J. Climate,* **2,** 1185–1195.

Kinter, J. L. III, J. Shukla, L. Marx, and E. K. Schneider, 1988: A simulation of the winter and summer circulations with the NMC global spectral model. *J. Atmos. Sci.,* **45,** 2486–2522.

Kitoh, A., 1994: Tropical influence on the South Pacific double jet variability. *Proc. NIPR Symp. Polar Meteorol. Glaciol.,* **8,** 34–45.

———, K. Yamazaki, and T. Tokioka, 1990: The double jet and semi-annual oscillations in the Southern Hemisphere simulated by the Meteorological Research Institute general circulation model. *J. Meteor. Soc. Japan,* **68,** 251–264.

Knutson, T. R., and S. Manabe, 1994: Impact of increased CO_2 on simulated ENSO-like phenomena. *Geophys. Res. Lett.,* **21,** 2295–2298.

———, and ———, 1995: Time-mean response over the tropical Pacific due to increased CO_2 in a coupled ocean–atmosphere model. *J. Climate,* **8,** 2181–2199.

Lau, N.-C., 1985: Modeling the seasonal dependence of the atmospheric response to observed El Niños in 1962–76. *Mon. Wea. Rev.,* **113,** 1970–1996.

———, S. G. H. Philander, and M. J. Nath, 1992: Simulation of ENSO-like phenomena with a low-resolution coupled GCM of the global ocean and atmosphere. *J. Climate,* **5,** 284–307.

Liang, X.-L., W.-C. Wang, and M. P. Dudek, 1995: Interannual climate variability and its change due to the greenhouse effect. *Global Planet. Change,* **10,** 217–238.

Lighthill, J., G. Holland, W. Gray, C. Landsea, G. Craig, J. Evans, Y. Kurihara, and C. Guard, 1994: Global climate change and tropical cyclones. *Bull. Amer. Meteor. Soc.,* **75,** 2147–2157.

Lindesay, J. A., and I. Smith, 1993: Modelling southern African rainfall responses to SST anomalies in the South Atlantic and Indian Oceans. Preprints, *Fourth Int. Conf. on Southern Hemisphere Meteorology and Oceanography,* Hobart, Australia, Amer. Meteor. Soc., 526–527.

Manabe, S., and R. J. Stouffer, 1988: Two stable equilibria of a coupled ocean–atmosphere model. *J. Climate,* **1,** 841–866.

———, and ———, 1996: Low-frequency variability of surface air temperature in a 1000-year integration of a coupled ocean–atmosphere model. *J. Climate,* **9,** 376–393.

———, ———, M. J. Spelman, and K. Bryan, 1991: Transient responses of a coupled ocean–atmosphere model to gradual changes of atmospheric CO_2. Part I: Annual mean response. *J. Climate,* **4,** 785–818.

———, M. J. Spelman, and R. J. Stouffer, 1992: Transient responses of a coupled ocean–atmosphere model to gradual changes of atmospheric CO_2. Part II: Seasonal response. *J. Climate,* **5,** 105–126.

Mason, S. J., and A. M. Joubert, 1997: Simulated changes in extreme rainfall over southern Africa. *Int. J. Climatol.,* **17,** 291–301.

McGregor, J. L., and K. Walsh, 1994: Climate change simulations of Tasmanian precipitation using multiple nesting. *J. Geophys. Res.,* **99,** 20 889–20 905.

———, ———, and J. J. Katzfey, 1993: Nested modelling for regional climate studies. *Modelling Change in Environmental Systems,* A. J. Jakeman, M. B. Beck, and M. J. McAleer, Eds., John Wiley and Sons, 367–386.

FIG. 10.54. Ratio of the interannual variances (see Fig. 10.38) of the lower-troposphere temperature from equivalent periods in the increased CO_2 and control integrations of the coupled model: (a) years 31–44, (b) years 44–57, (c) years 52–65, and (d) years 31–65. The same contouring and shading are applied as in Fig. 10.38 (solid contours indicate greater interannual variability in the increased CO_2 case). The zonal average of the ratio is shown to the right (Meehl et al. 1994).

Mearns, L. O., 1993: Implications of global warming for climate variability and the occurrence of extreme climate events. *Drought Assessment Management and Planning: Theory and Case Studies,* D. A. Wilhite, Eds., Kluwer, 109–130.

Mechoso, C., 1981: Topographic influences on the general circulation of the Southern Hemisphere: A numerical experiment. *Mon. Wea. Rev.,* **109,** 2131–2139.

Meehl, G. A., 1987: The annual cycle and interannual variability in the tropical Indian and Pacific Ocean regions. *Mon. Wea. Rev.,* **115,** 27–50.

———, 1989: The coupled ocean–atmosphere modeling problem in the tropical Pacific and Asian monsoon regions. *J. Climate,* **2,** 1146–1163.

———, 1990a: Development of global coupled ocean–atmosphere general circulation models. *Climate Dyn.,* **5,** 19–33.

———, 1990b: Seasonal cycle forcing of El Niño–Southern Oscillation in a global coupled ocean–atmosphere GCM. *J. Climate,* **3,** 72–98.

———, 1991: A reexamination of the mechanism of the semiannual oscillation in the Southern Hemisphere. *J. Climate,* **4,** 911–926.

———, 1992: Global coupled models: Atmosphere, ocean, sea ice. *Climate System Modeling,* K. Trenberth, Ed., Cambridge University Press, 555–581.

———, 1994: Coupled land–ocean–atmosphere processes and south Asian monsoon variability. *Science,* **266,** 263–267.

———, 1995: Global coupled general circulation models. *Bull. Amer. Meteor. Soc.,* **76,** 951–957.

———, 1996: Vulnerability of fresh water resources to climate change in the tropical Pacific region. *J. Water Air Soil Poll.,* **92,** 203–213.

———, 1997a: The south Asian monsoon and the tropospheric biennial oscillation. *J. Climate,* **10,** 1921–1943.

———, 1997b: Modification of surface fluxes in component models in global coupled models. *Climate Dyn.,* **14,** 1–15.

———, and B. A. Albrecht, 1988: Tropospheric temperatures and Southern Hemisphere circulation. *Mon. Wea. Rev.,* **116,** 953–960.

———, and W. M. Washington, 1990: CO_2 climate sensitivity and snow–sea-ice albedo parameterization in an atmospheric GCM coupled to a mixed-layer ocean model. *Clim. Change,* **16,** 283–306.

———, and ———, 1991: Response of a GCM with a hybrid convection scheme to a tropical Pacific sea surface temperature anomaly. *J. Climate,* **4,** 672–688.

———, and ———, 1995: Cloud albedo feedback and the super greenhouse effect in a global coupled GCM. *Climate Dyn.,* **11,** 399–411.

———, and ———, 1996: El Niño-like climate change in a model with increased atmospheric CO_2 concentrations. *Nature,* **382,** 56–60.

———, G. W. Branstator, and W. M. Washington, 1993a: Tropical Pacific interannual variability and CO_2 climate change. *J. Climate,* **6,** 42–63.

———, W. M. Washington, and T. R. Karl, 1993b: Low-frequency variability and CO_2 transient climate change. Part 1: Time-averaged differences. *Climate Dyn.,* **8,** 117–133.

———, M. Wheeler, and W. M. Washington, 1994: Low-frequency variability and CO_2 transient climate change. Part 3. Intermonthly and interannual variability. *Climate Dyn.,* **10,** 277–303.

———, G. J. Boer, C. Covey, M. Latif, and R. J. Stouffer, 1997: Intercomparison makes for a better climate model. *Eos,* **78,** 445–446, 451.

———, J. W. Hurrell, and H. van Loon, 1998: A modulation of the mechanism of the semiannual oscillation in the Southern Hemisphere. *Tellus,* **50A,** 442–450.

Mullan, A. B., and B. J. McAvaney, 1995: Validation of high latitude tropospheric circulation in the Southern Hemisphere (Subproject 9). *Proc. First Int. AMIP Scientific Conf.,* WCRP-92, WMO/TD-No. 732, World Climate Research Programme, 205–210.

Murphy, J. M., 1995: Transient response of the Hadley Centre coupled ocean–atmosphere model to increasing carbon dioxide. Part I: Control climate and flux correction. *J. Climate,* **8,** 36–56.

Nagai, T., T. Tokioka, M. Endoh, and Y. Kitamura, 1992: El Niño/Southern Oscillation simulated in an MRI atmosphere–ocean coupled general circulation model. *J. Climate,* **5,** 1202–1233.

Neelin, J. D., and H. A. Dijkstra, 1995: Ocean–atmospheric interaction and the tropical climatology. Part I: The dangers of flux-correction. *J. Climate,* **8,** 1325–1342.

———, and Coauthors, 1992: Tropical air–sea interaction in general circulation models. *Climate Dyn.,* **7,** 73–104.

Nicholls, N., 1991: Global warming, tropical cyclones, and ENSO. *Responding to the Threat of Global Warming,* ANL/EAIS/TM-17, Argonne National Laboratory, 2-19–2-36.

Palmer, T. N., and D. A. Mansfield, 1986a: A study of wintertime circulation anomalies during past El Niño events using a higher resolution general circulation model. I: Influence of a model climatology. *Quart. J. Roy. Meteor. Soc.,* **112,** 613–638.

———, and ———, 1986b: A study of wintertime circulation anomalies during past El Niño events using a higher resolution general circulation model. II: Variability of the seasonal mean response. *Quart. J. Roy. Meteor. Soc.,* **112,** 639–660.

Parrish, T. R., D. H. Bromwich, and R.-Y. Tzeng, 1994: On the role of the Antarctic continent in forcing large-scale circulations in the high southern latitudes. *J. Atmos. Sci.,* **51,** 3566–3579.

Pittock, A. B., A. M. Fowler, and P. H. Whetton, 1991: Probable changes in rainfall regimes due to the enhanced greenhouse effect. *Proc. Int. Hydrology and Water Resources Symp.,* Perth, Australia.

Rind, D., R. Goldberg, and R. Reudy, 1989: Change in climate variability in the 21st century. *Clim. Change,* **14,** 5–37.

Rowntree, P. R., 1972: The influence of tropical east Pacific Ocean temperature on the atmosphere. *Quart. J. Roy. Meteor. Soc.,* **98,** 290–321.

Ryan, B. F., D. A. Jones, and H. B. Gordon, 1992: The sensitivity of GCM models to the Australian monsoon equatorial shear line: Enhanced greenhouse scenario implications. *Climate Dyn.,* **7,** 173–180.

Schlesinger, M. E., 1984: Atmospheric general circulation model simulations of the modern Antarctic climate. *Environment of West Antarctica: Potential CO_2-Induced Changes,* National Academy Press, 155–196.

Shukla, J., and J. M. Wallace, 1983: Numerical simulation of the atmospheric response to equatorial Pacific SST anomalies. *J. Atmos. Sci.,* **40,** 1613–1630.

Simmonds, I., 1990a: A modelling study of winter circulation and precipitation anomalies associated with Australian region ocean temperatures. *Aust. Meteor. Mag.,* **38,** 151–161.

———, 1990b: Improvements in general circulation model performance in simulating antarctic climate. *Antarctic Sci.,* **2,** 287–300.

———, and W. F. Budd, 1991: Sensitivity of the Southern Hemisphere circulation to leads in antarctic pack ice. *Quart. J. Roy. Meteor. Soc.,* **117,** 1003–1024.

———, and X. Wu, 1993: Cyclone behaviour response to changes in winter Southern Hemisphere sea-ice concentration. *Quart. J. Roy. Meteor. Soc.,* **119,** 1121–1148.

———, G. Trigg, and R. Law, 1988: *The Climatology of the Melbourne University General Circulation Model.* University of Melbourne, Australia, Pub. No. 31, NTIS PB 88 227491, 62 pp.

Smith, I. N., M. Dix, and R. J. Allan, 1997: The effect of greenhouse SSTs on ENSO simulations with an AGCM. *J. Climate,* **10,** 342–352.

Sperber, K. R., S. Hameed, W. L. Gates, and G. L. Potter, 1987: Southern Oscillation simulated in a global climate model. *Nature,* **329,** 140–142.

Stouffer, R. J., S. Manabe, and K. Bryan, 1989: Interhemispheric asymmetry in climate response to a gradual increase of atmospheric CO_2. *Nature,* **342,** 660–662.

———, ———, and K. Ya. Vinnikov, 1994: Model assessment of the role of natural variability in recent global warming. *Nature,* **367,** 634–636.

Swanson, G. S., and K. E. Trenberth, 1982: Persistent anomaly statistics in the Southern Hemisphere. *Proc. Seventh Annual Climate Diagnostics Workshop,* Boulder, CO, National Oceanographic and Atmospheric Administration, 118–125.

Tett, S., 1995: Simulation of El Niño/Southern Oscillation-like variability in a global AOGCM and its response to CO_2 increase. *J. Climate,* **8,** 1473–1502.

Tokioka, T., A. Noda, A. Kitoh, Y. Nidaidou, S. Nakagawa, T. Motoi, and S. Yukimoto, 1995: A transient CO_2 experiment with the MRI CGCM—Quick report. *J. Meteor. Soc. Japan,* **73,** 817–826.

Trenberth, K. E., 1986: The signature of a blocking episode on the general circulation in the Southern Hemisphere. *J. Atmos. Sci.,* **43,** 2061–2069.

———, and K. C. Mo., 1985: Blocking in the Southern Hemisphere. *Mon. Wea. Rev.,* **113,** 3–21.

van Loon, H., 1967: The half-yearly oscillation in middle and high southern latitudes and the coreless winter. *J. Atmos. Sci.,* **24,** 472–486.

———, 1986: The characteristics of sea level pressure and sea surface temperature during the development of a warm event in the Southern Oscillation. *Namias Symposium,* Scripps Institution of Oceanography Reference Series 86-17, J. O. Roads, Ed., Scripps Institution of Oceanography, 160–173.

von Storch, J.-S., 1994: Interdecadal variability in a global coupled model. *Tellus,* **46A,** 419–432.

Walsh, K., and J. L. McGregor, 1995: January and July climate simulations over the Australian region using a limited-area model. *J. Climate,* **8,** 2387–2403.

Washington, W. M., and G. A. Meehl, 1989: Climate sensitivity due to increased CO_2: Experiments with a coupled atmosphere and ocean general circulation model. *Climate Dyn.,* **4,** 1–38.

———, and ———, 1996: High-latitude climate change in a global coupled ocean–atmosphere–sea ice model with increased atmospheric CO_2. *J. Geophys. Res.,* **101,** 12 795–12 801.

Whetton, P. H., 1997: Comment on "Global and terrestrial precipitation: A comparative assessment of existing climatologies" by D. R. Legates. *Int. J. Climatol.,* **17,** 163–170.

———, A. M. Fowler, M. R. Haylock, and A. B. Pittock, 1993: Implications of climate change due to the enhanced greenhouse effect on floods and droughts in Australia. *Clim. Change,* **25,** 289–317.

———, P. J. Rayner, A. B. Pittock, and M. R. Haylock, 1994: An assessment of possible climate change in the Australian region based on an intercomparison of general circulation modeling results. *J. Climate,* **7,** 441–463.

———, M. England, S. O'Farrell, I. Watterson, and B. Pittock, 1996a: Global comparison of the regional rainfall results of enhanced greenhouse coupled and mixed layer ocean experiments: Implications for climate change scenario development. *Clim. Change,* **33,** 497–519.

———, A. B. Mullan, and A. B. Pittock, 1996b: Climate change scenarios for Australia and New Zealand. *Greenhouse: Coping with Climate Change,* W. J. Bouma, G. I. Pearman, and M. R. Manning, Eds., CSIRO, 145–168.

———, A. B. Pittock, J. C. Labraga, A. B. Mullan, and A. Joubert, 1996c: Southern Hemisphere climate: Comparing models with reality. *Climate Change, People and Policy: Developing Southern Hemisphere Perspectives,* T. Giambelluca and A. Henderson-Sellers, Eds., John Wiley and Sons, 89–130.

Xu, J.-S., H. von Storch, and H. van Loon, 1990: The performance of four spectral GCMs in the Southern Hemisphere: The January and July climatology and the semiannual wave. *J. Climate,* **3,** 53–70.

Yang, S., and W. J. Gutowski Jr., 1994: GCM simulations of the three-dimensional propagation of stationary waves. *J. Climate,* **7,** 414–433.

Zebiak, S. E., and M. A. Cane, 1987: A model El Niño–Southern Oscillation. *Mon. Wea. Rev.,* **115,** 2262–2278.